PRESCOTT & DUNN'S
INDUSTRIAL MICROBIOLOGY
4TH EDITION

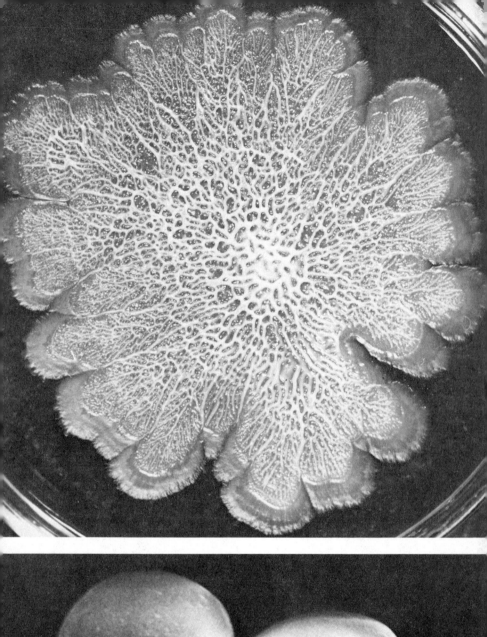

PRESCOTT & DUNN'S INDUSTRIAL MICROBIOLOGY

4TH EDITION

Edited by
Gerald Reed

Vice President
Amber Laboratories
Milwaukee, Wisconsin

AVI PUBLISHING COMPANY, INC.
Westport, Connecticut

Frontispiece Top: *Saccharomycopsis lipolytica giant colony*
　　　　　　　Courtesy of Dr. Kockova-Kratochvilova, Bratis-
　　　　　　　lava

Frontispiece Bottom: *Electron photomicrograph of Saccha-*
　　　　　　　romyces cerevisiae mother cell and daughter
　　　　　　　cell
　　　　　　　Courtesy of Dr. S.L. Chen, Universal Foods Cor-
　　　　　　　poration

Library of Congress Cataloging in Publication Data

Prescott, Samuel Cate, 1892–
　　Prescott & Dunn's industrial microbiology.

　　Includes index.
　　1. Industrial microbiology.　I.　Dunn, Cecil Gordon.
1904–　　　　　II.　Reed, Gerald.　II.　Title.　IV.
Title:　Prescott and Dunn's industrial microbiology.
[QR53.P67　　1982]　　　　660'.62　　　　81-7961
ISBN 0-87055-374-7　　　　　　　　　　AACR2

Printed in the United States of America by
The Saybrook Press, Inc., Saybrook, Connecticut

Preface

"It is the purpose of this volume to outline, in a concise but comprehensible manner, the fundamentals of industrial microbiology and to present descriptions of the more important processes within the field." This purpose from the preface of the *3rd Edition* of *Prescott & Dunn's Industrial Microbiology* has also been the guiding principle of the current *4th Edition*.

The intervening 20 years since publication of the *3rd Edition* have seen much progress and a great expansion of the field. This has required a restriction of the subject matter to food fermentations and food-related fermentations. It has also dictated the choice of an edited book over that of the monographs of the first three editions. Some of the wisdom and insight of Dr. Prescott and Dr. Dunn has undoubtedly been lost in the transition, but it is hoped that the expert knowledge of contributors to this volume will compensate for this loss.

The *4th Edition* has been written from the point of view of industrial practice. The scientific basis of the field has been outlined in individual chapters in sufficient detail to serve as a sound foundation for the industrial technology which is the principal subject matter of the book.

The full title of the earlier editions has been retained to honor Professors Prescott and Dunn, and the book is dedicated to Professor Dunn, who has provided guidance and counsel in the planning of this volume. May this book be a worthy successor to the preceding editions of *Prescott & Dunn's Industrial Microbiology*.

GERALD REED

September 1981

Dedication

For my friend and teacher
Cecil G. Dunn

Contributors

BENDA, DR. IRMGARD, Bayerische Landesanstalt für Weinbau und Gartenbau, Würzburg, Germany

BÖING, DR. J.T.P., Röhm GmbH, Darmstadt, Germany

BRANDT, MR. D.A., Vice President, Schenley Distillers, Cincinnati, OH

CHANDAN, DR. R.C., Professor, Michigan State University, East Lansing, MI

CHAUDHARY, DR. K., Professor, Department of Microbiology, Haryana Agric. University, Hissar (Haryana), India

DALBY, MR. DAVID K., Universal Foods Corp., Milwaukee, WI

EBNER, DR. HEINRICH, Heinrich Frings GmbH & Co. KG, Bonn, Germany

HAYMON, DR. L. WENDELL , Microlife Technics, Sarasota, FL

HELBERT, DR. J. RAYMOND, Miller Brewing Co., Milwaukee, WI

HESSELTINE, DR. C.W., U.S. Department of Agriculture, Peoria, IL

KAPOOR, MR. K.K., Assistant Professor, Department of Microbiology, Haryana Agric. University, Hissar (Haryana), India

MARTH, DR. E.H., Professor, University of Wisconsin, Madison, WI

MILLER, DR. MARTIN W., Professor, University of California, Davis, CA

NAKAYAMA, DR. K., Kyowa Hakko Kogyo Co., Tokyo, Japan

PONTE, DR. J.G., JR., Professor, Department of Grain Science and Industry, Kansas State University, Manhattan, KS

REED, DR. GERALD, Vice President, Amber Laboratories, Milwaukee, WI

TAURO, DR. P., Professor, Department of Microbiology, Haryana Agric. University, Hissar (Haryana), India

VAUGHN, DR. R.H., Professor Emeritus, University of California, Davis, CA

WANG, DR. H.L., U.S. Department of Agriculture, Peoria, IL

Contents

xi

     Gerald Reed   593

15   Enzyme Production, *J.T.P. Böing*   634

16   Citric Acid, *K.K. Kapoor, K. Chaudhary, and
     P. Tauro*   709

17   Amino Acids, *K. Nakayama*   748

18   Vinegar, *Heinrich Ebner*   802

19   Production of Fermentation Alcohol as a Fuel
     Source, *Gerald Reed*   835

     APPENDIX   861

     INDEX   867

Section I

General

1

Outline of Microbial Taxonomy, Metabolism, and Genetics

Gerald Reed

TAXONOMY

The classification of microorganisms is a complex and often perplexing enterprise. Our systems of classification are imposed upon the natural variety of microorganisms to serve our particular needs, and older systems of classification are abandoned if they cease to serve us well. Diverse criteria may be used to classify microorganisms: Morphology, means of propagation and sexual cycle, physiology, genetic or evolutionary relationships, habitat, or more pragmatic criteria such as industrial, agricultural or medical importance. In practice, classification is based on several criteria, and the resulting system, while useful, lacks clarity and uniformity.

A system of numerical taxonomy may be based on similarity matrices, or it may be based on the ratio of the bases guanine + cytosine/adenine + thymine (% GC) as a percentage of total DNA bases of the microorganism. For instance, for several genera of enterobacteria the % GC is between 50 and 60%, while for *Staphylococcus* it is between 30 and 40% and for *Lactobacillus* between 40 and 50%.

A basic distinction may be made between prokaryotic and eukaryotic cells. Eukaryotic cells are distinguished, as the name implies, by a distinct nucleus which is surrounded by a membrane. Nuclear DNA is associated with proteins and is present in definite structures, the chromosomes. Eukaryotic cells contain other structures such as mitochondria which are the loci of energy-producing metabolism. Photosynthetic eukaryotes contain separate chloroplasts which produce energy from sunlight. All eukaryotes require oxygen. During cell division, meiosis permits the transmission of genetic material to daughter cells.

In contrast, prokaryotes lack a well defined nucleus. The genetic material in the form of double stranded DNA is attached to the plasma membrane but is not separated from the cell content by its own membrane. Mitochondria are absent and the enzymes required for energy metabolism are distributed throughout the plasma. Photosynthesis where it occurs is also based on photosynthetic molecules associated with membranes. Prokaryotes may be obligate anaerobes, facultative anaerobes, or fully aerobic organisms. In many instances facultative anaerobes grow best in the presence of an atmosphere containing less than 20% oxygen.

Bacteria and the blue-green algae (Cyanobacteria) are prokaryotes. Fungi (including yeasts), algae, protists, and all plants and animals have eukaryotic cells. The above description indicates that prokaryotes are more primitive organisms and that their evolutionary development preceded that of the eukaryotes. It is thought that anaerobic bacteria developed about 3.5 billion years ago in the anoxic atmosphere of the earth. Cyanobacteria, which are largely responsible for the development of oxygen in the atmosphere by aerobic photosynthesis, may have appeared about 2 billion years ago.[1] Eukaryotes, which depend on the presence of oxygen in the atmosphere, developed about 1 to 1.5 billion years ago. The evolutionary development of microorganisms has been described in a most lucid manner by Schopf (1978) (see Fig. 1.1).

The following brief outline places some of the more important microbes into appropriate taxa. It is not meant as a description of the various subdivisions of microbes, nor is it in any sense complete. However, it may aid the reader in realizing taxonomic relationships, and it may serve as an aid to memory.

For a comprehensive and authoritative treatment of the subject of taxonomy the reader is referred to the following: Sneath (1957); Sokal and Sneath (1963); Thimann (1963); Hawker (1966); Lodder (1970); Moore-Landecker (1972); Mandelstam and McQuillan (1973); Colwell (1973); Laskin and Lechevalier (1973); Breed et al. (1957); Buchanan and Gibbons (1974); and Beuchat (1978).

Phycomycotina

Phycomycetes are a diverse group of lower, filamentous fungi. Most species are aquatic and have motile cells. However, the group of greatest industrial interest, the pin molds or bread molds (Zygomycetes), is terrestrial and has nonmotile zygospores. These include the genera *Rhizopus* and *Mucor*. *Rhizopus arrhizae* is one of the molds responsible for tempe fermentations. *R. stolonifer* is known as the black bread mold. *Mucor pusillus* and *M. miehei* are used for the production of microbial rennets, and *M. miehei* is useful for the production of an esterase which forms desirable cheese flavors.

Ascomycotina

The ascomycetes are characterized by a special sac-like structure (ascus) which contains the spores (ascospores). These are sexual spores resulting

---

[1]Photosynthetic bacteria other than Cyanobacteria carry out photosynthesis as a completely anaerobic process.

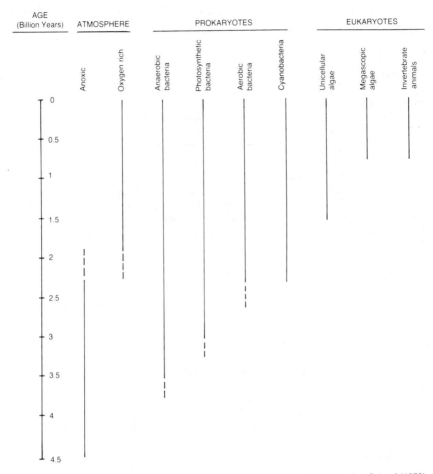

Simplified after Schopf (1978)

FIG. 1.1. EVOLUTION OF MICROBES

from the fusion of 2 nuclei followed by meiosis. The haploid ascospores, often 8 per ascus, contain 1 or several haploid nuclei. On germination, these spores develop into the typical, vegetative filaments which are called hyphae or mycelia. If the vegetative cells occur principally as single cells, they are generally designated as yeasts.

This is a very large class of microorganisms which includes many soil microbes, among them *Chaetomium*, a wood rotting fungus whose species are excellent producers of cellulases. It includes fungi living in plants or animals and in fresh and salt water. Many of these fungi are well known plant pathogens or food spoilage organisms responsible for the powdery mildew of fruits or the soft rot of vegetables. The red bread mold, *Neurospora sitophila*, is well known. *Endothia parasitica* is a commercial source of microbial rennet. The true morel mushroom, *Morchella esculenta*, belongs to the Ascomycotina.

Basidiomycotina

The basidiomycetes are the most highly evolved group of fungi. They are characterized by the fact that spores (basidiospores) are borne externally on a basidium. The latter is a structure in which nuclear fusion and reductive division occurs and which bears the externally located spores. This large group of organisms includes the mushrooms, puffballs, bracket fungi, rusts, smuts, jelly fungi, stinkhorns, and others.

The cultivated mushroom, *Agaricus bisporus*, carries 2 spores on the basidium, in contrast to the wild species, *A. campestris*, which carries 4 spores. Another mushroom, *Lentinus edodes*, is grown commercially in Japan on wooden logs. This "shiitake" mushroom has been one of the earliest sources of 5'-nucleotides (see Chapter 13).

Deuteromycotina

This subdivision of the fungi contains all organisms for which a true sexual stage (perfect stage) does not exist or has not been recognized. These Deuteromycetes or Fungi Imperfecti are difficult to classify and probably lack any common phylogenetic origin. Many of them have counterparts in the Ascomycetes. For instance, the red bread mold, *Neurospora sitophila*, (see under Ascomycotina), is known in its imperfect state as *Monilia sitophila*.

The genus *Aspergillus* is widely used for the production of organic acids and enzymes. It is a major contributor in the ripening of Oriental foods (see Chapter 12). *A. oryzae*, *A. niger*, and *A. awamori* are widely used species.

The genus *Penicillium*, which is used for the production of antibiotics, is also used for the production of organic acids. *P. roqueforti* and *P. camemberti* are used for the production of mold ripened cheeses. A species of *Botrytis*, *B. cinerea*, is responsible for the noble rot of grapes.

Yeasts

Yeasts are fungi which exist generally in the form of single cells and which reproduce by budding or by fission. Yeasts which form ascospores or basidiospores are properly classified with the sporogenous fungi, and asporogenous yeasts should be classified with the Fungi Imperfecti. It must be understood that grouping of these organisms as "yeasts" cuts across the classification of fungi which has been outlined above.

Some examples of ascomycetous and asporogenous yeasts follow. In some instances, equivalent species can be found in the 2 groups as shown in Table 1.1.

Ascomycetous Yeasts.— Species of *Saccharomyces* are used widely for the leavening of baked goods, and for the production of wine, beer, and distilled

TABLE 1.1. SOME ASCOSPORE-FORMING YEASTS AND THEIR ASEXUALLY EQUIVALENT SPECIES

Ascosporogenous Yeasts	Asexual Species
Saccharomyces exiguus	Torulopsis holmii
Saccharomyces telluris	Torulopsis bovina
Sacchromices rosei	Torulopsis stellata var. cambresi
Pichia kudriavzevi	Candida krusei
Pichia membranaefaciens	Candida valida
Pichia fermentans	Candida lambica
Kluyveromyces fragilis	Candida pseudotropicalis
Kluyveromyces lactis	Torulopsis sphaerica
Kluyveromyces marxianus	Candida macedoniensis
Dekkera bruxellensis	Brettanomyces bruxellensis
Dekkera intermedia	Brettanomyces intermedius
Debaryomyces hansenii	Torulopsis candida
Hanseniaspora valbyensis	Kloeckera apiculata
Hansenula capsulata	Torulopsis molischiana
Endomycopsis lipolytica	Candida lipolytica

Source: Reed and Peppler (1973).

beverages. In all of these commercial processes *S. cerevisiae* is used, with the exception of lager beer *(S. uvarum)*, some sourdoughs *(S. exiguus)*, and some wine fermentations *(S. uvarum, S. fermentati)*. *Kluyveromyces* species are used for the production of alcohol from dairy products *(K. fragilis)*. The following genera are present in many "natural" wine fermentations: *Hansenula, Pichia, Dekkera, Saccharomycodes* and *Schizosaccharomyces*. *Saccharomycopsis fibuligera* is used in the saccharification of starch.

Asporogenous Yeasts.—*Candida utilis* is used for the production of inactive dry yeast for food and feed. *C. lipolytica* and *C. tropicalis* are used for the same purpose. A *Brettanomyces* is used in the production of Belgian lambic beer. Species of *Kloeckera* and *Metschnikowia* occur in spontaneous wine fermentations.

This short enumeration gives some idea of the diversity of industrially useful yeasts. Other genera are listed in chapters dealing with specific industries.

Chemoautotrophic Bacteria

These organisms are called chemoautotrophic because of their ability to grow by oxidation of inorganic compounds. One of the groups, the hydrogen bacteria, can oxidize hydrogen. O_2 is the terminal electron acceptor. But the hydrogen bacteria are quite versatile in utilizing various carbon compounds, and show great similarity to heterotrophic organisms with similar nutritional requirements or capabilities.

Other groups of chemoautotrophic bacteria are the thiobacilli, which oxidize reduced sulfur compounds, and the nitrifying bacteria, which oxidize NH_3 or NO_2^-.[2] Some autotrophic species of the genera *Alcaligenes, Pseudomonas, Mycobacterium,* and *Nocardia* are mentioned in Chapter 13.

---

[2] *Hyphomicrobium*, which also can grow by utilizing ammonia or nitrate as the sole nitrogen source, is an obligate aerobic organism. Its preferred carbon sources are those with a single carbon atom, such as methanol or formate.

Phototrophic Bacteria

These purple, green, or brown bacteria carry out anaerobic photosynthesis, that is, without the formation of oxygen. This is in contrast to the Cyanobacteria, the algae, and the higher green plants, which produce oxygen during photosynthesis. The photosynthetic pigments are bacteriochlorophylls. All of them can fix molecular nitrogen. The rod-shaped genus *Rhodopseudomonas* which contains bacteriochlorophyll a and a carotenoid pigment is mentioned in Chapter 13.

Cyanobacteria

These important microorganisms have previously been included in the algae (blue-green algae or Cyanophyta), but they are prokaryotes which lack a nucleus. Their photosynthetic ability is based on the presence of chlorophyll a and phycobiliproteins. However, these are not present in a chloroplast but adhere to 2 membranes in the cell. The Cyanobacteria produce oxygen during photosynthesis, an ability they share with the algae and higher green plants. This clearly distinguishes them from the phototrophic bacteria which have already been mentioned.

Bacteria

The cells of these nonfilamentous and nonphotosynthetic organisms are relatively simple and quite small. Cocci have a diameter of about 0.5 to 1 μ; rods rarely exceed 1 μ in width but may attain a length of up to 20 μ. They are generally classified on the basis of their shape with subdivisions based on other morphological characteristics, on aerobic or anaerobic growth, on retention of the Gram stain, on physiological reactions, or on other criteria. Therefore, it is not surprising that schemes of classification by taxonomists differ appreciably depending on the weight given to one or other criteria used in classification.

Bacteria divide by fission, and some bacteria may produce asexual spores. These are formed within the bacterial cell (endospores), and they show greater resistance to heating than the bacterial cells. Cells shapes are distinguished as spherical (cocci), as rod-shaped (bacilli) or as curved rods. The cocci may be divided into diplococci (2 adhering cells), streptococci (a straight chain of cells), staphylococci (a sheet of cells), or sarcina (a three-dimensional arrangement or packet of cells).

The rods are divided into nonsporeforming rods, which include both Gram-positive and Gram-negative forms, and the sporeforming rods (bacillus). The sporeforming rods may be further divided into the aerobic genus *Bacillus* and the anaerobic genus *Clostridium*.

Curved rods are Gram-negative. They may be divided into those with a simple curved shape, the genus *Vibrio*, and those with a screw-shaped rod, the genus *Spirillum*.

Actinomycetes

These bacteria form mycelia. In contrast to fungal mycelia, the broadest filaments are quite thin with a diameter of less than 0.5 μ. The soil or oxidative forms include the genera *Streptomyces, Nocardia,* and *Mycobacterium.* The parasitic or fermentative forms include the genus *Actinomyces* proper.

In keeping with the purpose of this chapter the following table listing the more important genera of bacteria and Actinomycetes is highly abbreviated (Table 1.2) since the concept of a microbial species is difficult to define, and hence is imperfect and subject to revision.

Lactic Acid Bacteria

The trivial name "lactic acid bacteria" describes a number of bacteria, both cocci and rods, which are characterized by the production of lactic acid by fermentation of carbohydrates. They warrant special consideration because of their great importance in food fermentations. Apart from their well known use in the fermentation of dairy products (cheese, buttermilk, yoghurt, sour cream) they form the basis of the fermentation of sauerkraut, fermented pickles, fermented sausage, sourdough bread, and soda crackers, as well as the malo-lactic fermentation of wines. Homofermentative species produce lactic acid principally, while heterofermentative species produce lactic acid, ethanol, acetic acid, glycerol, mannitol, and CO_2 anaerobically from fermentable carbohydrates. Table 1.3 shows the number of species of 4 genera of lactic acid bacteria which have been identified (see also London 1976 and Stamer 1979).

Algae

These photosynthetic organisms resemble green plants, but they lack roots, stems, and leaves. They may be unicellular organisms or the cells may form branched or unbranched filaments. Algae are eukaryotes with a well defined nucleus. In contrast to photosynthetic bacteria, photosynthesis is localized in specific bodies, the chromatophores. Most algae multiply vegetatively or asexually by spore formation. But sexual reproduction is fairly frequent in grass-green, brown and red algae. It is rare in yellow-green and golden-brown algae.

Most marine algae are of macroscopic size. Some are used as food, for instance, *Porphyra,* which is consumed in the Orient. *Gelidium* is used for the extraction of agar, *Chondrus cryptus* (Irish moss) is the source of carrageenin, and the giant kelp of the Pacific coast of North America, *Macrocystis pyrifera,* is the source of alginates.

Some freshwater algae are of particular interest for the production of biomass (and single cell protein). These belong to the group Chlorophyta, green algae whose food reserve is starchy. Examples are the unicellular genus *Chlorella* and the genus *Scenedesmus,* which grows in nonfilamentous colonies. *Spirulina,* which is of equal interest, is a "blue green alga"; i.e., it belongs to the Cyanobacteria (see preceding).

TABLE 1.2.   BACTERIAL GENERA WHICH ARE OF IMPORTANCE IN THE FERMENTA-
TION INDUSTRY; INCLUDING REPRESENTATIVE SPECIES

Eubacteria

Rods

Rhizobium	rods, Gram −, aerobic, fix N_2 in symbiosis with legumes	*R. leguminosarum*
Escherichia	rods, Gram −, aerobic and facultatively anaerobic, indicator of fecal contamination, enzyme producer	*E. coli*
Salmonella	rods, Gram −, aerobic and facultatively anaerobic, food pathogen	*S. typhimurium*
Azotobacter	rods, Gram −, aerobic, fixes N_2 nonsymbiotically	*A. agilis*
Alkaligenes	rods, Gram −, some facultative anaerobes, litmus milk is turned strongly alkaline	*A. faecalis*
Pseudomonas	rods, Gram −, highly aerobic, motile, the pseudomonads have some auxotrophic members, e.g., "*Methanomonas*"	*P. fluorescens*
Acetobacter	rods, Gram −, highly aerobic, forms acetic acid	*A. aceti*
Brevibacterium	rods, Gram +, aerobic, reddish to brown pigment	*B. linens*
Cellulomonas	rods, Gram +, aerobic, decomposes cellulose	*C. thermocellum*
Corynebacterium	rods, Gram +, aerobic	*C. glutamicum*
Lactobacillus	rods, Gram +, anaerobic, homofermentative or heterofermentative producer of lactic acid	*L. casei*
Propionibacterium	rods, Gram +, anaerobic, produces propionic acid	*P. shermanii*
Bacillus	rods, Gram +, aerobic and facultatively anaerobic, sporeforming, highly diverse group	*B. subtilis*
Clostridium	rods, Gram +, sporeforming, anaerobic, cellulose fermenters are Gram −	*C. butyricum*

Cocci

Streptococcus	cocci, Gram +, microaerophilic, nonsporeforming, lactic acid producer	*S. lactis*
Pediococcus	cocci, Gram +, microaerophilic, nonsporeforming, lactic acid producer	*P. cerevisiae*
Leuconostoc	cocci, Gram +, microaerophilic, nonsporeforming, lactic acid producer	*L. citrovorum*

TABLE 1.2.   (Continued)

| *Staphylococcus* | cocci, Gram +, microaerophilic, non-sporeforming, pathogenic, and a food poisoning organism (enterotoxin) | *S. aureus* |

Actinomycetes

| *Streptomyces* | forms mycelium, aerobic, sporeforming, produces enzymes and antibiotics | *S. griseus* |

TABLE 1.3.   BIOCHEMICAL DIFFERENTIATION OF THE LACTIC ACID BACTERIA

Characteristic	Homofermentative	Heterofermentative
end product	mainly lactic acid	mainly lactic acid, CO_2
rods	*Lactobacillus* (16 species)	*Lactobacillus* (11)
cocci	*Pediococcus* (5)	*Leuconostoc* (6)
	Streptococcus (21)	
total number of species	42	17

Source: Buchanan and Gibbons (1974).

METABOLISM

Both prokaryotic and eukaryotic cells share a major metabolic pathway which leads from 1 molecule of glucose through several enzymatically catalyzed steps to pyruvic acid. This pathway requires no oxygen. In anaerobic fermentation processes, pyruvate is further broken down to end products which are excreted. For instance, in the alcoholic fermentation by yeasts, pyruvate is broken down to ethanol and carbon dioxide. In the bacterial lactic acid fermentation, pyruvate is broken down principally to lactic acid. In both instances, the net gain of available energy is stored in 2 moles of adenosine triphosphate (ATP).

In the process of respiration which is characteristic of eukaryotic cells and of aerobic bacteria, pyruvate is oxidized to water and carbon dioxide through the citric acid cycle. This results in a net gain of 36 moles of ATP per mole of metabolized glucose. Some of the metabolic pathways are described in the following chapters: The formation of ethanol and carbon dioxide through the glycolytic pathway and the citric acid cycle in Chapter 9 on "brewing"; the formation of lactic acid in Chapters 4 and 5 on cheese and other dairy products and Chapter 8 on wine; the formation of amino acids in Chapter 17.

Photosynthetic microorganisms are of lesser importance in the industrial production of foods. Some of these organisms are mentioned in Chapter 13 as potential producers of biomass. Several photosynthetic pathways of microorganisms are shown in Table 1.4.

TABLE 1.4. SOME CHARACTERISTICS OF PHOTOSYNTHETIC MICROBES

Characteristic	Green Plants and Algae	Cyanophyta[1] (Blue-green Algae)	Chlorobacteriaceae (Green Bacteria)	Thiorhodaceae (Purple Bacteria)
Cell organization	eukaryotic	prokaryotic	prokaryotic	prokaryotic
Photosynthetic O_2 evolution	yes	yes	no	no
Relation to O_2	aerobic	aerobic	anaerobic	anaerobic or facultatively aerobic
Major form of chlorophyll	chlorophyll	bacteriochlorophyll a	bacteriochlorophyll c or d	bacteriochlorophyll a or b

Source: Extensively modified after Lechevalier and Pramer (1971).
[1] More properly called Cyanobacteria.

GENETIC MODIFICATION

The genetics of microorganisms are not included in this volume in spite of the revolutionary changes which have taken place in this field. Hopwood (1979) has listed these discoveries in their chronological order as follows: (1) Conjugation by promiscuous sex factors (mediated by wild type plasmids); (2) *in vivo* rearrangements of transposable DNA elements (transposons); (3) recombination *in vivo* by restriction enzymes (restriction endonucleases); (4) *in vitro* recombinant DNA techniques (mainly in *Escherichia coli*); and (5) protoplast fusion as in the genera *Bacillus* and *Streptomyces*. But for most industrial microbiologists the traditional methods of hybridization (with eukaryotes) and mutation are the common methods of strain improvement. Treatment with mutagenic agents is still the most widely used tool for increasing genetic variability (Sermonti 1979). (See also Chapters 2, 3, and 17 of this book.)

The application of genetic modification of microorganisms has been widely practiced and has been most successful with hormones, biologically active peptides, antibiotics, amino acids, and enzymes. Bu'lock (1979) has pointed out that its application is rarer in fermentations producing large volume-low cost products, such as the production of biomass or alcohol, where "the whole process technology takes precedence over most other considerations." Therefore, it will not be surprising that the chapters of this book dealing with the production of enzymes and amino acids carry frequent references to the use of mutant strains, while other chapters do not. Demain (1976) has dealt lucidly with the principles governing genetic regulation of fermentation organisms for industrial use. The difficulties encountered in the genetic modification of Fungi Imperfecti through selection, mutation, and recombination have been addressed by Esser (1974); and a recent symposium has explored the genetics of microorganisms with particular emphasis on industrially important strains (Sebek and Laskin 1979).

REFERENCES

BEUCHAT, L.R.   1978.   Food and Beverage Mycology.   AVI Publishing Co., Westport, Conn.

BREED, R.S., MURRAY, E.G.D. and SMITH, N.R.   1957.   Bergey's Manual of Determinative Bacteriology, 7th Edition. Williams & Wilkins, Baltimore.

BUCHANAN, R.E. and GIBBONS, N.E.   1974.   Bergey's Manual of Determinative Bacteriology, 8th Edition. Williams & Wilkins, Baltimore.

BU'LOCK, J.D.   1979.   Process needs and the scope for genetic methods. *In* Genetics of Industrial Microorganisms. O.K. Sebek and A. I. Laskin (Editors). Am. Soc. Microbiol., Washington, D.C.

COLWELL, R.R.   1973.   Genetic and phenetic classification of bacteria. Adv. Appl. Microbiol. *16*, 137–175.

DEMAIN, A.L.   1976.   Industrial aspects of maintaining germ plasm and genetic diversity. *In* The Role of Culture Collections in the Era of Molecular Biology. R.R. Colwell (Editor). Am. Soc. Microbiol., Washington, D.C.

ESSER, K.   1974.   Some aspects of genetic research on fungi and their practical implication. Adv. Biochem. Eng. *3*, 360–387.

HAWKER, L.   1966.   Fungi: An Introduction.   Hutchinson and Co., London.

HOPWOOD, D.A.   1979.   The many faces of recombination. *In* Genetics of Industrial Microorganisms. O.K. Sebek and A.I. Laskin (Editors). Am. Soc. Microbiol., Washington, D.C.

LASKIN, A.I. and LECHEVALIER, H.A.   1973.   Handbook of Microbiology, Vol. 1.   CRC Press, Cleveland.

LECHEVALIER, H.A. and PRAMER, D.   1971.   The Microbes.   J.B. Lippincott Co., Philadelphia.

LODDER, J.   1970.   The Yeasts: A Taxonomic Study, 2nd Edition.   North Holland Publishing Co., Amsterdam.

LONDON, J.   1976.   The ecology and taxonomic status of the *Lactobacilli*. Annu. Rev. Microbiol. *30*, 279–301.

MANDELSTAM, J. and McQUILLAN, K.   1973.   Biochemistry of Bacterial Growth, 2nd Edition.   Blackwell Science Publishers, Oxford.

MOORE-LANDECKER, E.   1972.   Fundamentals of the Fungi. Prentice-Hall, Englewood Cliffs, N.J.

REED, G. and PEPPLER, H.J.   1973.   Yeast Technology.   AVI Publishing Co., Westport, Conn.

SCHOPF, A.   1978.   The evolution of the earliest cells. Sci. Am. *239* (3) 111–138.

SEBEK, O.K. and LASKIN, A.I.   1979.   Genetics of Industrial Microorganisms. Am. Soc. Microbiol., Washington, D.C.

SERMONTI, G.   1979.   Mutation and microbial breeding. *In* Genetics of Industrial Microorganisms. O.K. Sebek and A.I. Laskin (Editors). Am. Soc. Microbiol., Washington, D.C.

SNEATH, P.H.A.   1957.   Some thoughts on bacterial classification. J. Gen. Microbiol. *17* (1) 184–200.

SOKAL, R.R. and SNEATH, P.H.A.   1963.   Principles of Numerical Taxonomy.   W.H. Freeman and Co., San Francisco.

STAMER, J.R.   1979.   The lactic acid bacteria. Food Technol. *33* (11) 60–65.

THIMANN, K.V.   1963.   The Life of Bacteria, 2nd Edition. Macmillan Co., New York.

2

Yeasts

M.W. Miller

Yeasts, of all the various groups of microorganisms, have been the most intimately associated with man from the dawn of his existence. This association has been based primarily upon the ability of certain yeasts to rapidly and efficiently convert sugars into alcohol and carbon dioxide, thus effecting an alcoholic fermentation of sugary liquids such as fruit juices, grain extracts, and milk. It is very probable that yeasts are the oldest of cultivated plants. From models of a bakery and a brewery excavated at Thebes and dating from the 11th Dynasty (about 2000 B.C.), we have evidence that the use of yeasts by mankind was already well established. It is not unreasonable to believe from the organized state of brewing and baking exhibited by those models that man had been using yeasts for these purposes many years. In fact, evidence from archeologists shows that apparently all of the ancient civilizations utilized products of alcoholic fermentation very much as we do at the present time.

In the past century the welfare of man has been further benefited by the clarification of metabolic processes and basic biochemical reactions of living cells. This contribution in large part was due to the fact that yeasts, specifically *Saccharomyces cerevisiae*, were available in large quantities due to their use in baking and brewing. In fact, the origin of biochemical research started in 1897 when Hans and Eduard Büchner added cane sugar as a preservative to yeast extracts they had prepared for medical purposes and discovered that the cane sugar was rapidly fermented by the yeast extract. This first cell-free fermentation led to a series of studies about the nature of the intermediate metabolic steps in the fermentation of sugar to ethanol and carbon dioxide. Many famous biochemists, including Embden, Meyerhof, Parnas, Harden, and Young were associated with these studies. Besides the availability of *Saccharomyces cerevisiae* in large quantities, this yeast has a sexual cycle and lends itself to genetic analysis. Due to its single cell form of growth, it has other advantages for experimentation, such as a much higher metabolic rate (largely because the ratio of surface area to volume is greater than for cells growing in chains), a more rapid growth rate, and consequently the ability to bring about chemical changes very quickly.

While man had long used the fermentative capabilities of yeasts, the concept of yeasts *per se* may be considered to have had its beginning with the descriptions and drawings of yeast cells sent to the Royal Society in London in 1680 by Antonie van Leeuwenhoek. He observed these tiny "animalcules" in a droplet of fermenting beer by use of tiny, hand-ground and polished lenses which he made as a hobby. The fact that van Leeuwenhoek's lenses were capable of magnifying objects only 250–270 times their natural size emphasizes not only his extreme skill in grinding and polishing these lenses but also his unusual perceptiveness in being able to detect microorganisms as small as yeasts. Recognition of the significance of his findings was long delayed; in fact, it was nearly 150 years later before additional information on yeasts was sought.

Erxleben expressed the view in 1818 that yeasts consisted of living vegetative organisms responsible for fermentation. This view received little attention at that time but new interest was created by the development of a so-called vitalistic theory of fermentation by Cagniard de la Tour in France in 1835 and by Schwann and Kützing in 1837 in Germany. This vitalistic theory proposed that if yeasts are introduced into a sugar-containing solution, they use the sugar as a food and excrete the nonutilized parts as alcohol and carbon dioxide. This view was completely unacceptable to many chemists of that period who believed that chemical reactions rather than the activities of living cells explained the alcoholic fermentation of the sugars. Schwann termed yeasts "Zuckerpilz" meaning sugar fungus, from which the name *Saccharomyces* originates.

Studies by Pasteur, who was trained as a chemist and who had become involved with biological problems, finally proved that fermentation is due to the activities of living cells. His contributions to yeast microbiology and to our understanding of fermentations in general are presented especially in his "Études sur le Vin" ("Studies on Wine") in 1866 and "Études sur la Bière" (Studies on Beer") in 1876. In a later study he clarified the effect of oxygen on alcoholic fermentation, concluding that fermentation is a substitute for the respiratory processes and was considered the vital energy yielding process of yeasts living in the absence of oxygen. Further, his experiments tended to show that aeration suppressed fermentation, a phenomenon which was later studied more quantitatively by Meyerhof and is now referred to as "reaction of Pasteur-Meyerhof."

After van Leeuwenhoek's description of yeasts in 1680 as globular bodies which may be oval or spherical in shape, very little information was gained until after the studies of Pasteur when there was a period of rapid advancement in yeast morphology, taxonomy, and systematics. One of the many contributions of Pasteur to science was the introduction of methods for obtaining pure cultures, thus enabling one to make reliable morphological studies of yeasts. Emil Christian Hansen perfected these techniques and over a 30 year span carefully studied morphological and certain physiological characteristics of these microorganisms. Through these studies he was able to differentiate and to characterize many species of yeasts which are still recognized today.

Hansen developed the first comprehensive system of yeast taxonomy in 1896, a system which was greatly expanded by Guilliermond (1920, 1928) who contributed further details on the physiology, sexuality, and phylogenetic relationships among the yeasts. From 1931 to 1970, 5 monographs on yeast taxonomy emerged from the Technical University at Delft and were inspired directly or indirectly by the Dutch microbiologist Kluyver. The first was a complete and usable scheme of classification for sporulating yeasts prepared by Stelling-Dekker (1931), which was followed by 2 volumes on the nonsporeforming yeasts by Lodder (1934) and Diddens and Lodder (1942). In 1952 Lodder and Kreger-van Rij updated the studies of the previous authors. About the same time, Wickerham (1951) in the United States published on a number of new techniques and principles in the classification of yeasts. Later, through the cooperative efforts of 14 taxonomists in various countries of the world, under the editorship of J. Lodder (1970), the classification of the yeast was further revised and updated.

The interest in yeast taxonomy in recent years can thus be best illustrated by the fact that the 1952 monograph by Lodder and Kreger-van Rij represented the evaluation of 1317 strains of yeasts which were classified into 165 species with 17 varieties. In the 1970 taxonomic key edited by Lodder, the number of species had increased to 341. Four years later when Barnett and Pankhurst (1974) published an identification yeast key based primarily on the biochemical properties of yeasts, it included an additional 93 species which had been described between 1970 and 1973. New species are continually being added to the literature by many investigators in the field so that the total number of species today exceeds 500 classified in some 50 genera.

PROPERTIES OF YEASTS

What is a yeast? The origin of the word in many languages relates primarily to its ability to ferment. The English "yeast" and the Dutch "gist" are derived from the Greek term *zestos* which means boiled, a reference to the bubbling foam caused by the evolution of carbon dioxide. The German "hefe" and the French "levure" both have their origins in verbs meaning "to raise," again referring to the bubbling foam. Industrial terms such as "cultured, true, wild, top, or bottom yeasts" have little meaning from the botanical point of view. In fact, even from an industrial viewpoint they are confusing, for a yeast considered as a cultivated organism in one industry, for example, a brewery yeast, may well be considered to be a wild yeast by bakers. From a botanical point of view, yeasts are recognized to be a heterogeneous group. Botanists of the 19th century generally accepted the idea that yeasts belong to the plant kingdom. At present, many taxonomists are in agreement with the arrangement proposed by Ainsworth in which the fungi are treated as a separate kingdom and the yeasts are included in the division Eumycota.

Yeasts lack chlorophyll and are unable to manufacture by photosynthesis from inorganic substrates the organic compounds required for growth, as do higher plants, algae, and even some bacteria. Hence they must lead a saprophytic, or in a few instances a parasitic, lifestyle. Yeasts possess rather rigid, thick cell walls, have a well organized nucleus with a nuclear membrane (eukaryotic), and have no motile stages. The ability to form sexual spores within an ascus or produce them externally on a basidium places most yeasts in the subdivisions Ascomycotina and Basidiomycotina, respectively. Those yeasts in which a perfect (sexual) stage is not known are grouped together in the subdivision Deuteromycotina.

While the definition of yeasts varies somewhat according to author, they are generally defined as fungi which, in a stage of their life cycle, occur as single cells, reproducing commonly by budding or less frequently by fission. Generally organisms with plurinucleate cells or those producing black pigments or producing asexual spores borne on distinct aerial structures are excluded. The distinction of yeasts from related mycelial fungal forms is highly subjective, resulting in a number of transitional forms between yeasts and the more typical higher fungi. Taxonomic consideration of the yeasts relies heavily on morphological characteristics for genera. However, differentiation of species, because of the limitations of morphological criteria, relies very heavily upon physiological characteristics. Identification tests are standardized to enable investigators in different laboratories to compare their cultures with standard descriptions of type species and varieties.

Cellular Asexual Reproduction and Morphology

The early investigators described yeasts as being round to oval in appearance and noted that they divided by budding to form daughter cells. This description agrees with that of many yeasts. However, among the yeasts, the cell shapes and the means by which they reproduce are quite varied.

Budding.—The yeast thallus (vegetative body) in its simplest form is a single cell or perhaps one with a bud still attached as exemplified by *Saccharomyces cerevisiae*. Here the first bud is usually detached when the mother cell initiates a second bud. There are yeasts, however, in which the buds will remain attached so that the mother cell and first daughter cell may produce additional buds. In such instances, this will result in small to large clusters or chains of cells.

Pseudomycelia.—Similar to chain formation, a pseudomycelium (false mycelium) may be formed when, instead of the bud breaking away from the mother cell at maturity, it elongates and continues to bud in turn. In this manner chains of cells are formed, which in appearance resemble true mycelium (where cells are separated by a cross-wall or septum) but differ in the manner in which new cells arise (budding). Pseudomycelia vary in their complexity from very primitive, in which the number of cells are limited and where there is little or no differentiation among these cells, to other forms where cells comprising the main chain are rather elongated and the buds arise in clusters on the shoulder of these elongated stem cells. The side buds in turn may remain spherical, ovoidal, or may also elongate giving rise to further branching and complexity of form. While a particular yeast species may produce a characteristic pseudomycelial form, it is more commonly observed that several types of pseudomycelia are found within a single species. Consequently, while the type of pseudomycelium formed by yeast is of little value in classification, the ability or inability to form a pseudomycelium is used.

Budding, as has been mentioned before, represents the most common method of vegetative reproduction. Except for the species of a few genera, buds usually arise on the shoulders and at the ends of the long axis of the cells. This type of budding, referred to as "multilateral budding," is characteristic of *Saccharomyces* and most other ascomycetous yeasts. In the case of spherical cells, buds do not appear to be oriented to any particular area of the cell surface. In multilateral budding, only one bud is produced from a particular site on the cell. This is in contrast to some yeasts of basidiomycetous origin, which, while reproducing by budding, apparently produce a number of buds at the same site, although more than a single budding site on the cell surface has been observed. Special cases of budding are polarly oriented. *Hanseniaspora* and *Saccharomycodes* species bud repeatedly at one site on each of the tips of the cells. This type of budding causes the vegetative cell to assume a lemon shape and the yeasts are known as apiculates. The genus *Pityrosporum* is characterized by vegetative cells reproducing by repeated unipolar budding on a broad base. In one genus, *Trigonopsis*, yeast cells have a triangular shape with budding restricted to the 3 apices.

Fission.—Vegetative reproduction by fission is characteristic of 2 genera, *Endomyces* and *Schizosaccharomyces*. In these yeasts, reproduction is carried out by the formation of a cross-wall (septum) without a constriction of the original cell wall. When the process is complete, the new cell wall divides into 2 individual walls and the newly formed vegetative cells separate from each other. *Endomyces* also produces true mycelium and by most yeast taxonomists is not considered a yeast but rather a "yeast-related" genus. Its imperfect form is *Geotrichum*, the so-called "machinery mold."

Bud-fission.—There are a few yeasts in which asexual reproduction is intermediate between typical budding and fission. This so-called "bud-fission" results from a type of budding in which the base of the bud is very broad, somewhat like a bowling pin. Separation of the daughter cell from the mother is by the formation of a septum across the broad neck. *Saccharomycodes* and *Nadsonia* are genera exhibiting this type of reproduction. Fundamentally, bud-fission differs from typical budding only in the size of the septum. In budding it is so small that it appears that the bud is "pinched off " rather than distinctly separated by a septum.

Budding and Fission.—Some yeasts reproduce vegetatively by both budding and fission. Species of *Trichosporon* usually grow as mycelial strands (hyphae) by cross-wall formation. The strands can undergo disarticulation into individual vegetative cells called arthrospores, which, upon germination again, produce mycelium. Budding cells may also arise on the mycelial strands. In certain species of *Candida* and in *Saccharomycopsis (Endomycopsis)*, hyphae with cross-walls and budding cells are found. However, in these genera the hyphae do not disarticulate into arthrospores.

Clamp Connections.—A special type of mycelium is formed by some of the basidiomycetous yeasts (*Leucosporidium, Rhodosporidium,* and *Sporidiobolus*) in which clamp connections are formed between adjoining cells during the dikaryotic stage of their special life cycle. The clamp connection is a specialized mechanism which assures that 1 pair of compatible nuclei resides in each cell formed.

Cell Morphology

Spheroidal, globose, ovoidal, elongated, and cylindrical are descriptive terms for the general shape of many vegetative yeast cells. There are, however, certain yeasts which have highly characteristic cell shapes. The apiculate, due to the bipolar mode of budding, is characteristic of the species of *Hanseniaspora* and its imperfect genus *Kloeckera*, yeasts which are commonly isolated from the early stages of spontaneous fermentations or spoilage of fruits and similar raw materials. Yeasts of the genus *Dekkera* (imperfect = *Brettanomyces*) produce ogival cells. These are rather elongated cells rounded at one pole and somewhat pointed at the other. These yeasts are used in the production of ale in Europe and are also found in bottled wines and soft drinks as spoilage yeasts. The unipolar mode of asexual reproduction results in a characteristic "flask-like" cell shape characteristic of the genus *Pityrosporum* and is associated with skin disorders of warm-blooded animals. Tripolar budding results from the unique triangularly-shaped cells of *Trigonopsis*, originally isolated from beer in a Munich brewery and subsequently from grapes in South Africa. Also characteristic are the highly curved cells formed by *Cryptococcus cereanus*, a yeast found present in the fermenting juices of certain rotting cacti.

One must realize that the assignment of a particular cell shape as characteristic of a particular genus or species does not imply that every cell in a population will be that shape. It is true, however, that at some period in the ontogenetic development of yeast cells, the yeast cell will assume that form. An apiculate yeast, for example, begins its ontogenetic development as a small oval bud and will develop an apiculate shape only after separation from the mother cell and subsequent development of buds at the two cell poles. Upon aging, these cells assume a variety of shapes, most common of which would be an irregularly elongated cell.

The actual size of the individual yeast cells varies considerably and for a particular culture may be quite uniform, while in other species extreme heterogeneity in both size and shape is observed. For example, some yeast cells may be 2–3 μm in length, whereas others may attain lengths of 20–50 μm. Width of the cells is normally not as variable and measures normally between 1 and 10 μm. While a certain amount of variation is normal and is to be expected within a yeast culture, the cultural conditions and age of cells can exert a great deal of influence on the culture's morphological properties. Again, as stated before, descriptions of the various yeast species are based on results obtained with quite standardized conditions and media.

Microscopy has contributed a great amount to our knowledge of yeast cytology. Obviously, direct observation with a light microscope provided our first details. Staining techniques have helped to determine the location of specific cell components or surface areas. Transmission electron microscopy (TEM) of ultra-thin sections of yeast cells or of carbon-platinum replicas of freeze-fractured cells has revealed internal details at high magnifications. Lastly, by use of the scanning electron microscope (SEM) details of external vegetative cell and spore structures have been observed. This method has been particularly useful in gaining information on the topography of vegetative cells and sexual spores, as well as morphological details of asexual and sexual reproduction.

Because various yeast species may show significant differences in the ultrastructure and cellular organization, generalizations must be recognized as such. As with much of our earlier information on other aspects of yeast, *Saccharomyces cerevisiae* or bakers' yeast has been the subject of many of the cytological investigations. In the past two decades, using sophisticated equipment and techniques, investigators have used species belonging to some different genera to bring out particular structures lacking in other genera, all of which have multiplied our knowledge tremendously. It has also been shown that significant changes in cellular structures may be brought about by the cultural conditions under which the culture is grown prior to examination. The shape of the cell, absence or presence of vacuoles, inclusion bodies and lipid globules, as well as mitochondrial development and the extent of capsular polysaccharide formation can be strongly modified by the growing conditions and age of a culture.

Structure and Function of Cellular Components

In the following section specific information known about the structure and function of cellular components will be given. These include the cell wall (including extracellular capsule material in some species), the plasmalemma or cytoplasmic membrane, nucleus, mitochondria, ribosomes and microbodies, vacuoles, lipid globules, volutin or polyphosphate bodies, endoplasmic reticulum, and the cytoplasmic matrix itself.

Cell Wall.—With a regular light microscope the cell wall is observed as a distinct outline of the cell but does not reveal distinct features, appearing smooth and sometimes having slight irregularities but no other details. The wall is fairly rigid and is responsible for the particular shape which a yeast cell possesses. With the employment of electron microscopy and techniques by which isolated cell walls can be prepared, details of the physical characteristics have increased. In multilaterally budding yeasts, e.g., *Saccharomyces cerevisiae*, the most obvious structures to be found on a cell wall are the bud scars. The single birth scar resulting on the bud when separated from its mother is not particularly conspicuous but expands with the growth of the bud into a mature cell. Electron micrographs show that the birth scar area is similar to the rest of the cell wall area in that bud scars from the cell giving rise to buds can be located on the birth scar region of the cell itself. In the ascomycetous yeasts, the bud scars are quite characteristic in appearance, having a raised circular brim surrounding a depressed area of about 3 μm^2. Since successive buds are never formed at the same site in the ascomycetous yeasts, the number of bud scars on a cell is indicative of the reproductive capacity of the vegetative cell. Obviously, the number of buds formed per cell is limited. In a normal population which has reached the stationary phase of growth, most of the cells have either no bud scars or from 1 to 6 scars, while a very small fraction of the cell population will have as many as 12–15 bud scars. The limitation of the maximum number of scars on the oldest cells of the population is probably due to the exhaustion of a particular nutrient or to the crowding of cells in a particular medium.

Studies to determine the maximum number of buds that a cell can produce have shown that when successive buds are removed by the aid of a micromanipulator and when cell division is not limited by exhaustion of nutrients or crowding of cells, Saccharomyces cerevisiae can produce as few as 9 and as many as 43 buds per cell. The average number under these conditions was 24. Since under ideal conditions the yeast cell may duplicate itself within 1½ to 2 hr when vigorously growing, the reproductive capacity of these cells can be imagined. It has been found, however, as cells age, the generation time for a particular medium and set of conditions is extended and finally becomes as long as 6 hr per generation. In physical appearance, the cell wall of old cells becomes somewhat wrinkled in appearance in contrast to a young vigorous cell which has a smooth, turgid appearance except for the birth and bud scar structures.

Birth scars in the case of bipolar budding yeasts are only observed in the young, unbudded daughter cell. Yeasts which have this mode of reproduction have bud scars which are superimposed on each other and with repeated budding are characterized in appearance by a series of ring-like ridges on the polar extensions of the cells. Thus, the more buds a cell has produced, the longer the polar extensions become.

In yeasts of basidiomycetous origin, bud formation repeated at the same site often occurs. In such instances, the walls of successive buds arise each time under the original cell wall, giving rise to concentric collars which are particularly noticeable in TEM of ultra-thin sections cut longitudinally through the scar area. One bud scar is commonly observed, although 2 or more are not uncommon. In appearance the bud scar is different from that of the multilaterally budding yeasts in that the circular ridge is obscured and the center area either less depressed or not at all, giving rise to a pad-like scar structure.

Vegetative cells of the genus Schizosaccharomyces, which reproduces only by fission, exhibit scar rings at the point where the cross-walls divide after septum formation. The septum area elongates for the production of a new cell and a new cross-wall is formed at a point located distally from the original scar ring.

The filamentous, ascomycetous yeasts which grow by elongation of the hyphal tip do not show exterior scars. However, ultra-thin sections disclose that the cross-walls formed in these organisms are not normally solid plugs, but rather contain 1 to a number of pores, thus permitting protoplasmic continuity among all of the cells.

Chemical analysis and enzymatic degradation studies have led to our present understanding of the chemical composition of the cell wall of yeast. Again, most of the work has been done with species of *Saccharomyces* which we will first discuss and then contrast with information available about other species and genera. First, it is believed that the cell wall is composed of 3 layers which are not necessarily distinct from each other. First is an outer layer of glycoprotein which consists mainly of a phosphorylated mannan, then a middle layer of alkali-soluble β-glucan, and finally an inner layer of alkali-insoluble β-glucan. While analyses of cell walls often show variable amounts of lipid materials, their precise location is unknown and in fact may actually represent the lipid content of plasmalemmae not removed during cell breakage and cell wall purification. In addition to the yeast phosphorylated mannose and glucose polymers, a small amount (approximately 1%) of chitin, a linear polymer of N-acetylglucosamine, is present. Chitin of bakers' yeast is presently known to be restricted to the area of the bud scars. When the bud has fully developed it is first separated from the mother cell by a primary cross-wall of chitin which is subsequently covered with glucan and mannan prior to the separation of the mature bud from the mother. In the yeast-like genus *Endomyces* and in species of *Nadsonia, Rhodotorula, Cryptococcus,* and *Sporobolomyces,* recorded chitin contents are much higher, so it may well be an actual cell wall component. In contrast, in species of *Schizosaccharomyces* analyzed, the cell walls apparently do not contain chitin.

Capsular Materials.—Capsular materials, while not strictly a component of the cell wall, are produced extracellularly by representatives of several yeast genera. Capsular materials are generally classified as phosphomannans, β-linked mannans, heteropolysaccharides (contain more than one type of sugar), and finally a number of hydrophobic substances belonging to the sphingolipid type compounds.

Phosphomannans are produced extracellularly by certain species of *Hansenula, Pichia,* and *Pachysolen.* These viscous polymers are water soluble and form a sticky layer on the surface of the cells. They contain only D-mannose and phosphate, which is linked as a diester. The molar ratio of mannose to phosphate is characteristic of the particular species producing the capsular material, although it is known that this ratio can be affected by cultural conditions of growth. For example, in media in which the phosphate levels are suboptimal the mannose:phosphate ratio increases appreciably. In several species of the genus *Rhodotorula* and in *Torulopsis ingeniosa,* the chemical composition of the capsule material is a linear or slightly branched mannan in which the linkages between mannose units alternate between β-1,3 and β-1,4 linkages.

Heteropolysaccharides are identified as the capsular material of species of *Cryptococcus* . Since the antigenic properties of the capsular material of the pathogenic yeast *Cryptococcus neoformans* are of medical interest, the greatest amount of attention has been focused here. Mannose, xylose, and D-glucuronic acid account for the components of the heteropolysaccharide. Minor components which have been reported include D-galactose and acetyl groups. Other heteropolysaccharides have been identified from specific yeasts. *Lipomyces lipofer* produces a heteropolysaccharide consisting of D-mannose and D-glucuronic acid in a ratio of 2:1; *Candida bogorensis* contains glucuronic acid, fucose, rhamnose, mannose, and galactose.

A different group of compounds, tetraacetyl phytosphingosine and triacetyl dihydrosphingosine, which are complex hydrophobic compounds, appear to be responsible, at least in part, for the tendency of some yeasts (in particular, *Hansenula ciferrii*) to form surface growth in liquid medium.

Proteinaceous hair-like structures termed fimbriae have been reported on the cell surface of several basidiomycetous and ascomycetous yeasts. They range from 5 to 7 nm in diameter and from approximately 10 μm in the basidiomycetous yeasts to only 0.1 μm in length in ascomycetous yeast. It has been proposed from work with strains of *Saccharomyces* that these hair-like structures may be involved in cellular flocculation.

Plasmalemma (Cytoplasmic Membrane).—This membrane is between the cell wall layers and the cytoplasm and functions in the selective transport of nutrients from the medium into the cell and conversely protects the cell from the loss of low molecular weight compounds from the cytoplasm. During the growth of the cell it also represents the structure upon which the cell wall components are deposited. Physically the plasmalemma shows numerous deep invaginations. The outer surface is composed of particles which contain mannan-protein, and these particles may be involved in the formation of the mannan-protein complex of the cell wall. The plasmalemma is a so-called unit membrane that is a structure about 8 nm thick in which 2 electron-dense layers are separated by an electron-transparent layer. It is believed that the electron-dense layers represent proteins and the middle electron-transparent layer is lipids and phospholipids. The central layer is believed by investigators to consist of nonpolar groups while the borders adjacent are believed to contain proteins involved in the entry and exit of solutes and in some enzymatic action. For example, a magnesium dependent ATPase, active at neutral pH, is located in this area and is believed to play a role in energy dependent transport of certain solubles into the cell. Some carbohydrate in the form of mannan has also been identified in the central layer.

Cytoplasmic Matrix.—The ground substance or matrix in which various yeast structures such as the nucleus, vacuoles, etc., are located also contains large quantities of polyphosphates, glycolytic enzymes, ribosomes, reserve glycogen, and, in some yeasts, the reserve sugar trehalose. Some of the polyphosphates are highly polymerized and are a reservoir of high-energy phosphate utilizable in various metabolic processes such as sugar transport, synthesis of cell wall polysaccharides, etc. Glycogen, one of the main carbohydrate storage products, is a highly polymerized glucose molecule which accumulates primarily in a cell's stationary stage of growth when nitrogen is limiting and glucose is still available. Commercial bakers' yeast may have 12% glycogen on a dry weight basis. The other storage carbohydrate, trehalose, is a nonreducing disaccharide which is present from trace amounts to as high as 16% of the dry weight, depending upon the stage and condition of growth.

It is believed that the cytoplasmic matrix generally contains most hydrolytic enzymes, the enzymes of the glycolytic cycle, and also those of the pentose cycle. It must be recognized that it is difficult to determine which enzymes are truly soluble components in the cytoplasmic matrix and which ones are inadvertently detached from a cell structure by experimental study procedures.

The Nucleus.—While there is no general agreement among researchers that yeast nuclei have condensed chromosomes as is found in more advanced forms of life, genetic studies have shown 17 independently segregating groups of centromere-associated genes in cultures of diploid *Saccharomyces* species. This number probably represents the chromosome complement of the haploid stage. There is evidence that the number of chromosomes may vary with the species of yeast since *Hansenula holstii* has been shown genetically to contain 3 chromosomes while *H. anomala* has only 2 chromosomes.

While normally not directly observable by light microscopy, the nucleus of some yeasts can be seen by phase contrast when cells are grown on a medium containing 18–20% gelatin. The nucleus is composed of a nucleolus (an optically dense, crescent-shaped part) and a second more translucent portion which contains the chromatin material, about 90% of the cell's deoxyribonucleic acid (DNA), some RNA, and some polyphosphate compounds. Preparations stained with an acid fuchsin or iron hematoxylin will show the nucleolus, whereas aceto-orcein and Giemsa solution preferentially stain the chromatin material.

When viewed by TEM, ultrathin sections and freeze-fractured preparations of nonbudding cells show the nucleus to be more or less spheroidal. It is enveloped by a pair of unit membranes having numerous circular pores approximately 85 nm in diameter. The pores appear to be filled with small granules which may represent ribosomal subunits.

The nuclear envelope remains intact during cell division of ascomycetous yeasts, where mitosis results by elongation and constriction of the nucleus which normally takes place in the neck of the bud. Both the nucleolus and the chromatin-containing portion of the nucleus divide and are incorporated into the new daughter nuclei.

In basidiomycetous yeasts, however, mitosis is not intranuclear. The chromatin-containing portion of the nucleus appears to move into the bud before cell division, while the nucleolus remains in the mother cell, and, in contrast to the ascomycetous yeasts, there is a partial breakdown of the nuclear envelope, the chromatin material divides in the bud, and the nucleolus in the mother cell appears to disintegrate. A nuclear envelope reforms around the daughter nuclei after chromatin division along with newly formed nucleoli. One of the daughter nuclei moves back to the mother cell and cell division is completed. Under optimum conditions of growth, this sequence takes place in about 1½ hr.

Vacuoles.—The vacuoles are normally spherical in appearance, more transparent to a light beam than surrounding cytoplasmic material, and vary greatly in size. In actively dividing cells, there are usually numerous small vacuoles, whereas in older cells the number is reduced, often to 1 large vacuole. These cytoplasmic structures are particularly conspicuous in the stationary phase of growth when the cells are observed by phase contrast.

Transmission electron microscopy shows the vacuole to be surrounded by a single unit membrane, and when specimens are freeze-fractured, both the inner and outer surfaces of the yeast membranes appear to be covered with particles.

A study of the vacuolar contents has disclosed the presence of a number of hydrolytic enzymes including ribonuclease, esterase, and several proteases. It has also been found that a large fraction of the free amino acid pool of the yeast cell is stored in the vacuoles, as well as substantial concentrations of polymerized orthophosphate (volutin or polymetaphosphate). Certain purines or their derivatives of low solubility cause the formation of conspicuous crystals which are easily observed in the vacuoles of some yeasts and are referred to as dancing bodies due to active Brownian movement. It is thought that the autolysis of yeast cells begins with the breakdown of vacuoles under adverse conditions and the release of enzymes which then can attack cytoplasmic substrates.

Endoplasmic Reticulum.—A double membrane system similar to the endoplasmic reticulum (ER) of higher plants and animal cells has been shown to be present in yeast by TEM examination of ultra-thin sections and freeze-etched preparations. This system appears to be in close association with the plasmalemma and in some preparations to be actually connected to the outer nuclear membrane. Polyribosomes, which are centers of protein synthesis, are believed to be associated with the endoplasmic reticulum. The ER has also been implicated in the initiation of bud formation.

Mitochondria.—The primary function of mitochondria is that of oxidative energy conversion for the cell. Mitochondria are known to contain DNA, RNA, RNA-polymerase, and a number of the respiratory enzymes participating in the TCA cycle and electron transport systems. Under conditions of fermentation or under aerobic conditions where 5–10% glucose is present, the mitochondria degenerate to so-called promitochondria. These structures have poorly developed cristae, no longer synthesize cytochromes aa_3 and b, and the cell can no longer respire. Respiration is quickly regained by placing the cells in a medium containing a nonfermentable substrate such as glycerol and by aerating the culture.

The distribution of mitochondria in the cytoplasm seems to vary since in ultra-thin sections of cells of *Saccharomyces* they are found to be concentrated close to the plasmalemma, whereas in *Rhodotorula*, a nonfermentative yeast, they are randomly distributed throughout the cytoplasm. During budding, the mitochondria elongate and divide and are distributed between mother and daughter cells. The mitochondria have an inner membrane with numerous cristae which extend into the mitochondrial bodies and a fairly regular outer membrane. The membrane system has large amounts of lipids, phospholipids, and ergosterol.

Lipid Globules.—Yeasts have globules of lipid material stainable with Sudan Black or Sudan Red. While most yeast cells contain a small amount of these lipid materials, some yeasts, particularly when grown in a medium with a limiting nitrogen supply, can accumulate up to 50% of the dry weight as lipids. *Lipomyces starkeyi, Metschnikowia pulcherrima*, and *Rhodotorula glutinis* are notable fat-producing yeasts. The fat globules are highly refractile when observed by light microscopy. Electron microscopy has not revealed a membrane structure surrounding the globules.

Sexual Reproduction

Sexual reproduction constitutes a phase of the life cycle of the yeast, that is, an alternation of the haploid condition (1n set of chromosomes) and the diploid condition (2n). The events leading to the formation of sexual spores in yeasts are of great biological significance because they enable a yeast species to exploit fully the evolutionary processes such as hybridization selection, genetic recombination, and mutation, all of which are well known in higher forms of life and which can be exploited for the benefit of mankind. Since all of the yeasts are eukaryotic, the principles of reduction division or meiosis, which the diploid nucleus undergoes, are basically similar to those in higher forms of life.

Briefly, the fusion of 2 haploid nuclei results in a diploid nucleus which, through meiosis, is again reduced to the haploid number. Although yeast chromosomes are too small to be seen under the microscope, genetic evidence and the analogy with higher eukaryotes suggest that 2 members of each chromosomal pair have homologous genetic material. During meiosis, the homologous chromosomes become tightly paired, then form a 4-strand structure or tetrad which eventually results in the formation of 4 haploid nuclei each carrying 1 chromatid of the original tetrad. It is important that one realize that the haploid nuclei formed in this manner are not necessarily identical to those of the 2 cells, or nuclei, which gave rise to the original diploid. Genetic recombination, which occurs by the random assortment of different chromosomes and breakage of chromatids during the formation of the 4-stranded tetrad, and results in exchange of part of the chromatid (recombination), has been termed crossing over and has been demonstrated in yeast by geneticists.

Spores.—Four ascospores per ascus is common; however, there are species which characteristically form 1 or 2 spores per ascus. Even when 4 spores is the usual number, asci containing 1, 2, or 3 spores may be observed. In such cases the "extra" nuclei in the ascus are not incorporated and often disintegrate. In the few yeasts normally containing more than 4 spores per ascus, the tetrad nuclei usually undergo one more mitotic division so 8 spores may be formed in each ascus as in *Schizosaccharomyces octosporus*. In the case of additional supernumerary mitoses, multispored asci characteristic of some yeasts, such as *Kluyveromyces polysporus*, will result.

The process of spore delimitation within the cell has been studied in great detail in *Saccharomyces cerevisiae*. The original nuclear membrane of the diploid nucleus remains intact during the various stages of reduction division and results in a 4-lobed structure into which the 4 sets of chromatids are separated. Around each of the lobes a double membrane (prospore or fore-spore wall) is formed. The prospore wall thickens, causing the original nuclear membrane to break up into 4 individual portions. Part of the cytoplasm of the original cell becomes incorporated into each spore by the developing spore wall but a portion of the cytoplasmic matrix remains in the ascus and is referred to as epiplasm. This process, termed sporulation by "free cell formation," distinguishes it from spore formation by the appearance of cleavage planes (partitioning) characteristic of certain other fungi. The wall of the original cell, now an ascus, may be rapidly digested by endogenous enzymes and the spores liberated. In many yeasts, however, such as *Saccharomyces cerevisiae*, the cell wall does not autodigest, and the spores are liberated only after germination, swelling, and mechanical rupture of the cell wall. The sexual spores, whether released by mechanical forces or autolysis, can germinate, thus initiating another sexual cycle.

It should be emphasized that many yeasts require special growing conditions or sporulation conditions to pass through the sexual cycle, and if the specific conditions are not met, a yeast can continue to propagate indefinitely in the vegetative form. An excellent example is a yeast isolated in 1934 as an imperfect yeast which was not induced to sporulate until the proper conditions were known in 1970, a period of 36 years under laboratory culture without evidence of sporulation. Since definitive life cycles for basidiomycetous yeast were not known until the late 1960s, most of our information is derived from those yeasts which form ascospores as a sexual means of reproduction.

Yeast Sexuality

Various yeast species, depending upon the strain, are found to exhibit homothallism, heterothallism, and parasexuality.

Homothallism.—Homothallic yeasts are those which are capable of self-fertilization, that is, a single spore or cell is capable of carrying out a complete life cycle. Such yeasts are known in both the ascomycetous and the basidiomycetous yeasts.

Heterothallism.—Four mechanisms of heterothallism are known at the present time, the simplest being the biallelic, bipolar, sexual compatibility system in which species require mating types for sexual conjugation; the determinant for sex is located at 1 locus on a chromosome. The 2 compatible sex alleles, resulting in the 2 mating types, are generally termed a and α. Those yeasts which form ascospores and which are heterothallic have this system of compatibility.

In the basidiomycetous yeasts, however, 3 compatibility systems are known. Some species have a biallelic, bipolar, compatibility system similar to that found in the ascospore forming yeasts.

The second system has the sex determinant at 1 locus but there are 3 or more alleles. This system, known as multiallelic-bipolar, results in 3 or more mating types, for example, A_1, A_2, and A_3, etc. In this system, matings which will be fertile result from all paired combinations which involve different alleles such as $A_1 \times A_2$, $A_1 \times A_3$, and $A_2 \times A_3$.

The third and last system is termed tetrapolar and involves 2 unlinked loci and 2 allelic pairs. This system has 4 mating types, A_1B_1, A_1B_2, A_2B_1, and A_2B_2. Compatibility must be satisfied at both loci before fertile matings can occur; thus the genetic expression for the diploid would be $A_1A_2B_1B_2$. Matings between other types are not capable of completing the life cycle. Growth can occur but it is usually abortive and sporulation does not occur. Examples of this are particularly noticed when mating crosses are made and only locus A or B is satisfied, in which instance the mycelium formed is unhealthy and does not form clamp connections, which are indicative of compatible paired nuclei, nor does it form teliospores from which the sexual spores are formed.

While the occurrence of heterothallism in other fungi has been known for quite some time, the occurrence in yeasts was recognized only in 1943. It actually occurs quite commonly among the various groups of yeasts and its importance cannot be overemphasized. Due in part to the classic method of isolation for obtaining pure cultures, where one selects single well-isolated colonies on an isolation medium, many species have been described as lacking a sexual phase when in fact the isolation has been of only 1 mating type. Thus, the recognition of heterothallism has aided in the proper taxonomic designation of many of these organisms when the isolate has been mixed with compatible strains of a known species and the mating results in sporulation.

Further recognition should be made of the fact that the mating reaction itself among heterothallic yeasts is extremely variable in its strength. With some yeasts the mating reaction is so powerful that it is preceded by a visible agglutination when cells of compatible mating types are mixed. This agglutination is immediately followed by zygote formation, although interestingly enough, some yeasts with a very strong agglutination reaction are relatively low in the percentage of asci. At the other extreme, there are species where the mating reaction is so weak that it is virtually impossible to detect by microscopic examination. In such yeasts, the preparation of nutritionally deficient mutants for each mating type has proved valuable in determining the sexual ability of the culture in question. When weakly reacting mating types with individual nutritional deficiencies are mixed, an occasional zygote will be formed which, when the mixed suspension is plated onto a basic minimal medium, is capable of growth, whereas unmated strains are not. By this technique, one zygote can be detected, isolated, and studied from among the unmated cells of the originally mixed mating types.

It should be noted that while sexual differentiation in yeast strains exists, there is no implication that yeasts have reproductive structures to which the terms male and female can be applied. This term is normally reserved for highly differentiated male and female fusion cells, whereas the active structures in yeast are relatively unspecialized vegetative cells, ascospores, or sporidia. However, it is common to designate these structures as gametes, particularly if they represent different mating types. Their designation as a and α A_1, A_2, A_3, or A_1B_1, A_1B_2, etc., recognizes the inability to distinguish maleness or femaleness. In the case of homothallic vegetative cells which fuse prior to sporulation, such fusion may be considered to be a case of somatic conjugation.

Parasexuality.—Studies by various investigators have shown cases where yeasts and other fungi alternate between the haploid and diploid phases without the formation of sexual spores. This process, observed to occur in *Aspergillus nidulans* by Pontecorvo in 1954, is known as parasexuality. Subsequent to that time, similar observations have been made in other fungi, including some yeast species. This alternation of haploid and diploid phases is of particular interest since it is functional not only in fungi without known sexual stages, but also in yeasts where a sexual stage has been well established. There could be a distinct advantage to a yeast in which both haploid and diploid cells coexist. In the case of diploid cells where genes are paired, the presence of a recessive gene arising by mutation can spread, although its effect would be masked by the wild type, dominant gene. In this manner, one would have a stable phenotype along with genetic variation. In the case of a haploid cell, however, with only a single set of chromosomes, the mutation would not be masked by a dominant gene and would express itself rapidly. Thus, with both haploids and diploids coexisting, being genetically separated and being able to reproduce indefinitely in a vegetative phase, the yeast can have the advantages of haploidy or diploidy simultaneously.

Yeast Life Cycles

In discussing the life cycles of yeasts, those forming asci will be discussed separately from those forming teliospores or basidia.

Ascosporogenous Yeasts.—Ascosporogenous yeasts can be broadly subdivided into 2 classes, those in which the vegetative phase exists in the haploid condition and those in which the diplophase predominates. In the haploid class, such as the subgenus *Zygosaccharomyces*, the nucleus of the vegetative cell contains only a single set of chromosomes (1n). Thus, there must be a fusion of 2 nuclei to establish the diploid (2n) condition before reduction division can begin and subsequent sporulation can take place. In those yeasts which exist vegetatively primarily as diploid cells, such as *Saccharomyces cerevisiae*, the ascus can develop directly from the vegetative cell by reduction division of the diploid nucleus followed by spore formation. This general grouping of yeasts as haploid or diploid species is not a strict classification as there are certain yeasts in which haploid and diploid cells may occur side by side in the same culture. Thus, upon examination of a sporulating culture we may find asci which have arisen from a diploid cell and other asci which have obviously arisen from the conjugation of 2 haploid cells just prior to sporulation.

Haploid Vegetative Phase.—In ascosporogenous yeasts which characteristically exist as haploids in the vegetative phase, the diploid generation is usually limited to the zygote formed after fusion of 2 haploid cells and their nuclei. These yeasts normally have 1 of 3 types of life cycles.

Conjugation Tube.—In the first type of life cycle when conditions are favorable for sexual reproduction, the vegetative cells will form more or less distinct conjugation tubes. The tips of 2 conjugation tubes from 2 cells then grow together and plasmogamy (joining of the protoplasts) takes place. The nuclei of the 2 cells then approach each other and karyogamy (nuclear fusion) occurs, thus forming a diploid nucleus. In most instances, meiosis or reduction division immediately follows karyogamy and the resultant 4 nuclei produce ascospores in a dumbbell-shaped ascus (the original zygote). Spore distribution between the 2 portions of the dumbbell is random so that asci containing 2 spores in each half, 3 to 1 or even 4 to 0, can be observed.

Meiosis Bud.—The second type of life cycle exhibited by some haploid yeasts varies from the first in that the haploid vegetative cell produces a bud which, instead of separating in a normal fashion, stays attached to the mother cell, retaining a rather large connective opening. Nuclear division occurs (mitosis) and the 2 daughter nuclei move into the bud structure where karyogamy occurs. Meiosis also occurs in the bud resulting in 2 or 4 haploid nuclei. Because meiosis does occur in this bud-like structure, it has been termed the "meiosis bud." The haploid nuclei move back into the mother cell and ascospore formation takes place there. Haploid yeasts exhibiting this life cycle usually form only 1 or 2 spores per ascus and the extra nuclei presumably degenerate. While no septum has been observed to be formed across the bud opening of cells in the genus *Schwanniomyces*, electron microscopy studies have revealed that in a species of *Debaryomyces* the bud is first separated from the mother cell by a septum which then dissolves, allowing the 2 cell nuclei to fuse and the life cycle to be completed. From morphological similarities among the asci of these genera, it may be assumed that one or the other of these methods is functional in other genera where species form asci from a mother cell-daughter cell pairing.

Fusion Cell.—The third type of life cycle exhibited by haploid yeasts also involves the fusion of 2 "gametes," but the fusion cell does not become the ascus. The 2 nuclei move to a specialized structure which becomes the ascus. In the young ascus the nuclei undergo fusion, reduction division, and subsequent ascospore formation. Examples of this are found in species of *Eremascus* and *Nadsonia.*

Diploid Vegetative Phase.—Ascosporogenous yeasts which spend their vegetative days in the diploid condition produce their spores within the vegetative cells, which become the asci. For these yeasts there are 4 means known by which the diploid condition in the vegetative cell can arise from the haploid ascospore. These are:

(1) Two ascospores fused within the ascus before spores are dehisced. Fusion of the ascospores is followed by karyogamy of the 2 haploid, spore nuclei. Thus the first vegetative cell resulting from the conjugated spores is already a diploid cell. Such a life cycle occurs in *Saccharomycodes ludwigii* and also is found in certain strains of *Saccharomyces cerevisiae.*

(2) The ascospores germinate by budding off relatively small sized, haploid, vegetative cells. This continues for a very limited number of generations with fusion then occurring between pairs of spore-bud-cells. This results in diploid, larger-sized cells which then continue to propagate as the vegetative phase.

(3) This life cycle is a variation of number (2) where only 1 of the spores germinates, producing a haploid cell which then fuses with a yet ungerminated spore from the same ascus. Again, this results in a diploid giving rise to the vegetative phase. This means of diploidization is also found in certain strains of *Saccharomyces cerevisiae.*

(4) In some yeasts known to exist vegetatively in a diploid phase, no conjugation of ascospores or ascospore bud-cells has been found. It has, however, been shown that the haploid nucleus of an individual ascospore may divide mitotically so that the germinating spore has 2 haploid daughter nuclei which then fuse prior to the formation of the first bud. This diploid nucleus then divides mitotically so that the first bud coming from the single germinating ascospore already constitutes a diploid cell. This particular means of diploidization is obviously limited to homothallic yeasts. Species of *Hanseniaspora* reproduce in this fashion. There is also evidence that strains of *Saccharomyces chevalieri* also have this type of life cycle.

Basidiomycetous Yeasts.—Life cycles of the basidiomycetous yeasts also show variations but, in general, after fusion of 2 haploid (1n) yeast cells, there is a stage not found in the ascomycetous yeasts. This is the development of a dikaryotic condition (1n plus 1n) where the 2 nuclei after plasmogamy do not fuse in the zygote but rather are incorporated in pairs into the cells of a mycelium by a clamp connection. The clamp connection mechanism assures that 2 compatible nuclei are duplicated and propagated in subsequent cells of the mycelium. At the time of sporulation either thick-walled teliospores or thin-walled basidia are formed on the mycelium. After karyogamy and reduction division, sporidia are formed externally on the basidium or from a teliospore. The teliospore germinates into a tube-like promycelium upon which the sporidia are formed. The promycelium may be septate, with 4 cells being common, or aseptate. The sexual spores (sporidia) are not forceably discharged from the promycelium or basidia of these yeasts. They are formed by budding and further vegetative reproduction is also by this means.

In some teliospore-forming species, a self-sporulating, diploid phase is also known. It is believed that this phase is the result when a teliospore, after karyogamy has occurred, may fail to undergo reduction division so that the germinating buds are diploid. These diploid cells may reproduce asexually by budding or may develop into a uninucleate diploid mycelium without clamp connections. Teliospores formed on this type of mycelium may undergo reduction division so that the sporidia thus formed would partake in the usual haploid type of life cycle.

It has also been reported that diploid cells can give rise to a dikaryotic mycelium when the diploid nucleus undergoes a "somatic reduction," thus creating a dikaryon. Since many of the basidiomycetous yeasts have a tetrapolar compatibility system, fertile matings require the satisfaction of compatibility at 2 loci, whereas when this condition is met at only 1 of the 2 loci, the mycelium formed is unhealthy and does not form clamp connections or teliospores.

Sporulation

The determination of a yeast isolate's ability to form asco- or basidio-spores is of prime importance in determining its identification. This determination can be complicated by the fact that some yeasts can grow in a vegetative state for an indefinite number of generations and that very specific conditions must be met before the sexual cycle can be induced. Further, in heterothallic species, fertile matings and successful completion of the sexual cycle depend upon the mixing of compatible strains. Early workers were of the opinion that yeasts would sporulate only when conditions for growth became unfavorable. Today it is known that this is not necessarily true. Many yeasts freshly isolated from nature often sporulate very heavily on the relatively rich media commonly used for isolation. Further, yeasts cultured in a laboratory are propagated on much richer media than would normally be available to them in their natural habitat. However, many so-called domestic cultivated yeasts, for example, bakers' and brewers' yeasts, as well as many species commonly isolated from spoiled beverages and other food products, often sporulate poorly or not at all on media rich in nutrients. Some yeast species have a tendency to lose their ability to form sexual spores while others are not affected when they are held in laboratory culture for a period of years.

Innumerable media and various techniques have been used over the years for the purpose of inducing ascospore formation or to increase the percentage of sporulating cells. Yeast ascospores are somewhat more resistant to adverse conditions such as freezing, drying, and exposure to high temperatures and to harmful chemicals than are the vegetative cells. The heat resistance of the ascospore is only a few degrees (6°–12°C) greater than the vegetative cells under the same conditions. Thus, by heating a cell suspension for a short time at mildly elevated temperatures, such as 55°–60°C, vegetative cells can be killed but spores that might be present would still be viable.

Experience has shown that success in obtaining sporulation is best when the culture is well nourished and relatively young. In species of *Saccharomyces* it was found that young daughter cells which have not produced buds sporulate either very poorly or not at all. However, after producing one or more buds, sporulation occurs normally.

Fermentative (anaerobic) conditions do not promote sporulation. In *Saccharomyces cerevisiae* sporulation is enhanced when the vegetative cells are adapted for aerobic growth. However, it has been found that with some species of *Hanseniaspora* which produce 1 to 2 spheroidal spores, a reduced oxygen tension stimulates spore formation. When part of the vegetative cell inoculum is covered by a sterile cover slip, large numbers of asci may develop under the cover slip near the outer edge, but are infrequently found in the uncovered portion of the growth or far under the cover slip.

In the laboratory most yeasts will sporulate at room temperature (18°–25°C) although reduced temperatures (12°–15°C) are beneficial to some species of *Nadsonia* and *Metschnikowia*.

As one might expect from the diversity of the many yeast species, there is no one medium capable of inducing sporulation for all of the yeasts. Most media are adjusted so that the pH is in the neutral to slightly acid range (pH 5.8–7.0). Certain media have been found to be better for some species than others. For example, acetate agar is commonly used for achieving sporulation with species of *Saccharomyces*, although potato extract glucose agar (PDA) works better for some haploid species of this genus. Gorodkowa agar works well for most species of *Debaryomyces*; YM agar (yeast extract, malt extract maltose agar) induces sporulation of many species of *Pichia* and *Hansenula* and other yeasts. A medium based on vegetable juices (V-8), with or without added bakers' yeast, is also an excellent sporulation medium for species of a number of genera. In certain special cases a medium composed of the substrate from which the yeast was isolated may be used.

Since heterothallic yeasts require compatible mating types, cultures in which sporulation can not be induced and which would be classified in the so-called imperfect genera should be mixed with similar cultures in various combinations, or, when the mating types for a species are known, mixed with the same mating types and reobserved for sporulation. Many yeast isolates formerly believed to be asporogenous are properly classified into their perfect genera by these techniques.

Yeasts being observed for ability to sporulate should be checked frequently over a period of several weeks because the time required to sporulate as well as the number of asci formed varies a great deal among the yeast species and even within strains of the same species. Some can produce spores in 1-2 days, whereas other yeasts may not form asci for a week or two or longer. Frequent examination is necessary because with many yeasts the spores formed can also germinate on the same medium, and if observation were delayed or infrequent, one might not observe that sporulation had taken place.

Thus, although the ability or inability to produce sexual spores is not always easy to determine, the type of sexual spore formed is the primary criterion used by the taxonomist to place a yeast into 1 of 3 subdivisions of fungi. The spores formed exhibit a wide diversity in number per ascus, shape, size, surface markings, and color. Generally, most of these features are quite constant for a given species. For some genera, all species have a similar morphology, although in other genera, spore morphology differs among the species. Sporulation among the yeasts also has made possible genetic studies. However, from the standpoint of the yeast itself, it is the most important biological event in its life cycle and it governs to a large extent the evolutionary development of the various groups of the yeasts. Therefore, in spite of the time spent and difficulties encountered in determining a yeast's ability to sporulate, the importance is justified.

FERMENTATION

Man has used the ability of certain yeasts to carry out an alcoholic fermentation long before the responsible agent was recognized as a living thing. The everyday use of alcoholic beverages and panary products has also made large quantities of yeasts available, facilitating the work of the investigators elucidating the basic metabolic processes of living cells. The "cell free extract" of the Büchners in 1897 provided the means for studying alcoholic fermentation in the absence of living cells and thus initiated the many studies which elucidated the intermediate steps that occur in the transformation of glucose to ethanol and carbon dioxide. It must be kept in mind that "fermentative" yeasts also possess the ability to oxidatively utilize glucose as well as many other compounds which can not be fermented at all. The ability to carry out an alcoholic fermentation varies in vigor from species to species and in fact from strain to strain. Actually, there are many yeasts which are totally incapable of carrying out an alcoholic fermentation and thus are unable to derive energy or grow under anaerobic conditions. In fact, fermentation being "la vie sans air" (life without air) is not quite accurate. *Saccharomyces cerevisiae* under strictly anaerobic conditions will cease its fermentation even though a fermentable substrate is still present, thus causing a "stuck" fermentation. Apparently the presence of a small amount of oxygen is required for synthesis of certain vital compounds since supplementing a medium with ergosterol and oleic acid or with oleanoic acid will enable the fermentation to go to completion even if oxygen is absent.

Three basic rules regarding yeasts' capabilities to ferment were formulated many years ago by Kluyver. The first rule is that if a yeast is unable to ferment D-glucose, it can not ferment any other sugar. The second rule states that if a yeast can ferment D-glucose, then D-fructose and D-mannose will also be fermented. The third rule is that if a yeast can ferment maltose, it cannot ferment lactose, and vice versa. Exceptions are known for the third rule in that there are a few yeasts capable of fermenting both maltose and lactose. *Brettanomyces claussenii* is an example of a yeast able to ferment both disaccharides.

Respiration

Aerobic metabolism is affected by the level of oxygen present and also concentration of the sugars. For example, both with *Saccharomyces cerevisiae* and with *Candida utilis* the Embden-Meyerhof pathway accounts for approximately 90% of the glycolytic metabolism. Under aerobic conditions, however, the hexose monophosphate pathway is responsible for 6 to 30% glycolysis in *Saccharomyces cerevisiae* and for 3 to 50% in strains of *Candida utilis*. In contrast, species of *Rhodotorula* which are unable to ferment metabolize 60 to 80% of the glucose taken from the medium through the hexose monophosphate pathway.

The effect of substrate concentration can be illustrated for *S. cerevisiae* as follows. In media with glucose concentrations in excess of 5%, there is a total inhibition of synthesis of respiratory enzymes so that the yeast continues to ferment even though the medium is aerated. This phenomenon is known as the "glucose effect" or "Crabtree effect." Not all yeasts exhibit the Crabtree effect, e.g., species of *Kluyveromyces* and some haploid *Saccharomyces*. In contrast, a yeast (such as *Saccharomyces cerevisiae*) growing in low concentrations of glucose (0.1%) can shift from fermentation to respiration upon aeration of the medium. The decrease in fermentative ability upon aeration was first demonstrated by Pasteur (Pasteur effect) and industrially utilized by producers of bakers' yeast and of feed yeasts. As the level of the sugar decreases in an aerated growth medium, there is a corresponding increase in the activity of the enzymes of the tricarboxylic acid cycle (TCA) and increased activity of enzymes involved in the glyoxylate cycle and in the electron-transport system. Since the enzymes for the TCA and glyoxylate cycles are located in the mitochondrial fraction of the yeast, the observation that the presence of sugar inhibits mitochondrial formation is not unexpected.

The TCA cycle starts with pyruvate's being oxidized to acetyl coenzyme A instead of decarboxylation to acetaldehyde, which occurs during alcoholic fermentation. The β-oxidation of fatty acids also results in acetyl CoA. Acetyl CoA then condenses with oxaloacetate, giving rise to citrate. Subsequent reactions of the TCA cycle release 2 molecules of CO_2 per cycle and generate ATP by oxidative phosphorylation. Several of the intermediate compounds also may be used in the synthesis of amino acids and other cellular constituents.

The glyoxylate cycle or bypass has been more recently elucidated and was found to function for the replenishment of TCA cycle intermediates—especially L-malate and succinate removed from the cycle due to the synthesis of cellular constituents. The glyoxylate cycle also serves as the mechanism by which yeasts can grow on 2-carbon molecules such as ethanol and acetate. A biotin-dependent reaction producing oxaloacetate from pyruvate and CO_2 serves to supply additional oxaloacetate for incorporation into the TCA cycle.

The hexose monophosphate shunt pathway, also known as the pentose cycle, starts with glucose-6-phosphate which is oxidized to 6-phosphogluconate. This reaction is dependent upon the reduction of NADP to $NADPH_2$. Subsequently, 1 molecule of CO_2 is released and an additional molecule of $NADPH_2$ is formed. Aerobically, the cycle can result in the complete oxidation of glucose to $6CO_2$ and the formation of 12 molecules of $NADPH_2$. While the hexose monophosphate shunt does not generate high energy phosphate in the form of ATP, a portion of the $NADPH_2$ formed is oxidized in other metabolic pathways and in this way does contribute to ATP formation. The $NADPH_2$ is utilized in large part in synthetic reactions requiring reductions, such as the formation of lipids. The hexose monophosphate pathway also serves as a mechanism by which yeasts can respire pentoses and methyl pentose sugars if the yeasts have the proper pentose kinases and other enzymes that will convert the substrates into intermediates of the pentose cycle.

GROWTH REQUIREMENTS

Carbohydrates

In order to grow, yeasts require oxygen, proper temperature and pH, utilizable organic carbon and nitrogen sources, and various minerals; and some require vitamins and other growth factors. All yeasts are able to utilize D-glucose, D-fructose, and D-mannose, although a particular yeast may utilize other carbon sources more efficiently. Besides these 3 hexoses, another hexose, D-galactose, may be fermented, but this requires enzymatic adaptation by yeasts possessing the potential to ferment the sugar. Other monosaccharides and all L-sugars are unfermentable although a particular yeast may be able to utilize them by respiration.

In the fermentation of di-, tri-, or polysaccharides, the fermentation always goes through the hexose stage after enzymatic hydrolysis either at the cell surface or internally, depending upon the location of the specific enzymes. Consequently, if the particular hexose sugar is not fermented by the particular yeast, the di- or oligosaccharides are not fermented either. Enzymes for the hydrolysis of sucrose, raffinose, melibiose, starch, and inulin apparently are situated outside of the cell membrane (plasmalemma) although their precise location has still to be determined. Enzymes for the hydrolysis of other sugars such as maltose, lactose, cellobiose, and melezitose are located internally; the sugars are transported by specific permeases across the membrane and hydrolyzed within the cell. Fermentation of polysaccharides such as starch and inulin (a fructose-containing polysaccharide) is possible by relatively few yeasts, and then generally at reduced rates of fermentation. The ability to ferment pentose sugars or methyl pentoses is not found in any yeast, although it is not unusual for these compounds to be utilized by respiration. Thus, among the yeasts which are able to ferment, glucose (fructose, mannose), galactose, sucrose, maltose, raffinose, melibiose, lactose, and trehalose are the most commonly utilized sugars.

In brief, the breakdown of a sugar and its transformation to ethanol and carbon dioxide proceed in the following manner: D-glucose, D-fructose, and D-mannose are all transported across the cell membrane by a common transport system termed "facilitated diffusion." In addition, these 3 hexoses are all phosphorylated by the same enzyme, yeast hexokinase, to the corresponding hexose-6-phosphates. Both the rate of transportation and the rate of phosphorylation of these sugars vary depending upon the sugar and also on the species, in fact, even on the strain of yeast. Glucose-6- and mannose-6-phosphate are subsequently transformed to fructose-6-phosphate. The other fermentable hexose, D-galactose, when utilized by a yeast, is transported by a separate, induced transport system, and once in the cell it is phosphorylated by a specific enzyme and transformed by 3 additional enzymes before it can enter the glycolytic cycle.

Hydrolysis to hexose components is obviously required before di-, tri-, and oligosaccharides may be fermented. After the hexoses are phosphorylated to the hexose-6-phosphate, a reaction requiring adenosine triphosphate (ATP) and other enzymes converts fructose-6-phosphate into fructose-1,6-diphosphate using another molecule of ATP. The fructose-1,6-diphosphate is converted to pyruvic acid via the Embden-Meyerhof pathway of glycolysis. The net results are 2 molecules of pyruvate and the generation of 4 molecules of ATP, a net gain of 2 molecules of ATP. The 2 pyruvate molecules are subsequently decarboxylated yielding 2 molecules of CO_2 and 2 molecules of acetaldehyde. Finally, the 2 molecules of acetaldehyde are reduced to ethanol by alcohol dehydrogenase and the reduced form of the coenzyme nicotinic acid adenine dinucleotide ($NADH_2$), which had been formed during one of the intermediate steps of the Embden-Meyerhof pathway. The 2 molecules of ATP formed during the transformation of 1 molecule of hexose to 2 molecules of CO_2 and 2 of ethanol are used as a supply of energy required for cell growth and for synthesis of storage reserve products (glycogen and trehalose).

It should be noted that under conditions of no growth, for example, in a medium containing glucose but no nitrogen source, the yeast cells will still convert about 70% of the glucose to ethanol and CO_2 with the remainder going to reserve carbohydrates. When the glucose of the medium is depleted, it is felt that these reserve carbohydrates furnish the energy necessary for maintenance of the cell in replacement of the proteins and ribonucleic acids which are being constantly "recycled."

The concentration of ethanol produced depends not only on the conditions of fermentation, but perhaps even moreso upon the species and particular strain of a species. Suitable strains of Saccharomyces cerevisiae, for example, strains of wine yeast or distillers' yeast, can attain an ethanol concentration of 12 to 14% by volume with relative ease if a sufficient supply of fermentable sugar is supplied. As the ethanol concentration increases above this level, the rate of fermentation is greatly reduced and will eventually cease. Ethanol levels of 18 to 20% by volume have been achieved with select strains and specific fermentation conditions. In contrast, species of *Hanseniaspora* (and the asporogenous forms in the genus *Kloeckera*), which frequently are predominant yeasts in the early stages of natural fruit fermentations, are relatively sensitive to the presence of alcohol (about 3.5 to 6.0% alcohol).

Of industrial interest is the ability of some yeasts to ferment L-malic acid to ethanol and carbon dioxide. Some wine yeasts (strains of *Saccharomyces cerevisiae*) have this ability, but species of *Schizosaccharomyces* (*S. pombe* and *S. malidevorans*) are more efficient. This property is especially of interest in the fermentation of grape and fruit musts of low sugar content and high levels of malic acid.

Other Organic Carbon Sources

Yeasts can respire all sugars that they are capable of fermenting and in addition may respire a wide array of other organic compounds. In addition to the previously mentioned pentoses and methyl pentose sugars, compounds which yeasts can respire include organic acids, sugar alcohols, methanol, and ethanol, as well as some aromatic compounds and hydrocarbons. Only a limited number of yeasts can utilize the last 2 classes of compounds and then only a few of these. Hydrocarbons used are primarily limited to the n-alkanes, particularly those with 12 to 18 carbon skeletons. *Candida lipolytica, C. tropicalis,* and *C. maltosa* all grow well on these n-alkanes. A number of species of *Schwanniomyces, Metschnikowia, Debaryomyces,* and *Pichia* also can grow on these hydrocarbons. *Candida tropicalis* can grow on a few aromatic compounds as well, and strains of *Trichosporon cutaneum* can metabolize many more. The use of methanol as a potential substrate for the production of single cell protein has received the attention of a number of investigators. Some species which grow well on methanol are *Hansenula polymorpha, Pichia pastoris, Candida boidinii,* and a recently described *Torulopsis sonorensis,* which was found in necrotic cacti.

Nitrogen

Inorganic Sources.—The ability of yeasts to assimilate various compounds as a source of nitrogen also varies greatly among the yeasts. Among the inorganic sources of nitrogen, ammonium sulfate is utilizable by virtually all yeasts. In fact, most ammonium salts can supply nitrogen for the growth of yeasts, although some salts are more suitable than others. The ability to use nitrate is much more restricted and is, in fact, a criterion used in yeast classification. *Candida utilis* is an example of an industrially important yeast which utilizes nitrate. Bakers' and brewers' yeast (*Saccharomyces cerevisiae* strains) are unable to grow on nitrate. Nitrite is utilized by all of the yeasts which can use nitrate, although the toxicity of the nitrite ion varies among these yeasts, and caution must be used to assure that too high a concentration is not used. All species of the genera *Hansenula* and *Citeromyces* can utilize nitrate and nitrite, as can some species in genera such as *Rhodotorula, Torulopsis,* and *Candida.* Yeasts are also known that have the ability to use nitrite but not nitrate. Some species of *Debaryomyces* have this ability and are often isolated from cured meats and similar products containing this compound. Some investigators have published reports which claim that at least some *Rhodotorula* species can assimilate nitrogen from the air. However, this claim has been denied by others who have used an atmosphere containing radioactive nitrogen isotopes and who could not find any fixation of molecular nitrogen by the yeasts.

Amino Acids.—Amino acids also serve as a source of nitrogen for yeasts. While it was previously thought that yeasts could incorporate assimilated amino acids directly into proteins, it has more recently been shown that the utilization of a particular amino acid depends upon the ability of the particular yeast to deaminate the amino acid and to incorporate that nitrogen into other constituents of the cell. Glutamic and aspartic acids and their amines are easily deaminated or transaminated by most yeasts and serve as good sources of nitrogen. In general, yeasts take up only the L-form of the amino acid, although certain D-isomers can be assimilated. In general, ammonium sulfate is a better source of nitrogen than any one single amino acid. In certain yeasts the assimilation of a particular amino acid may be either blocked or stimulated by the presence of ammonium sulfate in the medium. Where mixtures of amino acids are present in addition to ammonium sulfate, the uptake of amino acids from the mixture is faster than the uptake of ammonium nitrogen. Studies have shown that di- and polypeptides can be assimilated by yeasts. In general, the growth of yeasts on peptides is inferior to that on amino acids or ammonium salts.

Other Organic Sources.—Of other compounds which can act as a source of nitrogen, urea has been found to be utilized by virtually all yeasts and is as good a source as is ammonium sulfate. It was found, however, for a yeast to grow well, the medium must also contain ample biotin. The ability to utilize purine and pyrimidine bases varies greatly among the yeasts, as does the growth response to the various bases. Bakers' and brewers' yeasts in general utilize very few whereas *Candida utilis* can assimilate a greater number of these nitrogen-containing compounds.

It has been shown for a number of yeasts that, during fermentation and growth, the yeasts actually excrete nitrogenous compounds into the medium. In many instances, these nitrogen-containing compounds are reabsorbed by the cell. Excreted compounds include amino acids, oligopeptides and nucleotides, components also to be found in the intracellular pool of nitrogenous compounds.

Vitamins

The requirement of yeasts for an exogenous source of vitamins varies widely, as some yeasts can synthesize all of their required vitamins, whereas other yeasts have multiple requirements. In recognition of requirement variability only the ability or inability to grow in a medium lacking vitamins is presently used in classification. With the exception of meso-inositol, the vitamins serve catalytic functions, normally as part of a particular coenzyme. Meso-inositol serves a structural function in membrane synthesis where it is incorporated into the phospholipids. The requirement for a particular vitamin in the medium is often qualified as to whether it is an absolute requirement, in which case the yeasts can not grow if the vitamin is not regularly supplied, regardless of time or condition of growth, or whether it is a relative requirement, meaning that the yeasts can grow very slowly by synthesizing the vitamin at a very reduced rate, but that the yeasts will grow much more vigorously if the vitamin is supplied in the medium. Yeasts having absolute requirements for a particular growth factor find use as a biological assay tool.

Biotin is the most commonly required vitamin to be supplemented in the medium whereas riboflavin and folic acid are apparently synthesized in sufficient quantities by all yeasts. Vitamin B_{12} is not known to be required or even synthesized. In some yeasts there is an interconversion between pyridoxine and thiamin so that a requirement for these can not be determined in the presence of either vitamin in an assay medium. In synthetic media the inclusion of 9 vitamins—biotin, pantothenic acid, folic acid, niacin, para-aminobenzoic acid, inositol, thiamin, riboflavin, and pyridoxine—is considered a complete supplementation. Adding high levels of vitamins, or in some cases their precursors, can "enrich" yeasts by taking advantage of their ability to concentrate vitamins, particularly the B complex, from a medium into the yeast cell.

Minerals

Analysis of the minerals found in yeast cells show about 50 elements to be present, most in extremely minute amounts. Trace and small amounts of elements such as iron, copper, zinc, cobalt, calcium, and magnesium are added to the medium as they are required by yeasts as enzyme activators, structural stabilizers, and components of proteins, pigments, etc. Phosphorus is normally supplied as dihydrophosphate; potassium facilitates its uptake. Monohydrophosphate is not taken up by the cell. Sulfur requirements are met by inorganic sulfates, and to some degree by other sulfur-containing compounds. Some yeasts can utilize the amino acid, methionine, or the tripeptide, glutathione, as sources of sulfur but most yeasts can not use cystine or cysteine. Sulfite can be used, but a number of yeasts are sensitive to bisulfite and to sulfurous acid. This sensitivity is used industrially for inhibiting "wild yeasts" in certain fermentations, notably wine.

pH and Temperature

Besides adequately supplying a yeast's nutritional needs, attention to proper pH values and temperature of incubation are essential for desired activities and growth of the yeast. Actually most yeasts are relatively tolerant of a wide range of pH values; most will grow at pH 2.8−3.0 if hydrochloric acid is used to adjust the pH of the medium. Lactic and acetic acids at low pH values are often inhibitory due to the high concentrations of the undissociated acid form which can pass through the cell membrane into the cytoplasm. Most yeasts will also grow well at neutrality and in slightly alkaline conditions (pH 8−8.5). Optimum pH for growth varies from 4.5 to 6.5 for most species. Temperature ranges for growth vary from relatively narrow, particularly for a few yeasts found associated with warm-blooded animals, to relatively broad ranges. Many yeasts can grow at 0°C or slightly below but the rate of growth is extremely slow; some have a maximum of 18°−20°C, whereas others can grow at 46°−47°C. Most species encountered grow from about 5° to 30°−37°C with an optimum approximating 25°C.

While optimum conditions are known for many yeasts, manipulation of nutrient concentrations, pH, temperature, and oxygen tension are used in industrial fermentations to best meet the desired purposes as economically as possible. Industrial fermenters further improve their processes by careful selection of yeast strains. Strain improvement by genetic selection or mutation often results in strains performing many-fold better than the original wild type.

The following general references to the technology of yeasts provide additional information: White (1954); Ingram (1955); Roman (1957); Cook (1958); Prescott and Dunn (1959); Reiff et al. (1960, 1962); Rainbow and Rose (1963); Peppler (1967); Rose and Harrison (1969, 1970, 1971); Lechevalier and Pramer (1971); Reed and Peppler (1973); Miller and Litsky (1976); Phaff et al. (1978).

REFERENCES

BARNETT, J.A. and PANKHURST, R.J.   1974.   A New Key to the Yeasts. North-Holland Publishing Co., Amsterdam-London.

COOK, A.H.   1958.   The Chemistry and Biology of Yeasts. Academic Press, New York.

DIDDENS, H.A. and LODDER, J.   1942.   The Anascosporogenous Yeasts, Part 2, 2nd Half. The Yeast Collection of the Centraalbureau voor Schimmelcultures.   North-Holland Publishing Co., Amsterdam. (Dutch)

GUILLIERMOND, A.   1920.   The Yeasts.   John Wiley & Sons, New York.

GUILLIERMOND, A.   1928.   Dichotomic Key for the Identification of Yeasts. Librairie le François, Paris.

INGRAM, M.   1955.   An Introduction to the Biology of Yeasts.   Sir Isaac Pitman & Sons, London.

LECHEVALIER, H.A. and PRAMER, D.   1971.   The Microbes.   J.P. Lippincott Co., Philadelphia.

LODDER, J. 1934. The Anascosporogenous Yeasts, Part 2, 1st Half. The Yeast Collection of the Centraalbureau voor Schimmelcultures. Royal Acad. Sci., Amsterdam, Afdeel. Natuurkd., 2nd Section, *32*, 1–256. (Dutch)

LODDER, J. 1970. The Yeasts, a Taxonomic Study, 2nd Edition. North-Holland Publishing Co., Amsterdam.

LODDER, J. and KREGER-VAN RIJ, N.J.W. 1952. The Yeasts, a Taxonomic Study. North-Holland Publishing Co., Amsterdam.

MILLER, B.M. and LITSKY, W. 1976. Industrial Microbiology. McGraw-Hill Book Co., New York.

PEPPLER, H.J. 1967. Microbial Technology. Reinhold Publishing Co., New York.

PHAFF, H.J., MILLER, M.W. and MRAK, E..M. 1978. The Life of Yeasts, 2nd Edition. Harvard Univ. Press, Cambridge, Mass.

PRESCOTT, C.S. and DUNN, C.G. 1959. Industrial Microbiology, 3rd Edition. McGraw-Hill Book Co., New York.

RAINBOW, C. and ROSE, A.H. 1963. Biochemistry of Industrial Microorganisms. Academic Press, New York.

REED, G. and PEPPLER, H.J. 1973. Yeast Technology. AVI Publishing Co., Westport, Conn.

REIFF, F. *et al.* 1960. Yeasts. I. Yeast Science. Verlag Hans Carl, Nurnberg, W. Germany.

REIFF, F. *et al.* 1962. Yeasts. II. Yeast Technology.Verlag Hans Carl, Nurnberg, W. Germany.

ROMAN, W. 1957. Yeasts. Dr. W. Junk, Publishers, The Hague, Netherlands.

ROSE, A.H. and HARRISON, J.S. 1969. The Yeasts. I. Biology of Yeasts. Academic Press, New York.

ROSE, A.H. and HARRISON, J.S. 1971. The Yeasts. II. Physiology and Biochemistry of Yeasts. Academic Press, New York.

ROSE, A.H. and HARRISON, J.S. 1970. The Yeasts. III. Yeast Technology. Academic Press, New York.

STELLING-DEKKER, N.M. 1931. The Sporogenous Yeasts. The Yeast Collection of the Centraalbureau voor Schimmelcultures. Drukkerij Holland, Amsterdam. (Dutch)

WHITE, J. 1954. Yeast Technology. Chapman Hall, London.

WICKERHAM, L.J. 1951. Taxonomy of yeasts. U.S. Dep. Agric. Tech. Bull. *1029.*

3

Pure Culture Methods

David K. Dalby

Microorganisms for the production of an economically important product are useful only if they can be maintained indefinitely in a pure and genetically stable form. This is also true for organisms used for biological assays and those which are investigated for their potential use in production. Responsibility for the care of these organisms should be in the hands of well trained and experienced individuals. Entrusting the care of cultures to engineers, chemists, etc., whose microbiological abilities are by training shallow, should be avoided whenever possible.

Industrial culture collections generally consist of two types of cultures: Those which are used frequently and those which are kept in reserve. The former are generally called "working stock cultures" and the latter "primary stock cultures." Working stocks are maintained on agar slants, in broth, or by any method which allows for quick access to active cells. Methods for maintaining working stocks are elementary, and detailed information can be found in most microbiology laboratory books and handbooks. Working stock cultures should be checked periodically for purity and their ability to carry out the desired process.

Primary stocks are cultures which are maintained in a state of low physiological activity and are used to replace working stocks whenever necessary. The methods used to maintain primary stocks should specify a minimum number of periodic transfers to reduce the chance of contamination or genetic change; and generally two separate methods should be utilized to ensure the safe maintenance of the primary stock cultures.

OBTAINING CULTURES

Culture Collections

Cultures and Patents.—The simplest method of obtaining cultures of microorganisms is to acquire subcultures from cultures previously isolated. Throughout the world there exist numerous culture collections ranging from very small private collections to large public collections which may maintain several thousand strains of microorganisms. Private collections are usually associated with specialized research groups in industries, universities, or institutes and are maintained solely for the use of these groups. Cultures in this type of collection are not usually distributed generally; however, specific requests from interested scientists are often honored.

The need to preserve and maintain microorganisms for industry, government, teaching, etc., has led to the organization of several large public culture collections. Well known collections of this type include the Commonwealth Mycological Institute (CMI), Agricultural Research Service Culture Collections (NRRL), American Type Culture Collection (ATCC), and the Centraalbureau voor Schimmelculturen (CBC). From an industrial point of view, the NRRL and ATCC deserve special consideration. These two organizations have been recognized by the United States Patent Office as official depositories of cultures, both domestic and foreign, which are an essential part of a patented process.

The American Type Culture Collection is a non-profit organization which preserves and distributes cultures of microorganisms and animal cells to all areas of scientific investigation and instruction. In addition to culture preservation, research is carried out in the areas of comparative microbiology, microbial systematics, and improved methods of characterization and preservation of cultures. This organization is supported by fees for cultures and services and by grants and contracts. ATCC has available two catalogues. One lists the approximately 18,000 strains of bacteria, bacteriophages, fungi, protozoa, and algae. The second lists animal cell lines, animal viruses, chlamydiae, and rickettsiae. These catalogues may be obtained for a fee from the American Type Culture Collection.

Development of the Agricultural Research Service Culture Collection began in the early 1900s and started primarily as a combination of a collection begun by Charles Thom and the microbial collection of the former USDA Bureau of Chemistry and Soils. The ARS collection has grown from its early existence by acquiring special collections, patent culture deposits, and other donations both solicited and unsolicited. Presently, the culture collection maintains over 17,000 strains of molds, bacteria, yeasts, and actinomycetes. The ARS does not publish a catalogue of organisms held in its collection, but may distribute free of charge cultures to scientists in industry and education both within and outside of the United States at the discretion of the curator of the collection. Policy regarding the deposit and distribution of cultures from the ARS culture collection is detailed in "Culture collections of microorganisms" (Iizuka and Hasegawa 1970), and "Sources and management of microorganisms for the development of a fermentation industry" (Hesseltine and Haynes 1974).

Noted World Collections.—Throughout the world there are many culture collections, and an extensive list has been compiled by Martin and Skerman (1972). No attempt will be made here to list all of the industrially important culture collections. However, it does seem proper to list a few collections where one may obtain cultures for industrial research (see Appendix Table A.1).

Shipping.—Each year thousands of cultures of microorganisms are shipped from country to country and within countries. Many laws and regulations exist governing the shipment of microorganisms; and it is important to be aware of these regulations. For the majority of microorganisms for the fermentation industry, little hazard to public health or agriculture is involved during shipment. However, shipment should conform to regulations by both the country of origin and the country of destination.

It is necessary to determine the pathogenicity of an organism in order to know which regulations affect its shipment. The standards used by the United States Government for evaluating the pathogenicity of microorganisms and for classifying etiological agents on the basis of hazard have been published (Anon. 1976). The following general criteria are used for this classification.

Class 1. Agents of none or minimal hazard under ordinary conditions of handling.

Class 2. Agents of ordinary potential hazard. This class includes agents which may produce disease of varying degrees of severity from accidental inoculation or other means of cutaneous penetration, but which can be contained by ordinary laboratory methods.

Class 3. Agents involving special hazard or agents obtained from outside of the United States which require a federal permit for importation unless they are specified for a higher classification. This class includes pathogens which require special conditions for containment.

Class 4. Agents that require the most stringent conditions for their containment because they are extremely hazardous to laboratory personnel or may cause serious epidemic disease. This class includes Class 3 agents from outside the United States when they are employed in entomological experiments or when other entomological experiments are conducted in the same laboratory area.

Class 5. Foreign animal pathogens that are excluded from the United States by law or whose entry is restricted by USDA administrative policy.

With the exception of industries involved in producing antisera, vaccines, pharmaceuticals, etc., the majority of organisms used by industry would fall into Class 1. Regulations regarding this class are limited primarily to the need to protect the United States mail from damage which could arise from broken shipping containers. Organisms that belong to Classes 2−5 require special permits and packaging procedures to ensure against the hazards which would result from improper handling of these more dangerous microorganisms.

The shipment of microorganisms which can be considered as plant pathogens is governed by two federal statutes, the Plant Quarantine Act of 1912 and the Federal Plant Pest Act of 1957. These acts prohibit the importation and movement of plant pests, pathogens, vectors, and articles that might harbor these organisms unless authorized by the U.S. Department of Agriculture. Authorization comes in the form of a permit, and the following guidelines are used in determining permit requirements.

Types of Organisms Requiring Permits:
(1) Foreign plant pests known to be injurious to crops grown in the United States.
(2) Domestic plant pests regulated by federal and state quarantines.
(3) Non-regulated domestic plant pests if shipment is into an area of the United States where the pests do not occur.
(4) Pests of noxious plants.

Types of Organisms Not Requiring Permits:
(1) Pure colonies of plant pest predators and parasites; pure cultures of plant pest pathogens.
(2) Non-pest organisms.

U.S. Government departments involved with regulating the shipment of microorganisms are: Agriculture; Commerce; Health, Education and Welfare (now Education, and Health and Human Services); Transportation; Treasury; and the U.S. Postal Service. Presently, rules and regulations are undergoing change and clarification, and the following should be contacted concerning present regulations and permits:

(1) For importation or interstate transport of agents which are animal pathogens:
Chief Staff Veterinarian
Organisms and Vectors
Veterinary Services, APHIS, USDA
Federal Building
Hyattsville, MD 20782

(2) For importation and interstate movement of agents which are human pathogens:
Center for Disease Control
Attn. Office of Biosafety
Atlanta, GA 30333

(3) For importation or interstate transportation of agents which are plant pests, pathogens, or vectors:
U.S. Dept. of Agriculture
Animal and Plant Health Inspection Service
Plant Protection and Quarantine
Federal Building
Hyattsville, MD 20782

Federal requirements for the packaging of both pathogenic and non-pathogenic materials may be found in the code of Federal Regulations 39CFR21.3 and 42CFR72.25. Information concerning the postal regulations of other countries can be found in Postal Service Publication 42. Examples of acceptable packaging methods are shown in Fig. 3.1 and 3.2, and sources for acceptable packaging are listed in Appendix A.2.

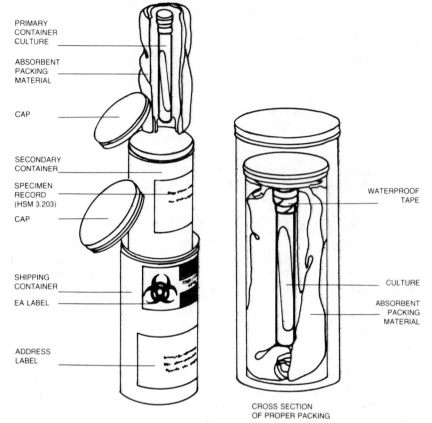

PRIMARY
CONTAINER
CULTURE

ABSORBENT
PACKING
MATERIAL

CAP

SECONDARY
CONTAINER

SPECIMEN
RECORD
(HSM 3.203)

CAP

SHIPPING
CONTAINER

EA LABEL

ADDRESS
LABEL

WATERPROOF
TAPE

CULTURE

ABSORBENT
PACKING
MATERIAL

CROSS SECTION
OF PROPER PACKING

Courtesy of Center for Disease Control, Atlanta, GA

FIG. 3.1.   DIAGRAM OF THE PACKAGING AND LABELING OF ETIOLOGICAL AGENTS
OF LESS THAN 50 ML VOLUME

Isolation

The ultimate sources of microorganisms for use in industrial processes
are found in nature. Soil, living or decaying plant and animal matter,
sewage sludge, etc., provide a wide spectrum of organisms suited to many
purposes. Pure cultures of microorganisms can be selected from such
sources by the use of several methods.

Plating Methods.—The use of media solidified with agar or some other
suitable jelling agent has been successful for many years in obtaining
cultures of many microorganisms. The process is very useful for yeasts,
bacteria, fungi, and other microorganisms which produce a distinct colony
on solidified media, and pure cultures can usually be obtained by using one
of the following variations.

Streak Plate.—A streak plate is prepared by dipping a sterile glass rod,
loop, or bent needle into a medium containing a suspension of cells of the
desired organism. The rod or wire is then used to make a series of parallel

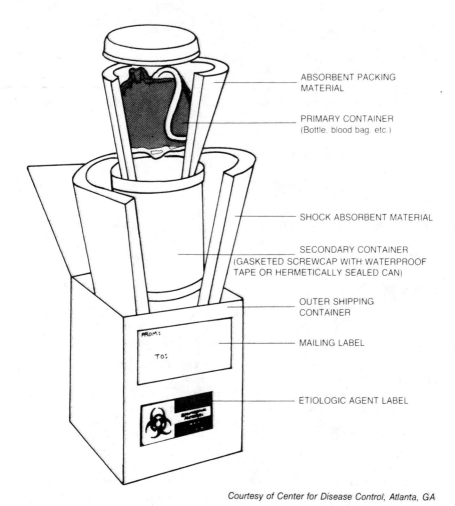

ABSORBENT PACKING MATERIAL

PRIMARY CONTAINER
(Bottle, blood bag, etc.)

SHOCK ABSORBENT MATERIAL

SECONDARY CONTAINER
(GASKETED SCREWCAP WITH WATERPROOF
TAPE OR HERMETICALLY SEALED CAN)

OUTER SHIPPING CONTAINER

MAILING LABEL

ETIOLOGIC AGENT LABEL

Courtesy of Center for Disease Control, Atlanta, GA

FIG. 3.2.   DIAGRAM OF THE PACKAGING AND LABELING OF ETIOLOGICAL AGENTS OF LESS THAN 500 ML VOLUME

non-overlapping streaks across the surface of the medium. As successive streaks are made the number of cells is diluted to such a point that the final streaks will usually yield separate and distinct colonies.

Pour Plate.—A second and very useful method is the pour plate. The process consists of adding a portion of the medium containing the desired cells into an agar medium which is still liquid but cooled to about 45°C. If the cell concentration is not known, a portion of the liquid agar may be serially diluted with additional agar medium and finally poured into sterile petri dishes. This results in the formation of a solidified medium with colonies scattered throughout the medium and not just on the surface. This allows for fairly easy separation.

A second series of isolations should be made from colonies of the streak or pour plate. This ensures that a pure culture is prepared.

Dilution.—This technique is widely used for obtaining pure cultures of many types of organisms. The process is carried out by evenly dispersing a sample of material in sterile water by use of a blender, homogenizer, or by vigorous shaking. The resulting suspension is then diluted to such a point that plating out a small portion of it will yield a number of colonies on each plate which can easily be separated and removed to fresh media. Variations of this process exist but lead to the same result (Smith 1969).

Antibiotics.—A very useful addition to the plating methods is the incorporation of substances which are inhibitory to certain groups of organisms. Some antibiotics, such as penicillin, are toxic for most prokaryotic organisms but not for eukaryotic organisms. Therefore, the use of penicillin is very helpful to reduce or eliminate the number of bacterial contaminants when yeasts or molds are isolated. Conversely, such compounds as nystatin can be used to inhibit eukaryotic organisms when prokaryotic organisms are being isolated.

Plating methods are extremely useful in the isolation of organisms from a natural habitat; however, it must be noted that microorganisms vary greatly in their nutritional, temperature, O_2, osmotic, pH, and other growth requirements. Therefore, it is necessary to be careful in selecting a medium which will provide a suitable environment for the microorganism.

Single Cell.—Some regard a culture as a pure culture only if the isolation is visually from a single cell or spore. It is necessary to have a single cell isolated for some types of work such as genetic investigations. However, the need for single cell isolation in the maintenance of cultures for industrial use is rather limited. If there is a need for single cell isolates, this can be fulfilled by the use of a micro-manipulator. The instrument is very helpful for work requiring the preparation of many single cell isolates, but it is also very expensive.

Aside from the use of a micro-manipulator there are several other methods available which require only equipment readily available in any laboratory. With fungi, for example, dilution plates can be prepared from spore suspensions and examinations made until germ tubes appear. A small portion of agar with one germ tube can then be removed to fresh medium.

Hypheal Tip.—Some fungi produce very few or no spores. For these organisms, isolates can be obtained by plating a single colony in the center of a petri dish. A tip from the radiating hyphae at the outer edge of a rapidly growing colony can then be cut and transferred to fresh medium.

Enrichment Culture.—This technique is a process by which a special environment is created which will allow for the selection of the desired organism, or group of organisms, from a mixture of many organisms. This may be accomplished by setting forth conditions which will allow the desired organism to outgrow the other organisms present or by inhibiting the growth of all organisms except those desired (Norris and Ribbons 1970; Stanier *et al.* 1976; Laskin and Lechevalier 1973).

Enrichment techniques may be carried out in liquid culture or on agar plates. An advantage to plating methods is the separation of colonies which allows for easy final isolation. Generally, liquid cultures are used to prepare a sufficient population from which isolation may be completed on solid media.

Enrichment techniques may employ alteration in the energy source, nitrogen source, carbon source, trace elements, pH, temperature, osmotic pressure, oxygen tension, source of inoculum, or any other parameter to select for the desired organism. For example, to select for ability to degrade hydrocarbons, a medium is prepared which contains only hydrocarbons as the source of energy. The medium can then be inoculated with soil, sewage sludge, etc., and incubated. After incubation, a portion of the medium is transferred to fresh hydrocarbon medium and allowed to incubate further. After several such transfers the majority of organisms remaining are capable of degrading hydrocarbons. Pure cultures may then be isolated from the liquid culture.

Genetic Alteration

Since the late 1930s the use of mutation as a method of strain improvement has been exploited in many industrial processes. Probably the most noted example of this type of work is the improvement in the production of penicillin. The basic procedure for producing mutants from microorganisms is relatively simple and involves exposing cells or spores to the action of a mutagen. The exposure may take place on the surface of agar in a petri plate, in a liquid culture, or by any method which allows contact between the mutagen and the cell. The process usually kills 99% of the exposed cells. The remaining living cells can then be isolated, subcultured, and checked for the desired result. Several commonly used mutagens are available and fall into two large groups, chemical and radiation.

Chemical.—A wide variety of chemical compounds are capable of producing mutations in microorganisms (Auerbach 1961; Norris and Ribbons 1970). The action of chemical mutagens may proceed in a number of ways. Some compounds cause the loss or addition of bases in DNA; some cause changes in the nucleic acids; while others are structurally closely related to nucleic acid bases and are incorporated into new strands of nucleic acid during replication. Chemical mutagens do not act equally well on different organisms. The best combination of mutagen and method of treatment is usually arrived at from published literature plus trial and error. It should also be noted that many of the chemical mutagens are carcinogenic and should be used with great care.

Nitrous Acid.—This compound is one of the easiest and least harmful of the chemical mutagens. Treatment is usually effected by suspending the cells in an acidic buffer and adding sodium nitrite. Mutagenesis is easily controlled by varying the length of time the cells are exposed.

Ethyl Methane Sulphonate (EMS), Ethyl Ethane Sulphonate (EES), and Diethyl Sulphate (DES).—These compounds belong to a larger group known as alkylating agents and give relatively high yields of mutants at high survival rates. Cells of the organism to be treated are usually suspended in a neutral buffer, and the mutagen is added to the suspension. The mutagenic reactions may be stopped by using sodium thiosulfate or by dilution during plating.

N-Methyl-N'-Nitro-N-Nitrosoguanidine (NTG).—Of the chemical mutagens, this compound is one of the most potent, and great care should be exercised during its use. NTG can produce a high yield of mutants coupled with a high survival rate under proper conditions. Large amounts of information on the conditions influencing the mutagenic and killing effects of NTG in *E. coli* and *Streptomyces coelicolor* have been reported by Adelberg *et al.* (1965) and Delic *et al.* (1970).

Base Analogues.—Compounds such as 5-bromouracil and 2-aminopurine have been used for mutagenesis. These compounds are often incorporated into the DNA in place of thymine and adenine, respectively. The mutagenic effect appears to be the result of incorrect base pairing in cell generations some time after the incorporation of the base analogue.

Acridine Compounds.—Compounds of this group have limited use on a routine basis. Acridines produce relatively few mutants in most systems but derivatives of acridines known as ICR (Institute of Cancer Research) compounds are known to produce a high yield of mutants in specific systems.

Radiations.—Several systems for mutagenesis have been developed using X-rays, γ-rays, neutrons, ultraviolet light, etc. Radiation systems are relatively convenient. If handled properly they may be used to produce a high yield of mutants in some systems.

Ionizing Radiations.—Ionizing radiations such as X-rays, γ-rays, neutrons, and other particles have been used successfully to produce mutations. This type of radiation is particularly useful whenever chemical agents or ultraviolet light is not effective. It should be noted that ionizing radiations often cause chromosomal breakage which may lead to undesirable structural changes such as translocations and inversions.

Ultraviolet Light.—The use of light between the wavelengths of 200 and 300 nm is a particularly easy and safe way of producing mutagenesis in microorganisms. Low pressure mercury vapor lamps emit UV light at a wavelength which is very efficient in producing mutations. Users of UV light should be aware of the possibility of photoreactivation upon exposure of treated cells to longer wavelengths of UV light and the visible spectrum. Of course, any user must also be aware of the damage which can be done to the unprotected eye.

Hybridization.—The use of genetic recombination in microorganisms is valuable for producing new strains of industrial importance. Genetic recombination in fungi and yeast can be accomplished by obtaining haploid strains from ascospores, basidiospores, etc., and then mating to produce new strains. With some lower fungi, the parasexual cycle has been utilized for strain improvement, and bacteria have been hybridized by transduction, conjugation, and transformation. The hybridization of yeasts and higher fungi requires 3 main steps: sporulation, spore isolation, and hybridization. Sporulation is controlled by many factors such as media composition, age of the culture, temperature, etc. These factors have been reviewed by Fogel and Mortimer (1971). Once sporulation has taken place, it is necessary to separate the asci from the nonsporulating cells. After the asci have been separated, the individual spores can then be obtained by one of several methods, such as ascus dissection. Hybrids can be prepared by mating spores, mating haploid cells, etc., to produce new strains.

Parasexuality was first described by Pontecorvo in 1952 and reviewed by him in 1956. Since that time this process has been found in many fungi, and utilized by some for the improvement of industrial strains. The parasexual cycle briefly consists of heterokaryotic mycelium in which fusion of nuclei takes place. These nuclei may be like or unlike. After fusion the diploid nuclei multiply along with the haploid nuclei; and finally genetic sorting out takes place during the production of uninucleate conidia. Successful utilization of the parasexual cycle for increased production of penicillin has been accomplished. However, its use as a common methodology has never materialized.

Genetic recombination in bacteria differs from that of the sexual process of eukaryotes (Notani and Setlow 1974; Curtiss 1969; Fennell 1960). In bacteria a portion of the genetic material of one cell (donor) is transferred to a second cell (recipient). The recipient cell, therefore, becomes diploid only in a section of its genetic complement. The process of recombination may proceed in 1 of 3 ways. The first, transformation, is accomplished by the adsorption of a piece of double-stranded DNA (which had been released into the medium by a donor cell) onto the surface of a recipient cell. One strand of the DNA is degraded and the remaining strand is incorporated into the genetic material of the recipient cell.

The most recently discovered process for genetic recombination, transduction, requires a bacteriophage. In transduction a portion of double-stranded DNA from the donor cell is carried to the recipient cell by a bacteriophage. When infection by the bacteriophage takes place, the genetic material carried from the donor to the recipient is also incorporated to form a partial zygote.

In the third process, conjugation, the genetic material is transferred to the recipient by direct contact. A single strand of DNA is transmitted. This portion of the genetic material may represent a major portion of the donor's genome.

MAINTAINING CULTURES

Slants and Broths

Use of slants and broth cultures for maintaining primary stock cultures is probably the simplest method available. For the majority of microorganisms, mature cultures grown on slants or in broth may be refrigerated and thereby preserved for some time, usually for a few months. However, use of slants and broths presents some serious shortcomings and, in general, it is the least desirable method available. It necessitates the frequent transfer of cultures which greatly increases the chance of contamination and undesirable genetic change. Many fungi, for example, grown under the artificial conditions of slants lose their ability to perform a desired biochemical process after several transfers. Bacteria, like fungi, may lose some of the desired properties such as virulence, antigenicity, or the ability to produce a particular metabolite.

Some of the problems just mentioned may be minimized by altering the media in which the microorganisms are subcultured. This helps to prevent the natural selection for strains which grow well in a particular medium. Much attention has been given to the composition of media which may be used successfully to culture microorganisms but the selection must be tailored to each particular situation (Booth 1971; Haynes et al. 1955; Norris and Ribbons 1970; Speck 1976).

Oil Overlay

Mineral oil may be used to overlay the mature growth of microorganisms on a slant (Buell and Westen 1947; Fennell 1960; Martin 1964). This process reduces water loss by evaporation, slows the exchange of gas, and allows subculturing without destruction of the primary stock culture. But use of the oil overlay method is subject to objections similar to those for the broth or slant method, such as loss of sporulation, loss of biochemical activity, etc. The technique is, however, particularly useful in maintaining fungi which do not produce spores.

Soil

A third and relatively simple method of preserving cultures is the dry spore stock on sterile soil (Collins and Lyne 1976). This method has been used successfully for a variety of microorganisms including fungi and bacteria. The first consideration in using soil should be to note whether the soil is to be used strictly as a carrier or also as a growth medium. If used as a carrier, abundant spores of the bacterial or fungal species must be prepared in advance. The mature spores may then be placed in sterile soil and the resulting preparation may be dried in air or under vacuum.

For many fungi, maintenance by use of soil stocks is very useful. Moist soil may be inoculated and the fungus may be allowed to grow until sporulation has been completed. This type of culture may be used both as primary stock and as a working stock culture.

Freezing

When the need for long time storage arises, it is necessary to reduce the metabolic activity of microbial cells to such a point that reproduction is halted. Presently two methods are employed to reach this goal. The biochemical activity of cells can be reduced to a state of "suspended animation" by reducing their temperature by the almost complete removal of water.

Freezing cells is a harsh process which can damage cells beyond repair. Methods have been developed which allow freezing with minimum damage to assure the recovery of live cells of many microorganisms (Muggleton 1962). A considerable amount of information has been published on the events that take place during the freezing process and many opinions have been advanced on the best method to be used. The following are a few generally accepted rules on the frozen storage of live cells.

(1) The temperature drop should be about 1°C per min to −20°C; then as rapid as possible to the storage temperature.
(2) Thawing should be as rapid as possible.
(3) Survival rates may be increased by the addition of glycerol, sugars, or other protective agents.
(4) Electrolytes should be kept at a minimum.

Liquid Nitrogen.—In recent years the use of liquid nitrogen for ultralow temperature storage of cells, including microbial cells, has become a very useful method for long term preservation. Assuming that cellular metabolism is completely stopped at temperatures of −130°C, organisms should be able to survive for an indefinite period of time provided they can withstand the cooling and thawing process. This process has been used successfully for a wide variety of cells including fungi, bacteriophages, protozoa, algae, mammalian cells, and bacteria (Norris and Ribbons 1970).

The overall process for ultralow temperature preservation is similar to that for the freezing process described earlier. A cell suspension protected by such materials as glycerol is cooled at a rate of about 1°C per min until a temperature of −35°C is reached. At this point the temperature is allowed to drop at an uncontrolled rate.

Some differences of opinion exist as to the method by which the frozen cells should be thawed. Several studies have been conducted. Generally the tubes are removed directly from the low temperature storage and placed in a 37°−40°C water bath and agitated for rapid thawing. Caution should be exercised during the thawing process. A cracked or otherwise faulty ampoule may have become contaminated with liquid nitrogen. As thawing proceeds the pressure created by the passage of this liquid nitrogen to the gas phase may cause an explosion.

The use of liquid nitrogen storage of microbial cells has several advantages: No subculturing of cells; non-spore-forming microbes such as basidiomycetes may be preserved; no changes in the biochemical mechanism or the genetic equipment of the cells; no contamination problems once the ampoules are stored. The disadvantages are: The initial cost of the equipment; the need for a constant supply of liquid nitrogen; and the difficulty in distributing cultures to other laboratories.

Lyophilization.—From the author's point of view, lyophilization is probably the most satisfactory method for long term preservation of those organisms which will withstand the process. For this reason lyophilization will be described in more detail than the previous methods.

Although it had been used earlier, the lyophilization process was developed into a valuable tool during the 1940s primarily by the Northern Regional Laboratory in Peoria, Ill. Later work at this institute turned this process into a very useful and highly reliable method for the preservation of many types of microorganisms (Fennell 1960; Collins and Lyne 1976). This process has many of the advantages of the liquid nitrogen process: It requires no subculturing of cells; there is no change in the biochemical reactions of the cells, and the cells are genetically stable; there are no contamination problems once the ampoules have been sealed; the finished ampoules can be easily shipped by mail; and the initial costs are much less than those for liquid nitrogen storage. The major disadvantage of the lyophilization process is simply that not all microorganisms can be preserved by this method.

Lyophilization or freeze-drying consists primarily of suspending propagative cells (bacteria or yeast cells, conidia, ascospores, etc.) in a protective medium, freezing, and the removal of water by sublimation under reduced pressure. The desiccated cultures are then sealed under vacuum and stored at low temperature (4°C). The majority of molds (penicillia, aspergilli, Mucorales) survive well, whereas *Pythicacea, Entomophtorales,* large spored fungi, and mycelial forms seldom survive the initial treatment. Many other single celled organisms such as yeast and bacteria also can be preserved by this method. If the organisms survive the initial treatment, they are likely to remain viable for a period of 20 years or more.

The spores or cells may be suspended in various protective media such as bovine serum or media containing sugars. However, skim milk is probably the most desirable medium, and it is more easily obtained. The microbial suspension is dispensed into small ampoules using sterile techniques. The ampoules are then quick frozen at about −35°C and placed under a vacuum of at least 200 μm of Hg. After drying, the ampoules are sealed under vacuum.

Ampoules for lyophilization may be prepared by cutting pyrex glass tubing into lengths of approximately 11 cm. One end of the tube is then sealed by rapidly rotating it in an oxygen-natural gas flame. Care must be used to seal the end completely. The other end is then fire polished to smooth the rough edges. Before lyophilization the tubes should be snugly stoppered with cotton and sterilized. If the cotton stoppers are too tight, the frozen cell suspension will melt during the process, resulting in an unusable preparation.

A suspension medium prepared from a mixture of 1 part fresh skim milk and 1 part distilled water should be sterilized. A cell suspension is prepared by introducing the sterile skim milk-water mixture into a fresh fungus culture about 2 days after sporulation has been completed. A highly concentrated suspension of fungal spores increases the chance of recovery of the culture after lyophilization. Approximately 3 drops of the cell suspension should be placed into each ampoule with a sterile Pasteur pipette. The excess cotton should be cut from the stopper leaving a stopper about 8–10 mm long which is pushed to within 10 mm of the medium. A small tag showing the culture identification on one side and the date on the other side is also placed inside the ampoule.

The preparation may be frozen in a beaker of ethanol with chips of dry ice at a temperature of $-35°$ to $-40°C$. It is important that the temperature not rise above $-35°C$ during freezing. Temperatures below $-35°C$ are usually not damaging to the culture, while temperatures above $-35°C$ may result in poor viability.

Once the preparations are frozen the tubes should be subjected to the vacuum. A vacuum of 200 μm Hg is usually satisfactory. The actual vacuum is not really critical. The important point is that the frozen preparation not melt during the process. Often it is necessary to maintain the frozen preparation in an ice bath for the first few minutes of drying. After the cultures are dry, the tubes are sealed under vacuum with an oxygen gas torch. Finished cultures may be stored at room temperature. However, it is advisable to store the ampoules in the dark since light can reduce viability; and storage in the refrigerator is recommended.

Several methods may be used to open the tubes, but sterility must be maintained: Thump the tube with the fingers to break up the pellet. Wipe the surface of the tube with a sterilizing solution (70% ethanol or Chlorox 1:1) after making a file scratch across the center of the cotton plug (Fig. 3.3). Apply a red hot glass rod to the scratch to crack the glass or simply apply

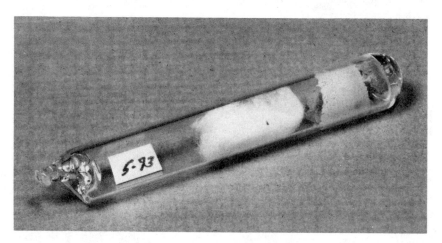

FIG. 3.3.   FREEZE-DRIED CULTURE IN VACUUM AMPOULE

The freeze-dried culture is seen on the right end of the tube; the cotton plug near the center; and the identifying label toward the left end, where the tube has been sealed.

pressure as one would to break glass tubing. Use care in opening the ampoules as the contents are under vacuum. Allow time for air, filtered by the plug, to seep into the ampoule. Otherwise, when the pointed end is snapped off the plug will be drawn to one end. Hasty opening may release live particles of the dried organism into the air of the laboratory. Since the cotton plug and the tube may contain spores or cells they should be autoclaved after use. If the tube does not have a cotton stopper make a file scratch on the sealed tube, sterilize the outer surface with ethanol or some other disinfectant, and break open inside a wrapping of sterile cotton.

Reculturing.—One of several methods may be used for reculturing. Regardless of the method, the resulting culture should be compared with a description of the culture originally lyophilized.

Method A: After opening the tubes, introduce a small volume of sterile water equal in volume to the pellet, and replace the cotton plug. Allow the ampoule to stand for 30 min, and then streak the suspension on agar medium. For bacteria, blood agar should be used since some bacteria require haemin for recultivation.

Method B: Prepare 50 ml of a suitable culture medium, place in a 250 ml Erlenmeyer flask, stopper, and sterilize. The contents of the ampoule may then be dumped into the flask. This method does not allow for the possibility that contamination may have occurred (as does the streaking method). Patience must be used in reculturing. Often several days to a week are required for growth to appear.

A medium suitable for the growth of each individual microorganism should be used for reculturing lyophilized cultures. Some investigations have shown that malt extract increases the chance of satisfactory results in reculturing fungi, and malt extract is generally included in reculturing media.

The main cause of failure to recover lyophilized cultures is poorly sealed tubes which lose their vacuum. A few days after lyophilization, the vacuum in the tubes should be checked. This can be done by using an induction coil (spark coil tester, high frequency coil, or generator, Tesla type). The tubes may be lined up side by side on a table top at room temperature. Touching the end of the tube with the discharge tip of the tester results in a white to purplish glow in the tube. Absence of this glow indicates a loss of vacuum.

Many factors influence the survival rates of microorganisms preserved by lyophilizaton. These have previously been reviewed. The most important generalization about lyophilization is simply this: If the culture withstands the original process of lyophilization, it is likely to survive for many years as long as the culture remains under vacuum.

MAINTAINING PURITY

Antibiotics

A wide variety of antibiotics and other chemicals are useful in obtaining pure cultures during isolation procedures and for maintaining purity in culture collections. Some common and easily available compounds and their applications are listed in Table 3.1.

TABLE 3.1.   ANTIBIOTICS FOR MAINTAINING CULTURE PURITY

Compound	Concentration (mg/liter)	Use
Rose Bengal	35	inhibits colony growth
Neomycin sulfate	20	inhibits *Actinomycetes*
Cycloheximide (Actidione)	2–200	inhibits yeast and some fungi
Na salt of Benzylpenicillin	40	inhibits bacteria
Griseofulvin	2–20	inhibits fungi

Mites

Invasion of fungal cultures by mites has long been a problem which is more easily prevented than cured. Mites feed on fungal spores and travel from one culture to another and spread contamination throughout a collection. A collection which is heavily infested by mites can usually be saved only by a reisolation program. This requires long time periods, and even then many cultures may be lost.

A catastrophe caused by an invasion of mites can usually be prevented by observing the following guidelines. Primary cultures should preferably be stored by a low risk method such as mineral overlay, lyophilization, or liquid nitrogen storage. Cultures which have completed sporulation should be stored under refrigeration. This will not prevent mite growth but will reduce the rate of growth and reproduction of mites so that an entire collection will not be lost before mites are detected. Samples of vegetation should not be placed in close proximity to the culture and collection storage area. Vegetation, particularly dried samples such as hay, are noted as carriers of mites and should not be allowed in the same room as culture collections.

Clean Room

Microbiologists find it important to have available a "transfer" or "clean" room. This should not be confused with a microbiological laboratory whose functions are primarily plate counts, assays, media preparation, and the like. A "transfer" or "clean" room is used in instances when critical work of culture transfer or culture preservation is done. Some brief examples are given.

Transfer Cultures.—In many instances an inoculum is needed for a large scale industrial process. This usually requires growing of the microorganism in several stages of ever increasing volume to obtain an inoculum large enough for the final commercial stage. This propagation often requires several transfers over a period of several days. If contamination is introduced at an early stage, the final product will suffer severely in quality.

Strain Research.—Research on strain selection often requires that a clean area be available that can be occupied by the investigator for an extended period of time without interruption by other individuals. This is necessary to minimize the danger of contamination.

Storage.—In some instances, for example for nutritional work, several days may be required for treated cultures to grow. It is often necessary to prepare and incubate such cultures in an area where little air contamination exists.

Extended Culture.—Some fermentation work requires a very long growth period. Just as for storage, an area with extremely low microbial contamination is required to complete such an investigation successfully.

Requirements of a Clean Room.—If a transfer room is to be used for critical work, it should have the following characteristics.

Limited Access.—It is important that during the process of transferring cultures no one enter or leave the room. This reduces air currents and the movement of air-borne bacterial and yeast cells and fungal spores.

Supervision and Responsibility by One Individual.—It is necessary to assign responsibility for such a critical area to one individual. He/she should be capable of caring for the area and should have the authority to maintain it in a proper way.

Clean Room Should Be Small.—The clean room should be small in size for proper cleaning and maintenance. Cleaning should be done frequently and thoroughly.

Filtered Air Supply.—A filtered air supply is essential to maintain a low level of air contamination after sanitation.

Cleanable Surfaces.—Surfaces must be able to withstand the action of harsh chemicals. Equipment which cannot be easily and completely cleaned should be excluded. A wide variety of agents are available for use as disinfectants on bench tops and other surfaces of a clean room. Ultraviolet light of a wavelength of 253.7 nm is an effective agent for destroying microorganisms in close proximity to the UV source. The value of ultraviolet radiation for disinfecting a large area such as a bench top is almost nil. After use, the source of a UV lamp will continue to glow long after it has ceased to emit germicidal radiation.

Ethanol, isopropanol, and other alcohols are often diluted to 70% with water and used as a general disinfectant. Use for bench top disinfecting is limited since some time is required for alcohols to become effective. Use for soaking of instruments is of value.

Formaldehyde is an effective agent against non-spore-forming organisms. It is available as a solution of about 40% formaldehyde in water (formalin), and it is used generally in concentrations of 0.1 to 0.5% formalin in water.

Hypochlorites are used as general germicides. They are effective against a wide range of microbes. A 1:1 mixture of household bleach is often used. Iodine in aqueous or alcoholic solution is a good general purpose disinfectant. It is effective against bacteria, spores, and viruses. A 1% solution in 70% ethanol is commonly used as a skin disinfectant.

Fumigation.—In some cases the air and bench contamination level will reach a point where fumigation of the entire room is required. For this reason the air of the clean room should be exhausted into a system not connected with an occupied area. Fumigation may be accomplished by several methods. Commonly, formalin is boiled to release the gas formaldehyde. Approximately 0.5 ml of formalin should be used for each 0.03 m^3 (1 ft^3) of air space.

IMPORTANCE OF PURE CULTURE METHODS

The food fermentations which are the subject of subsequent chapters are *not* pure culture fermentations. In some instances, they still depend on the mixed microflora of the raw material (e.g., sauerkraut or some sourdough fermentations). However, the major food fermentations depend on massive inoculation with microbial "seed" derived from a pure strain. The use of commercial bakers' yeast in doughs or the use of bacterial starter cultures in the production of cheese is characteristic of modern food fermentations. This massive inoculation is based on the selection of an appropriate strain and on its maintenance as a pure strain. This is true whether the microbial seed is commercially available (e.g., bakers' yeast) or whether it is produced "in house" (e.g., brewers' yeast). Therefore, the subject matter of the present chapter, "obtaining pure cultures" and "maintaining pure cultures" is fundamental to the commercial practice of food fermentations.

REFERENCES

ADELBERG, E.A., MANDEL, M. and CHEN, G.C.C. 1965. Optimal conditions for mutagenesis by N-methyl-N'-nitrosoguanidine in *Escherichia coli*. Biochem. Biophys. Res. Commun. *18*, 788–795.

ANON. 1976. Classification of etiologic agents on the basis of hazard. U.S. Public Health Serv. Center Dis. Control, Atlanta. Leaflet.

AUERBACH, C. 1961. Mutation and plant breeding. Chemicals and their effects. N.A.S.-N.R.C. Publ. *891*, 120–144.

BOOTH, C. 1971. Methods in Microbiology, Vol. 4. Academic Press, New York.

BUELL, C.B. and WESTEN, W.H. 1947. Application of mineral oil conservation method to maintaining collections of fungous [sic] cultures. Am. J. Bot. *34*, 555–561.

COLLINS, C.H. and LYNE, P.M. 1976. Microbiological Methods, 4th Edition. Butterworths, London.

CURL, E.A. 1958. Chemical exclusion of mites from laboratory fungal cultures. Plant Dis. Rep. *42* (9) 1026–1029.

CURTISS, R. 1969. Bacterial conjunction. Annu. Rev. Microbiol. *23*, 69–136.

DELIC, V., HOPWOOD, D.A. and FRIEND, E. 1970. Mutagenesis by N-methyl-N-nitro-N-nitrosoguanidine (NTG) in *Streptomyces caelicolor*. Mutat. Res. *9* (2) 167–182.

FENNELL, D.I. 1960. Conservation of fungus cultures. Bot. Rev. *26*, 79–141.

FOGEL, S. and MORTIMER, R.K. 1971. Recombination in yeast. Annu. Rev. Genet. *5*, 219–236.

HAYNES, W.C., WICKERHAM, L.J. and HESSELTINE, C.W. 1955. Maintenance of cultures of industrially important microorganisms. Appl. Microbiol. *3*, 361–368.

HESSELTINE, C.W. and HAYNES, W.C. 1974. Sources and management of microorganisms for the development of a fermentation industry. U.S. Dep. Agric., Agric. Res. Serv., Agric. Handb. *440*.

IIZUKA, H. and HASEGAWA, T. 1970. Culture collections of microorganisms. University Park Press, Baltimore.

LASKIN, A.I. and LECHEVALIER. H.A. 1973. Handbook of Microbiology, Vol. 1. Organismic Microbiology. CRC Press, Cleveland.

MARTIN, S.M. 1964. Conservation of microorganisms. Annu. Rev. Microbiol. *18*, 1–16.

MARTIN, S.M. and SKERMAN, V.B.D. 1972. World Directory of Collections of Cultures of Microorganisms. John Wiley & Sons, New York.

MUGGLETON, P.W. 1962. The preservation of cultures. Prog. Ind. Microbiol. *4*, 190–214.

NORRIS, J.R. and RIBBONS, D.W. 1970. Methods in Microbiology, Vol. 3A. Academic Press, New York.

NOTANI, N.K. and SETLOW, J.K. 1974. Mechanism of bacterial transformation and transfection. Prog. Nucleic Acid Res. Mol. Biol. *14*, 39–100.

PONTECORVO, G. 1956. The parasexual cycle in fungi. Annu. Rev. Microbiol. *10*, 393–400.

SMITH, G. 1969. An Introduction to Industrial Mycology, 6th Edition. Edward Arnold Publishing Co., London.

SPECK, M.L. 1976. Compendium of Methods for the Microbiological Examination of Foods. Am. Public Health Assoc., Washington, D.C.

STANIER, R.Y., ADELBERG, E. and INGRAHAM, J. 1976. The Microbial World. Prentice-Hall, Englewood Cliffs, N.J.

STEVENS, R.B. 1974. Mycology Guidebook. Univ. Washington Press, Seattle.

Section II

Food Fermentations

4

Cheese

Elmer H. Marth[1]

"Ein Dessert ohne Käse
ist wie ein Mädchen ohne Lächeln."
Brillat-Savarin (1755–1826)

Cheese usually is a concentrated form of two major components in milk, namely casein, the principal protein, and milkfat. Besides milk, selected bacteria, a milk-clotting agent, and sodium chloride are used to manufacture cheese. Variations in these basic constituents, use of additional ingredients, and variations in the environmental conditions surrounding the manufacture and subsequent ripening of cheese have given rise to hundreds of varieties of cheese. Extensive listings of varieties of cheese together with some information about their characteristics and manufacturing procedures are provided by Mair-Waldburg (1974A) and Walter and Hargrove (1969).

It is impossible to say when man first had experience with cheese or a cheese-like product. Mair-Waldburg (1974B) speculates that ancient man one day may have slaughtered a young animal after it nursed its mother. Upon opening its stomach, a cheese-like product was found that resulted from the action of rennin in the stomach on the milk. Another possibility, according to Mair-Waldburg (1974B), is that ancient man harvested the stomach of an animal and used it as a container for liquids (this is not too unlike the containers one sees at modern-day football games which are also used to hold certain liquids). When the freshly harvested stomach was used to hold milk, the combination of rennin, some lactic acid bacteria likely to be present, and a warm temperature again resulted in a cheese-like product. Mair-Waldburg (1974B) provides evidence to suggest that sheep were domesticated in Iran as early as 9000 years before Christ. It is not unreasonable to assume that some form of cheese was also known about the same time.

---

[1]Professor of Food Science and Bacteriology, Department of Food Science, University of Wisconsin-Madison, Madison, Wisconsin 53706.

It is not the purpose of this chapter to describe the history of cheese-making but only to indicate that the history of man and that of cheese-making are intimately interwoven. The story is fascinating, and some readers may desire more details. If so, they are advised to consult the discussions by Mair-Waldburg (1974B,C), Wekre (1974), Zehgruber (1974), and Marth (1953). Development of cheesemaking in the United States has been described by Price (1971).

This chapter will discuss cheese with emphasis on the microbiological aspects. This cannot be done without some attention being given to technological aspects because often they govern what microorganisms will do in cheese, and even whether the microorganisms will remain alive to achieve what is expected of them. To attain our goal, this chapter begins with a discussion of the principles of cheesemaking. This is followed by a description and discussion of several different varieties of cheese. The varieties to be considered were selected because they are representative of several different fermentative and technological processes and also because they are varieties of considerable economic importance. Finally, there is an extensive discussion of the public health aspects of cheese. Although currently in the United States cheese seldom causes these problems, the information was included because it is important in cheese microbiology and thus should be included in a chapter in a book whose major emphasis is microbiology. Furthermore, familiarity with this information may aid to keep cheese a safe food where it is now safe and may help to make it a safe food where it now is unsafe.

PRINCIPLES OF CHEESEMAKING

The basic steps in cheesemaking, as outlined by Kosikowski (1977), are: (1) setting milk (adding starter cultures and coagulant to prewarmed milk), (2) cutting the coagulum, (3) cooking the cut coagulum (curd), (4) removing whey from the curd, (5) allowing curd particles to "knit," (6) salting (this comes at different times in the procedure for different cheeses), (7) pressing, and (8) ripening of the finished cheese. The various facets of this general procedure will be considered in the following paragraphs.

Milk

Fresh milk obtained from healthy cows (or other animals that may serve as a source of milk) should be cooled rapidly and then promptly delivered to the cheese factory where it should be converted to cheese as soon as possible. The milk should be free from antibiotic residues, other chemical contaminants, and serious off-flavors. Furthermore, the raw milk should not have supported excessive growth of psychrotrophic bacteria because such growth can cause irregularities in the manufacture of cheese by affecting activity of lactic starter cultures (Cousin and Marth 1977A,B) and coagulation of milk by rennet extract (Cousin and Marth 1977C). The quality of cheese resulting from such milk also will be inferior to that of cheese made from milk without excessive growth of psychrotrophic bacteria (Cousin and Marth 1977D). The difficulties just mentioned occur because many species of psychrotrophic bacteria produce proteolytic enzymes that partially degrade the casein in milk (Cousin and Marth 1977E; DeBeukelar *et al.* 1977). After milk arrives at the factory, it is commonly clarified with a centrifuge to remove small extraneous particles and somatic cells. The milkfat content of the clarified milk may be adjusted, depending on the variety of cheese that is to be made. Some cheese is made from raw milk, but it is more common to use heat-treated (heat treatment less than that of pasteurization) or pasteurized milk. Heat-treated milk is sometimes preferred because the resultant cheese tends to be more flavorful than that made from pasteurized milk.

Tests to determine the microbiological quality, composition, and presence of unwanted chemical contaminants in milk are described in *Standard Methods for the Examination of Dairy Products* (Marth 1978). This book is revised periodically and the current edition should be consulted.

Starter Culture

One or several species of lactic acid bacteria are commonly added to pre-warmed milk. The small amount of acid produced by these bacteria early in the cheesemaking process (fermentation) facilitates subsequent clotting of milk by the coagulant. Furthermore, the starter bacteria become a part of the finished cheese where they contribute to processes that occur in the cheese during ripening.

The kind of cheese to be made determines which microorganisms to add to milk. For example, to make Cheddar cheese one would use *Streptococcus cremoris* and/or *Streptococcus lactis*, the so-called mesophilic lactic acid bacteria. In contrast, to make Swiss cheese one would use *Lactobacillus bulgaricus* and *Streptococcus thermophilus*, the so-called thermophilic lactic acid bacteria (the definition one chooses for thermophilic bacteria will determine if these bacteria are truly thermophilic). There are still other examples, but these two should suffice to illustrate the point.

Microorganisms other than lactic acid bacteria sometimes are added together with them when cheese is made. Examples are *Propionibacterium shermanii* for Swiss cheese or molds for blue or Camembert cheese. The functions of all starter cultures during ripening of cheese are discussed further when various varieties of cheese are considered later in this chapter.

Currently, concentrated frozen starter cultures can be purchased and added directly to milk in preparation for cheesemaking. Provided that the manufacturer's instructions are followed, this approach will result in greatest uniformity in culture activity from day to day and also will minimize, if not eliminate, problems caused by bacteriophage infections.

The alternative approach is to transfer the culture from one lot of a suitable medium to another until finally a volume of culture suitable for cheesemaking is obtained. When this is done, the problem of bacteriophage infection can be controlled through use of a phosphate-treated medium (Zottola and Marth 1966). Several commercial organizations market such a medium.

Regardless of how the culture is handled before it is added to milk, presence of antibiotics in milk (as a result of administering them to lactating dairy cows) will either retard activity and acid production by the culture or completely stop acid production if the starter culture is inactivated (Marth 1966). Loss of activity by the starter culture will lead to inferior cheese and may cause public health problems, as discussed later in this chapter. Presence of antibiotics is normally controlled by frequent testing of the raw milk from individual farms according to methods described in *Standard Methods for the Examination of Dairy Products* (Marth 1978).

Coagulant

A suitable coagulant is added to milk, usually a short time (e.g., 30 min) after the starter culture was added. The coagulant is an enzyme that splits colloidal casein into a carbohydrate-rich peptide fraction and the insoluble paracasein that precipitates in the presence of calcium ions.

Traditionally, rennet extract obtained from the fourth stomach of young calves has been used as the coagulant. A worldwide shortage of young calves has led to a shortage of rennet extract. Consequently, replacements such as blends of rennin and pepsin, rennet extracts from mature cows and coagulants of fungal origin have been developed. Fungi which produce coagulants for use in cheesemaking include *Mucor miehei, Mucor pusillus,* and *Endothia parasitica.* Much of the cheese currently made in the United States and marketed without extended aging is made with a coagulant of fungal origin.

Cutting the Coagulum

Rectangular frames with thin wires, horizontal on some and vertical on others, are used to cut the coagulated milk into cubes. Such cutting increases the surface area of the coagulum which facilitates its loss of whey. Cubes of coagulum also can be heated uniformly during the cooking process. Small cubes (e.g., 1.3 cm^3) lead to low-moisture cheese, whereas large cubes (e.g., 4–5 cm^3) lead to high-moisture cheese.

Cooking the Cut Coagulum

After the coagulum is cut, the cubes of coagulum (curds) suspended in whey are heated to a given temperature in a specific time (e.g., to 37°–38°C in 30 min for Cheddar cheese). This heating is accompanied by stirring of the curd-whey mixture and causes the cubes of curd to contract and thus express free whey. Cooking also serves to control acid production by the lactic starter culture, to suppress growth of some spoilage bacteria, to influence texture of the curd, and to aid in control of the amount of moisture in the finished cheese (Kosikowski 1977).

Draining Whey

After cooking is completed, whey is removed from the curd. This can be accomplished by draining whey from a vat that contains the whey-curd mixture, using appropriate precautions to prevent loss of curd. An alternative method, designated as dipping, involves scooping the curds from the whey-curd mixture. During the time needed for removal of curds from whey, some additional lactic acid is produced by the starter bacteria. The curd may be removed from the vat and placed in a form or mold, as in the manufacture of Camembert cheese. Alternatively, the curd may remain in the vat so some knitting of curd particles can occur, as in the manufacture of Cheddar cheese.

Knitting of Curds

This step allows for further production of lactic acid and for modification of the curd particles so they will adhere to each other and form a single mass of cheese. The characteristic texture of a given variety of cheese is partially determined by this process.

Salting of Curds

Salt (sodium chloride) is applied to curds in one of several ways. Dry salt may be sprinkled on loose curds as in the manufacture of Cheddar cheese or it may be rubbed onto the surface of freshly made cheese. Alternatively, freshly made cheese can be immersed in a nearly saturated aqueous solution of salt. Adding of salt contributes to the flavor, texture, and appearance of cheese; controls production of lactic acid; suppresses growth of spoilage microorganisms; and further reduces the amount of moisture in finished cheese (Kosikowski 1977).

Pressing of Curds

This step sometimes comes before salting (as with cheeses that are immersed in brine) or afterward (as with Cheddar cheese). Curds are placed into a form, sometimes called a hoop, and pressure is applied hydraulically or through use of weights. If cheese with an open texture is desired, external pressure may not be applied. Piling of curds in a vat, as in Cheddar cheese manufacture, is a form of pressing although hydraulic pressure is used later in the manufacture of this variety of cheese.

Pressing gives the cheese its characteristic shape and contributes to its compactness. Free whey is expressed and knitting of curd particles is completed during pressing. Use of vacuum chambers during or after pressing can aid in removing occluded air from cheese and thus give the product a closely knit body (Kosikowski 1977).

Ripening of Cheese

The finished cheese is placed in a room with controlled temperature and relative humidity (e.g., 4°C and 85% for Cheddar cheese) and is held there for several months to several years, depending on the variety of cheese and the extent of ripening that is desired.

Ripening allows for enzymatically-induced changes to occur in the protein and fat fractions of the cheese. These changes transform the freshly made cheese into one with desired and characteristic flavor, texture, aroma, and appearance (Kosikowski 1977).

Automation of Cheesemaking

Making of cheese was and for some cheesemakers continues to be largely a hand operation. However, increases in cost of labor and in amount of cheese a given factory desires to produce prompted development of machines which have automated various portions of the cheesemaking process. This also includes preparation of finished cheese for the retail market. It is not the purpose of this chapter to describe these machines, so the reader interested in this information should consult papers by Olson (1975) or Eisenreich (1974) or the book by Kosikowski (1977).

CHEDDAR CHEESE

Cheddar is the first of several varieties of cheese to be discussed in some detail. Next to be considered is Swiss cheese and this will be followed with information about surface-ripened and mold-ripened cheeses. Brief comments about process cheese will end the discussion of varieties of cheese.

Cheddar cheese is of English origin, having been developed in a dairying region surrounding the village of Cheddar which is situated at the foot of the Mendip Hills in Somerset (Chapman 1974). Church records indicate that cheese from this region was provided to King Henry II in 1170, but the procedure for making Cheddar cheese was not published until 1857 when it appeared in a research report of the Ayrshire Agricultural Association (Chapman 1974).

Cheddar cheese is made in cylindrical and block shapes. The size of these is variable with blocks as large as 290 kg having been made, but this is not common. Some Cheddar cheese is produced by filling curd into large steel barrels; this cheese probably will be used to make process cheese.

Manufacturing Process

The following steps (Kosikowski 1977) are involved in manufacturing Cheddar cheese.

(1) Whole milk is clarified and pasteurized (71.7°C for 16 sec) or heat-treated (an exposure to heat that is somewhat less than that of pasteurization). Sometimes raw milk is used.

(2) Milk to be made into cheese is warmed to 31°C and then inoculated with 0.5% (or more, depending on circumstances) of an active culture of *S. cremoris* and/or *S. lactis*. Annatto cheese color also may be added if a cheese with a yellow-orange color is desired.

(3) Inoculated milk is held at 31°C for 30 min to permit some bacterial activity, and then the coagulant is added. Coagulation of milk is completed in about 25 min.

(4) The coagulum is cut into cubes of curd and the curd-whey mixture with continuous agitation is warmed (cooked) to 38°C in 30 min.

(5) Whey is drained from the curds which remain in the vat, piled against the long sides of the vat, with a trench going down the center of the vat. The trenched curds are allowed to mat for about 15 min after drainage of whey is completed.

(6) The trenched curd is cut into blocks which are triangular in shape and about 10 cm in thickness. These slabs of curd, after resting on one side for 15 min, are turned and again allowed to rest. This may be repeated after which they are piled 2 high and again turned every 15 min. The slabs can be piled 3 high for the last 30 min. This process is called "cheddaring," and allows for development of acid and textural characteristics in the curd.

(7) The slabs of curd are fed into a curd mill which is suspended over the vat. This device cuts the slabs of curd into smaller pieces of curd which are ready for salting.

(8) Salt is applied and the curds are stirred for about 30 min. The salt is likely to be applied in 3 portions early during the stirring process.

(9) Curd is placed into a metal form or hoop of the desired shape and size. The hoop is outfitted with an appropriate cloth which covers all interior surfaces of the hoop. Hoops filled with curd are placed on a horizontal hydraulic press and pressure is applied for approximately 18 hr.

(10) Cheese is then covered with wax and moved to storage (2°–16°C, 85% relative humidity) for ripening. Rindless Cheddar, which is not waxed but is covered with a plastic film, also can be made.

(11) Ripening can be for as short a time as 3 months or as long as 1 year or more, depending on the intensity of flavor that is desired and the temperature at which ripening occurs.

(12) Ripened cheese can be cut into consumer-sized portions, packaged, and distributed. Alternatively, it can be used as an ingredient in process cheese or another food. Some ripened Cheddar cheese is grated and dried for use on or in other foods.

Changes During Ripening

Freshly made Cheddar cheese will contain appreciable numbers of the starter culture bacteria, *S. lactis* and/or *S. cremoris*. These bacteria begin to die within 2 to 8 weeks as the cheese ripens (Marth 1963). During this time the bacteria are likely to produce and/or liberate proteolytic and lipolytic enzymes that play a role in modifying the body and flavor of the cheese.

In contrast to the behavior of the lactic streptococci, lactobacilli in Cheddar cheese begin to grow after the cheese is several weeks old and attain a maximum population in cheese that is 3 to 6 months old. There may be a decrease in numbers after that, but large numbers of viable lactobacilli will survive for long times in Cheddar cheese. That lactobacilli can survive a long time was demonstrated by Minor *et al.* (1970) when they recovered bacteria similar to *Lactobacillus casei* from a cheese-like product that had spent 105 years submerged in Lake Michigan.

L. casei most commonly appears in Cheddar cheese although other lactobacilli, including *Lactobacillus plantarum, L. brevis, L. bulgaricus, L. helveticus, L. lactis, L. fermenti, L. acidophilus, L. arabinosus, L. pentosus, L. leichmannii,* and *L. delbrückii,* have been either isolated from Cheddar cheese or have been added to the cheese in attempts to improve its flavor (Marth 1963). Normally, lactobacilli are not added when Cheddar cheese is made, but these bacteria seem to reside within the cheese factory and so invariably are in the cheese. *L. casei* produces proteolytic and lipolytic enzymes that contribute to the ripening of cheese. Additionally, *L. casei* can produce hydrogen sulfide which, in small amounts, is needed for flavorful Cheddar cheese (Marth 1963).

As with the lactobacilli, micrococci are not commonly added in the manufacture of Cheddar cheese, but these bacteria constitute a major component of the microflora of the ripening cheese. Purposeful addition of certain micrococci when Cheddar cheese is made has resulted in rapid development of a desirable flavor. Beneficial micrococci have included strains of *Micrococcus freudenreichii, M. caseolyticus,* and *M. conglomeratus.* The beneficial effects of these bacteria have been attributed to their proteolytic activity.

Other bacteria also have been claimed to contribute to or improve the flavor of Cheddar cheese. Included are *Streptococcus faecalis, S. faecalis* var. *liquefaciens,* and *S. durans.* These bacteria are not commonly added when Cheddar cheese is made, but sometimes they might be in raw milk and thus get into the cheese if it were made from such milk.

As is true of most ripened cheeses, the flavor of Cheddar cheese is the result of the correct blend of numerous compounds, many of which are produced through microbial action when the cheese ripens. Major components of the flavor include carbonyl, nitrogenous, and sulfur compounds; fatty acids; alcohols; salt; water; and unmodified fractions of cheese (Marth 1963). The concentration of many of the components changes in the cheese during ripening.

Changes in the microflora and flavor during ripening are accompanied by changes in the body of the cheese. Freshly made Cheddar cheese has a firm and elastic body. A reduction in firmness and elasticity is apparent after 2 to 4 weeks of ripening at 7°C (Foster *et al.* 1957). This continues until a desirable, somewhat soft and smooth body develops. The body of a well-aged Cheddar (e.g., more than 1 year old) may be somewhat crumbly.

Defects

Abnormal fermentation or errors in the manufacture of Cheddar cheese can lead to defects in the finished product. The defects can be grouped into off-flavors, poor body, and public health concerns. Problems of a public health nature are described in considerable detail later in this chapter and will not be mentioned here.

Marth (1963) and Kosikowski (1977) indicate that the common off-flavors include bitterness, acid, rancid, unclean, fermented, and sulfide stinker. Defects in body and appearance mentioned by these authors include open texture, appearance of gas holes, and discoloration of the surface of cheese. Growth of mold can account for spoilage of this and other varieties of cheese.

Cheeses Similar to Cheddar

Slight variations in the procedure used to make Cheddar cheese have resulted in an assortment of varieties of cheese that resemble Cheddar. In England, such varieties include Cheshire, Derby, Lancashire, Caerphilly, Double Gloucester, and Dunlop (Chapman 1974). American variations include granular or stirred curd cheese, washed curd cheese, and Colby cheese.

SWISS CHEESE

Swiss or Emmentaler cheese is an example of a type of cheese which undergoes two fermentations, the lactic and propionic acid fermentations, and in which production of some gas is desirable because this causes formation of the characteristic holes or "eyes" in the body of the cheese. This kind of cheese is commonly called Emmentaler in Europe, where it originated in the Emme Valley, Canton of Bern, Switzerland, in the 15th century (Langsrud and Reinbold 1973A).

Traditionally, this cheese was made from raw milk, and that practice continues in some factories in Europe and elsewhere. The cheese can be made in the form of large wheels or millstones and having a firm rind. Individual wheels of cheese can weigh up to 100 kg. Now large amounts are made from pasteurized milk and in the form of rindless blocks that weigh from 36 to 41 kg. To facilitate cutting and packaging into consumer-sized units, some factories make larger blocks that weigh about 91 kg.

Manufacturing Process

Rindless block Swiss cheese is made by the procedure that follows. The outline is based on descriptions of the process by Reinbold (1972) and Kosikowski (1977).

(1) Fresh whole milk is clarified mechanically and then standardized to contain 3% milkfat. Such standardization can be done by adding an appropriate amount of skim milk to whole milk; the procedure is needed so finished cheese will contain the right amount of milkfat, which is 47 to 48% on a dry basis (e.g., fat in the dry matter).

(2) Milk is heated at about 68° to 70°C for 15 to 25 sec. This serves to destroy unwanted microorganisms that may be in the milk.

(3) Milk at about 32°C is inoculated with *Lactobacillus bulgaricus, Streptococcus thermophilus,* and *Propionibacterium shermanii. Lactobacillus helveticus* and *L. casei* have sometimes been used instead of *L. bulgaricus.* Propionibacteria other than *P. shermanii* also have been tested with some success. Inoculated milk is held at 32°C to allow for some growth by the added bacteria. Slight acid production that occurs aids in subsequent manufacturing operations.

(4) A suitable coagulant (e.g., rennet extract) is added to milk at 32°C, and milk is held at that temperature until a coagulum is formed.

(5) The coagulated milk is cut with a wire knife so small cubes of curd are produced. This allows for expulsion of whey.

(6) The curd, with the aid of an agitator, is slowly moved about in the whey for about 40 min. This further facilitates loss of whey and firms the curd.

(7) The curd is "cooked" by slowly raising the temperature of the curd-whey mixture to 50°–52°C over a 30–40 min period. After the desired temperature has been reached, the curd-whey mixture is stirred for 30–70 min. This process aids the curd in losing moisture and becoming firm, and controls bacterial activity.

(8) Curd and whey are pumped to a cheese-molding vat. Air should not be introduced during this process. Whey is drained from the curd.

(9) Curd in the molding vat is covered appropriately and then weights are placed on the curd. This is called "pressing," continues for 12–18 hr, and serves to expel gases and whey from the curd and to facilitate fusing of curd particles into a solid mass of cheese. Air, if present, interferes with this process.

(10) The mass of curd is cut into blocks of suitable size and the blocks are then placed into a solution containing at least 23% sodium chloride. Additional salt may be sprinkled on the top of each block of curd as it floats in the brine. Cheese remains in the brine for 1–2 days. During this time a rind develops on the cheese, salt is absorbed by the cheese, and bacterial activity is retarded because the cheese cools.

(11) Cheese is removed from the brine, allowed to dry for a day or less, and then is wrapped with a flexible, extensible, and fluid-proof wrapping material. Furthermore, the wrapping material should be impermeable to oxygen (to prevent mold growth on the cheese) and permeable to carbon dioxide so it can escape as it is produced during ripening of the cheese.

(12) Wrapped cheese is placed in a cold room at 7°–10°C for up to 10 days. This serves to stabilize the physical-chemical, enzymatic, and microbiological systems operative within the cheese at this time.

(13) Cheese is then moved to the warm room where it is held at 22°–24°C for 2 to 7 weeks. During this time flavor and eyes develop.

(14) Cheese is then refrigerated at 3° to 4°C until it is sold. Refrigerated storage arrests eye development, firms the cheese, inhibits bacterial growth, and controls development of some defects.

Changes During Ripening

Changes in Swiss cheese during manufacture and ripening have been described in considerable detail by Langsrud and Reinbold (1973B,C). The discussion to follow is based on their extensive review of the subject.

Growth of the starter bacteria begins while milk is being "ripened" and before the coagulant is added. Coagulation of milk serves to concentrate the bacteria in the coagulum, with only a small proportion remaining in whey. Growth of bacteria continues while the curd is being pressed. Brining and the concurrent drop in temperature of cheese retard further bacterial growth and there may be a reduction in numbers of *S. thermophilus* and *L. bulgaricus* during this time. Movement of cheese to the warm room is accompanied by growth of propionibacteria and a further reduction in numbers of *S. thermophilus* and *L. bulgaricus. L. casei* can grow during this time and is thought by some to be important in ripening of Swiss cheese.

Proteolysis takes place during ripening and the content of free amino acids in the cheese increases. Propionibacteria do not appreciably affect proteolysis. Most of the characteristic sweet flavor of Swiss cheese has been attributed to its proline content, which is greater than in other varieties of cheese. The ratio of proline to propionic acid is also believed to be important in the flavor of Swiss cheese. Furthermore, volatile carbonyl compounds contribute to the flavor.

The pH of Swiss cheese increases as the cheese progresses from pressing to the cold room, warm room, and final refrigerated storage. Immediately after pressing, cheese has a pH of 5.1 to 5.3. These values increase by 0.05 to 0.1 pH unit in the cold room. After holding in the warm room, the pH of the cheese is about 5.5 and after refrigerated storage, it is from 5.6 to 5.7.

Swiss cheese should contain eyes of proper size and form, and the eyes should be evenly distributed in the cheese. Normal eyes should be 1.3 to 2.5 cm in diameter and should have a distance of 2.5 to 7.6 cm between them.

Lipolysis, probably largely a result of bacterial metabolism, occurs in Swiss cheese and resulting fatty acids also contribute to the flavor of the finished product. Studies with Swiss and Emmentaler cheese have demonstrated that of the fatty acids, propionic and acetic are present in greatest amounts. Appreciable amounts of myristic, palmitic, stearic, and oleic acids also have been observed.

The flavor of Swiss cheese is not the result only of the compounds and processes that have just been mentioned. Instead, many other compounds (alcohols, aldehydes, esters, lactones, methyl ketones, etc.) have been found in Swiss cheese and most are there as a result of microbial activity. These compounds, even though often present in small amounts (ppm or less), contribute to the flavor of the finished cheese. The reader desiring more information on this aspect of Swiss cheese should consult the review by Langsrud and Reinbold (1973C).

Defects

Defects in Swiss cheese can result from incorrect manufacturing procedures, improper gas formation, or growth of unwanted microorganisms that can cause off-flavors and faulty eyes. Details on these defects have been given by Reinbold (1972) and Langsrud and Reinbold (1974).

Defects in eyes relate to distribution, number, size, shape, and interior appearance or condition. Some of these irregularities result from faulty manufacturing practices and others from an unsatisfactory fermentation in the cheese.

Defects also can occur in the body (dry, pasty, weak, brittle), color (bleached, pink ring, colored spots), size and shape (too small or large, bloated, uneven), and finish and appearance (dirty rind, surface splits, dried areas, soft rind, mold, rind rot) of the cheese. Microorganisms can contribute to some of these problems.

Cheeses Similar to Swiss

Several varieties of cheese that are similar to Swiss or Emmentaler have been developed in different countries. Some of these will be mentioned in the following paragraphs.

Jarlsberg is made in Norway and derives its name from the Jarlsberg region in southern Norway. The present form of this cheese was developed from 1956 to 1960 at the Norwegian Agricultural College. Milk having 3% milkfat is inoculated with mesophilic (rather than thermophilic) lactic acid bacteria and with propionic acid bacteria. The finished cheese contains about 47% fat in the dry matter.

Samsoe (Samsø) is produced in Denmark from milk containing 3.4% milkfat, and using ordinary (mesophilic) lactic starter cultures. Natural contamination serves as the source of propionic acid bacteria. Eyes in Samsoe cheese are limited in number and range in size from that of a pea to that of a cherry.

Iowa-style Swiss is somewhat similar to Samsoe although the starter cultures commonly associated with Swiss cheese are used.

Gruyère is made from raw cows' milk containing 3.4 to 3.7% milkfat, and using the regular Swiss cheese lactic starters. Propionibacteria are not added intentionally and so this cheese has only a few small eyes. A surface flora develops during ripening, resulting in a cheese that is more flavorful than is Swiss cheese. Gruyère cheese is produced primarily in the French-speaking region of Switzerland and in France.

Alpkäse or Bergkäse is made largely in the Alpine regions of West Germany, Switzerland, and Austria. Still other names are used for this cheese, particularly in Switzerland. Mesophilic and/or thermophilic lactic acid bacteria but no propionibacteria are added to whole or partially skimmed raw milk. The cheese has fewer eyes but more flavor than does Swiss cheese. Although the manufacturing procedure is similar to that for Emmentaler cheese, numerous local variations seem to exist.

SURFACE-RIPENED CHEESES

Some cheeses gain their principal sensory characteristics through the combined efforts of bacteria and yeasts that develop on the surface of the cheese during the ripening process. Examples of this type of cheese include brick and Limburger. Brick cheese will be discussed first and then some other surface-ripened cheeses will be mentioned.

Brick Cheese

Manufacturing Process.—Olson (1969) has described 2 methods that are generally used to make brick cheese. The following steps are involved in the first method.

(1) Pasteurized whole milk at 32°C is inoculated with *Streptococcus thermophilus*. Alternatively, a combination of *S. thermophilus* and *S. cremoris* or *Lactobacillus bulgaricus* may be used.

(2) After brief incubation, rennet or another suitable coagulant is added to the milk; the resultant curd is cut into 0.64 cm or 0.95 cm cubes and cooked at 38° to 45°C.

(3) After the curd is sufficiently firm, enough whey is drained so that about 2.5 cm remains above the curd surface.

(4) Curd and whey are dipped or pumped into rectangular hoops held on perforated screens.

(5) Hoops of curd are allowed to drain for 6 to 18 hr and are turned at intervals. Weights can be placed on the cheese during the draining process.

(6) Blocks of cheese are removed from hoops and immersed in brine containing 22% sodium chloride. Alternatively, salt can be applied to the exterior of the cheese.

(7) After 24 to 36 hr in brine, the cheese is removed and placed in a room at 15°C for 4 to 10 days. During this time the "smear" (growth of yeasts and bacteria) develops on the surface of the cheese.

(8) The smear is washed off, the cheese is waxed or packaged in plastic, and it is ripened for 4 to 8 weeks at about 4°C. More flavor can be obtained by leaving the smear intact for the entire ripening period.

The second or "sweet curd" method for making brick cheese employs *S. lactis* and/or *S. cremoris* as the starter culture and addition of water to the curd-whey slurry to control development of acid in the cheese. Some other modifications in manufacturing must be made to ensure that the minimum pH is 5.1 to 5.2 in 3-day-old cheese. Salting and ripening proceed as outlined above.

Brick cheese contains not more than 44% moisture and at least 50% of the total dry matter must be milkfat. A typical brick cheese is about 12 cm wide, 25 cm long, and 7.5 cm thick; it weighs about 2.25 kg.

Changes During Ripening.—Yeasts predominate in the surface microflora of brick cheese during the initial stages of ripening (Olson 1969). This is because of their ability to grow at the temperature and the relative humidity (approximately 95%) used for ripening as well as the low pH and high concentration of salt at the surface of the cheese. Depending on the water activity of the cheese, yeasts in one or more of the following genera may be present: *Debaryomyces, Rhodotorula, Trichosporon, Candida,* and *Torulopsis.*

Growth of yeasts serves to modify the surface of the cheese so that *Brevibacterium linens* and micrococci can grow. Yeasts accomplish this by metabolizing lactic acid and thus raising the pH of cheese at the surface above the minimum for growth of bacteria. Additionally, yeasts produce vitamins which may enhance growth of bacteria. Growth of yeasts also may contribute to the final flavor of brick cheese.

Brevibacterium linens and micrococci (*Micrococcus varians, M. caseolyticus,* and *M. freudenreichii*) develop after sufficient growth of yeasts has taken place. These bacteria release proteolytic enzymes that are largely responsible for producing the characteristic flavor of brick cheese that has had a surface smear develop during ripening.

Defects.—Olson (1969) described the major defects that can appear in brick and other surface-ripened cheeses. Included are defects in flavor, body and texture, and surface microflora.

Flavor defects include:

(1) Sour or acid—caused by excessive fermentation of lactose or inadequate washing of curd, or both. Too much acid retards development of surface microflora and too little acid results in fruity and gassy cheese.
(2) Bitterness—caused by abnormal protein degradation when the starter culture contains undesirable lactic acid bacteria such as *Streptococcus faecalis* var. *liquifaciens.*
(3) Flat—caused by insufficient growth of surface miroflora.
(4) Fruity and fermented—caused when the pH of cheese is high and the salt content is low so that anaerobic spore-forming bacteria can grow.

Defects in body and texture include:

(1) Corky—caused by inadequate acid development or excessive washing of curd, or both.
(2) Weak or pasty—caused by a combination of excessive moisture, too much or too little acid, and inadequate salt.
(3) Mealy—too much acid.
(4) Openness—caused by whey being trapped between firm cubes of curd; openings remain when whey drains from cheese.
(5) Gassiness—caused by growth of coliform bacteria, yeasts, certain strains of lactic acid bacteria, *Bacillus polymyxa,* or anaerobic spore-forming bacteria.
(6) Split cheese—caused by gas from anaerobic sporeformers.

Defects in the surface microflora include:

(1) Lack of growth—caused by low temperatures during ripening, too much salt in cheese, or drying of the surface of cheese.

(2) Mold growth—results when the surface smear fails to develop because the surface of the cheese is too dry.

Other Varieties of Surface-ripened Cheese

Limburger cheese has up to 50% moisture and is made by the procedures used for brick cheese. The initial ripening at 16°C and at a high relative humidity is longer than for brick cheese, thus allowing extensive growth of *B. linens*. Surface growth is not removed when the cheese is wrapped and moved to storage at 4° to 10°C. Extensive growth of *B. linens* on relatively small pieces of cheese accounts for the strong, pungent flavor and aroma of Limburger.

Port du Salut, Trappist, and Oka are wheel-shaped cheeses developed by Trappist monks and made by procedures somewhat similar to those for brick cheese. *Geotrichum* may appear in the surface microflora and contribute a distinctive flavor to the cheese.

Other varieties of surface-ripened cheese that are more common in Europe than in the United States include Saint Paulin, Bel Paese, Königkäse, Bella Alpina, Vittoria, Fleur des Alpes, Butter, and Tilsit. Liederkranz is a trade name for a surface-ripened cheese made in the United States by procedures similar to those for Limburger.

MOLD-RIPENED CHEESES

Certain molds are used as the major ripening agents of some cheeses. In some instances mold growth occurs throughout the cheese (e.g., blue cheese), whereas other times growth of mold appears only on the surface of the cheese (e.g., Camembert cheese). This section will discuss both types of cheese.

Blue Cheese

As just mentioned, this cheese is ripened primarily by growth and activity of mold throughout the cheese mass. Blue cheese was not made successfully in the United States until about 1918; information on appropriate procedures for making this cheese was not available earlier (Walter and Hargrove 1969). A blue cheese is about 19 cm in diameter and weighs about 2 to 2.3 kg; it is round with a flat top and bottom.

Manufacturing Process.—The following steps (Kosikowski 1966) are involved in producing blue cheese.

(1) Whole milk from cows is separated into cream and skim milk fractions, and the skim milk is pasteurized.

(2) Cream is bleached by adding benzoyl peroxide (maximum 0.002% of the weight of milk), pasteurized, and homogenized. Bleaching is done so that the finished cheese is white (except for mold growth) in color and homogenization increases the surface area of milkfat globules which facilitates lipolytic action that occurs during ripening of the cheese.

(3) Cream and skim milk are combined and 0.5% of an active lactic (*Streptococcus lactis* and/or *Streptococcus cremoris*) starter culture is added.

(4) Inoculated milk is held at 30°C for 1 hr to allow some acid production.

(5) A suitable coagulant is added, milk is allowed to coagulate, and the resultant curd is cut into cubes with 1.6 cm wire knives.

(6) Curds are allowed to remain in whey for about 1 hr while additional acid develops. Curds and whey are then heated to 33°C, held briefly, and whey is drained from the curds.

(7) Curd is trenched and inoculated with spores of *Penicillium roqueforti*; salt is also added and then the curd is stirred. Spores can be obtained in powdered form from commercial culture firms.

(8) Curd is placed into stainless steel blue cheese hoops.

(9) Hoops of curd are turned once every 15 min for 2 hr and then are allowed to drain overnight at 22°C.

(10) The next day curd is removed from the hoop and salt is applied to the surfaces of the cheese. Cheese is then stored at 16°C and 85% relative humidity and is salted once daily for 4 more days.

(11) After salting is completed, each flat surface of the cheese is pierced about 50 times with a suitable needle-like steel rod. This facilitates escape of carbon dioxide from the cheese and entrance of air so that growth of the mold is encouraged.

(12) Pierced cheese is then stored at 10° to 13°C and 95% relative humidity.

(13) After 1 month, surfaces of cheese are cleaned; cheese is wrapped in foil and then stored at about 2°C for 3 to 4 months to allow additional ripening.

Changes During Ripening.—The high acidity and the increasing amount of salt in the cheese cause a rapid demise of the lactic starter bacteria so that only a few viable cells remain in cheese that is 2 to 3 weeks old. Growth of *P. roqueforti* inside cheese becomes evident about 8 to 10 days after the cheese is pierced (Foster *et al.* 1957). Development of mold is maximal in 30 to 90 days; by then the mold has grown throughout the spaces between curd particles and along holes made when the cheese was pierced.

Penicillium roqueforti is primarily responsible for ripening blue cheese. Proteolytic enzymes from the mold act to soften the curd and thus to produce the desired body in the cheese (Marth 1974). Some components of blue cheese flavor may result from this proteolytic action. Perhaps more important, the mold produces water-soluble lipases which hydrolyze milkfat to free fatty acids. Included are caproic, caprylic, and capric acids which together with their salts are responsible for the sharp peppery flavor of blue cheese (Ernstrom and Wong 1974). *Penicillium roqueforti* forms hepta-none-2 from caprylic acid and this ketone is an important component of blue cheese flavor. Other ketones (pentanone-2 and nonanone-2) have been recovered from blue cheese and probably contribute to its flavor (Ernstrom and Wong 1974). Additionally, *P. roqueforti* reduces methyl ketones to form secondary alcohols (pentanol-2, heptanol-2, and nonanol-2) which also contribute to the flavor of the cheese (Jackson and Hussong 1958). The pH of blue cheese initially is at 4.5 to 4.7 and increases to 6.0 to 6.25 after 2 to 3 months of ripening.

Growth of microorganisms can occur on blue cheese after 2 to 3 weeks of ripening. This growth consists of yeasts, micrococci, and *Brevibacterium* spp. and contributes to the final flavor of the cheese. If cheese is waxed, microorganisms will not grow to produce the surface slime.

Defects.—Improper development of *P. roqueforti* can cause an assort-ment of defects. Too much growth can result in a musty, unclean flavor or in loss of the typical flavor, whereas too little growth is accompanied by defects in color, a body that is too firm, and insufficient flavor. Growth of unwanted molds can cause defects on the surface of blue cheese.

Cheeses Similar to Blue.—Roquefort is the original blue cheese. The designation "Roquefort" is applicable only to cheese made from ewes' milk in the Roquefort area of France. A similar product made elsewhere in France is called bleu cheese. A peculiarity of Roquefort cheese is the fact that it ripens in a network of caves and grottoes where cool moist air moves briskly, temperature never exceeds 10°C, and relative humidity remains at about 95% throughout the year (Walter and Hargrove 1969).

Gorgonzola is the principal blue-mold cheese of Italy where it is claimed to have been made in the Po Valley since 879 A.D. The English blue-mold cheese is Stilton, which has been made since about 1750. Stilton is milder than Roquefort or Gorgonzola. The texture of Stilton is sufficiently open so that the cheese usually does not need piercing to facilitate mold growth.

Camembert Cheese

Camembert is an example of cheese that is made with a mold developing only on the surface rather than throughout the mass of cheese as happens with blue cheese. Apparently cheese ripened through the action of mold growth on the surface has been produced in France for centuries. According to Kosikowski (1966), it was in 1791 that Marie Harel, who lived in the village of Camembert in Normandy, developed a product similar to the present Camembert cheese.

The typical Camembert cheese is about 11 cm in diameter, 2.5 to 3.8 cm thick, and weighs 225 to 250 g. The interior is light yellow and waxy, and creamy or almost fluid in consistency, depending on the degree of ripening (Walter and Hargrove 1969). The rind is a thin felt-like layer of mold mycelium and dried cheese. The mold is gray-white in color; sometimes bacterial growth occurs on the surface of the cheese and results in development of areas that are reddish-yellow in color.

Manufacturing Process.—The basic procedure for the manufacture of Camembert cheese employs the following steps:

(1) Pasteurized whole milk with about 3.5% milkfat is adjusted to about 32°C and is inoculated with 2% of an active lactic starter culture (*S. lactis* and/or *S. cremoris*) plus a sporulated culture of *Penicillium camemberti* (alternatively, spores of the mold can be applied to the surface of the cheese later in the manufacturing process).

(2) Annatto (yellow coloring) may be added to the milk.

(3) Inoculated milk is allowed to "ripen" for 15 to 30 min so that a titratable acid of 0.22% develops. Rennet extract (or other suitable coagulant) is added and the milk is stirred and then held quiescently until a firm curd develops.

(4) Curd is cut into cubes with 1.6 cm knives. Alternatively, uncut curd can be ladled into hoops.

(5) Curd is not cooked but is placed into open-ended, round, perforated, stainless steel molds or hoops. Filled hoops are allowed to drain for about 3 hr at about 22°C; no pressure is applied to the cheese during draining.

(6) Hoops of cheese are turned and draining continues. The turning process is repeated 3 to 4 times at 30 min intervals.

(7) Both flat sides of curd in hoops may now be inoculated by spraying the surface with a fine mist of *P. camemberti* spores suspended in water.

(8) After an hour, cheese is removed from hoops, placed on a drain table, and held at 22°C for 5 to 6 hr. Weights are generally not placed on the cheese.

(9) Dry salt is applied to the surface of the cheese which is then held overnight at about 22°C.

(10) Cheese is held for 1 or 2 weeks at 10° to 15°C and 95 to 98% relative humidity. It may be turned once during storage to facilitate uniform development of mold on the surface.

(11) Cheese is moved to storage at 4° to 10°C after being wrapped in foil. Storage under these conditions may be for several weeks before the cheese is packaged and moved into distribution channels. Final ripening occurs during distribution.

Camembert cheese should be consumed within 6 to 7 weeks after it is made. The process used to make Camembert cheese has been mechanized in the United States and Europe.

Changes During Ripening.—Foster *et al.* (1957) have outlined the major changes that occur when Camembert cheese ripens. Normally film yeasts and *Geotrichum* appear on the surface of the cheese within 3 to 4 days after it is placed in the "warm" room. Growth of *P. camemberti* is evident a few days later and becomes maximal when the cheese is 10 to 12 days old. Development of mold is followed by appearance of a reddish growth comprised of *Brevibacterium linens* and related pigmented, rod-shaped bacteria.

Film yeasts and *Geotrichum* are believed to ferment residual lactose at the surface of the cheese and also are thought to reduce acidity, thereby facilitating later growth of other organisms. Yeasts and *Geotrichum* do contribute to the flavor of Camembert cheese; however, excessive growth of these fungi leads to excessive softening of the rind and undesirably strong flavors in the cheese.

Development of *P. camemberti* is essential for production of the normal body and flavor of Camembert cheese. Pigmented bacteria *(Brevibacterium)* develop after fungi have reduced the acidity of the cheese at its surface. These bacteria also contribute to the flavor of the ripened cheese.

Defects.—Common defects associated with Camembert cheese include (a) early gas production during drainage of the cheese, and (b) growth of undesirable "wild" molds on the surface. The first defect can be minimized or eliminated by adequate sanitation and use of high quality milk. The second can be controlled by maintaining adequate humidity in the room where cheese is ripening; wild molds tend to develop on cheese when its surface becomes too dry for normal development of *P. camemberti*.

Late in 1971 at least 227 persons in 8 states of the United States became ill with acute gastroenteritis about 24 hr after consuming French Camembert or Brie cheese (Barnard and Callahan 1971; Schnurrenberger *et al.* 1971). The illness was attributed to the presence in cheese of enteropathogenic *Escherichia coli*. This problem is discussed in greater detail later in this chapter.

Cheeses Similar to Camembert.—Walter and Hargrove (1969) and Weigmann (1933) describe several kinds of cheese other than Camembert that are ripened largely through the activity of surface mold. Brie is probably the best known of these Camembert-like cheeses.

Brie is made in 3 sizes: (a) large—about 40 cm in diameter and 3.8 to 4.2 cm thick, about 2.7 kg; (b) medium—about 30 cm in diameter and somewhat thinner than the large size, about 1.6 kg; and (c) small—14 to 20 cm in diameter and 3.2 cm thick, 0.45 kg. The difference in size between Brie and Camembert causes differences in the ripening process of the two cheeses. This, together with variations in the manufacturing process, causes the flavor and aroma of Brie to differ from Camembert.

Coulommiers is a cheese similar to the small Brie but is ripened for less time. Another cheese similar to Brie is Monthery, which is made in two sizes roughly equivalent to the large-sized and medium-sized Brie. Monthery can be made from whole or partially skimmed milk.

Melum is similar to the small brie but has a firmer body and sharper flavor. This cheese also is designated as Brie de Melum. Other cheeses similar to Camembert and primarily produced in France include Olivet and Vendome. The latter is sometimes buried in ashes in a cool, moist cellar during ripening.

PROCESS CHEESE

James L. Kraft working by himself and Elmer E. Eldredge working for the Phenix Cheese Company were largely responsible for developing process cheese in the United States (Price and Bush 1974A). Kraft worked with Cheddar cheese and used sodium phosphate as the emulsifying salt, whereas Eldredge worked with Cheddar, Swiss, and Camembert cheese and used sodium citrate rather than phosphate. Eventually the Kraft and Phenix companies shared their patent rights; the initial patents were issued to Kraft in 1916 and to Eldredge in 1921 (Price and Bush 1974A). Today more than 50% of the cheese produced in the United States is converted to process cheese or a related product. Development of this industry has been described in detail by Price and Bush (1974A,B).

The Products

Process Cheese.—This product is made from a mixture of natural cheese, color, salt, and emulsifiers. The mixture is heated to 71°–80°C; the finished product is at pH 5.6–5.8 and contains about as much milkfat and moisture as does the original natural cheese (Kosikowski 1977).

Process Cheese Food.—This product is made from the same ingredients as process cheese, except that one or more of the following may be added: skim milk, whey, milk, cream, albumin, skim milk, and organic acids. The mixture is heated to 79°–85°C; the finished product is at pH 5.2–5.6 and contains not more than 44% water or less than 23% milkfat (Kosikowski 1977).

Process Cheese Spread.—This product contains the same ingredients as process cheese food, except that gums may be added to retain water. The mixture is heated to 88°–91°C; the finished product is at pH 5.2 or lower and contains between 44 and 60% water and not less than 20% milkfat (Kosikowski 1977).

Manufacturing Process Cheese and Related Products

Manufacture of process cheese begins with selection of the natural cheese to be used. An attempt is made to select cheese so that the finished product will be similar in flavor and other characteristics from day to day. A process cheese with desirable characteristics can be made by blending together 55% non-aged cheese, 35% medium-aged cheese and 10% fully-aged cheese (Kosikowski 1977).

After cheese has been selected, the surfaces are cleaned and trimmed and the cheese is ground with a grinder or other suitable device. The freshly ground cheese is mixed with the emulsifier and other ingredients, as appropriate. Emulsifiers are used to regulate the pH and produce a stable product which maintains its integrity during storage. Commonly used emulsifiers include sodium citrate, disodium phosphate, trisodium phosphate, sodium hexa-metaphosphate, and tetrasodium diphosphate.

The mixture of cheese and ingredients is then heated ("pasteurized") to the desired temperature in a suitable device. After heating (or "cooking"), the smooth molten cheese mass is packaged in suitable containers and cooled. Sometimes ribbons of process cheese are made which are then cut so that sliced process cheese is produced.

Microbiology of Process Cheese and Related Products

Process cheese and related products receive a heat treatment sufficient to inactivate vegetative cells of bacteria and molds, mold spores, and yeasts. Bacterial spores may survive the heat treatment. Sometimes spores of *Clostridium tyrobutyricum* or *Clostridium sporogenes* have germinated, grown, and produced gas in process cheese. This defect may occur after the product has been distributed to retail stores. The problem can be controlled by (1) using natural cheese with few anaerobic spores, (2) ensuring that the pH is not above 5.8, (3) ensuring that the sodium chloride content of the serum is 6−7%, and (4) holding process cheese, including its display in the store, at temperatures below 20°C (Kosikowski 1977). See the discussion on botulism later in this chapter for a description of how this problem relates to process cheese and related products.

Contamination of sliced process cheese with mold spores during packaging is possible and such contamination can result in mold growth on the product before it reaches the consumer. Appropriate hygienic measures can minimize or eliminate this problem.

PUBLIC HEALTH CONCERNS

When pathogen-free milk, an active lactic starter culture, and good hygienic practices are used to produce cheese, a safe food results. If the cheese is then handled and distributed in a sanitary manner, this food will be safe for consumption. When one considers the many millions of pounds of cheese that are produced annually worldwide and the few cases of reported illness attributed to this food, one must conclude that cheese has a remarkable record as a safe food.

In spite of this record, sometimes mistakes have been made in the steps leading from cow to milk to cheese to the consumer, and these mistakes have caused outbreaks of foodborne illness. Problems of greatest concern include staphylococcal food poisoning, salmonellosis, aflatoxicosis, gastroenteritis caused by enteropathogenic strains of *E. coli*, and presence in cheese of biologically active amines. All of these will be discussed in some detail, not because they are major problems, but because making the information available here may help interested persons in taking the steps needed to keep the problems under control. Furthermore, most discussions of cheese either ignore this information or treat it superficially, and hence the information may not be readily available to the person who is working with cheese.

Staphylococcal Food Poisoning

Staphylococcus aureus, a Gram-positive, coccus-shaped bacterium, can produce an enterotoxin (a protein with a molecular weight of approximately 28,000 to 34,000—there are several enterotoxins). This enterotoxin, when ingested, can cause such symptoms as nausea, vomiting, retching, and often diarrhea (Minor and Marth 1976). Recovery often follows in 24 hr, but several days may be required. The extent to which these symptoms may appear and the severity of illness are determined chiefly by the amount of toxin that was ingested and the susceptibility of the individual who becomes a victim of the disease (Minor and Marth 1976). Outbreaks of this disease have been caused by toxic cheese (Hendricks *et al.* 1959; Allen and Stovall 1960; Hausler *et al.* 1960; Zehren and Zehren 1968A,B).

Occurrence of Staphylococci in Milk.—Raw milk may become contaminated with staphylococci from several sources but the principal one is probably the mastitic bovine udder. Mastitis is often caused by *Staphylococcus aureus* and continues to be a major problem in dairy cattle. For example, in Wisconsin (U.S.A.) in 1974 at least 16% of all dairy cows suffered from mastitis (Anon. 1975); to be sure, not all of these cows were infected with *S. aureus*. Mastitis also seems to appear more regularly in large (91% of Wisconsin herds in 1974) than in small (38%) dairy herds. Since the trend is toward maintenance of ever larger herds on dairy farms, it is almost certain that mastitis will continue as a problem in animal health and also will result in the presence of *S. aureus* in some lots of raw milk.

The occurrence of staphylococci in raw milk has been verified by results of numerous surveys. Among the first surveys to do so was the one by Williams (1941). He studied 10 herds and found that more than 50% of the cows in these herds were shedding staphylococci in their milk. Staphylococcal counts in excess of 1000/ml were not uncommon and isolates obtained were generally coagulase-positive staphylococci. Williams (1941) did his work before antibiotics were used regularly to treat mastitis; hence the problem of antibiotic resistance in staphylococci of bovine origin and the increase in frequency of staphylococcal mastitis had not yet occurred. Twenty years after the work of Williams, samples of raw milk were tested by Clark and Nelson (1961). These investigators found that their samples contained 25 to 3300 coagulase-positive staphylococci/ml. Undoubtedly, present-day samples would yield similar results.

Occurrence of staphylococci in raw milk would be of less concern if enterotoxigenic strains were never present. Unfortunately, this is not true; data obtained by several researchers have demonstrated that some staphylococci from the bovine udder are enterotoxigenic. Bell and Veliz (1952) found that enterotoxin was produced by 25 of 35 cultures of staphylococci isolated from the udder. Enterotoxigenicity of staphylococci obtained from raw milk was determined by Casman (1965). He found that 4.2% of 190 staphylococcal cultures from raw milk were able to produce enterotoxin A and B. About 75% of the toxigenic strains produced enterotoxin A, whereas the remainder formed enterotoxin B. More recently Olson *et al.* (1970) tested 157 cultures of staphylococci from mastitic udders. Eleven of the cultures produced enterotoxin C, 11 others yielded enterotoxin D, 1 formed both C and D, and none produced enterotoxins A and B.

Data cited here are adequate to demonstrate that: (1) staphylococci are likely to be present in raw milk, and (2) some of the staphylococci in raw milk are likely to be enterotoxigenic.

Growth of Staphylococci in Milk.—The foregoing considerations would be of less consequence if staphylococci could not grow in raw or processed milk since growth is needed for synthesis of enterotoxins. In general, heated milks are excellent substrates, but raw milk is less favorable for staphylococcal growth.

Clark and Nelson (1961) held raw milk for 7 days at 4° and 10°C. Staphylococci failed to grow at 4°C but grew, although generally slowly, at 10°C. The initial number of staphylococci was not indicative of whether or not growth would occur during the holding period. Although the aerobic plate count of all test milks was low initially, the count was uniformly high after 4 days at 10°C. In one instance, marked growth of staphylococci was evident, which suggests that the kind of bacteria present and growing in milk may affect proliferation of staphylococci.

Both growth and enterotoxin production by staphylococci in milk were investigated by Donnelly et al. (1968). They inoculated low- and high-count raw milk with *S. aureus* and then held the milks at different temperatures. Data in Table 4.1 were obtained when 2 samples of low-count milk were inoculated with *S. aureus* and held at 30°C. It is again evident that behavior of staphylococci can differ in milks with similar initial numbers of bacteria. Although the indigenous flora grew in both samples of milk, staphylococcal growth was more pronounced in one than in the other milk. This more extensive growth by the staphylococcus resulted in enterotoxin production in one milk but not in the other. Similar results were obtained when incubation was at 25° and 35°C.

TABLE 4.1. GROWTH OF AND ENTEROTOXIN PRODUCTION BY *STAPHYLOCOCCUS AUREUS* IN RAW MILK AT 30°C[1]

| Time (hr) | Number/ml (× 10^6) | | | |
| | Trial 1 | | Trial 2 | |
	SPC[2]	S. aureus	SPC[2]	S. aureus
6	1	0.03	0.04	0.06
9	20	0.2	1	1
12	100	2	20	10
18	700	20	200	80[3]
24	1000	30	600	200[3]
36	3000	30	2000	400[3]
48	3000	30	2000	300[3]
72	4000	20	2000	300[3]
96	3000	9	800	200[3]

Source: Donnelly et al. (1968).
[1]Milk inoculated to contain 10^4 S. aureus/ml.
[2]SPC = Standard Plate Count: initial for Trial 1, 3×10^4/ml; for Trial 2, 10^4/ml.
[3]Enterotoxin detected.

The same investigators also determined the amount of time needed for enterotoxin production when staphylococci were added to low- and high-count milks. Enterotoxin production in low-count milk was more rapid at all temperatures when the largest number (10^6/ml) of staphylococci was added. When high-count milk was used, a large inoculum of staphylococci was needed for enterotoxin production, and then enterotoxin appeared only at 35°C.

The data that have just been described indicate that: (1) staphylococci can grow and produce enterotoxins in raw milk, (2) raw milk with few bacteria is more suitable for growth and synthesis of enterotoxin by staphylococci than is milk with many bacteria, and (3) there are differences among milks with few bacteria in their suitability for growth of staphylococci. These facts must be recognized as one strives to produce raw milk with ever fewer bacteria and as such milk is held for ever longer periods before it is processed. Enterotoxins, if formed at this time, would probably appear in a finished product.

Raw milk often is pasteurized or receives some other heat treatment before it is used to make cheese. Is such milk suitable for growth of and enterotoxin production by staphylococci? Donnelly et al. (1968) tested pasteurized milk and found that it supported growth and enterotoxin production by staphylococci. Pasteurized milk was even more suitable for enterotoxin production than was raw milk with few bacteria, particularly when the inoculum of staphylococci was small. When a larger inoculum of staphylococci was used, both pasteurized milk and raw milk with few bacteria were equally suitable for enterotoxin production. Tatini et al. (1971) did a similar experiment but gave milk several different heat treatments before staphylococci were added. Their data also indicate that heat-treated milk was more suitable for enterotoxin production than was the same milk in the raw state. The amount of heat, which ranged from that needed for pasteurization to that required for sterilization, had little apparent effect on the capacity of milk to support enterotoxin production.

From the foregoing discussion, it is evident that both raw and heated milk can support growth of and enterotoxin production by staphylococci. In spite of this, only very few reported outbreaks of staphylococcal food poisoning have been attributed to fluid milk and related products. (Minor and Marth 1972). There are good reasons why this is true. They include: (1) the great degree of control exerted in production, processing, and distribution of these products, (2) the extent of abuse needed to create a problem is greater than occurs because of the controls just mentioned, and (3) adventitious organisms can grow if milk is abused and thus they retard growth of staphylococci and also spoil the milk so that it would not be consumed.

Thermal Destruction of Staphylococci in Milk.—A heat treatment of some sort is commonly given to milk before cheese is made. Consequently, it is appropriate to briefly consider use of heat to inactivate staphylococci in milk.

According to data by Thomas et al. (1966), the D value (time in minutes at a given temperature to reduce the population by 90%) for S. aureus in skim milk at 60°C is 3.44 and at 65.6°C it is 0.28. The z value (degrees Fahrenheit needed for the thermal death curve to traverse one log cycle), according to Thomas *et al.* (1966) was 9.17. Heinemann (1957) tested the same strain of *S. aureus* that was used by Thomas *et al.* (1966). When the bacterium was heated in raw milk, it was completely inactivated after 80 min at 57.2°C, 24 min at 60°C, 6.8 min at 62.8°C, 1.9 min at 65.6°C, and 0.14 min at 71.7°C. Heinemann (1957) determined the z value to be 9.2, a figure that agrees well with that of Thomas *et al.* (1966).

Zottola *et al.* (1965,1969) isolated staphylococci from milk and cheese and then tested them for heat resistance in raw milk. When the test population exceeded 10^6/ml, 116 of 236 strains survived heating at 63.9°C for 21 sec; whereas when populations were below 10^6/ml, 108 of the strains survived exposure to 65.6°C for 21 sec. These heat treatments were used to simulate conditions which sometimes exist in cheesemaking when milk receives a subpasteurization treatment. This is done because many cheesemakers believe ripened cheese such as Cheddar is more flavorful when it is made from milk that has received a sub-pasteurization rather than a full pasteurization heat treatment.

Interactions Between Lactic Acid Bacteria and Staphylococci.—It was mentioned earlier in this section that use of an active lactic starter culture contributes to production of safe cheese. Why this is so will be evident from the discussion that follows.

Reiter *et al.* (1964) worked with a lactic starter culture that was to be used for cheesemaking. This culture inhibited growth of *S. aureus* in raw, pasteurized, and steamed milk. When lactic acid developed by the starter culture was neutralized as it was formed, inhibition was found to be a function of more than just pH. The quantity of lactic acid bacteria added to milk can influence growth, enterotoxin production, and even survivial of *S. aureus*. For example, Richardson and Divatia (1973) noted that adding 0.001 and 0.01% of a milk culture of lactic streptococci to milk containing approx. 10^6 *S. aureus*/ml served to appreciably retard growth of the staphylococcus. Lack of growth by the staphylococcus during the first 8 hr of incubation and subsequent inactivation were observed when the inoculum was 0.1, 0.5, or 1.0%. Enterotoxin appeared when the inoculum was 0.001 and 0.01% but not when any of the 3 larger inocula was used. When the number of staphylococci in milk was reduced, a smaller inoculum of lactic acid bacteria was needed to inhibit or inactivate *S. aureus*.

According to a report by Jezeski *et al.* (1967), an actively growing culture of *Streptococcus lactis* inhibited and sometimes inactivated *S. aureus* when both organisms were together in sterile or steamed skim milk. Enterotoxin was produced by *S. aureus* when it grew alone or in the presence of the streptococcus and a homologous bacteriophage. Absence of the bacteriophage enabled the lactic acid bacterium to prevent enterotoxin production by the staphylococcus. Haines and Harmon (1973) inoculated APT (All Purpose Tween) broth with *S. aureus* and *S. lactis* to determine when growth and enterotoxin production by *S. aureus* would be inhibited. They learned that: (1) no enterotoxin was produced and growth of *S. aureus* was retarded similarly when the initial pH values were 6.0, 6.5, and 7.0; (2) no enterotoxin was produced at either 25° or 30°C, but growth of *S. aureus* was inhibited somewhat more at 25° than 30°C; (3) enterotoxin was produced at 30°C when the number of *S. lactis* to *S. aureus* was 10:90, but not when it was 50:50 or 90:10; (4) different strains of *S. lactis* were equally inhibitory and prevented toxin production; and (5) different strains of *S. aureus* were inhibited in a similar way by *S. lactis*.

Further evidence for the inhibitory properties of lactic streptococci comes from a report by Gilliland and Speck (1972). They tested 6 different lactic streptococcus cultures for their ability to inhibit *S. aureus*. Inhibition of *S. aureus* was almost complete regardless of the starter culture used or the time, within limits, required for acid formation. These authors suggested that repression of staphylococci by lactic streptococci may involve production of antibiotics, hydrogen peroxide, and volatile fatty acids in addition to lactic acid. Of these, undoubtedly acid production is the single most important factor which causes the staphylococci to be inhibited.

Minor and Marth (1970) did a series of studies to more clearly define the role of acid in controlling growth of staphylococci in milk. They gradually added acid to milk which was inoculated with *S. aureus*. Acid was added over a 4, 8, or 12 hr period. When acid was gradually added over a 12 hr period, 90% reduction in growth of *S. aureus* was achieved if the final pH values were 5.2 for acetic, 4.9 for lactic, 4.7 for phosphoric and citric, and 4.6 for hydrochloric acid. To achieve a 99% reduction during a 12 hr period, the final pH value had to be 5.0 for acetic, 4.6 for lactic, 4.5 for citric, 4.1 for phosphoric, and 4.0 for hydrochloric acid. A final pH value of 3.3 was required for 99.9% reduction in growth with hydrochloric acid, whereas the same result was obtained at a final pH value of 4.9 with acetic acid. Correspondingly lower final pH values were required to inhibit growth within 8 and 4 hr periods.

Staphylococci in Cheese.—Although outbreaks of staphylococcal food poisoning have been associated with rennet-type cheese, this problem can be largely avoided if: (1) cheese is made from milk that was stored to preclude staphylococcal growth and then was given a heat treatment, such as pasteurization, that was adequate to inactivate staphylococci; (2) heated milk is not contaminated with staphylococci; and (3) an active starter culture that produces sufficient acid during cheesemaking is used. Since illness has been associated with cheese, several investigators have studied the behavior of *S. aureus* during the manufacture of different kinds of cheese.

Table 4.2 summarizes some of the data obtained by Tuckey *et al.* (1964) when they worked with Cheddar cheese. It is evident that some growth of staphylococci is possible even when cheese is made by a normal process; hence the need to inactivate staphylococci before the cheesemaking process begins. The initial increase (after 2 hr) is largely attributable to the concentration effect when curd is formed. Some growth continued during the manufacturing process except when the curd was salted. The detrimental effect of salt was limited, as might be expected, and growth of staphylococci again was evident by the time cheese was taken from press. Takahashi and Johns (1959) and Walker *et al.* (1961) made similar observations when they studied Cheddar and Colby cheese, respectively.

TABLE 4.2. BEHAVIOR OF *STAPHYLOCOCCUS AUREUS* DURING THE MANUFACTURE OF CHEDDAR CHEESE

	Number/g or ml $\times 10^5$	
Stage in Manufacturing Process	MSA[1]	S110[2]
Milk after inoculation (0 time)	0.84	0.89
Curd at 37.8°C (2 hr)	15.0	14.0
Whey at 37.8°C (2 hr)	0.03	0.01
Curd at draining (2 hr, 50 min)	19.0	20.0
Curd during cheddaring (4 hr)	25.0	37.0
Curd after salting (5 hr)	12.0	15.0
Curd after overnight pressing (26 hr)	25.0	23.0

Source: Tuckey *et al.* (1964).
[1]Mannitol salt agar.
[2]*Staphylococcus* Medium No. 110.

Additional work was done by Tatini *et al.* (1971) (Table 4.3). Their data indicate the importance of both the starter culture and the initial concentration of staphylococci in determining whether or not a cheese becomes toxic. These authors concluded that Colby cheese made normally would be toxic if it contained at least 15 million staphylococci/g. The value for Cheddar cheese was 28 million /g. If the starter culture failed to perform properly, then toxic cheese could result if 3 to 5 million staphylococci/g were present. Finally, it should be mentioned that the number of staphylococci declines as the cheese ripens. However, if they are present initially, it is not usual for them to persist in well-ripened cheese.

TABLE 4.3. GROWTH OF AND ENTEROTOXIN PRODUCTION BY *STAPHYLOCOCCUS AUREUS* DURING CHEDDAR CHEESE MANUFACTURE

Starter[1]	Number of *S. aureus*[2] ($\times 10^6$)		TA[3] at Milling (%)	pH Out of Press	Enterotoxin
	Initial	At Milling			
PR	0.06	6.4	0.61	5.35	—
PR	0.49	40	0.50	5.40	+[4]
WC	0.06	11	0.48	5.40	—
WC	0.28	40	0.51	5.35	—
C₂O	0.35	15	0.63	5.45	+
C₂O	4	210	0.55	5.30	+

Source: Tatini *et al.* (1971).
[1]PR and WC are mixed type multiple strain cultures. C_2O is an *S. lactis* single strain culture.
[2]*Staphylococcus* Medium No. 110.
[3]Titratable acidity.
[4]Enterotoxin present when whey was drained from cheese.

Aged Cheddar cheese is sometimes added to another food product to impart a desired flavor. Procedures to accelerate ripening of the cheese have been sought so that relatively fresh, and less expensive, cheese could be used for this purpose instead of the more costly aged product made by conventional processes. One method to prepare a liquid cheese product with a Cheddar-like flavor in fewer than 7 days has been described by Kristoffersen et al. (1967). Basically, the process involves making a slurry from 2 parts of 24 hr-old salted, unpressed Cheddar curd and 1 part of a solution of 5.2% sodium chloride. The slurry then is held at 30°C. Addition of 10–100 ppm of reduced glutathione enhances development of the Cheddar flavor. Gandhi and Richardson (1973) inoculated similar slurries with S. aureus and found that the bacterium could grow and produce enterotoxin in 24 to 48 hr at 32°C when the slurries contained 45 to 60% moisture. Treatment of the inoculated slurry with 0.5% hydrogen peroxide did not eliminate the staphylococci but addition of 0.2 or 0.3% sodium sorbate inhibited growth and thus prevented production of enterotoxin. The work of Gandhi and Richardson (1973) emphasizes the need to evaluate new processes for potential health hazards.

Tatini et al. (1973) also studied the potential for enterotoxin production by added staphylococci in the manufacture of brick, Swiss, Mozzarella, and blue cheese. No enterotoxin appeared in their experimental lots of Mozzarella or blue cheese. Environmental conditions in these two types of cheese were such that the staphylococci, even though some growth occurred, were unable to produce enterotoxin. In contrast to these observations, the investigators noted that staphylococcal growth and enterotoxin production did occur during the manufacture of brick and Swiss cheese. The kind of starter culture used, the initial population of staphylococci, and the population achieved by 7 hr after cheese was made appear to have determined whether or not the cheese was toxic. These authors evaluated several toxigenic strains of S. aureus when they experimented with Swiss cheese. As might be expected, when some strains were used, the cheese became toxic, whereas others failed to produce detectable enterotoxin even though growth was evident during the cheesemaking process. The authors concluded that staphylococcal populations of $3-8 \times 10^6/g$ could result in toxic Cheddar, Colby, brick, or Swiss cheese, depending on the kind and activity of the starter culture that was used and the strain of toxigenic S. aureus that was present.

Salmonellosis

This form of foodborne illness has an incubation period of from 3 to 72 hr, with most outbreaks occurring within 12 to 24 hr after food contaminated with the salmonellae has been ingested (Marth 1969). The principal symptoms of salmonellosis are nausea, vomiting, abdominal pain, and diarrhea that usually appears suddenly. Occurrence of these symptoms may be preceded by a headache and chills. Additional symptoms often associated with the disease include watery, greenish, foul-smelling stools; prostration; muscular weakness; faintness; a moderate fever; restlessness; twitching; and drowsiness.

Severity and duration of the disease vary with the amount of food (and hence salmonellae) consumed, the kind of Salmonella, and the resistance of the individual (Marth 1969). Intensity varies from slight discomfort and diarrhea to death in 2 to 6 days. Usually, symptoms persist for 2 to 3 days, followed by an uncomplicated recovery. In some instances, symptoms may linger for weeks or months. Some patients (0.2 to 5%) become carriers of the *Salmonella* organism which caused their infection. Mortality from gastroenteritis caused by salmonellosis is generally less than 1%.

Outbreaks of salmonellosis have been associated with consumption of different cheeses. The earlier literature is primarily concerned with typhoid fever (a form of salmonellosis), whereas more recent investigations indicate that salmonellae other than *Salmonella typhi* may be associated with cheese-induced illness.

Cheddar Cheeese.—Gauthier and Foley (1943) described an epidemic of typhoid fever in Canada during the autumn of 1941. Forty cases were involved and 6 deaths resulted. The only food common to all the patients was Cheddar cheese, made locally from raw milk and consumed when it was 10 days old. Although the factory where the cheese was made lacked adequate sanitation, the source of the epidemic was found to be a known typhoid carrier who, against orders from public health authorities, milked cows whose milk was used by the factory to produce Cheddar cheese.

Another outbreak of typhoid fever attributable to Cheddar cheese occurred in Quebec during February, 1944, and was described by Foley and Poisson (1945). The original source of infection was never traced, although it was found that the cheesemaker's wife had an active case. Nevertheless, the authors believe she was not responsible for infecting the cheese. Foley and Poisson (1945) also recommended use of a 3-month ripening period when Cheddar cheese is made from raw milk.

Out of 507 cases of typhoid fever in Alberta between 1936 and 1944, 111 were caused by Cheddar cheese, according to Menzies (1944). Samples of cheese from the last 3 outbreaks in 1944 were recovered and tested for *S. typhi*. The organism was found in 30-day-old cheese, but could not be recovered from 48- and 63-day-old cheese. As a consequence of this outbreak, Alberta halted sale of cheese made from raw milk unless the cheese was ripened for at least 3 months.

Survival of *S. typhi* in Cheddar cheese was studied by Ranta and Dolman (1941) and Campbell and Gibbard (1944). The former authors mixed *S. typhi* with Cheddar cheese and found that the organism survived for 1 month at 20°C. Inoculation of *S. typhi* onto the surface of cheese was accompanied by a similar survival at room temperature, a longer survival period at refrigerator temperature, and penetration of the organism to a depth of 4 to 5 cm into the cheese after 17 days.

Campbell and Gibbard (1944) inoculated milk with *S. typhi* and used it to make Cheddar cheese. All cheeses were ripened for 2 weeks at 14.4° to 15.6°C, after which 1 cheese from each duplicate set was transferred to storage at 4.4° to 5.6°C. At the lower temperature, 7 of 10 cheeses contained viable *S. typhi* cells for more than 10 months, whereas at the higher temperature the organism generally disappeared after 3 months of ripening. Size of inoculum and acidity of the cheese did not appear to affect the longevity of *S. typhi*.

Goepfert *et al.* (1968) investigated the behavior of *Salmonella typhimurium* during the manufacture and ripening of Cheddar cheese. Pasteurized milk was inoculated with *S. typhimurium* when the lactic starter culture was added. A slight increase in number of salmonellae occurred during the time between inoculation and cutting of the curd, followed by a rapid increase during the interval between cutting the curd and draining the whey. After accounting for concentration of cells through coagulation, an average of 3.5 generations of salmonellae developed during this period. Salting of the curd was associated with a reduction in the growth rate, and ripening of the cheese was accompanied by a decrease in the salmonellae population. Survival of *S. typhimurium* exceeded 12 weeks at a ripening temperature of 12.8°C and 16 weeks at 7.2°C. Limited tests demonstrated that acetate accumulating in ripening cheese may contribute to the demise of salmonellae.

Park *et al.* (1970B) made Cheddar cheese using a slow acid-producing strain of *S. lactis* to determine what might happen if salmonellae were present in cheese produced with an abnormal fermentation. Salmonellae grew rapidly during manufacture of cheese and limited additional growth occurred in cheese during the first week of ripening at 13°C, after which there was a gradual decrease in population of *S. typhimurium*. *S. typhimurium* survived during ripening of the low-acid cheese for up to 7 months at 13°C and 10 months at 7°C.

Park *et al.* (1970A) also made cold-pack cheese food that was contaminated with *S. typhimurium* and stored it at 4.4° and 12.8°C. Rapid decrease in number of salmonellae occurred during the first week of storage regardless of temperature or composition of the product. Viable salmonellae could not be recovered, after 3 weeks at 12.8°C or 5 weeks at 4.4°C, from cheese food adjusted to pH 5.0 with lactic acid and containing 0.24% potassium sorbate. Substituting sodium propionate for sorbate resulted in 14 and 16 weeks of survival by salmonellae when cheese food was held at 12.8° and 4.4°C, respectively. Partial or complete replacement of lactic acid by acetic acid was accompanied by somewhat longer survival of *S. typhimurium* than when only lactic acid was used. Elimination of added acid from the cheese food resulted in survival of *S. typhimurium* for 6 to 7 weeks when potassium sorbate was present, for 16 and 19 weeks when sodium propionate was used, and in excess of 27 weeks without any preservative. The ability of potassium sorbate to inactivate *S. typhimurium* was confirmed in another study by Park and Marth (1972B).

Effects of lactic acid bacteria on *S. typhimurium* were determined by Park and Marth (1972A). They observed that *S. cremoris, S. lactis*, and mixtures of the two, when added to milk in the amount of 0.25%, repressed growth but did not inactivate *S. typhimurium* during 18 hr of incubation at 21° or 30°C. Increasing the inoculum of the mixed lactic cultures to 1% resulted in inactivation of *S. typhimurium* at 30°C. When added at the level of 1%, *S. thermophilus* was more detrimental to *S. typhimurium* at 42°C than was *L. bulgaricus*. Mixtures of *L. bulgaricus* and *S. thermophilus*, when added to contaminated milks at levels of 1 and 5%, caused virtually complete inactivation of *S. typhimurium* during the interval between 8 and 18 hr of incubation at 42°C.

Colby Cheese.—A typhoid fever epidemic started in January 1944 in the northern part of Indiana and covered 18 to 20 counties (Rice 1944). Approximately 250 cases and 13 deaths were recorded in this outbreak. Thomasson (1944) noted that the carrier was never traced but illness was associated with consumption of Colby cheese. The cheese, made by a single dairy in the area, was produced from raw milk preheated to 32.2° to 37.8°C and was not allowed to ripen before sale.

Tucker *et al.* (1946) recorded an incident in which 384 cases of illness in Kentucky were caused by consumption of 12- to 14-day-old Colby cheese infected by *S. typhimurium.* Investigation revealed that a dead mouse had been removed from a 1000 gal. vat of milk used to produce the cheese. Tests on infected cheese demonstrated that *S. typhimurium* survived for 302 days during storage at 6.1° to 8.9°C.

Mold-ripened Cheeses.—Mocquot *et al.* (1963) made blue cheese from milk inoculated with 10^4 and 10^6 *Salmonella* sp. cells per ml and observed the behavior of the organisms during manufacture and ripening. The death rate of *Salmonella* was related to the pH value of 24-hr-old cheese and increased with a decrease in the pH. Survival and death of the organism were similar in the inner and outer portions of the cheese. The percentage survival of salmonellae in 6-day-old cheese was less than 0.01%.

Camembert cheese was responsible for an extensive outbreak of illness in Germany, with more than 6000 cases involved. According to Bonitz (1953) the cheese was infected with *S. bareilly*, and rennet was thought to be the original source of the contaminant. After conducting additional experiments, Bonitz (1957) changed his mind and postulated that infection probably entered the cheese via the glue used to fasten labels to the individual cheeses.

Other Cheese.—Occurrence of *Salmonella* in a variety of other ripened cheeses has been reported. Many of these cheeses are not common in the United States, but the information may be useful. In some instances investigators simply stated that salmonellae were recovered from cheese, without specifying the type of cheese being studied.

Bruhn *et al.* (1960) studied the survival of salmonellae in Samsoe cheese. This cheese has a pH value of 5.15 to 5.20 after 24 hr and contains 44 to 46% moisture. Samsoe cheese was ripened at 16° to 20°C for 5 to 6 weeks, after which it was held at 10° to 12°C for an additional 7 to 10 weeks. The authors noted that a 60 day period was necessary to achieve a 10,000-fold reduction in number of viable salmonellae. The death rate was less rapid at 10° to 12°C storage than at 16° to 20°C.

In May of 1944, an unusually high number of typhoid fever cases were reported in 4 counties of California. Investigation, according to Halverson (1944), showed that the sources of infection were Romano Dolce, Teleme, and high-moisture Jack cheeses, all made from unpasteurized milk. Over 90% of the affected persons had consumed one or several of the cheeses just mentioned. This outbreak provided the impetus for the state of California to pass laws controlling the manufacture and sale of cheese in that state.

Wahby and Roushdy (1955) studied the survival of *Salmonella enteritidis, S. typhi,* and *Salmonella paratyphi* B in Domiatti cheese, an Egyptian dairy product. When the cheese was held at 20° to 25°C, *S. enteritidis* survived for 17 days, *S. typhi* for 12 days, and *S. paratyphi* B for 27 days. Forty samples of Kareish cheese (an Arabian product) were examined chemically and bacteriologically by Moutsy and Nasr (1964). They observed that the cheese contained from 2.33 to 11.38% salt, 0.75 to 2.7% titratable acidity, 68 million to 6.3 billion bacteria per g, and 1000 to 100,000,000 coliforms per g. One sample yielded *S. typhimurium.*

The behavior of *S. typhimurium* and *S. enteritidis* in Kachkaval cheese (a hard cheese) was studied by Todorov (1966). He added 5 million to 350 million salmonellae per ml of milk, which was then made into Kachkaval cheese. Heating of the curd in a water bath at 71° to 72°C for 70 to 90 sec did not destroy the salmonellae. Their survival in the cheese ranged from 4 to 20 days, depending on the initial level of contamination.

Vizir (1940) studied the behavior of *Salmonella breslau,* a contaminant of Brynza cheese, in different media fortified with salt. At 12° to 16°C, the organism remained alive in milk with 5, 10, and 15% salt and in broth with 5 to 10% salt during the 45 days of the experiment. Lactic acid at a concentration of 0.9% or higher destroyed the organisms in 5 to 10 days, regardless of temperature. Zagaevskii (1963) noted that *S. typhimurium* and *Salmonella dublin* remained viable in Brynza cheese for up to 22 months.

An outbreak of typhoid fever attributable to a cheese made in a Norwegian home was reported by Hemmes (1942). A man ill with typhoid fever was a part of the household when the cheese in question was made. Studies on the aged cheese revealed that viable *S. typhi* were present after 40 but not 55 days.

Enteropathogenic *Escherichia coli*

Enteropathogenic *Escherichia coli* can be defined as any strain of *E. coli* having the potential to cause diarrheal disease. Enteropathogenic *E. coli* strains (EEC) have been divided into 2 groups, based on the type of disease produced. Those causing a disease with cholera-like symptoms (watery diarrhea leading to dehydration and shock) also produce enterotoxins, and thus are called toxigenic EEC. These strains have been implicated as the cause of "infantile diarrhea" and "traveler's diarrhea" (Dupont *et al.* 1971; Ryder *et al.* 1976). Those strains causing a *Shigella*-like illness (diarrhea with stools containing blood and mucus) are called invasive EEC because of their ability to penetrate the epithelial cells of the colonic mucosa. These strains do not produce an enterotoxin. Invasive EEC are associated with dysentery-like disease in people of all ages.

Incidence of *E. coli* and Coliforms in Cheese.—Presence of coliforms in cheese has been the subject of research for over 80 years. Early investigations were concerned with prevention of gassy defects in curd and cheese caused by coliform bacteria (Harrison 1905; Marshall 1900; Russell 1895). Since coliforms must reach numbers close to 10^7/g to cause gassiness in Cheddar cheese, cheese of normal appearance can still have substantial numbers of *E. coli* present (Yale and Marquardt 1943). Even with pasteurization, post-pasteurization contamination of milk with coliforms can be great enough to cause cheese to become gassy (Ernstrom 1954). Yale (1943) found that Cheddar cheese of high quality could contain up to 57,000 coliforms/g in the curd.

A general survey of coliform bacteria in Canadian pasteurized dairy products by Jones *et al.* (1967) showed that 18.7% of the coliforms isolated from these products were of intestinal origin. Three serotypes or 2% of the *E. coli* isolates were enteropathogenic serotypes. Lightbody (1962) found that 97% of Queensland Cheddar cheese contained coliforms after 2 to 3 weeks of aging. *E. coli* biotype I was found in 70% of the samples. Cheese samples with more than 10^6 coliforms/g were of poor quality. Some high grade cheese had more than 1000 coliforms/g. In further studies of Queensland Cheddar cheese, Dommet (1970) reported that improper pasteurization of milk, unsanitary equipment, and contaminated starter cultures were all responsible for coliform contamination of the cheese. In recent surveys of Canadian cheese varieties, Elliott and Millard (1976) noted that 15% of retail cheese samples contained over 1500 coliforms/g, and Collins-Thompson *et al.* (1977) found 18.1% of soft cheeses and 13.6% of semisoft cheeses exceeded 1600 total coliforms/g.

Recently Frank and Marth (1978) examined 106 samples of commercial cheese for the presence of fecal coliforms and EEC. Included in their survey were samples of Camembert, Brie, brick, Muenster, and Colby cheese. Of the samples tested, 58% contained fewer than 100 fecal coliforms/g, but 17% contained more than 10,000/g. No EEC serotypes were found in any of the cheese. A similar survey by Glatz and Brudvig (1980) also demonstrated the absence of EEC from commercial cheese.

Inhibition of *E. coli* by Lactic Acid Bacteria.—Frank and Marth (1977A,B) examined the effects of lactic acid bacteria on *E. coli* when both organisms were in skim milk. With no lactic acid bacteria present, the generation times of pathogenic and nonpathogenic strains of *E. coli* ranged from 28 to 35 min when incubation was at 32°C and from 66 to 109 min at 21°C. Addition of 0.25 or 2.0% of a commercial starter together with *E. coli* served to completely inhibit growth of *E. coli* in 6–9 hr of incubation at 32°C. At 21°C, *E. coli* often had difficulty initiating growth in the presence of the lactic acid bacteria. *S. cremoris* and *S. lactis* were equally inhibitory to *E. coli* at 32°C. At 21°C, *S. cremoris* was more inhibitory to *E. coli* than was *S. lactis*, but a commercial mixed-strain lactic starter culture was more inhibitory than was either of the pure cultures.

Behavior of EEC in Cheese.—During November and December 1971 at least 227 persons in 96 separate outbreaks in several states in the United States became ill with acute gastroenteritis about 24 hr after consuming imported French Camembert or Brie cheese (Barnard and Callahan 1971;

Schnurrenberger *et al.* 1971). *E. coli* of serogroup 0.124:B17 was isolated from stools of several patients and from samples of cheese believed to have caused the illness. This episode of foodborne illness prompted Park *et al.* (1973) and Frank *et al.* (1977) to study the fate of EEC during the manufacture and ripening of Camembert cheese. They observed the following: (1) growth of *E. coli* sometimes was minimal until after curd was cut and hooped, (2) populations of approximately 10^4 *E. coli*/g appeared in some cheeses 5–6 hr after the cheesemaking process began when milk initially contained about 10^2 *E. coli*/ml, (3) there was a demise of *E. coli* during ripening with some strains disappearing from cheese during the first 2 weeks and others surviving for 4 to 6 weeks, and (4) no growth of *E. coli* was observed in ripe cheese at a pH of 6.7 but rapid growth of *E. coli* occurred on the surface of the cheese.

The fate of EEC during the manufacture and ripening of brick cheese also was determined by Frank *et al.* (1978). Results differed from those obtained with Camembert cheese in that (1) somewhat larger populations of EEC developed initially during manufacture of brick cheese, (2) inactivation of EEC was slower in brick cheese with 10^3–10^4/g remaining after 7 weeks, and (3) growth of EEC on the surface of brick cheese was more limited.

Botulism

This serious and often fatal disease results after ingestion of a toxin produced by the anaerobic spore-forming *Clostridium botulinum*. In the United States, there have not been any known cases of botulism attributable to consumption of natural or process cheese (Kosikowski 1977). One outbreak of botulism in the United States in 1951 and another in Argentina in 1974 were associated with consumption of process cheese spread. This product also has an excellent safety record since there have been no known problems caused by it in nearly 30 years. Process cheese spread can meet legal standards if it has as much as 60% water and has a pH value of up to 5.2. At these limits, the product has less inhibitory potential than if the amount of water present and the pH were reduced.

Under the Good Manufacturing Practices for canned foods in the United States (Anon. 1977), process cheese spreads would be classified as a low-acid food requiring a heat process sufficient to destroy spores of *C. botulinum* unless it can be demonstrated that the spores cannot grow in the cheese spread. Since the heat treatment currently used is insufficient to destroy the spores of *C. botulinum*, the situation just described has prompted renewed interest in the behavior of *C. botulinum* in cheese spreads.

Kautter *et al.* (1979) inoculated jars of several varieties of process cheese spread (pH 5.05–6.32, water activity 0.930–0.953) with 24,000 spores of *C. botulinum* per jar. One variety, cheese-with-bacon, also received 460 spores per jar. When stored at 35°C, 46 of 50 jars of Limburger and 48 of 50 jars of cheese-with-bacon spread became toxic after 83 and 50 days, respectively. One jar of cheese-with-bacon spread that received the smaller inoculum became toxic during 6 months of storage at 35°C. No jars of 3 other varieties of cheese spread (Cheez Whiz, Old English, and

Roka Blue) became toxic when subjected to the same treatment.
Another series of experiments was done by Tanaka *et al.* (1979) in which
spores of *C. botulinum* (1000/g) were added to process cheese spread when it
was manufactured. Moisture content and pH of various batches ranged
from 49.7 to 59.2% and 5.80 to 6.28, respectively. Cheeses were incubated at
30°C for up to 48 weeks. No samples became toxic under these conditions
when the product contained 52 or 54% moisture and was made with sodium
phosphate as the emulsifier. Use of sodium citrate as the emulsifier resulted
in a product that remained toxin-free at 52% moisture and one that became
toxic at 54% moisture. Products with 58% moisture became gassy and toxic
regardless of the emulsifier that was used.

Aflatoxin

Aflatoxin is a collective term that refers to a group of highly toxic and
carcinogenic substances produced by the common molds *Aspergillus flavus*
and *Aspergillus parasiticus* during their growth on foods or feeds. Our
concern with aflatoxin and the toxigenic aspergilli is twofold. First, there is
the potential hazard to health of consuming even small amounts of aflatoxin. Second, the toxigenic aspergilli are widely distributed in nature, and
hence the likelihood is great that they will contaminate foods and feeds.

Aflatoxin sometimes can appear in milk, cheese, and other dairy products. That aflatoxin can appear in milk has been recognized since 1962
(Allcroft and Carnaghan 1962). In spite of this, little attention has been
given to this specific aspect of the aflatoxin problem in the United States.
Considerably more research on this subject has been done in West Europe,
particularly in the Federal Republic of Germany.

Two events have occurred recently in the United States which have
served to direct attention to the problem of aflatoxin in milk and milk
products. The first of these occurred in the southeastern states where corn
harvested in the fall of 1977 often contained appreciable amounts of aflatoxin. Feeding of such corn to dairy cattle resulted in aflatoxin in the milk
produced by the cows. This prompted the Food and Drug Administration to
establish a maximum of 0.5 part per billion (ppb) of aflatoxin M_1 (to be
described later) as allowable in fluid milk that enters interstate commerce.
The second event occurred in the summer of 1978 when dairy cows in
Arizona were fed contaminated cottonseed, and milk produced by the cows
contained aflatoxin M_1. A considerable amount of cheese also was thought
to have been made from the contaminated milk, but much of it was later
cleared through an intensive testing program by the firm that owned the
cheese. Some of the cottonseed that contained aflatoxin was shipped to
several other western states and this served to further aggravate the
problem. More extensive contamination of corn than in 1977 occurred in
1980 in the southeastern United States.

The Major Aflatoxins.—Although more than a dozen forms of aflatoxin
have been identified, only 5 major forms will be mentioned in this discussion. Some of their important characteristics are summarized in Table 4.4.

TABLE 4.4. CHARACTERISTICS OF THE AFLATOXINS

Aflatoxin	Molecular Formula	Molecular Weight	Melting Point (°C)	LD_{50} (mg/kg)	Relative Mutagenicity[1]
B_1	$C_{17}H_{12}O_6$	312	268°–269°	0.36	100
M_1	$C_{17}H_{12}O_7$	328	299°	ca. 0.36	3
G_1	$C_{17}H_{12}O_7$	328	244°–246°	0.78	3
B_2	$C_{17}H_{14}O_6$	314	286°–289°	1.69	0.2
G_2	$C_{17}H_{14}O_7$	330	237°–240°	2.45	0.1

[1]Results of Ames test (Wong and Hsieh 1976).

Aflatoxins B_1, G_1, B_2, and G_2 are the major forms produced by the molds, with B_1 and G_1 usually synthesized in largest amounts. The B-aflatoxins are given this designation because they have a bluish color under long-wave ultraviolet light, and the G-toxins are given their designation because they have a greenish-yellow color under the same type of light. The form of aflatoxin that is produced from B_1 by the cow and is excreted in milk is called M_1.

Chemically, aflatoxins are substituted coumarins that are relatively small molecules. Aflatoxins are *not* proteins as are many of the toxins produced by bacteria.

Of the aflatoxins, B_1 is most toxic and most carcinogenic. Aflatoxin M_1 is about as toxic but is considerably less carcinogenic than B_1. Aflatoxins G_1, B_2, and G_2 are less toxic and less carcinogenic than is B_1.

The aflatoxins are rather heat-stable but can be degraded by strong acidic or alkaline solutions, oxidizing agents, some molds, and a few bacteria (Marth and Doyle 1979). Recently Doyle and Marth (1978A,B,C,D,E) demonstrated that bisulfite as well as the molds that produced them can degrade aflatoxins.

Routes by Which Aflatoxin Gets into Milk and Cheese.—Aflatoxin can get into milk in 1 way and into milk products in 2 ways. Milk becomes contaminated only when cows consume feed that contains aflatoxin B_1, usually the major and always the most toxic form of aflatoxin. Some of the ingested aflatoxin B_1 is converted to M_1 by the liver of the cow and this form of aflatoxin is excreted in the milk. Products made from such milk will also contain aflatoxin M_1.

Growth of a toxigenic aspergillus on a dairy product such as cheese also can result in contamination of that product with one or several of the aflatoxins that are synthesized by the mold. It is possible for cheese to contain M_1, if made from contaminated milk, and also B_1 and other forms of aflatoxin if that same cheese subsequently supports growth of a toxigenic aspergillus. The routes of contamination as just described were first presented by Kiermeier *et al.* (1975).

Aflatoxin in Commercial Milk and Cheese.—Does aflatoxin M_1 really ever appear in commercial milk products? The answer is an unqualified "yes," although the incidence nationwide in the United States is likely to be small.

Tests on milk marketed late in 1977 in 4 southeastern states having contaminated corn showed that 4 to 8% of the samples contained 0.5 or more ppb of aflatoxin M_1. Lesser amounts appeared in an additional 38 to 72% of the samples (Table 4.5).

TABLE 4.5.   AFLATOXIN M_1 IN RETAIL MILK IN 4 STATES, 1977

State	No. of Samples	Samples (%) with Aflatoxin M_1 (ppb)			
		ND^1	T^2–0.2	0.3–0.4	≥0.5
Alabama	77	57	29	9	5
Georgia	75	20	51	21	8
S. Carolina	75	40	37	15	8
N. Carolina	75	29	51	16	4

Source: Stoloff (1980).
[1]None detected.
[2]Trace.

Recently a subcommittee of the Dairy Farm Methods Committee of the International Association of Milk, Food and Environmental Sanitarians addressed a series of questions about aflatoxin in milk to regulatory agencies in the various states in the United States (Termunde 1979). Of the regulatory agencies in 47 states that responded, (1) 13 indicated aflatoxin in milk is a problem in their states, (2) 18 indicated they test raw Grade A milk for aflatoxin, (3) 8 reported they found aflatoxin in retail milk, and (4) 1 each reported they had found aflatoxin in cottage cheese, cheese, buttermilk, yogurt, and ice cream.

Some data obtained in the Federal Republic of Germany (Table 4.6) indicate that aflatoxin M_1 was found in 34 to 82% of samples of milk, dried milk, yogurt, unripened cheese, Camembert cheese, hard cheese, and process cheese. It is apparent from these data and from those for the United States that aflatoxin M_1 does, in fact, sometimes occur in processed dairy products available to the consumer. The true incidence of aflatoxin M_1 in U.S. dairy products is probably low, and appearance of the toxin in finished products is likely to be limited largely to certain regions of the United States where cows are most likely to consume aflatoxin-contaminated feed.

TABLE 4.6.   AFLATOXIN M_1 IN GERMAN DAIRY PRODUCTS (1972–1974)

Product	No. of Samples	Positive Samples (%)	Aflatoxin M_1 Content (ppb)		
			Min	Max	Ave
Milk	260	45	0.05	0.33	0.07
Dried milk	41	73	0.20	2.00	0.50
Yogurt	54	82	0.05	0.47	0.20
Unripened cheese	80	34	0.10	0.51	0.23
Camembert cheese	65	51	0.10	0.73	0.31
Hard cheese	77	75	0.10	1.30	0.43
Process cheese	134	40	0.10	0.55	0.26

Source: Polzhofer (1977A).

Effects of Processing on Aflatoxin in Milk and Cheese.—Aflatoxin M_1 appears to be rather stable when exposed to the processing techniques that are used to manufacture dairy products. Hence, a substantial amount of the aflatoxin in milk would be expected to appear in products made from milk. Our information on what happens during processing is not complete, but some of the available data point to what can be expected.

Milk is commonly given some sort of heat treatment during conversion to a food for sale. Kiermeier and Mashaley (1977) exposed naturally and artificially contaminated milk to a series of heat treatments. In general, more aflatoxin M_1 was lost when milk was heated for minutes (15 or 30) than seconds (40) (Table 4.7). The loss of M_1 ranged from 6 to 41%. Although the results are somewhat variable, it is evident that heating of milk resulted in loss of some aflatoxin M_1.

TABLE 4.7.   LOSS OF AFLATOXIN M_1 BY MILK GIVEN VARIOUS HEAT TREATMENTS

Heat Treatment	Loss of Aflatoxin (%)
62°C for 30 min (added M_1)	35
74°C for 30 min (added M_1)	41
71°C for 40 sec (added M_1)	29
71°C for 40 sec (natural contaminant)	6–14
75°C for 40 sec (natural contaminant)	12
120°C for 15 min (added M_1)	22–28
120°C for 15 min (natural contaminant)	24–25

Source: Kiermeier and Mashaley (1977).

Distribution of aflatoxin M_1 between cheese and whey was studied by Stubblefield and Shannon (1974). According to their results, approximately 50% of the M_1 in milk appeared in most varieties of cheese. An exception was Ricotta with only 30% of the M_1 in milk appearing in the cheese.

TABLE 4.8.   AFLATOXINS B_1 + G_1 IN CHEDDAR CHEESE INOCULATED WITH ASPERGILLI

Mold and Cheese Sample	Aflatoxin B_1 + G_1 (µg/kg)	
	1 Week[1]	7 Weeks
A. parasiticus		
Top 0.64 cm layer	10,300	31,000
2nd 0.64 cm layer	9.6	624
3rd 0.64 cm layer	0	0
A. flavus		
Top 0.64 cm layer	17,300	3,920
2nd 0.64 cm layer	14.4	240
3rd 0.64 cm layer	0	0

Source: Lie and Marth (1967).
[1] After visible mold growth appeared.

The experiments just cited give us some insight into what happens to milk-borne aflatoxin M_1 during processing of the milk into products. Additional research is needed before we fully understand the fate of M_1 in a variety of dairy products, particularly in cheese during ripening.

Aflatoxin in Moldy Cheese.—At the beginning of this discussion it was indicated that growth of toxigenic aspergilli on a dairy product such as cheese could lead to the presence of aflatoxin in the food. Bullerman (1976)

and Bullerman and Olivigni (1974) isolated numerous molds from Swiss and Cheddar cheese. Although most of their isolates were molds in the genus *Penicillium*, nevertheless a few isolates from each type of cheese were aspergilli capable of producing aflatoxin.

What happens if toxigenic aspergilli are afforded the opportunity to grow? Lie and Marth (1967) explored this question by inoculating Cheddar cheese with spores of toxigenic aspergilli, allowing the mold to grow at room temperature and testing cheese for presence of aflatoxin. They found that both *A. flavus* and *A. parasiticus* could produce aflatoxin on cheese and that the aflatoxin penetrated into cheese to a distance of about 1.3 cm (Table 4.8). Later Shih and Marth (1972) observed that the toxigenic aspergilli also could produce aflatoxin on brick cheese and that the toxin penetrated into this cheese to a depth of nearly 2 cm.

Cheese containing aflatoxin B_1 was used by Kiermeier and Rumpf (1975) to manufacture pasteurized process cheese. They noted that only about 5% of the B_1 was lost during the manufacturing process. A similar observation was made by Polzhofer (1977B) when he made pasteurized process cheese from cheese that contained aflatoxin M_1.

Finally, it should be mentioned that ordinarily cheese is not well suited for production of aflatoxin by the toxigenic aspergilli. This is true for several reasons: (1) cheese lacks the carbohydrate needed by the mold for maximum production of aflatoxin, (2) cheese is commonly stored at temperatures below the minimum temperatures ($11°-13°C$) for aflatoxin production, and (3) other molds on cheese can easily outgrow the aspergilli, which often do poorly in a competitive environment. However, care must be exercised in handling cheese to prevent development of conditions which allow growth of the toxigenic aspergilli.

Amines

Ingestion of foods containing tyramine together with drugs of the monoamine oxidase inhibitor (MAOI) type can lead to hypertension attacks in some persons. Drugs of the MAOI type include iproniazid, nialamide, tranylcypromine, tranylcypromine sulfate, isocarboxazid, and phenelzine sulfate. Some of these continue to be used as antidepressants (Rice *et al.* 1976).

Tyramine acts by releasing norepinephrine from tissue stores, which in turn causes an increase in blood pressure (Rice *et al.* 1976). Tyramine has 1/20 to 1/50 of the ability of epinephrine to increase blood pressure. MAOI-type drugs increase the tissue stores of norepinephrine and thus potentiate the action of tyramine. Symptoms of a hypertensive crisis prompted by tyramine include high blood pressure, headache, fever, and sometimes perspiration and vomiting (Rice *et al.* 1976).

Histamine has been implicated in several outbreaks of food poisoning, including at least one that resulted from eating 2-year-old Gouda cheese (Doeglass *et al.* 1967). Symptoms of histamine poisoning include nausea, vomiting, facial flushing, intense headache, epigastric pain, burning sensation in the throat, dysphagia, thirst, swelling of the lips, and urticaria (Center Dis. Control 1973).

Tyramine and histamine are commonly found in fermented foods, including cheese. Values reported for different cheeses are listed in Table 4.9. Some lactic acid bacteria, as well as other bacteria likely to be in cheese, possess the enzymes needed to decarboxylate tyrosine or histidine and form tyramine and histamine, respectively (Rice et al. 1976). This could account for the presence of these compounds in cheese.

TABLE 4.9. AMOUNTS OF HISTAMINE AND TYRAMINE IN DIFFERENT KINDS OF CHEESE

Cheese	Histamine[1] (μg/g)	Tyramine (μg/g)
Blue	0–2300	27–1100
Boursault	0	110–1116
Brick	—	524
Brie	0	0–260
Camembert	0–480	20–2000
Cheddar	0–1300	0–1500
Colby	0–500	100–560
Edam	0	300–320
Gouda	0–850	20–670
Gruyère	—	516
Mozzarella	0	0–410
Parmesan	0–58	4–290
Processed	0	0–50
Provolone	10–525	38–150
Romano	0–161	80–238
Stilton	0	460–2170
Swiss	0	0–1800

Source: Rice et al. (1976).
[1]Dash = cheese was not tested; 0 = the amount present was less than the minimum that could be detected.

ACKNOWLEDGMENT

A contribution from the College of Agricultural and Life Sciences, University of Wisconsin-Madison, Madison, Wisconsin.

REFERENCES

ALLCROFT, R. and CARNAGHAN, R.B.A. 1962. Groundnut toxicity—*Aspergillus flavus* toxin (aflatoxin) in animal products: Preliminary communication. Vet. Rec. *74*, 863–864.

ALLEN, V.D. and STOVALL, W.D. 1960. Laboratory aspects of staphylococcal food poisoning from Colby cheese. J. Milk Food Technol. *23*, 271–274.

ANON. 1975. Wisconsin Dairy Facts. Wis. Dep. Agric., Madison.

ANON. 1977. Thermally processed low-acid foods in hermetically sealed containers. Code of Federal Regulations No. 21 (Revised Apr. 1), U.S. Gov. Printing Office, Washington, D.C.

BARNARD, R. and CALLAHAN, W. 1971. Follow-up on gastroenteritis attributed to French cheese. Morbidity Mortality Rep. *20*, 445.

BELL, W.B. and VELIZ, M.O. 1952. Production of enterotoxin by staphylococci recovered from the bovine mammary gland. Vet. Med. *47*, 321–322.

BONITZ, K. 1953. Epidemic of *Salmonella bareilly* caused by food poisoning in northern Germany. Dtsch. Med. Wochenschr. *78*, 1412–1413. (German)

BONITZ, K. 1957. On the epidemiology of *Salmonella bareilly* from infected cheese. Zentralbl. Bakteriol. I. Orig. *168*, 244–256. (German)

BRUHN, P.A., MØLLER-MADSEN, A., PEDERSEN, A.H. and JENSEN, H. 1960. Thermal-resistant pathogenic bacteria and their survival during aging of cheese. Beret. Forsoegsm. Kbh. *124*, 1–85. (Danish)

BULLERMAN, L.B. 1976. Examination of Swiss cheese for incidence of mycotoxin producing molds. J. Food Sci. *41*, 26–28.

BULLERMAN, L.B. and OLIVIGNI, F.J. 1974. Mycotoxin-producing potential of molds isolated from Cheddar cheese. J. Food Sci. *39*, 1166–1168.

CAMPBELL, A.G. and GIBBARD, J. 1944. The survival of *E. typhosa* in Cheddar cheese manufactured from infected raw milk. Can. J. Public Health *35*, 158–164.

CASMAN, E.P. 1965. Staphylococcal enterotoxin. Ann. N.Y. Acad. Sci. *128*, 124–131.

CENTER DIS. CONTROL. 1973. Follow-up on scrombroid fish poisoning in canned tuna fish—United States. Morbidity Mortality Week. Rep. *22*, 78.

CHAPMAN, H.R. 1974. Cheddar in the United Kingdom. *In* Handbook of Cheese. H. Mair-Waldburg (Editor). Volkswirtschaftlicher Verlag, GmbH., Kempten (Allgäu), West Germany. (German)

CLARK, W.S., JR. and NELSON, F.E. 1961. Multiplication of coagulase-positive staphylococci in grade A milk samples. J. Dairy Sci. *44*, 232–236.

COLLINS-THOMPSON, D.L., ERDMAN, I.E., MILLING, M.E., BURGNER, D.M., PURVIS, V.T., LOIT, A. and COULTER, R.M. 1977. Microbiological standards for cheese: Survey and viewpoint of the Canadian Health Protection Branch. J. Food Prot. *40*, 411–414.

COUSIN, M.A. and MARTH, E.H. 1977A. Lactic acid production by *Streptococcus lactis* and *Streptococcus cremoris* in milk precultured with psychrotrophic bacteria. J. Food Prot. *40*, 406–410.

COUSIN, M.A. and MARTH, E.H. 1977B. Lactic acid production by *Streptococcus thermophilus* and *Lactobacillus bulgaricus* in milk precultured with psychrotrophic bacteria. J. Food Prot. *40*, 475–479.

COUSIN, M.A. and MARTH, E.H. 1977C. Psychrotrophic bacteria cause changes in stability of milk to coagulation by rennet or heat. J. Dairy Sci. *60*, 1042–1047.

COUSIN, M.A. and MARTH, E.H. 1977D. Cheddar cheese made from milk that was precultured with psychrotrophic bacteria. J. Dairy Sci. *60*, 1048–1056.

COUSIN, M.A. and MARTH, E.H.   1977E.   Changes in milk proteins caused by psychrotrophic bacteria.   Milchwissenschaft *32*, 337–341.

DeBEUKELAR, N.J., COUSIN, M.A., BRADLEY, R.L., JR. and MARTH, E.H. 1977.   Modification of milk proteins by psychrotrophic bacteria.   J. Dairy Sci. *60*, 857–861.

DOEGLAS, H.M.G., HUISMAN, J. and NATER, J.P.   1967.   Histamine intoxication after cheese consumption. Lancet *ii*, 1361–1362.

DOMMET, T.W.   1970.   Studies on coliform organisms in Cheddar cheese. Aust. J. Dairy Technol. *25*, 54–61.

DONNELLY, C.B., LESLIE, J.E. and BLACK, L.A.   1968.   Production of enterotoxin A in milk.   Appl. Microbiol. *16*, 917–924.

DOYLE, M.P. and MARTH, E.H.   1978A.   Aflatoxin is degraded at different temperatures and pH values by mycelia of *Aspergillus parasiticus*.   Eur. J. Appl. Microbiol. Biotechnol. *6*, 95–100.

DOYLE, M.P. and MARTH, E.H.   1978B.   Aflatoxin is degraded by fragmented and intact mycelia of *Aspergillus parasiticus* grown 5 to 18 days with and without agitation.   J. Food Prot. *41*, 549–555.

DOYLE, M.P. and MARTH, E.H.   1978C.   Aflatoxin is degraded by mycelia of toxigenic and nontoxigenic strains of aspergilli grown on different substrates. Mycopathologia *63*, 145–153.

DOYLE, M.P. and MARTH, E.H.   1978D.   Bisulfite degrades aflatoxin: Effect of citric acid and methanol and possible mechanism of degradation.   J. Food Prot. *41*, 891–896.

DOYLE, M.P. and MARTH, E.H.   1978E.   Bisulfite degrades aflatoxin: Effect of temperature and concentration of bisulfite.   J. Food Prot. *41*, 774–780.

DUPONT, H.I., FORMAL, S.B., HORNECK, R.B., SNYDER, M.J., LIBONATI, J.P., SHEAHAN, D.G., la BRIE, E.H. and KALAS, J.P.   1971.   Pathogenesis of *Escherichia coli* diarrhea.   N. Engl. J. Med. *285*, 3–11.

EISENREICH, L.   1974.   Modern cheesemaking technique.   *In* Handbook of Cheese. H. Mair-Waldburg (Editor).   Volkswirtschaftlicher Verlag, GmbH., Kempten (Allgäu), West Germany. (German)

ELLIOTT, J.A. and MILLARD, G.E.   1976.   A comparison of two methods for counting coliforms in cheese.   Can. Inst. Food Sci. Technol. J. *9*, 95–97.

ERNSTROM, C.A.   1954.   An early gas defect in pasteurized milk Cheddar cheese.   Milk Prod. J. *45*, 21, 42.

ERNSTROM, C.A. and WONG, N.P.   1974.   Milk clotting enzymes and cheese chemistry.   *In* Fundamentals of Dairy Chemistry, 2nd Edition. B.H. Webb, A.H. Johnson and J.A. Alford (Editors).   AVI Publishing Co., Westport, Conn.

FOLEY, A.R. and POISSON, E.   1945.   A cheese-borne outbreak of typhoid fever, 1944.   Can. J. Public Health *36*, 116–118.

FOSTER, E.M., NELSON, F.E., SPECK, M.L., DOETSCH, R.N. and OLSON, J.C., JR.   1957.   Dairy Microbiology.   Prentice-Hall, Englewood Cliffs, N.J.

FRANK, J.F. and MARTH, E.H.   1977A.   Inhibition of enteropathogenic *Escherichia coli* by homofermentative lactic acid bacteria in skimmilk. I. Comparison of strains of *Escherichia coli*.   J. Food Prot. *40*, 749–753.

FRANK, J.F. and MARTH, E.H.   1977B.   Inhibition of enteropathogenic *Escherichia coli* by homofermentative lactic acid bacteria in skimmilk. II. Comparison of lactic acid bacteria and enumeration methods.   J. Food Prot. *40*, 754–759.

FRANK, J.F. and MARTH, E.H.   1978.   Survey of soft and semisoft cheese for presence of fecal coliforms and serotypes of enteropathogenic *Escherichia coli*.   J. Food Prot. *41*, 198–200.

FRANK, J.F., MARTH, E.H. and OLSON, N.F.   1977.   Survival of enteropathogenic and nonpathogenic *Escherichia coli* during the manufacture of Camembert cheese.   J. Food Prot. *40*, 835–842.

FRANK, J.F., MARTH, E.H. and OLSON, N.F.   1978.   Behavior of enteropathogenic *Escherichia coli* during manufacture and ripening of brick cheese.   J. Food Prot. *41*, 111–115.

GANDHI, N.R. and RICHARDSON, G.H.   1973.   Staphylococcal enterotoxin A development in Cheddar cheese slurries.   J. Dairy Sci. *56*, 1004–1010.

GAUTHIER, J. and FOLEY, A.R.   1943.   A cheese-borne outbreak of typhoid fever.   Can. J. Public Health *34*, 543–556.

GILLILAND, S.E. and SPECK, M.L.   1972.   Interactions of food starter cultures and foodborne pathogens: Lactic streptococci versus staphylococci and salmonellae.   J. Milk Food Technol. 35, 307–310.

GLATZ, B.A. and BRUDVIG, S.A.   1980.   Survey of commercially available cheese for enterotoxigenic *Escherichia coli*.   J. Food Prot. *43*, 395–398.

GOEPFERT, J.M., OLSON, N.F. and MARTH, E.H.   1968.   Behavior of *Salmonella typhimurium* during manufacture and curing of Cheddar cheese.   Appl. Microbiol. *16*, 862–866.

HAINES, W.C. and HARMON, L.G.   1973.   Effect of variations in conditions of incubation upon inhibition of *Staphylococcus aureus* by *Pediococcus cerevisiae* and *Streptococcus lactis*.   Appl. Microbiol. *25*, 169–172.

HALVERSON, W.L.   1944.   Preliminary report of typhoid fever due to unpasteurized cheese.   Calif. Health *1*, 171–173.

HARRISON, F.C.   1905.   Gas producing bacteria and their effect on milk and its products.   Ont. Agric. Coll. Bull. *141*.

HAUSLER, W.J., JR., BYERS, E.J., JR., SCARBOROUGH, L.C., JR. and HENDRICKS, S.L.   1960.   Staphylococcal food intoxication due to Cheddar cheese. II. Laboratory evaluation.   J. Milk Food Technol. *23*, 1–6.

HEINEMANN, B.   1957.   Growth and thermal destruction of *Micrococcus pyogenes* var. *aureus* in heated and raw milk.   J. Dairy Sci. *40*, 1585–1589.

HEMMES, G.D.   1942.   Transmission of typhoid through cheese.   Ned. Tijdschr. Geneesk. *86*, 3159–3161. (Dutch)

HENDRICKS, S.L., BELKNAP, R.A. and HAUSLER, W.J., JR.   1959. Staphylococcal food intoxication due to Cheddar cheese. I. Epidemiology.   J. Milk Food Technol. *22*, 313–317.

JACKSON, H.W. and HUSSONG, R.V. 1958. Secondary alcohols in blue cheese and their relation to methyl ketones. J. Dairy Sci. *41*, 920–924.

JEZESKI, J.J., TATINI, S.R., DEGARCIA, P.C. and OLSON, J.C., JR. 1967. Influence of *Streptococcus lactis* on growth and enterotoxin production by *Staphylococcus aureus*. Bacteriol. Proc. *1967*, 12.

JONES, G.A., GIBSON, D.L. and CHENG, K.J. 1967. Coliform bacteria in Canadian pasteurized dairy products. Can. J. Publ. Health *58*, 257–264.

KAUTTER, D.A., LILLY, T., JR., LYNT, R.K. and SOLOMON, H.M. 1979. Toxin production by *Clostridium botulinum* in shelf-stable pasteurized process cheese spreads. J. Food Prot. *42*, 784–786.

KIERMEIER, F. and MASHALEY, R. 1977. Influence of treatment of milk on the aflatoxin M_1 content of products made from such milk. Z. Lebensm. Unters. Forsch. *163*, 183–187. (German)

KIERMEIER, F., REINHARDT, V. and BEHRINGER, G. 1975. On the occurrence of aflatoxins in raw milk. Dtsch. Lebensm. Rundsch. *71* (1) 35–38. (German)

KIERMEIER, F. and RUMPF, S. 1975. On the fate of aflatoxin during the manufacture of process cheese. Z. Lebensm. Unters. Forsch. *157*, 211–216.

KOSIKOWSKI, F. 1966. Cheese and Fermented Milk Foods. Edwards Brothers, Ann Arbor, Mich.

KOSIKOWSKI, F. 1977. Cheese and Fermented Milk Foods, 2nd Edition. Edwards Brothers, Ann Arbor, Mich.

KRISTOFFERSEN, T., MIKOLAJCIK, E.M. and GOULD, I.A. 1967. Cheddar cheese flavor. IV. Directed and accelerated ripening process. J. Dairy Sci. *50*, 292–297.

LANGSRUD, T. and REINBOLD, G.W. 1973A. Flavor development and microbiology of Swiss cheese—A review. I. Milk quality and treatments. J. Milk Food Technol. *36*, 487–490.

LANGSRUD, T. and REINBOLD, G.W. 1973B. Flavor development and microbiology of Swiss cheese—A review. II. Starters, manufacturing processes and procedures. J. Milk Food Technol. *36*, 531–542.

LANGSRUD, T. and REINBOLD, G.W. 1973C. Flavor development and microbiology of Swiss cheese—A review. III. Ripening and flavor production. J. Milk Food Technol. *36*, 593–609.

LANGSRUD, T. and REINBOLD, G.W. 1974. Flavor development and microbiology of Swiss cheese—A review. IV. Defects. J. Milk Food Technol. *37*, 26–41.

LIE, J.L. and MARTH, E.H. 1967. Formation of aflatoxin in Cheddar cheese by *Aspergillus flavus* and *Aspergillus parasiticus*. J. Dairy Sci. *50*, 1708–1710.

LIGHTBODY, L.G. 1962. Coliform organisms in Queensland Cheddar cheese. Queensl. J. Agric. Sci. *19*, 305–307.

MAIR-WALDBURG, H. 1974A. Handbook of Cheese. Volkswirtschaftlicher Verlag, GmbH., Kempten (Allgäu), West Germany. (German)

MAIR-WALDBURG, H.   1974B.   On the history of cheesemaking in ancient times.   *In* Handbook of Cheese. H. Mair-Waldburg (Editor).   Volkswirtschaftlicher Verlag, GmbH., Kempten (Allgäu), West Germany.   (German)

MAIR-WALDBURG, H.   1974C.   Cheese in the literature.   *In* Handbook of Cheese. H. Mair-Waldburg (Editor).   Volkswirtschaftlicher Verlag, GmbH., Kempten (Allgäu), West Germany.   (German)

MARSHALL , C.E.   1900.   Gassy curd and cheese.   Mich. Agric. Coll. Bull. *183*.

MARTH, E.H.   1953.   The early history of cheesemaking.   Milk Prod. J. *44* (10) 30–31, 46–49.

MARTH, E.H.   1963.   Microbiological and chemical aspects of Cheddar cheese ripening. A review.   J. Dairy Sci. *46*, 869–890.

MARTH, E.H.   1966.   Antibiotics in foods—naturally occurring, developed and added.   Residue Rev. *12*, 65–161.

MARTH, E.H.   1969.   Salmonellae and salmonellosis associated with milk and milk products. A review.   J. Dairy Sci. *52*, 283–315.

MARTH, E.H.   1974.   Fermentations.   *In* Fundamentals of Dairy Chemistry, 2nd Edition. B.H. Webb, A.H. Johnson and J.A. Alford (Editors).   AVI Publishing Co., Westport, Conn.

MARTH, E.H.   1978.   Standard Methods for the Examination of Dairy Products, 14th Edition.   Am. Public Health Assoc., Washington, D.C.

MARTH, E.H. and DOYLE, M.P.   1979.   Update on molds: Degradation of aflatoxin.   Food Technol. *33* (1) 81–87.

MENZIES, D.B.   1944.   An outbreak of typhoid fever in Alberta traceable to infected Cheddar cheese.   Can. J. Public Health *35*, 431–438.

MINOR, T.E. and MARTH, E.H.   1970.   Growth of *Staphylococcus aureus* in acidified pasteurized milk.   J. Milk Food Technol. *33*, 516–520.

MINOR, T.E. and MARTH, E.H.   1972.   *Staphylococcus aureus* and staphylococcal food intoxications. A review. III. Staphylococci in dairy foods.   J. Milk Food Technol. *35*, 77–82.

MINOR, T.E. and MARTH, E.H.   1976.   Staphylococci and Their Significance in Foods.   Elsevier Scientific Publishing Co., Amsterdam.

MINOR, T.E., MARTH, E.H., OLSON, N.F., RICHARDSON, T., HANTKE, W.E., BRADLEY, R.L., JR. and CALBERT, H.E.   1970.   Microbiology and chemistry of 105-year old cheese.   J. Dairy Sci. *53*, 1795–1801.

MOCQUOT, G., LAFONT, P. and VASSAL, L.   1963.   Further observations on the survival of *Salmonella* in cheese.   Ann. Inst. Pasteur (Paris) *104*, 570–583.

MOUTSY, A.W. and NASR, S.   1964.   Studies on the sanitary condition of fresh Kareish cheese with special reference to the incidence of some food poisoning organisms.   J. Arab Vet. Med. Assoc. *24*, 99–106.

OLSON, J.C., JR., CASMAN, E.P., BAER, E.F. and STONE, J.E.   1970.   Enterotoxigenicity of *Staphylococcus aureus* cultures isolated from acute cases of bovine mastitis.   Appl. Microbiol. *20*, 605–607.

OLSON, N.F. 1969. Ripened Semisoft Cheese. Chas. Pfizer & Co., New York.

OLSON, N.F. 1975. Continuous cheesemaking for Cheddar and other ripened cheese. J. Dairy Sci. *58*, 1015–1021.

PARK, H.S. and MARTH, E.H. 1972A. Behavior of *Salmonella typhimurium* in skimmilk during fermentation by lactic acid bacteria. J. Milk Food Technol. *35*, 482–488.

PARK, H.S. and MARTH, E.H. 1972B. Inactivation of *Salmonella typhimurium* by sorbic acid. J. Milk Food Technol. *35*, 532–539.

PARK, H.S., MARTH, E.H. and OLSON, N.F. 1970A. Survival of *Salmonella typhimurium* in cold-pack cheese food during refrigerated storage. J. Milk Food Technol. *33*, 383–388.

PARK, H.S., MARTH, E.H., GOEPFERT, J.M. and OLSON, N.F. 1970B. The fate of *Salmonella typhimurium* in the manufacture and ripening of low-acid Cheddar cheese. J. Milk Food Technol. *33*, 280–284.

PARK, H.S., MARTH, E.H. and OLSON, N.F. 1973. Fate of enteropathogenic strains of *Escherichia coli* during the manufacture and ripening of Camembert cheese. J. Milk Food Technol. *36*, 543–546.

POLZHOFER, K. 1977A. Determination of aflatoxin in milk and milk products. Z. Lebensm. Unters. Forsch. *163*, 175–177. (German)

POLZHOFER, K. 1977B. Heat-stability of aflatoxin M_1. Z. Lebensm. Unters. Forsch. *164*, 80–81. (German)

PRICE, W.V. 1971. Fifty years of progress in the cheese industry. A review. J. Milk Food Technol. *34*, 329–346.

PRICE, W.V. and BUSH, M.G. 1974A. The process cheese industry in the United States: A review. I. Industrial growth and problems. J. Milk Food Technol. *37*, 135–152.

PRICE, W.V and BUSH, M.G. 1974B. The process cheese industry in the United States: A review. II. Research and development. J. Milk Food Technol. *37*, 179–198.

RANTA, L.E. and DOLMAN, C.E. 1941. Preliminary observations on the survival of *S. typhi* in Canadian Cheddar-type cheese. Can. J. Public Health *32*, 73–74.

REINBOLD, G.W. 1972. Swiss Cheese Varieties. Pfizer, New York.

REITER, B., FEWINS, B.G., FRYER, T.F. and SHARPE, M.E. 1964. Factors affecting the multiplication and survival of coagulase-positive staphylococci in Cheddar cheese. J. Dairy Res. *31*, 261–272.

RICE, T.B. 1944. Typhoid epidemic in Northern Indiana. Ind. Board Health Monthly Bull. *47*, 29, 42.

RICE, S.L., EITENMILLER, R.R. and KOEHLER, P.E. 1976. Biologically active amines in food: A review. J. Milk Food Technol. *39*, 353–358.

RICHARDSON, G.H. and DIVATIA, M.A. 1973. Lactic culture inocula required to inhibit staphylococci in sterile milk. J. Dairy Sci. *56*, 706–709.

RUSSELL, H.L.   1895.   Gas producing bacteria and the relation of the same to cheese.   Wis. Agric. Exp. Stn. 12th Annu. Rep., 139–150.

RYDER, R.W., WACHSMUTH, I.K., BUSTON, A.E. and BARRETT, F.F. 1976.   Infantile diarrhea produced by heat-stable enterotoxigenic *Escherichia coli*.   N. Engl. J. Med. *295*, 849–853.

SCHNURRENBERGER, L.W., BECK, R. and PATE, J.   1971.   Gastroenteritis attributed to imported French cheese.   Morbidity Mortality Rep. *20*, 427–428.

SHIH, C.N. and MARTH, E.H.   1972.   Experimental production of aflatoxin on brick cheese.   J. Milk Food Technol. *35*, 585–587.

STOLOFF, L.   1980.   Aflatoxin M_1 in perspective.   J. Food Prot. *43*, 226–230.

STUBBLEFIELD, R.D. and SHANNON, G.M.   1974.   Aflatoxin M_1: Analysis in dairy foods made from artificially contaminated milk.   J. Assoc. Off. Anal. Chem. *57*, 847–851.

TAKAHASHI, I. and JOHNS, C.K.   1959.   *Staphylococcus aureus* in Cheddar cheese.   J. Dairy Sci. *42*, 1032–1037.

TANAKA, N., GOEPFERT, J.M., TRAISMAN, E. and HOFFBECK, W.M. 1979.   A challenge of pasteurized process cheese spread with *Clostridium botulinum* spores.   J. Food Prot. *42*, 787–789.

TATINI, S.R., JEZESKI, J.J., MORRIS, H.A., OLSON, J.C., JR. and CASMAN, E.P.   1971.   Production of staphylococcal enterotoxin A in Cheddar and Colby cheese.   J. Dairy Sci. *54*, 815–825.

TATINI, S.R., WESALA, W.D., JEZESKI, J.J. and MORRIS, H.A.   1973. Production of staphylococcal enterotoxin A in blue, brick, Mozzarella, and Swiss cheeses.   J. Dairy Sci. *56*, 429–435.

TERMUNDE, D.E.   1979.   Interim report—farm methods committee, 1979–1980.   Intern. Assoc. Milk, Food Environ. Sanitarians, Ames, Iowa.

THOMAS, C.T., WHITE, J.C. and LONGREE, K.   1966.   Thermal resistance of salmonellae and staphylococci in foods.   Appl. Microbiol. *14*, 815–820.

THOMASSON, H.L.   1944.   A typhoid epidemic from cheese as experienced by a milk sanitarian.   Ind. Board Health Monthly Bull. *48*, 283, 295–296.

TODOROV, D.   1966.   Resistance of *Salmonella typhimurium* and *Salmonella enteritidis* in hard cheese (Kachkaval).   Proc. 17th Intern. Dairy Congr. *D*, Intern. Dairy Fed., 1966, Munich.

TUCKER, C.B., CAMERON, G.M., HENDERSON, M.P. and BEYER, M.R. 1946.   *Salmonella typhimurium* food infection from Colby cheese.   J. Am. Med. Assoc. *131*, 1119–1120.

TUCKEY, S.L., STILES, M.E., ORDAL, Z.J. and WITTER, L.D.   1964.   Relation of cheese-making operations to survival of *Staphylococcus aureus* in different varieties of cheese.   J. Dairy Sci. *47*, 604–611.

VIZIR, P.E.   1940.   The bacterial contamination of Brynza cheese, methods for counteracting it, and indexes for judging the cheese. V. The effect of degree of acidity and the salt concentration on the viability of *B. breslau.* Mikrobiol. Zh. Kiev. *7* (4) 65–78.

WAHBY, A.M. and ROUSHDY, A. 1955. Viability of enteric fever organisms in some Egyptian dairy products. Zentralbl. Veterinaermed. *2*, 57–65.

WALKER, G.C., HARMON, L.G. and STINE, C.M. 1961. Staphylococci in Colby cheese. J. Dairy Sci. *44*, 1272–1282.

WALTER, H.E. and HARGROVE, R.C. 1969. Cheese Varieties and Descriptions. U.S. Dep. Agric. Handb. *54.*

WEIGMANN, H. 1933. Handbook of Practical Cheesemaking, 4th Edition. Verlag Paul Parey, Berlin. (German)

WEKRE, E. 1974. On the history of cheesemaking in ancient times. *In* Handbook of Cheese. H. Mair-Waldburg (Editor). Volkswirtschaftlicher Verlag, GmbH., Kempten (Allgäu), West Germany. (German)

WILLIAMS, W.L. 1941. *Staphylococcus aureus* contamination of a grade "A" milk supply. J. Milk Technol. *4*, 311–313.

WONG, J.J.and HSIEH, D.P.H. 1976. Mutagenicity of aflatoxins related to their metabolism and carcinogenic potential. Proc. Natl. Acad. Sci. U.S.A. *73*, 2241–2244.

YALE, M.W. 1943. Significance of the coliform group of bacteria in American Cheddar cheese. J. Dairy Sci. *26*, 766–769.

YALE, M.W. and MARQUARDT, J.C. 1943. Coliform bacteria in Cheddar cheese. N.Y. Agric. Exp. Stn. Tech. Bull. *270.*

ZAGAEVSKII, L.S. 1963. The hygiene of milk production on farms infected with paratyphoid. Veterinariya (Moscow) *40*, 58–62.

ZEHGRUBER, K. 1974. On the history of cheese making in recent times. *In* Handbook of Cheese. H. Mair Waldburg (Editor). Volkswirtschaftlicher Verlag, GmbH., Kempten (Allgäu), West Germany. (German)

ZEHREN, V.L. and ZEHREN, V.F. 1968A. Examination of large quantities of cheese for staphylococcal enterotoxin A. J. Dairy Sci. *51*, 635–644.

ZEHREN, V.L. and ZEHREN, V.F. 1968B. Relation of acid development during cheese making to development of staphylococcal enterotoxin A. J. Dairy Sci. *51*, 645–649.

ZOTTOLA, E.A., AL-DULAIMI, A.N. and JEZESKI, J.J. 1965. Heat-resistance of *Staphylococcus aureus* isolated from milk and cheese. J. Dairy Sci. *48*, 774.

ZOTTOLA, E.A., JEZESKI, J.J. and AL-DULAIMI, A.N. 1969. Effect of short-time pasteurization treatments on the destruction of *Staphylococcus aureus* in milk for cheese manufacture. J. Dairy Sci. *52*, 1707–1714.

ZOTTOLA, E.A. and MARTH, E.H. 1966. Dry-blended phosphate-treated milk media for inhibition of bacteriophages active against lactic streptococci. J. Dairy Sci. *49*, 1343–1349.

5

Other Fermented Dairy Products

Ramesh C. Chandan

The 1975 per capita consumption of various fermented fluid milks in the world has been reported to be 38.5 liters (Haukka 1976), indicating that fermented dairy products constitute a vital part of the human diet in many parts of the world. The preservation and concomitant transformation of flavor and texture of milk by fermentation had been known centuries before the role of microorganisms in the process was appreciated. Historically, fermentation processes were based upon spontaneous souring of milk caused by inherent microflora. Modern processes affect milk fermentation under predictable, controllable, and exacting conditions to yield cultured dairy products of high nutritional and sanitary standards. This chapter deals with the microbiological and technological aspects of major fermented milks presently consumed in the United States, although brief reference to other fermented foods of greater importance in other parts of the world will be made. The chapter is organized to introduce the subject to the reader by touching briefly on the historical aspects, current trends in the consumption of relatively fresh fermented dairy foods (other than cheeses undergoing a long-term ripening process), standard specifications including legal aspects, manufacturing procedures of practical significance, and lastly certain nutritional and therapeutic aspects. The lactic cultures employed in the production of fermented milks are briefly described in terms of their application, and finally, a bibliography of key references is given.

The origin of fermented dairy products dates back to the dawn of civilization. The ancient Sanskrit scriptures of India, the Vedas, document the food value of dadhi, a fermented milk product similar to modern yogurt. Further evidence for the existence of soured milk as a food in early times is corroborated by the Bible. The historical, geographical, ecological, and dietetic patterns in various regions of the world are reflected in the diversity, variety, and types of fermented milks in vogue today (Table 5.1). These products are generally produced by the intense activity of lactic cultures. In addition, certain yeasts may be a part of the fermenting microflora yielding

TABLE 5.1. MAJOR FERMENTED DAIRY PRODUCTS CONSUMED IN VARIOUS REGIONS OF THE WORLD

Product Name	Major Country/Region	Kind of Milk Used
Sour cream or cultured cream, Smetana	United States, U.S.S.R., Central Europe	Cow
Cultured half and half	United States	Cow
Cultured buttermilk	United States	Cow
Ymer	Denmark	Cow
Taettmelk	Norway	Cow
Filmjolk	Sweden	Cow
"Long" milk	Scandinavia	Cow
Pitkapiima	Finland	Cow
Viili	Finland	Cow
Lactofil	Sweden	Cow
Acidophilus milk	United States, U.S.S.R.	Cow
Yakult	Japan	Cow
Yogurt, yoghurt, yoghaurt, yoghourt, yahourth, yaaurt, yourt, jugart, yaert, yaoert	United States, Europe, Asia	Cow, goat, sheep
Dough or abdoogh	Afghanistan and Iran	Cow, buffalo
Eyran	Turkey	Cow
Leben raib	Egypt	Cow, buffalo
Dahi	Indian subcontinent	Cow, buffalo
Mazurn	Armenia	Cow
Kisselo maleko	Balkans	Cow
Gioddu	Sardinia	Cow
Kefir	U.S.S.R.	Cow, goat, sheep
Koumiss	U.S.S.R.	Mare
Kurunga	Western Soviet Asia	Cow
Chal	Turkmenistan	Camel
Quarg	Germany	Cow
Cream cheese	United States, Europe	Cow
Cottage cheese	United States	Cow
Tvorog	U.S.S.R.	Cow

Sources: Campbell and Marshall (1975); Sandberg (1976); Babel (1976); Kosikowski (1977).

low levels of alcohol in the product. Furthermore, the use of milk of various animals adds another dimension to the variety of flavor, body and texture of fermented milk foods.

TRENDS IN CONSUMPTION

The total and per capita sales of various cultured dairy products in the United States are shown in Table 5.2. It has been estimated that approximately 0.9 billion kg (2 billion lb) of fluid cultured milk products and 0.45 billion kg (1 billion lb) of cottage cheese are manufactured annually. The

TABLE 5.2. TOTAL AND PER CAPITA SALES OF CERTAIN CULTURED DAIRY PRODUCTS IN THE UNITED STATES

Year	Yogurt TS[1]	PCS[2]	Cultured Buttermilk TS	PSC	Cultured Cream and Dips TS	PCS	Total Cottage Cheese TS	PCS
1955	17	0.11	1237	8.20	101	0.67	640	3.9
1960	44	0.26	1140	6.64	154	0.90	858	4.8
1965	61	0.32	1156	6.14	179	0.95	901	4.7
1970	172	0.86	1135	5.66	223	1.11	1044	5.2
1975	431	2.04	1018	4.83	360	1.71	984	4.6
1976	482	2.27	1035	4.87	355	1.67	1010	4.7
1977	534	2.49	1016	4.74	367	1.71	1017	4.7
1978	565	2.61	995	4.60	382	1.76	1023	4.7
1979[3]	567	2.59	948	4.34	401	1.83	1023	4.6

Source: Milk Ind. Found. (1980).
[1]Total sales (0.45 million kg or 1 million lb).
[2]Per capita sales (0.45 kg or 1 lb).
[3]Preliminary.

amount of yogurt is estimated at over 256 billion kg (565 million lb), giving per capita sales of approximately 1.2 kg (2.6 lb). Cultured buttermilk sales are of the order of 0.45 billion kg (1 billion lb) with per capita sales of 1.98 kg (4.3 lb). Cultured cream sales are over 181 million kg (401 million lb), which amount to per capita sales of 0.8 kg (1.8 lb). The cottage cheese sales of approximately 0.45 billion kg (1.00 billion lb) correspond to per capita sales of 2.1 kg (4.6 lb).

The trend in the consumption of cultured dairy products is apparent from Table 5.3. Yogurt, sour cream, and dips have registered considerable increases in sales during the last 15 years. Cultured buttermilk sales registered a decline of 25% during the period 1969–1979. The yogurt increase in sales of 204% in this period is particularly interesting. Although yogurt has been on sale since the turn of this century, its market was insignificant until about 1965. Since then the market has expanded dramatically. In 1979, yogurt sales increased only slightly. Kroger (1976) predicted sales of 1 billion dollars by 1986. The popularity of yogurt has increased in Europe as well. The yogurt boom is ascribed largely to its image of health and low-fat food. Certainly the advent of fruit-filled yogurt transferred its public image from a sour, tart, or ethnic food to a convenient, wholesome, and "natural" food. The potential for future growth of yogurt consumption in the United States may be indicated by comparison to the per capita consumption of yogurt in certain European countries. The data given in Table 5.4 indicate

TABLE 5.3. CHANGE IN PER CAPITA SALES OF CULTURED DAIRY PRODUCTS IN THE UNITED STATES 1969–1979

Product	% Change
Yogurt	+204
Sour cream and dips	+ 91
Cottage cheese	− 4
Cultured buttermilk	− 25

Source: Milk Ind. Found. (1980).

the yogurt consumption in some European countries ranges from 10 to 20 times the yogurt consumption in the United States. In view of the recent increase in popularity of yogurt, it is reasonable to expect product diversification in the future, leading to a family of products constituted of yogurt or derived from yogurt. Yogurt salad dressings, yogannaise, and frozen yogurt desserts have already appeared on the market. Soft-serve yogurt, hard-pack yogurt, and novelty items based upon yogurt have been well received by the consumer.

Cultured cream and party dips derived therefrom have also shown good growth (85%) in the last 10 years. These products have displayed increasing popularity despite the general decline in high-fat products. Cultured cream contains a minimum of 18% milk fat. Sour half and half, containing 10.5% milk fat, has been introduced to develop lower fat cultured cream type products.

Cottage cheese sales registered growth until 1972–1973. However, presently the trend has been slowed so that there is little or no growth. Low fat cottage cheese has shown consistent growth from 13 million kg (29 million lb) in 1960 to 62 million kg (136 million lb) in 1976 (Milk Ind. Found. 1980).

From a marketing standpoint, the culturing process imparts an added intrinsic property to dairy products since an extended shelf life is obtained at refrigeration temperatures of $5°-7°C$. The reduction in pH as a result of metabolic activity of the bacterial culture produces a variety of flavors and textures, as well as conditions hostile to the growth of pathogenic microorganisms. In addition, the cultures elaborate certain antibiotics or inhibitory materials of some significance in enhancing storage life, transportability, and safety in consumption of these products.

STARTER LACTIC CULTURES

Milk is the normal habitat of a number of lactic acid bacteria which may cause spontaneous souring. For sophisticated control, modern industrial processes utilize specially prepared lactic acid bacteria as starter cultures, or "starters," in the manufacture of fermented dairy products. A wealth of information on the starter cultures is available in the literature (Hammer and Babel 1957; Foster et al. 1958; Sellars 1967; Frazier 1967; Lloyd 1971;

TABLE 5.4.   YOGURT CONSUMPTION IN CERTAIN COUNTRIES IN 1973

Country	Annual per Capita Consumption (kg)	(lb)
Netherlands	13.5	30
Switzerland	9.9	22
Finland	7.6	17
France	7.2	16
United States	0.7	1.5

Source: Kosikowski (1977).

Vedamuthu 1976; Lawrence *et al.* 1976; Babel 1976; Kosikowski 1977). A starter consists of harmless microorganisms which, upon culturing in milk or milk-based mixes, impart desirable and predictable characteristics of flavor and texture attributable to a certain fermented milk product. A single-strain culture contains an individual strain of a bacterial species while a mixed/multistrain culture consists of a mixture of more than one strain or species. In the United States, Canada, and the Netherlands, a starter generally consists of mixed strains. A multistrain starter is considered to have an advantage over a single strain starter since fermentation will continue in the presence of a phage which specifically attacks one strain only. However, in mixed strain starters, a single strain may dominate at the expense of other constituent strains. A considerable number of plants are now using two mixed-strain starters (Collins 1977). In Australia and New Zealand, the preferred technology is to grow two single starters individually and to mix them prior to culturing of cheese vats. The advantage of this technique is that an individual strain, preselected for its performance and resistance to antibiotics and phage, is used for ensuring greater predictability and uniformity of culture effects.

For distribution of starters an earlier method involved shipping them as a liquid subculture. The liquid cultures are generally no longer distributed in commercial practice but, in rare cases, may be useful in distribution of cultures from a central laboratory to an operating plant. To prepare a liquid culture, the organisms are propagated in a suitable medium such as milk or whey and maintained in an active condition by periodic transfers. In general, a liquid culture contains about 10^9 organisms/ml of the starter. In addition to the continuous care required to maintain liquid cultures, it has been observed that repeated transfers may cause the culture to lose some of its critical characteristics. This problem is minimized by freeze-drying the cultures grown in milk and distributing the lyophilized culture in small vials. The freeze-dried cultures can be stored at room temperature for several years but the degree of viability of the organisms is very low. Reactivation of the lyophilized culture is necessary for proper performance. A study on the preparation of freeze-dried concentrates of the starters for direct vat inoculation has been reported by Sandine (1977). The storage stability appears to be enhanced by β-glycerol phosphate. Kilara *et al.* (1976) studied the effect of cryoprotective agents on freeze-drying and storage of lactic cultures. After 48 weeks, casitone yielded 15% survival of *Lactobacillus bulgaricus* and monosodium glutamate gave 36% survival of *Lactobacillus acidophilus*. Malt extract was found to give 30 and 22% survival for *Streptococcus thermophilus* and *Streptococcus lactis*, respectively.

Presently, the most common method for distribution of lactic cultures is in the form of frozen concentrates. The cultures are grown under optimum conditions in a fermentor, concentrated by centrifugation, and suspended in a suitable medium for maximum protection during freeze/thaw cycle and flash frozen in liquid nitrogen at −196°C. Culture concentrates contain approximately 10^{11} organisms/ml. The activity of a starter, frozen at

−196°C, is generally very satisfactory if it is thawed to 30°C and immediately inoculated into fresh substrate. High survival rates (95% viability) for starters can be achieved by this method. The use of frozen concentrated cultures in cultured dairy plants has eliminated the need for routine maintenance of a culture collection at the plant level. The culture concentrates are standardized for activity by the culture manufacturer. The concentrates may be designed for use in bulk starter preparation or for direct seeding in the culture vat. This development has virtually eliminated major incidents of contamination by bacteriophage or other undesirable organisms.

According to *Bergey's Manual* (Buchanan and Gibbons 1974), lactic acid bacteria generally used in fermented milks constitute the following two groups:

Part 14.   Gram-positive cocci
    Family II.   Streptococcaceae
    Genus I.   *Streptococcus*

    Species: 15   *Streptococcus thermophilus*
        20   *Streptococcus lactis*
        20a   *Streptococcus lactis* subsp. *diacetylactis*
        21   *Streptococcus cremoris*

    Genus II.   *Leuconostoc*
    Species: 5   *Leuconostoc cremoris* (formerly *Leuconostoc citrovorum*)

Part 16.   Gram-positive, asporogenous rod-shaped bacteria
    Family I.   Lactobacillaceae
    Genus I.   *Lactobacillus*

    Species: 5   *Lactobacillus bulgaricus*
        7   *Lactobacillus acidophilus*

Table 5.5 shows the microorganisms, types of fermentation, and fermentation times and temperatures used in manufacturing various cultured dairy products. In general, two distinct types of fermentation processes are involved. All the products result from lactic acid fermentations. In addition, kefir and koumiss utilize alcoholic fermentations by lactose-fermenting yeasts which produce up to 3% alcohol and CO_2 to impart effervescence. For buttermilk, sour cream, and cream cheese, it is customary to use a heterofermentative leuconostoc or *Streptococcus lactis* subsp. *diacetylactis* to generate flavor compounds (e.g., diacetyl) typical of such products.

The physiological characteristics of typical lactic acid bacteria used in cultured dairy products are presented in Table 5.6. These properties are helpful in the utilization and control of the various fermentation processes in the industry. The starters are graded according to their acid production, rate of growth, phage and antibiotic resistance, and ability to develop

TABLE 5.5.   FERMENTATION PROCESSES AND MICROORGANISMS ASSOCIATED WITH CERTAIN CULTURED MILK PRODUCTS

Product	Level of Acidity	Fermentation Type	Usual Fermentation Time and Temperature	Microorganisms
Yogurt	Moderate	Lactic acid	43°–45°C for 3 hr	*Lactobacillus bulgaricus* and *Streptococcus thermophilus*
Buttermilk and sour cream	Mild	Lactic acid	22°C for 18 hr	*Streptococcus lactis* subsp. *diacetylactis, S. lactis, S. cremoris, Leuconostoc cremoris*
Kefir and koumiss	Moderate	Lactic acid and alcoholic	15°–22°C for 24–36 hr	*Streptococcus lactis, S. cremoris,* lactose fermenting yeasts (*Torula, Candida*)
Acidophilus milk	High	Lactic acid	37°–40°C for 16–18 hr	*Lactobacillus acidophilus*
Bulgaricus milk (Bulgarian buttermilk)	High	Lactic acid	37°C for 10–12 hr	*Lactobacillus bulgaricus*
Cottage cheese	Mild	Lactic acid	22°C for 18 hr or 35°C for 5 hr	*Streptococcus lactis, S. cremoris*
Cream cheese	Mild	Lactic acid	22°C for 18 hr	*Streptococcus lactis, S. cremoris, Leuconostoc cremoris, S. lactis* subsp. *diacetylactis*

Sources: Chandan *et al.* (1969B); Kosikowski (1977).

TABLE 5.6. PHYSIOLOGICAL CHARACTERISTICS OF TYPICAL LACTIC ACID BACTERIA USED IN MANUFACTURING CULTURED DAIRY PRODUCTS

Characteristic	Streptococcus lactis	Streptococcus lactis subsp. diacetylactis	Streptococcus cremoris	Streptococcus thermophilus	Leuconostoc cremoris (citrovorum)	Lactobacillus bulgaricus	Lactobacillus acidophilus
Cell shape and configuration	Cocci, pairs, short chains	Cocci, pairs, short chains	Cocci, pairs, long chains	Cocci, pairs, long chains	Cocci, pairs, long chains	Medium or long rods, single, pairs, chains	Long rods, single, pairs, short chains
Homofermentative	+	+	+	+	−	+	+
Heterofermentative	−	−	−	−	+	−	−
Citric acid fermentation	−	+	−	−	+	−	−
CO_2 production	−	+	−	±	+	−	−
Diacetyl production	−	+	−	±	+	−	−
Acetoin production	−	+	−	−	+	−	−
Minimum growth temp, °C	8–10	8–10	8–10	20	4–10	22	20–22
Optimum growth temp, °C	28–32	28	22	40–45	20–25	40–45	37
Maximum growth temp, °C	40	40	37–39	50	37	52	45–48
Incubation temp, °C	21–30	22–28	22–30	40–45	22	42	37
Acid produced, % in milk	0.8–1.0	0.8–1.0	0.8–1.0	0.8–1.0	0.1–0.3	1.5–4.0	0.3–2.0
Salt tolerance, % max	4.0–6.5	4.0–6.5	4.0	2.0	6.5	2.0	6.5
Heat tolerance (60°C for 30 min)	±	±	±	++	±	+	−
Litmus milk reduction	++	++	++	+	±	++	±
Spore formation	−	−	−		−	−	±

Sources: Busch-Johannsen (1972); Davis (1975); Kosikowski (1977).

typical flavor and texture. The bacteria commonly used in cultured dairy foods are: for acid production, *Lactobacillus acidophilus, Streptococcus thermophilus, Streptococcus lactis,* and *Streptococcus cremoris;* for acid production and flavor development, *Lactobacillus bulgaricus, Streptococcus lactis* subsp. *diacetylactis;* and for flavor production, *Leuconostoc cremoris.* In some instances *Lactobacillus lactis, Lactobacillus helveticus, Leuconostoc dextranicum, Streptococcus durans,* and *Streptococcus faecalis* are used for acid and flavor production and *Propionibacterium shermanii* is used for flavor development. A reasonable degree of versatility in acid and flavor production may be achieved by combining different strains and species of these organisms.

The starters are generally purchased from commercial sources specializing in their production and marketing. Among the major culture suppliers are: Chr. Hansen's Laboratory, Milwaukee, WI; Marshall Div. of Miles Laboratories, Madison, WI; Microlife Technics, Sarasota, FL; Dairyland Food Labs., Waukesha, WI; Vivolac Cultures, Indianapolis, IN; Wiesby Laboratorium, Niebull, Germany; and Flora Donica, Odesse, Denmark. Pure cultures may be obtained from the American Type Culture Collection, Rockville, MD; Northern Regional Labs., Peoria, IL; the National Dairy Collection, National Institute of Research for Dairying, Reading, England; and the Institute for Fermentations, Osaka, Japan. Dairy research centers in Australia (Central Scientific and Industrial Research Organization), New Zealand (Dairy Research Institute), Holland (The Netherlands Instituut Voor Zuivelonderzoek), France (Jouy-en-Josas), and Switzerland (Liebefeld) maintain a good supply of lactic cultures.

Biochemical Basis of Culturing Dairy Products

The production of lactic acid, acetic acid, CO_2, diacetyl, and acetaldehyde from lactose and citric acid is of fundamental importance to the growth of lactic cultures and generation of characteristic flavor in products. The fermentation reactions take place in series (Walsh and Cogan 1973), employing both homofermentative and heterofermentative systems. Figure 5.1 summarizes, in a simplified way, the basic pathways of metabolism generally recognized in dairy microorganisms. Homofermentative lactobacilli and streptococci hydrolyze lactose into glucose and galactose. The formation of pyruvic acid follows the Embden-Meyerhof glycolytic pathway, hexose monophospate shunt pathway, Leloir pathway, and D-tagatose-6-phosphate pathway (Lawrence *et al.* 1976; Collins 1977). Gilliland *et al.* (1972) reported stimulation of lactic streptococci by the hydrolysis of lactose in milk to glucose and galactose by β-galactosidase. Using lactose and citric acid as primary substrates, significant quantities of typical metabolites are produced and accumulated in the fermented milks. Diacetyl imparts characteristic flavor to cultured butter, buttermilk, and sour cream. The production of diacetyl may be enhanced by environmental manipulation and by the selection of suitable strains of bacteria producing more diacetyl in preference to acetoin. Production of diacetyl generally increases below pH 5.5.

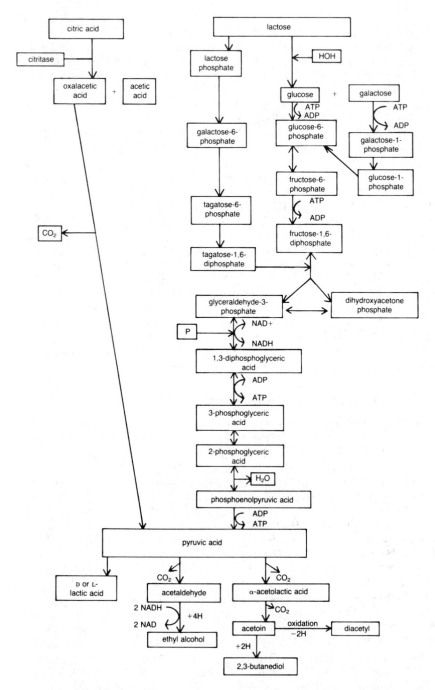

FIG. 5.1. FORMATION OF LACTIC ACID AND FLAVORING METABOLITES BY LACTIC CULTURES IN MILK

The fermentation profile of homofermentative lactobacilli is dependent upon the level of substrate and the degree of aeration (Collins 1977). Limiting concentrations of glucose and galactose result in production of lactate, acetate, and CO_2. Under aerobic conditions, cultures produce primarily acetic acid. However, many manufacturers enhance the acid-producing activity of starters by removal of oxygen from milk using a vacuum treatment.

Management and Preparation of Starters

The advent and industry-wide acceptance of frozen culture concentrates has simplified the management of cultures in most cultured dairy plants in the United States. However, working knowledge and employee training in lactic cultures are still advantageous in handling starters at the plant level. In countries where frozen culture concentrates are not yet fully developed or in case certain plants use proprietary strains and cultures in their plants, it is essential to develop and maintain appropriate microbiological expertise in the propagation, maintenance, and control of lactic starter cultures.

The starter is the most crucial component in the production of high quality fermented milks. Culture propagation should be conducted in a specified, secluded area of the plant where access of personnel is restricted. An effective sanitation program coupled with filtered air and positive pressure in the culture area, and preferably all the manufacturing areas including the packaging room, should significantly reduce the airborne contamination. Consequently, culture failure due to phage may be controlled, and extended shelf life of the product may be attained.

The media for culture propagation are generally composed of liquid skim milk or blends of cheese whey and nonfat dry milk dispersed in water so as to contain 9–10% milk solids. An improved whey-based phage inhibitory medium is available (Ausavanodom et al. 1977). Water, nonfat dry milk, whey, and other media ingredients must be free from substances inhibitory to the growth of the starter. Such inhibitory substances include sanitizing chemicals such as chlorine, iodine, and quaternary ammonium compounds as well as antibiotics and phages. Special media for optimum culture activity and phage resistance are available from commercial culture companies. The media generally contain demineralized whey, nonfat dry milk, phosphate, citrate, and growth factors present in yeast extracts. Phosphate acts as a sequestrant of Ca^{++} and thereby inhibits Ca^{++} dependent phage growth. Citrate provides a substrate for production of diacetyl and, along with phosphate, contributes to the buffering capacity. The powdered media are generally dispersed in water as such or blended with an equal weight of nonfat dry milk to attain 10–12% solids. Alternately, the media may be dispersed in liquid skim milk.

Heat treatment is necessary to destroy the contaminants in the medium and to alleviate unnecessary competition to the growth of a desirable lactic culture. In addition, heating the medium produces desirable nutrients by heat-induced reactions in milk constituents.

At present, cultures are purchased on a regular basis from commercial suppliers. They may be lyophilized or frozen concentrates shipped in liquid nitrogen or dry ice. For extended storage, the culture concentrate cans must be stored in liquid nitrogen. However, for relatively short storage periods of 4–6 weeks, the cultures may be stored in special freezers (at −40°C). The use of freezers offers an economical alternative if the turnover of cultures at the plant level is high and proper care in culture can rotation is taken (Lloyd 1975). The culture concentrates may be designed for bulk starter preparation or for direct inoculation into product mixes. The use of frozen culture concentrates eliminates the preparation of mother cultures and intermediate cultures. Figure 5.2 presents a flow diagram for the production of bulk starters in the manufacture of buttermilk, sour cream, cream cheese, and cottage cheese. The culture growth conditions are applicable to mixed strains of *Streptococcus lactis, S. cremoris, S. lactis* subsp. *diacetylactis,* and *Leuconostoc cremoris.* The procedure involves the use of nonfat dry milk as well as special media developed by several manufacturers for producing bulk starters.

Control of acid and flavor development in lactic starter cultures may be achieved by understanding their growth characteristics. By modifying the inoculation rate, incubation temperature, and time, it is possible to direct the fermentation, in a limited way, to fit the plant schedules. Care should be taken to preserve the balance of strains and organisms in the culture so that a symbiotic relationship is maintained.

Yogurt culture consists of two lactose-fermenting organisms, *Lactobacillus bulgaricus* (rod) and *Streptococcus thermophilus* (coccus). Culturing the two organisms together results in a symbiotic relationship since the growth rate and acid production by each organism are greater when grown together than in a single culture. Optimum growth temperatures for the rod and coccus are 45°C and 40°C, respectively. Depending upon incubation temperature, a differential in the ratio of rod to coccus would be evident. For full yogurt flavor development, a ratio of 1:1 is generally accepted as ideal. Upon repeated transfers, this ratio tends to change, depending upon the incubation temperature. A reasonable success in rectifying the balance has been achieved by varying the rate of inoculum, incubation time and temperature, acidity level in milk, and heat treatment of milk. Using a 2% inoculum and incubation at 44°C for 2.5 hr, proper balance of rod and coccus can be maintained in a yogurt culture. If the ratio is not 1:1, the streptococci may be increased by lowering incubation time or temperature. Conversely, the lactobacilli population may be encouraged by higher incubation time or temperature.

The yogurt culture may be obtained as a lyophilized culture or more commonly as a frozen concentrate. For the preparation of bulk starter, the procedures shown in Fig. 5.2 may be used with the exception of an incubation temperature of 44°C. Whole milk medium may be autoclaved at 121°C for 10 min, cooled to 44°C, inoculated with 2% inoculum, and incubated at 44°C until a pH of 4.6–4.7 is attained. Upon repeated transfers, optimum activity of the culture is obtained when the desired acidity is produced in 2.5

Using Frozen Culture

Special media for lactic cultures

Disperse in 378.5, 1136, or 1893 liters (100, 300, or 500 gal.) (depending upon the culture can size) of liquid skim milk or water to get 11.5% solids at 52°–57°C

Heat to 85°C for 40 min. Cool to 22°C. Inoculate 1 can of frozen culture concentrate. Incubate at 22°–25°C to a pH of 4.9, approx. 16–18 hr

Using Lyophilized Culture

Antibiotic-free skim milk or non-fat milk powder suspension

9% W/V

Dispense 750 ml milk into 24 1-liter bottles. Cap and heat at 85°C for 60 min. Cool and use up to 10 days. Add lyophilized culture to milk medium at 22°C. Incubate for 16–18 hr. May be subcultured 3 times before starting again with original lyophilized culture

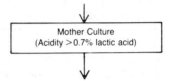

Mother Culture
(Acidity >0.7% lactic acid)

Maintain daily transfer by inoculating 1% mother culture in milk medium and incubating at 22°C for 16–18 hr. Heat 37.85 liters (10 gal.) milk in a 37.85 liter (10 gal.) can to 85°C for 30 min. Cool to 22°C and inoculate 0.5% mother culture. Mix for 2 min and incubate at 22°C for 16–18 hr

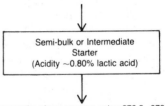

Semi-bulk or Intermediate Starter
(Acidity ~0.80% lactic acid)

Repeat the above process using 378.5–3785 liters (100–1000 gal.) of milk and 0.5–1.0% semi-bulk starter

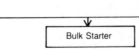

Bulk Starter

Store at 5°C up to 48 hr

FIG. 5.2.   FLOW DIAGRAM FOR SUGGESTED PRODUCTION OF BULK STARTER FOR USE IN THE MANUFACTURE OF BUTTERMILK, SOUR CREAM, CREAM CHEESE, AND COTTAGE CHEESE

hr. Mother cultures are propagated daily for a period of about 2–3 weeks after which a fresh lyophilized culture is activated. Different strains of frozen culture concentrates are rotated on a daily basis. General microscopic examination of the culture is recommended to ensure correct balance between rod and coccus.

New developments in starter technology have been summarized by Lawrence *et al.* (1976), Gilliland (1977), and Sandine (1977). For the preparation of concentrated cultures, batch processes are favored (Gilliland and Speck 1974). The literature reviews indicate that a pH of 6.0–6.3 is optimum for maximum starter cell production and the pH should be maintained by neutralization, preferably with NH_4OH. The addition of active catalase stimulates acid production by lactic culture during fermentation. The catalase destroys the inhibitor hydrogen peroxide elaborated by the culture during its growth. The frozen cultures display maximum viability and activity upon thawing when they have been stored at a temperature of −196°C. However, Stadhouders et al. (1971) showed that storage at −37°C is equally effective if the culture contains 7.5% lactose prior to freezing. The survival of the lactic culture during freezing at −17°C has been shown to be related to the fatty acid composition and glucose level in the cells. The amount of cellular capsular material also appears to be involved.

The application of genetic manipulation of lactic cultures in the technology of fermented dairy products is of particular interest in the future. Recent work of McKay and Baldwin (1975) suggests that the genes controlling proteolytic characteristics are located in plasmids. Lawrence et al. (1976) have reviewed the studies relative to lysogeny, transduction, plasmid analysis, modification-restriction system, and phage carrying state of lactic streptococci.

Continuous Starter Production

Lloyd (1971) and Kosikowski (1977) have reviewed the developments in this area. The design and use of equipment for continuous starter production provides a potential for mechanization and automation in the cheese and cultured dairy products industry. The concept of continuous starter production is based upon the basic method for continuous propagation of microorganisms (Kubitschek 1970; Stanier et al. 1976). The equipment consists of a large fermentation vessel with several sealable ports. Also, a provision is made for maintaining constant conditions of temperature, pH, oxygen tension, and pressure. Milk or a suitable medium is introduced from the inlet port and concomitantly the starter is pumped out. In this fashion the system functions as a feed and bleed operation. Adequate care is taken to ensure a balance between the output of bacteria in their growth phase and the inflow of nutrients. The acidity is maintained at a level low enough to prevent injury to the bacterial cells. A continuous culture system provides microorganisms in an exponential phase of growth in a constant stream using minimal substrate concentration. Accordingly, this system is valuable in studying regulation of synthesis, mutant selection, ecological behavior, and catabolism of the limiting substrates. In a chemostat, the flow rate (bleed rate) is coordinated with the growth rate of the culture. In a turbidostat, the culture density is measured by a turbidity-sensing device which regulates the flow rate.

Several workers have reported continuous systems for lactic starter growth. Wilkowske and Fouts (1958) found optimum rate of starter production at pH 5.3. Ashton et al. (1959) produced starter for direct inoculation into a cheese vat. Berridge (1966) devised a continuous system of *Streptococcus lactis* subsp. *diacetylactis* fermentation with the pH controlled by the flow rate. More acid was formed at pH 6.0 than at a lower pH. Continuous fermentation systems for milk products with an output of 500 liters/hr have been reported (Shmeleva and Jakovlev 1966). Robertson (1966) and Lattey (1968) explored the possibility of continuous starter systems at pH 5.0–6.3, and preservation of the culture by freezing at −196°C and storage at −75°C. Bulk frozen concentrates have been prepared using continuous fermentation in a whey medium containing yeast extract (Pont and Holloway 1968). Starter concentrates for direct vat inoculation have been demonstrated by continuous culturing of *Streptococcus lactis* C2 in trypsindigested whey fortified with yeast extract (Keogh 1970; Lloyd and Pont 1973). Pepsin-hydrolyzed whey medium reportedly gave better results (Berridge and Wilson 1970). Lewis (1967) reported maintenance of a yogurt culture continuously for a month in a chemostat. Details concerning a satisfactory pilot plant for continuous production of *Streptococcus lactis* have been described (Linklater and Griffin 1971).

The challenge in continuous culture technology is threefold. First, contamination with phage and other undesirable organisms has to be avoided during lengthy production runs. Also, the formation of undesirable mutants must be prevented. Secondly, problems in biological equilibrium among various strains or symbiotic relationships must be resolved in order to command industrial application. Thirdly, the development of continuous systems must be technologically and economically viable for cultured dairy plants to switch from frozen culture concentrates.

Starter Defects

During continuous growth the starter organisms may remain active and preserve their characteristics for some time. However, they may lose their activity rapidly depending on the compatibility of the species and strains. Also, activity is lost or changed due to the physical environment (Cowman and Speck 1965). In any case, change from the normal fermentation pattern is considered a defect. The common defects are:

1. Insufficient Acid Development.—This is one of the common defects in lactic cultures. Kosikowski (1977) defines a slow starter in terms of 1 ml culture which, upon inoculation into 10 ml of antimetabolite-free, heat-treated milk, produces less than 0.7% titratable acidity in 4 hr at 35°C. Factors contributing to a slow starter are:

(a) Composition of Milk.—Certain raw milks exert an inhibitory effect on many lactic starters, and this is attributed to various natural inhibitors including lactenins, lactoperoxidase, agglutenins, and lysozyme (Chandan *et al.* 1965; Randolph and Gould 1966). These inhibitors are present in all milk and show considerable variations with breed and season (Reiter and Moller-Madsen 1963; Reiter 1973). All these factors are heat-labile and their inhibitory property is arrested progressively on heating. When milk is

pasteurized at 72°C for 16 sec or autoclaved for 15 min, the natural inhibitors are completely destroyed. Further, the growth of starter cultures is stimulated in heated or autoclaved milk due to partial hydrolysis of casein, liberation of sulfhydryl groups, and formation of formate from lactose. Rapid acid production by lactic acid bacteria is observed in milk heated at 90°C for 1 hr, or 116°C for 15 min, or 121°C for 10 min. Autoclaving treatments are generally avoided for intermediate and bulk starter preparation because of the introduction of undesirable caramelized color and flavor in milk. However, flavor producing strains of *Leuconostoc cremoris* grow better in milk sterilized at 121°C for 15 min.

Recent trends in ultraheat or ultra high temperature treatment (UHT) of milk as a means of extending shelf life appear to have interesting implications for the cultured dairy product industry. Stone *et al.* (1975) demonstrated stimulation in growth of lactic starters in UHT milk. It appears that UHT milk is a better medium for culture growth than milk processed by batch or short-time pasteurization procedures.

Milk from mastitis-infected animals generally does not support the growth of lactic cultures. This effect is ascribed to the infection-induced changes in chemical composition of milk. For example, mastitis milk contains lower concentrations of lactose and unhydrolyzed protein and a higher chloride content and a higher pH than normal milk. Furthermore, a high leucocyte count in mastitis milk inhibits bacterial growth by phagocytic action (Sellars 1967; Babel 1976). Heat treatment restores the culture growth in mastitis milk.

Colostrum and late lactation milk contain nonspecific agglutenins which clump and precipitate sensitive strains of the starter. The agglutenins may possibly retard the rate of acid production by interfering with the transport of lactose and other nutrients (Sandine 1977).

Seasonal variation in the solids-not-fat fraction of milk affect the growth and the balance of strains in culture. Generally, a higher solids-not-fat level in milk favors the growth of lactic cultures. The ratio of leuconostocs to streptococci may be maintained by the addition of 0.02 M Mn^{++}/ml of the medium (DeMan and Galesloot 1962).

(b) Contaminating Microorganisms.—Prior degradation of milk constituents by contaminants affects the growth of lactic organisms. Preculturing of milk with psychrotrophic organisms enhanced acid production by *Streptococcus thermophilus*, *Lactobacillus bulgaricus*, *Streptococcus lactis*, and *Streptococcus cremoris* (Cousin and Marth 1977A,B). However, careful screening of milk for psychrotrophs is necessary for quality flavor production by lactic cultures.

(c) Antibiotics and Chemicals.—Various antibiotics gain entry into milk during antibiotic treatment of mastitis, and these inhibit acid production by bacteriostatic action, depending on the type of starter and the kind and amount of antibiotic involved. Concentrations as low as 0.005–0.05 International Units of penicillin, aureomycin, terramycin, or streptomycin per ml of milk are high enough to produce partial or full inhibition of the

starter. To overcome this difficulty many workers have suggested use of antibiotic resistant starter cultures and the use of additional materials like pancreatic extract or enzymes like penicillinase. However, a practical control is exercised by routine examination of milk and by avoiding use of milk showing a positive antibiotic test. In this regard, a rapid technique for antibiotic detection is particularly useful. The technique is based upon the inhibition of acid production by the assay organism *Bacillus stearothermophilus* var. *calidolactis* in the presence of extremely low antibiotic levels. The results are available in 2.5 hr.

Many sanitizing chemicals like quaternary ammonium compounds and chlorine compounds retard acid development by starter cultures. One to 5 parts per million of these sanitizing compounds are bactericidal to lactic cultures. Consequently, it is important to exert care and control in the use of sanitizers in the plant. Fatty acids (C-10 to C-16) also inhibit starters. These fatty acids may be due to partial hydrolysis of milk by lipases or they may be produced by lipolytic organisms. The fatty acids, particularly lauric, caprylic, and capric, lower the surface tension of milk to less than 40 dynes/cm. The inhibition of lactic cultures by free fatty acids is apparently related to the surface activity of the growth medium.

Avoiding the use of rancid milk is important not only from the standpoint of culture growth but more significantly because it would impart an objectionable flavor to the starter and the cultured dairy products derived therefrom.

Environmental pollutants such as telodrin, dieldrin, lindane, DDT, PCB, and PBB may be found occasionally in milk (Marth 1962; Murata *et al.* 1977). However, the insecticides have little or no effect on the growth or fermentation ability of lactic cultures (Kim and Harmon 1968). Murata *et al.* (1977) were able to culture milk containing PBB below 0.3 ppm on a fat basis. More work is needed to assess the impact of these pollutants on starters.

(d) Change in Fermentation Behavior.—After continuous use, the starters may change their fermentation activity and consequently produce lower amounts of lactic acid. This is attributed to genetic changes brought about by environmental factors (Citti *et al.* 1965).

When lyophilized or refrigerated organisms are used to prepare starters there may be a marked decrease in acid production. Some workers (Baumann and Reinbold 1966) have suggested the use of various chemicals like glycerine, azide, and sucrose to restore original acid-producing capacity. Acid production is stimulated by adding 0.25% pancreatic extract (Dahiya and Speck 1964). This stimulation is attributed to the various nucleic acid bases present in the added extract.

Strain dominance in mixed cultures, causing changes in the behavior of the composite cultures, is well documented (Sandine *et al.* 1972A). Results show that *Streptococcus lactis* subsp. *diacetylactis* tends to dominate *Streptococcus lactis* or *Streptococcus cremoris*. Different strains of *Streptococcus lactis* display major differences in domination during associative growth with *Streptococcus cremoris*.

(e) Phage Action.—Attack by bacteriophages is an important cause of slow acid production by lactic cultures. When the phage has reached a maximum level, all sensitive bacterial cells are infected and lysed within 30–40 min. When lysis occurs, acid production by the affected culture stops unless some resistant bacteria are present to carry on fermentation. Phages, strain-specific viruses, consist of a head (70 nm wide) and a tail (200 nm × 30 nm). The phage attacks streptococci as well as lactobacilli by attachment of the tail to the bacterial cell wall, followed by injection of DNA from the phage head into the cell. This is followed by synthesis of new phage particles and cell lysis which releases up to 200 new phage particles (Kosikowski 1977) to carry on further attack and lysis of the bacteria. If at least 50% of the fast acid-producing bacteria are phage resistant, a phage attack is not discernible (Collins 1977).

Phage control is effected in cultured milk plants by using 200–300 ppm chlorine on processing equipment and by fogging the culture rooms with 500–1000 ppm of chlorine. Heat treatment of milk (75°C for 30 min or 80°C for 20 sec) is considered adequate to inactivate various phages which attack lactic acid bacteria. By using combinations of proper procedures, such as sanitation, culture selection, and culture rotation, the probability of trouble with phage can be minimized. Use of phage-resistant and multiple-strain cultures is generally preferred.

The role of calcium in infection of starters by phage has been determined (Collins 1962; Henning *et al.* 1965). Since phages which attack lactic cultures require calcium for their activity, a medium with reduced calcium may be used to propagate starters. Calcium can be removed by ion exchange or by treating milk with various polyphosphates, phosphates, ammonium oxalate, and sodium salts of EDTA (Kadis and Babel 1962). Ammonium oxalate was rated better than other additives for removal of calcium, but phosphates are generally used because of their general acceptance as food additives.

In certain instances, acidity may be too high in relation to product standards. This problem is encountered in yogurt production. High acidity is usually associated with high incubation temperature, long incubation period, or excessive inoculum.

2. Insufficient or Abnormal Flavor Development.—Adequate production of lactic acid is essential for lowering the pH to a level where diacetyl and other compounds are formed in sufficient quantity. For good flavor, any factor interfering with proper acid development will retard or prevent adequate flavor development.

The culture may be incapable of producing adequate amounts of flavor due to a change in fermentation pattern induced by oxygen tension or due to a change in the balance of various bacterial cultures. Flavor can be improved by adding lactose, citric acid, and oxalacetate. The flavor can be enhanced by acidification of milk and inoculating with culture concentrates of *Leuconostoc cremoris* (Gilliland 1972).

The common flavor defects are maltiness, metallic flavor, methyl sulfide flavor, green flavor, fishy flavor, and fruity flavor. Maltiness is produced by the growth of *Streptococcus lactis* var. *maltigenes*. This organism is capable of surviving pasteurization (Leesment 1962). The malty flavor is caused by the production of 2-methyl propanal, 3-methyl butanal, 2-methyl propanol, and 3-methyl butanol (Morgan *et al.* 1966).

Metallic or puckery flavor is chiefly due to metallic contamination of the starter. Sometimes, overgrowth of leuconostocs also contributes to this development.

Green flavor in buttermilk is generally due to the production and accumulation of acetaldehyde in excessive amounts. In normal fermenting cultures, the acetaldehyde formed is metabolized by homofermenters to produce ethanol. In abnormal fermentations, heterofermenters produce acetaldehyde in excessive amounts which homofermenters are unable to metabolize (Keenan *et al.* 1966; Lindsay and Day 1965). A diacetyl:acetaldehyde ratio of 3:1 to 5:1 produces good buttermilk and sour cream flavor. If the ratio is 0.4:1, a green flavor is experienced by the taster. Harsh or yogurt flavor results when the ratio is 13:1 (Lindsay *et al.* 1965). Acetaldehyde is an essential component of yogurt flavor (Davis 1975). Approximately 25 ppm acetaldehyde is produced by symbiotic activity of a yogurt culture (Hamdan *et al.* 1971).

Fruity flavor is observed during some abnormal fermentations and is produced when excess acetaldehyde is converted to esters, ethyl hexanoate, and ethyl butyrate (Bills *et al.* 1965).

Fishy flavor is associated with the liberation of trimethylamine and related compounds by bacteria. Also, this flavor may be due to nonbacterial lipid oxidation (Forss 1964).

Methyl sulfide flavor is produced by *Aerobacter aerogenes* which may be a contaminant of the starter. This flavor is also referred to as cowy or feedy, or both (Toan *et al.* 1965).

3. Ropiness and Gassiness.—Ropiness or slimy milk is due to the change in the character of lactic streptococci or due to the dominance of the ropy strain of *Leuconostoc mesenteroides*. Psychrotrophic contaminants like *Alcaligenes visolactis, Aerobacter aerogenes,* and pseudomonads are often responsible for this defect (Hammer and Babel 1957).

Gassiness is due to the accumulation of gas during fermentation. A starter containing *Streptococcus lactis* subsp. *diacetylactis* may liberate large amounts of CO_2 gas (Collins 1977). Among contaminants, organisms of the *Escherichia-Enterobacter* group are chiefly responsible for this effect.

4. Bitterness.—This defect is due to limited proteolytic activity of some starter strains, and is commonly observed with Cheddar cheese cultures. Also, it can be attributed to the presence of proteolytic bacteria in the starter culture. *Streptococcus liquefaciens* and some sporeformers which survive normal heat treatment of the milk are the usual causes.

Flavor Production in Cultured Dairy Products

The characteristic flavor of cultured dairy products is produced by the activity of a lactic culture by metabolic transformation of milk constituents. In addition, certain flavoring compounds are produced in the culture itself. Table 5.7 gives an outline of the flavor components isolated from cultured dairy products. The list is only tentative and further work is necessary to understand the delicate balance between the known and unknown components. Meanwhile, starter distillates, produced by steam distillation of highly flavored butter cultures, are used to impart or enhance flavor in butter, sour cream, and creamed cottage cheese. Addition of starter distillates produces flavors resembling but not identical to cultured product flavor. The discrepancy is attributed to the loss of certain volatiles during distillation and imbalance in flavoring components resulting from distillation. Synthetic flavor formulation has been reported (Lindsay 1967). A typical formulation is presented in Table 5.8. Lactic acid was added only to butter (near pH 5.2). In the case of cultured cream and buttermilk, direct acidification with citric acid and δ-gluconolactone (Deane and Thomas 1964) was carried out to pH 4.6. One percent of salt improved the flavor of butter and cottage cheese. When these products were taste-evaluated by a panel, most people found no distinction and indicated no preference between the artificially flavored and naturally cultured product. Experienced judges, however, did distinguish natural buttermilk from the synthetic buttermilk. This subject is discussed further in the section on directly acidified products.

Shelf Life Extension by Lactic Cultures

Cultured milk products have been reported to contain antibacterial principles which apparently contribute to their increase in shelf life and control of food-borne pathogens (Speck 1972; Babel 1977). Evidence has been mounting on the occurrence of antibiotic materials in fermented milk products. Lactic acid, acetic acid, and hydrogen peroxide are additional factors inhibitory to spoilage organisms. *Pseudomonas fragi* is inhibited by a volatile fraction from a medium cultured with *Streptococcus diacetylactis* (Pinheiro *et al.* 1968). Similarly, *Salmonella typhimurium* and *Salmonella gallinarum* are inhibited by *Leuconostoc cremoris*. Several lactobacilli have been reported to repress the growth of *Staphylococcus aureus* and several species of *Pseudomonas* (Dahiya and Speck 1968; Price and Lee 1970). Actively growing starters, particularly *Streptococcus lactis*, inhibit the formation of staphylococcal enterotoxin A (Tatini *et al.* 1971; Haines and Harmon 1973). The occurrence of natural benzoic acid (up to 50 ppm) in cultured dairy products has been substantiated (Chandan *et al.* 1977). Filtrates of skim milk cultured with *Leuconostoc cremoris* inhibited, to a varying degree, *Staphylococcus aureus*, *Pseudomonas fluorescens*, *Escherichia coli*, *Pseudomonas fragi*, and *Enterobacter aerogenes* (Marth and Hussong 1963). Daly *et al.* (1972) determined the effect of *Streptococcus diacetyl-*

TABLE 5.7.   SOME FLAVORING COMPONENTS IN CULTURED DAIRY PRODUCTS

Component	Origin and Probable Significance	Reference
Present in Milk		
Free fatty acids—straight chain up to C_{18}	Rancid flavor, naturally present or derived from lipase action	Harper *et al.* (1961) Khatri and Day (1962) Magidman *et al.* (1962) Babel (1976)
4-*cis*-Heptenol	Cream flavor derived from isomeric C_{18} unsaturated fatty acids by autoxidation	Begemann and Koster (1964) DeJong and Vanden-Wal (1964)
δ-Lactones—C_8, C_{10}, C_{12}, C_{14}, and C_{16}	Butter flavor derived from hydroxy fatty acids by ring closure	Boldingh and Taylor (1962)
Methyl ketones, e.g., butanone-2, pentanone-2, and heptanone-2	Derived from β-keto acids of milk fat by thermal decarboxylation. Aged evaporated milk flavor	Lindsay *et al.* (1965) Parks *et al.* (1964) Wong and Patton (1962)
Dimethyl sulfide	Fresh milk flavor at 12 ppb level. Present in butter cultures, cultured and sweet cream butter, and smooths out diacetyl flavor	Patton *et al.* (1956) Day *et al.* (1964)
2-Furfural, 2-furfuryl alcohol, maltol, formic acid	Produced as a result of heat treatment of milk. Derived from lactose. Significance unknown	Jenness and Patton (1959)
Isovaleraldehyde	Fresh cream component, malty flavor	Wong (1963) Tobias (1976)
Diacetyl, acetone, and acetaldehyde	Fresh milk component	Jenness and Patton (1959) Tobias (1976)
Derived by the Action of Culture		
Lactic acid	Lactose, sharp sour flavor	Marth (1974)
Volatile acids, acetic, propionic, butyric, valeric acids	Citrate, lactose fat, sour flavor, and aroma. Acetic acid contributes significant culture flavor	Kandler (1961)
Aldehydes, ketones, alcohols, and esters	Acetone gives a cowy flavor	Lindsay (1967) Badings and Galesloot (1962)
(a) Acetaldehyde	Causes yogurt flavor defect in butter cultures but is essential component of yogurt flavor	Harvey (1960) Davis (1975)
(b) 3-Methyl butanal	Malty flavor defect	Jackson and Morgan (1954) MacLeod and Morgan (1958)
(c) 4-*cis*-Heptenal	Cream flavor	Tobias (1976)

TABLE 5.7. *(Continued)*

Component	Origin and Probable Significance	Reference
(d) Primary alcohols	Little contribution to flavor	Morgan *et al.* (1966)
(e) Methyl ketones, methyl and ethyl esters of aliphatic acids, and lactones	Isolated from cultured dairy products. Significance not certain	Winter *et al.* (1963)
Diacetyl	Butter flavor, buttermilk and sour cream flavor	Hammer and Babel (1957) Foster *et al.* (1958) Boldingh and Taylor (1962) Lindsay (1967)
Amino acids and peptides	Proteolysis products. Precursors of flavor compounds	Tobias (1976)

TABLE 5.8. FORMULATION OF A SYNTHETIC CULTURE FLAVOR

	Concentration, mg/kg Product			
Component	Buttermilk[1]	Cultured Cream[1]	Cottage Cheese[2]	Cultured Butter
Lactic acid	0	0	0	250.0
Acetic acid	1250.0	30.0	30.0	30.0
Acetaldehyde	0.5	0.2	0.2	0.2
Dimethyl sulfide	0.025	0.025	0.025	0.08
Diacetyl:				
Low level	1.0	1.0	1.0	0.5
High level	2.0	2.0	2.0	2.0
pH of the product	4.6	4.6	5.2	5.2

Source: Lindsay (1967).
[1]Acidified with citric acid and δ-gluconolactone (Deane and Thomas 1964).
[2]Normal dry curd.

actis on the growth of various organisms in milk. Inhibition ranged from 70 to 99.9% for *Pseudomonas fluorescens, P. fragi, P. viscosa, P. aeruginosa, Clostridium perfringens, Escherichia coli, Salmonella tennessee,* and *Vibrio parahemolyticus.* Evidently, the inhibitory effect is caused by an antimicrobial metabolite as well as low pH. Branen *et al.* (1975) found *Streptococcus diacetylactis* and *Leuconostoc cremoris* to inhibit *Pseudomonas putrefaciens, P. fragi,* and *P. fluorescens,* but wide variations were observed in different strains of the lactic cultures. Mather and Babel (1959C) reported inhibition of slime formation caused by *Pseudomonas fragi* and *Pseudomonas putrefaciens* when *Leuconostoc cremoris* was added to cottage cheese. In addition, the coliform organisms were inhibited. Elliker *et al.* (1964) reported extension of shelf life of cottage cheese by the use of *Streptococcus lactis* subsp. *diacetylactis* which inhibited the spoilage bacteria.

Two antibiotics produced by *Streptococcus lactis* and *Streptococcus cremoris* have been called nisin and diplococcin (Mattick and Hirsch 1949;

Babel 1977). Nisin has been isolated and found to possess variable inhibitory effects upon streptococci of groups A, B, E, F, G, K, M, and N, *Micrococcus lysodeikticus*, pneumococci, neisseria, some *Bacillus, Clostridium, Mycobacterium, Lactobacillus, Actinomyces, Erysipelothrix* species, and certain staphylococci (Mattick and Hirsch 1949; Shahani 1962).

The lactobacilli have been demonstrated to possess potent antimicrobial properties against a wide variety of spoilage and pathogenic organisms (Marth 1974). Inhibition of the growth of *Staphylococcus aureus, Pseudomonas aeruginosa, Sarcina lutea,* and *Escherichia coli* by lactobacilli has been observed *in vitro* (Polonskaya 1952; Grossowics *et al.* 1947). Sabine (1963), Tramer (1966), DeKlerk and Coetzee (1961), and Mikolajcik and Hamdan (1975) demonstrated an antibiotic effect of *Lactobacillus acidophilus* culture on a wide variety of microorganisms. Bryan (1965) reported inhibition of intestinal pathogenic and anaerobic putrefactive organisms by *L. acidophilus* and *L. bulgaricus.* Inhibitory factors elaborated by lactobacilli have been termed lactolin (Kodama 1952), lactobrevin (Kavasnikov and Sudenko 1967), lactocidin (Vincent *et al.* 1959), lactobacillin (Wheater *et al.* 1951), acidolin (Mikolajcik and Hamdan 1975), and acidophilin (Shahani *et al.* 1972, 1974, 1977). It is not yet clear whether these compounds are individually antimetabolites or are related to each other. The various microbial antimetabolites elaborated by the cultures appear to contribute to the total preservative potential and spectrum displayed by fermented dairy products.

GENERAL PRINCIPLES OF MANUFACTURE

Cultured dairy products are produced in various parts of the world from the milk of several species of mammals. The animals include: cow *(Bos taurus)*, water buffalo *(Bubalus bubalis)*, goat *(Capra hircus)*, sheep *(Ovis aries)*, mare *(Equus cabalus)*, and sow *(Sus scrofa)*. The composition of these milks as presented by Jenness (1974) is summarized in Table 5.9. Since the total solids in milk of various species range from 11.2 to 19.3%, the cultured products derived from them vary in consistency from a fluid to a custard-like gel. The range in casein content also contributes to the gel formation since upon souring this class of proteins coagulates at its isoelectric point of

TABLE 5.9.   COMPOSITION OF MILKS USED IN THE PREPARATION OF CULTURED DAIRY FOODS IN VARIOUS PARTS OF THE WORLD

Mammal	Fat %	Caseins %	Whey Proteins %	Lactose %	Ash %	Total Solids %
Cow	3.7	2.8	0.6	4.8	0.7	12.7
Water buffalo	7.4	3.2	0.6	4.8	0.8	17.2
Goat	4.5	2.5	0.4	4.1	0.8	13.2
Sheep	7.4	4.6	0.9	4.8	1.0	19.3
Mare	1.9	1.3	1.2	6.2	0.5	11.2
Sow	6.8	2.8	2.0	5.5	—	18.8

Source: Jenness (1974).

pH 4.6. The whey proteins are considerably denatured and insolubilized by heat treatments prior to culturing. The denatured whey proteins are also precipitated along with caseins to exert influence on the water binding capacity of the gel.

In the United States, bovine milk is practically the only milk employed in the industrial manufacture of cultured dairy products. Figure 5.3 shows the relationship among various forms of milk raw materials used in cultured dairy products. For optimum culture growth, the raw materials must be free from culture inhibitors like antibiotics, sanitizing chemicals, mastitis milk, colostrum, and rancid milk. Microbiological quality should be excellent for developing the delicate and clean flavor associated with top quality cultured dairy products. The raw materials generally include whole milk, skim milk, condensed skim milk, nonfat dry milk, and cream. In addition, other food materials like sweeteners, stabilizers, flavors, fruit preparations, and salt are required as components of certain cultured dairy foods. These materials are blended together in proportions to obtain a standardized mix conforming to the particular product to be manufactured. The typical chemical composition of dairy ingredients and cultured dairy products is shown in Table 5.10.

A cultured dairy product plant requires a special design to minimize contamination of the products with phage and spoilage organisms. Filtered air is useful in this regard. In some locations, water may require special treatment including filtration to control chemical and microbiological contamination of cottage cheese curd during washing of the curd. The plant is generally equipped with a receiving room to receive, meter or weigh, and store milk and other raw materials. In addition, a culture propagation room

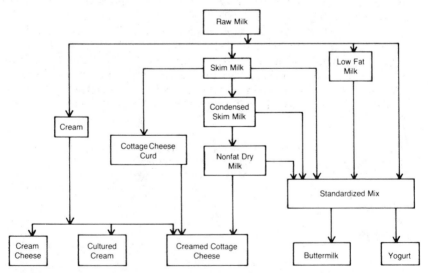

FIG. 5.3.   DAIRY INGREDIENTS AND THEIR DERIVATIVES USED IN CULTURED DAIRY FOODS

TABLE 5.10. TYPICAL COMPOSITION OF DAIRY INGREDIENTS AND CULTURED DAIRY PRODUCTS

Product	Moisture %	Fat %	Protein %	Ash %	Lactose %	pH
Whole milk	87.4	3.7	3.5	0.7	4.9	6.6
Skim milk	90.6	0.1	3.6	0.7	5.0	6.6
Condensed skim milk (30%)	70.3	0.5	8.8	2.0	12.7	6.5
Condensed skim milk (40%)	60.4	0.5	11.7	2.7	16.9	6.5
Nonfat dry milk	3.0	0.8	35.9	8.0	52.3	6.6
Cream, plastic	18.2	80.0	0.7	0.1	1.0	6.6
Cream, whipping, heavy	57.3	36.8	2.2	0.5	3.2	6.6
Cream, whipping, light	62.9	30.5	2.5	0.5	3.6	6.6
Cream, light	73.0	19.3	2.9	0.6	4.2	6.6
Half and half	81.0	10.6	3.2	0.7	4.5	6.6
Yogurt, plain	85.0	1.5	5.3	1.0	7.0	4.3
Yogurt, fruit flavored	75.3	1.2	4.0	0.9	18.7[1]	4.3
Cultured buttermilk	90.1	0.9	3.3	0.9	4.8	4.5
Cultured cream	73.5	18.0	3.2	0.6	4.3	4.2
Sour half and half	80.1	12.0	2.9	0.7	4.3	4.2
Acidophilus milk	88.0	3.3	3.6	0.6	3.0	3.9
Kefir	89.5	1.5	3.5	0.6	4.5	4.6
Koumiss	90.0	1.1	2.1	0.4	4.8	4.4
Cottage cheese, creamed	79.0	4.5	12.5	0.8	2.7	5.0
Cottage cheese, low-fat	82.5	1.0	12.4	1.4	2.7	5.0
Cream cheese	51.0	37.0	8.8	1.2	2.0	4.8

Sources: Webb et al. (1974); Kosikowski (1977); and U.S. Dep. Agric. (1977).
[1]Total sugar.

along with a control laboratory, a dry storage area, a refrigerated storage area, a processing room, and a packaging room form the backbone of the plant. The processing room contains equipment for standardizing and separating milk, pasteurizing and heating, and homogenizing along with the necessary pipelines, fittings, pumps, valves, and controls. For cottage cheese manufacture, the equipment consists of cheese vats with mechanical agitation devices, cutting knives, whey draining, curd washing, and creaming systems. A control laboratory is generally set aside where culture preparation, process control, product composition, and shelf life tests may be carried out to ensure adherence to regulatory and company standards. Also, a quality control program is established by laboratory personnel. A utility room is required for maintenance and engineering services needed by the plant. The refrigerated storage area is used for holding fruit, finished products, and other heat-labile materials. A dry storage area at ambient temperature is primarily utilized for temperature-stable raw materials and packaging supplies.

The sequence of stages of processing milk in a dairy plant is described by Jones and Harper (1976). Table 5.11 presents an outline and summary of the steps and the salient features of each step.

Milk is commonly stored in silos which are large vertical tanks with a capacity up to 100,000 liters. A silo consists of an inner tank made of

TABLE 5.11. SEQUENCE OF PROCESSING STAGES IN THE MANUFACTURE OF CULTURED DAIRY PRODUCTS

Step	Salient Feature
1. Milk procurement from the farm	Sanitary production of milk from healthy cows is necessary. For microbiological control, refrigerated bulk milk tanks should cool to 10°C in 1 hr and <5°C in 2 hr. Avoid unnecessary agitation to prevent lipolytic deterioration of milk flavor. Milk pickup is in insulated tanks at 48 hr intervals.
2. Milk reception and storage in manufacturing plant	Temperature of raw milk at this stage should not exceed 10°C. Insulated or refrigerated storage up to 72 hr helps in raw material and process flow management. Quality of milk is checked and controlled.
3. Centrifugal clarification and separation	Leucocytes and sediment are removed. Milk is separated into cream and skim milk or standardized to desired fat level at 5° or 32°C.
4. Mix preparation	Various ingredients to secure desired formulation are blended together at 50°C in a mix tank equipped with powder funnel and an agitation system.
5. Heat treatment	Using plate heat exchangers with regeneration system, milk is heated to temperatures of 85°–95°C for 10–40 min, well above pasteurization treatment. Heating of milk kills contaminating and competitive microorganisms, produces growth factors by break-down of milk proteins, generates microaerophilic conditions for growth of lactic organisms, and creates desirable body and texture in the cultured dairy products.
6. Homogenization	Mix is passed through extremely small orifice at pressure of 10.3–13.7 kPa, causing extensive physico-chemical changes in the colloidal characteristics of milk. Consequently, creaming during incubation and storage of cultured dairy products is prevented. The stabilizers and other components of a mix are thoroughly dispersed for optimum textural effects.
7. Inoculation and incubation	The homogenized mix is cooled to an optimum growth temperature, depending upon the culture used. Inoculation is generally at the rate of 0.5–5% and the optimum temperature is maintained throughout incubation period to achieve a desired pH. Quiescent incubation is necessary for product texture and body development.
8. Cooling, fruit incorporation, and packaging	The coagulated product is cooled down to 5°–22°C, depending upon the product. Using fruit feeder or flavor tank, the desired level of fruit and flavor is incorporated. The blended product is then packaged.
9. Storage and distribution	Storage at 5°C for 24–48 hr imparts in several products desirable body and texture. Low temperature ensures desirable shelf life by slowing down physical, chemical, and microbiological degradation.

Sources: Jones and Harper (1976); Davis (1975); Babel (1976); Kosikowski (1977).

stainless steel containing 18% chromium, 8% nickel, and less than 0.07% carbon. Acid and salt resistance in the steel is attained by incorporating 3% molybdenum. To minimize corrosion, this construction material is used for the storage of acidic products as well as those products containing added salt. The stainless steel tank is usually covered with 50−100 mm of insulation material which in turn is surrounded by an outer shell of stainless or painted mild steel or aluminum. The silo tanks generally have an agitation system (60−80 RPM), spray balls mounted in the center for cleaning in place (CIP), an air vent, and a manhole. The air vent must be kept open during cleaning with hot cleaning solutions. This precaution is necessary to prevent a sudden development of vacuum in the tank and consequent collapse of the inner tank upon rinsing with cold water.

For reconstitution of dry powders, viz., nonfat dry milk, sweeteners, and stabilizers, the use of a powder funnel and recirculation loop, or a special blender is convenient.

The common pasteurization equipment consists of a vat, plate, triple-tube, scraped, or swept surface heat exchanger. Vat pasteurization is conducted at 63°C with a minimum holding time of 30 min. This temperature is raised to 66°C in the presence of sweeteners in the mix. For an HTST system, the equivalent temperature-time combination is 73°C for 15 sec or in the presence of sweeteners, 75°C for 15 sec. An UHT system employs temperatures greater than 90°C and as high as 148°C for 2 sec. Alternatively, culinary steam may be used directly by injection or infusion to raise the temperature to 77°−94°C, but allowance must be made for an increase in water content of the mix due to steam condensation in this process. In some plants, steam volatiles are continuously removed by vacuum evaporation to remove certain undesirable odors (feed, onion, garlic) associated with milk.

The homogenizer is a high pressure pump forcing the mix through extremely small orifices. It includes a bypass for safety of operation. The process is usually conducted by applying pressure in two stages. The first stage pressure, ranging from 6.8 to 13.7 kPa, reduces the average milk fat globule diameter size from approx 4 μm (range 0.1 to 16 μm) to less than 1 μm. The second stage uses 3.4 kPa and is designed to break the clusters of fat globules apart with the objective of inhibiting creaming in milk. Culturing vats for the production of fermented dairy products are generally designed with a cone bottom to facilitate draining of relatively viscous fluids after incubation.

For temperature maintenance during the incubation period, the culturing vat is provided with a jacket for circulating hot or cold water or steam located adjacent to the inner vat containing the mix. This jacket is usually insulated and covered with an outermost surface made of stainless steel or aluminum or painted steel. The vat is equipped with a heavy-duty, multispeed agitation system, a manhole containing a sight glass, and appropriate spray balls for CIP cleaning. The agitator is often of swept surface type for optimum agitation of relatively viscous cultured dairy products. For efficient cooling after culturing, plate or triple-tube heat exchangers are used.

The culturing vat is designed only for temperature maintenance. Therefore, efficient use of energy requires that the mix not be heat treated in the culturing vat.

Most plants attempt to synchronize the packaging lines with the termination of the incubation period. Generally, textural defects in cultured products are caused by excessive shear during pumping or agitation. Therefore, positive drive pumps are preferred over centrifugal pumps for moving the product after culturing or ripening. For incorporation of fruit, it is advantageous to use a fruit feeder system adapted from the frozen dessert industry (Arbuckle 1977). Various packaging machines of suitable speeds are available to package various kinds and sizes of cultured dairy products.

CULTURED BUTTERMILK

Cultured buttermilk is obtained from pasteurized skim milk or part skim milk cultured with lactic and aroma producing organisms. *Streptococcus lactis* and/or *Streptococcus cremoris, Streptococcus lactis* subsp. *diacetylactis* and *Leuconostoc cremoris* are frequent cultures. The term buttermilk is also used for a phospholipid-rich fluid fraction obtained as a by-product during the churning of cream in butter manufacture. However, cultured buttermilk is a viscous, cultured, fluid milk, containing a characteristic pleasing aroma and flavor.

Cultured buttermilk is usually produced in dairy plants processing milk and other fluid dairy products by a process similar to the flow sheet given in Fig. 5.4. It is packaged in traditional milk cartons.

The processes used in the manufacture of cultured buttermilk include pasteurization, homogenization, and culturing systems. The ingredients are skim milk, low-fat milk, cream, condensed skim milk, nonfat dry milk, culture, and salt. The addition of 0.20–0.25% sodium citrate to milk provides a precursor to enhance flavor production by the culture. Butter flakes are also incorporated in certain markets. Milk fat level ranges from 0.5 to 1.8%; however, the proposed federal buttermilk standards (Anon. 1977) call for a maximum milk fat content of 0.5% and a minimum of 8.25% milk solids-not-fat. The standards also require a minimum titratable acidity of 0.5% calculated as lactic acid. Characteristic flavor ingredients, nutritive carbohydrate sweeteners, stabilizers, and coloring which does not simulate the color of milk fat, may also be used. Optionally, a minimum of 2000 International Units of vitamin A and 400 International Units of vitamin D per 0.95 liter (1 qt) of buttermilk may be added. The standards also define the products obtained from milk of various fat concentrations. Cultured milk shall be obtained from milk containing not less than 3.25% milk fat. Cultured low-fat milk shall contain not less than 0.5 nor more than 2.0% milk fat. For standardizing the milk solids-not-fat level to 8.25%, in addition to the traditional dairy ingredients, any milk-derived ingredient may be used. However, this ingredient should not decrease the protein:milk solids-not-fat ratio and the protein efficiency ratio of the resulting mixture. Butter granules or flakes may be added to buttermilk either by churning 18–20% fat cream or by spraying melted butter oil on chilled buttermilk.

Cultured Buttermilk

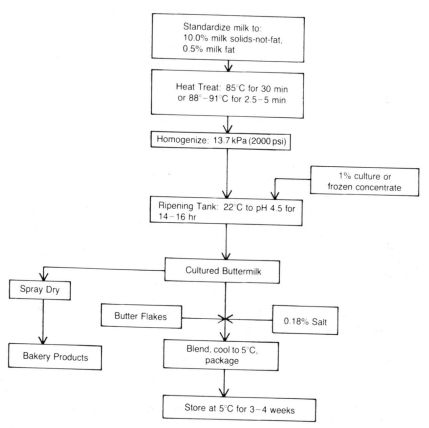

FIG. 5.4.   FLOW SHEET DIAGRAM FOR THE MANUFACTURE OF CULTURED BUTTER-MILK

Under refrigeration, the keeping quality of cultured buttermilk is extended to 3–4 weeks. Wheying off may occur but can be avoided by using a suitable stabilizer and proper processing conditions. To increase the popularity of buttermilk, several attempts have been made to market fruit flavored cultured milk (Kosikowski 1977). However, commercial application has not been realized to any appreciable level.

The quality of buttermilk in the various regions of the country is not consistent. Kosikowski (1977), Babel (1976), and Emmons and Tuckey (1967) reported various factors causing quality problems. A proper buttermilk flavor is produced by maintaining high standards of sanitation, an active starter and a uniform temperature of 22°C. It is possible to improve the buttermilk flavor by the addition of citric acid, sodium citrate, sodium chloride, and/or cream. Thin consistency, excessive CO_2 production, unclean or putrid odor, and bitter taste may result from contamination of the culture, improper pasteurization, or poor sanitary practices in the plant. A certain amount of CO_2 makes an important contribution to the desirable flavor of buttermilk (Collins 1977). Metallic flavor can be avoided by the use of stainless steel or glass-lined equipment. Development of too much acidity (>0.85% lactic acid) is generally checked by decreasing the amount of inoculum, by rapid cooling after ripening, or by blending pasteurized skim milk. Sharp acid flavor may be moderated by the addition of 0.1–0.2% sodium chloride. The fermentation period is important in flavor production and retention. Improper incubation times cause poor flavor in buttermilk, especially when citric acid is in limiting quantities.

Buttermilk possesses a characteristic fluidity. The viscosity of buttermilk is directly related to acidity. Some strains of *Streptococcus cremoris* produce thick-bodied buttermilk because of the tendency of the bacterial cells to form chains. Certain contaminants like *Alcaligenes viscolactis* may produce a ropy condition. The chemical composition of milk also influences the viscosity of the product; a higher total solids content produces a viscous product and a lower total solids content results in a thin-bodied buttermilk. A solids-not-fat content of 9% in skim milk is optimum for viscosity. Agitation of the product at an acidity of 0.85% lactic acid produces a good body (Kosikowski 1977). Ten percent reconstituted nonfat dry milk produces a very acceptable product (Babel 1976).

Excessive agitation, high storage temperature, and contamination with proteolytic organisms yield a relatively thinner buttermilk. The microbiological quality of milk is therefore critical in developing the flavor and the body of buttermilk. For this reason, the use of return milk for buttermilk manufacture is precluded. Gelatin and fat may be incorporated to improve the body of buttermilk. Considerable foam may be produced in buttermilk if air is pumped into the product as a result of violent agitation or leaky pumps. For standardizing the viscosity of buttermilk produced from batch to batch, a viscometer may be used.

Buttermilk Defects and Possible Causes

1. *Flat, "Green".*—Acidity too low, low setting temperature, low solids milk, lack of flavor organisms, short incubation period.
2. *High Acid, Sour.*—Setting temperature too high, incubation time too long, lack of fat, inoculation too heavy, poor refrigeration of finished product.
3. *Biting on the Tongue, "Carbonated".*—Too many flavor organisms, overripening, contaminated culture.
4. *Metallic.*—Contact with copper or iron, rusty equipment, high acidity and age.
5. *Oxidized.*—Metal contamination, use of oxidized dairy products.
6. *Rancid.*—Mixing homogenized milk with raw dairy products, use of rancid milk or cream.
7. *Whey Off.*—Excesssive agitation when warm, breaking at too low acidity, adding skimmed milk to reduce viscosity, low solids milk, high

storage temperature, high setting temperature, too low or too high treatment of skim milk.

8. *Lumpiness.*—High acidity, high setting temperature, poor agitation during breaking, low heat treatment of milk.

9. *Thick, Heavy Body.*—High acidity, ropy culture, solids too high, high incubation temperature, long incubation time, high heat treatment of milk, homogenization of high fat milk.

10. *Thin, Light Body.*—Inactive culture, low setting temperature, low heat treatment of milk, low acidity, low solids, too much agitation during breaking.

11. *Incorporated Air, Foamy.*—Agitation too vigorous, pumping with centrifugal pump, leaky pumps, or valves.

CULTURED CREAM

Cultured cream or sour cream is manufactured by ripening pasteurized cream of 18% fat content with lactic and aroma-producing bacteria. This product resembles cultured buttermilk in terms of culturing procedure. However, in consistency it is an acid gel containing butter-like aromatic flavor. The federal standards stipulate a minimum titratable acidity of 0.5% lactic acid, a minimum fat of 18%, and a maximum stabilizer concentration of 0.1% (U.S. Dep. Agric. 1977). Additives permitted are standardizing flavors, stabilizers, and enzymes.

Cultured cream is used as a topping on vegetables, salads, fish, meats, and fruits, as a filling in cakes, and in soups and cookery in place of buttermilk or sweet cream. It can be dehydrated by spray drying and used as an ingredient wherever its flavor is needed.

Manufacturing principles for cultured cream are outlined in Fig. 5.5. The manufacturing method shown is a general procedure. Modifications in this process are available to suit certain purposes. Homogenization twice at 17.2 kPa and at 71°C has been used to produce a very thick product (Guthrie 1963). One might manufacture cultured cream in individual packs which may be filled before ripening or soon after the ripening stage, followed by cooling. A heavy-bodied product is formed on setting. Factors affecting viscosity of cultured cream are: (a) acidity, (b) mechanical agitation, (c) heat treatment, (d) solids-not-fat content, (e) rennet addition, and (f) homogenization. It is considered desirable to supplement cream containing solids-not-fat less than 6.8% with milk solids to increase viscosity. The HTST pasteurization produces a thin product as compared to the long hold, vat pasteurization method. Rennet addition in small quantities (0.5 ml single-strength to 37.85 liters of cream) aids in thickening the product. As in buttermilk, flavor may be improved by incorporating into cream 0.15–0.20% sodium citrate, which is metabolized by *Streptococcus lactis* subsp. *diacetylactis* and *Leuconostoc cremoris* to produce more aroma compounds (diacetyl and volatile acids).

Guthrie (1963) reported that differences in the smoothness of cultured cream are related to differences in lactic cultures. Ropiness characteristic of a lactic culture enhances the smoothness and to some extent the viscosity of cultured cream. The body of the product appears to be independent of the culture used.

The hot-pack process ensures long shelf life by destroying the microorganisms and the enzymes present in the finished product. Packaging in a plastic or metal container with a hermetically sealed lid further ensures prevention of recontamination by microorganisms as well as protection from oxidative deterioration of milk fat in the finished product.

Cream used in the manufacture of cultured cream should be fresh with a relatively low bacterial count. During cream separation from milk, the bacteria tend to concentrate in the lighter phase, cream, thereby enhancing its vulnerability to spoilage. Pasteurization at 74°C for 30 min or at 85°C for 1 min is satisfactory from a bacteriological standpoint (Babel 1976).

Artificial Cultured Cream

This product may be defined as cultured cream in which part or all of the milk fat has been replaced by other oils or fats. Such a product is sold commercially in the United States. It appears to have advantages over conventional cultured cream in terms of price and caloric value. In this regard, a suitable product may be manufactured using a process identical to that for cultured cream with the exception of the starting material. An artificial cream may be prepared by emulsifying a suitable fat in either skim milk or in suspensions of casein compounds or soybean protein products. A suitable emulsifier, stabilizer, flavor, and color may be incorporated in the starting mix.

Sour Cream Dip

Party dips based on sour cream are made by blending appropriate seasoning bases in cultured cream. By packaging under refrigerated conditions, the product has shelf life of 2–3 weeks under refrigerated storage. However, for a shelf-life of 3–4 months, the process shown in Fig. 5.5 is used. To build extra body and stability, 1–2% nonfat dry milk and 0.8–1.0% stabilizer are incorporated at 80°C. The mixture is pasteurized by holding for 10 min, and homogenized at 17.2 kPa to resuspend and smooth the product. The seasonings are blended at this stage while the mix is still at 80°C, followed by hot-packing in sealed containers. Upon cooling and storage at 5°C, partial vacuum inside the container assists in the prevention of oxidative deterioration to yield an extended shelf life of 3–4 months.

Sour Half and Half

To satisfy the consumer demand for relatively low-fat and less expensive cultured cream substitutes, sour or cultured half and half has been developed commercially. It is manufactured from a mix containing a minimum of

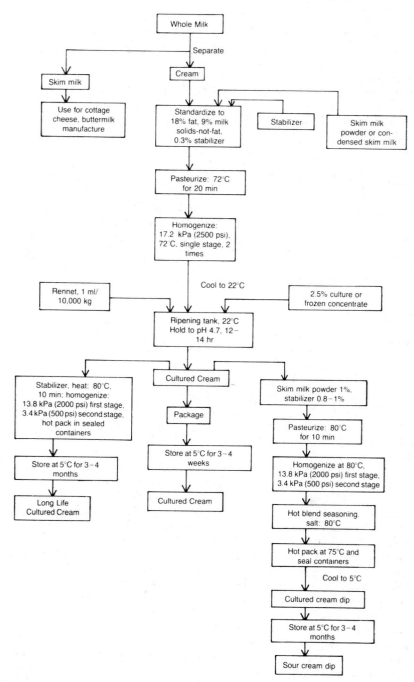

FIG. 5.5.   DIAGRAMMATIC SCHEME FOR CULTURED CREAM AND DIP MANUFACTURE

10.5 and a maximum of 18% milk fat. To compensate for reduction in solids due to lower milk fat level, it is customary to increase the milk solids-not-fat level to 10–12%. Accordingly, a satisfactory body, viscosity, and texture are obtained.

The manufacturing principles for cultured half and half are similar to those for cultured cream except for the formulation of the initial mix. Skim milk, cream, and other dairy ingredients are blended to a standard of 10.5% fat and 10–12% milk solids-not-fat. This mix is then processed along the same lines as cultured cream (Fig. 5.5).

Salad Dressing

Using sour half and half as a base, appropriate flavor bases may be blended to produce a distinctive creamy salad dressing. This dressing contains about 50–75% less calories than conventional salad dressings, but has comparable flavor and texture. The reduction in calories is primarily due to the lower fat level of 10.5% in sour half and half dressing as compared to 30–80% oil in regular dressing.

YOGURT

The manufacture of yogurt has recently been reviewed by Humphreys and Plunkett (1969), Robinson and Tamine (1975), Kroger (1976), and Kosikowski (1977).

Yogurt Ingredients and Rheological Aspects

Dairy Ingredients.—Yogurt is generally made from a mix standardized from whole, partially defatted milk, condensed skim milk, cream, and nonfat dry milk. Alternatively, milk may be partly concentrated by removal of 15–20% water in a vacuum pan. Supplementation of milk solids-not-fat with nonfat dry milk is the preferred industrial procedure. All dairy raw materials should be selected for high bacteriological quality. Ingredients containing mastitis milk and rancid milk should be avoided. Also, milk partially fermented by contaminating organisms and milk containing antibiotic and sanitizing chemical residues cannot be used for yogurt production. The procurement of all ingredients should be based upon specifications and standards which are checked and maintained with a systematic sampling and testing program by the quality control laboratory. Since yogurt is a manufactured product, it is likely to have variations according to the quality standards established by marketing considerations. Nonetheless, it is extremely important to standardize and control the day-to-day product in order to meet consumer expectations and regulatory obligations associated with a certain brand or label.

The milk fat levels in yogurt range from 1.0 to 3.25%. The proposed federal standards of identity (Anon. 1977) define the product in 3 categories.

The product containing a minimum of 3.25% milk fat is termed yogurt. Low-fat yogurt contains not less than 0.5% and not more than 2% milk fat. The product containing less than 0.5% milk fat is labeled as nonfat yogurt. In all the categories of yogurt, a milk solids-not-fat minimum of 8.25% and a titratable acidity minimum of 0.5% lactic acid are stipulated.

The ingredients defined are: cream, milk, partially skimmed milk, skim milk alone, or in combination. Concentrated skim milk, nonfat dry milk, or other milk derived ingredients may be used to standardize milk solids-not-fat content of a mix. Presumably, the milk derived ingredients include casein, sodium and calcium caseinates, whey, and whey protein concentrates alone or in combination. The use of milk derived ingredients is conditional in that the ratio of protein to total nonfat solids of the food and the protein efficiency ratio of all protein present should not be diminished by their use. Additives permitted are nutritive carbohydrate sweeteners, coloring, stabilizer, and fruit preparations for flavoring yogurt. The culture is specified as a mixture of *Lactobacillus bulgaricus* and *Streptococcus thermophilus*.

Sweeteners.—Nutritive carbohydrates used in yogurt manufacture are similar to the sweeteners used in ice cream and other frozen desserts described by Arbuckle (1977). Sucrose is the major sweetener used in yogurt production. Sometimes, corn sweeteners and honey may also be used. The level of sucrose in yogurt mix appears to affect the production of lactic acid and flavor by yogurt culture. Bills et al. (1972) reported a decrease in acetaldehyde production at 8% or higher concentration of sucrose. Sucrose may be added in a dry, granulated, free-flowing, crystalline form or as a liquid sugar containing 67% sucrose. Liquid sugar is preferred for its handling convenience in large operations. However, storage capability in sugar tanks along with heaters, pumps, strainers, and meters is required. The corn sweeteners, primarily glucose, usually enter yogurt via the processed fruit flavor in which they are extensively used for their flavor enhancing characteristics. Up to 6% corn syrup solids are used in frozen yogurt. Non-nutritive sweeteners (e.g., Ca-saccharin) have been used along with maltol to produce a product containing about 50% of the calories of normal sweetened yogurt.

Lactase has been suggested for hydrolysis of lactose to a sweeter mixture of glucose and galactose in yogurt, thereby reducing the level of sucrose required to achieve a constant degree of sweetness (Kosikowski and Wierzbicki 1971; Engel 1973). Goodenough and Kleyn (1976) investigated qualitative and quantitative changes in yogurt during its manufacture. They reported average lactose concentration of 8.5% in yogurt mix. Upon fermentation, the lactose level dropped to 5.75% with a concomitant increase of 1.20% galactose. Glucose was detected only in trace quantities. Commercial yogurts had an average of 4.06% lactose, 1.85% galactose, 0.05% glucose, and pH of 4.15.

Stabilizers.—The primary purpose of using a stabilizer in yogurt is to produce smoothness in body and texture, impart gel structure, and reduce wheying off or syneresis. The stabilizer increases shelf life and provides a reasonable degree of uniformity of the product. Stabilizers function through their ability for form gel structures in water, thereby leaving less free water for syneresis. In addition, some stabilizers complex with casein. A good yogurt stabilizer should not impart any flavor, should be effective at low pH values, and should be easily dispersed in the normal working temperatures in a dairy plant. The stabilizers generally used in yogurt are gelatin, vegetable gums like carboxymethyl cellulose, locust bean and guar, and seaweed gums like alginates and carrageenans (Hall 1975).

Gelatin is derived by irreversible hydrolysis of the proteins collagen and ossein. It is used at a level of 0.3–0.5% to get a smooth shiny appearance in refrigerated yogurt. Gelatin is a good stabilizer for frozen yogurt. The term Bloom refers to the gel strength as determined by a Bloom gelometer under standard conditions. Gelatin of a Bloom strength of 225 or 250 is commonly used. The gelatin level should be geared to the consistency standards for yogurt. Amounts above 0.35% tend to give yogurt of relatively high milk solids a curdy appearance upon stirring. At temperatures below 10°C, the yogurt acquires a pudding-like consistency. Gelatin tends to degrade during processing at ultrahigh temperatures and its activity is temperature dependent. The yogurt gel is considerably weakened by a rise in temperature. Furthermore, being an animal product, gelatin is generally not acceptable in Kosher yogurts.

The seaweed gums impart a desirable viscosity as well as gel structure to yogurt. Algin and sodium alginate are derived from giant sea kelp. Carrageenan is made from Irish moss and compares with 250 Bloom gelatin in stabilizing value. These stabilizers are heat stable and promote stabilization of the yogurt gel by complex formation with Ca^{++} and casein.

Among the seed gums, locust bean gum or carob gum is derived from the seeds of a leguminous tree. Carob gum is quite effective at low pH levels. Guar gum is also obtained from seeds and is a good stabilizer for yogurt. Guar gum is readily soluble in cold water and is not affected by high temperatures used in the pasteurization of yogurt mix. Carboxymethyl cellulose is a cellulose product and is effective at high processing temperatures.

The stabilizer system used in yogurt mix preparations is generally a combination of various vegetable stabilizers to which gelatin may or may not be added. Their ratios as well as the final concentration (generally 0.5–0.7%) in the product are carefully controlled to get desirable effects. Other stabilizers reportedly used are agar and pectin (Humphreys and Plunkett 1969). $CaCl_2$ may be useful in controlling whey separation (Pette and Lolkema 1951).

For detailed descriptions of various industrial gums, the reader is referred to Whistler (1973).

Fruit Preparations for Flavoring Yogurt.—The fruit preparations for blending in yogurt are specially designed to meet the marketing requirements for different types of yogurt. They are generally present at levels of 15–20% in the final product (Craven 1975). A majority of the fruits contain natural flavors. The following types of yogurts are marketed in the United States.

1. Fruit-on-bottom or Eastern Sundae Style Yogurt.—In this type, 59 ml (2 oz) of fruit preserves are layered at the bottom followed by 177 ml (6 oz) of inoculated yogurt mix on the top. No flavor or sweetener is added to the yogurt layer. After placing lids on the cups, incubation and setting of the yogurt takes place in the cups. When a desirable pH of 4.2–4.4 is attained, the cups are placed in refrigerated rooms for rapid cooling. For consumption, the fruit and yogurt layers are mixed by the consumer. Fruit preserves have a standard of identity. A fruit preserve consists of 55% sugar and a minimum of 45% fruit which is cooked until the final soluble solids content is 68% or higher (65% in the case of certain fruits) (Gross 1974). Frozen fruits and juices are the usual raw materials. Commercial pectin, 150 grade, is normally utilized at a level of 0.5% in preserves and the pH is adjusted to 3.0–3.5 with a food-grade acid, viz., citric, during manufacturing of the preserves.

2. Western Sundae Style Yogurt.—In this type, fruit preserves or special fruit preparations may form the bottom layer. The top layer consists of yogurt containing sweeteners, with the flavor and color indicative of the fruit on the bottom. The flavor level is usually 2–4% in the top layer. In other respects, this yogurt is identical to Eastern Sundae Style.

3. Swiss-style Yogurt.—Also known as Continental Style, French Style, and Stirred yogurt, the fruit preparation is thoroughly blended in yogurt after culturing. Stabilizers are necessary in this form of yogurt unless milk solids-not-fat levels are relatively high (14–16%). In this style, cups are filled with a blended mixture of yogurt and fruit. Upon refrigerated storage for 48 hr, the clot is reformed to exhibit a fine body and texture. Overstabilized yogurt possesses a solid-like consistency and lacks a refreshing character. Yogurt should not be so thin that it can be drunk. It should melt in the mouth without chewing.

Flavors and certified colors are usually added to the fruit-for-yogurt for improved eye appeal and better flavor profile. The fruit base should meet the following requirements. It should (a) exhibit true color and flavor of the fruit when blended with yogurt, and (b) be easily dispersible in yogurt without causing texture defects, phase separation, or syneresis. The pH of the fruit base should be compatible with the yogurt pH. The fruit should have a low yeast and mold population in order to prevent spoilage and to extend shelf life. Fruit preserves do not necessarily meet all these requirements, especially of flavor, sugar level, consistency, and pH. Accordingly, special fruit bases of the following composition are designed for use in stirred yogurt.

	%
Fruit	17–41
Sugar	22–40
Corn syrup solids	10–24
Modified food starch	3.5–5.0

	%
Fruit flavor, artificial	0.1
or natural	1.25
Color	0.01 or to specifications
Potassium sorbate	0.1
Citric acid to pH 3.7–4.2	

CaCl$_2$ and certain food-grade phosphates are also used in several fruit preparations. The soluble solids range from 60 to 65% and viscosity is standardized to 5 ± 1.5 Bostwick units (cm), 30 sec reading at 24°C. Standard plate counts on the fruit bases are generally less than 500/g. Coliform count and yeast and mold count are less than 10/g. The fruit flavors vary in popularity in different parts of the country and during different times of the year. In general, more popular fruits are: strawberry, raspberry, blueberry, peach, cherry, orange, lemon, purple plum, boysenberry, spiced apple, apricot, and pineapple. Blends of these fruits are also popular. Fruits used in yogurt base manufacture may be frozen, canned, dried, or combinations thereof. Among the frozen fruits are strawberry, raspberry, blueberry, apple, peach, orange, lemon, cherry, purple plum, blackberry, and cranberry. Canned fruits are: pineapple, peach, mandarin orange, lemon, purple plum, and maraschino cherry. The dried fruit category includes apricot, apple, and prune. Fruit juices and syrups are also incorporated in the bases. Sugar in the fruit base functions in protecting fruit flavor against loss by volatilization and oxidation. It also balances the fruit and the yogurt flavor. The pH control of the base is important for fruit color retention. The color of yogurt should represent the fruit color in intensity, hue, and shade. The base should be stored under refrigeration to obtain optimum flavor and to extend the shelf life.

Yogurt Starter and Its Contribution to Texture and Flavor.—The starter is a critical ingredient in yogurt manufacture. Tramer (1973) and Davis (1975) have discussed practical aspects of yogurt culture. Freedom from contaminants, vigorous growth in yogurt mix, good flavor, body, and texture production and a reasonable resistance to phages and antibiotics are primary requirements of a yogurt culture. Equal cell numbers of *Lactobacillus bulgaricus* and *Streptococcus thermophilus* are desirable for flavor and texture production. The lactobacilli grow first, liberating the amino acids glycine and histidine, and stimulating the growth of streptococci (Bautista *et al.* 1966). Tramer (1973) demonstrated differences in the comparative ability to produce acid by various strains of yogurt cultures in commercially autoclaved versus heat-treated (95°C, 30 min) milk. The rate of acid production by yogurt culture should be synchronized with plant production schedules. Using frozen culture concentrates, incubation periods of 5 hr at 45°C, 11 hr at 32°C, or 14–16 hr at 29°–30°C are required for yogurt acid development. Using bulk starters at 1% inoculum level, the period is 2.5–3.0 hr at 45°C, 8–10 hr at 32°C, or 14–16 hr at 29°–30°C (Yeager 1975). Stone *et al.* (1975) reported that milk, UHT pasteurized at temperatures of 115.6°–157.2°C with a holding time of 0.02 sec, exhibited higher starter activity in comparison with vat-pasteurized milk.

The production of flavor by yogurt cultures is a function of time as well as the sugar content of yogurt mix. Gorner et al. (1968) reported that acetaldehyde production in yogurt takes place predominantly in the first 1–2 hr of incubation. Eventually, 23–55 ppm of acetaldehyde are found in yogurt. Hamdan et al. (1971) reported acetaldehyde levels of 22–26 ppm in their cultures at the fifth hour of incubation, and the acetaldehyde level declined in later stages of incubation. Yogurt flavor is typically ascribed to the formation of lactic acid, acetaldehyde, acetic acid, and diacetyl.

The milk coagulum during yogurt production results from the drop in pH due to the activity of the yogurt culture. The streptococci are responsible for lowering the pH of a yogurt mix to 5.0–5.5 and the lactobacilli are primarily responsible for further lowering of the pH to 3.8–4.4. Attempts have been made to improve the viscosity and to prevent syneresis of yogurt by including a slime producing strain of Streptococcus filant or Streptococcus lactis var. hollandicus (Galesloot and Hassing 1968; Busch-Johannsen et al. 1971; Tramer 1973). The texture of yogurt tends to be coarse or grainy if it is allowed to develop firmness prior to stirring or if it is disturbed at pH values higher than 4.6. Rennet addition to yogurt mix, excessive whey solids, and incomplete blending of mix ingredients are additional causes of a coarse texture. Homogenization treatment and high fat content tend to favor smooth texture. Gassiness in yogurt may be attributed to defects in starters or contamination with sporeforming Bacillus species, coliform, or yeasts, producing excessive CO_2 and hydrogen. In comparison with plate heat exchangers, cooling with tube type heat exchangers causes less damage to yogurt structure (Mann 1973). Further, loss of viscosity of yogurt may be minimized by well designed booster pumps, metering units, and valves involved in yogurt packaging.

The pH of yogurt during storage continues to drop. Gavin (1966) noticed a drop from 4.62 to 4.15 in 6 days of storage at 4°C. High temperatures of storage accelerate the drop in pH. Most yogurt manufacturers incorporate 5–7% sucrose in their yogurt mix prior to culturing. Tramer (1973) reported that various strains of yogurt cultures responded differently to various levels of sucrose. No inhibition of culture activity was noticed up to 5.5% sucrose concentrations. At higher sucrose levels, acid production by some yogurt cultures was partially inhibited. This effect was primarily ascribed to a stress on lactobacilli and was related to a total solids level in yogurt. Total solids, consisting of milk solids and sweeteners, above 22% level were inhibitory to Lactobacillus bulgaricus. The inhibition of yogurt culture is also caused by antibiotic residues in milk. Mocquot and Hurel (1970) reported that both Streptococcus thermophilus and Lactobacillus bulgaricus are affected by 0.005 IU penicillin/ml, 0.066 μg aureomycin/ml and 0.38 IU streptomycin/ml. Streptococcus thermophilus is exceedingly sensitive to penicillin. It is affected at 0.01 IU/ml and acid production ceases at 0.03 IU/ml. Phages are not a practical threat to yogurt making if using frozen cultures and their proper rotation are practiced along with high sanitation standards in the plant. However, phages for yogurt cultures have been isolated (Reinbold and Reddy 1973; Kosikowski 1977). Hypochlorites and quaternary ammonium compounds also inhibit yogurt cultures (Bouchez and van Belleghem 1971).

Heat treatment at 85°C for 30 min or equivalent is an important step in manufacture. The heat treatment (a) produces a relatively sterile medium for the exclusive growth of the starter, (b) removes air from the medium to produce a more conducive medium for microaerophilic lactic cultures to grow, (c) effects thermal breakdown of milk constituents, especially proteins, releasing peptones, sulfhydryl groups which provide nutrition and anaerobic conditions for the starter, and (d) denatures and coagulates milk albumins and globulins which enhance the viscosity and produce custard-like consistency in the product. Homogenization also aids in texture development and additionally it alleviates the surface creaming and wheying off problems. Ionic salt balance in milk is also involved in the wheying off problem.

Manufacturing Procedures

Plain Yogurt.—The steps involved in the manufacture of set-type and stirred-type plain yogurts are shown in Fig. 5.6. Plain yogurt normally contains no added sugar or flavors in order to offer the consumer natural yogurt flavor for consumption as such or an option of flavoring with other food materials of the consumer's choice. In addition, it may be used for cooking or for salad preparation with fresh fruits or grated vegetables. The fat content may be standardized to the levels preferred by the market. Also, the size of package may be geared to the market demand. Wax coated cups as well as plastic cups and lids are the chief packaging materials used in the industry.

Fruit Flavored Yogurt.—A general manufacturing outline for both Sundae Style and Swiss yogurts is presented in Fig. 5.7. Several variations of this procedure exist in the industry. Fruit incorporation is conveniently effected by the use of a fruit feeder at a 15–20% level. Prior to packaging, the stirred-yogurt texture can be made smoother by pumping it through a valve or a stainless steel screen.

The incubation times and temperatures are coordinated with the plant schedules. Incubation temperatures lower than 40°C in general tend to impart a slimy or sticky appearance to yogurt. *Lactobacillus acidophilus* culture concentrates may be incorporated into yogurt after incubation to make acidophilus yogurt (Vedamuthu 1974), but the viability of the *L. acidophilus* cultures may be rather limited (Speck 1977; Gilliland and Speck 1977).

Wilcox (1971) reviewed the processes for making yogurt with polyunsaturated corn oil, instant yogurt, and yogurt enriched with vitamin C.

Post-culturing Heat Treatment.—The shelf life of yogurt may be extended by heating yogurt after culturing to inactivate the culture and the

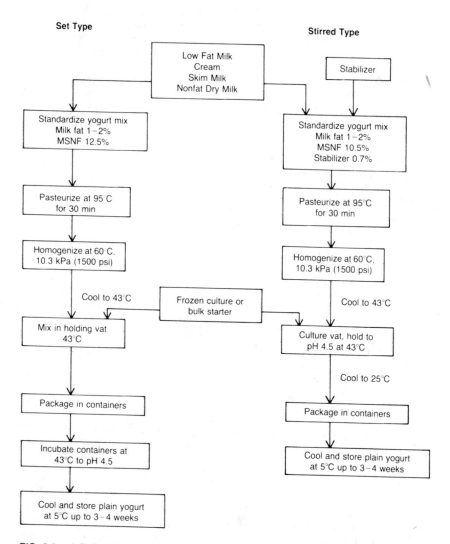

FIG. 5.6.   A FLOWSHEET OUTLINE FOR THE MANUFACTURE OF SET AND STIRRED
TYPE PLAIN LOW FAT YOGURTS

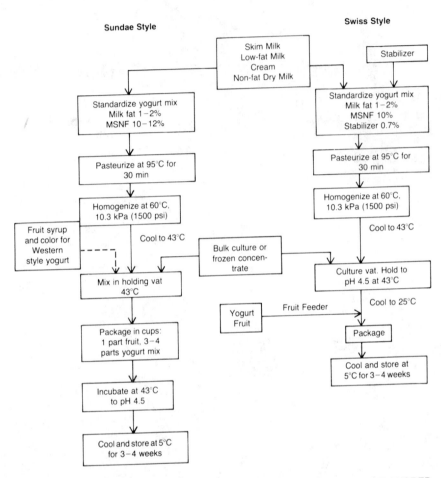

FIG. 5.7. A FLOWSHEET OUTLINE FOR THE MANUFACTURE OF FRUIT FLAVORED LOW FAT YOGURT

constituent enzymes (Mann 1973). Heating to 60°–65°C stabilizes the product so the yogurt shelf life will be 6–8 weeks at 12°C. However, this treatment destroys the "live" nature of yogurt, which may be a desirable consumer attribute to retain. The proposed federal standards permit the thermal destruction of viable organisms with the objective of shelf life extension, but the parenthetical phrase "heat treated after culturing" must show on the package following the yogurt labeling. The post-ripening heat treatment may be designed to accomplish two objectives: (1) to ensure destruction of starter bacteria, contaminating organisms, and enzymes, and (2) to preserve or redevelop the texture and body of the yogurt by appropriate stabilizer and homogenization processes, similar to hot-pack cultured cream (Fig. 5.5).

Frozen Yogurt

Soft-serve yogurt, hard-pack yogurt, and novelty items based upon yogurt are relatively new products getting an enthusiastic response by the consumer. Push-ups, frozen yogurt on a stick, skippy cups, and tetrapaks are being manufactured and marketed.

The frozen yogurt base mix may be manufactured in a cultured dairy plant and shipped to a soft-serve operator or an ice cream plant. Alternatively, the mix may be prepared and frozen in an ice cream plant. The following formulation is generally used: 1.5–2.0% milk fat, 13–15% milk solids-not-fat, 0.15–0.20% gelatin (250 Bloom), 7–10% sucrose, and 4–5% corn syrup solids (24–26 DE). These ingredients (except one-half of the sugar) are standardized in a blend tank and pasteurized at 88°C for 40 min. The mix is then homogenized at 58°–63°C at 10.3 kPa, and cooled to 44°C. Yogurt culture is then inoculated and incubation of the mix is continued until pH 3.9 is attained. The yogurt mix is then cooled to 25°C and the remaining sugar as well as fruit is blended. Special fruit preparations designed for frozen yogurt are used at a level of 15–20%. This mix is then frozen in an ice cream freezer at 50–60% overrun, packaged, and hardened similar to ice cream. To obtain a soft-serve product, a soft-serve freezer is used at a draw temperature of −8°C.

Yogurt Quality Control

Quality control programs for yogurt manufacture include control of product flavor, body and texture, color, process, and composition. The flavor defects are generally described as too intense, too weak, or unnatural. The sweetness level may be excessive, weak, or may exhibit corn syrup flavor. The flavor may be too tart, weak, or atypical. The ingredients used may impart undesirable flavors like stale, metallic, old ingredients, oxidized, rancid, or unclean. Lack of control in processing procedures may cause overcooked, caramelized, or too tart flavors. Proper control of processing parameters and ingredient quality ensures good flavor.

In hard-pack frozen yogurt, a coarse and icy texture may be caused by the formation of ice crystals due to fluctuations in storage temperature. Sandiness may be due to lactose crystals resulting from too high levels of milk solids or whey solids. A soggy or gummy defect is caused by too high milk solids-not-fat level or too high sugar content. A weak body results from too high overrun and insufficient total solids.

Color defects may be caused by the lack of intensity or authenticity of hue and shade. Proper blending of fruit purees and yogurt mix is necessary for uniformity of color. The compositional control tests are: fat, moisture, pH, and overrun (for frozen yogurt), and microscopic examination of yogurt culture to ensure a ratio of 1:1 in *Lactobacillus bulgaricus* and *Streptococcus thermophilus*. Good microbiological quality of all ingredients is necessary. As a guideline, raw milk and cream should contain fewer than 500,000 and 800,000 bacteria/ml, respectively. The total bacteria count of pasteurized fluid dairy products should not exceed 50,000 counts/g or ml. Coliform counts for pasteurized products should not exceed 10/g or ml. Also, periodic checks for yeast and mold counts in fruit preparations are useful.

ACIDOPHILUS MILK

Acidophilus milk is produced by fermentation of milk with *Lactobacillus acidophilus*. This product has been considered to possess therapeutic value, particularly following antibiotic therapy (Sandine *et al.* 1972B; Speck 1975), and is therefore prescribed by dieticians and physicians for digestive disorders. Its therapeutic properties are attributed to the ability of the organisms to become transplanted in the intestinal tract. Lactose (present in the fermented milk), in general, is a slowly assimilable disaccharide and is unchanged by the digestive process until it reaches the intestine where it is metabolized by *Lactobacillus acidophilus* to produce an acidic condition. This condition discourages the growth and proliferation of gas-forming putrefactive organisms in the gut. Based on this reasoning, efforts have been made to produce a product containing a live culture of the strains that can be transplanted in the intestines. The manufacturing principle resembles buttermilk manufacture and is shown in Fig. 5.8.

Lactobacillus acidophilus is a slow fermenting organism and to get a reasonably fast growth rate it is customary to ultraheat milk (98°C for 30 min or 145°C for 2–3 sec) before inoculation. For viability of the cells, it is preferable to terminate the incubation period when the acidity reaches 0.65%. Continued incubation may result in 1% or more acidity. Acidophilus milk should be cooled to 5°C immediately. Fresh acidophilus milk contains in excess of 500 million cells/ml, although the viable count decreases rapidly with storage. The product obtained by fermentation may be too sour for many palates. For moderating the sour taste, acidophilus milk may be mixed with cultured buttermilk. Alternatively, souring is controlled by using a mixed culture. Bioghurt is a trade name of a milk product fermented with *Lactobacillus acidophilus* and *Streptococcus lactis*. It belongs to a family of dairy products fermented by a mixture of organisms which are nonsymbiotic, and nonspecific. One of the cultures is generally *Lactobacillus acidophilus*.

Several combinations of mixed starters are possible and one might employ two or more of the organisms generally utilized in the manufacture of cultured dairy products (Table 5.5). *Lactobacillus (Bifidobacterium) bifidus* may also be utilized. The manufacturing process would be similar to yogurt manufacture with modifications according to the cultural conditions of the fermenting organisms. Apparently, it is possible to incorporate different levels of acidities and flavor by using the cultures in various combinations. This area of cultured dairy products needs further development.

Acidophilus Milk

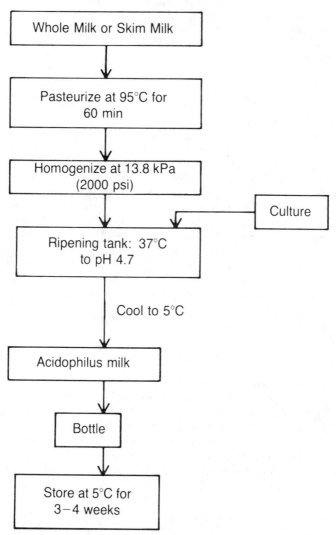

FIG. 5.8.   FLOW SHEET DIAGRAM FOR ACIDOPHILUS MILK MANUFACTURE

Consumption of milk containing live *Lactobacillus acidophilus* cells has been facilitated by the recent introduction of uncultured low-fat milk inoculated with about 500 million acidophilus cells per ml (Speck 1975). An acidophilus cell concentrate is prepared by centrifugation, suspended in regular refrigerated milk, and packaged. The product does not ferment as long as it is held under refrigeration during distribution and storage, permitting the consumers to enjoy the sweet flavor of milk as well as ingest live *Lactobacillus acidophilus* cells.

Campbell and Marshall (1975) mentioned a Japanese milk product containing an extract of *Chlorella* algae, vitamin C, and viable cells of *Lactobacillus* (Shirota strain) in excess of 10^8/ml. It is reported to have a pleasant, mildly sour flavor and thin consistency. *Lactobacillus casei* has also been employed as a culture in fermented milk drinks.

KEFIR

Kefir belongs to the class of acid and alcoholic fermented milks. Kefir is very popular in the U.S.S.R., with an annual per capita consumption of 4.5 kg (10 lb) (Kosikowski 1977). It is produced by fermentation with kefir grains. The grains are white to yellow in color, and insoluble in water and common solvents. However, when added to milk, their size expands by imbibing water and their color changes to white. According to Kosikowski (1977) the grains contain 24% slimy polysaccharide secreted by the culture. The symbiotic microflora contains yeasts (*Saccharomyces kefir* and *Torula kefir*), lactobacilli (*Lactobacillus caucasicus*), leuconostocs, and lactic streptococci. The product is foamy and fizzy due to its CO_2 content. An outline of kefir manufacturing principles is shown in Fig. 5.9.

Kefir grains are generally recovered and used repeatedly. For recovery the grains are sieved off after fermentation. They are either suspended in cold water and stored at 4°C or dried in cheesecloth at room temperature for 36–48 hr and stored dry at 4°C. The grains retain their activity for more than a year when stored dry and cold, but in water suspension they lose their activity after a week.

KOUMISS

Koumiss is similar to kefir except that mares' milk is used in its manufacture. Russians use koumiss for the treatment of pulmonary tuberculosis in doses of 1.4 liters (3 pints) per day for 2 months. The product is alcoholic and produces slight intoxication. The alcohol content of koumiss may vary from 1 to 2.5%, depending upon whether the product is of the weak or strong type. Corresponding to the alcoholic content, the titratable acidity varies from 0.7 to 1.8% lactic acid. Mares' milk is low in casein content and does not curdle like cows' milk. Accordingly, koumiss is a grayish-white wholesome drink.

The starter for koumiss consists of *Lactobacillus bulgaricus* and a lactose-fermenting yeast, *Torulopsis holmii*. Lactic acid, ethanol, and CO_2 are the major products giving koumiss a sour alcoholic flavor and fizzy appearance. In certain cases of slow fermentation, horse flesh or tendon or some vegetable matter is added. Presumably, this practice provides the microbial flora needed for fermentation. A detailed description of koumiss production is provided by Kosikowski (1977).

The manufacturing principles used in koumiss production are shown in Fig. 5.10. In view of the production of CO_2 in the product, appreciable pressure is developed in the capped bottles. In general, koumiss is quickly marketed. To satisfy the rising consumer demand in the U.S.S.R., koumiss is now being made from skimmed cows' milk.

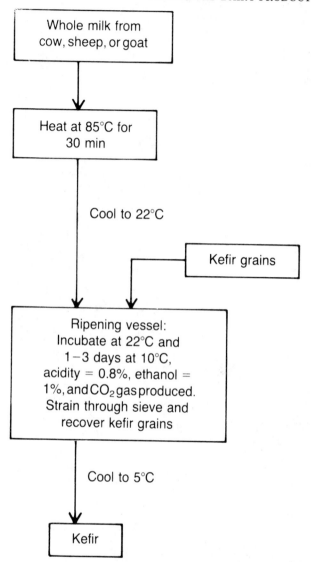

FIG. 5.9.   AN OUTLINE OF THE MANUFACTURING PRINCI-PLE FOR KEFIR PRODUCTION

COTTAGE CHEESE

Cottage cheese belongs to the class of natural, unripened, soft cheeses. It differs from cream cheese in that it has a considerably lower fat content, which gives it a popular place in low-calorie diets.

Cottage cheese is made from pasteurized, skim milk, or reconstituted skim milk powder. Coagulation is accomplished by lactic streptococci, and a very small amount of rennet or coagulator may or may not be added. After

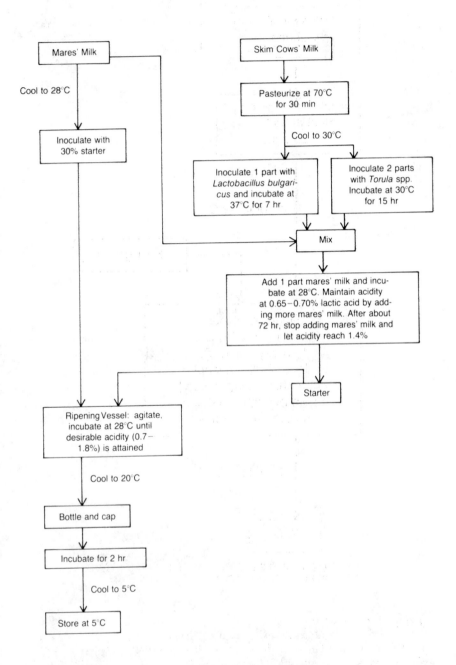

FIG. 5.10. PRINCIPLES OF KOUMISS MANUFACTURE ACCORDING TO KOSIKOWSKI (1977)

curdling, the curd is cut into cubes using 1.3, 1.6, or 1.9 cm wire cheese knives. It is marketed as small curd and large curd cottage cheese. The curd is cooked, washed, and mixed with salt and cream dressing.

Federal standards define creamed cottage cheese as the product containing moisture not exceeding 80% and milk fat not less than 4%. The low-fat product may contain 0.5–2.0% milk fat. Cottage cheese dry curd contains less than 0.5% fat. In addition, silicone antifoams and sorbate preservatives may be used. It may be marketed in flavored form by incorporating little bits of pimientos, chives, or pineapple.

The product is commonly used as a major ingredient of salads. In an ideal creamed cottage cheese, the basic flavor should be similar to fresh, clean, cultured milk or cream. Cottage cheese should have a mild acid, light salty taste, with the flavor and aroma of a good lactic butter culture. The body and texture should be uniform, smooth and meaty, not too firm and not too soft and pasty. It should be of uniformly sized particles (regardless of style or cut of curd) and have a natural color. Creamed cottage cheese should have a uniform layer of cream around the curd particles with a minimum of free cream. Any excess cream should be of a thick consistency and not whey-like or watery.

Manufacturing Methods

Figure 5.11 illustrates the manufacturing principles involved in cottage cheese production. The methods for making cottage cheese can be classified into 3 groups, based upon the length of time allowed for coagulation.

Short Set Method

This method requires 4.5–5.5 hr for coagulation and proper acid development; 5 to 7% lactic culture is used along with a setting temperature of 31°–32°C. This method is designed to be completed, except for packaging, in an 8 hr day. Labor is utilized in the afternoon for cooking, washing, and cleanup.

Raw skim milk is pasteurized at 73°C for 16.5 sec. It is desirable to avoid using any previously pasteurized skim milk as a minimum heat treatment is required to avoid curd weakness. After pasteurization, the skim milk is cooled promptly and pumped to the vat where the proper setting temperature of 32°C is obtained.

Setting.—The titratable acidity of the pasteurized skim milk is determined in order to judge the rate of subsequent acid development. The bulk culture is added to the vat at the rate of 5 to 7%. The skim milk is then stirred thoroughly every 30 min for 1.5 hr. The temperature should be kept at 32°C. The acidity test is run after the culture is mixed with the skim milk and again after 1.5 hr. During this time, if the culture is producing acid normally, the acidity should increase by 0.05 to 0.07%. If the acidity has not increased by 0.05%, 1% more culture is added for each 0.01% increment below 0.05%.

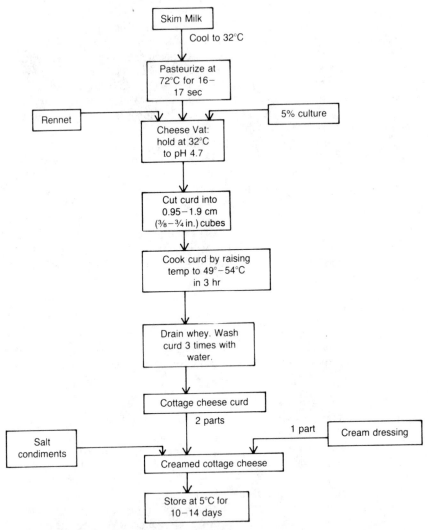

FIG. 5.11.   FLOW DIAGRAM ILLUSTRATING THE PRINCIPLES OF SHORT SET COT-
TAGE CHEESE MANUFACTURING PROCEDURE

The next step is to add rennet or coagulator diluted with 40 volumes of
cold water before it is added to the vat. Rennet may be used at the rate of 1
ml per 454 kg of skim milk for large curd and medium curd, and 0.3−0.5 ml
per 454 kg for small curd. A commercial coagulator, if used, should be used
according to the manufacturer's instructions for the type of cheese being
made. The coagulator or rennet is mixed thoroughly with the skim milk and
the vat is covered for 2 hr for consummation of coagulation.

Cutting.—Two hours after the coagulator is added, curd formation is checked by one of the methods described later. When the curd is ready, the first cut, along the length of the vat, is made with the horizontal knife. It is followed by the second cut lengthwise with the vertical knife and the third cut crosswise of the vat with the same vertical knife. Care should be taken to avoid overlapping of cuts.

Cooking.—After cutting, the curd is held without agitation for 10 min to allow the cut surfaces of the cubes to firm slightly. The water in the jacket of the vat is raised 12°C above the vat temperature. The curd is pushed very gently from the sides of the vat toward the center with a stainless steel stirrer. The vat is stirred every 5 to 10 min. Jacket temperature is maintained 12°–18°C above the vat temperature in order to raise the temperature of the vat content to 49°–54°C in 1 to 1.5 hr. A mechanical agitator on slow speed may be used after the temperature reaches 38°–41°C, followed by a gradual increase in agitation speed to prevent matting. Small curd cheese requires more agitation and cooks out faster than large curd cheese. If the curd is not sufficiently firm when the temperature reaches 49°–54°C, the curd is held at that temperature with agitation until firm. The firmness of large, intermediate, and small size curd can best be judged by chilling a small amount of curd in cold water (5°C). The curd is adequately cooked when a handful of water-chilled curd springs apart after being squeezed together with moderate pressure. Alternatively, the curd is ready for washing when you get a reading of 4–5 on a Lundstedt meter which is designed to measure the firmness of the curd.

Washing.—After the curd is cooked, the jacket of the vat is drained and whey removed until the curd begins to surface. Wash water, acidified and chlorinated, is then added, equal to the volume of whey removed, to reduce the temperature in the vat to 27°–30°C. After 10 min agitation, the wash water is drained. Washing is repeated 2 more times to reduce temperature in the vat to 13°–16°C and then to 5°C or lower. After the last wash, the curd is trenched and permitted to drain for 30–60 min before creaming or removal to storage. Alternatively, the water-curd slurry is pumped into a special blender for drainage. A new development in curd washing uses a vertical curd washing system designed to effect conservation of wash water. Consequently, a lower BOD in the plant effluent is achieved.

Creaming.—The drained curd is creamed in the vat or in a special blender or by weighing the curd into cans containing a previously weighed amount of dressing described subsequently. The most desirable method is to mix the dressing with the curd in the vat or in a blender and then transfer by gravity or by special pumps directly to the packaging machines.

Medium Set Method

This method works well with a 24 hr operation in which milk is received one day and cottage cheese from that milk is marketed the following day.

This method is essentially the same as the Short Set Method except that 2–4% lactic culture is used at a setting temperature of 27°–28°C and 8–10 hr are required for proper acid development and coagulation.

Long Set Method

This method requires the following modifications of the Short Set Method previously described. The setting temperature should be 21°–22°C and 0.1–1.5% lactic culture is required for a 12–16 hr setting time. Less culture is required and the work load can be readily distributed over a normal work day. Rennet or coagulator should be added just after the addition of the culture. The cooking should be slower at the start and faster at the finish. Total cooking time is usually 30–60 min longer than for the Short Set Method.

Water Supply

Water used in making cottage cheese is in direct contact with the product and hence is just as important in determining flavor and keeping quality as any other ingredient. The water should be from an approved drinking water supply and should be free of sediment and off-flavors, and practically free of sulfur, sodium carbonate, sodium bicarbonate, and sodium chloride. Water used for the reconstitution of nonfat dry milk and for cottage cheese curd should be adjusted to pH 4.0–4.75 with food-grade phosphoric, lactic, or citric acid. All water which comes in contact with the product should be chlorinated with 10–20 ppm chlorine with a contact time of at least 10 min. Water containing phenols or phenol-like compounds may develop chlorophenol (medicinal) flavor when chlorinated and should be treated or filtered before chlorination. For example, activated carbon may be used to remove the phenolic materials.

Lactic Culture

Bulk lactic culture should be prepared according to the method outlined in Fig. 5.1. The culture (*Streptococcus lactis* or *Streptococcus cremoris* and low levels of *Leuconostoc cremoris* or *Streptococcus lactis* subsp. *diacetylactis*) should possess a clean, smooth, and fine flavor as well as good activity.

Methods for Determining Proper Cutting Time

Three methods can be used for determining the proper time to cut cottage cheese curd:

1. **Titratable Acidity of Whey.**—When the curd is firm, a sample of the whey is drawn from at least 10 cm (4 in.) below the surface using a plastic pipette. The sample is filtered to get clear whey. The titratable acidity of the

clear whey should be checked periodically until it is 0.34 to 0.36% higher than the initial titratable acidity of the skim milk. For skim milk fortified with nonfat dry milk to 10, 11, or 12% solids, the acidity of the whey at cutting should be 0.55, 0.60, or 0.65%, respectively. Cutting at low acidity yields a sweeter cheese which cooks out faster but may tend to mat during cooking. Cutting at high acidity yields a more acidic cheese which is brittle and more difficult to cook out but tends to mat less during cooking.

2. pH Determination.—Cottage cheese may also be cut when the pH of the curd is in the range of 4.6–4.7.

3. Acid Coagulation Test.—The acid coagulation or A-C test can also be used to determine the proper time for cutting. This test can be used alone or with one of the other tests to determine the correct titratable acidity or pH at which to cut curd from any particular milk supply. The steps in the A-C test are as follows:

(1) Proceed through the regular manufacturing steps and then, just before adding rennet, remove 200 ml of the seeded skim milk into a stainless steel beaker.
(2) Add rennet to the vat but not to the beaker of skim milk.
(3) Suspend the beaker of skim milk in the vat to ensure the same temperature and rate of acid development.
(4) When the skim milk in the vat has coagulated, check the sample in the beaker for evidence of coagulation.
(5) When a gel can be detected, cut with 1 or 2 strokes of a small knife or spatula.
(6) Repeat cutting at intervals of 5 min until fine lines of whey appear within 1 min after cutting. This is the A-C endpoint and the curd is ready for cutting.

Cottage Cheese Dressing

A properly prepared cottage cheese dressing adds a significant shelf life to the finished product and enhances flavor and appearance of the cheese. A certain level of viscosity, obtained by the use of a suitable stabilizer, is necessary to properly bind the free flowing cream and whey in the cheese.

For the preparation of the dressing, a standardized blend is prepared to contain 12.5% fat, 8.5% milk solids-not-fat, 2.7% salt, and 0.25% cottage cheese stabilizer. For low-fat cottage cheese, the composition is 3.0% fat, 15.0% milk solids-not-fat, 2.7% salt, and 0.25% stabilizer. Before heat is applied (or at a temperature below 32°C), the stabilizer is added. The amount of stabilizer added will largely govern the viscosity of the finished product. It is advisable to start at a low point and, if a more viscous product is desirable, slightly increase the amount of stabilizer. It is preferable to mix stabilizer, salt, and milk solids before addition to the dressing mix in a pasteurizing vat. A practical way of introducing the dry ingredients is through a funnel attached to a centrifugal pump. If this is not feasible, the vat agitator is started at high speed and the dry ingredients are slowly sprinkled into a partially filled vat. The agitator is kept at high speed and the product is heated to a temperature of 75°–77°C and held for 30 min. The dressing is homogenized at 13.8 kPa, single stage, at 57°C and cooled to below 4°C. Fresh dressing is made every 2 days. Approximately 1 part of cream dressing mixed with 2 parts of dry cheese curd will yield a finished creamed cottage cheese of 4% milk fat. Similarly, this ratio will yield a low-fat product of 1% milk fat.

The production of desirable aroma and flavor in cottage cheese may be augmented by the use of certain cultures in culturing skim milk or cream dressing (Babel 1976; Kosikowski 1977). Mather and Babel (1959A) reported 2.3 ppm diacetyl and 55.7 ppm of acetoin in skim milk coagulum. After cutting, whey removal, and washing, the curd retained only 1 ppm diacetyl and 29 ppm of acetoin. Work of Babel and Mather (1961) and Lundstedt and Fogg (1962) has resulted in development of processes for flavor enhancement utilizing the flavor producing characteristics of *Leuconostoc cremoris* and *Streptococcus lactis* subsp. *diacetylactis*, respectively. These organisms are added to cottage cheese via cream dressing. The growth of *Leuconostoc cremoris* produces certain metabolites inhibitory to *Pseudomonas fragi*, *Pseudomonas putrefaciens*, and coliform organisms (Mather and Babel 1959B). The spoilage of cottage cheese is considerably controlled by reducing the pH of creamed cottage cheese below 5.0 (Harmon and Smith 1956). This effect is ascribed to the retardation of the growth rate of bacteria from several genera, including *Pseudomonas*, *Achromobacter*, and *Alcaligenes*, which are involved in proteolytic and lipolytic degradation of cottage cheese.

Yield

Cottage cheese yield from a vat of skim milk is commonly expressed as the kg of curd per 100 kg of skim milk. Alternatively, the yield may be expressed as the kg of curd per kg of solids in the skim milk. Occasionally, it is expressed as the kg of curd per kg of casein in the milk. A satisfactory yield of large cottage cheese curd (prior to creaming) made with the conventional method from skim milk of 9% solids, is 15.5 kg per 100 kg of skim milk or 1.72 kg of curd per kg of solids in the skim milk. The curd can be increased by about 1.80 kg for every kg of dry milk solids fortified into skim milk. A fortified skim milk containing 12% solids should yield 21.6 kg of curd per 100 kg of the starting material.

The yield of cottage cheese changes as the solids or casein content of the skim milk varies with the breed of cows, season of year, and stage of lactation. Also, the moisture content of a curd, normally 80%, may vary due to the cooking and draining methods used prior to creaming. Furthermore, overpasteurization may result in a weak and fragile curd subject to shattering and loss during cooking and draining. Excessive acidity at the cutting step may result in curd losses due to shattering. Proper care in cutting, cooking, and washing can reduce mechanical losses considerably. Some lactic cultures have been found to coagulate milk more firmly than others and give better cottage cheese yields.

Defects

Common defects and possible causes in cottage cheese have been discussed by Angevine (1972) and Lundstedt (1972, 1974). They are enumerated here as follows with a description of probable causes:

1. *Acid Flavor.*—Too high an acidity before cutting; failure to expel whey; insufficient or improper washing; poor quality skim milk and/or starters.

2. *Bitter and Unclean Flavor.*—Contamination by unclean wash water, equipment, utensils, undesirable bacteria during or after manufacture; use of poor quality skim milk, starter, and/or cream for creaming.

3. *Weak, Soft Curd.*—Excessive heat treatment of skim milk before setting, or of nonfat dry milk during manufacture; too high an acidity at the time of cutting; insufficient heating during cooking; too much coagulator along with too high an acidity.

4. *Shattered Curd.*—Excessive heat treatment of skim milk before setting; excessive preheat treatment of skim milk during dry milk manufacture; too much acid at cutting; improper cutting and stirring; undesirable bacteria producing gas; chilling curd too fast.

5. *Tough, Rubbery Curd.*—Insufficient acid development at cutting; cooking too rapidly or too much heat during cooking; excessive loss of moisture after manufacture.

6. *Floating or Gassy Curd.*—Water supply high in carbonate content; contaminated or poor starter cultures; insanitary equipment; gas-producing cultures.

CREAM CHEESE

Cream cheese contains at least 33% milkfat and not more than 55% moisture. Ingredients used are: cream, milk, skim milk, condensed milk, and nonfat dry milk. Cream cheese is manufactured by a cooked curd process similar to cottage cheese. Depending upon the plant schedule, the process may be the long set or short set type. Figure 5.12 shows the principles of cream cheese manufacture. After the coagulum is formed, it is stirred and heated gradually to 54°C by direct addition of hot water or by indirect heating through a jacket. Whey separation may be conducted mechanically by specially designed centrifugal separators. Hot packing significantly extends the shelf life of the product. Product deterioration is generally due to the growth of yeasts and molds.

DIRECT ACIDIFICATION PROCESSES

Direct acidification is considered as an alternate method for the production of soured dairy products. In recent years increasing atten-

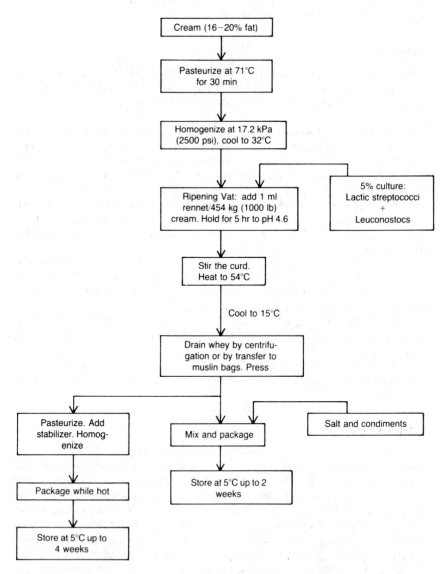

FIG. 5.12. FLOW DIAGRAM ILLUSTRATING THE PRINCIPLES OF CREAM CHEESE MANUFACTURE

tion has been paid to the development of imitation cultured dairy products using this technique. This process eliminates the need for acid production by lactic starter cultures and an acid condition is generated in the product by the addition of a food-grade acid. Stabilizers and artificial flavors are incorporated to simulate the body and flavor of a cultured dairy product. The base of directly acidified products may consist of dairy ingredients or may be an emulsified vegetable oil stabilized with sodium caseinate. The acids used are citric, phosphoric, gluconic, or lactic. Lactic acid is sparingly used because it is relatively expensive. Glucono-delta-lactone hydrolyzes slowly in milk to form gluconic acid, simulating somewhat the rate of acid production by a lactic culture. The flavors containing diacetyl are generally derived from culture distillates. Among the stabilizers and emulsifiers are: vegetable gums, starch, carrageenan, gelatin, partial glycerides, caseinates, and sodium phosphate.

The advantages claimed for direct acidification over conventional bacteriological souring include: (a) the process is easier to control; since lactic cultures are not involved, none of the problems associated with their usage (e.g., phage, antibiotic residues in milk) are encountered; (b) the process is quicker and more economical in terms of manpower; (c) it eliminates the need for expensive culturing vessels and is more economical on floor space; and (d) products of readily reproducible flavor can be made.

Buttermilk and sour cream are presently produced on a limited scale by direct acidification processes. According to the federal standards, they must be labeled as "acidified" products. In the case of cottage cheese the label includes "directly set" or "curd set by direct acidification."

Several developments have been recorded in the area of cottage cheese making by direct acidification. Deane and Hammond (1960) developed a process based on the principle of acidifying milk with neutral cyclic esters which hydrolyze in aqueous media to form organic acids. The most suitable acidifying agent for this purpose is D-gluconodelta-lactone, a neutral inner ester of gluconic acid. In milk the gluconodelta-lactone slowly hydrolyzes to form gluconic acid which reduces the pH of milk to the desired level of 4.5–4.7. The rate of hydrolysis of the glucono-delta-lactone is controlled by temperature. Higher temperature accelerates the rate. For economic reasons, this process is not commercialized. Little (1968) patented a process using hydrochloric acid or phosphoric acid for lowering the pH of skim milk to 4.6 at 4°C and inducing coagulation with rennet. After cutting the curd, the cooking process is started. Corbin (1971) patented a process using a food-grade acid to lower the pH of skim milk to 5.0, followed by acidification with D-glucono-delta-lactone. Wakeman's patent (1972) utilizes direct acidification of milk at low temperature and induces the curd formation by quiescent heating. Kosikowski (1977) reviewed the use of acid whey powder as an acidulant in cultured dairy products. In addition, cottage cheese has been made from ultrafiltered skim milk, obtained by fractionation of milk with special membranes. Continuous and automated processes for cottage cheese making are reviewed by Ernstrom and Kale (1975). They reported that the cultures contribute not only the acid but also a factor involved in developing desirable body and texture to cottage cheese curd. They recommended a combination of pre-culturing milk to pH 5.5 – 5.7 with starter and direct acidification to achieve desirable moisture retention and firmness in cheese curd.

The cultured dairy products differ from acidified products in several constituents. They contain starter culture cells (several million/g), high enzymatic activity, and products derived by their action on lactose, milk fat, and proteins. Furthermore, they contain antibiotics produced by the starters. The significance of these differences in nutrition, if any, is not fully known.

THERAPEUTIC AND NUTRITIONAL VALUE

Metchnikoff (1908) postulated that *Lactobacillus bulgaricus* possesses therapeutic value exercised by suppressing toxin production of putrefactive bacteria in the human intestine. This conclusion was based on his study of the inhabitants of the Balkans who consumed a rather large quantity of fermented milk (Bulgarian buttermilk) in their diets and displayed extraordinary vigor and longevity. In the U.S.S.R., koumiss is presently used for the treatment of pulmonary tuberculosis. In general, the health effects of fermented foods are controversial. Data in support and against the therapeutic claims are available in the literature and final resolution must be based upon future scientific investigations.

Hargrove and Alford (1977) studied the growth rate of experimental rats fed milk, acidified milks, and several fermented milks. They reported a nutritional superiority for yogurt in comparison with milk, acidified milk, cultured buttermilk, and acidophilus milk.

Since certain strains of *Lactobacillus acidophilus* can be transplanted to the human gut, attempts have been made to utilize the beneficial effects of this organism. Acidophilus culture is available for human consumption as a liquid concentrate or as dry granules and tablets (Storrs and Stern 1967). In addition, milk is fermented with *Lactobacillus acidophilus* to get a tart drink which has a limited consumer appeal. Alternatively, *L. acidophilus* cell concentrates are added to cold, pasteurized milk to furnish a source of concentrated cells for the consumer. The culture concentrate may also be added to yogurt which then acts as a carrier of the acidophilus culture, but the viability of acidophilus may be limited under these conditions (Speck 1977; Gilliland and Speck 1977). Sandine *et al.* (1972B) and Speck (1975) have reviewed the role of lactic acid bacteria in human health. The use of lactobacilli in intestinal therapy as well as the maintenance of balance in the intestinal microflora is stressed. Ingestion of viable *Lactobacillus acidophilus* cells of the order of $10^8 - 10^9$ per day is evidently satisfactory for the maintenance of proper balance. More work is necessary to determine the relationship between the intake of cultured milks and the ecology of intestinal microflora in the context of therapeutic and nutritional benefits.

The presence of antibiotics and antitumor activity in fermented dairy products is corroborated by several studies (Marth 1974; Babel 1977). Minor and Marth (1972) reported that yogurt and cultured buttermilk did not support the growth or viability of *Staphylococcus aureus*. The lactobacilli elaborate antibiotic-like materials, called lactolin, lactobrevin, lactocidin, lactobacillin, acidolin and acidophilin by various workers, which are effective against a wide variety of pathogenic organisms. It is not yet clear whether the antimicrobial compounds are individual antimetabolites or are related to each other. Nevertheless, it is possible that these antibiotic materials might contribute to the therapeutic value associated with acidophilus milk and yogurt by a favorable alteration in the ecological balance of the intestinal microflora.

The lactic streptococci and leuconostocs are also associated with the antibiosis effect *in vitro*. Nisin and diplococcin are products of the metabolic activity of *Streptococcus cremoris* and *Streptococcus lactis*, respectively. Similarly, antimicrobial compounds produced by *Streptococcus lactis* subsp. *diacetylactis* and *Leuconostoc cremoris* extend the shelf life of cottage cheese. Their contribution to the health properties, if any, is presently unknown.

Reddy *et al.* (1973) investigated the inhibitory effect of yogurt on Ehrlich ascites tumor cell proliferation. The mice given yogurt displayed 28% inhibition of tumor cell proliferation. The antitumor activity of the extracts of media fermented by *Lactobacillus bulgaricus* was reported earlier by Bogdanov *et al.* (1962). More recently, Mitchell and Kenworthy (1976) reported results on the effect of an antimicrobial agent of *Lactobacillus bulgaricus* upon the enterotoxic activity of *Escherichia coli*. They found anti-enterotoxic activity in the cells of *Lactobacillus bulgaricus*. This evidence suggests that certain lactobacilli may prevent or ameliorate coliform associated diarrheas. According to Marth (1974) some cultured dairy products display antibiotic activity. Acidophilus milk and yogurt inhibit *Escherichia coli* and *Mycobacterium tuberculosis*. Koumiss and kefir are bacteriostatic or bactericidal to *Escherichia coli*, *Staphylococcus aureus*, *Bacillus subtilis*, *Bacillus cereus*, and other organisms. An excellent review on this subject is available (Sandine *et al.* 1972B).

The nutrient content of various cultured dairy foods is shown in Table 5.12. These products are exceptional sources of high quality protein, calcium, phosphorus, potassium, and riboflavin in the human diet. In addition, the microflora of yogurt has appreciable lactase activity and may thereby contribute to the digestion of lactose in lactose-intolerant persons (Goodenough and Kleyn 1976; Kilara and Shahani 1976), although directly acidified yogurt has no lactase activity. While comparing the B complex vitamins in cultured and acidified yogurt, Reddy *et al.* (1976B) found that acidified yogurt contained slightly higher levels of B

TABLE 5.12.   NUTRIENT CONTENTS OF CULTURED DAIRY PRODUCTS IN 100 G PORTIONS

Nutrient	Cultured Buttermilk	Cultured Cream	Cultured Half and Half	Low-fat Yogurt Plain	Low-fat Yogurt Fruit Flavored	Cottage Cheese Creamed	Cottage Cheese Low-fat	Cream Cheese
Calories	40	214	135	63	99	103	72	349
Protein, g	3.31	3.16	2.94	5.25	3.98	12.49	12.39	7.55
Fat, g	0.88	20.96	12.00	1.55	1.15	4.51	1.02	34.87
Carbohydrate, g	4.79	4.27	4.26	7.04	18.64	2.68	2.72	2.66
Ca, mg	116	116	104	183	138	60	61	80
Fe, mg	0.05	0.06	0.07	0.08	0.06	0.14	0.14	1.2
Mg, mg	11	11	10	17	13	5	5	6
P, mg	89	85	95	144	109	132	134	104
K, mg	151	144	129	234	177	84	86	119
Na, mg	105	53	40	70	53	405	406	296
Zn, mg	0.42	0.27	0.50	0.89	0.67	0.37	0.38	0.54
Ascorbic acid, mg	0.98	0.86	0.86	0.80	0.60	Trace	Trace	0
Thiamin, mg	0.034	0.035	0.035	0.044	0.034	0.021	0.021	0.017
Riboflavin, mg	0.154	0.149	0.149	0.214	0.162	0.163	0.165	0.197
Niacin, mg	0.058	0.067	0.067	0.114	0.086	0.126	0.128	0.101
Pantothenic acid, mg	0.275	0.360	0.363	0.591	0.446	0.213	0.215	0.271
Vitamin B_6, mg	0.034	0.016	0.016	0.049	0.037	0.067	0.068	0.047
Folacin, µg	—	11	11	11	8	12	12	13
Vitamin B_{12}, µg	0.219	0.300	0.300	0.562	0.426	0.623	0.633	0.424
Vitamin A, IU	33	790	452	66	49	163	37	1,427

Source: U.S. Dep. Agric. (1976).

vitamins indicating that some milk B vitamins are utilized by the culture during yogurt production. Another effect of the yogurt culture is to precipitate the casein of milk into an easily digestible form (Davis 1975).

The enzymes of lactic cultures are known to degrade proteins and lipids of milk, partially predigesting these nutrients. The lipolytic and proteolytic activities of various lactic cultures were studied by Chandan et al. (1969A,C) and Searles et al. (1970). Specific lipase activities (net μmole free fatty acids produced in 10% tributyrin emulsion/hr at 37°C/ml cellular DNA) of the lactobacilli were all below 1.0, while those of streptococci, leuconostocs, and propionibacteria ranged from 2.3 to 33. Specific proteolytic activities (μg crystalline trypsin equivalents/ml cellular DNA) of leuconostocs and streptococci were all below 0.10, while the activities of the propionibacteria and lactobacilli ranged from 0.2 to 3.2. Lactobacillus bulgaricus cells possess an appreciable proteolytic activity (Argyle et al. 1976B). At the optimum pH of 5.8–5.9, casein hydrolysis was approximately 4 times greater when compared to whey proteins. Poznanski et al. (1965) demonstrated the cooperative breakdown of casein by a proteinase of Lactobacillus bulgaricus and a peptidase of Streptococcus thermophilus. Chandan et al. (1970) and Argyle et al. (1976A) showed slow irreversible aggregation of whey proteins in milks acidified directly or by the action of lactic cultures. These studies show the relative importance of lactobacilli in protein hydrolysis and of streptococci and leuconostocs in fat hydrolysis in cultured dairy products. The hydrolytic and protein aggregation effects may contribute to the physical and nutritional properties of cultured dairy products.

Cultured dairy foods contain large populations of various starter organisms, and it is reasonable to conjecture some contribution of single-cell proteins of these cultures to the overall nutrient profile of the cultured product. Reddy et al. (1976A) fermented cheese whey with Lactobacillus bulgaricus and reported approximately 7% of the crude protein originated from cells of the lactobacillus. Erdman et al. (1977) investigated the amino acid profiles of protein from lactobacilli. They found the protein to contain essential amino acids in concentrations equal to or greater than the FAO (Food and Agric. Organ. U.N.) reference protein, indicating that the lactobacilli may be a good protein source for human consumption. The digestibility of the cell protein of Lactobacillus bulgaricus was found to be 89.2% compared with 98.5% for casein.

REFERENCES

ANGEVINE, N.C.   1972.   Quality control of Cottage cheese and cultured products.   Tech. Bull. Am. Cultured Dairy Products Inst., Washington, D.C., Oct. 10.

ANON.   1977.   Cultured and acidified buttermilk, yogurts, cultured and acidified milks, and eggnog. Proposal to establish new identity standards. Fed. Reg. 42 (112) 29919–29925.

ARBUCKLE, W.S. 1977. Ice Cream, 3rd Edition. AVI Publishing Co., Westport, Conn.

ARGYLE, P.J., JONES, N., CHANDAN, R.C. and GORDON, J.F. 1976A. Aggregation of whey proteins during storage of acidified milk. J. Dairy Res. *43*, 45–51.

ARGYLE, P.J., MATHISON, G.E. and CHANDAN, R.C. 1976B. Production of cell-bound proteinase by *Lactobacillus bulgaricus* and its location in the bacterial cell. J. Appl. Bacteriol. *41*, 175–184.

ASHTON, T.R., AIREY, F.K., LEEDHAM, L.F. and GOMM, R.J. 1959. Continuous cheese starter manufacture. Proc. 15th Intern. Dairy Congr., 1959, London *2*, 605–609.

AUSAVANODOM, N., WHITE, R.S., YOUNG, G. and RICHARDSON, G.H. 1977. Lactic bulk culture system utilizing whey-based bacteriophage inhibitory medium and pH control. II. Reduction of phosphate requirements under pH control. J. Dairy Sci. *60*, 1245–1251.

BABEL, F.J. 1976. Technology of dairy products manufactured with selected microorganisms. *In* Dairy Technology and Engineering. W.J. Harper and C.W. Hall (Editors). AVI Publishing Co., Westport, Conn.

BABEL, F.J. 1977. Antibiosis by lactic bacteria. J. Dairy Sci. *60*, 815–821.

BABEL, F.J. and MATHER, D.W. 1961. Creamed Cottage cheese. U.S. Pat. 2,971,847. Feb. 14.

BADINGS, H.T. and GALESLOOT, T.E. 1962. Studies on the flavor of different types of butter starters with reference to the defect "yoghurt flavor" in butter. Proc. 16th Intern. Dairy Congr., 1962, Copenhagen *B*, 199–206.

BAUMANN, D.P. and REINBOLD, G.W. 1966. Freezing of lactic cultures. J. Dairy Sci. *49*, 259–264.

BAUTISTA, E.S., DAHIYA, R.S. and SPECK, M.L. 1966. Identification of compounds causing symbiotic growth of *Streptococcus thermophilus* and *Lactobacillus bulgaricus* in milk. J. Dairy Res. *33*, 299–307.

BEGEMANN, P.H. and KOSTER, J.C. 1964. Components of butterfat. 4-*cis*-Heptenal—A cream-flavored component of butter. Nature *202*, 552–553.

BERRIDGE, N.J. 1966. A note on the self-regulating capability of a starter organism in continuous culture. J. Soc. Dairy Technol. *19*, 232–233.

BERRIDGE, N.J. and WILSON, P. 1970. Continuous starter production using whey. 18th Intern. Dairy Congr., 1970, Sydney *1E*, 232.

BILLS, D.D., MORGAN, M.E., LIBBEY, L.M. and DAY, E.A. 1965. Identification of compounds responsible for fruity flavor defect of experimental Cheddar cheese. J. Dairy Sci. *48*, 765. (abstract)

BILLS, D.D., YANG, C.S., MORGAN, M.E. and BODYFELT, F.W. 1972. Effect of sucrose on the production of acetaldehyde and acids by yoghourt culture bacteria. J. Dairy Sci. *55*, 1570–1573.

BOGDANOV, I., POPKHRISTOV, P. and MARINOV, L. 1962. Anticancer effect of antibioticum bulgaricum on sarcoma 180 and on solid form of Ehrlich carcinoma. Abstr. 8th Intern. Cancer Congr., 1962, Moscow.

BOLDINGH, J. and TAYLOR, R.J. 1962. Trace constituents of butterfat. Nature *194*, 909–913.

BOUCHEZ,D. and VAN BELLEGHEM, M. 1971. Causes for noncoagulation of milk in the yoghourt test. Rev. Agric. Brussels *24*, 671–683.

BRANEN, A.L., GO, H.C. and GENSKE, R.P. 1975. Purification and properties of antimicrobial substances produced by *Streptococcus diacetylactis* and *Leuconostoc citrovorum*. J. Food Sci. *40*, 446–450.

BRYAN, A.H. 1965. Lactobacilli for enteric infections. Drug Cosmet. Ind. *96*, 474–476, 585, 588–589.

BUCHANAN, R.E. and GIBBONS, N.E. 1974. Bergey's Manual of Determinative Bacteriology, 8th Edition. Williams and Wilkins, Baltimore.

BUSCH-JOHANNSEN, G. 1972. New ways of using cultures in dairy plants. Dtsch. Molk. Ztg. *93*, 1196–1201.

BUSCH-JOHANNSEN, G., GROW, K.D. and KAPOOR, J. 1971. Yoghurt production through slow acidification at 30°C. Dtsch. Molk. Ztg. *92*, 1460–1463.

CAMPBELL, J.R. and MARSHALL, R.T. 1975. The Science of Providing Milk for Man. McGraw-Hill, New York.

CHANDAN, R.C., ARGYLE, P.J. and JONES, N. 1969A. Proteolytic activity of lactic cultures. J. Dairy Sci. *52*, 894. (abstract)

CHANDAN, R.C., ARGYLE, P.J., JONES, N. and GORDON, J.F. 1970. Changes in bovine whey proteins during storage of acidified milks. Proc. 18th Intern. Dairy Congr., 1970, Sydney *1E*, 415.

CHANDAN, R.C., GORDON, J.F. and MORRISON, A. 1977. Natural benzoate content of dairy products. Milchwissenschaft *32*, 534–537.

CHANDAN, R.C., GORDON, J.F. and WALKER, D.A. 1969B. Dairy fermentation processes. Process Biochem. *4*, 13–22.

CHANDAN, R.C., PARRY, R.M., JR. and SHAHANI, K.M. 1965. Purification and some properties of bovine milk lysozyme. Biochim. Biophys. Acta *110*, 389–398.

CHANDAN, R.C., SEARLES, M.A. and FINCH, J. 1969C. Lipase activity of lactic cultures. J. Dairy Sci. *52*, 894. (abstract)

CITTI, J.E., SANDINE, W.E. and ELLIKER, P.R. 1965. Comparison of slow and fast acid-producing *Streptococcus lactis*. J. Dairy Sci. *48*, 14–18.

COLLINS, E.B. 1962. Behavior and use of lactic streptococci and their bacteriophages. J. Dairy Sci. *45*, 552–558.

COLLINS, E.B. 1972. Biosynthesis of flavor compounds by microorganisms. J. Dairy Sci. *55*, 1022–1028.

COLLINS, E.B. 1977. Influence of medium and temperature on end products and growth. J. Dairy Sci. *60*, 799–804.

CORBIN, E.A., JR. 1971. Cheese manufacture. U.S. Pat. 3,620,768. Nov. 16.

COUSIN, M.A. and MARTH, E.H. 1977A. Lactic acid production by *Strepto-*

coccus lactis and *Streptococcus cremoris* in milk precultured with psychrotrophic bacteria.   J. Food Prot. *40*, 406–410.

COUSIN, M.A. and MARTH, E.H.   1977B.   Lactic acid production by *Streptococcus thermophilus* and *Lactobacillus bulgaricus* in milk precultured with psychrotrophic bacteria.   J. Food Prot. *40*, 475–479.

COWMAN, R.A. and SPECK, M.L.   1965.   Activity of lactic streptococci following storage at refrigeration temperatures.   J. Dairy Sci. *48*, 1441–1444.

CRAVEN, K.   1975.   What's new with yoghurt fruits?   Cultured Dairy Prod. J. *10*, 16–18.

DAHIYA, R.S. and SPECK, M.L.   1964.   Growth of streptococcus starter culture in milk fortified with nucleic acid derivatives.   J. Dairy Sci. *47*, 374–377.

DAHIYA, R.S. and SPECK, M.L.   1968.   Hydrogen peroxide formation by lactobacilli and its effect on *Staphylococcus aureus*.   J. Dairy Sci. *51*, 1568–1572.

DALY, C., SANDINE, W.E. and ELLIKER, P.R.   1972.   Interactions of food starter cultures and food borne pathogens: *Streptococcus diacetylactis* versus food pathogens.   J. Milk Food Technol. *35*, 349–357.

DAVIS, J.G.   1975.   The microbiology of yoghourt.   *In* Lactic Acid Bacteria in Beverages and Food.   J.G. Carr, C.V. Cutting and G.C. Whiting (Editors).   Academic Press, New York.

DAY, E.A., LINDSAY, R.C. and FORSS, D.A.   1964.   Dimethyl sulphide and the flavor of butter.   J. Dairy Sci. *47*, 197–199.

DEANE, D.D. and HAMMOND, E.G.   1960.   Coagulation of milk for cheesemaking by ester hydrolysis.   J. Dairy Sci. *43*, 1421–1429.

DEANE, D.D. and THOMAS, W.R.   1964.   Use of chemical compounds to replace lactic cultures in the manufacture of sour cream.   J. Dairy Sci. *47*, 684. (abstract)

DeJONG, L. and VANDENWAL, H.   1964.   Identification of some isolinoleic acids occurring in butterfat.   Nature *202*, 553–555.

DeKLERK, H.C. and COETZEE, J.N.   1961.   Antibiosis among lactobacilli.   Nature *192*, 340–341.

DeMAN, J.C. and GALESLOOT, T.E.   1962.   The effect of the addition of manganese to milk upon the growth of starter bacteria.   Neth. Milk Dairy J. *16*, 1–23.

ELLIKER, P.R., SANDINE, W.E., HAUSER, B.A. and MOSELEY, W.K. 1964.   Influence of culturing Cottage cheese dressing with different organisms on flavor and keeping quality.   J. Dairy Sci. *47*, 680. (abstract)

EMMONS, D.B. and TUCKEY, S.L.   1967.   Cottage cheese and other cultured milk products.   *In* Pfizer Cheese Monographs, Vol. 3.   Chas. Pfizer Co., New York.

ENGEL, W.G.   1973.   The use of lactase to sweeten yogurt without increasing calories.   Cultured Dairy Prod. J. *8*, 6–7.

ERDMAN, M.D., BERGEN, W.G. and REDDY, C.A. 1977. Amino acid profiles and presumptive nutritional assessment of single-cell protein from certain lactobacilli. Appl. Environ. Microbiol. 33, 901–905.

ERNSTROM, C.A. and KALE, C.G. 1975. Continuous manufacture of Cottage and other uncured cheese varieties. J. Dairy Sci. 58, 1008–1014.

FORSS, D.A. 1964. Fishy flavor in dairy products. J. Dairy Sci. 47, 245–250.

FOSTER, E.M., NELSON, F.E., SPECK, M.L., DOETSCH, R.N. and OLSON, J.C. 1958. Dairy Microbiology. Macmillan and Co., London.

FRAZIER, W.C. 1967. Food Microbiology, 2nd Edition. McGraw-Hill, New York.

GALESLOOT, T.E. and HASSING, F. 1968. Manufacture of stirred yogurt of high viscosity. Officieel Orgaan K. ned. Koninklijke Netherlandse Zuivelbond 60 (18) 488–490. [Dairy Sci. Abstr. (1960) 30, 370. (abstract 2260)].

GALESLOOT, T.E., HASSING, F. and VERINGA, H.A. 1968. Symbiosis of yogurt. (I) Stimulation of *Lactobacillus bulgaricus* by a factor produced by *Streptococcus thermophilus*. Neth. Milk Dairy J. 22, 50–63.

GAVIN, M. 1966. Combined effect of temperature and acidity on the keeping quality of yoghurt. Milchwissenschaft 21, 85–87.

GILLILAND, S.E. 1972. Flavor intensification with concentrated cultures. J. Dairy Sci. 55, 1028–1031.

GILLILAND, S.E. 1977. Preparation and storage of concentrated cultures of lactic streptococci. J. Dairy Sci. 60, 805–809.

GILLILAND, S.E. and SPECK, M.L. 1974. Frozen concentrated cultures of lactic starter bacteria. A review. J. Milk Food Technol. 37, 107–111.

GILLILAND, S.E. and SPECK, M.L. 1977. Instability of *Lactobacillus acidophilus* in yogurt. J. Dairy Sci. 60, 1394–1398.

GILLILAND, S.E., SPECK, M.L. and WOODARD, J.R., JR. 1972. Stimulation of lactic streptococci in milk by β-galactosidase. Appl. Microbiol. 23, 21–25.

GOODENOUGH, E.R. and KLEYN, D.A. 1976. Qualitative and quantitative changes in carbohydrates during the manufacture of yogurt. J. Dairy Sci. 59, 45–47.

GORNER, F., POLO, V. and BERTAN, M. 1968. Changes in content of volatile substances in yoghurt during ripening. Milchwissenschaft 23, 94–100.

GROSS, D.R. 1974. Fruit preserves and jellies. *In* Encyclopedia of Food Technology, Vol. 2. A.H. Johnson and M.S. Peterson (Editors). AVI Publishing Co., Westport, Conn.

GROSSOWICKS, N., KAPLAN, D. and SCHNEERSON, S. 1947. Production of antibiotic substances by a lactobacillus. Proc. 5th Intern. Congr. Microbiol., 1947.

GUTHRIE, E.S. 1963. Further studies of the body of cultured cream. Cornell Univ. Agric. Exp. Stn. Bull. *986*.

HAINES, W.C. and HARMON, L.G.   1973.   Effect of selected lactic acid bacteria on growth of *Staphylococcus aureus* and production of enterotoxin. Appl. Microbiol. *25*, 436–441.

HALL, T.A.   1975.   Yogurt formulation with attention to stabilizer systems. Cultured Dairy Prod. J. *10*, 12–14.

HAMDAN, I.Y., KUNSMAN, J.E., JR. and DEANE, D.D.   1971.   Acetaldehyde production by combined yoghurt cultures.   J. Dairy Sci. *54*, 1080–1082.

HAMMER, B.W. and BABEL, F.J.   1957.   Dairy Bacteriology.   John Wiley & Sons, New York.

HARGROVE, R.E. and ALFORD, J.A.   1977.   Nutritional superiority of yogurt as compared to other fermented and non-fermented milks.   J. Dairy Sci. *60* (Supplement I) 34. (abstract)

HARMON, L.G. and SMITH C.K.   1956.   The influence of microbiological populations on the shelf life of creamed Cottage cheese.   Mich. State Univ. Agric. Exp. Stn. Q. Bull. *38*.

HARPER, W.J., GOULD, I.A. and HANKINSON, C.L.   1961.   Observation on the free volatile acids in milk.   J. Dairy Sci. *44*, 1764–1765.

HARVEY, R.J.   1960.   Production of acetone and acetaldehyde by lactic streptococci.   J. Dairy Res. *27*, 41–45.

HAUKKA, J.   1976.   The Technology and Manufacture of Fermented Milks. Valio Finnish Cooperative Dairies Association, Helsinki, Finland.

HENNING, D.R., SANDINE, W.E., ELLIKER, P.R. and HAYS, H.A.   1965. Studies with a bacteriophage inhibitory medium.   I.   Inhibition of phage and growth of single strain lactic streptococci and leuconostoc.   J. Milk Food Technol. *23*, 273–277.

HUMPHREYS, C.L. and PLUNKETT, M.   1969.   Yoghourt: a review of its manufacture.   Dairy Sci. Abstr. *31*, 607–622.

JACKSON, H.W. and MORGAN, M.E.   1954.   Identity and origin of the malty aroma substance from milk cultures of *Streptococcus lactis* var. *maltigenes*.   J. Dairy Sci. *37*, 1316–1324.

JENNESS, R.   1974.   The composition of milk.   *In* Lactation: A Comprehensive Treatise. Part III.   Nutrition and Biochemistry of Milk/Maintenance.   B.L. Larson and V.R. Smith (Editors).   Academic Press, New York.

JENNESS, R. and PATTON, S.   1959.   Principles of Dairy Chemistry.   John Wiley & Sons, New York.

JONES, V.A. and HARPER, W.J.   1976.   General processes for fluid milks. *In* Dairy Technology and Engineering.   W.J. Harper and C.W. Hall (Editors).   AVI Publishing Co., Westport, Conn.

KADIS, V.W. and BABEL, F.J.   1962.   Effectiveness of ammonium oxalate, salts of ethylenediaminetetraacetic acid and sodium tripolyphosphate in limiting bacteriophage development in milk.   J. Dairy Sci. *45*, 486–491.

KANDLER, O.   1961.   Metabolism of starter organisms.   Milchwissenschaft *16*, 523–531.

KAVASNIKOV, E.I. and SUDENKO, V.I. 1967. Antibiotic properties of *Lactobacillus brevis.* Mikrobiol. Zh. Kiev 29, 2.146 [*Cited* Dairy Sci. Abstr. (1967) 29, 3972].

KEENAN, T.W., LINDSAY, R.C., MORGAN, M.E. and DAY, E.A. 1966. Acetaldehyde production by single-strain lactic streptococci. J. Dairy Sci. 49, 10–14.

KEOGH, B.P. 1970. Survival and activity of frozen starter cultures for cheese manufacture. Appl. Microbiol. 19, 928–931.

KHATRI, L.L. and DAY, E.A. 1962. Analysis of free fatty acids of fresh milk fats and ripened cream butter. J. Dairy Sci. 45, 660. (abstract)

KILARA, A. and SHAHANI, K.M. 1976. Lactase activity of cultured and acidified dairy products. J. Dairy Sci. 59, 2031–2035.

KILARA, A., SHAHANI, K.M., and DAS, N.K. 1976. Effect of cryoprotective agents on freeze drying and storage of lactic cultures. Cultured Dairy Prod. J. 11, 8–11.

KIM, S.C. and HARMON, L.G. 1968. Effect of insecticide residues on growth and fermentation ability of lactic culture organisms. J. Milk Food Technol. 31, 97–100.

KODOMA, R. 1952. Studies on lactic acid bacteria. II. Lactolin, a new antibiotic substance produced by lactic acid bacteria. J. Antibiot. 5, 72–74.

KOSIKOWSKI, F.V. 1977. Cheese and Fermented Milk Foods, 2nd Edition. Edwards Bros., Ann Arbor, Mich.

KOSIKOWSKI, F.V. and WIERZBICKI, L.E. 1971. Low-lactose yogurt from microbial lactase (β-D-galactosidase) applications. J. Dairy Sci. 54, 764. (abstract)

KROGER, M. 1976. Quality of yogurt. J. Dairy Sci. 59, 344–350.

KUBITSCHEK, H.E. 1970. Introduction to Research with Continuous Cultures. Prentice-Hall, Englewood Cliffs, N.J.

LATTEY, J.M. 1968. Studies on ultra-deep frozen cheese starters. N.Z. J. Dairy Technol. 3, 35–41.

LAWRENCE, R.C., THOMAS, T.D. and TERZAGHI, B.E. 1976. Review of the progress of dairy science. Cheese starters. J. Dairy Res. 43, 141–193.

LEESMENT, H. 1962. Bacteria in starter producing malty flavor. Proc. Intern. Dairy Congr., 1962, Copenhagen 1962B, 209–216.

LEWIS, P.M. 1967. A note on the continuous flow culture of mixed populations of lactobacilli and streptococci. J. Appl. Bacteriol. 30, 406–409.

LINDSAY, R.C. 1967. Cultured dairy products. *In* Symposium on Foods: The Chemistry and Physiology of Flavors. H.W. Schultz (Editor). AVI Publishing Co., Westport, Conn.

LINDSAY, R.C. and DAY, E.A. 1965. Rapid quantitative method for determination of acetaldehyde in lactic starter cultures. J. Dairy Sci. 48, 665–669.

LINDSAY, R.C., DAY, E.A. and SANDINE, W.E. 1965. Green flavor defect in lactic starter cultures. J. Dairy Sci. 48, 863–869.

LINKLATER, P.M. and GRIFFIN, C.J.   1971.   The design and operation of a continuous milk fermentor.   J. Dairy Res. *38*, 127–136.

LITTLE, L.L.   1968.   Process for making cheese by coagulating milk at low temperature.   U.S. Pat. 3,406,076. Oct. 15.

LLOYD, G.T.   1971.   New development in starter technology.   Dairy Sci. Abstr. *33*, 411–416.

LLOYD, G.T.   1975.   The production of concentrated starters by batch culture.   II. Studies on the optimum storage temperature.   Aust. J. Dairy Technol. *30*, 107–108.

LLOYD, G.T. and PONT, E.G.   1973.   The production of concentrated starters by batch culture.   Aust. J. Dairy Technol. *28*, 104–108.

LUNDSTEDT, E.   1972.   A new and superior type of Cottage cheese.   Cultured Dairy Prod. J. *7* (1) 8–10.

LUNDSTEDT, E.   1974.   Reflections on the dairy industry.   Cultured Dairy Prod. J. *9* (1) 14–15, 24.

LUNDSTEDT, E. and FOGG, W.B.   1962.   Citrated whey starters. II. Gradual formation of flavor and aroma in creamed Cottage cheese after the addition of small quantities of citrated Cottage cheese whey cultures of *Streptococcus diacetylactis*.   J. Dairy Sci. *45*, 1327–1331.

MacLEOD, P. and MORGAN, M.E.   1958.   Differences in the ability of lactic streptococci to form aldehydes from certain amino acids.   J. Dairy Sci. *41*, 908–913.

MAGIDMAN, P., HERB, S.F., BARFORD, R.A. and RIEMENSCHNEIDER, R.W.   1962.   Fatty acids of cow's milk. A technique employed in supplementing gas-liquid chromatography for identification of fatty acids.   J. Am. Oil Chem. Soc. *39*, 137–142.

MANN, E.J.   1973.   Digest of world literature. Dairy Ind. *38* (Aug.) 386–387; (Sept.) 432–433.

MARTH, E.H.   1962.   Chlorinated hydrocarbons deposited in biological materials. II. Animal and animal products.   J. Milk Food Technol. *25*, 72–77.

MARTH, E.H.   1974.   Fermentations. *In* Fundamentals of Dairy Chemistry. B.H. Webb and A.H. Johnson (Editors).   AVI Publishing Co., Westport, Conn.

MARTH, E.H. and HUSSONG, R.V.   1963.   Effect of skimmilks cultured with different strains of *Leuconostoc citrovorum* on growth of some bacteria and yeasts.   J. Dairy Sci. *46*, 1033–1037.

MATHER, D.W. and BABEL, F.J.   1959A.   Studies on the flavor of creamed Cottage cheese.   J. Dairy Sci. *42*, 809–815.

MATHER, D.W. and BABEL, F.J.   1959B.   A method for standardizing the diacetyl content of creamed Cottage cheese.   J. Dairy Sci. *42*, 1045–1056.

MATHER, D.W. and BABEL, F.J.   1959C.   Inhibition of certain types of bacterial spoilage in creamed Cottage cheese by the use of a creaming mixture prepared with *Streptococcus citrovorus*.   J. Dairy Sci. *42*, 1917–1926.

MATTICK, A.T.R. and HIRSCH, A.   1949.   The streptococci and antibiotics. 12th Intern. Dairy Congr. *2*, 546–550.

McKAY, L.L. and BALDWIN, K.A.   1975.   Plasmid distribution and evidence for a proteinase plasmid in *Streptococcus lactis* C$_2$. Appl. Microbiol. *29*, 546–548.

METCHNIKOFF, E.   1908.   The Prolongation of Life. G.P. Putnam's Sons, New York.

MIKOLAJCIK, E.M. and HAMDAN, I.Y.   1975.   *Lactobacillus acidophilus*. II. Antimicrobial agent. Cultured Dairy Prod. J. *10*, 10, 12–14, 16, 18, 20.

MILK IND. FOUND.   1980.   Milk Facts.   Milk Industry Foundation, Washington, D.C.

MINOR, T.E. and MARTH, E.H.   1972.   Fate of *Staphylococcus aureus* in cultured buttermilk, sour cream and yogurt during storage.   J. Milk Food Technol. *35*, 302–306.

MITCHELL, I.DeG. and KENWORTHY, R.   1976.   Investigations on a metabolite from *Lactobacillus bulgaricus* which neutralizes the effect of enterotoxin from *Escherichia coli* pathogenic for pigs.   J. Appl. Bacteriol. *41*, 163–174.

MOCQUOT, G. and HUREL, C.   1970.   The selection and use of some microorganisms for the manufacture of fermented and acidified milk products. J. Soc. Dairy Technol. *23*, 130–146.

MORGAN, M.E., LINDSAY, C., LIBBEY, L.M. and PEREIRA, R.L.   1966. Identity of additional aroma constituents in milk cultures of *Streptococcus lactis* var. *maltigenes*.   J. Dairy Sci. *49*, 15–18.

MURATA, T., ZABIK, M.E. and ZABIK, M.J.   1977.   Polybrominated biphenyls in raw milk and processed dairy products.   J. Dairy Sci. *60*, 516–520.

PARKS, D.W., KEENEY, M., KATZ, I. and SCHWARTZ, D.P.   1964.   Isolation and characterization of the methyl ketone precursors in butterfat. J. Lipid Res. *5* (2) 232–235.

PATTON, S., FORSS, D.A. and DAY, E.A.   1956.   Methyl sulfide and the flavor of milk.   J. Dairy Sci. *39*, 1469–1470.

PETTE, J.W. and LOLKEMA, J.   1951.   Yoghourt. V. Firmness and whey separation of milk yoghurt. Neth. Milk Dairy J. *5*, 27–45.

PINHEIRO, A.J., LISKA, B.J. and PARMALEE, C.E.   1968.   Properties of substances inhibitory to *Pseudomonas fragi* produced by *Streptococcus citrovorus* and *Streptococcus diacetilactis*.   J. Dairy Sci. *51*, 183–187.

POLONSKAYA, M.S.   1952.   Antibiotic substance produced by *Lactobacillus acidophilus*. Mikrobiologiya *21* (3) 303–310 [Dairy Sci. Abstr. (1953) *15* (7) 562].

PONT, E.G. and HOLLOWAY, G.L.   1968.   A new approach to the production of cheese starter. Aust. J. Dairy Technol. *23*, 22–29.

POZNANSKI, S., LENOIR, J. and MOCQUOT, G.   1965.   The proteolysis of casein under the action of certain bacterial endoenzymes. Lait *45*, 3–26. (French)

PRICE, R.J. and LEE, J.S.   1970.   Inhibition of Pseudomonas species by hydrogen peroxide producing lactobacilli.   J. Milk Food Technol. *33*, 13–18.

RANDOLPH, H.E. and GOULD, I.A.   1966.   Effect of the inherent properties of milk on the production of acid by selected lactic cultures.   J. Dairy Sci. *49*, 254–258.

REDDY, C.A., HENDERSON, H.E. and ERDMAN, M.D.   1976A.   Bacterial fermentation of cheese whey for production of ruminant feed supplement rich in crude protein.   Appl. Environ. Microbiol. *32*, 769–776.

REDDY, G.V., SHAHANI, K.M. and BANERJEE, M.R.   1973.   Inhibitory effect of yogurt on Ehrlich ascites tumor-cell proliferation.   J. Natl. Cancer Inst. *50*, 815–817.

REDDY, K.P., SHAHANI, K.M. and KULKARNI, S.M.   1976B.   B-complex vitamins in cultured and acidified yogurt.   J. Dairy Sci. *59*, 191–195.

REINBOLD, G.W. and REDDY, M.G.   1973.   Bacteriophage for *Streptococcus thermophilus*. Dairy Ind. *38*, 413–416.

REITER, B.   1973.   Some thoughts on cheese starters.   J. Soc. Dairy Technol. *26*, 3–15.

REITER, B. and MOLLER-MADSEN, A.   1963.   Reviews of the progress of dairy science—cheese and butter starters.   J. Dairy Res. *30*, 419–449.

ROBERTSON, P.S.   1966.   Recent developments affecting the Cheddar cheese making process.   J. Dairy Res. *33*, 343–369.

ROBINSON, R.K. and TAMINE, A.Y.   1975.   Yogurt—a review of the product and its manufacture.   J. Soc. Dairy Technol. *28*, 149–163.

SABINE, D.B.   1963.   An antibiotic-like effect of *Lactobacillus acidophilus*. Nature *199*, 811.

SANDBERG, L.   1976.   Introduction to different fermented milk varieties. *In* The Technology and Manufacture of Fermented Milks.   Valio Finnish Cooperative Dairies Assoc., Helsinki, Finland.

SANDINE, W.E.   1977.   New techniques in handling lactic cultures to enhance their performance.   J. Dairy Sci. *60*, 822–828.

SANDINE, W.E., DALY, C., ELLIKER, P.R. and VEDAMUTHU, E.R. 1972A.   Causes and control of culture-related flavor defects in cultured dairy products.   J. Dairy Sci. *55*, 1031–1039.

SANDINE, W.E., MURALIDHARA, K.S., ELLIKER, P.R. and ENGLAND, D.C.   1972B.   Lactic acid bacteria in food and health. A review with special reference to enteropathogenic *Escherichia coli* as well as certain enteric disease and their treatment with antibiotics and lactobacilli.   J. Milk Food Technol. *35*, 691–702.

SEARLES, M.A., ARGYLE, P.J., CHANDAN, R.C. and GORDON, J.F.   1970. Lipolytic and proteolytic activities of lactic acid bacteria.   Proc. Intern. Dairy Congr., 1970, Sydney *IE*, 111.

SELLARS, R.L.   1967.   Bacterial starter cultures. *In* Microbial Technology. H.J. Peppler (Editor).   Reinhold Publishing Corp., New York.

SHAHANI, K.M.   1962.   Inhibitory effect of nisin upon various organisms. J.

Dairy Sci. *45*, 827–832.

SHAHANI, K.M., REDDY, G.V. and JOE, A.M. 1974. Nutritional and therapeutic aspects of cultured dairy products. Proc. 19th Intern. Dairy Congr., 1974, New Delhi *Ie*, 569–570.

SHAHANI, K.M., VAKIL, J.R. and CHANDAN, R.C. 1972. Antibiotic acidophilin and process of preparing the same. U.S. Pat. 3,689,640. Sept. 5.

SHAHANI, K.M., VAKIL, J.R. and KILARA, A. 1977. Natural antibiotic activity of *Lactobacillus acidophilus* and *bulgaricus*. II. Isolation of acidophilin from *L. acidophilus*. Cultured Dairy Prod. J. *12*, 8–11.

SHMELEVA, L. and JAKOVLEV (YAKOVLEV), D. 1966. Continuous culture of lactic acid bacteria. 17th Intern. Dairy Congr., 1966, Munich *C*, 367–374.

SPECK, M.L. 1972. Control of food-borne pathogens by starter cultures. J. Dairy Sci. *55*, 1019–1022.

SPECK, M.L. 1975. Interactions among lactobacilli and man. J. Dairy Sci. *59*, 338–343.

SPECK, M.L. 1977. Acidophilus food products. Soc. Ind. Microbiol. *27*, 44. (abstract)

STADHOUDERS, J. 1974. Dairy starter cultures. Milchwissenschaft *29*, 329–337.

STADHOUDERS, J., HUP, G. and JANSEN, L.A. 1971. A study of the optimum conditions of freezing and storing concentrated mesophilic starters. Neth. Milk Dairy J. *25*, 229–239.

STANIER, R.Y., ADELBERG, E.A. and INGRAHAM, J.L. 1976. The Microbial World, 4th Edition. Prentice-Hall, Englewood Cliffs, N.J.

STONE, W.K., LARGE, P.M. and THOMAS, W.C. 1975. Ultra high temperature pasteurization increases starter activity. Cultured Dairy Prod. J. *10*, 11–12.

STORRS, A.B. and STERN, R.M. 1967. *Lactobacillus acidophilus* concentrates. *In* Microbial Technology. H.J. Peppler (Editor). Reinhold Publishing Corp., New York.

TATINI, S.R., JEZESKI, J.J., MORRIS, H.A., OLSON, J.C. and CASMAN, E.P. 1971. Production of staphylococcal enterotoxin A in Cheddar and Colby cheeses. J. Dairy Sci. *54*, 815–825.

TOAN, T.T., BASETTE, R. and CLAYDON, T.J. 1965. Methyl sulfide production by *Aerobacter aerogenes* in milk. J. Dairy Sci. *48*, 1174–1178.

TOBIAS, J. 1976. Organoleptic properties of dairy products. *In* Dairy Technology and Engineering. W.J. Harper and C.W. Hall (Editors). AVI Publishing Co., Westport, Conn.

TRAMER, J. 1966. Inhibitory effect of *Lactobacillus acidophilus*. Nature *211*, 204–205.

TRAMER, J. 1973. Yoghourt cultures. J. Soc. Dairy Technol. *26*, 16–21.

U.S. DEP. AGRIC. 1976. Composition of foods. U.S. Dep. Agric., Agric. Handb. *81*.

U.S. DEP. AGRIC. 1977. Federal and state standards for the composition of milk products (and certain non-milk fat products). U.S. Dep. Agric. Handb. *51*.

VEDAMUTHU, E.R. 1974. Cultures of buttermilk, sour cream and yoghurt with special comments on acidophilus yogurt. Cultured Dairy Prod. J. *9*, 16−21.

VEDAMUTHU, E.R. 1976. Getting the most out of your starter. Cultured Dairy Prod. J. *10*, 16−20.

VINCENT, J.G., VEOMETT, R.C. and RILEY, R.F. 1959. Antibacterial activity associated with *Lactobacillus acidophilus*. J. Bacteriol. *78*, 477−484.

WAKEMAN, A.H. 1972. Preparing cheese curd. U.S. Pat. 3,645,751. Feb. 29.

WALSH, B. and COGAN, T.M. 1973. Diacetyl, acetoin and acetaldehyde production by mixed species lactic starter cultures. Appl. Microbiol. *26*, 820−825.

WEBB, B.H., JOHNSON, A.H. and ALFORD, J.A. 1974. Fundamentals of Dairy Chemistry, 2nd Edition. AVI Publishing Co., Westport, Conn.

WHEATER, D.M., HIRSCH, A. and MATTICK, A.T.R. 1951. Lactobacillin —an antibiotic from lactobacilli. Nature *168*, 659.

WHISTLER, R.L. 1973. Industrial Gums. Academic Press, New York and London.

WILCOX, G. 1971. Eggs, Cheese and Yogurt Processing. Noyes Data Corp., Park Ridge, N.J.

WILKOWSKE, H.H. and FOUTS, E.L. 1958. Continuous and automatic propagation of dairy cultures. J. Dairy Sci. *41*, 49−56.

WINTER, M., STOLL, M., WARNOFF, E.W., GREUTER, F. and BUCHI, G. 1963. Volatile carbonyl constituents of dairy butter. J. Food Sci. *28*, 554−561.

WONG, N.P. 1963. A comparison of the volatile compounds of fresh and decomposed cream by gas chromatography. J. Dairy Sci. *46*, 571−573.

WONG, N.P. and PATTON, S. 1962. Identification of some volatile compounds related to the flavor of milk and cream. J. Dairy Sci. *45*, 724−728.

YEAGER, C. 1975. Yogurt processing methods. Cultured Dairy Prod. J. *10*, 10.

6

Lactic Acid Fermentation of Cabbage, Cucumbers, Olives, and Other Produce

Reese H. Vaughn

The preparation of foods by lactic acid fermentation (pickling) has been one of the important methods of food conservation for untold centuries and, until the development of canning, mechanical dehydrating, and freezing processes during the past 170 years, pickling together with sun drying and natural refrigerating and freezing constituted the major method for preserving foods for extended periods of time. Now, with the development of more and more sophisticated techniques for preservation, food habits have changed, particularly in the United States. At present, pickled foods such as cucumbers, olives, various peppers, and other vegetables are used as adjuncts which serve as appetizers or, as with sauerkraut, are consumed as a substantial part of the meal.

The more important products produced in whole or in part by lactic acid fermentation in salt brine are pickles, sauerkraut, and olives. Lesser amounts of other vegetables including carrots, cauliflower, celery, okra, onions, sweet and hot peppers, and green tomatoes are also fermented, most being used in mixed pickle products.

The production data shown in Tables 6.1, 6.2, and 6.3 and the apparent consumption data found in Table 6.4 indicate the relative importance of sauerkraut, pickles, and olives in the food industry of the United States.

Cucumbers constitute the largest single crop grown specifically for pickling. The majority of this production comes from Michigan, Wisconsin, and California, although cucumbers for pickling are grown in other states where the climatic conditions are more or less favorable and there is a constant demand for them by the pickling industry.

Sauerkraut is the only product other than cucumbers fermented in large quantity in the United States. Its production is localized in the states bordering the Great Lakes, including New York, Pennsylvania, Ohio, Mich-

TABLE 6.1.   PRODUCTION OF CABBAGE FOR SAUERKRAUT IN THE UNITED STATES

Year	Amount Harvested		Production		Yield		Value		Total Dollars (Thousands)
	ha	Acres	MT	Tons	Per ha (MT)	Per Acre (Tons)	Dollars Per MT	Dollars Per Ton	
1966	4304	10,760	161,604	179,560	37.55	16.69	22.00	20.00	3596
1967	5712	14,280	246,060	273,400	43.09	19.15	18.70	17.00	4653
1968	5024	12,560	208,665	231,850	41.54	18.46	19.80	18.00	4179
1969	5112	12,780	200,250	222,500	39.17	17.41	20.90	19.00	4237
1970	5192	12,980	239,490	266,100	46.12	20.50	19.47	17.70	4707
1971	4584	11,460	211,455	234,950	46.12	20.50	19.25	17.50	4111
1972	4320	10,800	178,290	198,100	41.26	18.34	23.54	21.40	4245
1973	5216	13,040	197,235	219,150	37.82	16.81	27.28	24.80	5424
1974	5564	13,910	253,305	281,450	45.52	20.23	34.10	31.00	8720
1975	4724	11,810	215,775	239,750	45.67	20.30	34.54	31.40	7525
1976	4588	11,470	208,845	232,050	45.51	20.23	34.32	31.20	7235
1977	4152	10,380	207,450	230,500	49.97	22.21	33.55	30.50	7022

Source: U.S. Dep. Agric. (1979).

TABLE 6.2.   PRODUCTION OF CUCUMBERS FOR PICKLES IN THE UNITED STATES

Year	Amount Harvested		Production		Yield		Value		Total Dollars (Thousands)
	ha	Acres	MT	Tons	Per ha (MT)	Per Acre (Tons)	Dollars Per MT	Dollars Per Ton	
1966	52,408	131,020	484,020	537,800	9.22	4.10	89.76	81.60	43,866
1967	62,480	156,200	539,415	599,350	8.64	3.84	100.32	91.20	54,662
1968	59,328	148,320	508,014	564,460	8.57	3.81	101.31	92.10	52,003
1969	54,008	135,020	465,300	517,000	8.62	3.83	101.09	91.90	47,534
1970	53,432	133,580	529,920	588,800	9.92	4.41	103.51	94.10	55,391
1971	51,040	127,600	506,790	563,100	9.92	4.41	102.52	93.20	52,461
1972	51,532	128,830	514,035	571,150	9.97	4.43	103.40	94.00	53,660
1973	50,372	125,930	538,920	598,800	10.71	4.76	109.23	99.30	59,448
1974	52,804	132,010	537,300	597,000	10.17	4.52	144.10	131.00	77,954
1975	56,068	140,170	606,825	674,250	10.96	4.87	141.90	129.00	86,730
1976	51,352	128,380	570,477	633,800	11.12	4.94	138.60	126.00	79,751
1977	49,596	123,990	565,346	628,100	11.40	5.07	138.60	126.00	79,244

Source: U.S. Dep. Agric. (1979).

TABLE 6.3.   PRODUCTION OF CALIFORNIA OLIVES

Year	Bearing		Production		Yield		Value		
	Hectarage	Acreage	MT	Tons	Per ha (MT)	Per Acre (Tons)	Dollars (Average) Per MT	Per Ton	Total Dollars (Thousands)
1966	10,642	26,604	56,700	63,000	5.33	2.37	265.10	241.00	15,183
1967	10,816	27,041	12,600	14,000	1.17	0.52	421.30	383.00	5,362
1968	10,800	26,999	77,400	86,000	7.18	3.19	404.80	368.00	31,648
1969	10,932	27,330	63,000	70,000	5.76	2.56	360.80	328.00	22,960
1970	11,035	27,588	46,800	52,000	4.23	1.88	271.70	247.00	12,844
1971	11,072	27,681	49,500	55,000	4.48	1.99	162.80	148.00	8,140
1972	11,327	28,317	21,780	24,200	1.91	0.85	456.50	415.00	10,043
1973	11,424	28,560	63,000	70,000	5.51	2.45	438.90	399.00	27,930
1974	11,396	28,491	52,650	58,500	4.61	2.05	477.40	434.00	25,389
1975	11,632	29,080	60,300	67,000	5.31	2.36	369.60	336.00	22,512
1976	12,314	30,784	72,000	80,000	5.85	2.60	363.00	330.00	26,400
1977	13,328	33,320	38,700	43,000	2.90	1.29	398.20	362.00	15,566

Source: Olive Admin. Comm. (1977–1978); U.S. Dep. Agric. (1978).

TABLE 6.4. PER CAPITA CONSUMPTION OF PICKLES, SAUERKRAUT AND OLIVES [KG-(LB) PROCESSED WEIGHT]

	Apparent Consumption, kg (lb) per Person					
	Pickles		Sauerkraut		Olives	
Year	kg	lb	kg	lb	kg	lb
1920	0.54	1.2	0.32	0.8		0.3
1930	0.82	1.8	1.04	2.3		0.5
1940	1.00	2.2	0.95	2.1		0.7
1950	1.50	3.3	0.86	1.9		0.8
1960	2.04	4.5	0.68	1.5	0.36 (0.13)[1]	0.8 (0.36)[1]
1966	3.00	6.6	0.64	1.4	0.36 (0.17)	0.8 (0.40)
1968	3.50	7.7	0.72	1.6	0.32 (0.10)	0.7 (0.25)
1969	3.50	7.7	0.64	1.4	0.54 (0.14)	1.2 (0.36)
1971	3.45	7.6	0.72	1.6	0.41 (0.16)	0.9 (0.41)
1972	3.54	7.8	0.64	1.4	0.32 (0.17)	0.7 (0.44)
1973	3.63	8.0	0.64	1.4	0.32 (0.13)	0.7 (0.33)
1974	3.58	7.9	0.68	1.5	0.41 (0.16)	0.9 (0.43)
1975	3.54	7.8	0.59	1.3	0.36 (0.15)	0.8 (0.40)
1976	3.82	8.4	0.64	1.4	0.45 (0.22)	1.0 (0.48)
1977	3.63	8.0	0.64	1.4	0.45 (0.23)	1.0 (0.50)

Source: U.S. Dep. Agric. (1968, 1977).
[1]Figures in (   ) represent California canned ripe olives, taken from data of Olive Admin. Comm. (1977–1978).

igan, Indiana, Wisconsin, and also in Colorado and Washington where the climate favors the growing of cabbage and there is an established demand for it for kraut production.

Commercial olive production in the United States is confined almost entirely to the Sacramento and San Joaquin valleys of California. The only other producing areas in North America are Arizona and Baja California, Mexico, where comparatively small hectarages are grown. Olives will grow in other areas of the United States and Mexico but climatic conditions are not right for the economical setting of fruit. The olive tree, to be productive, requires winter-chilling but at the same time is very susceptible to freezing.

THE PRODUCTION OF SAUERKRAUT

The use of cabbage (Brassica oleracea) as a food antedates known recorded history. Sauerkraut, a product resulting from the lactic acid fermentation of shredded cabbage, is literally acid (sour) cabbage. The antecedents of sauerkraut differed considerably from that prepared at present. At first the cabbage leaves were dressed with sour wine or vinegar. Later the cabbage was broken or cut into pieces, packed into containers, and covered with verjuice (the juice expressed from immature apples or grapes), sour wine, or vinegar. Gradually the acid liquids were replaced by salt and a spontaneous fermentation resulted. One may speculate that sauerkraut manufacture comparable to the method used today developed during the period of 1550 to 1750 A.D. although cabbage has been known and used commonly for about 4000 years. Those readers particularly interested in the historical evolution of the sauerkraut fermentation should consult Pederson (1960, 1979) and Pederson and Albury (1969).

Originally sauerkraut was made only in the home because it provided a means for utilizing fresh cabbage which otherwise would spoil before it could be used. Now the commercial production of sauerkraut has become an important food industry. Even so, a significant quantity is still produced in the home, particularly in rural and suburban areas where home vegetable gardens still exist.

Cabbage varieties best suited for growth in the major production areas are used. Early, midseason, and late types are grown. Varieties formerly used such as Early Flat Dutch, Late Flat Dutch, Early Jersey Wakefield, and others have been replaced in part by new cultivars which have been bred to be well-adapted to mechanical harvesting and at the same time inherently contain less water, thus reducing the generation of in-plant liquid wastes (see Stamer 1975). Mild-flavored, sweet, solid, white-headed cabbage is the choice as it makes a superior kraut.

Preparation for Fermentation

Properly matured sound heads of cabbage are first trimmed to remove the outer green, broken or dirty leaves. The cores are cut mechanically by a reversing corer that leaves the core in the head. Then the cabbage is sliced by power-driven, rotary, adjustable knives into long shreds as fine as 0.16 to 0.08 cm (1/16 to 1/32 in.) in thickness. In general, long, finely cut shreds are preferred, but the thickness is determined by the judgment of the manufacturer. The shredded cabbage (known also as slaw) is then conveyed by belts or by carts to the vats or tanks for salting and fermentation.

Salt plays a primary role in the making of sauerkraut and the concentrations used are carefully controlled. According to the legal standard of identity the concentration of salt must not be less than 2%, nor more than 3%. As a result most producers use a concentration in the range of 2.25 to 2.5% of salt. Salt is required for several reasons. It extracts water from the shredded cabbage by osmosis, thus forming the fermentation brine. It suppresses the growth of some undesirable bacteria which might cause deterioration of the product and, at the same time, makes conditions favorable for the desirable lactic acid bacteria. Salt also contributes to the flavor of the finished sauerkraut by yielding a proper salt-acid ratio (balance) if the cabbage is properly salted. The use of too little salt causes softening of the tissue and produces a product lacking in flavor. Too much salt interferes with the natural sequence of lactic acid bacteria, delays fermentation and, depending on the amount of oversalting, may produce a product with a sharp, bitter taste, cause darkening of color, or favor growth of pink yeasts.

Uniform distribution of salt throughout the mass of shredded cabbage cannot be overemphasized. In some factories the slaw is weighed on conveyor belt lines and the desired amount of salt is sprinkled on the shreds by means of a suitable proportioner as it moves along the conveyor to the vat. In other plants hand-carts are used to carry the shredded cabbage to the vat. Some prefer to salt the weighed cabbage in each cart. Others transport the slaw in carts which are weighed occasionally to check the capacity. The shreds are then dumped into the vat, distributed by forks, and then salted with a specific weight of salt.

The variations of salt concentrations in the brines covering kraut have been thoroughly investigated by Pederson and Albury (1969) and discussed by Pederson (1975, 1979). No mention of recirculation of the brines to gain uniformity in concentration of salt was noted. It would seem that this method of ensuring uniform salt distribution in sauerkraut brines would be as effective as it is in the olive industry. Only small alterations in tank or vat design would be required to make it possible to completely recirculate the brine, pumping from the bottom and discharging at the surface.

Brine begins to form once the shreds are salted, and the tank is closed once it has been filled to the proper level. Formerly, the slaw was covered with a thick layer of outer leaves and then fitted with a wood cover (head) which was heavily weighted. Within a few hours the brine had formed and the fermentation had started. The head then was fixed in position in much the same manner as with pickle or olive tanks. Now, however, a sheet plastic cover is used. This cover is much larger in area than the top of the vat or tank itself. The plastic sheeting is placed firmly against the top of the shredded cabbage with the edges draped over the sides of the container to form an open bag. Then enough water or preferably salt brine is placed in this bag so that the weight of the liquid added forces the cabbage shreds down into the brine until the brine covers the surface of the uppermost shreds. Unless the shreds are completely covered with brine, undesirable discoloration together with undesirable flavor changes will occur. This newer method of covering and weighting provides nearly anaerobic conditions, particularly after fermentation becomes acid and quantities of carbon dioxide are produced. Precautions to avoid pinholes or tears in the plastic is mandatory if aerobic yeast growth is to be avoided. With the old method of closure film forming yeasts always were a problem and if the scum was not removed at intervals a yeasty flavor was imparted to the kraut. *Pichia membranaefaciens* yeast strains, in particular, voraciously oxidize lactic acid contained in salt brines. Other genera also may be involved and besides destroying acid also contribute to yeasty flavor.

By the time the tank or vat is filled with the salted shreds and weighted, brine has formed and a fermentation has started in a sequence of bacterial species responsible for the lactic acid fermentation.

Microbiology of the Sauerkraut Fermentation

Although the lactic acid fermentation was described by Pasteur in 1858 and much work had been done in the intervening years with various lactic bacteria from cabbage and cucumber fermentations, it was not established that a definite sequence of bacterial species of lactic acid bacteria were responsible for the fermentation of either vegetable until 1930 when Pederson first described the lactic acid bacteria he observed in fermenting sauerkraut. Pederson found that the fermentation was initiated by the species *Leuconostoc mesenteroides*. This species was followed by gas-forming rods and finally by non-gas-forming rods and cocci. Since 1930 additional studies by Pederson and Albury (1954, 1969) have firmly established the importance of *Leuconostoc mesenteroides* in initiating the lactic fermentation of sauerkraut. Also they more closely identified the species and sequence of the other lactic acid bacteria involved. Now it is accepted that the kraut fermentation is initiated by *Leuconostoc mesenteroides*, a heterofermentative species, whose early growth is more rapid than other lactic acid bacteria and is active over a wide range of temperatures and salt concentrations. It produces acids and carbon dioxide that rapidly lower the pH, thus inhibiting the activity of undesirable microorganisms and enzymes that may soften the shredded cabbage. The carbon dioxide replaces air and creates an anaerobic condition favorable to prevention of oxidation of ascorbic acid and the natural color of the cabbage. Also carbon dioxide stimulates the growth of many lactic acid bacteria. It also may be that this species provides growth factors needed by the more fastidious types found in the fermentation.

While this initial fermentation is developing, the heterofermentative species *Lactobacillus brevis* and the homofermentative species *Lactobacillus plantarum* and sometimes *Pediococcus cerevisiae* begin to grow rapidly and contribute to the major end products including lactic acid, carbon dioxide, ethanol, and acetic acid. Minor end products also appear. These are a variety of additional volatile compounds produced by the various bacteria responsible for the fermentation, by auto-chemical reactions, or the intrinsic enzymes of the fermenting cabbage itself. Hrdlicka *et al.* (1967) reported the formation of diacetyl and acetaldehyde, the primary carbonyls formed during cabbage fermentation. Volatile sulfur compounds are major flavor components of fresh cabbage according to Bailey *et al.* (1961) and Clapp *et al.* (1959) and also of sauerkraut. However, according to Lee *et al.* (1976), the major portion of the volatiles of sauerkraut is accounted for by acetal, isoamyl alcohol, n-hexanol, ethyl lactate, *cis*-hex-3-ene-1-ol, and allyl isothiocyanate. Of these, only the latter two have been identified as major constituents of fresh cabbage. These latter authors concluded that although these two compounds define the character of cabbage products (kraut) they do not contribute significantly to the determination of its quality. They further believe that the fresh and fruity odor of such compounds as ethyl butyrate, isoamyl acetate, n-hexyl acetate, and mesityl oxide are probably more important in determining the acceptability of sauerkraut.

Temperature is a controlling factor in the sequence of desirable bacteria in the sauerkraut fermentation at a salt concentration of 2.25%. At the optimum of 18.3°C (65°F) or lower the quality of the sauerkraut is generally superior in flavor, color, and ascorbic acid content because the heterofermentative lactic acid bacteria exert a greater effect.

According to Pederson and Albury (1969) an average temperature of about 18°C (65°F) with a salt concentration of 2.25% may be considered normal in the kraut-producing areas of the United States. At (or near) this temperature, fermentation is initiated by *Leuconostoc mesenteroides* and continued by *Lactobacillus brevis* and *Lactobacillus plantarum*, the latter species being most active in the final stages of fermentation. Under these conditions a final total acidity of 1.7 to 2.3% acid (calculated as lactic acid) is formed, and the ratio of volatile to nonvolatile acid (acetic/lactic) is about 1 to 4. The fermentation is completed in 1 to 2 months, more or less, depending upon the quantity of fermentable materials, concentration of salt, and fluctuations in temperature. At higher temperatures, as would be expected, they found that the rate of acid production was faster. For example, at 23°C (73.4°F) a brine acidity of 1.0 to 1.5% (calculated as lactic acid) may be observed in 8 to 10 days and the sauerkraut may be completely fermented in about 1 month. At a still higher temperature of 32°C (89.6°F), the production of acid generally is very rapid with acid production of 1.8 to 2.0% being obtained in 8 to 10 days. As the temperature increased, they observed a change in the sequence of lactic acid bacteria. First, the growth of *Leuconostoc mesenteroides* was retarded and *Lactobacillus brevis* and *Lactobacillus plantarum* dominated the fermentation. At higher temperatures the kraut fermentation became essentially a homofermentation dominated by *Lactobacillus plantarum* and *Pediococcus cerevisiae*. As a result, the quality attributes of flavor and aroma deteriorated and the kraut was reminiscent of acidified cabbage because of the large quantity of lactic acid and little acetic acid produced by the homofermentative species. They also observed that sauerkraut fermented at higher temperatures would darken readily and, therefore, should be canned as quickly as possible after the fermentation was completed.

An extremely important observation they made was that kraut could be successfully fermented even when started at the low temperature of 7.5°C (45.5°F). *Leuconostoc mesenteroides* can grow at lower temperatures than the other lactic acid bacteria involved in the fermentation. At this low temperature (7.5°C or 45.5°F) an acidity of 0.4% (as lactic acid) is produced in about 10 days and 0.8 to 0.9% in less than a month. This amount of acidity coupled with saturation of the mass of kraut and brine with carbon dioxide is sufficient to provide the conditions necessary for preservation and later completion of the fermentation provided that anaerobiosis is maintained throughout the period of latency. When the kraut mass warms enough, the fermentation then is completed by the lactic acid bacteria of the genera *Lactobacillus* and *Pediococcus*, known to grow poorly if at all at 7.5°C (45.5°F). Thus, it may require 6 months or more before the fermentation is completed. Such kraut is generally of superior quality because it remains cool and is not subjected to high temperature during fermentation. In good commercial practice this variation in temperature permits the processor to maintain a supply of new, completely fermented sauerkraut throughout most of the year.

Precedent for the recommendation by Pederson and Albury that sauerkraut be fermented at not over 18.3°C (65°F) had already been recorded by Parmele *et al.* (1927), Marten *et al.* (1929), and others.

Defects and Spoilage of Sauerkraut

Abnormalities of sauerkraut, although varied, with few exceptions can be and generally have been avoided by application of scientific knowledge already available to the industry. For example, the simple expedient of providing anaerobiosis has eliminated most of the problems involving discoloration (auto-chemical oxidation), loss of acidity caused by growth of molds and yeasts, off-flavors and odors (yeasty and rancid) caused by excessive aerobic growth of molds and yeasts, slimy, softened kraut caused by pectolytic activity of these same molds and yeasts, and pink kraut caused by aerobic growth of asporogenous yeasts, presumably members of the genus *Rhodotorula*. [See Peterson and Fred (1923) and Pederson and Kelly (1938).]

Stamer *et al.* (1973) described the induction of red color in white cabbage juice by *L. brevis* while studying the effects of pH on the growth rates of the 5 species of lactic acid bacteria commonly associated with the kraut fermentation. *L. brevis* was the only species which produced such color formation in white cabbage juice and did so only when the juice was buffered with either calcium carbonate or sodium hydroxide. No color development occurred when the pH of the juice (3.9) was not adjusted or when the pH of the juice was raised to 5.5 and the juice sterilized by filtration before it was reincubated. Therefore, red color formation was caused by *L. brevis* and did not arise as the result of chemical or inherent enzymatic reactions of the juice.

It remains to be seen whether this interesting phenomenon will be observed in industrial kraut fermentations. Since color induction by *L. brevis* was found to be pH dependent it seems unlikely to be found in normal kraut fermentations but could easily result from accidental addition of alkali to the shredded cabbage during salting.

Slimy or ropy kraut has been observed for many years. It is generally caused by dextran formation induced by *Leuconostoc mesenteroides* and is transitory in nature. This species prefers to ferment fructose rather than glucose. Therefore, in the fermentation of sucrose, the fructose is fermented leaving the glucose which interacts to form the slimy, ropy, water-insoluble dextrans. These vary from an almost solid, gelatinous mass to a ropy slime surrounding the bacterial cells. These variations are easily demonstrated by growing *L. mesenteroides* in a 10% sucrose solution containing adequate accessory nutrients. The fermenting kraut may become very slimy during the intermediate stage of fermentation but with additional time the dextrans are utilized by other lactic acid bacteria. Thus, it is imperative to distinguish between dextran induced slimy kraut and permanently slimy kraut caused by pectolytic activity. The former condition certainly is not a defect but should be considered a normal step in a natural progression.

THE PREPARATION OF PICKLES

The cucumber *(Cucumis sativus)*, one of the oldest vegetables cultivated by man, is thought to have had its origin in Asia, perhaps India, more than

3000 years ago. It is popular both as a fresh and as a pickled vegetable and is grown widely in temperate climates although originally of semitropical origin. Successful culture of this vegetable is dependent upon avoidance of frost and drought and the control of microbial pathogens and insect pests.

Cucumbers for pickling must be grown from varieties known to have regular form, firm texture, and good pickling characteristics.

Formerly, the common pickling cucumber varieties recommended by various authorities included the Chicago pickling, Boston pickling, Jersey pickling, National pickling, Heinz pickling, Fordhook pickling, Snow's perfection, Packer, and various other strains of lesser importance (Seaton et al. 1936; Jones and Etchells 1943). These varieties, all open-pollinated or monoecious plants, which bear both male and female blossoms, although still available, are being replaced largely by hybrids developed to be used for once-over mechanical harvest. These new hybrid varieties are called "gynoecious" because they have a preponderance of female flowers but are not 100% female. Now, however, most gynoecious hybrid seed must have a pollinator added. These new cultivars often have greater vigor and uniformity than the open-pollinated ones formerly grown. In addition, several of the hybrids are early maturing so they can be used to advantage in harvest scheduling.

Hybrids are used for once-over mechanical harvest and also have given good performance for hand harvesting. Michigan, North Carolina, and California, in that order, lead in total production of pickling cucumbers. California leads in yields per ha (acre) and has done so for many years. In 1976, California produced 33.3 MT per ha (14.8 tons per acre) for a total of 63,900 MT (71,000 tons). At present, Michigan is the only state to have committed itself heavily to machine harvesting of cucumbers, having harvested over 95% of its crop by machine in recent years.

According to Sims and Zahara (1978), the desirable characteristics of a variety suitable for once-over machine harvesting are: (1) relatively small vine with length not over 76 cm (30 in.); (2) relatively short internodes to obtain the maximum number of fruit-setting points; (3) concentrated fruit set and even maturity; (4) early maturity; (5) tendency for fruit to remain on the vine until removed by harvester; (6) resistance to skin and internal damage; (7) late yellowing of fruit if variety is black spined. The use of white-spined hybrids has increased in recent years because their fruits do not turn yellow at the blossom end and maintain their greenness longer, thus providing a more uniform color; (8) uniform shape with a minimum of deformities produced under stress; blocky ends are preferable over pointed ones; (9) a thick wall, small seed cavity, and slow seed development; and (10) blossoms readily fall off fruit. Obviously, the same desirable criteria are valid for hand harvested cucumbers.

Pickling cucumbers are harvested while still immature. Fully grown (ripe) ones are undesirable for pickling because they become too large, change color and shape, are full of mature seeds, and are too soft for most commercial uses. Whether harvested by hand or by machine, care must be taken in picking and transporting the cucumbers to avoid undue bruising and crushing. It should be mandatory to deliver the cucumbers to the salting station or factory as soon as possible after harvest to prevent deterioration. Too long a holding time prior to brining allows the cucumbers to "sweat." This condition promotes the growth of undesirable softening organisms which may cause spoilage early in the brine fermentation before the pH becomes inhibitory to the pectolytic or cellulolytic microorganisms, which are nearly always found to be present on the cucumbers at time of harvest.

To minimize spoilage during fermentation, it is important to remove all unsound, decomposed, broken, or crushed cucumbers. Sorting to remove all crushed or broken, defective, and distorted cucumbers (wilt, rot, crooks, nubbins, etc.) should be done before brining to minimize spoilage during fermentation. Sorting is followed by size grading unless the cucumbers are to be fermented field run. Mechanical graders are used to separate the cucumbers into 4 or more sizes. Final size grading and sorting is done after fermentation.

Three types of pickled cucumbers are made. They include fresh pack (also called fresh cure, home style, and other names) which, at most, are held in salt brine for only as long as 2 days, then packed in jars or cans and pasteurized; salt stock pickles from which a variety of processed products are made; and fermented dill pickles. The two latter kinds undergo a complete lactic acid fermentation whereas the fresh pack pickles undergo, at best, a marginal fermentation unless held in brine for 24 hr or more (see Pederson 1979).

It is estimated that 40 to 50% of the annual harvest of cucumbers is made directly into fresh pack or pasteurized products including whole dills, dill spears, dill chips, sweet slices, etc. The remainder of the crop is converted into fermented salt stock pickles or fermented dills by lactic acid fermentation. The cured salt stock is desalted and processed into various staple pickle products including sweet and sour pickles, mixed pickles, processed dills, sliced pickles, relishes, etc.

Brining Techniques for Salt Stock

There are 2 general methods for preparing salt stock pickles for fermentation: dry salting and brining.

Dry Salting.—The dry salting procedure is not used extensively for cucumbers at present because of its tendency to yield soft, flabby, shriveled pickles that do not fill out properly when processed. However, dry salting is used for other produce, especially cauliflower, red bell and pimiento peppers, salt-cured ripe olives, and is the procedure of choice for sauerkraut.

For cucumbers, dry salting is done after first adding salt brine to cover the bottom of the tank to a depth of at least 30.5 cm (12 in.) to form a cushion, thus preventing bruising, breaking, or crushing the fresh cucumbers when they are dumped into the tank. Dry salt is added at the rate of about 22.5 kg (50 lb) for every 450 kg (1000 lb) of small cucumbers and 29.25 kg (65 lb) for every 450 kg (1000 lb) of large cucumbers. When full, the tank is covered with a circular, slatted, wooden head until there is room for about 15 cm (6 in.) of brine above the cover. The slatted head is then secured with heavy cross timbers held at the ends with clamps. For convenience in handling, the slatted cover may be 2 semicircular pieces for large tanks or even 3 pieces when the largest diameter tanks are involved. If the brine formed by osmosis does not cover the cucumbers or cover when the tank is closed, 40° salometer brine is added to the desired level. The brine should be recirculated a day or two after the tank is filled in order to equalize the concentration of salt throughout the brine. On long storage the brine may be increased slowly until it is about 60° salometer.

In the brining industry the concentration of salt is expressed in degrees salometer which is % saturation of NaCl by weight. A saturated solution of pure sodium chloride (100° salometer) contains 26.359 g at 15.5°C (60°F). Thus, a salometer reading of 10° is equal to 2.64% NaCl by weight (rounded to the nearest tenth). Salt hydrometers are calibrated so that readings will cover several ranges of salt: low, medium, and high. Hydrometers also are available that are calibrated in % salt by weight.

Brine Salting.—Most picklers use the brine salting technique for fermenting cucumbers rather than the dry salting procedure just described. A "low" or a "high" brine process may be used. The low brine has a salt concentration of 25° to 30° salometer, whereas the "high" brine contains 40° or more salt by hydrometer.

The cucumbers are handled in the same manner as described previously for the dry salting process except that brine is used to cover the produce. The tanks are headed in the same way if they are of wood or concrete construction.

Recently, molded plastic and fiberglass tanks have been found useful for replacement of the wood or concrete containers lost by attrition. These plastic and fiberglass containers have several distinct advantages. They are not subject to the usual biological degradation of wood or chemical corrosion of concrete; they do not have to be maintained during the off season, as do wooden ones, to keep them from developing leaks which sometimes require extensive coopering to repair. The drain valves are plastic (polyvinylchloride), as are all other piping, so metal corrosion and resultant contamination of the cucumbers is eliminated. The greatest advantage of these newer containers, however, is that, if they are properly designed, the closures are nearly airtight so that former problems with loss of acidity caused by growth of aerobic yeasts is greatly reduced. With the use of plastic sheeting to cover the brine in open tanks, the problem of control of film yeasts in cucumber fermentations has, in recent years, been reduced to a minimum. Sheet plastic (polyethylene) may be used in the same manner as described previously with sauerkraut or as done with cucumbers and olives held in open tanks in California. In the latter case, the plastic film is floated on the surface of the brine over the false head and secured to the inside of the tank with pliable wood slats nailed so that the plastic is held in place at the surface of the brine. This arrangement will provide nearly anaerobic conditions unless the plastic has imperfections or the slats are improperly placed. Complete, or nearly complete, anaerobiosis can be attained by using "Sealtite" (a wax used widely in the wine industry) to seal the plastic cover to the sides of the tank.

Microbiology of the Cucumber Fermentation

Once the tank of cucumbers has been filled, the cover secured, and brine added, there is a rapid development of microorganisms in the brine. In general, no attempt is made to control the microbial populations of the brines so the cucumbers undergo a "spontaneous" fermentation. The natural controls of the microbial population of the fermenting cucumbers include the concentration of salt in the brine, the temperature of the brine, the availability of fermentable materials, and the relative numbers and types of microorganisms present on the cucumbers and in the brine at the start of the fermentation. The rapidity of the fermentation is directly related to the temperature of the brine and its concentration of salt.

The initial brine strength will vary, depending upon the individual pickling company. In the past, higher concentrations of salt were used because high salt levels were believed to retard spoilage. Now, with increasing frequency, cucumbers are fermented in brines in the 5–8% NaCl range. At this concentration of salt, the sequence of species of lactic acid bacteria approximate that already described for the sauerkraut fermentation with the exception that the species of *Leuconostoc* never predominate the initial stages of the fermentation, even at 5% salt. At 8% salt these species may not be detected at all. The other lactic acid bacteria, *Pediococcus cerevisiae*, *Lactobacillus brevis*, and *Lactobacillus plantarum*, occur in most, if not all, fermentations made in the range of 5 to 8% salt. *Pediococcus cerevisiae* is less salt resistant so sometimes it is absent in the brine at the higher concentration (8%). The same is true with *Lactobacillus brevis*.

During the primary stage of fermentation, a great many unrelated bacteria, yeasts, and molds have been isolated. All are widely distributed in nature and, at the beginning of the fermentation, far outnumber the desirable lactic acid bacteria in uncontrolled fermentations. The primary stage of fermentation, therefore, is the most important phase of the pickling process. If for any reason the fermentation does not proceed normally during this period, any of the unessential microorganisms may become predominant and contribute to spoilage. The primary stage normally lasts 2 or 3 days, exceptionally as long as 7 days, or even more. During this period, the numbers of lactic acid bacteria increase rapidly, both fermenting and oxidizing yeasts increase significantly, and the extraneous and undesirable forms decrease rapidly and may disappear entirely. At the same time, a steady increase in total acidity and a corresponding decrease in the pH of the brine is observed.

In low-salt brine stabilized at about 5% NaCl, a mixture of the low-acid-tolerant species of *Leuconostoc* and the high-acid-tolerant species of *Lacto-*

bacillus and *Pediococcus* predominate in the intermediate stage of fermentation. If the fermentation is normal, the extraneous and undesirable bacteria have completely disappeared by the end of 10 to 14 days, although yeasts are still present in significant numbers. There is a further increase in total acidity and the pH value has also decreased more.

The data of Etchells and Jones (1943), shown in Fig. 6.1, aptly demonstrate the changes in populations of coliform bacteria, acid-forming organisms, and yeasts found in natural fermentations of brined cucumbers in salt concentrations of 20°, 40°, and 60° salometer, respectively. It is seen that the coliform bacteria and other Gram-negative species are readily inhibited in brine fermentations having 40° salometer salt or less because of prompt development of acid by the lactic acid bacteria. However, in brines

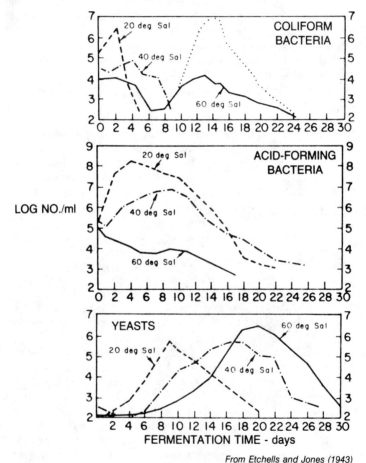

From Etchells and Jones (1943)

FIG. 6.1.   EFFECTS OF BRINE STRENGTH ON THE PREDOMINATING MICROBIAL GROUPS IN NATURAL FERMENTATIONS OF BRINED CUCUMBERS

at 60° salometer salt, the lactic bacteria and the coliform salt-resistant type of *Aerobacter* (represented by the dotted line in Fig. 6.1) and yeasts compete for the fermentable materials and may produce large quantities of gas (CO_2 and H_2) and cause a high percentage of hollow stock (bloaters). It is interesting that the salt-resistant type of *Aerobacter* also occurs in olive fermentations and has been identified by Foda and Vaughn (1950) as an indole-positive type of *Aerobacter aerogenes*.

Species of *Leuconostoc* are fostered by low temperatures (7.2°–10°C or 45°–50°F) and low salt concentrations (2.5–3.7%) according to Pederson and Albury (1950, 1954). Etchells *et al.* (1975) believe that this species would not normally be encountered in commercially brined salt stock cucumber fermentations at 6–8% salt and 24°–29°C (75°–85°F). There nearly always are exceptions that "prove the rule," however. One of the present author's first encounters (unpublished) with species of *Leuconostoc* involved approximately 150 barrels (190 liter or 50 gal. capacity) of refrigerated dill pickles being produced for the delicatessen market. Most of the barrels of fermenting cucumbers had a very slimy brine. The brine temperatures varied from 6° to 8°C (43° to 47°F) and the salt concentrations were in the range of 4–6% NaCl. On examination, it was determined that the predominating bacteria belonged to the species *Leuconostoc mesenteroides*. In retrospect, it is believed that the low brine temperature countered the effect of the salt concentration to permit *L. mesenteroides* to dominate the brines of the refrigerated dill pickles, even at 6% NaCl. Even so, it is agreed that *Leuconostoc mesenteroides* is not an integral part of the lactic acid bacteria populations of normal salt stock fermentations in the range of 5 to 8% salt and always is absent in brines having more salt.

Pediococcus cerevisiae, Lactobacillus brevis, and *Lactobacillus plantarum* are responsible for the final stage and the completion of the lactic acid buildup in uncontrolled fermentations in brines containing salt stabilized in the range at 5 to 8%. All 3 species are found when the cucumbers are fermented at less than about 8% salt. However, the activity of *P. cerevisiae* is severely restricted at this salt concentration and ceases to proliferate when the pH falls to about 3.7. This leaves only the 2 species of *Lactobacillus* left to complete the fermentation (see Etchells *et al.* 1975). At the end, the total acidity may reach as high as 0.9%, calculated as lactic acid, and have a pH value as low as 3.3, providing oxidative yeasts are held in check by anaerobiosis.

Fermented Dill Pickles

Genuine dill pickles differ from the other well known dill pickles (fresh cure and processed dills) because they are the product of bacterial fermentation in a dill flavored, spiced, salt brine. They owe their distinctive flavor and aroma to the products of fermentation of the lactic acid bacteria and to the blending of flavor and aroma of dill herb and spices that were added to the brine.

The larger sizes of cucumbers are generally used for preparing fermented dill pickles. They are washed and placed in suitable containers, together with the requisite amount of dill weed (generally cured in vinegar, salt brine) and dill spices, and brined.

Dill pickles are generally fermented in a low-salt brine of 5% or even less NaCl, but some use up to 7 or 8% brine. The use of vinegar to help retard the growth of undesirable microorganisms by decreasing the pH value of the brine is a common practice. The optimum temperature for the fermentation is between 21° and 26.7°C (70° and 80°F). The fermentation is usually active for 3 to 4 weeks and an additional curing period of 3 to 4 weeks is considered essential. During this period of 6 to 8 weeks, the flesh of the pickles becomes entirely translucent. The brine contains about 0.5 to 1.2% total acidity calculated as lactic acid. In addition, there is a small amount of volatile acid (acetic), ethanol, and other minor products produced by the lactic acid bacteria and yeasts present during the fermentation. The pH values range from about 3.3 to 3.5 if the pickles have been held under nearly complete anaerobiosis. Otherwise, if oxidative yeasts persist, the pH values will be higher due to the loss of acid utilized by them.

Formerly, it was the almost universal practice to use 190 liter (50 gal.) barrels for the fermentation, although a few picklers did use small, wooden tanks. Production now must be considered a bulk fermentation, for most fermented dills are made in wood tanks or plastic or fiberglass receptacles containing 0.9 MT (1 ton) or more of cucumbers.

"Overnight," "Refrigerated," or "Icebox" dill pickles are similar to genuine dills with the important exception that they are stored at a low temperature (38°–45°F) where a slow lactic acid fermentation produces (at the end of 6 months), a total acidity of only about 0.3 to 0.6% calculated as lactic acid. These refrigerated pickles retain some of the fresh cucumber flavor and are highly prized as a food adjunct. However, they are so perishable they must be kept under refrigeration until sold for consumption. Consequently, their availability is not widespread and many people have never had the pleasure of eating them.

Bulk fermentation has become possible because of increased understanding of the need for anaerobic conditions in the fermentation. Wood tanks are fitted with a plastic film and the plastic and fiberglass containers are designed to provide anaerobiosis.

At the beginning of either type of dill fermentation, the original microbial population includes a wide variety of unrelated bacteria, yeasts, and molds. They all are widely distributed in nature and far outnumber the lactic acid bacteria at the start. However, if the fermentation proceeds in a normal fashion, the lactic acid bacteria soon predominate.

The sequence of lactic acid bacteria already described for salt stock pickles also is found in the fermenting brines of the dill pickles just described. However, because of the lower salt concentration, and in the case of the refrigerated dills, the lower temperature, *Leuconostoc mesenteroides* plays a more important role. The work described by Pederson and Ward (1949) and Pederson and Albury (1950) substantiate the initiation of the fermentation by *Leuconostoc mesenteroides* at lower temperatures and lower salt concentrations.

Once the lactic fermentation has an appreciable start, then the other species, *Pediococcus cerevisiae*, *Lactobacillus brevis*, and *Lactobacillus plantarum* begin to dominate the fermentation. The fermentation then is completed by the 2 species of *Lactobacillus*, *L. brevis* and *L. plantarum*.

Genuine dill pickles may be marketed in bulk in plastic containers of various sizes, or, as is done by some picklers, packed in glass, covered with an acidified brine, closed, and pasteurized at 74°C (165°F) for 15 min (*center of jar temperature*) and cooled rapidly to 37.8°C (100°F) or less (see Fabian and Wickerham 1935; Jones *et al.* 1941). So far as is known, the refrigerated dills are always sold in bulk and held under refrigeration until consumed, because of their extreme susceptibility to spoilage.

Deterioration of Pickles

Extensive study has been made of the spoilage of cucumbers during fermentation, curing, and storage. Most of the deterioration is caused by the activity of microorganisms, either by the elaboration of deteriorative enzymes or as the result of copious production of gaseous endproducts (carbon dioxide and hydrogen). Chemical defects are generally confined to metallic contamination or unanticipated alteration of flavor and aroma by the use of specific chemicals or undefined congenerics used for spicing purposes. Auto-chemical and physico-chemical reactions also have occurred.

The most damaging defect caused by microorganisms is tissue destruction resulting from cellulolytic or pectinolytic enzymes elaborated by a variety of organisms. Tissue destruction and loss of texture or firmness generally means nearly total economic loss to the pickler. Gaseous deterioration, resulting in the production of internal cavities or distorted stock caused by excessive gas pressure is another common spoilage caused by microorganisms. This defect is known as "bloater" or "floater" spoilage and is shown in Fig. 6.2. Affected pickles may have lens-shaped internal cavities or the locules may be slightly separated. In severe cases the locules become completely separated and the flesh in each locule is compressed so that the interior is completely hollow and the shape then is reminiscent of a balloon. The salt stock pickles damaged by destructive gas pressure generally may be salvaged by diverting them to relish type products. However, bloater spoilage of dill pickles may mean an economic loss at the present time.

Softening

Softening occurs when microorganisms are capable of elaborating pectinolytic or cellulolytic enzymes under the conditions of salinity, acidity, etc., which exist in pickle brines. Softening is a progressive spoilage which occurs most frequently soon after the cucumbers are brined for production of dill or salt stock pickles. The skin of the cucumber is attacked first, usually at the blossom end. In a short time, the entire skin may be affected, become

DEGREE OF BLOATING

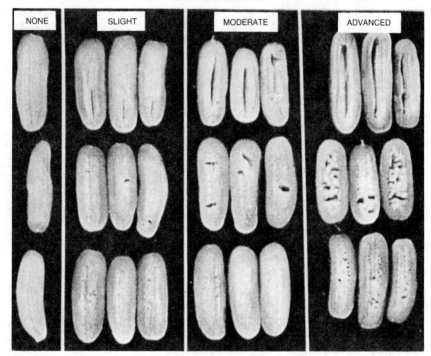

FIG. 6.2.   THREE PRINCIPAL TYPES OF BLOATER DAMAGE FOUND IN PICKLES, SHOWING DIFFERENT DEGREES OF DAMAGE

slippery, and be easily removed. This characteristic first manifestation of softening has given rise to the descriptive terms "slips" or "slippery" pickles in the industry. "Mushy" pickles result when the softening progresses into the deeper layers of cells in the pickles and more and more pectic materials, present in the middle lamella separating the individual cells of the cucumber, are attacked. It is interesting that the form of the pickle may appear normal, but when pressure is applied, it turns to a mushy consistency.

Three kinds of pectolytic enzymes are produced by the bacteria. Pectin methylesterase (pectinesterase) splits off methyl groups from the pectin molecule leaving pectic acid. Polygalacturonase degrades pectic acid leaving saturated digalacturonic acid or higher oligouronides. Polygalacturonic acid *trans*-eliminase splits pectic acid leaving unsaturated digalacturonic acid or higher unsaturated oligouronides as the major end products.

A variety of bacteria, yeasts, and molds are known to produce pectolytic enzymes. The Gram-positive types of *Bacillus* include the following species: *B. subtilis, B. pumilus, B. polymyxa, B. macerans,* and *B. stearothermophilus*; all have been studied and their pectolytic enzymes well described by Vaughn and his associates (see Nortje and Vaughn 1953; Nagel and Vaughn 1961; Davé and Vaughn 1971; and Karbassi and Vaughn 1979).

The Gram-negative types, involving several genera and including *Achromobacter*, *Aerobacter*, *Aeromonas*, *Escherichia*, *Erwinia*, and *Paracolobactrum*, also have been found to contain strains having pectolytic activity (see King and Vaughn 1961; Hsu and Vaughn 1969; and Vaughn *et al.* 1969B).

All of the bacterial species and genera listed in the preceding paragraph have been shown to degrade pickles, making them first slippery and then mushy in texture when tested *in vitro* with sterilized cucumbers. The limiting pH for the activity of the bacterial pectolytic enzymes is in the range of 5.0 to 5.5. It is concluded, therefore, that representative pectinolytic species of the preceding genera of bacteria may cause softening of cucumbers, either salt stock or dills, if:

(1) The pectinolytic bacteria predominated the microbial populations of the cucumbers and their brines.

(2) The pH of the brined cucumbers was in the desirable range for softening—pH 5.5 or above. It is known that some of the softening bacteria can raise the pH of the brines unless the initial values are inhibitory (Vaughn *et al.* 1954).

(3) The desirable lactic fermentation is retarded or arrested in some manner so that the pH values of the brined cucumbers remain relatively high for several days (see Fabian and Bryan 1932).

(4) The brine concentration is in the range of 5 to 8% NaCl.

A variety of different yeasts and molds also have the ability to decompose pectinous substances. The first reports of pectolytic activity by yeasts apparently were made by Cruess and Douglas (1936) and Roelofsen (1936) in July and October, respectively. However, it was not firmly established until 1951 that yeasts did possess pectolytic activity. The work of Luh and Phaff with *Saccharomyces fragilis* was reported then, and later confirmed by Roelofsen (1953) in a reiteration of his 1936 publication, originally published in an obscure journal in the Dutch language. A few other yeasts are known to be pectinolytic. Vaughn *et al.* (1969A) described 3 species of *Rhodotorula* that produced polygalacturonase and, again, in 1972 described 2 other species of *Saccharomyces* that could decompose pectinous material. All of the yeasts produced polygacturonases that were active in the acid range well below the pH value 5.5 thought to be limiting for the bacterial enzymes. Salt concentrations above 5% become limiting for the growth of some of the yeast strains. This may be one reason why, in the past, more yeasts causing softening have not been recovered from cucumber brines. So far as is known, yeasts produce only polygalacturonase.

A variety of molds produce softening enzymes including both cellulolytic and pectolytic types. The molds are known to produce pectinesterase, polygalacturonase, and pectin-*trans*-eliminase. These enzymes have been carefully described by Phaff (1947) who worked with *Penicillium chrysogenum* and by Edstrom and Phaff (1964) who used *Aspergillus fonsecaeus* to describe pectin-*trans*-eliminase purification and properties.

The fungi include representatives of the genera *Alternaria, Aspergillus, Cladosporium, Dematium, Fusarium, Geotrichum, Mucor, Myrothecium, Paecilomyces, Penicillium, Phoma,* and *Trichoderma.* Etchells *et al.* (1955) demonstrated that molds grow and secrete the softening enzymes into the cucumber flowers. The introduction of the fresh or dried flowers containing the enzymes, together with the cucumbers to which they are attached, provides the spoilage factor when the brine is added. Tanks filled with small cucumbers retaining a high percentage of flowers or with experimentally added flowers possess high enzyme activity, and the pickles usually become soft or inferior in firmness. Losses caused by contaminated flowers can be greatly reduced by draining the original cover brine and replacing it with new brine. This apparently reduces the amount of softening enzyme so that softening becomes negligible or nil (Etchells *et al.* 1958).

Draining the cover brine had been a satisfactory method for control of the softening problem since 1954 until recent state and federal regulations concerning disposal of salt severely restricted its use. Studies then were directed toward a search for inhibitors of pectolytic and cellulolytic enzymes. This search, involving isolation of an inhibitor from various plant sources, was successful (Bell and Etchells 1958; Bell *et al.* 1962, 1965B). The forage crop, *Sericea lespedeza* was a particularly good source of the inhibitor and Bell *et al.* (1965A) reported that 50–100 ppm of a crude extract from this source would block softening of no. 1 size cucumbers. Unfortunately, there has been difficulty in obtaining approval for use of the inhibitor under commercial conditions so Etchells *et al.* (1975) have turned to studies involving reclaiming the salt brines for recycling. (It seems to this author, at least, that the simplest method for reclaiming the brine would be the use of heat to inactivate the enzymes, followed by precipitation, flocculation, and filtration procedures, etc., commonly used for purification of water supplies.)

Gaseous Spoilage

Gas-producing microorganisms, now known to cause gaseous deterioration of pickles, represent a number of genera of yeasts and bacteria. Although it was suspected earlier, the first substantial evidence that gaseous spoilage was caused by microorganisms was presented by Veldhuis and Etchells in 1939. They found that hydrogen was produced in significant quantities in the fermentations at 60° Salometer brine and in some, but not all, fermentations at lower concentrations of salt. They also isolated, but did not identify, an organism which produced significant amounts of hydrogen (probably an *Aerobacter*—author's comment). Somewhat later Jones *et al.* (1941) and Etchells and Jones (1941) suggested that gaseous fermentation by unidentified yeasts was the cause of "floater" spoilage. Still later Etchells *et al.* (1945) presented evidence that coliform bacteria of the genus *Aerobacter* caused the formation of hydrogen in pickle fermentations.

Yeasts active in natural pickle fermentations are undesirable, either because they cause copious gas formation and consequently increase bloater formation, or because they utilize lactic acid and, if uncontrolled, cause a rise in pH to the point where spoilage bacteria can renew their activities. The fermenting (subsurface) yeasts have been identified as belonging to the genera *Brettanomyces*, *Hansenula*, *Saccharomyces*, and *Torulopsis* by Etchells *et al.* (1961). The oxidative (surface or film) yeasts have been identified as belonging to the genera *Candida*, *Endomycopsis*, *Debaryomyces*, and *Zygosaccharomyces* by Etchells and Bell (1950). Mrak and Bonar (1939) considered the genus *Debaryomyces* to be responsible for film formation on salt stock pickles.

Lactobacillus brevis, a gas-forming lactic acid bacterium, was shown by Etchells and Bell (1956) to cause floater formation *in vitro*. More recent work by Etchells and his associates has unravelled most of the unknown factors in the complete explanation of "bloater" spoilage. They found that when unheated large-sized cucumbers are brined, serious bloating occurs even when *Lactobacillus plantarum* is the fermenting species (Etchells *et al.* 1973). It was found that when unheated cucumbers are brined, respiration of the fruit liberates enough carbon dioxide into the brine so that, when combined with the small amount produced by *L. plantarum*, it is sufficient to cause bloater damage (Fleming *et al.* 1973B).

Other gas-forming bacteria found in the initial stage of the cucumber fermentation are known to cause *in vitro* spoilage in pasteurized 5% brine cucumbers. These include the *Bacillus polymyxa-macerans* group of aerobic bacilli and the gas-forming pseudomonad, *Aeromonas liquefaciens* (Vaughn 1953).

Thus, it is seen that all of the gas-forming microorganisms found in pickle brines, desirable as well as undesirable, may at one time or another be responsible for producing enough carbon dioxide and/or hydrogen to cause gaseous spoilage.

Nitrogen purging of the fermenting brine is used to reduce undesirable levels of carbon dioxide that otherwise might result in bloater formation (Fleming *et al.* 1973B).

Chemical and Physical Deterioration

The main cause of nonbiological chemical deterioration of pickles is the direct addition of undesirable chemicals to the brines or the pickles themselves. The changes taking place may affect the appearance and color or the flavor and aroma of the pickles.

Contamination with copper and iron formerly was the most common type of chemical deterioration. Vinegar or lactic acid which was used for preparation of brines or liquors for pickles formerly was frequently heavily contaminated with copper and iron. Other metallic contamination with these two chemicals resulted from contact of corrodible equipment (valves, pipe lines, etc.) made from alloys containing copper and iron with the comparatively corrosive brines used for pickling. Stainless steel and plastic components have eliminated this problem.

Copper replaces magnesium in the chlorophyll and pheophytin contained in the pickles and causes them to turn an unnatural, artificial green to blue green color. Only 5 to 10 ppm of copper in the brine or liquor is enough to spoil the pickles.

Iron is involved in the blackening of pickle brines and pickles. Black brine in pickle fermentations, instead of always involving sulfate or reduction of protein sulfur compounds, in all probability frequently is the reaction of iron with polyphenolic compounds to form complex iron "tannates" which are black, bluish-black, or greenish-black in color. Oxygen is required to oxidize iron to the ferric state and to keep this reaction progressing. This complex reaction starts at the surface and extends downward as the oxygen penetrates. Iron sulfide, however, will form from reduced (ferrous) iron, so sulfureted pickle brines are nearly uniformly black from the top to the bottom of the brine and, presumably, have the rotten egg odor which identifies H_2S.

Rahn (1913) was one of the first to investigate the problem of black brines in pickles. He found that the black color resulted from the production of iron sulfide caused by bacterial reduction of sulfate in gypsum ($CaSO_4$) to H_2S, which then reacted with iron to produce the black iron sulfide. Fabian *et al.* (1932) confirmed the work of Rahn and extended it to include a discussion of the possible liberation of hydrogen sulfide in the course of protein decomposition by bacteria. Although hydrogen sulfide production was well documented, the bacteria responsible were not identified.

Bacterial blackening is supposedly caused by *Bacillus nigrificans*, related to *B. mesentericus* (Fabian and Nienhuis 1934).

Pure Culture Fermentation Studies

Etchells *et al.* (1964) reported the results of their study of pure culture fermentation of brined cucumbers, an investigation which was of great importance to the pickling industry. This work led to the concept for the development of "In-Container" or "Ready to Eat" pickled products of a variety of kinds. The processes for production of the ready to eat dill pickles and other products were described in a pubic service patent by Etchells *et al.* (1968B). The process involved heat-shocking and aseptic packing of the product into sanitized containers, followed by covering with a brine pasteurized at 76.7°C (170°F) and cooled to 4.4°C (40°F) and inoculation with pure cultures of lactic acid bacteria which resulted in a controlled fermentation. However, in attempting to adapt the pure culture concept to tanks (2.25 to 13.5 MT or 2.5 to 15 tons capacity) used for bulk fermentation of cucumbers, many problems were encountered economically which could not be resolved. However, a bulk fermentation procedure has since been described (Etchells *et al.* 1973, 1975) which is practical and if the essential steps are followed will ensure against losses by spoilage organisms of whatever kind.

The Controlled Fermentation of Cucumbers

There are a number of steps in the process developed by Etchells and associates which are new. Some others have been practiced in the industry for at least 50 years. Nonetheless, all of the steps aid in advancing our knowledge of the science and technology of cucumber fermentation.

Since it was impractical to destroy the contaminating organisms with heat, as was done in the pure culture process described above, thorough washing and in-container chlorination have been used to reduce the initial contamination by microorganisms. Chlorine (about 80 ppm) is added to a 25° salometer brine and used as a cover brine. The chlorinated cover brine is carefully acidified with acetic acid (food grade) or its equivalent of 200 grain vinegar. The cover brine is chlorinated again, about half a day after adding the cover brine.

Salt is added to the cover brine to maintain the original brine strength, which otherwise would be diluted by the water content (approximately 95%) of the cucumbers. The amount of salt added to the cover brine at the outset, and after 1 or 2 days, will depend upon the size of the cucumbers being brined.

After the initial salt addition has become equilibrated, but before the second salt addition and around 3 or 4 hr before the addition of the starter culture(s), sodium acetate is added to buffer the cover brine. This buffering is done to ensure that all of the fermentable sugars will have been utilized during the active stage of the lactic acid culture. Otherwise, acid-tolerant, fermenting yeasts could later ferment the residual sugar and produce levels of carbon dioxide perhaps sufficient to cause some gaseous spoilage.

Inoculations may be made with the 2 species of homofermentative lactic acid bacteria; *P. cerevisiae* and *L. plantarum* or else *L. plantarum* alone. It is most important that the starter cultures selected be able to give maximum performance under the conditions optimum for fermentation. They must grow well at 23.9°−29.4°C (75°−85°F), not be retarded by 6 to 8% salt, and produce a minimum amount of carbon dioxide. The strains of *P. cerevisiae* must not inhibit *L. plantarum* if the 2 species are to be used together in starting the fermentation.

Nitrogen purging is started as soon as the tank of cucumbers is headed, brined, and acidified. The type and rate of purging will depend upon schedules for different sizes of cucumbers and container capacities. Etchells *et al.* (1975) report, "It has been found repeatedly that restricting a buildup of carbon dioxide in the brine during the entire fermentation, using the controlled process, prevents bloater formation." Figure 6.3 shows the various changes taking place in the controlled fermentation of cucumbers brined in bulk, described by Etchells *et al.* (1975).

THE FERMENTATION OF OLIVES

Although the origin of the art of preparing table olives by lactic acid fermentation is lost in antiquity, the history of the table olive industry in California has been documented. The first olive trees in the state were

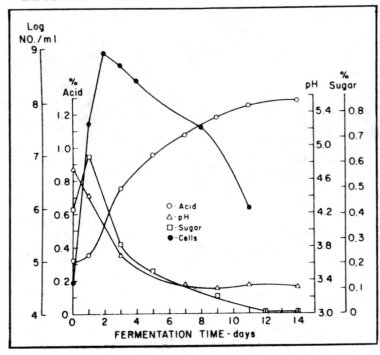

From Etchells et al. (1975)

FIG. 6.3. CONTROLLED FERMENTATION OF CUCUMBERS BRINED IN BULK. EQUILIBRATED BRINE STRENGTH DURING FERMENTATION, 6.4% NACL; INCUBATION TEMPERATURE 27°C

grown from seed said to have been planted at the Mission San Diego in 1769. The seeds were brought from San Blas, Mexico, by Don José de Galvez during an expedition to rediscover the port of Monterey. The seedlings, thus obtained by selection, are the source of the present mission variety of *Olea europaea* (LeLong 1890).

Although olives were used for oil production in the California missions as early as 1780, the first production of olive oil outside of the missions did not occur until 1871. The olive was not planted extensively until about 1860, but by 1870 was showing promise of becoming of some importance to California agriculture. Between 1870 and 1900 many olive varieties had been introduced from Europe and Africa and much effort had been made in testing them, principally for oil production, since oil was the major product of the industry.

Pickling of olives for table use, practiced as an art in the missions, on the farms, and in the homes for many years, was of little commercial value until about 1900. (Directions for pickling ripe and green olives are to be found in the early agricultural literature of California.)

For some reason, the art of pickling "ripe" olives was studied more extensively than that of "green" olives. According to Cruess (1958), it was independently discovered about 1900 by Professor Bioletti of the University of California and Mrs. Freda Ehmann, a commercial packer, that ripe olives, after a preliminary treatment with lye (NaOH) to destroy the bitter principle, could be canned and preserved by heat in much the same manner as other foods. This original development, coupled with application of modern science and technology have, somehow, combined to make the California canned ripe olive the major product of the industry.

Formerly, early in this century, oil was still a major product of the olive industry and, until about 1960, still accounted for an appreciable quantity of fruit in each of the yearly product-disposition figures. In years of high yields the olives destined for oil were in an approximately 1 to 1 ratio with olives canned. However, because of increased harvest costs in California and other competitive factors from abroad, oil production now has become largely a salvage operation, and tons of olives used for oil approximate those used for producing Spanish-type green olives, another salvage product. (Data supporting these statements are available from the Olive Administrative Committee, Fresno, California 93728.)

The disposition of olives for products in approximate order of importance in California comprise black-ripe and green-ripe olives (whole and pitted), Spanish-type, Sicilian-type, Greek-type including brined and salt-cured fruits, and oil.

A portion of the black-ripe olives may be prepared as sliced (cross-section rings), chopped, or segmented (longitudinal) into 4 or more pieces per olive. Canned ripe olives are the major products of the industry and account for 70% or more of all of the olives harvested. All of the green-ripe canned olives are processed and canned at harvest time (called direct or fresh cure by the industry). Fresh cured olives, although not subjected to a lactic acid fermentation, also have spoilage problems which will be discussed later.

The remainder of the harvested olives, of necessity, have to be stored in salt brine prior to processing. They also undergo a lactic acid fermentation. Thus, there are 4 brine fermentations, including "storage," Sicilian-type, Spanish-type, and Greek-type brined olives.

The "storage" and Sicilian-type fermentations may be considered to be identical for each variety of olive because the fruits are placed directly in brine without lye treatment, whereas the Spanish-type olives are treated with lye to destroy most of the bitterness, washed to remove some of the alkali, and then brined. The brine-cured Greek-type olives are placed in a high salt brine which may not undergo a lactic fermentation, but a fermentation caused by salt tolerant yeasts. Salt-cured Greek-type olives are cured (desiccated) with coarse salt, so they do not undergo a fermentation in the strict sense. Therefore, they will not be considered here.

Commercial Varieties

There are 5 main varieties of olives grown in bearing hectarage (acreage) in California. The Mission olive, formerly the main olive of the industry has, during the past 25 years or so, declined to last place in importance because of its tendency to produce small fruits and its propensity for late maturity. The main varieties grown now, in commercial importance, are the Manzanilla, Sevillano, and Ascolano in the order named. Lesser varieties also are grown, including the Barouni which now is seldom used for canned ripe olives but is utilized in all sizes for Spanish-type olive production. Is is also sold on the fresh-produce market for home pickling as are some of the Northern California Sevillano fruits and other varieties, at the discretion of the fresh-market shippers. All except the Mission olive are of foreign origin: the Manzanilla and Sevillano olives were imported from Spain, the Ascolano from Italy, and the Barouni from Tunisia.

Each variety is distinctive in size, shape, and general utility. The Sevillano olives are the largest, followed in order by the Ascolano, Barouni, Manzanilla, and Mission varieties. The Manzanilla variety most nearly approaches the requirements of an all-purpose olive, since it may be used for the production of California ripe olives, green fermented olives, or olive oil. The other varieties have characteristics which limit their usefulness in one way or another.

Harvesting

Most of the olives used for ripe or green pickling are harvested starting between the middle or latter part of September and ending at about Thanksgiving time (late November) if the crop is large. The time will depend on the variety of fruit, the locality, the growing season, climatic conditions, and other factors. A sharp frost means termination of picking olives for processing. Frosted olives generally are salvaged for oil production. It is desirable to remove all, or nearly all, of the fruits produced in a season to help ensure a good set of fruit the following year.

Because canned black-and-green-ripe olives are the main products of the California olive industry, harvesting practices favor the selection of fruit most desirable for production of these two commodities. Therefore, when the majority of the fruit in an orchard has reached the desired maturity, the harvest is started and continued until all, or nearly all, of the olives have been gathered.

Mature (tree-ripened) olives do not have the final dark purple to jet black color at the time of harvest for pickling, but are at the color turning stage, from green to straw yellow or, at most, light cherry red. More highly colored fruit may be overripe. Overripe fruit, which is highly colored, generally deteriorates during the lye pickling process, and consequently is diverted to oil production, or if the texture still is quite firm, may be used for the preparation of the Greek-type brine-cured or salt-cured olive specialties.

California canned black-ripe olives (generally uniformly black in color after canning) are made from fruit ranging in color from green to not more than cherry red at harvest. The California green-ripe or "home-cured" olives (yellowish-green to light greenish-brown, frequently mottled color after canning) are made from fresh olives having a green to not more than cherry red color. California fermented green olives (Sicilian- and Spanish-types) are produced from fruits ranging from green to straw yellow in color.

Harvesting, formerly all done by hand, is slowly becoming mechanized. Machine tree-shakers are being used in newer groves where the trees have been pruned in a fashion to promote removal of olives by shaking. However, in old groves the trees were pruned in such a manner that machine removal was impractical because of failure to remove enough fruit, or the trees were infirm and there was much damage due to limb breakage. Studies also have been made with chemical spray treatment to facilitate machine harvest of the olives [see Fridley *et al.* (1971) and Hartmann and Opitz (1977) for more details on mechanical harvesting of olives]. Unfortunately, fresh olives are quite easily bruised and the amount of bruising is accentuated with mechanical harvesting, but fortunately, the bruises do not affect the appearance of the processed black-ripe olives or Greek-type olives. However, the bruised spots do persist in the processed green-ripe and the fermented green Sicilian-type and Spanish-type olives.

In the interest of economy, hand labor is reduced as much as possible. Harvested fruit whether hand- or machine-harvested is placed in bulk-bins for transportation to the pickling plants by truck or other mechanized equipment. These bins, widely used in the food industry, are designed to be handled by forklift trucks. They are approximately 4 ft^2 and 2 ft deep, have a 2-port pallet entry, and, if filled completely, hold 453.6 kg (1000 lb) or more of fruit.

Sweating of the fruits held in bins can be a problem. To avoid possible fruit damage caused by sweating, the olives should be put through the various plant operations in as short a time as possible; preferably all bins of olives should have been delivered to the plant, dumped onto the transfer lines, put through the destemmer, and sorted and size graded in not more than 24 hr.

In years of high yields of olives, some packers store fruit under refrigeration for varying lengths of time. The olives are cold-susceptible so they should not be stored below 7.2°C (45°F) in order to avoid an undesirable taste in the processed fruit. Freshly harvested olives can be held under refrigeration without undue mold development if the refrigerated storage is properly operated. On longer holding, molding becomes a problem in the center of the mass of olives, and the loss increases with the increase in storage time.

Olives can be held in a light brine (5% NaCl) under refrigeration for a period of at least 6 months without loss resulting from microbial activity if the containers are anaerobic or nearly airtight at the least.

Once the harvested olives have been destemmed, sorted, and size-graded, they are ready to be processed, stored in salt brine for future processing, or made into fermented Sicilian- or Spanish-type green olives.

The Fermentation of Storage and Sicilian-type Olives

The "storage" and Sicilian-type fermentations are considered identical for each variety of olive because the fruits are placed directly in brine without lye treatment to destroy the bitterness of the olives. If there are no cherry red colored olives in a tank, the disposition of the fruit can be for Sicilian-type olives, or the fruit may be processed into canned black-ripe olives, according to the economic demand.

The open fermentation tanks are filled and headed in much the same manner as already described for cucumbers. The newer fiberglass tanks are filled with olives, brined, and the cover locked in place without the use of a false head to hold the olives submerged in the brine.

The salt concentration of the brine will vary according to the variety of olives and the final disposition to be made of the tank of olives. Sevillano and Ascolano olives are subject to salt shrivel so it is customary to use a lower concentration of salt (4 to 5% NaCl) if either variety is to be sold as Sicilian-type olives. Otherwise the brine strength will range between 5 and 8% salt because, in the processing of black-ripe olives, salt shrivel is reduced to a minimum by use of a needling machine to puncture the skin of the fruits. This facilitates osmotic exchange between the needled olives and the processing solution, generally water, or, at most, 10° to 12° salometer salt brine. The shrivel is markedly reduced or eliminated by this manipulation.

The olives, whatever their final disposition, undergo a lactic acid fermentation. The amount of total acidity produced is quite variable but will usually range between 0.2 and 0.7% (calculated as grams lactic acid per 100 ml of brine). The development of acidity in (holding) storage brines varies widely because of a number of factors. The Sevillano variety does not fresh cure well, so the majority of processors hold this variety in salt brine for a minimum of 30 days before putting the fruit through the black-ripe, canned olive process. This holding period, obviously, does not favor an acidity in the upper range, for olives ferment slowly at the ambient temperatures prevailing in the brines, especially during the final weeks of the harvest. If held in open tanks, the acidity may be lost by oxidative yeasts, which decompose the acid developed. This is quite an important factor when open tanks of such olives are held in shaded or covered areas during the colder months when the oxidative yeasts and molds are still more active than the desirable lactic acid bacteria. The tendency now is to use anaerobic methods to prevent this loss of acidity, either by use of plastic film to cover the surface of the brine or by the use of polyethylene or fiberglass containers designed to minimize the access of air to the fermenting olives.

The changes in the microbiological populations observed in Sicilian-type or holding solution fermentations are quite similar, and the olives are subject to the same microbial spoilage problems. The lactic acid bacteria found in these fermentations are of the same kinds already described for the cabbage (sauerkraut) and cucumber (pickle) fermentations.

Detailed studies made, but never published (Vaughn *et al.* 1945–1951, 1952–1954), indicate that there is a sequence of microorganisms. The initial stage of fermentation is dominated by organisms other than the lactic acid bacteria. In 4 or 5 days, if the brine strength is not more than 5% NaCl, a population of lactic acid bacteria begins to appear. *Leuconostoc mesenteroides*, *Pediococcus cerevisiae*, and *Lactobacillus plantarum* have been found, but *L. mesenteroides* never dominates the fermentation, as it does with cabbage during the initial stage of fermentation. *P. cerevisiae* and *L. plantarum*, or the latter alone, are always found. The initial population of extraneous organisms, mainly coliform bacteria and bacilli, disappear gradually and, if the fermentation is normal, no longer can be found after 10 to 14 days of fermentation. Fermenting yeasts appear in the first 1 to 2 weeks and continue throughout the fermentation. Some of these yeasts are acid-formers, but their contribution to the increase in total acidity of the olive fermentations may be offset by the decomposition of the acidity by oxidative yeasts, which also may become established if conditions are favorable for their growth.

The total acidity developed during the fermentation of holding solution olives destined for future processing varies according to the time the fruit is held in brine. The total acidity may be about 0.1% (total acidity calculated as lactic, g/100 ml) when held for minimum of 30 days as is required for the Sevillano variety or, if the olives are held longer, and the oxidative yeasts and molds are controlled, the total acidity may be as high as 0.6% as lactic acid.

Spoilage problems involving these fermentations will be discussed later.

The Fermentation of Spanish-type Olives

The first extensive commercial-scale experiments made to study this fermentation were reported by Cruess (1930). From about 1935 until 1955 there was a gradual increase in the quantity of Spanish-type olives fermented in California. All sizes of fruit of the Manzanilla, Sevillano, and Barouni varieties were fermented as Spanish-type green olives, the Barouni variety in particular. However, rising production costs, and the adoption of the industry-wide marketing agreement limiting the sizes of each variety that may be canned in California made it unprofitable to continue making Spanish-type olives from all sizes except for the Barouni olives, which were exempt because of their unsuitability for canning. At present, the production of Spanish-type olives is strictly a salvage operation, and the noncanning sizes of all varieties are utilized.

Also, for economic reasons, the use of the traditional 190 liter (50 gal.) barrels for fermentation has been discontinued. The development of plastics fostered the replacement of the wood barrel with rigid plastic, bottle-shaped containers of 1514 liter (400 gal.) capacity made of polyethylene, with polyvinyl chloride fittings for draining. These bottles have the advantage of holding the equivalent of 8 barrels of olives, being mobile with the aid of lift trucks, even when filled with olives and brine. Another advantage is that use of the bottles permits lye treatment, washing (leaching), and brining in the same container as was done earlier by using large redwood tanks holding the equivalent of about 55 barrels of olives (Ball *et al.* 1950). Also there is a considerable economy realized because the bottles require no maintenance upkeep when not in use.

The bottles are filled with olives of the appropriate size, transported to the lye treatment and washing area, and the olives are covered with a lye solution. The concentration of lye used for treating green olives to hydrolyze the bitter glucoside, oleuropein, varies between 0.9 and 1.25% as used in California. Stronger lye solutions must be used very carefully, for they frequently cause softening and blistering, as well as undesirable skin sloughing with all varieties. Fruit treated during the early part of the season (September 15 to October 15) is prone to blistering when a gaseous fermentation starts because the skin has separated from the flesh due to lye temperature and, if the skin remains intact, gas formation causes a blister to form. This form of blister formation, commonly known as "fish-eye" spoilage, will be discussed in more detail subsequently.

Cooling of the lye solution and/or olives is indicated. The practice in the San Joaquin Valley has been to use block ice or a heat-exchanger to cool the lye solutions during the early part of the season when the ambient temperatures still remain high during the day.

The lye is allowed to penetrate about ½ to ¾ of the way to the pits of all varieties treated with the exception of the Barouni. This variety must be treated to the pit, for the flesh, unless so treated, will become an undesirable reddish purple color in the area not exposed to lye and the color will intensify as the acidity of the brine increases. This color change is thought to be caused by a leuco-anthocyanin present in the flesh of the fruit.

The time required for the desired lye penetration varies according to the concentration and temperature of the lye solution and the variety, size, maturity, and temperature of the fruit. However, an attempt is made to maintain a schedule to complete the penetration in 12 to 14 hr.

Formerly, the removal of residual lye by washing and leaching with water was carried out by changing the leaching water every 3 to 6 hr during the day. The interval at night might reach 10 hr between changes of water. The washing-leaching extended for 24 to 48 hr before the olives were brined. By 1943 (Vaughn et al. 1943), the trend was to shorten the washing-leaching period in order to minimize the graying of the color of the fruit. Now, because the number of changes of leaching water have been reduced to 3 or 4 changes in 24 hr, the olives are quite alkaline when brined.

Formerly, the industry used 2 different concentrations of salt when brining olives for the Spanish-type fermentation (Vaughn et al. 1943). The Sevillano variety is susceptible to salt shrivel in brine having more than 5% NaCl. Therefore, the processors started this variety in a low concentration of salt (4.0 to 5%), and then increased the salt in the brine slowly to 7−8% over an indefinite interval.

At present, the majority of all varieties used in the industry are brined with about 10% salt solution which is acidified with enough lactic acid to neutralize the residual lye remaining after the washing-leaching period. Most processors add sufficient acid to the brine to lower the pH of the stabilized olives and brine to a value of between 4.5 and 5.0.

Sometimes glucose (crystalline corn sugar) is added at the time of brining. If this is done, the sugar is added at the rate of 0.45 kg (1 lb) (approximately) per 189.3 liters (50 gal.) of olives and brine. However, the limitation of loss of fermentable materials by use of a much shorter, less rigorous washing and leaching before brining has largely eliminated the need for addition of sugar during the fermentation, especially when the plastic bottle is used.

In normal fermentation of green olives of the Spanish-type, the initial stage of the pickling process is the most important phase. During this stage, lasting up to 14 days, if the brines are not acidified, the original contaminating population of Gram-negative and Gram-positive bacteria are eliminated as the result of acid production, both by themselves and by the developing population of lactic acid bacteria.

Formerly, *Leuconostoc mesenteroides* and *Streptococcus faecalis* were always found in low salt fermentations of the Sevillano variety. *L. mesenteroides*, which dominated the latter phase of the primary stage, as well as the early part of the secondary stage of fermentation, disappeared from the population within 3 or 4 weeks. *Lactobacillus plantarum*, a non-gas-forming species, dominated the latter part of the intermediate stage as well as the final stage of fermentation. This species is known to persist in olive brines for more than a year. *L. brevis*, a gas-forming type, is found in the latter phases of the intermediate stage and is present in appreciable numbers during the final stage of fermentation. This latter species never approaches the population levels produced by *L. plantarum*. Furthermore, as shown by Vaughn *et al.* (1943), *L. brevis* was never found in the high salt fermentations conducted with the Manzanilla variety. Table 6.5 summarizes the predominating bacterial population trends described, and also shows the accompanying changes in the acidity and pH values.

Pediococci were not found in any of the fermentations described by Vaughn *et al.* (1943). However, in later studies (Ball *et al.* 1950; Vaughn and Martin 1971) it was found that *Pediococcus cerevisiae* may be isolated from olive brines during the last phases of the initial stage of fermentation and the first part of the intermediate stage of fermentation, then it declines rapidly. Not all brines of Spanish-type olives contain *P. cerevisiae*. Also, since all varieties of olives are started in 10% salt brines now, the gas-forming species *L. brevis* also may not be found in the fermentations. Now, the only lactic acid bacterium sure to be found in all fermentations is *L. plantarum*.

Control of Spanish-type Fermentations

The olive fermentation is very slow in comparison to either the sauerkraut or pickle fermentation. Data in the literature indicate that either the sauerkraut or pickle fermentation will have produced maximum total acidity before the olive fermentation has passed through the initial stage of fermentation [consult Pederson (1979); Etchells *et al.* (1975); and Vaughn *et al.* (1943)]. Olives brined for Spanish-type fermentations in late September and early October produce more acid because of more favorable ambient temperatures. The olives brined from the middle of October until the end of the harvest ferment more slowly and may become dormant soon after the start of fermentation because of the approach of winter and the resultant

TABLE 6.5.   GROSS FLORAL AND CHEMICAL CHANGES IN BRINE FROM A SEVILLANO FERMENTATION

Time in Days	Approximate Numbers of Microorganisms per ml of Brine		Total Acidity, Grams Lactic Acid per 100 ml Brine	pH	Grams NaCl per 100 ml Brine	Most Abundant Bacteria or Yeasts Isolated During the Stages of Fermentation
	Gram-positive	Gram-negative				
Primary Stage						
0	43	254	0.0	8.20	6.25	Gram-negative bacteria *Aerobacter* and *Pseudomonas*
1	1,500	3,050	0.014	6.80	—	
2	6,890,000	—	0.037	6.35	—	Gram-positive bacteria *Streptococcus* and *Leuconostoc*
3	120,500,000	13,500,000	0.054	5.75	3.04	
5	390,500,000	228,500,000	0.108	5.20	—	A few yeasts
7	237,000,000	27,600,000	0.153	5.00	—	
Intermediate Stage						
9	707,000,000	22,000,000	0.153	4.65	—	Gram-negative bacteria *Aerobacter*
11	—	—	0.234	4.30	—	
12	410,000,000	2,950,000	0.198	4.50	3.28	Gram-positive bacteria *Leuconostoc* and *Lactobacillus*
15	34,000,000	0	0.243	4.35	3.33	
18	126,000,000	0	0.297	4.40	3.51	No yeasts
21	140,000,000	0	0.333	4.35	3.63	
Final Stage						
28	14,000,000	0	0.423	4.22	3.69	Gram-positive bacteria *Lactobacillus plantarum*
35	3,500,000	0	0.405	4.20	3.69	*Lactobacillus brevis*
42	4,500,000	0	0.414	4.05	3.86	No Gram-negative bacteria or yeasts
56		0	0.445	4.20	3.91	
77	1,500,000	0	0.432	4.20	3.83	—
196	700,000	0	0.414	4.10	4.09	—
365	2,650,000	0	0.514	4.50	4.09	—

Source: Vaughn *et al.* (1943).

drop in temperature. Therefore, it is obvious that temperature control is necessary to prevent dormancy of the fermentations.

The fermentations of all varieties of Spanish-type olives brined in California may be accelerated by proper incubation. Commercial scale experiments made by Cruess (1930) indicated that an average temperature range of 21° to 24°C (70° to 75°F) was satisfactory for acceleration of acid production without impairing the quality of the fermented olives. Additional studies by Vaughn *et al.* (1943) showed that the optimum temperature for maximum acid production of pure cultures of *Leuconostoc mesenteroides* and *Lactobacillus plantarum* was 30°C (86°F), and for *L. brevis* was 34°C (93.2°F).

Temperatures much above 32.2°C (90°F) have been observed to be undesirable for pickling. Therefore, the temperature recommended by Cruess can be followed with assurance. A temperature range of 23.9° to 30°C (75° to 86°F) is commonly used for incubation at present, but only to the extent that fermented Spanish-type olives are required for early delivery.

Fermenting olives should not be incubated until the potential spoilage bacteria, which probably are always present at the beginning of the fermentation, have been eliminated. With normal uncontrolled fermentations, the time required for disappearance of the undesirable bacteria varies from 1 to as much as 3 weeks, depending upon the activity of the lactic acid bacteria and the availability of fermentable material. Those processors who acidify the initial brine may, if desirable, initiate incubation at the start of the fermentation.

Fermentation of Spanish-type olives may also be accelerated by use of starter cultures and addition of fermentable sugar. Cruess (1930) recommended the use of starters of normal brine. Later (1937), Cruess suggested the use of pure cultures of lactic acid bacteria for ensuring start of fermentation. Pure culture starter inoculations were used extensively in California from 1937 until about 1955, particularly with the Manzanilla variety. *Lactobacillus plantarum* was the species of choice to use for starters. Details of preparation and use of the starters in the industry have been described by Vaughn *et al.* (1943). Now, although sophisticated pure culture starters are available for commercial culture laboratories, for economic reasons, when inoculation is indicated, normal brine is used to reseed the suspect fermentation.

Cruess (1930) was the first to recognize the need for addition of supplementary fermentable sugar to obtain satisfactory acid formation in Manzanilla olives and in Sevillano brines which did not otherwise develop enough acidity. Supplementary glucose (from corn) is used extensively, especially with the Barouni, Manzanilla, and Mission varieties. The sugar is commonly added at the rate of 0.45 to 0.9 kg (1 to 2 lb) for each 189.3 liters (50 gal.) of olives and brine. Unless acidification of the initial brine is practiced, the processor waits from 1 to 3 weeks before adding the sugar. For the best olives and avoidance of malodorous spoilage, the fermentations are controlled to a pH of at least 4.0 and, preferably, 3.8.

Acidification of the brine was first suggested by Cruess (1930) as means of preventing potential loss of fruit caused by abnormal fermentation. It already had been reported by Fornachon *et al.* (1940) that *Lactobacillus brevis* and other gas-forming lactobacilli had an optimum pH in the range of 5.0 to 6.0 for growth and decomposition of different carbon compounds. This latter study was considered to be good evidence for the possible need to acidify all olive brines. Therefore, additional studies, for the most part under commercial conditions, were made by Vaughn *et al.* (1943). They found that when acidification of the brine, with either acetic or lactic acid, was accompanied by use of a starter culture of *L. plantarum* and addition of supplementary glucose, satisfactory fermentations were obtained. Acidification apparently functions to spark the activity of the lactic acid bacteria as well as to eliminate undesirable spoilage organisms.

Perhaps the most important control of all for protecting olives in fermentation, storage, Sicilian- or Spanish-type, is the maintenance of anaerobic or nearly anaerobic conditions. Control of oxidative molds and yeasts is mandatory (Balatsouras and Vaughn 1958; Mrak *et al.* 1956; Vaughn *et al.* 1969A) but only recently became recognized as such by the industry as a whole.

The advent of plastic film and molded plastic containers during the 1960s was responsible for important changes in the fermentation of green olives in California. The use of pliable plastic film to cover the brines in the open bulk fermentation tanks (redwood) for the first time permitted an airtight closure that, once in place, needed little maintenance. This was a major advance because it virtually eliminated oxidative molds and yeasts as sources of potential spoilage. (Plastic films are also used on the open redwood tanks used for storage and Sicilian-type fermentation for the same reason.)

When the industry was using 189.3 liter (50 gal.) barrels exclusively, the first attempt to restrict brine surface exposure to the air was by use of cellar bungs commonly used in the wine industry in the early 1940s. However, these did not prevent spillage caused by diurnal expansion and contraction of the brine, so during warm weather it was necessary to add fresh brine to the barrels daily if the film yeasts were to be held to a minimum.

Expansion bungs were devised to minimize spillage. The first such bungs were improvised by cutting the bottoms out of 0.95 liter (1 qt) size carbonated beverage or whiskey bottles and inserting the necks in rubber stoppers of a size suitable to fit the side opening of the barrels. This device controlled expansion and contraction of the brine without spillage. However, breakage was a problem. Finally, a plastic bung was developed. This more durable bung, 15.2 cm (6 in.) in diameter, provided a constant air surface to total volume ratio of 182.5 cm^2 (28.3 $in.^2$) to 189.3 liters (50 gal.) of olives and brine.

The opening in the removable top of the plastic bottle used for fermentation of Spanish-type green olives has the same diameter as that of the plastic expansion bung just described. The advantage of the plastic bottle is quite obvious—182.5 cm^2 (28.3 $in.^2$) exposed air surface to 1514 liters (400 gal.) of olives and brine. The amount of brine exposed to the air can be reduced further by floating a circular disk of slightly less diameter on the surface of the brine in the top opening. Exposure to air can be entirely eliminated by floating a 2.5 cm (1 in.) layer of paraffin, microcrystalline wax, or "vaspar," a 1 to 1 mixture of paraffin and vaseline. Results reported by Vaughn and Martin (1971) have shown a significant increase in acid production of olives fermented in the plastic bottles as compared with those olives fermented in 189.3 liter (50 gal.) barrels or in 757 liter (200 gal.) plastic liners held in wooden shells which also have been used in the industry.

By a combination of control measures, it is possible and also practical to complete the Spanish-type green olive fermentation in 3 to 4 weeks under ideal commercial conditions (see Fig. 6.4). To accomplish this it is necessary to use control measures which require maintenance of a brine temperature of 24° to 30°C (75° to 86°F); adjustment of the salt concentration of the acidified brine to 5 to 7% (W/V); to ensure the presence of desirable lactic acid bacteria in the brine (use of a starter, if indicated); and the addition of supplementary sugar to the brine to ensure that an acidity of at least 0.8% (total acid calculated as grams lactic acid per 100 ml brine) is produced. Maintenance of nearly anaerobic conditions is mandatory (Ball *et al.* 1950; Vaughn and Martin 1971).

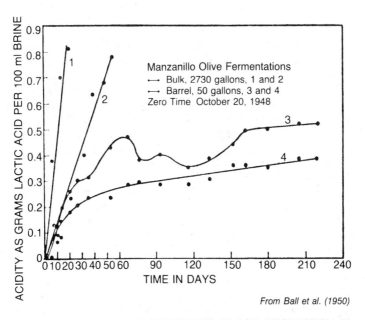

From Ball et al. (1950)

FIG. 6.4. SHOWING THE DIFFERENCE IN ACID PRODUCTION IN CONTROLLED BULK FERMENTATIONS COMPARED WITH CONVENTIONAL BARREL FERMENTATIONS

Spoilage Problems

Bacteria, yeasts, and molds may cause spoilage of olives at any time after the harvest and the final packaging of the olives. All olives, processed black- and green-ripe olives, as well as the brined olives that undergo the lactic acid fermentation, are subject to microbial attack.

The commonest and, consequently, best-known spoilage types caused by microorganisms are gassy deterioration, malodorous fermentation, and tissue softening. Of these, gassy fermentation and softening occur most frequently and under the widest variety of conditions; in all stages of the lactic fermentation as well as in olives undergoing "direct cure" for canning.

Gassy, "Floater," or "Fish-eye" Spoilage.—This abnormality is characterized by the development of blisters resulting from the accumulation of gases which cause separation of the skin from the flesh of the olives and by the formation of fissures or gas pockets which may extend to the pits of the fruit, as shown in Fig. 6.5. (Gas pockets are never found when blisters caused by too concentrated or too warm lye treatment solutions are observed so this chemical deterioration should not be confused with the biological one under consideration.)

Through the extensive studies of Cruess and Guthier (1923), Alvarez (1926), Tracy (1934), and Vaughn and his students (Foda and Vaughn 1950; Vaughn et al. 1943; West et al. 1941), it is well established that the coliform bacteria are chiefly responsible for blister and gas pocket formation. All of the species of coliform bacteria have been implicated in gassy spoilage except *Escherichia coli* (see Foda and Vaughn 1950).

Cultures of *Bacillus polymyxa* and *B. macerans* also cause gassy deterioration of olives (Gililland and Vaughn 1943; Vaughn 1954). These species also may cause softening of olives. The gas-forming pseudomonad *Aeromonas liquefaciens* also may form gas pockets in olives but, like the bacilli, is equally important as a softening organism (Vaughn et al. 1969B). Saccharolytic species of the anaerobes of the genus *Clostridium* also cause a violent gassy fermentation in olives as shown by Gililland and Vaughn (1943). These anaerobes may also be involved in softening of olives but are more important for the malodorous spoilage they cause (butyric fermentation and zapatera).

It was formerly the contention of the author that only those bacteria which produce hydrogen were dangerous. There always are exceptions. Recently, Vaughn et al. (1972) associated yeasts of the genera *Saccharomyces* and *Hansenula* with gassy fermentation and softening in olives. These yeasts produced typical gas blisters but did not cause fissure formation. Two of the species, *Saccharomyces kluyveri* and *S. oleaginosus*, also caused severe softening of olives. The cultures of *Hansenula* were not pectolytic.

Control measures, for the most part, include sanitation, regulated control of the fermentation by reducing the pH by acidification, ensuring a population of desirable lactic acid bacteria, and, in the case of processed olives for canning, use of pasteurization. Control measures are discussed in more detail by Vaughn (1954).

FIG. 6.5A. GAS FORMA-
TION IN OLIVES

Blister formation commonly
called "fish eye" spoilage.

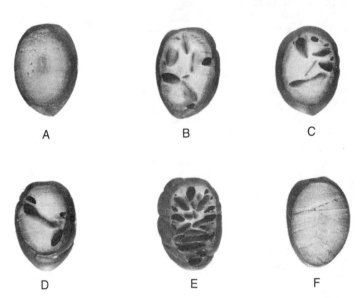

From Gililland and Vaughn (1943)

FIG. 6.5B. GAS POCKET FORMATION BY INOCULATION OF OLIVES WITH
PURE CULTURES

A—Uninoculated control. B—*Aerobacter aerogenes.* C—*Escherichia coli.* D—
Aerobacillus polymyxa. E—*Clostridium butyricum.* F—*Saccharomyces cerevisiae.*
All olives are of the Mission variety; sterilized by intermittent steaming and incu-
bated before inoculation.

Malodorous Fermentations.—There are 3 extremely malodorous fermentations, caused by bacteria, which develop in olives. They are the butyric
acid fermentation, hydrogen sulfide fermentation, and zapatera spoilage.

The butyric acid fermentation has been associated with olives since
Hayne and Colby (1895) first recorded olive spoilage by the "butyric fer-

ment." This abnormal fermentation is characterized by its butyric acid or rancid butter odor during the initial stages of the development of this malodorous abnormality. However, as the spoilage progresses, the odor intensifies and finally results in a very malodorous stench.

Gililland and Vaughn (1943) first isolated pure cultures of anaerobic, spore-forming butyric acid bacteria from samples of butyric spoiled olives. All of the cultures were found to be of the saccharolytic but not proteolytic types of the genus *Clostridium*. Most of the cultures were closely related to or identical with the species *Clostridium butyricum*. Sevillano olives formerly were prone to develop butyric fermentations because the salt concentration was kept lower in order to avoid salt shrivel. However, if the salt concentration is kept in the range now used by the industry (7 to 8% NaCl W/V) the butyric fermentation cannot develop. It always occurred during the initial stage of fermentation in storage, Sicilian- and Spanish-type olives and generally affecting Sevillano olives, but the other varieties as well, if the salt was in the range of 5% and the pH value was at 4.5 or above. At present, olives affected by the butyric fermentation are rarely found.

Hydrogen sulfide fermentation of olives is characterized by the identifying odor of H_2S gas. At first the odor may be slight, but as the fermentation progresses the odor intensifies and is reminiscent of the smell of rotten eggs. Black brines may occur in this fermentation if sufficient contaminating ferrous iron is present to cause the formation of the black iron sulfide. Black brines have been observed by the author in both storage and Sicilian-type olive brines. Chalky white brines have also been observed to occur in Spanish-type olives when contaminating zinc from galvanized pipelines, buckets, and barrel hoops got into the brine and formed the whitish zinc sulfide. The majority of hydrogen sulfide fermentation brines show neither kind of sulfide formation but are clear or have a microbial turbidity.

Early published reports on the occurrence of hydrogen sulfide fermentation in brined olives is lacking, so there is no authenticated record of the first hydrogen sulfide fermentation occurring in California olive brines. The author first observed such fermentations and recognized them as hydrogen sulfide fermentations during the 1937 harvest season in the upper Sacramento Valley. In all probability, this fermentation has occurred for as long as olives have been brined in California or elsewhere.

From 1937 on, periodic attempts were made to isolate sulfate reducing bacteria from sulfureted brines. "Mineral" autotrophic enrichments could be carried through several transfers but eventually failed. It was not until the study made by Levin and Vaughn (1966) that it was recognized that the "mineral" media used in the earlier studies lacked essential nutrients now recognized as necessary for perpetuating the growth of most, if not all, strains of sulfate-reducing vibrios.

Levin and Vaughn associated the halophilic *Desulfovibrio aestuarii* with the hydrogen sulfide fermentation of fermenting storage and Sicilian-type olives involving the Sevillano variety. Control of the sulfate-reducing *D. aestuarii* is accomplished by acidification to a pH value below 5.5 either by direct addition of acid or control of the fermentation. Salt concentration is of no control value because some of the bacteria grow in the presence of 12 to 14% NaCl (W/V).

Avoidance of hydrogen sulfide fermentation is effective if open tanks of fermenting olives are pumped over to recirculate the brine after each heavy rain to ensure an inhibitory pH in the upper layers of the tank.

Sulfureted olives may be salvaged by replacing the brine and then aerating violently to oxidize the remaining hydrogen sulfide. More than one change of brine and aeration may be necessary. Once the odor is depleted then the brine must have the pH value adjusted to a safe level below 5.5.

"Zapatera," another malodorous fermentation of olives, apparently was first described by Cruess (1924), who had observed the abnormality in spanish green olives while on a visit to Spain. This spoilage occurs in all types of brined olives and, to the author's knowledge, is found in all olive growing areas of the world where olives are brined. The off-odor associated with zapatera at first is described as "cheesy" or "sagey," but as the spoilage progresses the "cheesy" odor disappears, and as the odor further intensifies, it develops into an unmistakably characteristic foul, fecal-like stench.

Under the conditions that prevail in California, zapatera spoilage, unlike the butyric fermentation already described, occurs when the desirable lactic acid fermentation is allowed to stop before the pH of the brine has decreased to a value of 4.5 or a little less. At the start of the spoilage the pH of the infected brine increases while the titratable acidity decreases. Continuous loss in acidity is observed as the spoilage progresses (Ball 1938; Cruess 1941; Vaughn et al. 1943).

The implication of bacteria as the cause of zapatera at first was confused. Smyth (1927), the first to investigate the problem, concluded that the spoilage was caused by one or more of a group of spore-forming, proteolytic, facultative rods normally present in the soils of Andalusia, an olive growing area in Spain. Soriano and Soriano (1946) and Soriano (1955) claimed that zapatera spoilage in Argentina was caused by the sulfate-reducing bacterium *Desulfovibrio desulfuricans*. However, hydrogen sulfide odor is either not detectable or, at most, is not pronounced in the case of zapatera, either in California olives or olives from other parts of the world where olives are grown. Therefore, it was doubtful that these investigators had isolated the bacterium capable of causing the spoilage.

Because of the previous inability to isolate bacteria capable of causing zapatera, the author and his associates decided to investigate the acidic constituents of normal and zapatera fermented brines. It was found by Delmouzos et al. (1953) that normal fermented olives contained acetic, lactic, and some succinic acid, whereas zapatera brines also contained formic, propionic, butyric, valeric, isovaleric, caproic, and caprylic acids. This information suggested the association of species of the genus *Clostridium* with zapatera spoilage, especially since previous investigation by Bhat and Barker (1947) and Tabachnick and Vaughn (1948) had demonstrated lactate utilization to be accentuated by the presence of acetate. Further study by Kawatomari and Vaughn (1956) associated a number of species of *Clostridium* with zapatera spoilage. Two saccharolytic, proteolytic species, *C.*

bifermentans and *C. sporogenes*, predominated among the cultures isolated and studied.

However, the association of species of *Clostridium* with zapatera spoilage did not give complete explanation of the etiology of the spoilage. Propionic acid is one of the abnormal acids found in the spoiled brines. *Clostridium propionicum* was not found in the study made by Kawatomari and Vaughn so an additional investigation was made to determine whether species of *Propionibacterium* also might be involved in the spoilage. Plastourgos and Vaughn (1957) found species of *Propionibacterium* including representative isolates of *P. pentosaceum* and *P. zeae* to be abundant in zapatera brines. All produce propionic acid from lactate in culture media and in olive brines. It is obvious, therefore, that the propionic acid bacteria are the cause of the cheesy odor that occurs early in zapatera spoiled olives. It is also obvious that the spoilage is caused by the participation of species of at least 2 genera of bacteria, *Clostridium* and *Propionibacterium*, and is but a manifestation of nature's attempt to cause mineralization of the olives.

Control to prevent zapatera spoilage involves direction of the lactic fermentation until the pH value is at least at 4.0, and preferably 3.8 or below.

Zapatera spoiled olives should not be sold because once the spoilage develops, the odors cannot be removed. Nevertheless, zapatera olives still are found in commercial channels throughout the world.

Softening Spoilage

The softening of olives is usually caused by the activity of pectolytic microorganisms. Bacteria, molds, and yeasts all have been incriminated in the problem. However, the form and texture of fresh olives may be changed by use of too concentrated sodium hydroxide solutions, by frosting, or by heating. Therefore, it is not always easy to determine whether the softening is of microbial origin or whether it results from chemical or physical mistreatment of the fruit.

Softening manifests itself in 1 to 3 characteristic changes in the appearance of the olives. These changes are known in the industry by the very descriptive terms "soft stem end," "nail head," and "sloughing," respectively. These types of softening have been observed in all kinds of pickled olives and all varieties are susceptible.

Soft stem end spoilage is cause by a variety of bacteria, molds, and yeasts which are known to produce pectolytic enzymes which cause softening of olives. Olives are mechanically destemmed on arrival at the processing plant or in the field at harvest when gathered from the trees. This leaves interior tissues unprotected by stem tissue, so enzymes enter through the unprotected area and cause the splitting of the pectic materials into fragments. The pectolytic attack in turn softens the stem end area and causes a puckering of the stem end. As more tissue is attacked, the softened area expands until the entire flesh of the olive has been affected and becomes soft and, many times, quite mushy.

The bacteria known to be involved in soft stem end spoilage include spore-forming species of *Bacillus* (Nortje and Vaughn 1953) and *Clostridium* (Vaughn 1953), and the Gram-negative bacteria, including representatives of the genera *Aerobacter, Aeromonas, Achromobacter, Escherichia*, and *Paracolobactrum* (Vaughn *et al.* 1969B).

Molds (fungi) responsible for stem end softening of olives include *Aspergillus, Fusarium, Geotrichum, Paecilomyces*, and *Penicillium* (Balatsouras and Vaughn 1958).

Yeasts involved in soft stem end spoilage of olives include oxidative, pink yeasts of the genus *Rhodotorula* (Vaughn *et al.* 1969B) and fermenting yeasts of the genus *Saccharomyces* (Vaughn *et al.* 1972).

The bacteria all produce pectolytic enzymes that are active at a pH range of about 6.0 to 10.0 and have an optimum pH at about 8.0 to 8.5. It is obvious, therefore, that the bacterial enzymes will not be a factor in softening unless, for some reason, the fermentation is abnormally slow. On the other hand, they are very damaging in the process of preparation of olives for black- or green-ripe canned olives during the final stage of washing prior to canning. The bacterial induced soft stem end spoilage is controlled by acidification and/or control of the lactic acid fermentation and by pasteurization in the case of the ripe olive process.

The fungal and yeast enzymes on the other hand are active at pH values in the range commonly found for storage, Sicilian- and Spanish-type olives (about 3.8) and still are active at neutrality. It is certain, therefore, that the molds and yeasts are responsible for the softening of olives in brine at pH 5.5 and below.

The bacteria generally produce two pectolytic enzymes: pectin methyl esterase, which demethylates the pectin molecule to pectic acid, and either endo- or exo-polygalacturonic acid *trans*-eliminase, which causes degradation of the polygalacturonate to either unsaturated digalacturonic acid or unsaturated trigalacturonic acid.

The yeasts known to be pectinolytic produce polygalacturonase, which degrades pectin to saturated digalacturonic acid. One mold, *Aspergillus fonsecaeus*, studied by Edstrom and Phaff (1964) produces pectin *trans*-eliminase, and the major end products of pectin degradation include unsaturated and methylated galacturonates. Most of the other molds produce polygalacturonase and the major end product is monogalacturonic acid.

Since the molds are aerobic, the maintenance of anaerobic conditions will control their activity. Yeast control to prevent such softening is not so simple. The experience of the author with fermenting yeast spoilage would lead one to believe that the best control measure is to direct the lactic acid fermentation to completion. The yeast episode described by Vaughn *et al.* (1972) was the first encounter with fermenting, pectolytic yeasts in 35 years of experience with olives by the author.

Nail head spoilage is characterized by the formation of a concave depression under the skin of the olive which causes the skin to depress into the concave area which generally is nearly devoid of tissue. The depressions are from about 0.32 to 0.64 cm (⅛ to ¼ in.) in diameter. The spoilage occurs infrequently, and the cause is not perfectly known. The author has reproduced a type of nail head spoilage by bruising but not penetrating the skin of fresh olives, then fermenting them in brine and making the olives into canned black ripe olives. The resultant spoilage was identical to that observed with commercial samples (Vaughn 1956). It may be that nail head spoilage is a purely physical defect due to a combination of bruising and resultant physiological deterioration of the bruised area.

Sloughing spoilage of olives occurs most frequently with olives undergoing the black-ripe process for canned olives (Vaughn et al. 1969B). However, the Gram-negative pectolytic bacteria described in 1969 did not cause rupture of the skin and sloughing of the flesh, although the olives became soft and mushy. A search was made for free-living, cellulase-producing bacteria from olives undergoing sloughing. Cellulolytic bacteria were associated with sloughing spoilage of California ripe olives by Patel and Vaughn (1973). The most active cellulolytic bacterium was identified as *Cellulomonas flavigena*. Other cellulolytic bacteria studied included species of *Xanthomonas, Aerobacter, Escherichia, Kurthia, Micrococcus*, and *Alcaligenes*. All cultures produced skin rupture and sloughing of the flesh of sterile olives to a degree but the cellulomonad was the most active. Control is simple. Pasteurization may be used or the washing cycle can be reduced from 4 to not more than 3 days.

Species of *Fusarium* isolated from storage and Sicilian-type brines by Balatsouras and Vaughn (1958) also were found to be cellulolytic, as well as pectinolytic. It is conceivable, therefore, that they might cause some sloughing involving skin rupture, at least, which would occur in storage olives during ripe processing. Control of the fusaria would involve strict control of anaerobiosis in the fermentations.

Other Abnormalities

Two other abnormalities sometimes occur in olives. They are "yeast spot" formation in green olives and refermentation of bottled green olives. "Yeast spot" formation is a very common defect associated with fermented green olives in California. This abnormality is characterized by the formation of raised white spots (pimples or pustules) directly under the pores (stomata) of the olive, with a colony of microorganisms developing between the epidermis and the underlying tissues. The Sevillano variety is prone to this abnormality. It is not confined to California but is also observed in other places where olives are fermented. "Yeast spots" also have been observed on pickled cucumbers and fermented green tomatoes. Although commonly called "yeast spots" throughout the industry, Vaughn et al. (1953) found that most, if not all, of the pustules contained lactobacilli. These bacteria were identified as *Lactobacillus plantarum* and *L. brevis* and other species closely related to the latter one. The cause of such colonization is not known, nor is any control available.

Since such olives are considered unsightly by the brokers, some loss of value of the pickled fruit is experienced, although the affected olives are perfectly normal and healthful in other respects.

Refermentation of glass packed olives sometimes occurs with accompanying gas formation and unsightly sedimentation caused by the growth of microorganisms. The cause of this refermentation is a reservoir of fermentable material still left in the olives, or the unwitting addition of fermentable sugar contained in stuffing (pimiento and almonds in particular) for pitted olives, or in various spices and other flavoring constituents added when the olives are packed in glass.

Most refermentation is caused by lactobacilli and, sometimes, a mixture of these bacteria and fermenting yeasts. To control refermentation effectively, it must be determined that all of the fermentable constituents in the olives and other added materials have been decomposed. It is common practice to referment pitted olives stuffed with pimiento or nuts for at least a month prior to finally packing the olives in glass. Pasteurization [heating to a center temperature of 85° to 88°C (185° to 190°F)] has been used. Some packers use sorbic acid (potassium sorbate) as a preservative, and vacuum seals are preferred.

Refermented olives are salvageable but the cost of packing and repacking is doubled because control was not practiced before the olives were packed the first time.

Newer Developments

Recent studies have been directed toward increasing the rate of fermentation, particularly with the Manzanilla variety. Attention has been directed toward the use of heat and studies on antibacterial substances present in olives.

Heat received the first attention. Samish *et al.* (1966) reported a manyfold increase in the rate of the lactic acid fermentation of green olives following a short prebrining treatment with hot alkali which, in effect, caused lye peeling of the fruits. With this method, the rate of fermentation was stimulated when the concentration and/or the temperature of the alkali was increased (Samish *et al.* 1968).

Etchells *et al.* (1966) found that a pre-brining heat treatment in water, after application of lye to destroy part of the bitterness of the olives, increased the fermentability of the heat-shocked olives over those receiving lye treatment alone. As a result, Etchells and co-workers concluded that the fermentability of the different varieties depended more on their inhibitor content than any other single property they might have. Confirmatory results with such heat treatment studies have been reported by Borbolla y Alcala *et al.* (1971).

The notion of a microbial inhibitor in olives had been mentioned by Vaughn (1954) who speculated that oleuropein might eventually be shown to be inhibitory to the lactic acid bacteria. However, until Fleming and Etchells (1967) reported the occurrence of an inhibitor of lactic acid bacteria in green olives, there was no concrete evidence that such inhibitors might exist. They found the inhibitor was soluble in water and ethanol, heat stable in water at 100°C, but was labile to alkaline conditions. Unfortunately, the results reported were done with frozen olives—never pickled in California because of the adverse effects on texture caused by freezing.

Shortly thereafter, Juven et al. (1968A,B) reported that the bitter glucoside of green olives was found to inhibit the growth of lactic acid bacteria. In 1969, Fleming et al. described an inhibitor of lactic acid bacteria which was a phenolic compound devoid of acid-hydrolyzable reducing sugar but with a bitter taste. In 1970 Juven and Henis reported the identification of the glucoside oleuropein and its aglycone as the main antibacterial components of green olives. It appears from the literature that the compounds investigated by the Israeli and United States groups were probably identical. This assumption is supported by the more detailed report by Fleming et al. (1973A) showing that two hydrolysis products of oleuropein, the aglycone and elenoic acid, inhibited the growth of lactic acid bacteria but oleuropein itself had no effect.

While the two hydrolysis products of oleuropein appear to be inhibiting to the lactic acid bacteria in vitro, there must be some reservation with regard to their inhibitory role in Spanish-type green olive fermentations. The stability of the antimicrobial factor in alkali is poor (see Fleming and Etchells 1967; Juven and Henis 1970; and Juven et al. 1968B). It is common practice in California to leach lye-treated green olives with 3 or 4 changes of water in 24 hr and then add an unacidified brine. When so brined the olives are quite alkaline and remain so for some time. Under such conditions it is felt that the inhibitory aglycone would be subject to further decomposition.

Also there is disagreement concerning the inhibitory activity of oleuropein. Juven et al. (1968A) reported that it inhibited the growth of different bacteria, including Lactobacillus plantarum and Leuconostoc mesenteroides. No such activity for the glucoside was recorded by Fleming et al. (1973A). Recently, Garrido-Fernandez and Vaughn (1978) made a detailed in vitro study to assess the possible utilization of oleuropein as a major source of carbon by various microorganisms associated with fermentation of olives. It was found that both the desirable lactic acid bacteria and spoilage organisms did use oleuropein as a source of carbon, many without a significant delay in growth. It was found that increase in oleuropein from 0.2 to 0.4% (W/V) had little effect on the spoilage organisms, delaying somewhat, but not preventing, eventual growth by any of the lactic acid bacteria tested.

Another complicating factor is that the epidermis (skin or peel) of the olive is known to have a low permeability for water and gasses as shown by Duran and Tamayo (1964). The permeability of the skin is greatly increased when subjected to a hot alkali treatment, when heat shocked after normal lye treatment, or frozen. These chemical and physical methods for increasing the rate of fermentations of green olives are all subject to the same criticism—they all may adversely affect the texture of the olives so treated.

Green Olive Processing Without Fermentation

A method of salt-free storage of olives combining acidulated water (lactic and acetic acids), the food preservative sodium benzoate, and anaerobiosis was described by Vaughn *et al.* (1969C). This method is being used to store olives for ripe processing formerly stored in brine. The use of salt-free storage has eliminated the problem of salt shrivel of Ascolano and Sevillano olives and improves the texture and flavor of Sevillano olives so stored. For reasons unknown the Mission variety does not store well for extended periods (6 months) but becomes softened and grainy in texture after 1 to 2 months of storage.

The method was developed to alleviate the problems of disposal of salt brines, a result of environmental protection legislation. An additional possibility exists. Vaughn and Martin (1967–1968) observed that green olives lose much of their bitterness when stored for several months in this salt-free environment. The olives, when repacked with the proper salt brine to balance the acidity of the fruits, have very good organoleptic properties. On the basis of "blind" tasting they have passed as green fermented olives.

REFERENCES

ALVAREZ, R.S.   1926.   A causative factor of "floaters" during the curing of olives. J. Bacteriol. *12*, 359–365.

ANON.   1956.   Processing of Green Olives. Sponsored by Juan de la Cierva Cordoriu. Superior Council of Scientific Investigations, Madrid. (Spanish)

BAILEY, S.D., BAZINET, M.L., DRISCOLL, J.L. and McCarthy, A.I.   1961. The volatile sulfur components of cabbage. J. Food Sci. *26*, 163–170.

BALATSOURAS, G.D. and VAUGHN, R.H.   1958.   Some fungi that might cause softening of storage olives. Food Res. *23*, 235–243.

BALL, R.N.   1938.   Proc. 17th Annu. Tech. Conf. Calif. Olive Assoc., 1938, Riverside, Calif. (mimeograph), 30.

BALL, R.N., van DELLEN, E., JAQUITH, J.B., VAUGHN, R.H., TABACH-NICK, J. and WEDDING, G.T.   1950.   Experimental bulk fermentation of California green olives. Food Technol. *4*, 30–32.

BELL, T.A. and ETCHELLS, J.L.   1958.   Pectinase inhibitor in grape leaves. Bot. Gaz. *119*, 192–196.

BELL, T.A., ETCHELLS, J.L., SINGLETON, J.L. and SMART, W.W.G., JR. 1965A.   Inhibition of pectinolytic and cellulolytic enzymes in cucumber fermentations by sericea. J. Food Sci. *30*, 233–239.

BELL, T.A., ETCHELLS, J.L. and SMART, W.W.G., JR.   1965B.   Pectinase and cellulase enzyme inhibitor from sericea and certain other plants. Bot. Gaz. *126*, 40–45.

BELL, T.A., ETCHELLS, J.L., WILLIAMS, C.F. and PORTER, W.L.   1962. Inhibition of pectinase and cellulase by certain plants. Bot. Gaz. *123*, 220–223.

BHAT, J.V. and BARKER, H.A.   1947.   *Clostridium lacto-acetophilum* nov. spec.; and the role of acetic acid in the butyric fermentation of lactate. J. Bacteriol. *54*, 381–391.

BORBOLLA y ALCALA, J.J. DE LA, GONZALEZ-CANCHO, F. and GON-ZALEZ-PELLISO, F. 1971. Green olives and color change in brine. I. Grasas Aceites Seville 22, 455–460. (Spanish)

CLAPP, R.C., LONG, L., JR., DATEO, G.P., BISSETT, F.H. and HASSEL-STROM, T. 1959. The volatile isothiocyanates in fresh cabbage. J. Am. Chem. Soc. 81, 6278–6281.

CRUESS, W.V. 1924. Olive pickling in Mediterranean countries. Calif. State Agric. Exp. Stn. Circ. 278.

CRUESS, W.V. 1930. Pickling green olives. Calif. State Agric. Exp. Stn. Bull. 498.

CRUESS, W.V. 1937. Use of starters for green olive fermentations. Fruit Prod. J. 17 (1) 1–12.

CRUESS, W.V. 1941. Olive products. Ind. Eng. Chem. 33, 300–303.

CRUESS, W.V. 1958. Commercial Fruit and Vegetable Products, 4th Edition. McGraw-Hill Book Co., New York.

CRUESS, W.V. and DOUGLAS, H.C. 1936. An interesting spoilage of Sicilian olives. Fruit Prod. J. 15, 334.

CRUESS, W.V. and GUTHIER, E.H. 1923. Bacterial decomposition of olives during pickling. Calif. State Agric. Exp. Stn. Bull. 368.

DAVÉ, B.A. and VAUGHN, R.H. 1971. Purification and properties of a polygalacturonic acid trans-eliminase produced by Bacillus pumilus. J. Bacteriol. 108, 166–174.

DELMOUZOS, J.G., STADTMAN, F.H. and VAUGHN, R.H. 1953. Malodorous fermentation acidic constituents of zapatera olives. J. Agric. Food Chem. 1, 333–334.

DURAN, G.M. and TAMAYO, A.I. 1964. Study on the histological structure of the fruit of Olea europaea L. Grasas Aceites Seville 15, 72. (Spanish)

EDSTROM, R.D. and PHAFF, H.J. 1964. Purification and certain properties of pectin trans-eliminase from Aspergillus fonsecaeus. J. Biol. Chem. 239, 2403–2408.

ETCHELLS, J.L. and BELL, T.A. 1950. Film yeasts on commercial cucumber brines. Food Technol. 4, 77–83.

ETCHELLS, J.L. and BELL, T.A. 1956. Bloater formation by gas-forming lactic acid bacteria in cucumber fermentation. Bacteriol. Proc. 1956, 28.

ETCHELLS, J.L., BELL, T.A. and COSTILOW, R.N. 1968B. Pure culture fermentation process for pickled cucumbers. U.S. Pat. 3,403,032. Sept. 24.

ETCHELLS, J.L., BELL, T.A., FLEMING, H.P., KELLING, R.E. and THOMP-SON, R.L. 1973. Suggested procedure for the controlled fermentation of commercially brined pickling cucumbers—the use of starter cultures and reduction of carbon dioxide accumulation. Pickle Pak Sci. 3, 4–14.

ETCHELLS, J.L., BELL, T.A. and JONES, I.D. 1955. Cucumber blossoms in salt stock mean soft pickles. N.C. Agric. Exp. Stn. Res. Farming 13, 1–4, 14–15.

ETCHELLS, J.L., BELL, T.A., MONROE, R.J., MASLEY, P.M. and DEMAIN, A.L. 1958. Populations of softening enzyme activity of filamentous fungi on flowers, ovaries, and fruit of pickling cucumbers. Appl. Microbiol. *6*, 427–440.

ETCHELLS, J.L., BORG, A.F. and BELL, T.A. 1961. Influence of sorbic acid on populations and species of yeasts occurring in cucumber fermentations. Appl. Microbiol. *9*, 145–149.

ETCHELLS, J.L., BORG, A.F. and BELL, T.A. 1968A. Bloater formation by gas-forming lactic acid bacteria in cucumber fermentations. Appl. Microbiol. *16*, 1029–1035.

ETCHELLS, J.L., BORG, A.F., KITTEL, I.D., BELL, T.A. and FLEMING, H.P. 1966. Pure culture fermentation of green olives. Appl. Microbiol. *14*, 1027–1041.

ETCHELLS, J.L., COSTILOW, R.N., ANDERSON, T.E. and BELL, T.A. 1964. Pure culture fermentation of brined cucumbers. Appl. Microbiol. *12*, 523–535.

ETCHELLS, J.L., FABIAN, F.W. and JONES, I.D. 1945. The *Aerobacter* fermentation of cucumbers during salting. Mich. State Agric. Exp. Stn. Tech. Bull. *200*.

ETCHELLS, J.L., FLEMING, H.P. and BELL, T.A. 1975. V.2. Factors influencing the growth of lactic acid bacteria during the fermentation of brined cucumbers. *In* Lactic Acid Bacteria in Beverages and Food. J.G. Carr, C.V. Cutting and G.C. Whiting (Editors). Academic Press, London, New York, San Francisco.

ETCHELLS, J.L. and JONES, I.D. 1941. An occurrence of bloaters during the finishing of sweet pickles. Fruit Prod. J. *20*, 370, 381.

ETCHELLS, J.L. and JONES, I.D. 1943. Bacteriological changes in cucumber fermentation. Food Ind. *15*, 54–56.

FABIAN, F.W. and BRYAN, C.S. 1932. Experimental work on cucumber fermentation. I. The influence of sodium chloride on the biochemical and bacterial activities in cucumber fermentations. Mich. State Agric. Exp. Stn. Tech. Bull. *126*.

FABIAN, F.W., BRYAN, C.S. and ETCHELLS, J.L. 1932. Experimental work on cucumber fermentation. V. Studies on cucumber pickle blackening. Mich. State Agric. Exp. Stn. Tech. Bull. *126*.

FABIAN, F.W. and NIENHUIS, A.L. 1934. Experimental work on cucumber fermentation. VII. *Bacillus nigrificans* n. sp. as a cause of pickle blackening. Mich. State Agric. Exp. Stn. Tech. Bull. *140*.

FABIAN, F.W. and WICKERHAM, L.J. 1935. Experimental work on cucumber fermentation. VIII. Genuine dill pickles—a biochemical and bacteriological study of the curing process. Mich. State Agric. Exp. Stn. Bull. *146*.

FLEMING, H.P. and ETCHELLS, J.L. 1967. Occurrence of an inhibitor of lactic acid bacteria in green olives. Appl. Microbiol. *15*, 1178–1184.

FLEMING, H.P., WALTER, W.M., JR. and ETCHELLS, J.L. 1969. Isolation of a bacterial inhibitor from green olives. Appl. Microbiol. *18*, 856–860.

FLEMING, H.P., WALTER, W.M., JR. and ETCHELLS, J.L.   1973A.   Antimicrobial properties of oleuropein and products of its hydrolysis from green olives. Appl. Microbiol. *26*, 777–782.

FLEMING, H.P., THOMPSON, R.L., ETCHELLS, J.L., BELL, T.A. and KELLING, R.E.   1973B.   Bloater formation in brined cucumbers fermented by *Lactobacillus plantarum*. J. Food Sci. *38*, 499–503.

FLEMING, H.P., THOMPSON, R.L., ETCHELLS, J.L., KELLING, R.E. and BELL, T.A.   1973C.   Carbon dioxide production in the fermentation of brined cucumbers. J. Food Sci. *38*, 504–506.

FODA, I.O. and VAUGHN, R.H.   1950.   Salt tolerance in the genus *Aerobacter*. Food Technol. *4*, 182–188.

FORNACHON, J.C.M., DOUGLAS, H.C. and VAUGHN, R.H.   1940.   The pH requirements of some heterofermentative species of *Lactobacillus*. J. Bacteriol. *40*, 649–655.

FRED, E.B. and PETERSON, W.H.   1922.   The production of pink sauerkraut by yeasts. J. Bacteriol. *7*, 257–269.

FRIDLEY, R.B., HARTMANN, H.T., MEHLSCHAU, J.J., CHEN, P. and WHISLER, J.   1971.   Olive harvest mechanization in California. Calif. State Agric. Exp. Stn. Bull. *885*.

GARRIDO-FERNANDEZ, A. and VAUGHN, R.H.   1978.   Utilization of oleuropein by microorganisms associated with olive fermentation. Can. J. Microbiol. *24*, 680–684.

GILILLAND, J.R. and VAUGHN, R.H.   1943.   Characteristics of butyric acid bacteria from olives. J. Bacteriol. *46*, 315–322.

HARTMANN, H.T. and OPITZ, K.W.   1977.   Olive production in California. Univ. Calif. Div. Agric. Sci. Leaflet *2474*.

HAYNE, A.P. and COLBY, G.E.   1895.   Olives. Calif. Agric. Exp. Stn. Rep. *1895*. (Appendix to report for 1894–1895)

HRDLICKA, J., CURDA, D. and PAVELKA, J.   1967.   Volatile carbonyl compounds during fermentation of cabbage. Sb. Vys. Sk. Chem. Technol. Praze Potraviny *E*(15) 51. (Original not seen—cited by Lee *et al.*, 1974)

HSU, E.J. and VAUGHN, R.H.   1969.   Production and catabolite repression of the constitutive polygalacturonic acid *trans*-eliminase of *Aeromonas liquefaciens*. J. Bacteriol. *98*, 172–181.

JONES, I.D. and ETCHELLS, J.L.   1943.   Physical and chemical changes in cucumber fermentations. Food Ind. *15*, 62–64.

JONES, I.D., ETCHELLS, J.L., VELDHUIS, M.K. and VEERHOFF, O.   1941.   Pasteurization of genuine dill pickles. Fruit Prod. J. *20* (10) 304–305, 316, 325.

JUVEN, B. and HENIS, Y.   1970.   Studies on the antimicrobial activity of olive phenolic compounds. J. Appl. Bacteriol. *33*, 721–732.

JUVEN, B., SAMISH, Z. and HENIS, Y.   1968A.   Identification of oleuropein as a natural inhibitor of lactic fermentation of green olives. Isr. J. Agric. Res. *18*, 137–138.

JUVEN, B., SAMISH, Z., HENIS, Y. and JACOBY, G. 1968B. Mechanism of enhancement of lactic acid fermentation of green olives by alkali and heat treatments. J. Appl. Bacteriol. *31*, 200–207.

KARBASSI, A. and VAUGHN, R.H. 1980. Purification and properties of polygalacturonic acid *trans-eliminase* from *Bacillus stearothermophilus*. Can. J. Microbiol. *26*, 377–384.

KAWATOMARI, T. and VAUGHN, R.H. 1956. Species of *Clostridium* associated with zapatera spoilage of olives. Food Res. *21*, 481–490.

KING, A.D., JR. and VAUGHN, R.H. 1961. Media for detecting pectolytic gram-negative bacteria associated with the softening of cucumbers, olives, and other plant tissues. J. Food Sci. *26*, 635–643.

LEE, C.Y., ACREE, T.E., BUTTS, R.M. and STAMER, J.R. 1976. Flavor constituents of fermented cabbage. Intern. Union Food Sci. Technol., 1974, Madrid. Proc. 4th Intern. Congr. Food Sci. Technol. *1*, 175–178.

LeLONG, B.M. 1890. The mission olive. Calif. State Bd. Hort. Annu. Rep. *1890*. 185–189.

LEVIN, R.E. and VAUGHN, R.H. 1966. *Desulfovibrio aestuarii*, the causative agent of hydrogen sulfide spoilage of fermenting olive brines. J. Food Sci. *31*, 768–772.

LUH, B.S. and PHAFF, H.J. 1951. Studies on polygalacturonase of certain yeasts. Arch. Biochem. Biophys. *33*, 212–227.

MARTEN, E.A., PETERSON, W.H., FRED, E.B. and VAUGHN, W.E. 1929. Relation of temperature of fermentation to quality of sauerkraut. J. Agric. Res. *39*, 285–292.

MRAK, E.M. and BONAR, L. 1939. Film yeasts from pickle brines. Zentralbl. Bakteriol. Parasitenk. Abt. 2 *100*, 289–294.

MRAK, E.M., VAUGHN, R.H., MILLER, M.W. and PHAFF, H.J. 1956. Yeasts occurring in brines during the fermentation and storage of green olives. Food Technol. *10*, 416–419.

NAGEL, C.W. and VAUGHN, R.H. 1961. The characteristics of a polygalacturonase produced by *Bacillus polymyxa*. Arch. Biochem. Biophys. *93*, 344–352.

NORTJE, B.K. and VAUGHN, R.H. 1953. The pectolytic activity of species of the genus *Bacillus*: Qualitative studies with *Bacillus subtilis* and *Bacillus pumilus* in relation to softening of olives and pickles. Food Res. *18*, 57–69.

OLIVE ADMIN. COMM. 1977–1978. California Olive Industry Statistics 1977–78. Olive Administrative Committee, Fresno, Calif.

PARMELE, H.B., FRED, E.B., PETERSON, W.H., McCONKIE, J.E. and VAUGHN, W.E. 1927. Relation of temperature to rate and type of fermentation and to quality of commercial sauerkraut. J. Agric. Res. *35*, 1021–1038.

PASTEUR, L. 1858. Account of lactic fermentation. Mem. Soc. Imp. Sci. Agric. Lille Ser. 2 *5*, 13–26. (French)

PATEL, I.B. and VAUGHN, R.H. 1973. Cellulolytic bacteria associated with sloughing spoilage of California ripe olives. Appl. Microbiol. *25*, 62–69.

PEDERSON, C.S.   1930.   Floral changes in the fermentation of sauerkraut. N.Y. Agric. Exp. Stn. Geneva Tech. Bull. *168*.

PEDERSON, C.S.   1960.   Sauerkraut. *In* Advances in Food Research, Vol. 10. C.O. Chichester, E.M. Mrak and G.F. Stewart (Editors).   Academic Press, New York.

PEDERSON, C.S.   1975.   Pickles and sauerkraut. *In* Commercial Vegetable Processing. B.S. Luh and J.G Woodroof (Editors).   AVI Publishing Co., Westport, Conn.

PEDERSON, C.S.   1979.   Microbiology of Food Fermentations, 2nd Edition. AVI Publishing Co., Westport, Conn.

PEDERSON, C.S. and ALBURY, M.N.   1950.   The effect of temperature upon bacteriological and chemical changes in fermenting cucumbers. N.Y. State Agric. Exp. Stn. Bull. *744*.

PEDERSON, C.S. and ALBURY, M.N.   1954.   The influence of salt and temperature on the microflora of sauerkraut fermentation. Food Technol. *8*, 1–5.

PEDERSON, C.S. and ALBURY, M.N.   1969.   The sauerkraut fermentation. N.Y. State Agric. Exp. Stn. Bull. *824*.

PEDERSON, C.S. and KELLY, C.D.   1938.   Development of pink color in sauerkraut. Food Res. *3*, 583–588.

PEDERSON, C.S. and WARD, L.   1949.   The effect of salt upon the bacteriological and chemical changes in fermenting cucumbers. N.Y. State Agric. Exp. Stn. Bull. *273*.

PETERSON, W.H. and FRED, E.B.   1923.   An abnormal fermentation of sauerkraut. Zentralbl. Bakteriol. Parasitenk. Abt. 2 *58*, 199–204.

PHAFF, H.J.   1947.   The production of exocellular pectic enzymes by *Penicillium chrysogenum*. I. On the formation and adaptive nature of polygalacturonase and pectinesterase. Arch. Biochem. *13*, 67–81.

PLASTOURGOS, S. and VAUGHN, R.H.   1957.   Species of *Propionibacterium* associated with zapatera spoilage of olives. Appl. Microbiol. *5*, 267–271.

RAHN, O.   1913.   Bacteriological studies on brine pickles. Canner Dried Fruit Packer *37* (20) 44; (21) 43.

ROELOFSEN, P.A.   1936.   Protopectinase-forming yeasts. Rep. 16th Meet. Assoc. Exp. Stn. Personnel, 1936, Djember, Java. (Dutch)

ROELOFSEN, P.A.   1953.   Polygalacturonase activity in yeast, *Neurospora* and tomato extract. Biochem. Biophys. Acta *10*, 410–413.

SAMISH, Z., COHEN, S. and LUDIN, A.   1966.   Method for the preservation of olives. Isr. Pat. Applic. *24907*.

SAMISH, Z., COHEN, S. and LUDIN, A.   1968.   Progress of lactic acid fermentations of green olives as affected by peel. Food Technol. *22*, 1009–1012.

SEATON, H.L., HUTSON, R. and MUNCIE, J.H.   1936.   The production of cucumbers for pickling purposes. Mich. Agric. Exp. Stn. Spec. Bull. *273*.

SIMS, W.L. and ZAHARA, M.B.   1978.   Growing pickling cucumbers for mechanical harvesting. Univ. Calif. Coop. Ext. Serv. Leafl. *2677*.

SMYTH, H.F.   1927.   A bacteriologic study of the Spanish green olive. J. Bacteriol. *13*, 56.

SORIANO, S.   1955.   Bacteria causing "zapatera" of olives and their control. 1st Natl. Conf. Olive Culture, 1955. Minist. Agric. Ganad., Buenos Aires. (Spanish)

SORIANO, S. and SORIANO, A.M. DE.   1946.   Microbiological study of olives preserved in brine. Rev. Assoc. Argent. Dietol. *4* (13) 132. (Spanish)

STAMER, J.R.   1975.   V. 1. Recent developments in the fermentation of sauerkraut. *In* Lactic Acid Bacteria in Beverages and Food. J.G. Carr, C.V. Cutting and G.C. Whiting (Editors). Academic Press, London, New York, San Francisco.

STAMER, J.R., HRAZDINA, G. and STOYLA, B.O.   1973.   Induction of red color formation in cabbage juice by *Lactobacillus brevis* and its relationship to pink sauerkraut. Appl. Microbiol. *26*, 161–166.

TABACHNICK, J. and VAUGHN, R.H.   1948.   Characteristics of tartrate-fermenting species of *Clostridium*. J. Bacteriol. *56*, 435–443.

TRACY, R.L.   1934.   Spoilage of olives by colon bacilli. J. Bacteriol. *28*, 249–265.

U.S. DEP. AGRIC.   1968.   Food Consumption, Prices, Expenditures. U.S. Dep. Agric. Econ. Res. Serv., Agric. Econ. Rep. *138*, 70, 78.

U.S. DEP. AGRIC.   1977.   Food Consumption, Prices and Expenditures. U.S. Dep. Agric. Econ. Stat. Coop. Serv., Agric. Econ. Rep. *138*, 16, 21. (supplement)

U.S. DEP. AGRIC.   1978.   Agricultural Statistics 1978. U.S. Govt. Printing Office, Washington, D.C.

U.S. DEP. AGRIC.   1979.   Agricultural Statistics, 1979. U.S. Govt. Printing Office, Washington, D.C.

VAUGHN, R.H. *et al.*   1945–1951, 1952–1954.   Unpublished data. Berkeley and Davis, Calif.

VAUGHN, R.H.   1953.   Unpublished data. Davis, Calif.

VAUGHN, R.H.   1954.   Lactic acid fermentation of cucumbers, sauerkraut, and olives. *In* Industrial Fermentations. L.A. Underkofler and R.J. Hickey (Editors).   Chemical Publishing Co., New York.

VAUGHN, R.H.   1956.   Unpublished data. Davis, Calif.

VAUGHN, R.H., DOUGLAS, H.C. and GILILLAND, J.R.   1943.   Production of Spanish-type green olives. Calif. State Agric. Exp. Stn. Bull. *678*.

VAUGHN, R.H., JAKUBCZYK, T., MacMILLAN, J.D., HIGGINS, T.E., DAVÉ, B.A. and CRAMPTON, V.M.   1969A.   Some pink yeasts associated with softening of olives. Appl. Microbiol. *18*, 771–775.

VAUGHN, R.H., KING, A.D., JR., NAGEL, C.W., NG, H., LEVIN, R.E., MacMILLAN, J.D. and YORK, G.K., II. 1969B. Gram negative bacteria associated with sloughing, a softening of California Ripe Olives. J. Food Sci. *34*, 224–227.

VAUGHN, R.H., LEVINSON, J.H., NAGEL, C.W. and KRUMPERMAN, P.H. 1954. Sources and types of aerobic microorganisms associated with the softening of fermenting cucumbers. Food Res. *19*, 494–502.

VAUGHN, R.H. and MARTIN, M.H. 1967–1968. Unpublished data. Madera, Calif.

VAUGHN, R.H. and MARTIN, M.H. 1971. Bulk fermentation of Spanish type green olives in California. Inf. Oleicoles Intern., Madrid, Oct.–Dec. 1971, *56–57*, 209–217. (French)

VAUGHN, R.H., MARTIN, M.H., STEVENSON, K.E., JOHNSON, M.G. and CRAMPTON, V.M. 1969C. Salt-free storage of olives and other produce for future processing. Food Technol. *23*, 124–126.

VAUGHN, R.H., STEVENSON, K.E., DAVÉ, B.A. and PARK, H.C. 1972. Fermenting yeasts associated with softening and gas-pocket formation in olives. Appl. Microbiol. *23*, 316–320.

VAUGHN, R.H., WON, W.D., SPENCER, F.B., PAPPAGIANIS, D., FODA, I.O. and KRUMPERMAN, P.H. 1953. *Lactobacillus plantarum* the cause of "yeast spots" on olives. Appl. Microbiol. *1*, 82–85.

VELDHUIS, M.K. and ETCHELLS, J.L. 1939. Gaseous products of cucumber fermentations. Food Res. *4*, 621–630.

WEST, N.S., GILILLAND, J.R. and VAUGHN, R.H. 1941. Characteristics of coliform bacteria from olives. J. Bacteriol. *41*, 341–352.

7

Fermented Sausage

L. Wendell Haymon

The process of fermenting sausage was probably one of the earliest forms of meat processing. Salting fresh meat and curing fresh meat by drying were man's earliest attempts at food preservation. Sausage manufacturing probably began before written history. The first mention of sausage manufacturing in written history was in the 9th century, B.C., when it was mentioned in Homer's *Odyssey*. The sausage was called *oryae*. The play *Orya* by Epicharmus about 500 B.C. mentions *oryae*. The word *salami* was probably coined from the product made in Salamis, Cyprus, a city destroyed in 449 B.C. (Pederson 1979). Sausages eaten by the Babylonians, Greeks, and Romans were no doubt fermented and dried meat products. Brested (1938) stated that Caesar's legions in Gaul consumed dry sausages. Descriptions of the process of making sausages confirm that many types of dry sausages were eaten by the Babylonians, Greeks, and Romans.

Pederson (1979) has written an excellent review of the history of sausage uses and manufacturing. The various regions of the Mediterranean developed characteristic sausages, as shown by the salamis known as Genoa, Milano, and Lombardi (Anon. 1938). The Mediterranean countries consumed a highly seasoned, non-smoked product classified as the Latin type. The Northern European countries developed a form of the Roman product, but slightly spiced, heavily smoked, moist, and higher in salt content. This product is often referred to as the Germanic type. In the colder areas of Europe sausage was made in the winter months, stored, and aged until the summer; hence it was called summer sausage. There is little doubt that these sausage products were heavily inoculated with indigenous flora prompting the growth of lactic acid bacteria, yeasts, and mold in and on the surface of the sausage. Surface growth of mold on sausage yields a unique product in several areas of the world.

Early in the 20th century, bacteria were discovered to be responsible for (1) lactic acid production and (2) nitrate reduction in sausages. The contribution of bacteria to sausage production was considered comparable to the changes brought about in the manufacture of cheeses. Jensen and Paddock

(1940) used various species of lactic acid bacteria to standardize and improve the character of sausages, and were the first to be issued a U.S. patent for sausage fermentation. They showed in their patent that several species of *Lactobacillus* could be used as starter cultures. Jensen (1942) found that the pleasant acid tangy character of Thuringer style sausage is formed by several species of *Lactobacillus* and *Leuconostoc*. He summarized his work by saying (1) chance inoculation by indigenous bacteria is not economical, and (2) processes for semi-dry sausages could be shortened.

The European process for making sausage required the use of nitrate instead of the common use of nitrite. Niinivaara (1955) isolated a culture from meat, *Micrococcus aurantiacus* (M-53), which reduces nitrate to nitrite and shows inhibition of other meat flora. A sampling of the microbial population of fermented sausage is shown in Table 7.1. The sausages were sampled from the marketplace and analyzed for lactic acid bacteria. The predominant species was *Lactobacillus plantarum*. The colonies spotted on Baird-Parker medium were typically coagulase-negative micrococci or staphylococci. Sausages containing high levels of *Pediococcus* were fermented with starter cultures.

Commercial Cultures

Deibel and Niven (1957) reported use of the bacterial species, *Pediococcus cerevisiae*, as a starter culture for the semi-dry summer sausage. *P. cerevisiae* is a true lactic acid organism belonging to the Streptococcaceae family (Deibel *et al.* 1961A) and differs substantially from other Gram-positive cocci of this family. Its optimum temperature is 43°C, and it grows well in 5–7% saline medium. Everson *et al.* (1970) stated that its classification as *P. cerevisiae* was erroneous, and it was reclassified as *P. acidilactici*.

Deibel *et al.* (1961B) showed that the ultimate aim of using a starter culture is to gain greater control of the fermentation. They developed lyophilized cultures of *P. cerevisiae* which performed satisfactorily under United States manufacturing conditions, and the use of *P. cerevisiae* was approved by the U.S. Dep. of Agriculture. This culture was first marketed under the name "Accel" (Anon. 1958). The first commercial starter culture in Europe was the "Bactoferment" (*M. aurantiacus*) after Niinivaara (1955). It was augmented by a culture called "Duploferment," a mixture of *L. plantarum* and *M. aurantiacus*, in 1966.

In 1971 Rothchild and Olsen were issued a patent describing the process of making sausage using frozen concentrates of *P. cerevisiae*. In 1976 a U.S. patent was issued to Olsen and Rothchild for a frozen, concentrated bacterial product which maintains viability at −18°C for long periods of storage. This product is sold under the trade name "Lactacel."

Figure 7.1 shows the growth of lactic acid starter culture in fermenting meats. The pH decreases faster with increasing temperature, and the optimum temperature for fermentation with *P. cerevisiae* is 43°–45°C. The action of the homofermentative organism produces an environment which

TABLE 7.1.   MICROBIOLOGICAL POPULATION OF FERMENTED SAUSAGES (LOG NUMBER)

Sausage Type	TPC[1]	APT[2] (20°C)	APT[2] (30°C)	Azide Agar	Tomato Juice Agar[2]	LBS[2,3]	TGY + Sucrose[2,3]	Baird-Parker[4]
1. Genoa	4.6435	5.4150	5.6435	4.5563	4.4914	4.6532	<1.0000	<1.0000
2. Lebanon	7.5315	7.1761	7.0792	7.4472	7.2788	6.8451	5.3222	<1.0000
3. Hard salami	7.5911	7.2304	7.1761	7.5682	3.8751	2.6990	6.3010	<1.0000
4. Genoa	7.8921	7.5051	7.5798	7.7782	7.5185	7.1761	5.6990	2.8451CN[5]
5. Hard salami	7.5563	7.4771	6.9031	7.5051	7.0792	5.5513	5.3802	3.2553CN
6. Pepperoni	6.9345	7.2788	7.2553	7.2553	5.1461	4.5051	6.2304	2.4777CN
7. Lebanon	2.4150	3.2788	3.3424	3.2788	3.3802	3.2041	<1.0000	<1.0000
8. Hard salami	4.6021	6.1461	7.0414	7.2304	6.9638	3.5315	<1.0000	<1.0000
9. Genoa	6.7324	6.6721	6.8921	7.2041	5.2304	2.3010	6.4150	<1.0000
10. Novarra	8.7761	7.9731	7.9191	7.7924	8.0000	7.9823	<1.0000	2.6990CN
11. Hard salami	5.7243	6.6628	6.5315	7.4914	6.8451	<1.0000	<1.0000	<1.0000
12. Lebanon	7.7709	7.8325	7.7324	7.8976	7.8261	7.5441	5.1761	<1.0000
13. Genoa	7.8451	7.5315	7.7243	7.7993	7.5315	4.5911	6.3424	<1.0000
14. Hard salami	5.7404	5.3423	5.5185	5.7076	6.6990	4.4771	<1.0000	<1.0000
15. Genoa	8.4472	8.5563	6.4624	8.0792	8.5319	7.9777	6.3010	<1.0000
16. Pepperoni	5.6532	3.2304	6.7853	6.7559	6.0414	4.4472	<1.0000	<1.0000
17. Genoa	7.2304	7.0414	7.1761	7.1461	7.2553	3.0000	6.0792	<1.0000
18. Hard salami	6.2788	5.5563	6.1461	6.1761	6.1139	3.5441	3.7782	<1.0000
19. Pepperoni	5.4314	4.4472	4.3802	5.9294	3.4771	<1.0000	<1.0000	<1.0000
20. Genoa	7.3222	7.1761	7.0792	7.0414	7.1139	7.1761	5.8261	2.8921CN
21. Hard salami	6.6335	6.6335	6.6021	7.1761	6.6532	6.8129	5.8062	2.6721CN
22. Hard salami	5.4150	5.2553	5.3979	5.0792	5.3617	3.3802	4.7993	2.4150CN
23. Thuringer	5.6812	4.7243	4.7324	5.3010	4.7160	5.3802	4.3424	2.4771CN

Source: Haymon (1978).

[1] Total Plate Count at 25°C for 3 days.
[2] Incubation for 3 days on All Purpose Tween medium or Lactobacillus Selective medium.
[3] Incubated at 30°C. (TGY is tryptose, glucose, yeast extract medium.)
[4] Incubated for 48 hr at 35°C.
[5] CN = Coagulase-negative.

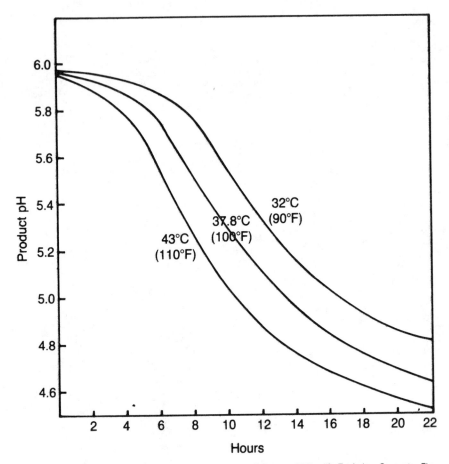

Courtesy of Microlife Technics, Sarasota, Fla.

FIG. 7.1. EFFECT OF TIME AND TEMPERATURE ON THE pH OF FERMENTING MEATS INOCULATED WITH A COMMERCIAL LACTIC ACID STARTER CULTURE

is excellent for drying (around pH 5.2 to 5.3), and the action of salt and sugars with the lactic acid produces a safe and marketable summer sausage.

Action of Sausage Organisms

The native chorizo-type sausage in the Philippines was studied by Sison (1967). He showed the presence of several strains of *Micrococcus*, but the fermentation was due to *P. cerevisiae*, either indigenous or added. *Leuconostoc mesenteroides* and a *Streptococcus* strain closely related to *S. faecalis* were isolated. *Lactobacillus brevis* was also mentioned as a potential starter culture organism.

The major activity of the lactic acid bacteria is the conversion of sugars, usually glucose or sucrose, to lactic acid by the Embden-Meyerhof pathway. Deketelaere *et al.* (1974) found that lactic acid was the major acid formed during the fermentation of carbohydrates, with minor amounts of acetic acid, generally about 10 to 1 of lactic to acetic acid. They also found small quantities of butyric and propionic acids. *Leuconostoc mesenteroides* and *Lactobacillus brevis* convert slightly less than 50% of the sugar fermented to lactic acid, and a similar amount to ethanol, acetic acid, and carbon dioxide (Rogosa 1974). *Streptococcus lactis* subspecies *diacetylactis* produces diacetyl and acetoin which impart a nutty flavor and aroma to some sausages.

Palumbo and Smith (1977) have shown that micrococci dominate the surface of the stuffed sausage during the early stages of the fermentation. These micrococci are killed at pH 5.5 and are not found in the sausages after heat treatments of drying.

The major functions of the micrococci during the fermentation are the reduction of nitrate to nitrite and the production of catalase. Lactic acid bacteria rarely reduce nitrate to nitrite. Staphylococci actively reduce nitrate to nitrite, particularly *S. aureus, S. xylosus,* and *S. epidermidis* (Haymon 1978). Micrococci are indigenous to all animal species and each species carries a specific flora of micrococci. Excessive nitrite levels in sausage have been noted when using high nitrate concentrations with micrococci, producing a defect in the sausage known as "nitrite burn" (Deibel and Evans 1957; Bacus and Deibel 1972). The micrococci are also lipolytic, and they produce lipase during the early stages of the fermentation. Cantoni *et al.* (1967) confirmed the action of micrococci on pork fat. There was a dramatic increase in free fatty acids, volatile fatty acids, and carbonyl compounds after 28 days of drying. Tjaberg *et al.* (1969) had shown that lactic acid bacteria produce varying amounts of hydrogen peroxide. The micrococci produce catalase which effectively destroys the hydrogen peroxide produced at the surface of the sausage. Haymon and Acton (1978) have summarized the role of micrococci in the flavor development of fermented sausages.

Demeyer *et al.* (1974) showed increases in free fatty acids and carbonyl compounds in Belgian salami during drying. Triglycerides were partially degraded to free fatty acids and the unsaturated free fatty acids were degraded to carbonyl compounds. The hydrolytic activity was attributed to lipase action of micrococci.

Inhibition of Staphylococci

Another function of the action of lactic acid bacteria in the production of sausages is the inhibition of *Staphylococcus aureus*. In the last 15 years several incidences of food poisoning due to the consumption of fermented meats were reported to the National Center for Disease Control (Anon.

1971A,B,C, 1975). The inhibition or suppression of *Staphylococcus aureus* growth also suppresses Enterotoxin A production.

Genigeorgis (1976) has reviewed the competitive aspects of staphylococci with other bacteria in foods. Streptococci and *P. cerevisiae* were the most inhibitory to growth of staphylococci and enterotoxin production. *Leuconostoc citrovorum* and lactobacilli do not inhibit growth and only slightly inhibit production of Enterotoxin B (Haines and Harmon 1973).

The inhibition of staphylococci in sausage is more pronounced as the ratio of lactics to staphylococci increases and as the temperature of fermentation decreases. Higher temperatures (38.9°C is optimum for *S. aureus*) and brine concentration tend to favor the growth of mesophilic, salt tolerant staphylococci (Peterson *et al.* 1964). The beneficial effect of lactic acid starter cultures in inhibiting or suppressing staphylococcal growth and enterotoxin production in fermented sausages has been demonstrated (Barbes and Deibel 1972; Daly *et al.* 1973; Metaxopoulas 1976; Niskanen and Nurmi 1976).

Haymon and Gryczka (1978) have shown that *P. cerevisiae* limits staphylococcal growth to less than 3 log cycles using 18 hr old broth cultures of *S. aureus*. Indigenous staphylococci in sausage were limited to less than 2 log cycles of growth at optimum temperatures by *P. cerevisiae* and *Micrococcus varians*.

Economics

The traditional process of fermenting sausages consisted of grinding the meats in some fashion, mixing in sodium chloride and sodium or potassium nitrate, blending in spices and seasonings, and transferring this mixture to curing pans. The meat mixture was tightly packed in layers about 15 to 20 cm (6 to 8 in.) deep and held at 40°C for 48 to 72 hr. During this curing period, nitrate reducing bacteria converted some of the nitrate to nitrite, a substance required for the red cured meat pigmentation reaction. After stuffing, the sausages were held in a "green" room for 12 to 48 hr at 10°–15°C. Following the greening the product was moved either to a smoking pit or to a smokehouse for heating and smoking. For dry sausages the green room was followed by drying in a room at ambient temperatures for up to 120 days.

This process for summer sausages has been reduced to 12–24 hr of total processing time since the introduction of starter cultures of *P. cerevisiae*. Dry sausages are cured or ripened in 24–48 hr with starter cultures instead of the 7–14 days by the traditional process. Because the sausages are acidified to near their isoelectric point, where they have their lowest solubility, they can be dried more easily at that pH. Drying times for dry sausages such as pepperoni and Genoa have decreased to 25–40 days depending on the caliber of the sausage and on trichinae certification procedures.

SUMMARY

The total microbiological population of fermented sausage has changed from chance or random contamination to one of safe and economical production schedules. The fermented sausages in the supermarket are safe from botulism poison and staphylococcus food poisoning, and are wholesome, well defined, processed meat products.

REFERENCES

ANON. 1938. Foreign style sausage. *In* Sausage and Meat Specifications, The Packer's Encyclopedia, Part 3. National Provisioner, Chicago.

ANON. 1958. Technical service Accel for the production of thuringer, summer sausage, cervelat, Lebanon bologna, and pork roll. Merck & Co., Rahway, N.J.

ANON. 1971A. Gastroenteritis associated with salami. Morbid. Mortal. Weekly *20*, 253, 258.

ANON. 1971B. Gastroenteritis associated with Genoa salami. Morbid. Mortal. Weekly *20*, 261–266.

ANON. 1971C. Gastroenteritis attributed to Hormel San Remo Stick Genoa salami. Morbid. Mortal. Weekly *20*, 370.

ANON. 1975. Staphylococcal food poisoning associated with Italian dry salami. Morbid. Mortal. Weekly *24*, 374, 379.

BACUS, J.N. and DEIBEL, R.H. 1972. "Nitrite burn" in fermented sausage. J. Appl. Microbiol. *24*, 405–408.

BARBES, L.E. and DEIBEL, R.H. 1972. Effect of pH and oxygen tension on staphylococcal growth and enterotoxin formation in fermented sausage. Appl. Microbiol. *24*, 891–898.

BRESTED, J.H. 1938. The Conquest of Civilization. Harper and Row, New York.

CANTONI, C., MOLNAR, M.R., RENON, P. and GIOLITTI, G. 1967. Micrococci in pork fat. J. Appl. Bacteriol. *30*, 190–196.

DALY, C., LA CHANCE, M., SANDINE, W.E. and ELLIKER, P.R. 1973. Control of *Staphylococcus aureus* in sausage by starter cultures and chemical acidulation. J. Food Sci. *38*, 426–430.

DEIBEL, R.H. and EVANS, J.B. 1957. "Nitrite burn" in cured meat products, particularly in fermented sausages. Am. Meat Inst. Found. Bull. *32*.

DEIBEL, R.H. and NIVEN, C.F., JR. 1957. *Pediococcus cerevisiae*: a starter culture for summer sausages. Bacteriol. Proc. Gen. Meet., Soc. Am. Bacteriol., 1957, Detroit.

DEIBEL, R.H., NIVEN, C.F., JR. and WILSON, G.D. 1961A. Microbiology of meat curing. III. Some microbiological and related technological aspects in the manufacture of fermented sausages. Appl. Microbiol. *9*, 156–161.

DEIBEL, R.H., WILSON, G.D. and NIVEN, C.F., JR.   1961B.   Microbiology of meat curing. IV. A lyophilized *Pediococcus cerevisiae* starter culture for fermented sausage. Appl. Microbiol. *9*, 239–243.

DEKETELAERE, A., DEMEYER, D., VANDERCKHOVE, P. and VERVAEKE, I.   1974.   Stoichiometry of carbohydrate fermentation during dry sausage ripening. J. Food Sci. *39*, 297–300.

DEMEYER, D., HOOZEE, J. and MESDOM, H.   1974.   Specificity of lipolysis during sausage ripening. J. Food Sci. *30*, 293–296.

EVERSON, C.W., DANNER, W.E. and HAMMES, P.A.   1970.   Bacterial starter cultures in sausage products. J. Agric. Food Chem. *18*, 570–571.

GENIGEORGIS, C.A.   1976.   Quality control for fermented meats. J. Am. Vet. Med. Assoc. *169*, 1220–1228.

HAINES, W.C. and HARMON, L.C.   1973.   Effect of selected lactic acid bacteria growth on *S. aureus* and production of enterotoxin. Appl. Microbiol. *25*, 436–441.

HAYMON, L.W.   1978.   Unpublished data. Microlife Technics, Sarasota, Fla.

HAYMON, L.W. and ACTON, J.C.   1978.   Flavors from lipids by microbiological action. *In* Lipids as a Source of Flavor. M.K. Supran (Editor). Am. Chem. Soc. Symp. Ser. *75*, Washington, D.C.

HAYMON, L.W. and GRYCZKA, A.J.   1978.   Inhibition of *Staphylococcus aureus* and enterotoxin A production in fully dried sausages using frozen commercial starter cultures. Unpublished data. Sarasota, Fla.

JENSEN, L.B.   1942.   Microbiology of Meats. Garrard Press, Champaign, Ill.

JENSEN, L.B. and PADDOCK, L.   1940.   Sausage treatment with *Lactobacilli*. U.S. Pat. 2,225,783. Dec. 24.

METAXOPOULAS, J.   1976.   Effect of processing parameters of commercially manufactured Italian style dry sausage on the growth of *Staphylococcus aureus*. Ph.D. Thesis. Univ. California, Davis.

NIINIVAARA, F.P.   1955.   The influence of pure cultures of bacteria on the maturing and reddening of raw sausage. Acta Agral. Fennica *85*, 95–101. (German)

NISKANEN, A. and NURMI, E.   1976.   Effect of starter culture on staphylococcal enterotoxin and thermonuclease production in dry sausage. Appl. Microbiol. *31*, 11–20.

OLSEN, R.H. and ROTHCHILD, H.   1976.   Bacterial product useful for making sausage. U.S. Pat. 3,960,664. June 1.

PALUMBO, S.A. and SMITH, J.L.   1977.   Chemical and microbiological changes during sausage fermentation and ripening. *In* Enzymes in Food and Beverage Processing. R.L. Ory and A.J. St. Angelo (Editors). Am. Chem. Soc. Symp. Ser. *47*, Washington, D.C.

PEDERSON, C.S.   1979.   Fermented sausage. *In* Microbiology of Food Fermentations, 2nd Edition.   AVI Publishing Co., Westport, Conn.

PETERSON, A.C., BLACK, J.J. and GUNDERSON, M.F.   1964.   Staphylococci in competition. III. Influence of pH and salt on staphylococcal growth in mixed populations. Appl. Microbiol. *12*, 70–76.

ROGOSA, M. 1974. Part 16; Family I, Genus I, Lactobacillus. *In* Bergey's Manual of Determinative Bacteriology, 8th Edition. R.E. Buchanan and N.E. Gibbons (Editors). Williams & Wilkins Co., Baltimore.

ROTHCHILD, H. and OLSEN, R.H. 1971. Process for making sausage. U.S. Pat. 3,561,977. Feb. 9.

SISON, E.C. 1967. Microbiology and technology of native fermented sausage. M.A. Thesis. Univ. Philippines Coll. Agric., Laguna.

TJABERG, T.B., HANGAM, M. and NURMI, E. 1969. Studies on discoloration of Norwegian salami sausage. Proc. 15th Meet. Eur. Meat Res. Workers, Aug. 17–24, 1969, Helsinki.

8

Bakery Foods[1]

J.G. Ponte, Jr.
and Gerald Reed

The history of bread baking which has been briefly mentioned in the chapter on the production of bakers' yeast will not be treated. The reader is referred to the excellent books by Jacob (1944) or Darby *et al.* (1977) as well as to the references given at the end of this chapter.

Baked foods are produced and consumed in most of the countries of the world. Considerable variation exists in the type of baked foods that are made from country to country, and often among regions within a given country. Some of the ways in which bread can vary are summarized as follows:

Leavening. Breads may be unleavened, such as chappatti; leavened with yeast, such as pan breads; chemically leavened, such as Irish soda bread; or leavened with bacteria, such as salt rising bread.

Formulation. Many breads around the world are made with lean formulas comprising flour,water, yeast, and salt, while some breads, such as certain premium breads made in the United States, also contain considerable amounts of other ingredients such as fat, eggs, and milk.

Shape and Size. A virtually endless variety of shapes and sizes have evolved; some breads are baked in pans while many are baked on the oven hearth.

Specific Volume. This factor can vary from the low-density, high-volume white pan breads made in North America to the dense, dark breads of Central Europe.

---

[1]Contribution No. 80-56-B, Dep. of Grain Science and Industry, Kansas Agric. Exp. Stn., Manhattan, KS 66506.

Crust Characteristics. Some breads, such as Vienna loaves, have thin crispy crusts, while others, such as German pumpernickel breads, have thick crusts. Colors range from quite light to dark colors.

Crumb Characteristics. Continuous mix breads in the United States have a very fine, uniform crumb structure, while other breads such as French baguettes have coarse, irregular structures. Some Middle East breads have little crumb structure at all, being comprised largely of crust because of the high baking temperatures that are used.

Means of producing these baked foods also vary considerably. Large, highly-mechanized bakeries account for much of the production in developed countries. In less developed countries, bakeries are typically quite small, and operations are conducted with the help of little, or at times no, mechanical equipment.

A common denominator in many of these countries is that baked foods, bread in particular, has traditionally been an important factor in human nutrition. In some countries, bread consumption accounts for a large part of the diet.

Aykroyd and Doughty (1970) compared the consumption of wheat in many countries. Part of their data are summarized in Table 8.1.

TABLE 8.1. CONSUMPTION OF WHEAT AS PERCENTAGE OF TOTAL CALORIES IN DIFFERENT COUNTRIES

Country	Total Calories from Wheat %	Country	Total Calories from Wheat %
Turkey	58	United Kingdom	22
Afghanistan	55	Finland	20
Jordan	50	Canada	19
Greece	48	Germany, Federal Republic	18
Chile	44	United States	17
Iraq	42	Peru	16
Romania	38	Costa Rica	12
Israel	37	Japan	10
France	30	Brazil	9
Ireland	29	Ceylon	9
Uruguay	28	Philippines	5
Switzerland	24	Uganda	1

Source: Aykroyd and Doughty (1970).

These figures indicate that in some countries over half of the total calories consumed come from wheat. Many of the developed countries appear to consume roughly 17 to 30% of their calories from wheat.

The trend in consumption of wheat products in developed countries is illustrated in Fig. 8.1. The general trend is downward, but exceptions include Italy and Japan.

While wheat is consumed in many forms (gruels, pasta products, stews, etc.) much of the wheat for human consumption is processed into flour which is then made into baked foods. This is especially true in the developed

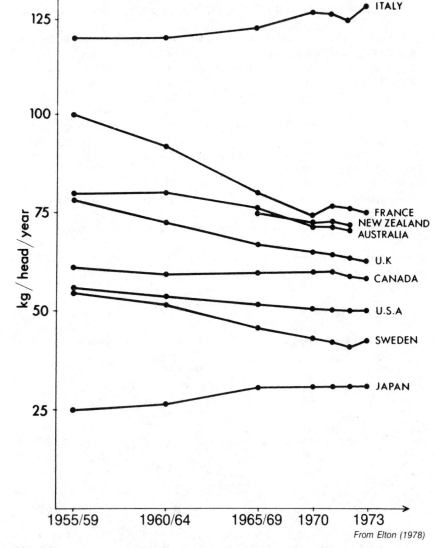

FIG. 8.1.   CONSUMPTION OF WHEAT BY HUMANS IN SOME TYPICAL DEVELOPED COUNTRIES

countries. For example, in the United Kingdom during 1974–1975, about 5 MT of wheat are milled yearly to give 3.6 MT of flour (Elton 1978); the by-products such as bran and wheat germ are utilized as animal feed. Of this white flour, 67% is used in commercial breadmaking, 11% in biscuit manufacture, 8% for household use, and 14% for miscellaneous usage.

Developed countries as well as many less developed countries have well organized baking industries that today produce much of the baked foods that are consumed. Data on the size and scope of the baking industry for one country, the United States, will be summarized here.

Table 8.2 indicates that over $14 billion worth of baked foods were produced in the United States during 1978. A total of about 240,000 workers were employed to produce these foods.

TABLE 8.2. UNITED STATES BAKERY FOOD SALES
(In millions of dollars)

	1977	1978	1979[1]
Industry (SIC2051): Bread, Cake and Related Products			
Value of shipments	10,400	11,076	12,006
Value added	5,563	5,874	—
Total employment (000)	201	201	—
Production workers (000)	108	108	—
Industry (SIC2052): Cookies and Crackers			
Value of shipments	2,935	3,170	3,450
Value added	1,639	1,714	—
Total employment (000)	39	40	—
Production workers (000)	32	32	—

Source: Anon. (1979).
[1]Forecast

The distribution of baked foods is shown in Table 8.3. Roughly 52% of the value of all baked foods is made up of bread and rolls. White bread dominates the total bread market, accounting for 66% of all the bread and rolls

TABLE 8.3. DISTRIBUTION OF BAKED FOODS, UNITED STATES, 1976

Type of Product	U.S. Value $
White bread	4,489,160
Other bread	1,396,170
Rolls, bread type	878,570
Total bread, rolls	6,763,900
Crackers, biscuits, cookies	2,550,660
Pretzels	215,690
Total crackers, cookies	2,766,350
Cakes and pastry	1,661,100
Toaster pastries	72,960
Doughnuts	774,930
Pies	428,260
Total sweet goods	2,937,250
Frozen cake, pies, pastries	508,310
Total Bakery Foods, 1976	12,975,810

Source: Anon. (1978A).

sold. The category of "other breads" is made up of wheat, rye, etc., and makes up about 21% of the total bread and roll market. However, it may be noted that the other, or variety, bread market is expanding substantially, in part due to expanded consumer interest in "natural" foods. It is expected that by 1980 about 30% of the bread market will be taken up by variety breads (Mrdza 1978).

As with other major industries, there is a trend toward fewer but larger manufacturing establishments in the United States baking industry. From 1958 to 1972 there was a drop from about 6000 to about 3300 bakeries employing 20 or more persons (Anon. 1975). Most of the baked products are made in large wholesale bakeries; there were approximately 1100 wholesale plants having 50 or more employees in 1977. During that year, the largest 20 companies had sales of $4.7 billion of baked foods made in 342 plants; 5 companies had sales of $2.7 billion for products made in 207 plants (Anon. 1978B).

As previously noted, cereal products, largely in the form of bread, have an important role in the diet of many countries, including developed countries, although some of these are experiencing a slow decline in cereal consumption.

Senti (1971) summarized the contribution of grain products to the United States diet, using as his basis reports from over 7500 households in 1965. Figure 8.2 shows that grain products provided 40% of thiamin intake, 31% of the iron, 26% of the food energy, 20% of the protein, and 19% of the riboflavin. The more important types of grain products consumed in the United States are given in Fig. 8.3 from a study in 1965. These values indicate that bread and other bakery products comprise a large part of the grain products eaten. Some differences in consumption are noted between northern and southern households. Consumers in the South appear to purchase somewhat less bread, but more of other bakery products, flour, and mixes, and cornmeal and grits compared to northern households.

Enrichment plays a major role in the nutrient contribution of bread and other cereal foods. During the early 1940s, the baking and milling industries in cooperation with the medical profession and government agencies voluntarily adopted an enrichment program to restore B vitamins and iron to bread. The enrichment program has continued to the present, and it is generally believed that enrichment has had a positive impact on the well-being of the country. Today virtually no incidence of beriberi or pellagra can be found, in marked contrast to the situation prior to initiation of enrichment. A major part of this improvement is credited to the enrichment program.

Present United States enrichment levels for 0.45 kg (1 lb) bread call for 1.8 mg thiamin, 1.1 mg riboflavin, 15.0 mg niacin, and 12.5 mg iron. Calcium may optionally be added to a level of 600 mg per 0.45 kg.

Enrichment in other countries varies considerably (Aykroyd and Doughty 1970). Canada, Chile, El Salvador, and Mexico practice enrichment. In Europe, the United Kingdom and Sweden and Denmark enrich bread, but other countries do not; France essentially prohibits the practice. Japan enriches flour for its national school lunch program, and also markets "special enriched" wheat products.

GRAIN PRODUCTS

AS PART OF FOOD DOLLAR AND SOURCE OF NUTRIENTS

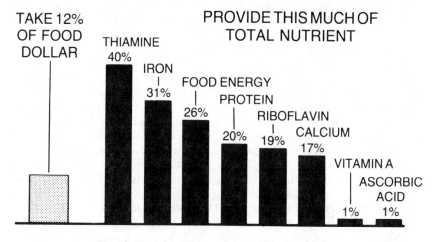

U.S. HOUSEHOLDS, 1 WEEK IN SPRING, 1965

U.S. DEPARTMENT OF AGRICULTURE

NEG. ARS. 5943-69 (4)   AGRICULTURAL RESEARCH SERVICE

From Senti (1971)

FIG. 8.2.   ECONOMY OF GRAIN PRODUCTS AS A SOURCE OF NUTRIENTS

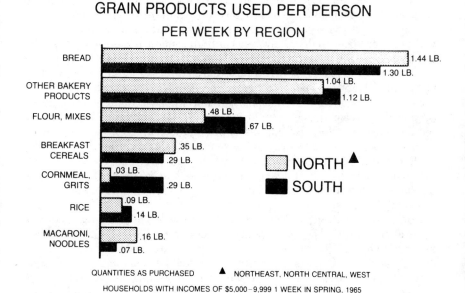

From Senti (1971)

FIG. 8.3.   CONSUMPTION OF VARIOUS TYPES OF GRAIN PRODUCTS BY REGION IN SPRING OF 1965

YEAST COMPARED TO OTHER LEAVENING SYSTEMS

The porous structure of baked foods is due to the leavening action of a gas, carbon dioxide (CO_2), which is evolved either by fermentation or by the reaction of inorganic chemicals. It should be noted, however, that this CO_2 does not generate new gas cells in dough. Rather, the CO_2 derived from leavening diffuses into cells already present in the dough system; these primary cells arise from air bubbles occluded during dough mixing or from air adsorbed on flour particles (Baker and Mize 1941).

Yeast fermentation under the anaerobic conditions of the dough forms 2 moles of ethanol and 2 moles of CO_2 from 1 mole of glucose. This means that fermentation of 180 mg of glucose (1 mMole) results in the formation of 88 mg of CO_2 (2mMole) which occupies a volume of about 480 ml at atmospheric pressure. The metabolic reactions of yeast are relatively slow in comparison with chemical reactions. One can calculate that a dough containing 3% compressed yeast will produce enough carbon dioxide gas during 1 hr of proofing to yield bread of specific volume of 7. This assumes that all of the evolved gas is retained in the dough and a good oven spring.

Chemical leavening is carried out with baking powder, a mixture of sodium bicarbonate and acid salts (usually phosphates or pyrophosphates); 336 mg of $NaHCO_3$ (2 mMole) produce 88 mg of CO_2 (2 mMole) which occupy a volume of 480 ml at atmospheric pressure. The reaction is almost instantaneous and has to be slowed down by coating the inorganic acid salts with fat. The coating is chosen in such a manner that a part of the reaction takes place on the bench and most of it in the oven.

Baked goods produced by fermentation with yeast are usually preferred because of the desirable flavor which results from the reaction of fermentation by-products with other dough ingredients. But the leavening power of yeasts is very poor in dough systems of high osmotic pressure, that is with low moisture or with high salt or sugar content. Therefore, breads, rolls, and buns are usually leavened with yeast, and cakes, cookies, and biscuits with chemical leavening. There are some borderline areas in which either of these two leavening systems is used. For instance, doughnuts may be produced as "yeast-raised" doughnuts or as "cake doughnuts" (leavened by baking powder). This also applies to pizza crusts which may be produced with either leavening system, or to some coffee cakes. Flat dough pieces such as pancakes are leavened with baking powder.

In some instances, both leavening systems may be used. This applies to products such as doughnuts or pizza crusts which have already been mentioned as borderline cases.

Not all yeast-raised products are baked in ovens. The yeast-raised dumplings of central Europe, the "steamed" yeast raised bread of the Chinese, and English muffins (grilled on a hot surface) are some exceptions.

With very flat doughs, leavening may also be obtained by the steam generated in the doughs during baking. This is true of soda crackers, which are laminated during processing, and of some of the flat breads of the Mideast.

METHODS OF BREAD PRODUCTION

The type of bread most widely produced and consumed in the United States today is white pan bread. This bread is made by a number of procedures, but the most common methods utilized are: (1) sponge dough, (2) straight dough, (3) continuous mix, and (4) liquid ferment.

Sponge Dough

The sponge dough process was introduced during the 1920s and is today the predominant breadmaking method used by the baking industry. The popularity of the sponge dough method is due, in part, to the greater processing tolerance of the procedure and to the flavor of the bread, generally regarded as superior compared to breads made by other methods.

A schematic diagram of the sponge dough process is illustrated in Fig. 8.4, and Table 8.4 summarizes a typical formula used in white pan bread production.

The "sponge," comprising about 65% of the total flour plus a portion of the total dough water, yeast, and "yeast food," is first mixed. This mixing period is relatively brief, and merely aims at uniformly combining the sponge ingredients. The sponge is then discharged into a trough, where it will undergo a fermentation period of some 4.5 hr in a controlled environment.

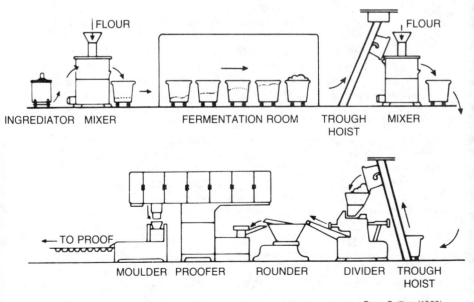

From Seiling (1969)

FIG. 8.4.   SCHEMATIC DIAGRAM OF SPONGE DOUGH PROCESS

TABLE 8.4.   WHITE PAN BREAD FORMULATION—SPONGE AND STRAIGHT DOUGH

	Sponge Dough[1]		Straight Dough[1]
	Sponge	Dough	
Flour	65.0	35.0	100
Water (variable)	40.0	25.0	65
Yeast	2.5	—	3.0
Yeast food	0.2–0.5	—	0.2–0.5
Salt	—	2.25	2.25
Sweetener (solids basis)	—	8–10	8–10
Fat	—	3.0	3.0
Dairy product	—	0.5–3.0	0.5–3.0
Crumb softener	—	0.2–0.5	0.2–0.5
Rope and mold inhibitor	—	0.125	0.125
Dough improver	—	0–0.5	0–0.5
Enrichment	—	as needed	as needed

Source: Cotton and Ponte (1974).
[1]Ingredients based on 100 parts flour.

From a starting temperature of about 25°C, the final temperature will increase by approximately 6°C due to the exothermic reactions brought about by yeast activity. Sponge volume will also increase by a factor of 4 or 5 as a consequence of carbon dioxide production during fermentation. Complex biochemical and physical changes take place during the fermentation period and these will be discussed in subsequent sections of this chapter.

At the end of sponge fermentation, the sponge is transferred into a dough mixer. The balance of the flour is also placed into the mixer, along with the water and remaining ingredients. The mixer is operated first slowly to incorporate and blend these components, then the mixer is speeded up (typically the mixer arms rotate at about 72 rpm) until the dough is completely mixed and properly "developed." At this point, the dough has been transformed from a sticky, wet-appearing mixture into a smooth, cohesive dough, characterized by a glossy sheen. This change occurs because of the unique properties of wheat flour. Upon the addition of water and the input of energy, wheat proteins and lipids form gluten. Gluten comprises the continuous phase of dough, and possesses film-forming and gas-retaining properties. As Saccharomyces cerevisiae evolves carbon dioxide, this gas diffuses into previously-formed gas vesicles; the dough, due to the unique nature of gluten, is able to retain this gas and is thereby leavened.

The mixed dough is placed in troughs and allowed to rest for 20 to 30 min. During this period, the dough recovers from mechanical stress; it relaxes, and is better able to undergo the remaining processing stages.

Dividing the dough is the next stage. During this stage, the dough is cut into pieces of desired weight by a machine that volumetrically divides the pieces and discharges them onto a moving belt. The dough pieces are conveyed to a rounder, where the rough-appearing pieces are forced along a metal sleeve such that the pieces become rounded and have a smooth, dry skin. In this condition, the dough pieces retain more carbon dioxide and are less sticky.

The dividing and rounding operations impose a certain amount of stress to the dough pieces, with the result that they are somewhat degassed and unpliable. To compensate for this effect, the dough pieces upon leaving the rounder are accorded another rest period, or intermediate "proof," of some 8 to 12 min. This takes place in tray-type conveyors enclosed within cabinets of different configurations.

From the intermediate proofer, the dough pieces are conveyed to molding machines, which transform the more-or-less round pieces of dough into cylinders. Molders perform their functions with a series of rollers which sequentially squeeze the dough piece into a sheet, curl the sheet into a cylinder, and finally roll and seal the cylinder. Automatic molders feed the dough cylinders into bread pans.

Pans containing the dough pieces are placed in fermentation units called proof boxes for the last fermentation period prior to baking. The environment in these units is typically maintained at 35° to 43°C at a relative humidity of 80 to 95%. The dough pieces expand in the pans to a desired volume, a process usually requiring approximately 60 min. The proofed loaves are then placed in an oven for baking. Gas within the dough fabric expands and the "oven spring" is produced. Steam and alcohol vapors also contribute to this expansion. Enzymes are active until the bread reaches about 75°C. At this temperature the starch gelatinizes and the dough structure is "set." When the bread surface temperature reaches 130° to 140°C, sugars and soluble proteins react chemically to produce an attractive crust color. The center of the loaf does not exceed 100°C (Ponte *et al.* 1963).

Remaining stages in the breadmaking process include cooling of the baked bread, slicing, wrapping, and distribution to stores for sale to the consumer.

Straight Dough

All of the ingredients for the straight dough (see Table 8.4) are combined and mixed in one operation. After mixing, the straight dough is given a bulk fermentation period of 2 to 4 hr and is then processed in a manner similar to that described for the doughs in the sponge dough process.

Straight doughs require less labor, time, and equipment compared to sponge doughs, but, as noted earlier, exhibit less tolerance to processing variations. This factor, plus the characteristically blander flavor of straight dough bread, has led over the years to the commercial preference of sponge doughs over straight doughs. Today, straight doughs are used mostly by smaller bakeries or for specialty bread production by larger bakers.

Continuous Mix

Continuous breadmaking technology was introduced in the United States during the 1950s. In common with other industries, the baking industry sought to lower production costs and increase product uniformity by means of continuous processing. The Do-Maker process (Baker 1954) and the Amflow method (Anon. 1958) are the continuous mix processes used in the United States.

Figure 8.5 and Table 8.5 show, respectively, a flow diagram and a formula for the Amflow procedure. This method and the Do-Maker method share essential features.

Initially a ferment or brew (stage 1) is made (with or without some flour in the case of the Do-Maker), and this is fermented in a holding tank for about 1 hr. A second stage (water, salt, and sugar) is added to the system, and fermentation continues for a total of about 2.5 hr. At the end of fermentation the brew is pumped to a premixer; cooling of the brew is achieved by means

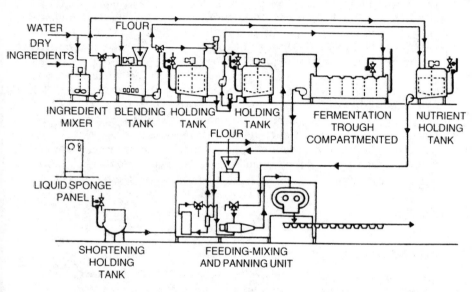

From Seiling (1969)

FIG. 8.5. SCHEMATIC DIAGRAM OF CONTINUOUS MIX BREADMAKING (AMFLOW METHOD)

TABLE 8.5. CONTINUOUS-MIX PROCESS FORMULA FOR FLOUR BREW

Formula[1]		Stage 1[1]	Stage 2[1]	Dough Stage[1]
100.0	Flour	30	—	70
67.0	Water	56	4.0	7.0
3.0	Yeast	3.0	—	—
0.5	Yeast food	0.5	—	—
2.0	Salt	—	2.0	—
6.0	Sugar	—	1.0	5.0
3.0	Milk	—	—	3.0
0.1	MCP	0.1	—	—
0.1	Mold inhibitor	—	—	0.1
3.0	Shortening	—	—	3.0
	Oxidation/variable			

Source: Trum (1964).
[1]Ingredients based on 100 parts flour.

of a heat exchanger. The remaining flour and water, sugar, melted fat, oxidant, and other optional ingredients are also fed into the premixer. After the premixer combines all of these ingredients, the loosely mixed mass is pumped to a developing chamber. This chamber contains 2 counter rotating arc impellers. As the dough passes through the chamber it is under pressure and is subjected to intense mechanical energy. Under these conditions the dough is quickly developed and is then extruded into pans from a slit in the bottom of the chamber by a system of intermittently actuated knives. The dough from the continuous mix developer is much softer and warmer (about 38°C) compared to sponge doughs. The dough pieces are proofed and baked in a manner similar to that of sponge doughs.

Continuous mix breads differ from "conventional" breads. They have a much finer, more fragile crumb structure. These factors coupled with a flavor that some consumers feel is less desirable than that of sponge dough bread have limited the acceptance of continuous mix bread in some markets.

Liquid Pre-ferment

Procedures for making bread employing liquid or semi-liquid ferments have been known for many years. Advances in equipment design stemming from continuous mix technology made the utilization of liquid ferment breadmaking systems more feasible. Such systems combine standard equipment to mix and process doughs with pumps and tanks to handle liquid brews. Advantages often cited for liquid ferment systems compared to traditional breadmaking procedures include (1) savings in plant space, (2) labor savings, (3) more processing flexibility, and (4) improved sanitation.

A flow diagram and representative formula for a liquid ferment process are shown in Fig. 8.6 and Table 8.6, respectively.

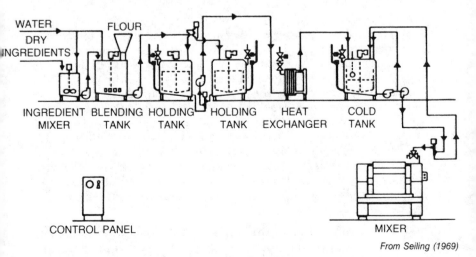

From Seiling (1969)

FIG. 8.6.   SCHEMATIC DIAGRAM OF LIQUID-FERMENT PROCESS

TABLE 8.6.   BREAD FORMULA MADE WITH FLOUR-CONTAINING FERMENT

Ingredient	Formula[1]	Liquid Sponge Formula[1]	Materials Added at Mix[1]
Flour	100	50.02	49.98
Water	66	55.77	10.23
Yeast	3	3	
Yeast food	0.625	0.625	
Salt	2.25	0.5	1.75
Sugar	8	1.5	6.5
Milk	3		
Mold inhibitor	0.125	0.125	
Fat	3		3
Emulsifier	0.5		0.5
Total	186.500	111.540	74.960

Source: Euverard (1967).
[1]Ingredients based on 100 parts flour.

In this system, a liquid ferment or brew is prepared and allowed to undergo fermentation much as described for continuous mix operations. After fermentation, the brew is pumped to a dough mixer (same type as used in conventional sponge dough processing), along with remaining ingredients. The dough is mixed and then subjected to the same processing steps as outlined for the dough in the sponge dough process.

FUNCTIONS OF YEAST IN BREADMAKING

Yeast obviously has an irreplaceable role in breadmaking. At least 3 major functions of yeast in breadmaking can be recognized: (1) leavening, (2) flavor development, and (3) dough maturing.

Leavening

Yeast gives off carbon dioxide gas due to its metabolism of simple sugars. In breadmaking, *Saccharomyces cerevisiae* utilizes sugars derived from flour, and/or sugars added as a component of the dough formula.

During fermentation in the sponge of the traditional sponge dough process, the only source of fermentable carbohydrate for the yeast is that obtained from the flour. Early in the fermentation carbon dioxide is evolved rapidly as *S. cerevisiae* utilizes the available free sugars from the flour (Cooper and Reed 1968). A sharp drop-off in the rate of carbon dioxide production occurs after about 1 hr or so when the free sugars are depleted. At this point the yeast adapts to the fermentation of maltose, and gas production again rises. Maltose becomes available as a result of amylase hydrolysis of a part of the flour starch that has been mechanically damaged during milling. At the end of about 3 hr this source of fermentable sugar is exhausted and gas production drops off.

Added Sweetener.—During the dough stage of sponge dough processing and in straight doughs or other breadmaking systems where all ingredients are present, *S. cerevisiae* will utilize sweetener added to the dough as an ingredient. Glucose in commercial dextrose, and glucose and fructose contained in some corn sweetener systems will be directly utilized by yeast. Added sucrose is almost immediately hydrolyzed into the constituent monosaccharides glucose and fructose due to the action of invertase. The yeast will then ferment both simple sugars, but at differing rates; glucose is preferred and is fermented at a faster rate than is fructose. Bread made with added sucrose typically will contain more fructose as a residual sugar than glucose. The relative rates of utilization of fructose, glucose, and maltose in dough are shown in Fig. 8.7.

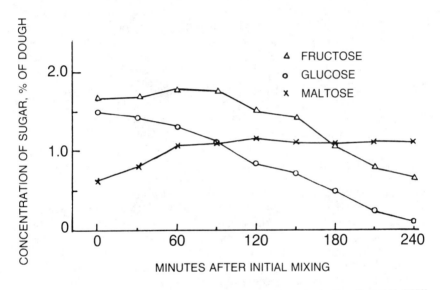

From Tang et al. (1972)

FIG. 8.7. CHANGES IN SUGAR CONCENTRATIONS IN FERMENTING DOUGH

Added Amylases.—Amylases are known to randomly split starch molecules into smaller units of varying size (α-amylase) and progressively release maltose from the terminal portion of starch molecules (β-amylase). These enzymes are important in providing fermentable sugars for yeast growth and gas production, and to improve handling and other properties of doughs. Amylase activity enhances the external surface color, volume, and keeping quality of bakery products.

Before wheat was harvested by heavily mechanized methods, uncontrolled germination of wheat and barley during wet crop years led to the accumulation of high levels of α-amylase. Ungerminated wheat and barley contain low levels of α-amylase. In contrast, β-amylase activity is substantial and changes little upon seed germination. In order to elevate enzyme activity in flours milled from mechanically harvested cereal grains, malted wheat and barley have been traditionally added. More recently the addition of fungal α-amylase to flours at the mill or bakery has been practiced. The enzyme is derived from *Aspergillus oryzae*, *A. niger*, or *A. awamori* and is available in the form of diluted powders or water-dispersible tablets.

The influence and effects of fungal α-amylase supplementation on the development of desired quality of bread, rolls, buns, and crackers are numerous (Barrett 1975). Hydrolysis of dextrins yields maltose and glucose which sustains the fermentation rate and also produces additional sugar in the finished product. Fungal enzyme mixtures contain higher levels of glucoamylase than do cereal and bacterial amylase preparations. This is important where sugars are not added to the formula and where the yeast has low maltase activity. α-Amylase activity lowers dough viscosity and affects dough softness. The relative fineness of the cells comprising bread crumb (i.e., the "grain") and other yeast-leavened baked foods is influenced by the amount and type of α-amylase present in dough.

Consequences of Carbon Dioxide Production.—Evolution of carbon dioxide into the dough by the action of *S. cerevisiae* leads to the porosity of bread, a characteristic that undoubtedly has been a major factor in the acceptance of bread by many nations down through the years. This porosity is due to the carbon dioxide produced by the yeast and to the ability of the dough to retain the gas. The latter ability is related to the film-forming properties of dough made with wheat flour. Gluten, the protein complex of dough, forms a continuous structure in which are embedded starch granules; the gluten can be stretched into viscoelastic films that form the walls of gas cells. Carbon dioxide diffuses into these cells during the course of fermentation. The gas cells undergo subdivision during various dough-processing steps and eventually form the basic cellular or porous structure of bread.

As observed earlier, the classic investigations of Baker and Mize (1941) indicate that carbon dioxide generated during yeast fermentation does not in itself produce gas cells. Yeast-generated carbon dioxide diffuses into cells already formed from two sources, viz., air bubbles occluded in the dough during mixing and air adsorbed on flour particles (one-fifth of the volume of flour is entrapped air).

Flavor Development

Fresh bread has a pleasing and appealing flavor that in part is responsible for its universal acceptance as a human food. Yet the flavor is subtle and difficult to characterize in spite of considerable effort to do so. A comprehensive review of this subject has recently been compiled (Maga 1974).

Bread flavor is derived from two main sources: yeast fermentation and crust browning.

Yeast Fermentation.—The contribution of the flavor of bakers' yeast to bread flavor has rarely been considered, and is certainly not a major factor. Nevertheless, it is known that doughs made with high concentrations of yeast yield baked products with "yeasty" or slightly "cooked" flavors (Wölm and Rothe 1974). Höhn et al. (1975) have found that thiamin and thiamin diphosphate are the major flavor compounds in yeast which may contribute to an undesirable flavor of the bread.

The characteristic flavor of yeast raised bread arises from yeast fermentation and subsequent reaction of fermentation products with other dough compounds during baking. The taste and odor of fermenting dough resemble that of fermenting beer wort and fermenting wine, and the flavor compounds which have been isolated are qualitatively the same. During the baking some of these flavor compounds escape and others react with amino acids and other compounds of the dough to yield the characteristic flavor of bread. The fermentation by-products formed during yeast fermentation fall largely into the classes of organic esters and acids, alcohols, and carbonyl compounds. Attempts to synthesize bread flavor by combining compounds found in doughs, liquid ferments, or bread have not been successful.

Many observers feel that traditional methods of breadmaking involving substantial periods of bulk fermentation yield bread with the most desirable flavor. A number of studies have been done showing that certain organic compounds do increase in concentration as a function of fermentation time (Maga 1974). Countering this view, however, is the work of other investigators (Collyer 1966; Kilborn and Tipples 1968) which has shown that taste panel members could not distinguish flavor differences in bread made with or without bulk fermentation. These findings would indicate that flavor compounds formed during baking and during proofing were sufficient to yield satisfactory bread flavor.

It should be noted here that some of the organic compounds formed during fermentation may arise from bacterial action (Robinson et al. 1958). Lactic acid bacteria found in dough presumably are associated with yeast; commercial bakers' yeast normally contains a low number of contaminating microorganisms. In addition to S. cerevisiae, other yeasts may be responsible for characteristic flavors of certain breads (Ng 1976).

The above discussion indicates that the exact role of yeast fermentation in bread flavor development is not well understood. Yet, it would seem clear that yeast fermentation by-products must play an important role in bread flavor. Even if traditional bulk fermentation is not used in breadmaking, the loaves must be proofed. During this period S. cerevisiae proliferates and produces compounds which undoubtedly contribute to flavor, either directly or as flavor precursors. Breakdown products of flour protein undoubtedly play an important role in flavor and color development. Yeast proteolytic enzymes modify peptones and polypeptides for growth; however, a portion of these products react with sugars to impart desirable flavors upon baking.

Proteolytic enzymes are in fact added to doughs on a large commercial scale, and it is estimated that two-thirds of the white bread baked in the United States is treated with enzymes derived from Aspergillus oryzae (Barrett 1975). Proteolysis during the fermentation stage results in shortened protein chains which realign into sheets of protein film. As a result, less time is required before the point of maximum extensibility is reached.

Proper levels of fungal proteinase improve handling and machining properties of dough and yield bread loaves having increased volume and better symmetry.

Fungal and bacterial proteinases mellow the gluten during fermentation to yield the proper balance of extensibility and strength in cracker dough. These doughs can be rolled out very thin without tearing and lie flat in the oven without bubbling or curling at the edges.

Crust Browning.—The extent of crust browning is influenced substantially by previous activities of *S. cerevisiae* in the dough, as discussed above. The importance of crust browning in flavor development can be shown by removing the crust of freshly baked bread and storing the bread until it has cooled, or by baking bread in a microwave oven where crust browning does not occur. Under these conditions the product will lack the flavor usually associated with bread. Presumably, therefore, a part of bread flavor is formed in the crust during baking and then diffuses into the crumb where it becomes absorbed. Over 100 flavor compounds have been found in bread (Maga 1974).

Table 8.7 lists some of the compounds reportedly produced during fermentation and/or baking (Magoffin and Hoseney 1974).

TABLE 8.7. COMPOUNDS REPORTEDLY PRODUCED DURING FERMENTATION AND/OR BAKING

Organic Acids		Alcohols	Aldehydes and Ketones	Carbonyl Compounds
Butyric	Acetic	Ethanol	Acetaldehyde	Furfural
Succinic	Lactic	n-Propanol	Formaldehyde	Methional
Propionic	Formic	Isobutanol	Isovaleraldehyde	Glyoxal
n-Butyric	Valeric	Amyl alcohol	n-Valeraldehyde	3-Methyl butanal
Isobutyric	Caproic	Isoamyl alcohol	2-Methyl butanol	2-Methyl butanal
Isovaleric	Caprylic	2,3-Butanediol	n-Hexaldehyde	Hydroxymethyl
Heptanoic	Isocaproic	β-Phenylethyl	Acetone	furfural
Pelargonic	Capric	alcohol	Propionaldehyde	
Pyruvic	Lauric		Isobutyraldehyde	
Palmitic	Myristic		Methyl ethyl ketone	
Crotonic	Hydrocinnamic		2-Butanone	
Itaconic	Benzylic		Diacetyl	
Levulinic			Acetoin	

Source: Magoffin and Hoseney (1974).

Dough Maturing

Yet another function of yeast is to cause many changes in dough that are often summarized as dough "maturing" or "ripening." A properly matured dough is one that exhibits optimum rheological properties (optimum balance of extensibility and elasticity), such that it may be machined well and will lead to bread with desirable volume and crumb characteristics. Some of the reactions leading to dough maturing are considered below.

Alcohol and carbon dioxide, among other products, are derived from yeast fermentation. Alcohol is water miscible and, since appreciable amounts are formed, it therefore influences the colloidal nature of the flour proteins and alters the interfacial tension within dough. Some of the carbon dioxide dissolves in the aqueous phase of the dough and forms weakly ionizable carbonic acid, which lowers the pH of the system. Carbon dioxide also distends the dough as the gas expands, thereby contributing mechanical work into the dough system.

Ammonia from ammonium sulfate and ammonium chloride added to the dough as "yeast foods" are assimilated by *S. cerevisiae*, causing a liberation of sulfuric and hydrochloric acids. These acids along with carbonic acid further lower the pH which, in turn, significantly influences gluten hydration and swelling, the reaction rate of enzymes in dough, oxidation-reduction reactions, and various chemical reactions (Magoffin and Hoseney 1974).

S. cerevisiae also produces reductases, which affect dough rheological properties by acting through intermediate substrates found in dough (Reed and Peppler 1973).

EFFECT OF INGREDIENTS AND PROCESSING ON YEAST PERFORMANCE

Recent studies have dealt with a larger number of variables which affect yeast activity in doughs (Tscheuschner *et al.* 1974; Flückiger 1974). These variables are treated separately because this permits a better understanding of the manner in which they influence yeast performance. In general, fermentation activity is increased by higher yeast concentrations, by higher fermentation temperatures, and by the addition of sugars up to 4–6% based on the weight of flour. Fermentation activity is decreased by sugar concentrations above 6%, by increased salt concentrations, by pH values below 4.5, and by the addition of mold inhibitors.

The use of oxidants and other dough conditioners affects the elasticity of doughs and the permeability of doughs to carbon dioxide gas. This affects the amount of CO_2 retained in the dough and, consequently, the leavening effect of the yeast. But oxidants and most dough conditioners have little or no effect on the fermentation activity of yeast per se, and, therefore, they will not be discussed below.

It may be noted in passing that recent work by Bell *et al.* (1977) indicates that dough permeability appears to be related to the well-known improving effect of fat in dough. During the early baking stage, when the loaf is rapidly expanding, fat-containing doughs exhibit more carbon dioxide retention than doughs made with no added fat.

Fermentable Sugars

Under the anaerobic conditions prevailing in a dough, yeast ferments sugars to ethanol and carbon dioxide. These sugars are the monosaccharides glucose and fructose and the disaccharides sucrose and maltose. Lactose is not fermented by bakers' yeast. Starches and dextrins are not fermented by yeast but may serve as sources of fermentable sugars if they are hydrolyzed by amylases. Flour contains from about 0.3 to 0.5% of fermentable sugars. In traditional doughs which consisted of water, flour, yeast, and salt and in today's lean doughs, the rate of gas production follows a double humped curve as shown in Fig. 8.8. The relatively high rate of gas production at 30 min represents the fermentation of sugars as they pre-exist in the flour. The second increase in gassing rate occurs after 60–90 min and corresponds to the liberation of maltose from the starch of the flour by amylases. The final drop of the rate after 2½ hr reflects the exhaustion of the supply of fermentable sugars.

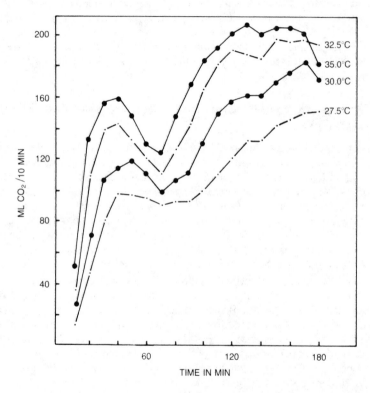

From Harbrecht and Kautzmann (1967)

FIG. 8.8.   GASSING RATE CURVES OF DOUGHS AS A FUNCTION OF TEMPERATURE

Flour contains both α- and β-amylase but the concentration of α-amylase is quite low and limits the formation of maltose. Therefore, malt, which contains sufficient α-amylase, is generally added to flour before delivery to the baker. Fungal α-amylase may be used. At times, the baker may further supplement his doughs with either malt or fungal amylase, in the production of certain items. The rate of hydrolysis of raw starch in flour is quite slow, and only so-called damaged starch granules can be hydrolyzed enzymatically. The amount of damaged starch accounts for about 5–8% of the weight of the flour. This means that the total amount of sugar ultimately available for fermentation is limited unless additional sugars are added to the dough.

In lean doughs maltose is the principal fermentable sugar. Therefore, it is important to use a strain of bakers' yeast with good "malto-zymase" activity. This enzyme complex has been thought of as a group of enzymes capable of hydrolyzing maltose to glucose and of fermenting glucose via the glycolytic pathway. It is now apparent that yeasts contain sufficient internal maltase (glucosidase) to hydrolyze maltose quickly. At present it is believed that transport of maltose into the yeast cell is the limiting step in maltose fermentation, and the presence of an active transport mechanism catalyzed by a "maltose permease" has been assumed. Some yeast strains contain the required enzyme system constitutively. Others have to be adapted to the fermentation of maltose. While the fermentation of maltose has been of great practical significance in the past, it has lost importance because of the addition of fermentable sugars to doughs. Even lean doughs contain from 0.5 to 2% added sugar in the United States.

Only a few investigations have been carried out on the actual levels of sugars in fermenting doughs. Figure 8.9 shows the sugar levels of a dough made from a liquid pre-ferment at the beginning of the pre-ferment period,

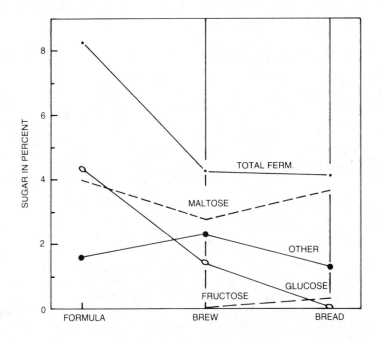

From Piekarz (1963)

FIG. 8.9. DISAPPEARANCE OF SUGARS IN A PRE-FERMENT DOUGH SYSTEM CONTAINING CORN SYRUP

at the end of the pre-ferment period, and in the final bread. The original source of sugars was a corn syrup which contributed 8% fermentable sugar (3.9% maltose and 4.1% glucose). Glucose is rapidly fermented throughout the fermentation period. Maltose is fermented slowly in the pre-ferment. In the dough the level of maltose actually increases because the rate of maltose formation from starch is greater than the rate of fermentation. Therefore, the final bread contains hardly any glucose but almost 4% maltose (Piekarz 1963). The fate of sugars in a straight dough has been demonstrated by Koch *et al.* (1954). In his tests, 1 g of yeast solids fermented about 1.2 g of sugar per hour. The compressed yeasts available in the United States ferment about 2.5 g of sugar per g of yeast solids per hour in straight doughs and lean doughs, and about 1 g of sugar in sweet doughs (see also Table 8.8; Reed 1974A).

Residual bread sugars in laboratory-produced sponge dough bread made with several different sweetener types are shown in Table 8.9. These previously unpublished data (Ponte *et al.* 1970) indicate, again, that where both fructose and glucose are present in dough, glucose is more rapidly fermented. Maltose levels are low (0.7 to 0.9%) if no maltose is added as a part of the sweetener system. The residual lactose, of course, is derived from the usage of dairy ingredients in the formulation. Taste panel comparisons of these breads suggest that the first 4 breads (made with either 10.0% dextrose, 6.7% sucrose, 10.5% high fructose corn syrup, or 95 D.E. corn syrup) had approximately the same level of residual sweetness.

Effect of pH and Temperature

The activity of bakers' yeast is almost constant over a pH range from 4 to 7. This is also the range for various doughs used in the industry with the exception of sourdoughs. Below a pH of 4 activity drops sharply, and above a pH of 7 the dropoff is gradual. The relative insensitivity of yeast to a 300-fold range of hydrogen ion concentrations is due to the fact that the internal pH of the yeast cell is maintained fairly constant over the entire range. The pH near the center of the cell is approximately 5.8 but it differs for different structures within the cell. Gassing rates at various pH levels have been determined by Franz (1961), Seeley and Ziegler (1962), and Garver *et al.* (1966).

In liquid pre-ferments which do not contain flour or nonfat dry milk solids, the pH drops during fermentation because of the production of carbon dioxide and organic acids by yeasts and lactic acid bacteria. Buffer salts must be added to such pre-ferments to keep the pH above 4.5.

Bakery sponges are usually set at a temperature of 24°–26°C and there is a rise of 3°–4°C during the sponge fermentation. Dough temperatures are generally somewhat higher and in the case of continuously mixed doughs may reach 35°C. Almost all bakery fermentations are carried out within this range of 25°–35°C. These temperatures are convenient for bakery operations and the exact temperatures are chosen to produce doughs with suitable elasticity and handling characteristics and to permit opti-

TABLE 8.8.   COMMON PARAMETERS FOR COMMERCIAL USE OF YEAST IN INDUSTRY

	Bread Dough	Beer		Wine	Whisky	Sake
		Lager[1]	Ale			
Raw Material	Flour Sugar	Wort	Wort	Grape or Fruit Juice	Cereal Mash	Steamed Rice
Time of fermentation	1–3 hr	8–10 days	2–6 days	5–10 days	3 days	21–28 days
Temp. of fermentation	30°–35°C	10°C	20°C	15°–27°C	35°C	7°–15°C
pH	5.2–4.7	5.2–4.2	5.2–4.0	3.5	4.9–4.0	—
Final EtOH conc. (%)	2–3	4	4	11–13	7–9	18
Cells[2], start	275	6–10	6–10	5–10	5–10	
end	300	30–70	30–70	50–150	50–150	100
Gas production rate[3]	18–35	16–25	16–25	25–32	—	—

Source: Reed (1974A).
[1]For primary fermentation.
[2]Expressed in million cells/g or /ml.
[3]Expressed as mM EtOH/hr/g yeast solids.

TABLE 8.9.   RESIDUAL BREAD SUGARS AS A FUNCTION OF SWEETENER TYPE

Sweetener	Amount Added to Dough[1]	% Residual Sugar, Dry Solids Basis			
		Fructose	Dextrose	Maltose	Lactose
Dextrose	10.0	T	5.9	0.9	0.2
Sucrose	6.7	2.6	1.7	0.7	0.2
High fructose corn syrup	10.5	2.5	1.8	1.4	0.2
Corn syrup, 70 D.E.	15.0	0.2	3.1	5.2	0.2
Corn syrup, 95 D.E.	13.4	0.1	4.9	1.5	0.3

Source: Ponte et al. (1970).
[1]Sweeteners added on as-is basis, based on weight of flour. Bread was produced by sponge dough method, utilizing the same formula except for sweetener variation (compensation was made for water in syrups).

mum bread quality. These temperatures do not provide optimum gas production rates.

There have been only a few exacting studies on the effect of temperature on gas production rates. Available data indicate that the fermentation rate increases by a factor of 1.5 to 2 for a 10°C increase in temperature (Garver et al. 1966; Seeley and Ziegler 1962). Figure 8.8 shows that there is a 50% increase in the rate of fermentation if the temperature is raised from 27.5°C to 32.5°C. The rate of CO_2 evolution in straight doughs increased twofold from 20° to 27°C (Flückiger 1974).

During the early stages of baking the volume of the loaf increases considerably. This is due to thermal expansion of the entrapped gas of the dough, to the formation of additional CO_2 because of its decreased solubility in the dough water, and to the production of additional CO_2 by fermentation. It is difficult to estimate how much of this so-called "oven spring" is due to fermentation. There is about a 10 min period in the oven before the center of the loaf reaches a temperature of 55°C, the temperature at which yeast cells are killed rapidly. Figure 8.10 shows the rate of kill at temperatures of 48°, 50° and 52°C. There is a considerable lag (also temperature dependent) before the curve follows first order reaction kinetics (van Uden 1971).

Osmotic Pressure

Yeast fermentation is strongly inhibited at high osmotic pressures in doughs. The major contributors to osmotic pressure in doughs are salt and sugars. At salt concentrations up to 1.5% there is little inhibition in doughs, but at concentrations of 2–2.5% which are common in bread doughs there was considerable inhibition (Schulz 1962). With sucrose, glucose, maltose, and fructose, inhibition becomes apparent at concentrations exceeding 4–5% (Schulz 1965).

Yeasts with a high invertase activity are more inhibited by high sucrose concentrations than yeasts with low invertase activity. This is probably due

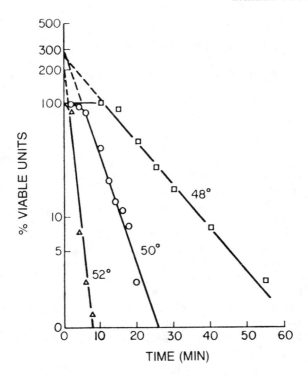

After van Uden et al. (1971)

FIG. 8.10. SEMILOGARITHMIC SURVIVAL CURVES OF
SACCHAROMYCES CEREVISIAE EXPOSED TO VARIOUS
TEMPERATURES

to the increase in osmotic pressure when sucrose is hydrolyzed by the enzyme (Sato *et al.* 1961; Thorn and Reed 1959; Lesaffre 1978).

Yeasts vary greatly in their tolerance to high osmotic pressure. This tolerance is a function of the strain but it also depends to a considerable extent on the conditions under which the yeast is grown. Good osmotolerance is particularly important for yeast raised sweet goods which may contain 20–25% of sugar based on the weight of the flour. Table 8.10 shows that for bakers' compressed yeast the rate of gas production was only about 35% of that in lean doughs.

TABLE 8.10. GASSING POWER OF BAKERS' YEASTS

| | mM CO_2 per hr per g Yeast Solids | | |
	Regular Doughs	Lean Doughs	Sweet Doughs
Bakers' compressed yeast	24	27	9.5
Bakers' active dry yeast	16	16	14

Fermentation Inhibitors

Ethanol is a strong inhibitor of yeast growth and yeast fermentation. At ethanol levels exceeding 4% (weight per volume) there is some inhibition of the rate of ethanol formation and carbon dioxide evolution. For each g of sugar that has been fermented about 0.45 g of ethanol is formed. Sponge doughs and straight doughs have been reported to contain 3 and 1.5% ethanol, respectively (Kosmina 1966), and liquid pre-ferments from 1.5 to 1.75% (Bottomley 1961). At such levels the inhibiting effect of ethanol is minimal. It is not negligible in concentrated pre-ferments. Cole *et al.* (1962) reported the presence of 1.8, 3.3, and 6.8% ethanol (by vol.) in pre-ferments containing 3.2, 6.6, and 11.9% sucrose, respectively. Most of the ethanol formed by fermentation is driven off during baking so that a freshly baked loaf may contain no more than 0.8% ethanol based on the weight of flour (Wiseblatt 1960).

Mold inhibitors are commonly added to commercial white bread. This is particularly important if the bread is sliced before wrapping since additional surface area is exposed to air. Propionates are the most widely used inhibitors, and levels of 0.3% of sodium or calcium propionate are frequently used. Other suitable mold inhibitors are sodium diacetate and vinegar. There are considerable differences in the degree of inhibition reported by various authors because other variables such as pH may affect the degree of yeast inhibition. Schulz (1967) reported the highest rate of inhibition and also found that maltose fermentation was more strongly inhibited than the fermentation of sucrose or glucose. His tests had been done with lean formulae. At a level of 0.25% propionate, an inhibition of the rate of fermentation by 20% can be expected.

Effect of Yeast Nutrients

The effect of vitamins, minerals, nitrogenous materials, and carbon sources on yeast growth have already been treated in the preceding chapter. For doughs which undergo very short fermentation periods the addition of extra nutrients is not required. But for normal fermentation periods the addition of a readily assimilated nitrogen source is useful although there is little growth of yeast during the fermentation.

Such nitrogen and additional minerals are usually added in the form of "yeast foods." Such yeast foods contain not only yeast nutrients but also oxidants and sometimes salts which adjust the pH of the dough. Yeast foods generally contain about 10% of either ammonium chloride or ammonium sulfate as a source of nitrogen, potassium bromate and/or iodate as oxidants, and monocalcium phosphate for pH adjustment if the pH of the water is alkaline. Flour, salt, and calcium sulfate are often used as fillers. A normal level of addition of such a yeast food is 0.5% based on the weight of flour.

Addition of minerals is rarely required. If very soft water is used for the makeup, the addition of calcium salts is desirable. Schultz *et al.* (1942) in their classical paper found that the stimulating effect of flour on fermenting activity was due to thiamin. This is important for liquid pre-ferments which contain no flour and which consequently require thiamin. For this reason sufficient thiamin is usually added during the production of bakers' yeast to supply at least 50 μg of thiamin per g of yeast solids to the compressed yeast.

The oxidants and the monocalcium phosphate do not affect yeast activity directly. However, they serve to improve gas retention in doughs and hence improve the leavening effect of the yeast. This is the rationale for the combined use of yeast nutrients and oxidants. For a more detailed discussion of yeast foods see also Reed (1972).

FORMS OF YEAST USED IN BAKING

Determination of the Activity of Bakers' Yeast

A laboratory bake test is accurate in the sense that it reflects performance of the yeast in bakery operations. In one such test straight doughs yielding 2 0.45-kg (1-lb) loaves were produced (Peppler 1972). In such tests one measures either the time required for a dough to proof to a given height or one measures the volume of the bread for a given proof time. Such tests may also be carried out on a smaller scale, for instance, with "pup" loaves requiring only 100 g of flour for each test (Anon 1957). There are two basic problems with this type of test. First, the bake test is imprecise. Its results depend on the skill of the operator, on the type of flour used, on the temperature of the bake shop, and on the proper functioning of mixers, proof cabinets, molders, and ovens. The second problem regards the great variety of uses of yeast in a bakery. A particular bake test reflects accurately only that operation which it imitates on a laboratory scale. But bakers produce bread from straight doughs, sponge doughs, lean doughs, sweet doughs, doughs made with liquid pre-ferments, or various combinations of these. Therefore, it is almost impossible to reflect all of the variations in bakers' processes in one or several simple bake tests.

The alternative is a simple determination of the fermenting power of the yeast by measuring the amount of carbon dioxide evolved in a given time period. This can be done in solutions of various sugars in a simple fermentometer such as the one described by Schultz et al. (1942). Such simple gas measuring devices are still in use and a suitable arrangement for determining the activity of a wine yeast has recently been described (Reed and Chen 1978). With such tests one can determine the amount of carbon dioxide produced by bubbling the gas through an alkaline solution and by back-titration. Or, one can measure the amount of gas volumetrically at atmospheric pressures or one measures the pressure of the gas in a defined volume. These simple tests have a serious drawback. The osmotic pressure in doughs is very much greater than in simple sugar solutions, and yeasts whose activity is greatly reduced by higher osmotic pressures vary greatly in their osmotolerance.

For these reasons it is advisable to measure carbon dioxide evolution in actual doughs. Since total gas evolution is measured and not the amount of gas remaining in the dough it is not essential that doughs be mixed to a given degree of elasticity. Such dough pieces weighing from 10 to 100 g can be introduced into hermetically sealed cups with pressure meters. More frequently they are placed into instruments which measure the volume of total evolved gas at normal atmospheric pressure. A suitable instrument is the S.J.A. fermentograph which facilitates the measurement by an automatic chart of gas evolution over a given time period. Figure 8.11 shows such a Fermentograph curve.

Doughs may also be prepared and inserted into pressure cups. The gassing power is then expressed as mm Hg pressure. Shogren *et al.* (1977) used such a system with a conventional straight dough containing flour, 100%; skim milk solids, 4%; sugar, 6%; salt, 1.5%; malt, 0.25%; compressed yeast, 3%; and 20 ppm of $KBrO_3$. All percentage values are expressed as percentages of the flour used as is common in formulations of baked goods. Figure 8.12 shows the results of such gassing power tests when the percentage of water was varied from 40 to 200% (based on flour). A normal percentage of water would be in the 60–70% range. Gassing power increases greatly at higher levels of absorption (percentage of water). While the authors explain this on the basis of additional nutrients leached from the flour, it is more likely that the reduced osmotic pressure of the doughs at higher concentrations of water accounts for the increased gassing power of the yeast.

An interesting method which involves the preparation of doughs but avoids the variability of flour has been developed by Schulz (1972). This starch dough method as modified by Brümmer (1977) requires the mixing of a dough consisting of 400 ml of water, 500 g of cornstarch, 15 g carob flour, 25 g sucrose or any other sugar, and either 12.5 g of compressed yeast or 3 g of active dry yeast. Dough pieces weighing 400 g are placed into 2 liter measuring cylinders and the volume is measured every 15 min for a period of 150 min.

Not all of the carbon dioxide evolved by the fermentation of sugars remains entrapped in the dough. A certain fraction of the gas escapes and does not serve as a leavening gas. The amount of gas that escapes from the dough depends on the strength of the flour and on proper development of the dough. For this reason any method which measures total gas development by a yeast is valid only if one assumes the yeast does not affect the permeability of the dough membrane for carbon dioxide gas. This assumption is probably justified for compressed yeast; it is not always justified for active dry yeast if leached yeast solids (mainly glutathione) affect the rheology of the dough.

The results of baking tests are often expressed in terms of proof minutes, and the results of gassing power tests in terms of ml of CO_2 evolved. These are arbitrary expressions and do not permit a comparison of the data obtained in different laboratories. It is more meaningful to express yeast activity on the basis of the millimoles of CO_2 evolved per hour and per gram of yeast solids. Such values will generally vary between 10 and 25 mM of CO_2/hr/g yeast solids, depending on the type of yeast and the particular dough composition (see also Reed 1974A). Typical values for U.S. com-

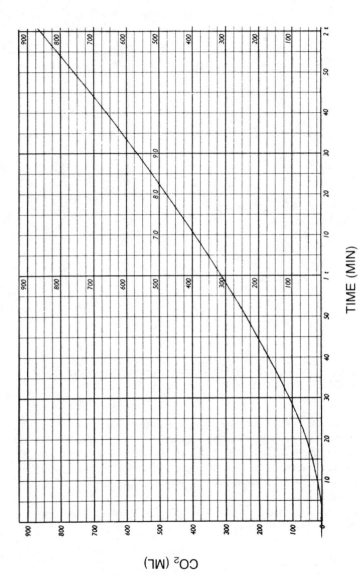

TIME (MIN)

Courtesy of Universal Foods Corp.

FIG. 8.11. FERMENTOGRAPH GASSING POWER CURVE OF A STRAIGHT DOUGH (AUTOMATICALLY RE-CORDED).

The straight dough piece contained 2.5 g of yeast solids. During the first hour, the rate of CO_2 development was 5.6 mM per g of yeast solids, and during the second hour, 9.8 mM per g of yeast solids.

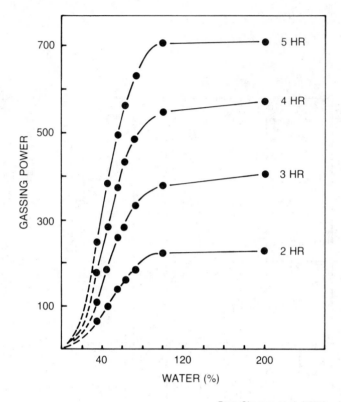

From Shogren et al. (1977)

FIG. 8.12.   GASSING POWER (MM HG) VS. PERCENTAGE WA-
TER IN DOUGHS WITH 2.75% YEAST, AND FOR FERMENTATION
PERIODS OF 2, 3, 4, AND 5 HR

pressed yeast and active dry yeast for 3 types of dough systems are shown in
Table 8.10. One can relate these values to the amount of sugar fermented
by a given amount of yeast. Ten mM of CO_2 are evolved by the fermentation
of 0.9 g of glucose. Therefore, 1 g of yeast solids which leads to the production
of 10 mM of CO_2 will have fermented 0.9 g of this sugar (compressed yeast
contains 30% solids).

The stability of compressed yeast and active dry yeasts has already been
treated in the preceding chapter.

Compressed Yeast

The cut kilograms (lb) of compressed yeast are wrapped in wax paper and
packaged 22.7 kg (50 lb) to a case. The cases are shipped by refrigerated
trucks either directly to the bakery or are held refrigerated in distribution
centers for later delivery to bakers. In the United States, deliveries to

bakers are made every other day, twice weekly, or once a week depending to some extent on the distance from the yeast factory. While bakers request shipment of the freshest yeast it is probably more important that the yeast be properly cooled before shipment and that it be shipped to the bakery and stored in the bakery so that its temperature does not exceed $5°-8°C$. Refrigeration of yeast in the bakery is particularly important with regard to yeast that may have been brought to the mixer floor and that may not have been used during a shift. If this yeast is allowed to warm up, it may not be possible to cool it down again because respiration of the yeast at temperatures above $20°C$ makes it difficult to do so effectively.

For larger bakeries and especially for bakeries using liquid pre-ferments, the yeast press cake is usually crumbled at the yeast factory and packed in 22.7 kg (50 lb) bags with polyethylene inner liners. The same precautions have to be used with regard to this crumbled yeast as to the compressed yeast cakes.

In the past it has been customary to disperse compressed yeast in buckets with water before adding it to the mixer. This is not necessary and compressed yeast cakes can be added directly to the flour in high speed mixers. For the preparation of liquid pre-ferments it is customary to suspend the yeast and other minor ingredients in a mixing tank from which it is pumped to the pre-ferment tank. In some larger bakeries yeast slurry tanks have been installed. Yeast is suspended in an equal weight of water. The slurry is kept at $5°-10°C$ with slight agitation to prevent settling of the yeast. From the slurry tank the desired amount of yeast can be pumped directly to the mixer (Masselli 1959; Buckheit 1971).

The above conditions for refrigeration of compressed yeast apply particularly to the United States where yeast is often shipped over distances up to 1500 km, and where a very fast fermenting yeast is required. In some other countries yeast of a different strain and of somewhat lower nitrogen content can be shipped without refrigeration.

For use by consumers and for sale through grocery stores, compressed yeast is packaged in 18 g and 56 g weights, wrapped either in aluminum foil or in wax paper. This yeast generally has a lower nitrogen content and about 10% of starch has been added. Both of these measures assure a better shelf life which exceeds several weeks. Nevertheless, the development of mold on yeast cakes is a problem when turnover in the stores is slow. For consumer use compressed yeast has been largely replaced by active dry yeast.

Active Dry Yeast

In general, active dry yeast has not replaced compressed yeast in wholesale bakeries. This is readily apparent from the figures shown in Table 8.10. These indicate that for a given amount of yeast solids, active dry yeasts ferment more slowly than compressed yeasts in regular doughs and lean doughs. However, there is some use of active dry yeast in sweet doughs reflecting the good fermenting activity at higher sugar levels.

In the United States, active dry yeast is shipped to bakers in fiber drums with polyethylene liners. It has a useful storage life of up to 3 months at ambient temperatures, and up to 6 months if it is kept refrigerated. For export and for storage for prolonged periods the yeast is packaged in 11.3 kg (25 lb) cans which are flushed with nitrogen gas to replace the air atmosphere. For smaller 0.9 kg (2 lb) cans it is simpler to apply a vacuum. In either case the cans must be hermetically sealed. Alternatively, the yeast may be packed in flexible packaging material under vacuum or a carbon dioxide atmosphere. All of these yeasts have a useful shelf life of at least 1 year provided the seal is not broken.

In the United States, active dry yeast has largely replaced compressed yeast for sale to institutions such as restaurants, schools, prisons, etc., and for sale to consumers through grocery stores. Active dry yeast is also preferred in countries where a hot climate or lack of refrigerated facilities make it difficult to distribute compressed yeast satisfactorily. Air lift dried "instant" yeasts which have recently come on the market are packed in flexible, hermetically sealed pouches or bags. They have a relatively high fermenting activity which is intermediate between that shown for compressed yeast and active dry yeast in Table 8.10. These yeasts have the same excellent stability as long as they are protected by an inert atmosphere. They have very poor stability once the seal has been broken and air has been admitted.

Consumer active dry yeast is usually packaged in smaller aluminum foil pouches (7 g per pouch). These pouches are flushed with nitrogen gas and heat sealed. The storage life is also at least 1 year. However, these packages are subject to mechanical shock during shipment and particularly during handling in grocery stores. The development of small leaks in the sealing area is not uncommon. This results in an accelerated loss in yeast activity. For this reason the reliability of the hermetic seal is more important than differences in the original fermenting activity of the yeast.

For use in baking active dry yeast is generally rehydrated in water of about 30°–40°C before addition to the mixer or to the pre-ferment tank. A rehydration period of 5 min is adequate to obtain good dispersion and rehydration. If the yeast is finely ground it can be added directly to the flour in the mixer. The air lift dried "instant" yeasts are particularly suitable for direct addition to the flour without prior rehydration (Greup 1974).

Doughs made with active dry yeast are slacker, more extensible, and more relaxed than doughs made with compressed yeast. This is due to the leaching of a reducing compound, glutathione (GSH), into the water used for rehydration. Direct addition of the yeast to the flour prior to addition of the dough water minimizes this effect but does not eliminate it. Ponte et al. (1960) has shown that the slackening effect of ADY is indeed due to GSH by separating rehydrated ADY from the rehydration water. If the rehydration water is discarded there is no slackening effect. However, this separation is not practical for commercial operations. As the result of the presence of GSH in rehydrated ADY the mixing time of doughs is reduced by about 25%. The slackening effect is beneficial for doughs made from very strong flours, for pizza doughs, bun doughs, and some sweet doughs where a well-relaxed dough is desired. In many other systems the slackening effect is not desired and in that case it can be counteracted by increased oxidation. Figure 8.13 shows the oxidation requirements of straight doughs for compressed yeast and active dry yeast.

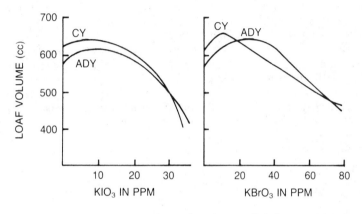

From Ponte et al. (1960)

FIG. 8.13. EFFECT OF POTASSIUM IODATE AND POTASSIUM BRO-MATE ON LOAF VOLUME OF BREAD MADE WITH COMPRESSED YEAST (CY) AND ACTIVE DRY YEAST (ADY) BY THE STRAIGHT DOUGH METHOD

Special Active Dry Yeast Preparations

Throughout the past 50 years there have been numerous attempts to dry compressed yeast together with such materials as starches, flour, inorganic salts, and others. In general these have not been successful. In some cases an attempt has been made to include other dough ingredients with ADY products. For instance, Distiller's Co., Ltd. (1976) has patented a product which contains ADY, an edible oil, L-cysteine, and azodicarbonamide. The latter two compounds are dough conditioners. Hartmeier (1976) used malto-dextrin as a carrier for the drying of compressed yeast and added grape or fruit syrups prior to drying.

Normally, ADY contains no additives. For ADY products of moisture values below 6%, an emulsifier such as sorbitan-monostearate is generally added to facilitate rehydration and to minimize the leaching phenomenon (Langejan 1974). A "protected ADY" of improved stability can be obtained by adding the emulsifier and about 0.1% of an antioxidant, butylated hy-droxyanisole, to ADY of low moisture (Chen *et al.* 1966). Such yeasts are commercially available and used in some instances as consumer yeasts.

Concentration of Yeasts in Doughs

Bakers' compressed yeast contains between 25 and 35 $\times$ 10^9 cells per g. The total number of cells depends, of course, on the size of the cell, and for smaller cells the number of cells per g is higher. In a normal sponge dough the number of yeast cells is about 300 to 400 $\times$ 10^6 cells per g. There is little or no multiplication of cells during the 3 to 4 hr sponge period but the number of budding cells increases from 30 to 50% (Thorn and Ross 1960). This is true only because a large number of cells has been added in the form of bakers' compressed yeast. For small concentrations of yeast, growth is considerable during long fermentation periods (overnight).

The actual concentration of compressed yeast used commercially varies with the type of dough system and with the desired proof time. Generally proof times are between 45 and 60 min. Finney *et al.* (1976) used a straight dough procedure. They determined optimum bread quality by varying total fermentation time (exclusive of proof time) for varying yeast concentrations. Their results are shown in Fig. 8.14. The general shape of the curve is

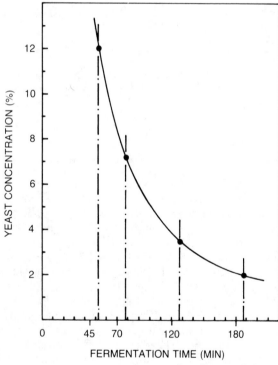

FIG. 8.14. YEAST CONCENTRATIONS AND FERMENTATION TIMES REQUIRED TO PRODUCE OPTIMUM BREADS WITH THE STANDARD COMMERCIAL FLOUR

From Finney et al. (1976)

similar to the early work by Fisher and Halton (1937). It is apparent that the effective fermentation time cannot be reduced beyond a certain limit no matter how much compressed yeast is used, that is, if one wishes to produce bread of excellent quality. When the fermentation time was decreased from 180 to 70 min, the yeast concentration had to be increased from 2 to 7.2% and the requirement for bromate addition was tripled. Proof time decreased from 55 to 21.5 min.

Table 8.11 shows the levels of compressed yeast customarily used in various dough systems and for the production of various baked goods. In some instances the range of concentrations used in practice is quite small. This is particularly true for sponge dough breads and breads made with continuous mix processes. For various sweet doughs and frozen, unbaked doughs the range is quite large, reflecting variations in dough composition and in processing conditions. The interdependence of some of these variables is shown in Table 8.12 which indicates the requirement for larger yeast concentrations for liquid pre-ferments with lesser amounts of flour in the pre-ferment.

TABLE 8.11. CUSTOMARY LEVELS OF COMPRESSED YEAST IN DOUGHS (BASED ON WEIGHT OF FLOUR)

Type of Baked Goods	Yeast (%)	Author
Breads		
Sponge and dough	2.5	Uhrich (1975)
Continuous mix bread	2.5–3.5	Uhrich (1975)
Specialty breads[1]	2.5–4.0	Ford (1978)
Whole wheat	2.75	Cain (1966)
High-fiber breads	3–5	Dubois (1978)
No-time doughs	3–4	Shirley (1977)
No-time doughs	10	Flückiger (1974)
Sour rye bread	1.2	Slaten (1960)
Buns and rolls		
Hard rolls, straight or sponge	2	Feinberg (1963)
Hamburger buns	4	Wolfe (1963)
English muffins	2–8	Pfefer (1976)
Sweet goods		
Yeast raised doughnuts	4–7	Braden (1976)
Danish, coffee cakes, etc.	5–10	Meigs (1968)
Specialties		
Partially baked rolls and buns	2	Wolfe (1963)
Frozen, unbaked doughs	5–5.5	Fuhrmann (1978)
Pretzels	0.5	Reed (1974B)
Soda crackers	0.2	Sugihara (1977)
Traditional breads		
Habab (Syria)	1.5	Tweed (1979)
Creola (Cuba)	1.0	Tweed (1979)
Barbery (Syria)	0.25	Tweed (1979)

[1]Breads from mixed grains.

TABLE 8.12.   BASIC DIFFERENCE BETWEEN SPONGE-DOUGH PROCESS AND LIQUID FERMENT PROCESS

	Sponge Dough	Liquid Pre-ferments		
Flour levels	50–100%	40–70%	15–40%	0–15%
Sugar in pre-ferment	none required	up to 0.5%	up to 2%	2–3%
Set temperature	23°–27°C	24°–27°C	27°–30°C	28°–31°C
	74°–80°F	76°–80°F	80°–86°F	82°–88°F
Fermentation time	3–5 hr	2–3 hr	2 hr or less	2 hr or less
Yeast percentage	2½%	gradual increase to 3.5%		
Yeast nutrient and oxidation	15 ppm	20–50 ppm	25–60 ppm	35–75 ppm
Buffers, enzymes, acidity additives, mix reducing agents	none required	gradual increase to high levels		
Dough mixing time	normal	+10%	+40%	+60%
Dough temperatures	normal—27°C (80°F)	gradual increase to 35°C (95°F)		
Dough floor time	normal—20 min	gradual increase in time		

Source: Uhrich (1975).

USE OF YEAST IN SPECIAL DOUGH SYSTEMS
Short Time Doughs

Such doughs are often used when it is important to reduce the overall time required for bread processing. Short time doughs may yield bread in about 2 hr, as opposed to 7–8 hr for conventional sponge doughs. Retail bakers and food service operators utilize short time doughs to avoid night and early morning working hours, and to reduce labor costs. Large wholesale bakers do not employ short time doughs in their operations. Sometimes such doughs are incorrectly called no-time doughs.

Baked goods of reasonable quality can be obtained with short time doughs, but higher levels of yeast and oxidants, warmer dough temperatures and increased dough mixing are required, and the use of cysteine or other agents to relax the dough is sometimes advisable. The more important problems with short time doughs include decreased product shelf life and poorer processing tolerance. Decreased shelf life is not as serious a problem with retail bakers or food service operations as it is with wholesale bakers. Lessened processing tolerance (i.e., temperature, dough elasticity, timing, etc.) is acceptable to the small baker, but not to the large, heavily mechanized baker. It is often difficult to achieve proper proof height for short time doughs in the normal 55–60 min anticipated by the baker.

Finney *et al.* (1976) were able to obtain bread of equal quality for the conditions shown in Table 8.13. The interrelationship among fermentation time, proof time, and oxidation requirement is quite apparent. The principles which follow from this relationship have been well established. That is, a decrease in fermentation time calls for an increase in yeast concentration and a drastic decrease in proof time. Oxidation requirements are greatly increased but the absolute values depend very much on the type of flour used.

TABLE 8.13.   CONDITIONS FOR THE PRODUCTION OF QUALITY STRAIGHT-DOUGH BREAD BY SHORT-TIME DOUGH PROCEDURES

Fermentation Time min	Yeast Concentration %	$KBrO_3$ Concentration[1] ppm	Proof Time min
180	2	0–30	55
120	3.5	0–45	36.5
70	7.2	0–90	21.5
45	12.0	0–180	12

Source: Finney et al. (1976).
[1]The range indicates varying bromate requirements of 7 hard red spring and hard red winter wheat flour.

It is interesting to compare the results of this laboratory investigation with actual bakery practice as reported by Shirley (1977). Preparing short time doughs with floor times of 20 to 30 min it was necessary to increase mixing time, to use protease or cysteine (a reducing compound) to obtain full development of the dough, and to increase yeast levels. Some additional steps had to be taken which could not be properly brought out in the laboratory procedures which have been discussed. Because of the very short fermentation time, the pH did not drop sufficiently and vinegar was added. Levels of sugar had to be reduced to prevent excessive browning; and the level of salt was reduced to reduce excessive proof times.

Frozen Doughs

Yeast leavened, unbaked doughs may be preserved by freezing, either in the form of small dough slabs or in the form of formed rolls or loaves. The products are later thawed, proofed, and baked. This freezing process is used by some bakeries to ensure a supply of doughs for bake-off on weekends or holiday periods. Frozen storage for this purpose is rarely longer than 1 week. A major market for frozen doughs is "in-store" bakeries and institutions, which bake bread on the premises but do not wish to operate the heavy mixing and make-up equipment of a bakery (Drake 1970). For this application the production of frozen doughs presents no major problem since they can be stored at $-25°$ to $30°C$ for $2-4$ weeks without appreciable loss in bake activity.

There is also a consumer market for frozen bread and roll doughs. This requires a shelf life of several months and deterioration of yeast activity during frozen storage is a serious problem. This deterioration is caused by a loss of yeast viability and by changes in the structure of the dough. Loss of yeast viability is by far the more important cause of deterioration. This results in prolonged proof times and inferior internal and external characteristics of the baked goods. The market for frozen, unbaked goods has been discussed by Vetter (1979).

Compressed yeast can be frozen and kept at temperatures between $-25°$ and $-30°C$ for several months without appreciable loss in viability and bake activity (Godkin and Cathcart 1949). Mazur and Schmidt (1968) froze yeast with extremely fast freezing rates. Freezing in fractions of a second or freezing to the temperature of liquid nitrogen ($-76°C$) is harmful to yeast survival.

In the production of frozen baked goods the loss in yeast viability does not occur during freezing but throughout the period of frozen storage. Figure 8.15 shows the loss of viability in straight doughs as a function of yeast concentration and fermentation time prior to freezing. It is clear that longer fermentation periods lead to greater damage of the yeast cells. This damage can be due to (1) an increased sensitivity of cells in a state of high metabolic activity, (2) the effect of soluble dough constituents (sugar, salt, etc.), or (3) the effect of the products of yeast metabolism, that is, CO_2, ethanol, or other fermentation by-products.

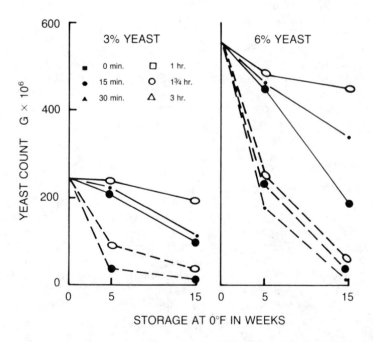

From Kline and Sugihara (1968)

FIG. 8.15. CHANGES IN VIABLE YEAST COUNT OF STRAIGHT DOUGHS DURING FROZEN STORAGE AT $-17°C$.
Effect of initial yeast level and fermentation time prior to molding and freezing. (The ordinate shows the yeast count in millions of cells per g of dough.)

Most authors are inclined to see the cause of yeast damage in the heightened susceptibility of the yeast itself. This seems to be a reasonable assumption (Merritt 1960) although no experimental evidence is available to support this point of view. The second hypothesis has not been tested. Hsu *et al.* (1979A,B) have dealt with the third hypothesis. They have shown that the volatile fraction of liquid ferments is a major factor in producing yeast damage. But not all of the damage could be attributed to the 2.5% of ethanol which had been formed prior to freezing. It must also be remembered that a concentration of 2.5% of ethanol has a small but demonstrable effect in inhibiting fermentation.

Too little attention has been paid to the effect of freezing rate, thawing rate, to the temperature of frozen storage, or to possible fluctuations in the temperature during frozen storage. Hsu *et al.* (1979B) have reported that freezing at different temperatures causes different levels of damage. Proof times of the frozen doughs were 72, 71, and 132 min for freezing at $-10°$, $-20°$, and $-40°C$, respectively. Freezing at $-78°C$ resulted in doughs which could not be proofed in 6 hr.

In assessing the applicability of laboratory results to commercial operations one has to keep in mind that some authors have worked with storage periods of only 2–4 weeks (Lorenz and Bechtel 1964, 1965; Hsu *et al.* 1979A,B), while others have used periods of frozen storage of several months (Kline and Sugihara 1968; Sugihara and Kline 1968).

Regardless of the various hypotheses which have been proposed, the recommendations for the conduct of commercial operations are quite consistent. There are: a high level of yeast (5–6%); high levels of oxidation (30–40 ppm bromate); cool doughs (18°C after mixing); and rapid conveying of the dough slabs or formed loaves into the freezer. Lorenz (1974) has summarized these recommendations as well as the earlier literature.

Complete Bakery Mixes

Institutional bakeries and some wholesale bakeries use so-called "bakery mixes" as principal dough ingredients. Such mixes contain sugar, shortening, salt, all of the minor dough ingredients, and part or all of the flour required. They do not contain the yeast. Bakery mixes are particularly suited to the production of sweet goods, such as doughnuts. They generally require only the addition of yeast and water.

For the formulation of "complete" mixes, finely ground ADY may be added. Such mixes are stable for a limited time; generally for 2–4 weeks. A much more stable complete mix can be obtained by use of a "protected ADY." In addition moisture pickup by the yeast from the flour must be prevented by use of a low moisture flour. With flour whose moisture content had been reduced to 9–10% the shelf life of complete mixes could be increased to 3 months, and for 8% moisture flour it could be increased to 1 year (Chen *et al.* 1966). Better stability of complete mixes may also be obtained by packaging in an inert atmosphere. Complete mixes are available in the United States for the wholesale market, and in Japan for the wholesale and consumer market. There is also a consumer market for such mixes in the United Kingdom.

BACTERIAL PROCESSES IN THE PRODUCTION OF BAKED GOODS

Sourdoughs

Sourdough bread in the United States is usually made from a mixture of rye and wheat flour containing up to 80% of wheat flour. The acid taste may be obtained by adding mixtures of lactic and acetic acids. Such breads are usually leavened with yeast, and the action of lactic acid bacteria (if it occurs) is only incidental. In Europe and for some American rye sour breads the use of "sourdough starters" is preferred because it gives the full flavor which depends on the microbial souring action. The use of such rye sour cultures predates recorded history. It is relatively easy to produce such starter by mixing rye flour and water in equal proportions and permitting it to ferment for several days at 30°C. This forms the basis of the culture which can then be maintained by further additions of rye flour and water.

Sourdough starter cultures are also available commercially and these can be used to inoculate the doughs. Such cultures usually contain mixtures of lactic acid bacteria. Spicher and Shröder (1978) classified the organisms as subgroups of the genus *Lactobacillus* as follows: *Thermobacterium (L. acidophilus)*, *Streptobacterium (L. casei, L. plantarum, L. farciminis, L. alimentarius)*, and *Betabacterium (L. brevis, L. buchneri, L. fermentum, L. fructivorans)*. All of the strains isolated by these authors fermented maltose, about three-fourths fermented fructose, one-half fermented sucrose, and about one-fourth fermented lactose. Ordinary bakers' yeast *(S. cerevisiae)* usually contains 10,000 to several million lactic acid forming bacteria per g, and its use will result in souring of the dough if the fermentation is carried out for prolonged periods of time and at relatively elevated temperatures.

The lactic acid bacteria usually found in sour rye fermentations are *L. plantarum, L. brevis,* and *L. fermenti.* Homofermentative bacteria producing only lactic acid are usually thermophiles and do not grow at the relatively low temperatures of dough fermentations. Homofermentative bacteria producing only acetic acid also do not find a favorable climate in the anaerobic doughs. Therefore, lactic and acetic acids are mainly produced by heterofermentative bacteria found in doughs (Sugihara 1977).

Today bakers' yeast is frequently added to sourdoughs to increase the leavening activity of the system. In spontaneous sourdoughs (to which no bakers' yeast had been added) Spicher and Schöllhammer (1977) found 4 types of yeasts. For 3 of these groups the yeasts could be identified as *S. cerevisiae, Pichia saitoi,* and *Trichosporon penicillatum,* respectively. Yeasts in one of the groups could not be identified. The number of yeast cells in spontaneous sourdoughs was about 1.3×10^9 per g and that of bacterial cells from 4 to 6×10^9 per g. The growth of yeasts and bacteria in such doughs is illustrated in Fig. 8.16 together with the concomitant drop in pH

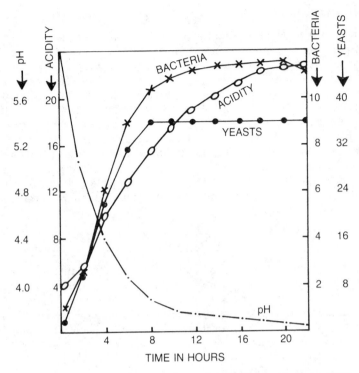

FIG. 8.16. MULTIPLICATION OF YEAST AND BACTERIA AND pH OF RYE SOURDOUGH

From Rohrlich and Stegeman (1958)

and the increase in titrable acidity. For a description of sourdough processes see also Schulz (1962).

A white sourdough bread produced only in the western United States is the San Francisco sourdough bread. It is of special interest because it is still produced by propagation with sponges from the previous day's production. The characteristic yeast of this process is *Torulopsis holmii* and the bacterium producing both lactic and acetic acids has been named *Lactobacillus sanfrancisco* (Sugihara *et al.* 1971). The sponges contain only water, flour, and the microorganisms and require about 8 hr at 25°C for full development. The final pH may be as low as 3.9. About 20% of the starter sponge is used in the dough (based on the weight of dough flour).

The Italian Pannetone is a traditional Christmas fruitcake which is also made from starter sponges. The characteristic yeast is identical with that of San Francisco sourdough. The dominant lactic acid producer was identified as *L. brevis* (Sugihara 1977).

Soda Crackers

There is a considerable similarity between the bacteriological events in sourdough sponges and in cracker dough sponges. Cracker sponges ferment for about 18 hr before the doughs are made up, and a portion of the cracker sponge is often kept for inoculation of the next sponge. Very small percentages of bakers' yeast are used (0.125 to 0.2%). The yeasts multiply rapidly during the first 10 to 15 hr of the sponge fermentation to reach 40×10^6 cells per g. This alcoholic fermentation is followed by a bacterial fermentation which results in the production of lactic acid, a further drop in pH, and a growth to about 75×10^6 cells per g (Micka 1955; Jensen 1974). When the cracker sponges are mixed with additional flour, soda is added to raise the pH and the dough is sheeted after a 4 hr dough fermentation. The leavening action is largely due to formation of steam during the short baking period in a hot oven.

GENERAL LITERATURE REFERENCES

The following book chapters deal with the use of bakers' yeast: Schulz 1962; Harrison 1963; Burrows 1970; Reed and Peppler 1973. The patent literature has been reviewed by de Renzo (1975). General texts dealing at least in part with the science of baking have been published by Matz 1972; Pomeranz and Shellenberger 1971; Pyler 1973; Pomeranz 1971; Sultan 1976; and Jenkins 1975.

REFERENCES

ANON. 1957. Cereal Laboratory Methods, 6th Edition. Am. Assoc. Cereal Chem., St. Paul, Minn.

ANON. 1958. The newest of the continuous dough-making systems. Baker's Dig. *32* (6) 49–52.

ANON. 1975. Census of manufacturers report on wholesale baking. Milling Baking News *54* (1) 15.

ANON. 1978A. Bakery trends '78. Bakery Prod. Marketing *13* (6) 96–124.

ANON. 1978B. United States Industrial Outlook. U.S. Dep. Commer., Bur. Domestic Commerce, GPO, Washington, D.C.

ANON. 1979. Commerce study sees baking growth. Milling Baking News *57* (53) 56–60.

AYKROYD, W.R. and DOUGHTY, J. 1970. Wheat in Human Nutrition. Food Agric. Organ., Rome.

BAKER, J.C. 1954. Continuous processing of bread. Proc. Am. Soc. Bakery Eng., Chicago, 65–79.

BAKER, J.C. and MIZE, M.D. 1941. The origin of the gas cell in bread dough. Cereal Chem. *18*, 19–34.

BARRETT, F.F. 1975. Enzyme uses in the milling and baking industries. *In* Enzymes in Food Processing. G. Reed (Editor). Academic Press, New York.

BELL, B.M. *et al.*   1977.   Physical aspects of the improvement of dough by fat. Food Chem. *2*, 57−70.

BOTTOMLEY, R.A.   1961.   Some observations on the continuous process of bread making. Food Technol. *15*, 423−428.

BRADEN, B.W., JR.   1976.   Yeast raised doughnuts. Proc. Am. Soc. Bakery Eng., Chicago, 127−132.

BRÜMMER, J.M.   1977.   Method for the determination of CO_2 developing from saccharose, maltose, and other types of sugar. Proc. 5th Intern. Ferment. Symp., 1976, W. Berlin. Comm. Ferment. Intern. Union Pure Appl. Chem. (IUPAC). (abstract)

BUCKHEIT, J.T.   1971.   Yeast—its contolled handling in the bakery. Baker's Dig. *45* (1) 46−49, 60.

BURROWS, S.   1970.   Baker's yeast. *In* The Yeasts, Vol. 3. A.H. Rose and J.S. Harrison (Editors).   Academic Press, New York.

CAIN, E.O.   1966.   Formulation and production of wheat bread.   Proc. Am. Soc. Bakery Eng., Chicago, 63−69.

CHEN, S.L., COOPER, E.J. and GUTMANIS, F.   1966.   Active dry yeast. I. Protection against oxidative deterioration during storage. Food Technol. *20*, 1585−1589.

COLE, E.W., HALE, W.S. and PENCE, J.W.   1962.   The effect of processing variables on the alcohol, carbonyl, and organic acid contents of pre-ferments for bread baking. Cereal Chem. *39*, 114−122.

COLLYER, D.   1966.   Fermentation products in bread flavor and aroma. J. Sci. Food Agric. *17*, 440−445.

COOPER, E.J. and REED, G.   1968.   Yeast fermentation: effects of temperature, pH, ethanol, sugars, salt and osmotic pressure. Baker's Dig. *42* (6) 22−24, 28−29, 63.

COTTON, R.H. and PONTE, J.G., JR.   1974.   Baking industry. *In* Wheat: Production and Utilization. G.E. Inglett (Editor).   AVI Publishing Co., Westport, Conn.

DARBY, W., GRIVETTI, L. and GHALIOUNGUI, P.   1977.   Food: Gift of Osiris. Academic Press, London.

DE RENZO, D.J.   1975.   Bakery Products Yeast Leavened. Food Technol. Rev., Vol. 20. Noyes Data Corp., Park Ridge, N.J.

DISTILLER'S CO., LTD.   1976.   Active dry yeast. Brit. Pat. 1,451,793. Oct. 6.

DRAKE, E.   1970.   Frozen bakery goods: a growth market. Baking Ind. J. *3* (2) 36, 38, 41.

DUBOIS, D.K.   1978.   The practical application of fiber materials in bread production. Proc. Am. Soc. Bakery Eng., Chicago, 47−54.

ELTON, G.A.H.   1978.   The role of cereals in the diet of the developed world. *In* Cereals 78: Better Nutrition for the Worlds Millions. Y. Pomeranz (Editor). Am. Assoc. Cereal Chem., St. Paul, Minn.

EUVERARD, M.R.   1967.   Liquid ferment systems for conventional dough processing. Baker's Dig. *41* (5) 124−129.

FEINBERG, A.J.   1963.   Factors governing successful hard roll production. Proc. Am. Soc. Bakery Eng., Chicago, 109–117.

FINNEY, P.L. *et al.*   1976.   Short time baking systems. I. Interdependence of yeast concentration, fermentation time, proof time, and oxidation requirement. Cereal Chem. *53*, 126–134.

FISHER, E.A. and HALTON, P.   1937.   Studies on test baking. I. The technique and some factors affecting fermentation. Cereal Chem. *14*, 349–372.

FLÜCKIGER, R.   1974.   Yeast quality requirements for different baking processes. Getreide Mehl Brot *28* (9) 230–233.

FORD, K.W.   1978.   Specialty breads. Update and new trends. Proc. Am. Soc. Bakery Eng., Chicago, 61–64.

FRANZ, B.   1961.   Kinetics of the alcoholic fermentation during the propagation of bakers' yeast. Nahrung *5*, 458–481.

FUHRMANN, D.F.   1978.   Frozen dough products. Proc. Am. Soc. Bakery Eng., Chicago, 90–95.

GARVER, J.C., NAVARINI, I. and SWANSON, A.M.   1966.   Factors influencing the activation of bakers' yeast. Cereal Sci. Today *11*, 410–418.

GODKIN, W.J. and CATHCART, W.H.   1949.   Fermentation activity and survival of yeast in frozen fermented and unfermented doughs. Food Technol. *3*, 139–146.

GREUP, D.H.   1974.   Active dry baker's yeast. Getreide Mehl Brot *28* (10) 259–263.

HARBRECHT, A. and KAUTZMANN, R.   1967.   Comparative investigation of the determination of the leavening power of bakers' yeasts. Branntweinwirtschaft *107*, 21–23. (German)

HARRISON, J.S.   1963.   Bakers' yeast. *In* Biochemistry of Industrial Microorganisms. C. Rainbow and A.H. Rose (Editors). Academic Press, New York.

HARTMEIER, W.   1976.   Instant soluble dry product from must or other fruit juices and active yeast. Ger. Pat. Applic. 25 15 360. Oct. 14.

HÖHN, E., SOLMS, J. and ROTH, H.R.   1975.   The aroma substances of yeast. II. Sensory evaluation of thiamin and thiamindiphosphate. Lebensm.-Wiss. + Technol. *8*, 212–216.

HSU, K.H., HOSENEY, R.C. and SEIB, P.A.   1979A.   Frozen dough. I. Factors affecting stability of yeasted dough. Cereal Chem. *56* (5) 419–424.

HSU, K.H., HOSENEY, R.C. and SEIB, P.A.   1979B.   Frozen dough. II. Effects of freezing and storing conditions on the stability of yeasted dough. Cereal Chem. *56* (5) 424–426.

JACOB, H.E.   1944.   Six Thousand Years of Bread.   Doubleday Doran and Co., Garden City, N.Y.

JENKINS, S.   1975.   Bakery Technology. Lester and Orpen, Toronto.

JENSEN, O.G.   1974.   Biscuit and cracker technology. *In* Encyclopedia of Food Technology, Vol. 2. A.H. Johnson and M.S. Peterson (Editors).   AVI Publishing Co., Westport, Conn.

KILBORN, R.H. and TIPPLES, K.H.   1968.   Sponge and dough type bread from mechanically developed doughs. Cereal Sci. Today *13*, 25–28, 30.

KLINE, L. and SUGIHARA, T.F.   1968.   Factors affecting the stability of frozen bread doughs. I. Prepared by the straight dough method. Baker's Dig. *42* (5) 44–46, 48–50.

KOCH, R.B., SMITH, F. and GEDDES, W.F.   1954.   The fate of sugars in bread doughs and synthetic solutions undergoing fermentation with bakers' yeast. Cereal Chem. *31*, 55–72.

KOSMINA, N.P.   1966.   *Cited by* Rohrlich, M. and Bruckner, G. *In* Das Getreide. Paul Parey Verlag, Hamburg, W. Germany.

LANGEJAN, A.   1974.   Dried baker's yeast. U.S. Pat. 3,843,800. Oct. 22.

LESAFFRE ET CIE, PARIS.   1978.   New cultures of baker's yeast, procedure to obtain them, and yeasts obtained from these cultures. Belg. Pat. 862.191. June 22.

LORENZ, K.   1974.   Frozen dough. Baker's Dig. *48* (2) 18–19, 22, 30.

LORENZ, K. and BECHTEL, W.G.   1964.   Frozen bread dough. Baker's Dig. *38* (6) 59–63.

LORENZ, K. and BECHTEL, W.G.   1965.   Frozen dough: Effect of bromate level on white bread. Baker's Dig. *39* (4) 53, 56–59.

MAGA, J.A.   1974.   Bread flavor. CRC Crit. Rev. Food Technol. *5*, 55–142.

MAGOFFIN, C.D. and HOSENEY, A.C.   1974.   A review of fermentation. Baker's Dig. *48* (6) 22–23, 26–27.

MASSELLI, J.A.   1959.   The fundamentals of brew fermentation. Proc. Am. Soc. Bakery Eng., Chicago, 160–167.

MATZ, S.A.   1972.   Bakery Technology and Engineering, 2nd Edition.   AVI Publishing Co., Westport, Conn.

MAZUR, P. and SCHMIDT, J.J.   1968.   Interactions of cooling velocity, temperature, and warming velocity on the survival of frozen and thawed yeast. Cryobiology *5* (1) 1–17.

MEIGS, H.T.   1968.   Sweet doughs. Am. Soc. Bakery Eng. Bull. *186*. July.

MERRITT, P.P.   1960.   The effect of preparation on the stability and performance of frozen, unbaked, yeast-leavened doughs. Baker's Dig. *34* (4) 57–58.

MICKA, J.   1955.   Bacterial aspects of soda cracker fermentation. Cereal Chem. *32*, 125–131.

MRDZA, G.   1978.   Trends for specialty breads. Cereal Foods World *23* (11) 635–639.

NG, H.   1976.   Growth requirements of San Francisco sour dough yeasts and baker's yeast. Appl. Environ. Microbiol. *31*, 395–398.

PEPPLER, H.J.   1972.   Yeast. *In* Bakery Technology and Engineering, 2nd Edition. S.A. Matz (Editor).   AVI Publishing Co., Westport, Conn.

PFEFER, D.N.   1976.   English muffins. Proc. Am. Soc. Bakery Eng., Chicago, 51–56.

PIEKARZ, E.R.   1963.   Evaluation of sugars in ferment systems. Proc. Am. Soc. Bakery Eng., Chicago, 118–126.

POMERANZ, Y.   1971.   Wheat Chemistry and Technology. Am. Assoc. Cereal Chem., St. Paul, Minn.

POMERANZ, Y. and SHELLENBERGER, J.A.   1971.   Bread Science and Technology.   AVI Publishing Co., Westport, Conn.

PONTE, J.G., JR., DE STEFANIS, V.A. and TITCOMB, S.T.   1970.   Unpublished data.   ITT Continental Baking Co., Rye, N.Y.

PONTE, J.G., JR., GLASS, R.L. and GEDDES, W.F.   1960.   Studies on the behavior of active dry yeast in bread making. Cereal Chem. 37, 263–279.

PONTE, J.G., JR. et al.   1963.   Some effects of oven temperature and malted barley flour level on breadmaking. Baker's Dig. 37 (3) 44–48.

PYLER, E.J.   1973.   Baking Science and Technology.   Seibel Publishing Co., Chicago.

REED, G.   1972.   Yeast food. Baker's Dig. 46 (6) 16–17, 60.

REED, G.   1974A.   Comparison on the use of commercial yeasts. Process Biochem. 9 (9) 11–12, 32.

REED, G.   1974B.   Pretzels. In Encyclopedia of Food Technology, Vol. 2. A.H. Johnson and M.S. Peterson (Editors).   AVI Publishing Co., Westport, Conn.

REED, G.   1975.   Fermentation defined. Proc. Am. Soc. Bakery Eng., Chicago, 35–42.

REED, G. and CHEN, S.L.   1978.   Evaluating commercial active dry wine yeasts by fermentation activity. Am. J. Enol. Vitic. 29 (3) 165–168.

REED, G. and PEPPLER, H.J.   1973.   Yeast Technology.   AVI Publishing Co., Westport, Conn.

ROBINSON, R.J., LORD, T.H., JOHNSON, J.A. and MILLER, B.S.   1958. The aerobic microbiological population of pre-ferments and the use of selected bacteria for flavor production. Cereal Chem. 35, 295–305.

ROHRLICH, M. and STEGEMAN, J.   1958.   Multiplication of sour dough organisms and acid formation during dough fermentations in a discontinuous process. Brot Gebaeck 12, 41–63.

SATO, T., TSUMURA, N., TANAKA, Y., OKADA, T. and KOYANAGI, Y. 1961.   Comparison of practical characters of bakers' yeasts of the world. Nippon Jozo Kyokai Zasshi 57 (1) 75–80; (2) 67–76.

SCHULZ, A.   1962.   Baker's yeast and baking. In The Yeasts, Vol. 2. F. Reiff et al. (Editors). Verlag Hans Carl, Nurnberg, W. Germany. (German)

SCHULZ, A.   1965.   Investigations of the leavening activity of bakers' yeasts. Brot Gebaeck 19 (4) 61–65.

SCHULZ, A.   1967.   Investigations of the leavening activity of commercial German yeasts. Brot Gebaeck 21 (6) 115.

SCHULZ, A.   1972.   Determination of fermentative power of yeast from a national and international point of view. Ber. Getreidechem.-Tag. Detmold 7 (146) 83–86.

SCHULZ, A.S., ATKIN, L. and FREY, C.N.   1942.   Determination of vitamin B₁ by yeast fermentation method. Ind. Eng. Chem. (Anal. Ed.) *14*, 35−39.

SEELEY, R.D. and ZIEGLER, H.F.   1962.   Yeast. Some aspects of its fermentative behavior. Baker's Dig. *36* (4) 48−52.

SEILING, S.   1969.   Equipment demands of changing production requirement. Baker's Dig. *43* (5) 54−59.

SENTI, F.R.   1971.   Impact of grain production fortification on the nutrient content of the U.S. diet. Cereal Sci. Today *16* (3) 92−102.

SHIRLEY, E.H.   1977.   The Canadian concept of no-time dough systems. Proc. Am. Soc. Bakery Eng., Chicago, 36−43.

SHOGREN, M.D., FINNEY, K.F. and RUBENTHALER, G.L.   1977.   Note on the determination of gas production. Cereal Chem. *54* (3) 665−668.

SLATEN, H.P.   1960.   Production of sour rye bread. Proc. Am. Soc. Bakery Eng., Chicago, 60−67.

SPICHER, G. and SCHÖLLHAMMER, K.   1977.   Comparative studies on the yeasts of "pure culture" and "spontaneous" sour doughs. Getreide Mehl Brot *31* (8) 215−222.

SPICHER, G. and SCHRÖDER, R.   1978.   The microflora of sour dough. IV. The species of rod-shaped lactic acid bacteria of the genus *Lactobacillus* occurring in sour doughs. Z. Lebensm.-Unters.-Forsch. *167*, 342−354.

SUGIHARA, T.F.   1977.   Non-traditional fermentations in the production of baked goods. Baker's Dig. *51* (5) 76−80, 142.

SUGIHARA, T.F. and KLINE, L.   1968.   Factors affecting the stability of frozen bread doughs. II. Prepared by the sponge and dough method. Baker's Dig. *42* (5) 51−54, 69.

SUGIHARA, T.F., KLINE, L. and MILLER, N.W.   1971.   Microorganisms of the San Francisco Sour Dough Bread Process. I. Yeasts responsible for the leavening action. Appl. Microbiol. *21*, 456−458.

SULTAN, W.J.   1976.   Practical Baking, 3rd Edition.   AVI Publishing Co., Westport, Conn.

TANG, R.T. *et al.*   1972.   Quantitative changes in various sugar concentrations during breadmaking. Baker's Dig. *46* (4) 48−55.

THORN, A.J. and REED, G.   1959.   Production and baking techniques for active dry yeast. Cereal Sci. Today *4*, 198−201.

THORN, A.J. and ROSS, J.W.   1960.   Determination of yeast growth in doughs. Cereal Chem. *37*, 415−421.

TRUM, G.W.   1964.   The AMF continuous bread pilot plant. Cereal Sci. Today *9*, 248−254.

TSCHEUSCHNER, H.D., HEINICKEL, U. and QUENDT, H.   1974.   Effects of material and process parameters on gas formation in yeast leavened doughs. Bäcker Konditor *22* (1) 7−11.

TWEED, A.R.   1979.   The production of traditional breads—Iran, Cuba, Syria, Japan. Proc. Am. Soc. Bakery Eng., Chicago, 38−49.

UHRICH, M.   1975.   Formulation of liquid pre-ferments. Proc. Am. Soc. Bakery Eng., Chicago, 42–52.

VAN UDEN, N.   1971.   Kinetics and energetics of yeast growth. *In* The Yeasts, Vol. 2. A.H. Rose and J.S. Harrison (Editors). Academic Press, New York.

VETTER, J.L.   1979.   Frozen, unbaked bread dough: Past, present, future. Cereal Foods World *24*, 42–43.

WISEBLATT, L.   1960.   The volatile organic acids found in dough, oven gases and bread. Cereal Chem. *37*, 734–739.

WOLFE, R.T.   1963.   Production of buns and bun type products. Proc. Am. Soc. Bakery Eng., Chicago, 92–99.

WÖLM, G. and ROTHE, M.   1974.   Short time doughs with wheat bread and flavor effects. III. Effects of yeast concentration. Nahrung *18* (2) 165–170.

9

Wine and Brandy

I. Benda

WINE PRODUCING AREAS AND VOLUME OF WINE PRODUCTION

Wine is the end product of the complete or partial alcoholic fermentation of the juice of fresh grapes. Dessert wines as well as sparkling wines, which are discussed at the end of the chapter, are also considered wines.

The culture of vines (*Vitis* sp.) is carried out in the temperate, subtropical zones of the earth. At present, about 10 million ha are in cultivation; and about 3/4 of this total is accounted for by the classical wine producing countries of the Mediterranean basin. The largest wine producing countries are France with 73 million hl and Italy with 65.9 million hl annually. Total world production is 293 million hl per year. Distribution of this production over the continents of the world is shown in Table 9.1.

TABLE 9.1.   WINE PRODUCTION IN THE WORLD

	1972–1976		1977		1978	
	1000 hl	%	1000 hl	%	1000 hl	%
World, total	319,457	100	289,088	100	292,831	100
Europe	254,418	79.6	222,839	77.1	229,198	78.3
North America	14,348	4.5	17,434	6.0	17,773	6.1
South America	33,041	10.3	33,502	11.6	30,635	10.5
Africa	12,456	3.9	9,113	3.2	9,372	3.2
Asia	1,763	0.6	1,993	0.7	2,160	0.7
Oceania (Australia and New Zealand)	3,431	1.1	4,180	1.4	3,693	1.2

Source: Abbreviated and modified from Mauron (1979).

Total wine consumption in 1978 was roughly 286 million hl as shown in Table 9.2. Italy, France, and Portugal show by far the largest per capita consumption, being in the neighborhood of 100 liters per year.

TABLE 9.2.   WINE CONSUMPTION IN THE WORLD

	1972–1976		1977		1978	
	1000 hl	%	1000 hl	%	1000 hl	%
World, total	282,918	100	282,625	100	286,442	100
Europe	231,962	82.0	228,226	79.6	228,360	79.7
North America	14,978	5.3	16,516	5.8	17,752	6.2
South America	28,085	9.9	32,911	11.5	31,675	11.1
Africa	5,292	1.9	5,281	1.9	5,272	1.9
Asia	1,001	0.3	1,008	0.4	1,059	0.3
Oceania (Australia and New Zealand)	1,590	0.6	2,212	0.8	2,324	0.8

Source: Abbreviated and modified from Mauron (1979).

North America in 1978 produced 17.8 million hl of wine which accounts for about 5% of world production; and it consumed over 17.7 million hl with a per capita consumption of 6.5 liters per year. Undoubtedly, wine production and consumption in the United States are expanding rapidly. The major production area is California which produces 85% of the wine, and which is also known as the principal area of the United States for the cultivation of *Vitis vinifera*. The vineyards of California are located mainly in the coastal regions north and south of San Francisco, in the central valley, and east of Los Angeles. Wine is produced mainly from German and French grape varieties.

In contrast with California the other wine producing areas (Oregon, Washington, Michigan, Arkansas, New York) cultivate varieties of *Vitis labrusca*, for instance the Concord variety. The latter variety contributes a "foxy" flavor note to wines which has been the cause of some controversy. Varieties of *Vitis rotundifolia* are grown in Virginia, North Carolina, South Carolina, and Georgia but are of lesser importance for total wine production. In addition, so-called French hybrids are grown in the Eastern United States. These vines are interspecific crosses of European *Vitis vinifera* grapes with native, wild American grapes. Large-scale cultivation of *Vitis vinifera* grapes in the northern and eastern portions of the United States has not been practical because of the severe winter frosts. However, the cultivation of *Vitis vinifera* in some of these areas is not without interest, and attempts to introduce such vines continue.

The cultivation of interspecific grape varieties in Europe has decreased considerably during recent years. The countries of the European Common Market produce about 45% of the world production of wine and have uniform and strongly enforced regulations for wine production. Since 1970 quality wine has been allowed to be produced only from grapes of the *Vitis vinifera* varieties.

Table 9.3 summarizes wine production in the United States. These figures do not agree fully with those for the production of wine grapes since in some areas, for instance in Illinois, wine is produced from grapes grown in other states.

TABLE 9.3.   WINE PRODUCTION IN THE UNITED STATES
(3785 Liters or 1000 Gal.)

	1974	1975	1976	1977	1978
California	322,194	329,401	330,399	369,091	373,425
New York	36,666	36,664	32,930[1]	33,772[1]	35,833[1]
Illinois	7,420	7,836	7,366[1]	6,659[1]	6,347[1]
Georgia	947	1,600	not available		
Virginia	2,088	1,552	not available		
Washington	906	1,310	not available		
South Carolina	1,471	1,194	not available		
Ohio	566	1,102	not available		
Michigan	1,378	1,049	not available		
Arkansas	728	788	not available		
Other states	1,844	1,475	not available		
Total	376,208	383,971	379,492	417,940	426,773

Source: Bureau of Alcohol, Tobacco and Firearms, U.S. Dep. Treasury; Anon. (1980).
[1]Estimated.
Crop year is July 1 to June 30.

ALCOHOLIC FERMENTATION

Introduction

Grape juice, the raw product for the preparation of wine, is highly variable as it depends on the conditions of growth of grape vines, moreso than any other medium that is used in the fermentation industry. Climatic and edaphic factors of the vineyards as well as methods of cultivation affect the composition of the fermentation medium. In particular the grape variety affects the suitability and quality of the grape juice. Climatic conditions which change from year to year are the principal reason why the fermentation of grape juice results in wines of constantly changing quality. In the final analysis this prevents the complete industrialization of wine production, and consequently, the production of wine is connected with serious problems and risks.

Grape juice contains the following major constituents: Water, carbo-hydrate (glucose, fructose, pentose, pectin), nitrogenous compounds as proteins and protein split products, acids (tartaric and malic), minerals (the potassium, calcium, and magnesium salts of phosphoric, sulfuric, hydrochloric, and silicic acids), tannins, pigments, vitamins, enzymes, and aroma compounds.

Sugar

Sugars and nitrogen compounds which are formed during the growth and maturation of grapes are the principal substrates for the subse-quent fermentation. The content of fermentable sugars in grape juice, glucose plus fructose, depends on the maturity of the grape and can vary between 120 and 250 g per liter (Vogt et al. 1974). The pentoses, principally arabinose and rhamnose, which occur in the order of about 1 g per liter in must, cannot be fermented by yeasts. The two hexoses, glucose and fructose, occur in variable ratios, but generally in the ratio of 1:1. In California wine grapes of the 1958 vintage, the ratio of glucose to fructose varied from 0.80 to 1.07 (Amerine and Thoukis 1958) as shown in Table 9.4. These authors did not find a shift of this ratio with increasing maturity of the grapes while Kliewer (1967) concluded that increasing maturity caused a shift to relatively higher fructose concen-trations.

Any yeast which is capable of fermenting sugars will ferment the monoses glucose, fructose, and mannose. However, yeast species show certain preferences. Most yeast species occurring in must prefer glu-cose, that is, they ferment glucose faster. Other species do not show such a preference or are fructophilic. This characteristic is species dependent (Peynaud and Domercq 1955A) and will be discussed in more detail below. Saccharomyces cerevisiae belongs to the glucophilic species and, therefore, unfermented residual sugar may contain relatively higher percentages of fructose (Ough and Amerine 1963). Some yeast strains of the Sauterne area ferment fructose faster in media which contain equimolar concentrations of the two sugars (Gottschalk 1946; Sols 1956). Gottschalk found that such yeasts change to a glucophilic behav-ior after lyophilization. This leads to the conclusion that the preference of intact Sauterne yeasts for fructose is due to the preferential perme-ability of its cell wall for this sugar. Finally, Dittrich (1977) points out that grape juice with very high sugar concentrations in which only osmophilic yeasts survive (S. rouxii and S. bailii) (see Bambalow 1970; Peynaud and Domercq 1955A) shows a stronger fermentation of fruc-tose. For this reason Minárik et al. (1978) recommend the use of selected glucophilic yeasts for the fermentation of sweet wines of the Tokay type.

TABLE 9.4. GLUCOSE AND FRUCTOSE CONTENTS AND RATIOS OF VARIOUS WINE GRAPE VARIETIES

Variety	Date Harvested	°Balling	Total Acid g Tartaric/100 ml	pH	Glucose g/100 ml	Fructose g/100 ml	Glucose/Fructose
Cabernet Sauvignon	Sept. 3	21.0	0.86	3.42	9.44	9.69	0.98
Delaware	Sept. 3	21.8	0.72	3.48	9.95	10.87	0.92
Gewürztraminer	Sept. 3	22.3	0.73	3.74	10.19	10.78	0.95
Folle Blanche	Oct. 16	18.3	0.81	3.30	8.31	9.22	0.90
Gamay Beaujolais	Aug. 22	20.0	0.94	3.32	9.47	8.86	1.07
Mission	Sept. 23	22.2	0.68	3.64	10.10	12.51	0.81
Palomino	Oct. 23	16.4	0.42	3.89	6.32	7.94	0.80
Sauvignon blanc	Sept. 3	20.0	1.01	3.38	8.97	10.12	0.89
Sylvaner	Sept. 3	21.1	0.95	3.48	9.38	10.95	0.86
Valdepeñas	Sept. 3	20.8	0.81	3.52	10.19	12.22	0.83
White Riesling	Sept. 3	21.2	0.76	3.35	9.63	11.54	0.83

Source: Amerine and Thoukis (1958).

Nitrogen

The concentration of protein and protein split products in grape juice varies with the location and the extent of fertilization of the soil, and the nitrogen content of must is between 0.2 and 1.4 g per liter. Amerine *et al.* (1980) found an average nitrogen content of 0.6 g per liter. Yeasts readily assimilate amino acids but proteins can be hydrolyzed and used for cell growth. Between 50 and 70% of the total nitrogen of musts can be assimilated by yeasts (Tarantola 1955). The quantity of nitrogenous substances is entirely adequate for a vigorous fermentation, and under normal circumstances there is even an excess.

However, a strong reduction of the assimilable nitrogen is found in the juice of grapes which have been infected with the "noble rot" fungus, *Botrytis cinerea*, during maturation. Such juices are highly prized in some European growing areas for the production of high quality wines. The fungus reduces the protein and amino acid nitrogen of the grapes strongly. The content of free amino acids in such musts may be lowered to 50% of its original value (Dittrich and Sponholz 1975). The member countries of the European Economic Community permit the addition of up to 30 g per hl of ammonium phosphate or ammonium sulfate to provide additional nitrogen in readily assimilable form.

pH Value

The pH of grape must varies between 3.0 and 3.9 because of its content of acids, mainly tartaric and malic. The total acid content is between 5 and 15 g per liter, expressed as tartaric acid. Microbiologically the low pH can be considered a favorable and selective factor. Most bacteria with the exception of the acetic acid bacteria and the lactic acid bacteria prefer a neutral to slightly alkaline medium and do not develop at the low pH of must. Therefore, the susceptibility of must to infection is greatly limited. But for yeasts a pH range between 3 and 6 is most favorable for growth and fermentation activity. Beer yeasts and distillers' yeasts can tolerate pH values as low as 2.7 fairly well (Trautwein and Wassermann 1931; Neish and Blackwood 1951), but a change in pH can affect the formation of fermentation by-products. For instance, at higher pH values the concentration of glycerine is increased. There is also a positive correlation between the pH of the must and the formation of pyruvic acid (Rankine 1967A). Ribéreau-Gayon *et al.* (1975B) have also found good toleration of highly acid musts by wine yeasts. However, even within the range from pH 3 to 4 there is a noticeable effect on lag phase and fermentation activity. At higher pH values the lag phase is reduced and fermentation activity increased (Ough 1966A,B). Finally, the effect of pH on growth of yeasts and fermentation activity depends also on the concentration of sugar and ethanol (Neish and Blackwood 1951; Trautwein and Wassermann 1931).

Temperature

Temperature affects yeasts and consequently the course of the wine fermentation considerably. Many publications shed some light on the rather complex interactions in the practice of wine making. Fundamentally it must be stressed that temperature optima as well as maximum and minimum temperatures which can be tolerated are distinctly different for yeast respiration, fermentation, cell growth, and alcohol tolerance. Further it is well known that yeast species and the strains of a particular species may differ considerably with regard to the effect of temperature. The effect of temperature is also dependent in part on the composition of the fermentation medium. One also has to keep in mind that the killing of yeast cells at high temperatures which results in a stuck fermentation is coupled with the danger of infection by thermophilic microorganisms. In connection with this problem one must consider not only the temperature of the cellar and that of the fermenting must, but the development of heat through the exothermic fermentation. On the other hand, low temperatures which prevail in northern areas may produce problems by delaying the onset of fermentation. Apart from the optima for cell growth and the physiological activity of yeasts one must consider problems due to retention or loss of alcohol and aroma substances at low or at high temperatures, respectively.

The fermentative metabolism of yeast cells can be carried out within a rather large temperature range. Maximal values for *S. cerevisiae* are near 40°–45°C or higher (White and Munns 1951) and minimal temperatures approach 0°C (Osterwalder 1934; Saller 1955). Within the 15° to 35°C range the rule is that the higher the temperature the faster the fermentation and the faster its cessation. The lower the temperature the higher the yeast cell count at the end of fermentation. From this it follows that the amount of residual sugar at the end of fermentation is inversely proportional to the fermentation temperature. Independent of this the fermentation intensity diminishes again above a certain temperature which is generally above 35°C. At very low temperatures as used by Saller (1955) the multiplication of cells decreases again as could be expected. However, the emphasis on fermentative metabolism and a reduced evaporation of ethanol lead to higher final ethanol values.

The temperature sensitivity of the yeast is also affected by the ethanol formed during fermentation. This negative effect which affects yeast growth and secondarily affects ethanol yield is greater the higher the temperature (see Table 9.5).

Temperature also affects the formation of by-products. With higher temperatures within the 15° to 35°C range the concentration of glycerin, acetoin, butanediol-2,3 and acetaldehyde increase (Lafon 1955; Rankine and Bridson 1971; see Table 9.6). But contrary findings regarding the formation of glycerol at higher temperatures have also been reported (Brockmann and Stier 1948). For acetic and other volatile acids, concentrations were higher at 25°C than at either 15° or 35°C (Lafon 1955;

TABLE 9.5. INFLUENCE OF TEMPERATURE ON ETHANOL PRODUCTION BY 8 YEASTS IN GRAPE JUICE CONTAINING 23% SUGAR
Means of 4 Replicates.

Strain Number	Ethanol Production (% V/V) At 15°C	At 25°C	At 35°C	Mean
316	12.75	12.38	2.25	9.13
268	13.23	12.95	2.42	9.53
173	12.95	12.91	4.09	9.99
271	13.19	12.84	6.73	10.92
342	13.16	12.70	6.73	10.86
278	13.19	13.01	7.88	11.36
350	13.23	13.10	8.54	11.62
348	13.16	12.85	8.63	11.55
Mean	13.11	12.84	5.91	10.62
Sig. diff.[1] between strains	0.20	0.28	1.16	—

Source: Rankine (1953).
[1] At 0.1% probability level.

TABLE 9.6. EFFECT OF THE TEMPERATURE ON THE FORMATION OF SECONDARY FERMENTATION PRODUCTS

Fermentation Product	Temperature 15°C	20°C	25°C	30°C	35°C
Sugar, fermented, in g/liter	220	191	198	175	139
Ethanol, in vol. %	12.6	11.5	11.9	10.5	7.8
Glycerin, in mM/liter	62	69	68	73	78
Acetic acid, in mM/liter	11.8	15.3	15.5	14.1	14.3
Succinic acid, in mM/liter	6.5	6.1	5.4	5.2	5.3
Acetoin, in mM/liter	1.1	1.1	1.5	1.3	2.2
Butanediol-2,3, in mM/liter	5.3	5.7	6.4	6.4	7.4
Acetaldehyde, in mM/liter	0.6	0.7	1.1	1.2	1.2

Source: Abbreviated from Lafon (1955).

(Lafon 1955; Rankine 1953). Higher temperatures also lead to higher concentrations of pyruvic acid and 2-ketoglutaric acid. For instance the concentration of pyruvic acid is three times higher at 35° than at 20°C (Lafon-Lafourcade and Peynaud 1966).

Finally, the formation of higher alcohols is also temperature dependent with a maximum at 20°C (Dittrich 1977; see Table 9.7). At higher temperatures the formation of higher alcohols decreases and at 35°C it is only about ¼ of the value at 20°C (Peynaud and Guimberteau 1962). Similar

TABLE 9.7.   FORMATION OF HIGHER ALCOHOLS AS A FUNCTION OF TEMPERATURE
In mg/liter.

	Propanol-1	2-Methylpropanol (Isobutanol)	Amyl Alcohols	Hexanol	2-Phenylethanol
10°C	33	52	80	4	25
20°C	42	62	90	4	28

Source: Dittrich (1977).
Must from grapes of the variety Müller-Thurgau, harvested 1970.

observations have been made for the 10° to 33°C range. Here the highest concentrations of isoamyl alcohol, active amyl alcohol, and isobutanol were formed at 24°C while the formation of n-propanol was minimal at that temperature (Ough et al. 1966).

The proper choice of a temperature for the practical fermentation of wine depends on a number of factors so that actual fermentation temperatures vary within a rather wide range, from about 10° to 30°C. The outdoor temperature in various wine producing areas differs greatly so that it is sometimes necessary to cool the must so that its temperature does not exceed 30° to 35°C during fermentation. In northern countries the must is sometimes heated in order to shorten the lag period.

For the production of white wine in Germany temperatures between 18° and 20°C have been recommended by Troost (1972) while lower temperatures of 10° to 16°C have been recommended for California (Amerine et al. 1980). The advantages of a lower fermentation temperature are the fresher and more fruity character of the wine, the formation of more ethanol and smaller losses of ethanol, as well as a reduction in the danger of bacterial infections and, consequently, a lesser danger of producing volatile acidity. For these reasons low fermentation temperatures have already been recommended by Saller (1955) and Tchelistcheff (1948).

For red wines one chooses a temperature between 22° and 30°C as long as the grapes are fermented on the skins in order to increase color extraction. A richer aroma has also been reported by Amerine and Ough (1957) for such higher temperatures. Further information on the effect of temperature in wine production can be found in the extensive study of Ough and Amerine (1966) and the investigations of Cantarelli (1966).

Sugar Concentration

Some musts have very high sugar concentrations, for instance, the must made from specially selected berries (Auslese). For such musts the high osmotic pressure has a negative effect on yeast cells since both growth of yeasts and fermentation activity are lowered. This osmotic effect can be observed at sugar concentrations exceeding 300 g per liter. At higher sugar levels alcohol formation decreases as shown in Table 9.8. The inhibition is further increased by the ethanol which is formed during the

TABLE 9.8.   EFFECT OF SUGAR CONCENTRATION ON ALCOHOL FORMATION

Sugar Concentration of Must in g/liter	Ethanol Concentration Formed in 2 Months, in Vol. %
370	8.6
420	6.3
470	5.9
550	3.4
750	0

Source: Ribéreau-Gayon et al. (1975B).

fermentation. In high quality musts from selected berries which have been infected with *Botrytis cinerea* the inhibition of yeast growth is not just due to a lack of nitrogenous materials but also to the very high sugar concentration (Dittrich 1977).

The tolerance of high sugar concentrations differs for various yeast species. Within the genus *Saccharomyces*, the species *S. bailii* var. *osmophilus, S. rouxii,* but also *S. italicus,* are considered as osmotolerant or osmophilic (Munitis *et al.* 1976). Further information is available in Schröder (1960) and Windisch (1969). A disagreeable side effect of the fermentation of highly concentrated musts is the increased formation of acetic acid by yeast (Dittrich 1977).

Carbon Dioxide

Fundamental studies on the effect of carbon dioxide in cellar practice have been carried out by Geiss (1952) and Schmitthenner (1949). Based on their work a process for the fermentation under CO_2 pressure has been developed in Germany. This process is not widely practiced at present but a process for the storage of slightly fermented sweet must under CO_2 pressure has assumed significance in recent years. High CO_2 concentrations in must inhibit yeast growth. At 15 g per liter CO_2 (about 7.2 atm at 15°C) yeast growth ceases (see also Kunkee and Ough 1966). However, sugar fermentation is carried out by yeast cells up to a pressure of 14 atm. Finally, at about 30 atm CO_2 pressure the yeast cells are killed. It is, therefore, possible to control the fermentation at CO_2 concentrations of less than 15 g per liter. Yeast growth is inhibited and the fermentation can also be kept at a certain limited rate. The limitation of yeast metabolism under CO_2 pressure saves sugar, and higher alcohol yields can be obtained than under normal conditions of fermentation (Troost 1972). An additional advantage for certain markets is the possibility of retaining higher levels of residual sugars. The formation of higher concentrations of volatile acids is a disadvantage of fermentations under CO_2 pressure. Also, the metabolism of lactic acid bacteria is not inhibited which may result in a malo-lactic fermentation and its undesirable consequences (Koch *et al.* 1953). These observations have recently been confirmed by Mayer (1974). They clearly indicate that multiplication of *Leuconostoc oenos* in wine is stimulated by higher CO_2 concentrations.

In California the fermentation under CO_2 pressure has not been practical. Amerine and Ough (1957) concluded that their negative results are due to increased production of volatile acids by bacteria, a condition which is favored by the higher pH values of California musts.

In alcoholic media the inhibition of yeast growth due to CO_2 pressure is increased. At relatively low CO_2 concentrations of 0.6 to 1.8 g per liter, growth is inhibited. However, this effect depends also on the original yeast cell counts in the medium, and this is also true at higher CO_2 concentrations (Haubs *et al.* 1974). In addition, the effect of CO_2 pressure on yeast depends on pH. Inhibition is increased at lower pH values (Kunkee and Ough 1966).

Sulfur Dioxide

The effect of SO_2 (sulfurous acid) on the course of the fermentation must be discussed here although its effect has much broader enological significance. Basically one must distinguish between the biological and the chemical effect of SO_2. The biological effect comprises an inhibition of undesirable microorganisms of the must which affect the course of the fermentation and the quality of the wine negatively. This group of microorganisms includes the acetic acid bacteria and in certain cases the lactic acid bacteria, as well as some yeast species naturally present in must.

The chemical effect of SO_2 is desired to bind acetaldehyde which is formed during the fermentation and which has undesirable organoleptic properties. Further, SO_2 is added to prevent oxidative reactions caused as the result of enzymic reaction or of a strictly chemical nature, although SO_2 is less effective in inhibiting strictly chemical oxidation.

From a biological point of view the use of SO_2 is of great importance in fighting acetic acid bacteria. These aerobic organisms can multiply in warm weather on injured or fungus-infected berries on the vines. Additional growth of acetic acid bacteria occurs if the harvested grapes are kept for a longer period prior to fermentation, particularly in a warm climate. These bacteria are feared because of their production of acetic acid and acetic acid esters which lower the quality of the wine and may make it completely unpalatable. Acetic acid also inhibits yeast fermentation, particularly in conjunction with ethanol. A concentration of 1 g per liter of acetic acid can inhibit the rate of fermentation and at 3 g per liter fermentation can cease under certain circumstances (Schanderl 1959). *Saccharomyces cerevisiae* is particularly sensitive to the presence of acetic acid while other species, such as *Saccharomycodes ludwigii* and *Schizosaccharomyces pombe*, are less sensitive.

For musts which are low in total acidity it is often desirable to use SO_2 to inhibit malo-lactic fermentation by lactic acid bacteria to prevent a further diminution of the acid level. Finally, it is desirable to suppress certain yeast species such as *Kloeckera apiculata* and *Metschnikowia pulcherrima* (formerly *Candida pulcherrima*) with SO_2 to retain good wine quality. For instance, *Kloeckera apiculata* can produce more than 1 g per liter of volatile acids (Benda 1970; Sponholz and Dittrich 1974; see also Table 9.9). In addition both *K. apiculata* and *M. pulcherrima* produce higher concentrations of pyruvic and 2-ketoglutaric acid, which is a disadvantage because it leads to the binding of larger amounts of SO_2 (Sponholz and Dittrich 1974). Apiculate yeasts are normally present in musts in high numbers and may inhibit the development of *S. cerevisiae* (Schulle 1953). This leads to a slower start of the fermentation proper and to increased SO_2 requirements in processing.

TABLE 9.9.   PRODUCTION OF VOLATILE ACIDS BY IMPORTANT WINE YEASTS

Yeast Species	Volatile Acids in g/liter (as Acetic Acid)
Kloeckera apiculata (strain 1)	1.09–1.13
Kloeckera apiculata (strain 4)	1.01–1.04
Metschnikowia pulcherrima (strain 5)	0.13–0.14
Torulopsis stellata (strain 3)	1.08–1.28
Saccharomyces rosei (strain 6)	0.11–0.14
Saccharomyces uvarum (strain 8)	0.51–0.67
Saccharomyces uvarum (strain 10)	0.42–0.45
Saccharomyces cerevisiae (strain 2)	0.50–0.77
Saccharomyces cerevisiae (strain 9)	0.34

Source: Benda (1970).

SO_2 has also a direct effect on the course of the fermentation by *Saccharomyces* yeasts. It delays the onset of fermentation, and the lag period is longer the greater the amount of H_2SO_3 added to the must. For instance the addition of 100 mg per liter delayed the onset of fermentation by 3 days, while the addition of 200 mg per liter delayed it by 20 days (Schanderl 1959). The metabolism of yeast is inhibited only by "free" SO_2 (Ingram 1948, 1958). As soon as "free" sulfurous acid is bound in the form of sulfonates, the original inhibition ceases and fermentation proceeds normally since the sulfonates hardly affect yeast metabolism (Rehm *et al.* 1965). Sulfonates are mainly formed with acetaldehyde, pyruvic acid, and 2-ketoglutaric acid (Burroughs and Sparks 1973; Kielhöfer 1958; Kielhöfer and Würdig 1960A,B,C). Rehm *et al.* (1965) could not exclude the possibility that sulfonates increase the inhibiting effect of free sulfurous acid.

The toxic effect of SO_2 is due to dissolved, molecular SO_2. In the medium there is a dynamic equilibrium of the various forms of free SO_2, namely SO_2 proper, HSO_3^- and $SO_3^=$, which is dependent on pH. Molecular SO_2 is taken up by the yeast cells and the degree of toxicity is determined by the concentration of molecular SO_2 (Macris and Markakis 1974). Manometric measurements with *S. bayanus* have been carried out at low pH values by Bréchot *et al.* (1969). The effect of SO_2 led to an inhibition of anaerobic fermentation by 30–40%, an inhibition of respiration by 40–80%, and an increase in aerobic fermentation by 30–60%. In the absence of SO_2 the Pasteur effect is normal; in its presence the Pasteur effect is decreased or negative.

The effect of SO_2 on the yeast cells varies from species to species; and within any species there is a difference in sensitivity among strains. Some yeast species, for instance *Saccharomycodes ludwigii* and *S. bailii* (synonym: *S. acidifaciens*), show a greater tolerance (Schanderl 1962). The rate of cell kill by SO_2 for various yeasts has been determined by Macris and Markakis (1974). Its effect is also dependent on pH of the medium to which it is inversely proportional (Ingram 1948). It also depends on the ethanol concentration, and therefore the inhibiting effect is greater in wine than in must.

The effect of free and bound SO_2 on wine bacteria has been reported by Lafon-Lafourcade and Peynaud (1974). The bound form has a 5 to 10 times lesser effect than the free form. However, in some wines bacterial decomposition of acids could be inhibited by 80 to 100 mg per liter of bound SO_2.

In practice the requirement for SO_2 addition depends on the presence of compounds which bind H_2SO_3. These are present in various concentrations in must and their formation during fermentation depends on processing conditions and to a certain extent on the presence of microorganisms which participate in the fermentation. The addition of SO_2 causes an increased production of aldehyde by yeasts. Therefore, it is not desirable to add SO_2 during the fermentation in order to prevent an increase in the SO_2 requirement (Kielhöfer and Würdig 1960A,B,C). Berries infected with *Botrytis cinerea* require higher amounts of SO_2 since some metabolites of the fungus and the accompanying acetic acid bacteria, such as 2-ketogluconic acid, 2,5-diketogluconic acid, 5-ketofructose, and xyloson bind SO_2 (Peynaud and Sapis 1975; Ribéreau-Gayon 1973). A higher requirement for SO_2 for wines from *Botrytis* infected grapes is also caused by a lack of thiamin in the must. During the fermentation the yeast is not able to synthesize enough thiamin to decarboxylate the keto acids, pyruvic acid and 2-ketoglutaric acid (Dittrich *et al.* 1974). The requirement of young wines for SO_2 is higher after a slow fermentation than after a vigorous fermentation, since slow fermentations lead to the accumulation of SO_2 binding substances in the medium (Dittrich 1977; Zürn 1976).

Finally, it should be mentioned that several yeast strains produce higher concentrations of SO_2 during the fermentation by reduction of the sulfate of the must (Eschenbruch 1974; Mayer and Dufour 1973; Würdig and Schlotter 1971). Under commercial conditions *Saccharomyces* strains can produce up to 130 mg SO_2 per liter. Dittrich and Staudenmayer (1968) report that from 162 investigated yeast strains, 6 strains produced more than 50 mg SO_2 per liter, of which 1 strain produced 103 mg per liter, and another even 155 mg per liter. The property of SO_2 formation is constitutive. It is not species specific but strain specific (Dott *et al.* 1976; Minárik 1975). For yeast strains producing much SO_2 it can be assumed that there is faulty regulation of the biosynthesis of enzymes responsible for the assimilatory metabolism of sulfur compounds (Bonish and Eschenbruch 1976; Heinzel and Trüper 1976). Further information on the sulfate and sulfite metabolism of *S. cerevisiae* is found in Dott *et al.* (1977), Eschenbruch *et al.* (1973), and Eschenbruch and Bonish (1976A,B).

YEASTS

In contrast with media used in other branches of the fermentation industry, grape must carries with it the yeast flora which is required for the alcoholic fermentation. During the past two decades numerous investigations have dealt with the yeast flora of must and wine, and specifically with ecological and physiological questions. In many wineries in the United States, Canada, South Africa, and Australia, pure yeast cultures are used commercially. However, in the majority of the wine producing countries the spontaneous fermentation by naturally occurring yeasts is still largely practiced. In the following section the more important results of these investigations will be summarized and discussed.

Ecology

Grape berries are the habitat of various yeast species while still on the vine. The decisive factor for the yeast propagation is the sugar-containing grape juice which reaches the surface of the berries from the interior. Belin (1972, 1977) investigated various parts of grapes with scanning electron microscopy. He found relatively small numbers of yeast cells. Normally yeast cells occur at the upper portion of the pedicel on the lenticels as well as on those portions of the grape skin which surround the stomata. Larger numbers of cells are found where the surface of the grape skin is cut or torn and where the juice of the berries exudes. It is known that the number of yeast cells increases greatly if the berries have been injured. Apart from this the number of cells increases with increasing maturity of the grapes.

Grape juice which has been obtained aseptically from berries contains yeast cells which number in the order of 10^3 to 10^4 cells per ml (Peynaud and Domercq 1953; see also Table 9.10). After crushing the number of cells jumps rapidly and the freshly pressed must contains from 10^5 to 10^6 yeast cells per ml. These results suggest that the major source of inoculation of the must with yeasts is the cellar equipment. Yeast cells can multiply well on the surface of this equipment which has been wetted with grape juice. Comparable results with cell counts have been reported by Barnett *et al.* (1972).

Yeasts are transferred to grapes by insects. These insects are mainly bees, wasps, and species of the genus *Drosophila*. The insects are strongly attracted by volatile aroma compounds which are developed during the maturation of grapes, and they feed on the juice of the berries. Extensive investigations by Benda (1960) have clearly shown that wind is not respon-

TABLE 9.10.   YEAST CELL COUNTS AT VARIOUS STAGES OF THE GRAPE HARVEST
Number of Cells per ml of Must.

Sample Number	Aseptically Harvested Grapes	Grapes in Collecting Buckets[1]	Must After Pressing
1	<1	5	800
2	30	—	1360
3	26	—	—
4	1	25	—
5	120	—	—
6	160	23	—
7	2	40	660
8	50	10	6400
9	4	280	460
10	65	—	—
11	<1	2	—
12	1	4	1800
13	1	15	1500

Source: Peynaud and Domercq (1953).
[1]Harvested in the usual manner.

sible for the transfer of yeast cells to grape berries. Di Menna (1955) and
Parle and Di Menna (1966) have obtained identical, negative results, but
one does isolate live yeasts from the body surface and from the intestinal
tract of captured insects. Wine yeasts which are normally found on grapes
were isolated from the intestines of bees and wasps. Such yeasts were *K.
apiculata, M. pulcherrima,* and in smaller numbers *S. cerevisiae* and *Torulopsis stellata.* It was interesting that live yeasts could still be found in the
middle and end intestines after the bees had been in winter rest for 1 1/2
months. Bees in winter rest do not defecate until the first "cleaning" flight
in spring (Mooser 1958).

Similar results were reported by Stević (1962). The yeasts he found in
bees and wasps were mostly species of *Torulopsis,* but *Candida, Pichia,
Saccharomyces,* and other genera were also represented. The yeasts can be
found on the bees during the time of active flight as well as during the
winter rest. About 65–70% of the bees still are carriers of live yeast at the
end of the winter rest period. In spring these yeasts are transferred to the
nectar of flowers and to ripening fruits; and rapid yeast multiplication can
begin again in nature.

The relationship of *Drosophila* species to this cycle of yeasts is also
interesting. American workers have shown that yeasts are an important
constituent of the diet of adult flies and of larvae. Numerous yeast species
could be isolated from the crop of the fruit flies (Dobzhansky *et al.* 1956).

Fruit flies are regularly present on fermenting substances (McKenzie
1974). They are attracted by various metabolites of yeasts in a specific way
(Wolf and Benda 1965). The specific reaction of *Drosophila melanogaster*
toward such metabolites permits a distinction not only among species but
also among strains of yeasts (Benda and Wolf 1965; Wolf and Benda 1965,
1967).

The currently held opinion that yeasts survive the winter in the soil
cannot be maintained on the basis of newer investigations. While some

authors isolated wine yeasts from soil (Capriotti 1955; Di Menna 1959; Švejcar 1969), others could find no yeasts or only small numbers of live yeasts (Benda 1960; Parle and Di Menna 1966; Stević 1962); furthermore, a transfer of yeasts from the soil to berries is rather unlikely; and finally, the number of live yeast cells on uninjured grape berries is rather small. The investigations of Mooser (1958) and Stević (1962) which have been mentioned indicate that bees play a dominant role in the distribution of yeasts in nature and in their survival during winter.

So far it has not been possible to prove that the yeast flora depends on climate although some findings point in that direction. For instance, Castelli (1957) found that the asporogenous species *K. apiculata* and *K. corticis* decrease numerically as one moves from northern Italy to the south. In contrast the sporogenous species of *Hanseniaspora*, mainly *H. guilliermondii*, are found exclusively in southern Italy and in Sicily. Benda (1962A) found a certain dependence on climate. During the cool and rainy season of 1960, a greater number of *M. pulcherrima* and a lesser number of *K. apiculata* and *Aureobasidium pullulans* could be found in comparison with the dry and warm season of 1959.

Numerous reports deal with the yeast flora of musts (Benda 1964; Castelli 1957; Domercq 1956; Minárik 1966; Mrak and McClung 1940; Van Zyl and Du Plessis 1961). An enumeration of all yeast species which have so far been isolated from grapes, must, and wine is available from Kunkee and Goswell (1977). It is worthwhile noting that the reports of individual authors from various wine producing regions give parallel results. Basically, the greatest number of yeast species in musts does not belong to the *Saccharomyces* species, which are preferred for wine fermentations. Only about 1–10% of the yeasts present in the original must belong to *Saccharomyces* if one obtains the must in a sterile manner. The picture changes in favor of larger numbers of *Saccharomyces* yeasts as soon as the appearance of juice on the grapes attracts insects and leads to a faster multiplication of yeasts. Finally, one has to consider that in practice larger numbers of germs of *Saccharomyces* yeasts are transferred to the crushed grapes from vats and presses in the winery. Therefore, *Saccharomyces* yeasts can constitute a higher percentage of total yeasts at the beginning of the fermentation.

The earliest representatives of yeasts in must are principally those belonging to *K. apiculata*. These have been found in all wine growing regions. A close relative of this species, *K. corticis* (synonym *K. magna*), is found in central Italy. Its sporogenous form, *Hanseniaspora valbyensis* (synonym *H. guilliermondii*) can be found under exceptional conditions in southern Italy and Sicily. Castelli (1957) ascribes this finding to climatic conditions. In some wine producing areas larger numbers of *M. pulcherrima* have also been found, mainly in Germany, Czechoslovakia, and Spain. The latter species is rarely found in France. The species *T. stellata (T. bacillaris)* which is characteristic of German musts should also be mentioned. Typical film forming yeasts such as *Hansenula anomala, Pichia membranaefaciens, P. fermentans, Candida krusei,* and *C. vini,* as well as representatives of *Debaryomyces* and *Brettanomyces* and other yeasts, are either absent from musts or play only a minor role. Apart from the yeasts which have already been mentioned, *Aureobasidium pullulans* has been found on grapes and in must in German wine producing areas. This fungus cannot be readily classified taxonomically. According to Cooke (1962) it belongs to the so-called "black" yeasts; and according to Von Arx (1968) the fungus is a transition form between the imperfect yeasts of Cryptococcaceae and Sporobolomycetaceae and the Moniliales. In the system of Ainsworth (1973) it is classed with the Deuteromycotina. The hyphae of *A. pullulans* form many yeast-like budding blastoconidiae. Older colonies are remarkable because of their brown or black-brown hyphae (see Fig. 9.1 and 9.2).

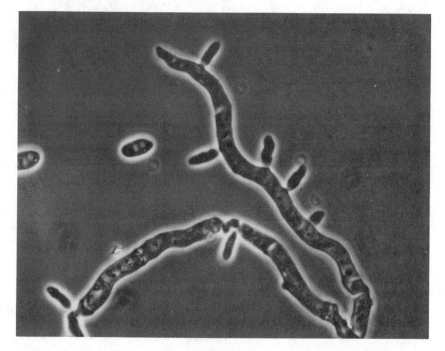

FIG. 9.1.   *AUREOBASIDIUM PULLULANS* (MYCELIUM WITH BLASTOSPORES)

With the start of the fermentation the picture of the yeast flora shifts rapidly. This shift is caused by the addition of SO_2 to the crushed grapes or must, and by the increasing alcohol concentration which favors the stronger fermenting and more alcohol tolerant yeasts (see Fig. 9.3 and Table 9.11).

Generally the fermentation begins with the weakly fermenting yeasts, and the main fermentation and the end of the fermentation are carried out by the strongly fermenting true wine yeasts of the genus *Saccharomyces*.

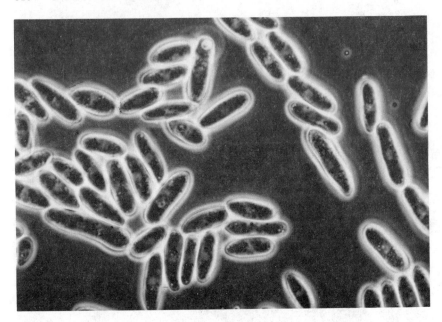

FIG. 9.2.   *AUREOBASIDIUM PULLULANS* (BLASTOSPORES)

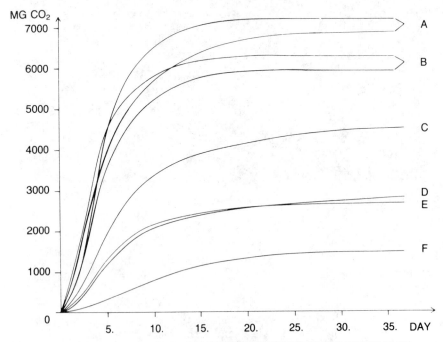

FIG. 9.3.   FERMENTATION ACTIVITY OF YEASTS OCCURRING IN GRAPE MUST
A—*Saccharomyces cerevisiae*. B—*Saccharomyces uvarum*. C—*Saccharomyces rosei*. D—*Torulopsis stellata*. E—*Kloeckera apiculata*. F—*Metschnikowia pulcherrima*.

TABLE 9.11.   THE NATURAL YEAST FLORA OF MUST

Start of the Fermentation	Main Fermentation	End of the Fermentation
[1]*Kloeckera apiculata*	[1]*Saccharomyces cerevisiae*	[1]*Saccharomyces cerevisiae*
[1]*Metschnikowia pulcherrima*	[1]*Saccharomyces uvarum*	*Saccharomyces bayanus*
[1]*Torulopsis stellata*	*Saccharomyces bayanus*	
Kloeckera corticis	*Saccharomyces chevalieri*	
Candida krusei	*Saccharomyces delbrueckii*	
Candida vini	*Saccharomyces fermentati*	
Hansenula anomala	*Saccharomyces florentinus*	
Hansenula subpelliculosa	[2]*Saccharomyces rosei*	
Pichia fermentans	*Saccharomyces rouxii*	
	Kluyveromyces veronae	

Source: Modified from Benda (1962A).
Yeast species were isolated from grape musts of *Franconia* during the course of the fermentation.
[1]Predominant species during the different stages of the fermentation.
[2]*S. rosei* tolerates relatively small concentrations of ethanol and appears between the start of the fermentation and the main fermentation.

These are of prime importance for the production of wine. The number of *Saccharomyces* species found in fermenting grape musts is very large, but *S. cerevisiae* comprises the major portion in all wine producing areas. *S. uvarum* and *S. oviformis* are strongly represented but occur in changing percentages in the various regions. Table 9.12 summarizes the occurrence of *Saccharomyces* species in fermenting musts.

Taxonomy and Physiology

All yeast species which are important for the production of wine or at least those which occur frequently in fermenting musts will be discussed in this section. Species of the genus *Saccharomyces* are undoubtedly the most important yeasts. They are able to produce high concentrations of ethanol in spontaneously fermenting musts and they are responsible for the main and end fermentation. If pure culture yeasts are added to the must these yeasts are responsible for all phases of the fermentation. Nevertheless, one should not underestimate the other yeast genera which may affect the quality of the end product, even though this effect has not been fully clarified. Some physiological data are available for these other genera, and some interesting observations have been made with regard to their effect on the aroma of the wine. Relatively little is known about the mutual effect of the various yeast species and strains within a population although some interesting details have been observed.

Basically one has to distinguish between yeast species which participate in the fermentation and those which may become spoilage organisms after

TABLE 9.12. OCCURRENCE OF SPECIES OF THE GENUS *SACCHAROMYCES* IN FERMENTING MUST

Always[1] or Usually[2] Present	Frequently Present	Rarely Present
S. cerevisiae[1]	*S. bailii* var. *bailii*	*S. bisporus* var. *mellis*
S. uvarum[2]	*S. bayanus*	*S. capensis*
	S. chevalieri	*S. coreanus*
	S. exiguus	*S. delbrueckii*
	S. heterogenicus	*S. fermentati*
	S. italicus	*S. florentinus*
	S. rosei	*S. globosus*
		S. kluyveri
		S. microellipsoides
		S. rouxii
		S. transvalensis

Comprehensive listing of *Saccharomyces* species occurring in wine producing districts which have been studied.

the end of the fermentation. It is not pertinent that some spoilage yeasts are already present in the must—usually in small numbers—since they do not normally multiply during the fermentation, unless they are species which cause after-fermentations in the bottle. Spoilage yeasts will be discussed in a separate section.

In this section following, individual species will be described with regard to enologically important properties. For the determination of species the reader is referred to the standard texts by Lodder and Kreger van Rij (1952) and Lodder (1970), and for a discussion of their physiology from an enological point of view to Domercq (1956), Ribéreau-Gayon and Peynaud (1960), and Schanderl (1959).

With regard to the description of individual yeast species it must be pointed out that the strains of a species may differ greatly, mainly in physiological characteristics. This is important for the selection of pure culture yeasts.

Several species of *Saccharomyces* are the more important yeasts for winemaking as has already been mentioned, and they constitute the true wine yeasts in the narrower sense. All species of this genus ferment well. Table 9.13 indicates that a large number of species may participate in the fermentation. But this number can be narrowed to a smaller group which is of real importance. Within this group *S. cerevisiae* is the principal yeast. Table 9.13 shows some physiological properties of *Saccharomyces* species which are of importance in wine making. The data shown are either of value for the taxonomy of these yeasts or they are of enological importance.

TABLE 9.13.   PHYSIOLOGICAL CHARACTERISTICS OF SACCHAROMYCES SPECIES IMPORTANT IN WINE MAKING

Yeast Species	Sugar Assimilation					Sugar Fermentation						Ethanol[1] % by Vol.	Volatile Acids[1] as Acetic Acid in g/liter
	Glu-cose	Galac-tose	Su-crose	Mal-tose	Lac-tose	Glu-cose	Galac-tose	Su-crose	Mal-tose	Lac-tose	Raffinose		
S. bailii	+	− or +	− or +	−	−	+	−	− or +	−	−	− or 1/3		
strain "acidifaciens"												8.0–12.6	0.67–3.23
strain "elegans"												8.1– 9.7	0.56–2.17
S. bayanus	+	−	+	+	−	+	−	+	+	−	1/3		
strain "bayanus"												11.7–12.9	1.02–1.31
strain "oviformis"												11.9–18.4	0.29–2.17
S. cerevisiae	+	+	+	+	−	+	+	+	+	−	1/3	8.0–16.8	1.00–1.87
S. chevalieri	+	+	+	−	−	+	+	+	−	−	1/3	11.2–18.3	0.48–2.17
S. exiguus	+	+	+	−	−	+	+	+	+	−	1/3	12.6–14.8	0.63–1.10
S. italicus	+	+	+	− or +	−	+	+	+	+	−	−	12.6–17.0	1.46–2.09
S. rosei	+	−	+	+	−	+	−	+	−	−	1/3	10.1–13.0	0.00–0.24
S. rouxii	+	+ or −	− or +	− or +	−	+	−	− or +	+	−	−	13.0–14.9	0.77–1.14
S. uvarum	+	+	+	+	−	+	+	+	+	−	3/3		
strain "uvarum"												7.2– 9.9	0.52–0.75
strain "carlsbergensis"												7.3–13.0	0.89–2.10

Source: Data for sugar assimilation and fermentation from Lodder (1970), and for ethanol and volatile acids from Ribéreau-Gayon and Peynaud (1960) and Melas-Joannidis *et al.* (1958).
[1]Under aerobic conditions.

Saccharomyces cerevisiae.—(Synonyms: *S. cerevisiae* var. *ellipsoideus*, *S. ellipsoideus*, *S. vini.*) As a wine yeast, this important species is known as *S. cerevisiae* var. *ellipsoideus*. But recent revisions of the taxonomy of yeasts have eliminated it as a separate variety (Van der Walt 1970B).

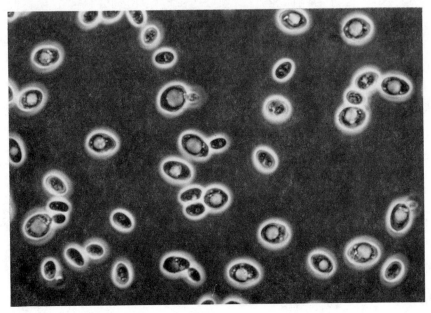

FIG. 9.4.   *SACCHAROMYCES CEREVISIAE*

Many strains of this species have been isolated and propagated but most of these are presumably not that different from each other. But the literature describes several strains which show quantitative differences in one or several physiological characteristics. Schanderl (1959) reports on the selection of strains of *S. cerevisiae* with the following characteristics:

High ethanol production of 18–20% by vol.

Cold resistance (ethanol formation of 8–12% by vol. at temperatures of 4° to 10°C; "cold fermenting yeasts"

Alcohol tolerance (normal fermenting power at ethanol concentrations of 8–12% by vol. in a sugar-containing medium; yeasts used for secondary alcoholic fermentation)

Resistance against SO_2 ("sulfite yeasts")

Osmotolerance (alcohol formation of 10–13% by vol. in a medium with an original sugar concentration of 30%; "Trockenbeerenauslese yeasts")

High alcohol formation with fast sedimentation and the formation of flocculent deposits ("Champagne yeasts")

Flor or film formation in the presence of air after the alcoholic fermentation, combined with rapid formation of the specific compounds of the
sherry bouquet ("Sherry yeasts")

It is not known to which extent the described properties are true, constant
characteristics of these strains. But it is likely that they are true strain
characteristics because the yeasts have been kept for many years in industrial collections. Other characteristics are probably also genetically fixed
properties, such as the formation of hydrogen sulfide, quantitative differences in the formation of acetaldehyde, glycerin, and volatile acids (Rankine 1953), and in volatile aroma substances formed by yeasts. Strains of *S.
cerevisiae,* among others, can also be differentiated by *Drosophila melanogaster* on the basis of different aroma substances formed by yeast metabolism (Benda and Wolf 1965; Wolf and Benda 1965).

S. cerevisiae can be found in all wine producing regions. The species is
characterized by its ability to form relatively high concentrations of ethanol
which is in the range of 8–17% by vol. (Ribéreau-Gayon and Peynaud
1960). Rankine (1953) used 44 strains of this species. With a sugar content
of the must of 23% he found a range of 8–15% by vol. Only 23 of the strains
tested formed more than 14% by vol. of ethanol. Only small cell counts can
be obtained with healthy grapes, but in fermenting must the yeast multi-
plies rapidly and displaces other yeast species. It becomes dominant because of its ability to form much ethanol and because of its tolerance to
ethanol. In many instances *S. cerevisiae* is the only yeast responsible for the
final stages of the fermentation.

S. cerevisiae has assumed great importance as a pure culture yeast for use
as a "yeast starter." Its relatively high resistance to ethanol, however, can
also lead to undesirable after-fermentations in wines containing residual
sugar, if there is a reinfection with the yeast during bottling (Domercq
1956; Van der Walt and Van Kerken 1958).

Saccharomyces bayanus.—(Synonyms: *S. oviformis, S. pastorianus, S.
beticus, S. cheriensis.*) During recent revisions of the taxonomy of yeast
classification the former species "oviformis" and "pastorianus" have been
included with *S. bayanus* because of the lack of sharp criteria for physiological or morphological differences (Van der Walt 1970B). In addition, the
former species *S. beticus* and *S. cheriensis* whose names are so frequently
found in the literature have been included. There are some interesting
differences among these groups, however, although they cannot be differentiated in a systematic manner. For instance, the strain "pastorianus" is
used in brewing because of its ability to ferment maltotriose well (among
other reasons). The strains "beticus" and "cheriensis" are important for
the production of sherry as film forming yeasts (Iñigo Leal and Arroyo
Varela 1964). They form films or mats on the surface of wines with higher
alcohol concentrations, while changing from a fermentative to a respiratory
metabolism, and they form much acetaldehyde and esters. The types "bayanus" and "oviformis" are frequently used for the production of table wines.

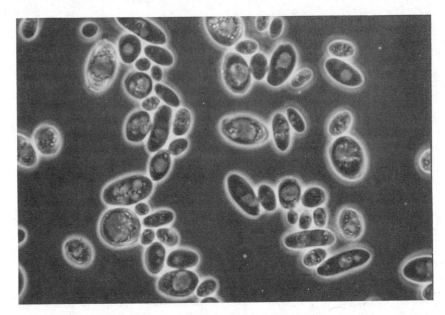

FIG. 9.5.   *SACCHAROMYCES BAYANUS* (STRAIN "OVIFORMIS")

The most important physiological difference is their ability to form ethanol in different concentrations. "Bayanus" strains (3 strains tested) gave on the average 10% by vol. of ethanol aerobically and 12.3% anaerobically, while "oviformis" strains (86 strains tested) gave 12.3 and 17.6%, respectively. "Bayanus" strains show a more or less pronounced heterotrophy with growth factors; "oviformis" strains do not show this (Ribéreau-Gayon and Peynaud 1960).

Ecological investigations in various wine producing regions have shown that "bayanus" strains are strongly represented in several Italian regions (Castelli 1957). They are less frequently found in France, where they are mainly present in the Médoc (Domercq 1956). In other areas their appearance is still less frequent. The occurrence of "oviformis" strains is also limited. Czechoslovakia, where they appear more often, is an exception. A high percentage of "oviformis" yeasts is found in Tokay wines from selected grapes (Beerenauslese), i.e., in the so-called Asszu-wines (Minárik 1963). In these wines they accounted for up to 78% of the yeasts. Four strains were capable of forming more than 19% by vol. of ethanol. This is the reason why these strains are found mainly at the end of the fermentation period. Because of their ability to form high concentrations of ethanol they are also used as pure culture "yeast starters," mainly for the production of champagne. They are largely responsible, just as the *S. cerevisiae* yeasts, for the clouding of wines with residual sugar (Peynaud and Domercq 1959; Scheffer and Mrak 1951; Van der Walt and Van Kerken 1958).

Saccharomyces uvarum.—(Synonym: *S. carlsbergensis.*) The species *S. uvarum* includes the formerly separate species *S. carlsbergensis* (Van der Walt 1970B). The literature on the flora of wines and musts mentions both names, although the designation *uvarum* predominates. *S. uvarum* has more elongated cells, a difference which has not been considered sufficient to retain 2 separate species. Both types show the same pattern for assimilation and fermentation of sugars, and there are scarcely any differences in

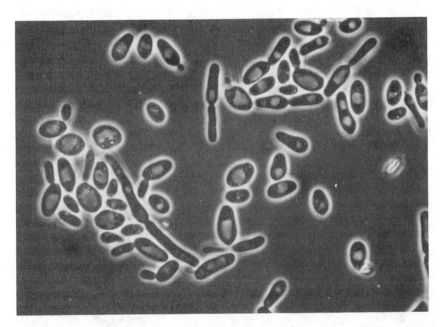

FIG. 9.6.   *SACCHAROMYCES UVARUM*

the formation of ethanol and fermentation by-products (Ribéreau-Gayon and Peynaud 1960). However, strains used in the brewing industry seem to ferment maltotriose more readily than those found in wine (Van der Walt 1970B). In the brewing industry strains of *S. uvarum* and *S. carlsbergensis* have always played a large role, and strains of *S. carlsbergensis* are important as so-called bottom fermenting yeasts. The strain "uvarum" is the most characteristic yeast in apple wine (Beech and Carr 1977).

Both strains of *S. uvarum* which have been mentioned can frequently be found in must. It is distributed over most of the wine producing regions and occurs quite frequently in dry Tokay wines (Minárik 1963). In the Franconia region of Germany it also occurs in large numbers (after *S. cerevisiae*) (Benda 1962A). Nevertheless, it cannot be considered an important yeast of the microflora of grape must. Normally, this species occurs during the main phase of the fermentation, and it is superseded in the final phase by *S. cerevisiae* and *S. bayanus*.

Its ability to form ethanol is intermediate. Under aerobic conditions the yeast developed 8.6% ethanol by vol. (6 strains of "uvarum") and 10.0% ethanol by vol. (14 strains of "carlsbergensis"), on the average, and under anaerobic conditions 7.1 and 7.4%, respectively (Ribéreau-Gayon and Peynaud 1960). Its enological role has also been described by Malan and Tarantola (1956).

Sometimes *S. uvarum* has caused clouding of wines (Ribéreau-Gayon *et al.* 1975B; Scheffer and Mrak 1951; Van der Walt and Van Kerken 1958), but it cannot be really considered to be a spoilage yeast.

Saccharomyces rosei.—(Synonym: *Torulaspora rosei.*) *S. rosei* is a typical representative of the yeast flora of grape must, just as *S. cerevisiae* and *S. uvarum* are. It is widely distributed although its proportionate share of total counts is not large. It can be found in most wine producing areas, but only in Italy with sufficient frequency (Castelli 1957) and in Japan (Yokotsuka 1954).

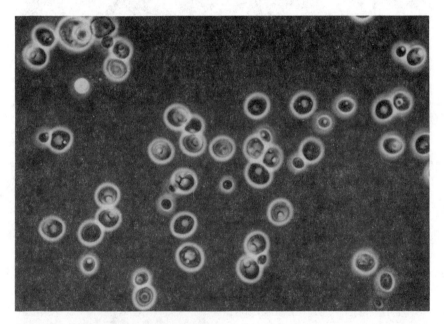

FIG. 9.7.   *SACCHAROMYCES ROSEI*

S. rosei can generally be found at the beginning of the main fermentation phase. It cannot be found among the yeasts responsible for the final phase of the fermentation. The ability to form alcohol aerobically was between 10.1 and 13.0% by vol. (34 strains tested) and anaerobically between 7.8 and

12.8% (Ribéreau-Gayon and Peynaud 1960). The authors comment that individual strains of *S. rosei* have widely different capacities for the formation of ethanol with average values between 7 and 11% by vol. Minárik (1966), who tested 29 strains, obtained similar results with ethanol percentages between 5 and 10% by vol.; most of the strains reached only between 6 and 8% by vol. The ability of *S. rosei* to form ethanol and to tolerate ethanol is weak to intermediate.

It is interesting that this *Saccharomyces* species produces few volatile acids. The strains investigated by Ribéreau-Gayon and Peynaud (1960) formed between 0.00 and 0.29 g per liter, with an average of 0.04 g per liter. The ability to form ethyl acetate is also slight. This property was of sufficient interest to try pure culture fermentations with *S. rosei* (Castelli 1967; Toledo and Teixera 1955). After 2 to 4 days of fermentation with *S. rosei*, one has to re-inoculate with a more strongly fermenting yeast, for instance, *S. cerevisiae*. However, one has to work cleanly since infection of the must with *S. cerevisiae* or some other strongly fermenting yeast at the beginning of the fermentation leads to the rapid disappearance of *S. rosei*. Wines obtained with *S. rosei* and *S. cerevisiae* by the mentioned authors were considered to be neutral and clean.

Saccharomyces italicus.—(Synonym: *S. steineri.*) The two strains "italicus" and "steineri" which have been isolated from grape must have now been combined into the single species, *S. italicus*, since the idea that they

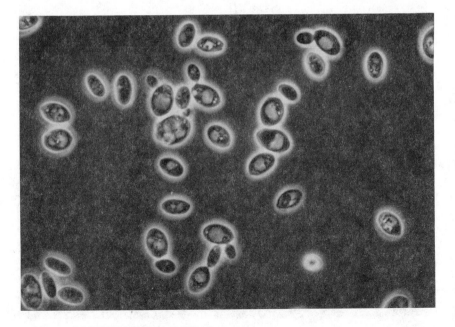

FIG. 9.8. *SACCHAROMYCES ITALICUS*

differ in their ability to ferment sucrose could not be maintained (Van der Walt 1970B). Both strains ferment sucrose if they are brought into a suitable medium; however, sucrose is fermented more slowly than glucose (Ribéreau-Gayon and Peynaud 1960).

S. italicus can be found in several European and non-European wine producing areas, but it does not occur frequently. Therefore, its enological significance is limited. In southern Italy it occurs with greater frequency: it could be found in 42% of the musts (Capriotti 1954). A more frequent occurrence has also been reported for the musts of Porto (Marques Gomes and De Castro Reis 1961–1962).

The ability to form ethanol is on the average similar to that of *S. cerevisiae*, but fermentation in musts with higher sugar concentrations is slower (Ribéreau-Gayon and Peynaud 1960). It has no remarkable characteristics with regard to the formation of fermentation by-products.

Saccharomyces chevalieri.—(Synonyms: *S. fructuum, S. mangini.*) This yeast has also been described in the literature under the designation *S. fructuum* and *S. mangini* as a representative of the flora of grape musts. In the meantime these yeasts have been found to be identical with *S. chevalieri* (Lodder and Kreger van Rij 1952; Lodder 1970). It has been reported to be a "metastable" species. With several strains a change from nonmaltose fermenters to maltose fermenters could be observed. This would permit classification as *S. cerevisiae* (Van der Walt 1970B).

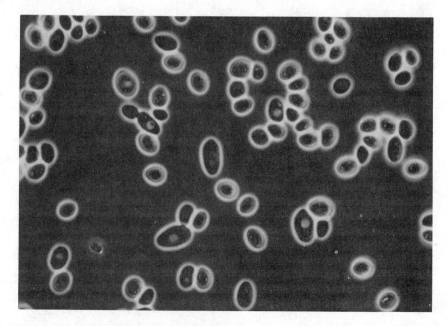

FIG. 9.9. *SACCHAROMYCES CHEVALIERI*

S. chevalieri has been identified mostly in France, Spain, and Italy (Domercq 1956; Iñigo Leal *et al*. 1963; Malan 1951, 1953, 1954, 1955). Its physiological properties which are of importance in wine making are similar to those of *S. cerevisiae* (Ribéreau-Gayon and Peynaud 1960).

Saccharomyces rouxii.—*S. rouxii* appears in the early literature under several synonyms; but mainly as the genus *Zygosaccharomyces*. It is characterized by its ability to withstand high osmotic pressure. *S. rouxii* which had been isolated from fermenting, preserved fruits was able to ferment in a medium containing 0.5% yeast extract and 70% sucrose (Mossel 1951). It is closely related to other osmophilic yeasts, such as *S. bisporus* var. *mellis* and *S. bailii* var. *osmophilus*. Lack of stability with regard to the assimilation and fermentation of maltose indicates a certain transition among these species (Van der Walt 1970B). According to Peynaud and Domercq (1955A), *S. rouxii* is fructophilic.

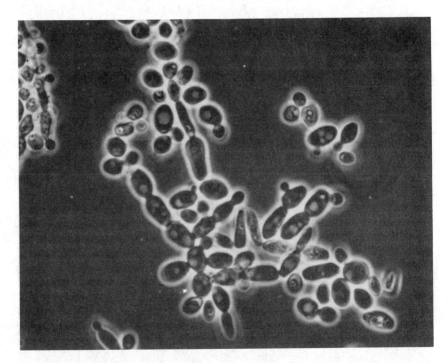

FIG. 9.10.   *SACCHAROMYCES ROUXII*

S. rouxii has frequently been isolated from musts with high sugar concentrations, among others from musts of grapes infected with *Botrytis cinerea*. But cell counts of this species are generally low (Capriotti 1965; Kroemer and Krumbholz 1931; Melas-Joannidis *et al.* 1958). It occurred repeatedly in high sugar musts, and occasionally among film forming yeasts in wines in the region of West Andalusia (Somavilla *et al.* 1975; Iñigo Leal and Arroyo Varela 1964). Under aerobic conditions it forms 13.3% ethanol by vol. and it forms 0.09 g of volatile acids per liter (Domercq 1956). This is confirmed by the work of Melas-Joannidis *et al.* (1958) who found that their strains produced from 13.0 to 14.9% ethanol by vol. However, the latter authors reported the formation of higher concentrations of volatile acids, namely, 0.77 to 1.14 g per liter.

Saccharomyces bailii.—(Synonyms: *S. acidifaciens; S. elegans.*) These formerly separate species have now been classified into one species, *S. bailii* (Van der Walt 1970B). Observed differences in the assimilation of some sugars were not sufficiently clear cut to permit the maintenance of separate species. The same is true of morphological characteristics. However, *S. bailii* could be subdivided into the two varieties *bailii* and *osmophilus* which differ in their tolerance of high osmotic pressure. Both *S. bailii* var. *bailii* and *S. bailii* var. *osmophilus* grow on a 50% (w/w) glucose-yeast extract agar; the variety *osmophilus* still on 60% (w/w) glucose-yeast extract agar. The latter variety forms pieces of skin or a continuous skin if it is grown on malt extract.

S. bailii has often been found in media containing high concentrations of sugar, ethanol, SO$_2$, or acetic acid. This specific tolerance must be due to the

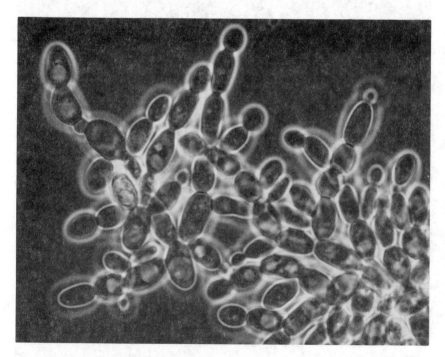

FIG. 9.11. *SACCHAROMYCES BAILII* VAR. *BAILII*

property of its cell membrane (Van der Walt 1970B). The alcohol tolerance of this yeast is greater than its ability to produce alcohol (Ribéreau-Gayon and Peynaud 1960). Strains which have been described as *S. acidifaciens* and *S. elegans* in the literature are fructophilic, similar to *S. rouxii* (Peynaud and Domercq 1955A).

S. bailii has been isolated variously from grape must in Italy, France, Spain, and Czechoslovakia (Castelli and Del Giudice 1955; Castelli and Iñigo Leal 1957; Minárik 1966; Peynaud and Domercq 1955B). Its effect on the fermentation of musts is small. It ferments slowly and forms on the average no more than 9% ethanol by vol. (Ribéreau-Gayon and Peynaud 1960), but its tolerance of high concentrations of sugar and SO_2 is of interest. Several strains could be isolated from botrytized Sauterne grapes with high sugar concentrations (Dubourg 1896). Forty strains of *S. bailii* could be isolated from spontaneously fermenting grape juice concentrates with more than 65% sugar. The rate of fermentation of such concentrates varied with the original sugar and acid concentration (Bambalow 1970). *S. bailii* frequently causes after-fermentations and clouding of wine in tanks and in bottled wine because of the mentioned tolerance (Domercq 1956; Rankine and Pilone 1974; Schanderl 1962; Van der Walt and Van Kerken 1958). It also occurs as contaminant on the equipment of cellars (Peynaud and Domercq 1959).

Saccharomyces exiguus.—This yeast species has been isolated from fermenting musts (Capriotti 1965; Melas-Joannidis *et al.* 1958). However, it

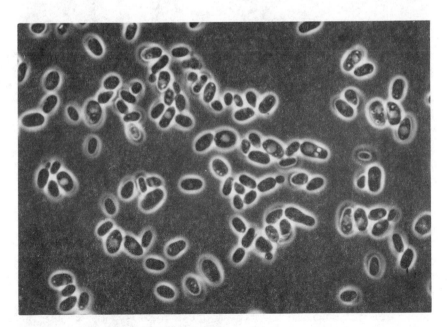

FIG. 9.12.   *SACCHAROMYCES EXIGUUS*

does not occur frequently. Its most frequent occurrence was in the Mancha region of Spain where it could be found in 17% of the investigated musts (Castelli and Iñigo Leal 1957). It forms 12.6–14.8% ethanol by vol. and 0.63–1.10 g of volatile acid per liter (6 strains tested) (Melas-Joannidis *et al.* 1958).

Kloeckera apiculata.—This asporogenous yeast is characterized by bipolar budding, and it is, therefore, easily recognized with the microscope. It is the imperfect form of the sporogenous species *Hanseniaspora valbyensis* and *H. uvarum*. It has frequently been shown that *H. valbyensis* loses the ability to form spores after prolonged culture in the laboratory, although this differs from strain to strain. Phaff (1970) assumes that spore formation has often been overlooked in perfect forms which led to their designation as *K. apiculata* in the literature.

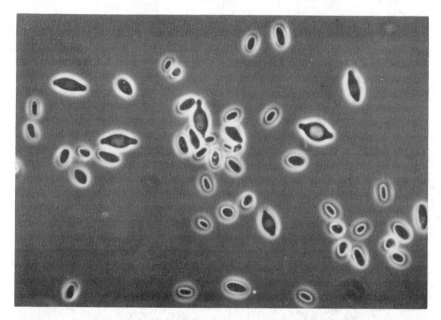

FIG. 9.13.   *KLOECKERA APICULATA*

Together with *S. cerevisiae* the species *K. apiculata* is the most widely distributed and the most strongly represented yeast species in all wine producing areas. The wine producing area of Haryana in India may be an exception (Relan and Vyas 1971). *K. apiculata* occurs in grape musts mainly before the fermentation and in its early phases in large numbers. The proportion of this yeast in the original flora of the must depends on several factors, such as the maturity and soundness of the grapes, the presence of insects, and contamination by cellar equipment, as well as on the treatment of the crushed grapes or the must, for instance addition of SO_2. Benda (1962A) harvested exclusively ripe but uninjured grapes under aseptic conditions in the wine producing region of Würzburg, Germany. The unsulfited musts from these grapes showed almost exclusively the presence of *K. apiculata,* or the presence of this species in association with smaller or larger numbers of *Metschnikowia pulcherrima* (syn.: *Candida pulcherrima*), or with smaller numbers of *Torulopsis stellata* (syn.: *Torulopsis bacillaris*).

Apart from *K. apiculata,* grape musts frequently contain other species of the genus although in smaller numbers. These are *K. javanica* var. *javanica* (syn.: *K. jensenii*) and *K. africana* (Domercq 1956; Minárik 1966; Verona 1951), as well as *K. corticis* (Benda 1970). A comprehensive review of the microflora of Italian musts indicates the frequent occurrence of *K. apiculata* and an equally frequent occurrence of its sporogenous form *Hanseniaspora valbyensis* (syn.: *Hanseniaspora guilliermondii*), and a less frequent occurrence of *K. corticis* (Castelli 1957). It is interesting that the sporogenous form of *K. apiculata, Hanseniaspora valbyensis,* occurs exclusively in southern Italy and in Israel which is probably due to climatic conditions.

K. apiculata is very sensitive to SO_2. The addition of 150 mg SO_2 per liter of must eliminates this yeast, while smaller doses show a correspondingly smaller effect (Ohara *et al.* 1959). The fungicidal effect also depends on the number of yeast cells in the must (Beech and Carr 1977). If *K. apiculata* is present in large numbers, then the effect is somewhat reduced. Therefore, the activity of *K. apiculata* cells is limited by the customary sulfiting of the crushed grapes or the must. Nevertheless, this species together with *M. pulcherrima* and *T. stellata* and the originally present *Saccharomyces* cells initiates the fermentation, provided pure culture yeasts are not inoculated. *K. apiculata* ceases its growth and its fermenting activity once the relatively low alcohol concentration of 4–5% by vol. has been reached. It is then completely superseded by the more strongly fermenting *Saccharomyces* yeasts.

K. apiculata strains which have been tested by Domercq produced from 3.7 to 6.4% ethanol by vol. under aerobic conditions. The formation of volatile acids varied from 0.7 to 2.6 g per liter, generally in the neighborhood of 1 g per liter. The yeast also forms high concentrations of ethyl acetate (125–374 mg/liter) (Ribéreau-Gayon and Peynaud 1960) as well as amyl acetate (Sponholz and Dittrich 1974). These esters produce a foreign fruity odor and taste which can make the wine unpalatable at higher concentrations (see also Castor 1954).

The yeast also produces higher concentrations of pyruvate which is important because it binds sulfite. Further, in comparison with *S. cerevisiae* it produces relatively large amounts of phenyl ethyl alcohol which in reasonable concentrations has a positive effect on palatability. This alcohol together with possible derivatives can contribute to the rounding out of the aroma and taste of the wine. Another positive effect is the production of larger concentrations of glycerol, which contributes to the greater "fullness" of the wine (Sponholz and Dittrich 1974).

The presence of *K. apiculata* has a positive effect on the fermentation of other yeasts and on the completeness of the fermentation (Schulle 1953). This author obtained the best results if inoculation of the must was carried out with cells of *S. cerevisiae* and *K. apiculata* in a ratio of 24:1 to 5:1. But fermentation by *S. cerevisiae* was inhibited when the number of *K. apiculata* cells was larger than that of *S. cerevisiae* cells. He attributed the activating or inhibiting effect of *K. apiculata* to its metabolites.

In a must which contained *S. cerevisiae* yeasts, the amount of ethanol formed diminished with higher proportions of *K. apiculata* cells (Tarantola 1945–1946). A pure culture of *S. cerevisiae* produced 14.1% ethanol by vol. Equal numbers of *S. cerevisiae* and *K. apiculata* cells (1:1) at the beginning of the fermentation produced 13.88% ethanol by vol.; 3 times the number of *K. apiculata* cells (1:3) produced 13.58%, and 9 times the number (1:9) produced 13.47%.

Synergistic effects of yeast species have also been observed with other combinations. In a must with 21.2% sugar which had been inoculated with *Hanseniaspora uvarum* and *Saccharomyces* species in the ratio of 5:1, the number of *Saccharomyces* yeasts increased during the fermentation to larger numbers than in a must inoculated with a pure culture of *Saccharomyces*. Conversely the presence of *Saccharomyces* yeasts inhibited the growth of *Hanseniaspora uvarum*. In all of these instances the amount of ethanol formed was the same.

Metschnikowia pulcherrima.—(Synonym: *Candida pulcherrima*.) The earlier literature referred to this asporogenous yeast as *C. pulcherrima*, but

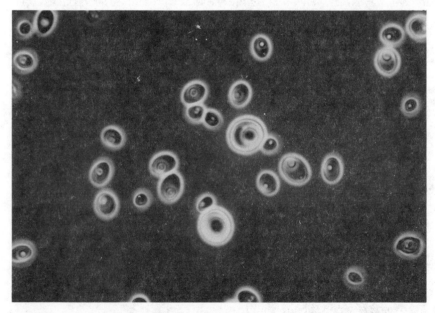

FIG. 9.14.   *METSCHNIKOWIA PULCHERRIMA*

the presence of needle-like spores has now been proven (Pitt and Miller 1968). The same is true for a yeast, *Torulopsis burgeffiana*, which had earlier been isolated from grapes (Benda 1962B). Both types are identical and have now been included in the newly formed species, *M. pulcherrima* (Miller and Van Uden 1970).

M. pulcherrima is remarkable in pure culture by the strong refraction of light by older cells, which contain a large sphere of lipid material. Besides, several strains form the red dye pulcherrimin.

The yeast is a typical representative of the yeast flora in some wine growing regions; in others it is rare or absent (Benda 1964; Castelli 1957; Minárik 1964). It may occur in large numbers in must, together with *K. apiculata*, and participate in the initial phase of the fermentation. It has a more oxidative metabolism. It forms no more than 4% by vol. of ethanol. Its production of volatile acids is small (Benda 1970), but the production of ethyl acetate and fusel oils, mainly isobutanol and isoamyl alcohol, is very large (Sponholz and Dittrich 1974). It also forms relatively high concentrations of compounds which bind SO_2. The quantity of by-products formed by this yeast depends on the degree to which it participates in the fermentation. Therefore, it may affect the quality of the wine favorably or unfavorably (Wagener and Wagener 1968).

The number of *M. pulcherrima* cells is already reduced by the addition of SO_2 and by competition with *Saccharomyces* species in the early stages of the fermentation. Therefore, its contribution to the formation of fermentation by-products should not be overrated.

Torulopsis stellata.—(Synonym: *Torulopsis bacillaris.*) The formerly separate species *T. bacillaris* has now been incorporated into the species *T. stellata* (Van Uden and Vidal-Leiria 1970). Both strains can be distinguished morphologically which caused Kroemer and Krumbholz (1931) to place them into different species, but they are identical in the diagnostically important physiological properties. Both strains tolerate high sugar concentrations but differ in their fermentation activity. *T. stellata* is an osmophilic yeast. The optimum temperature for fermentation is 25°C, but the yeast can still ferment well in the 30°−35°C range (Peynaud and Domercq 1955B).

T. stellata occurs in some wine growing areas only sporadically. It is frequently found in France (Domercq 1956; Sapis-Domercq and Guittard 1976). In Germany it occurs frequently in the Franconia wine producing areas as part of the grape flora (Benda 1962A). It occurs very often in musts from *Botrytis* infected grapes (Kroemer and Krumbholz 1931; Minárik *et al.* 1978; Le Roux *et al.* 1973). Kroemer and Krumbholz report that some of the strains which they isolated grow best on must containing 300−400 g sugar per liter, and that growth ceases only with sugar concentrations of 900 g per liter.

The yeast is found during the early stages of the fermentation, and it is slowly eliminated during the main fermentation. Generally it produces between 7 and 10% ethanol by vol., but concentrations as high as 12.5% have been reported. The production of volatile acids is higher than 1 g per

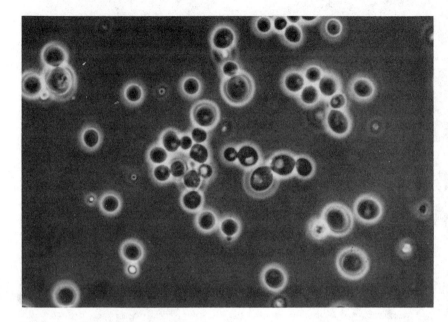

FIG. 9.15.   *TORULOPSIS STELLATA*

liter; that of ethyl acetate is slight (Benda 1970; Domercq 1956; Ribéreau-Gayon and Peynaud 1960). It tolerates SO_2 better than *K. apiculata*. Some of the fungicides which are used to prevent *Botrytis* infection, and which inhibit *Saccharomyces* species, do not inhibit *T. stellata* and lead to a positive selection of this yeast in must (Minárik and Rágala 1975).

Schizosaccharomyces pombe.—The yeasts of the genus *Schizosaccharomyces* do not truly belong to the yeast flora of must and wine, although they occur occasionally on grapes or in must. Only in the Western part of Sicily do they occur normally on grapes and form part of the yeast flora (Florenzano *et al.* 1977). These yeasts will be discussed here because of several attempts to use them in the production of wine for the degradation of malic acid.

Yeasts of the genus *Schizosaccharomyces* grow vegetatively by cell division instead of by budding. In this respect they differ from all other yeasts. Among the 4 species of the genus *Schizosaccharomyces*, the species *S. pombe* is of greatest interest because of its ability to produce ethanol in concentrations similar to those produced by *S. cerevisiae* and because of its ability to degrade L-malic acid. Under anaerobic conditions this degradation leads to the formation of ethanol and CO_2 (Dittrich 1963A; Mayer and Temperli 1963). The enzyme pathways have been investigated by Temperli *et al.* (1965) and Flesch and Holbach (1970). There are 2 possible paths based on 2 separate enzymes which split malic acid, the malic enzyme (1), and the malate dehydrogenase enzyme (2):

(1) $COOH\text{-}CH_2\text{-}CHOH\text{-}COOH + NAD^+ \xleftarrow{\text{malate enzyme}} CH_3\text{-}CO\text{-}COOH + CO_2 + NADH + H^+$
   malic acid   pyruvic acid

$CH_3\text{-}CO\text{-}COOH \xleftarrow{\text{pyruvate decarboxylase}} CH_3\text{-}CHO + CO_2$
pyruvic acid   acetaldehyde

$CH_3\text{-}CHO + NADH + H^+ \xleftarrow{\text{alcohol dehydrogenase}} CH_3\text{-}CH_2\text{-}OH + NAD^+$
acetaldehyde   ethanol

(2) $COOH\text{-}CH_2CHOH\text{-}COOH + NAD^+ \xleftarrow{\text{malate dehydrogenase}} COOH\text{-}CH_2\text{-}CO\text{-}COOH + NADH + H^+$
   malic acid   oxalacetic acid

$COOH\text{-}CH_2\text{-}CO\text{-}COOH \xleftarrow{\text{oxalacetate decarboxylase}} CH_3\text{-}CO\text{-}COOH + CO_2$
oxalacetic acid   pyruvic acid

$CH_3\text{-}CO\text{-}COOH \xleftarrow{\text{pyruvate decarboxylase}} CH_3\text{-}CHO + CO_2$
pyruvic acid   acetaldehyde

$CH_3\text{-}CHO + NADH + H^+ \xleftarrow{\text{alcohol dehydrogenase}} CH_3\text{-}CH_2\text{-}OH + NAD^+$
acetaldehyde   ethanol

Three strains of *Schizosaccharomyces* produced 10–12% by vol. of ethanol under anaerobic conditions, and 13.2–15.1% under aerobic conditions. But the rate of fermentation of sugars is slower than with *Saccharomyces* species (Peynaud *et al.* 1964). Formation of volatile acids is slight (Dittrich 1963B), and so is the production of higher alcohols (Parfait and Jouret 1975). The yeast may be considered to be osmophilic, since growth on a yeast extract agar containing 50% of glucose by weight was good (Slooff 1970). The yeast also tolerates acetic acid quite well. The presence of 10 g acetic acid per liter reduced the rate of fermentation only slightly (Dittrich 1977). It also tolerates low pH values well (Yang 1973A). *S. cerevisiae* fermentations were strongly inhibited at a pH of 2.8 and ceased at a pH of 2.5, while *Schizosaccharomyces* remained active at these pH values.

Finally, *S. pombe* has a high tolerance for SO_2. It could be isolated from a must with 415 mg of total SO_2 per liter and 130 mg free SO_2 (Brugirard and Roques 1972). The presence of 300 mg SO_2 per liter inhibits fermentation only slightly. In a mixed culture with *S. cerevisiae* and in a must containing 150–300 mg SO_2 per liter, *S. pombe* was not superseded by *S. cerevisiae* and fermentation of L-malic acid was complete (Yang 1975). Minárik and Navara (1967) did not confirm the high tolerance of *S. pombe* for SO_2. *S. pombe* is a very efficient fermenter of malic acid. In a sterile must L-malic acid is completely fermented, however, at a rate about 1/10 that of the sugar fermentation (Peynaud *et al.* 1964). The yeast tolerates high acid concentrations. In a must with a total acid concentration of 67.5 g per liter, a

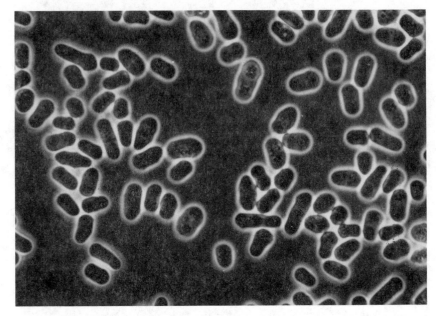

FIG. 9.16.   *SCHIZOSACCHAROMYCES POMBE*

reduction by 15.2 g per liter resulted from fermentation with *S. pombe* (see Table 9.14).

The reduction of acids in musts with high acid concentrations by *S. pombe* has been repeatedly attempted (Benda and Schmitt 1966, 1969; Bidan *et al.* 1974; Castelli and Haznedari 1971; Gallander 1977; Peynaud *et al.* 1964; Yang 1973B). There is no problem with the fermentation of sterile musts.

TABLE 9.14.   FERMENTATION OF MALIC ACID BY *SCHIZOSACCHAROMYCES POMBE* IN GRAPE MUST
Total Malic Acid Concentration in g/liter.

Before Fermentation	After Fermentation
10.0	1.7
15.3	5.1
20.6	8.3
30.2	15.7
40.4	24.0
49.0	32.6
59.4	46.0
67.5	52.3

Source: Dittrich (1964).

However, in mixtures with *Saccharomyces* species, growth of *S. pombe* is inhibited, particularly in cool musts, since the optimum growth temperature for *S. pombe* is about 30°C (Peynaud and Sudraud 1964). Depending on the relative proportions of the 2 species and depending also on the temperature, growth of *S. pombe* is arrested sooner or later. This has been studied in some detail with mixed cultures of *S. malidevorans* and *S. cerevisiae* (Rankine 1966). The data shown in Table 9.15 indicate the rapid disappearance of *Schizosaccharomyces* cells with increased growth by *Saccharomyces* cells.

TABLE 9.15.   EFFECT OF MIXED YEAST CULTURES ON THE FERMENTATION OF MALIC ACID IN FILTER-STERILIZED GRAPE JUICE AT 25°C[1]

	Schizosaccharomyces Cells[2]		
Inoculated, %	At the Midpoint of the Fermentation, %	At the End of the Fermentation, %	Malic Acid Fermented
0	0	0	21
10	2	3	47
50	4	7	87
90	16	29	96
100	100	100	100

Source: Abbreviated and modified from Rankine (1966).
[1]*Schizosaccharomyces malidevorans* and *Saccharomyces oviformis* cells were used.
[2]*Schizosaccharomyces* cells as percentage of the total yeast population.

Most of the experiments yielded wines with unsatisfactory organoleptic properties, although Benda and Schmitt (1969) showed that there were relatively large differences among individual strains of *S. pombe*.

Attempts have also been made to achieve a reduction of acid by a secondary fermentation (Amati and Minguzzi 1975; Benda 1974; Haznedari 1976). Such tests showed that the extent of acid reduction depends not only on the strain of yeast used but also on the ethanol content of the wine. Snow and Gallander (1979) fermented with *Schizosaccharomyces pombe* to achieve a rapid reduction in malic acid, removed the yeast, and fermented with *S. cerevisiae* to dryness.

Pure Culture Yeasts

Enological Considerations.—It has already been mentioned that the spontaneous fermentation of musts is carried out by the yeast flora of the grapes as well as by yeasts from the processing equipment of the cellar. Various technological steps during the processing of the must which reduce insoluble solids consequently reduce the number of yeast cells. The addition of sulfur dioxide has a selective effect by killing or inhibiting non-*Saccharomyces* yeasts. Sulfiting is designed to eliminate acetic acid bacteria and undesirable yeasts. Nevertheless, variable numbers of non-*Saccharomyces* yeasts survive the application of customary concentrations of SO_2. These yeasts participate in the early stages of the fermentation although sulfiting favors the predominance of the desired *Saccharomyces* yeasts.

Stimulated by the use of pure culture methods in brewing, the wine industry has also introduced the use of pure cultures from specially selected and grown strains. These are generally used with musts whose natural flora has been reduced technologically, but which are certainly not sterile; or, they are used with sterile musts in which the natural flora has been killed by pasteurization. A true, pure culture fermentation can be carried out only in the latter case.

The advantages of pure culture fermentations are obvious. They help in eliminating the risk associated with spontaneous fermentations. Such risks can be due to several circumstances, such as defects of the grapes, a low temperature at the start of the fermentation, residual fungicides, and others. Besides, it is possible to ameliorate various problems of the wine fermentation by using yeasts with particular physiological characteristics. A selectivity in this respect is quite feasible as has already been mentioned in the paragraphs dealing with *S. cerevisiae*. Rankine (1955, 1968A) has dealt with this problem extensively.

Opinions differ with regard to the effect of pure culture fermentations on the organoleptic quality of the wine. It is, however, certain that the choice of a particular yeast strain does not markedly improve the quality of the wine. It is the quality of the grapes which is decisive for the quality of the wine that can be achieved. In general, yeast strains with the greatest ability to produce ethanol contribute the least to wine aroma (Malan 1956). This applies specifically to *Saccharomyces*. It can be concluded that differences among strains of *Saccharomyces* species will not be great, particularly since the aroma components of the grape overlay those of the fermentation by-products. Ribéreau-Gayon and Peynaud (1960) expressed the same opinion by saying that all strains of *S. cerevisiae* produce wine of nearly equal character. Similar conclusions have been drawn by Castor (cited by Amerine 1980), but further investigations of this question should be carried out since there are not enough data available on organoleptic evaluations which have been subjected to the required statistical analysis.

Conditions are quite different if one compares different *Saccharomyces* species, and particularly if fermentations are carried out with strains from different genera (Rankine 1968A, 1977; Ribéreau-Gayon and Peynaud 1960). Rankine checked the formation of higher alcohols by various *Saccharomyces* species, and determined the threshold of perception with a taste panel. He found that in several instances the concentrations of higher alcohol exceeded the threshold values, which was particularly true for isoamyl alcohol. This alcohol assumes, therefore, great significance for the wine aroma. But other compounds also play an important part, such as the volatile acids, ethyl acetate, and hydrogen sulfide, whose formation may be different for different species of *Saccharomyces*.

Experiments and practical experience in wine making have demonstrated that wines made by spontaneous fermentation can differ to a greater or lesser extent from those produced by fermentation with pure yeast cultures. Different metabolites of the different yeast species can affect the aroma of the wine positively or negatively. Wines produced by spontaneous fermentation have a more complex distribution of aroma components than wines fermented with *S. cerevisiae*. Values for the concentration of some higher alcohol, as isoamyl alcohol, isobutanol, 2-phenyl ethanol, and active isoamyl alcohol, are higher for wines which have undergone spontaneous fermentation (Wucherpfennig and Bretthauer 1970). Sponholz and Dittrich (1974) have obtained similar results (see Table 9.16) which show that *K. apiculata* and *M. pulcherrima* might be responsible yeasts which are prevalent during the earlier stages of the fermentation. This is also true

TABLE 9.16.   ANALYTICAL COMPARISON OF WINES OBTAINED BY SPONTANEOUS FERMENTATION AND BY FERMENTATION WITH PURE CULTURE YEASTS

By-products of the Alcoholic Fermentation	Spontaneous Fermentation	Pure Culture Yeast Fermentation
Glycerin, g/liter	4.4	3.8
Pyruvate, mg/liter	18	16
Ketoglutarate, mg/liter	40	33
n-Propanol, mg/liter	18	15
Isobutanol, mg/liter	21	8
Optically active isoamyl alcohol, mg/liter	32	17
Isoamyl alcohol, mg/liter	106	77
2-Phenyl ethanol, mg/liter	29	6

Source: Abbreviated from Dittrich (1977).
The pure culture yeast fermentation was carried out with a 1% inoculum of a liquid culture.

for the formation of esters, mainly of ethyl acetate, by individual yeast species. In comparison with *Saccharomyces* yeasts, *K. apiculata* and *M. pulcherrima* produce 10 times as much ethyl acetate (Ribéreau-Gayon and Peynaud 1960; Sponholz and Dittrich 1974). The formation of volatile acids by *Kloeckera* species has been investigated under a variety of conditions. *K. apiculata* produces much higher levels of volatile acids than *S. cerevisiae*. This is still apparent if the initial ratio of *S. cerevisiae* to *K. apiculata* cells in the fermenting must is 1:1 (Tarantola 1945–1946). The formation of volatile acids also differs greatly among various species of *Saccharomyces* (see Table 9.13). *S. rosei* produces very low concentrations of volatile acids (Castelli 1967). Therefore, its use in wine fermentations has been suggested by this author. He recommends inoculation of the must with *S. rosei* and after 4 to 5 days inoculation with a *Saccharomyces* species producing higher concentrations of ethanol. But the must should be sterile and sterility should be maintained during processing, since even slight infection with a more active *Saccharomyces* species inhibits *S. rosei*.

The question of the organoleptic quality of wines produced with pure culture yeast continues to be a matter for controversy. There is no doubt that pure culture fermentations produce wines with a cleaner flavor. On the other hand critics of the pure culture fermentation object that it produces wines with a less rounded and less complex flavor. This is certainly due to the presence of some fermentation by-products. For instance, Sponholz and Dittrich (1974) point to the higher alcohols and specifically to phenyl ethanol, which in rather small concentrations have a positive effect on wine aroma. It must be added that wines resulting from spontaneous fermentations may have higher concentrations of glycerol (Rankine and Bridson 1971; Ribéreau-Gayon and Peynaud 1960; Sponholz and Dittrich 1974).

Two additional points have to be made in connection with the effect of pure yeast culture fermentations on wine aroma and flavor. First, some yeast strains produce high and exceptionally high concentrations of pyruvic acid and 2-ketoglutaric acid, dependent on differences in the carboxylase content of the cells; as well as high amounts of acetaldehyde (Ponade 1973; Rankine 1967A, 1968B, 1972). These compounds play an important part as binding sites for sulfite. Apart from that a slow fermentation can lead to the accumulation of high concentrations of acetaldehydes which also increase the requirement for sulfite (Dittrich *et al.* 1973; Zürn 1976; Zürn and Perscheid 1977) (see also Table 9.17).

TABLE 9.17. ACETALDEHYDE CONCENTRATIONS AND THE RESULTING SO_2 CONCENTRATIONS IN EXPERIMENTAL WINES WITH AND WITHOUT PURE CULTURE YEASTS

	Spontaneous Fermentation (mg/liter)	Liquid Pure Yeast[1] (mg/liter)	Active Dry Yeast, No. 1[2] (mg/liter)	Active Dry Yeast, No. 2[2] (mg/liter)
Start of the fermentation	delayed	delayed	rapid	rapid
Acetaldehyde	143	105	35	38
Free SO_2	30	29	33	36
Total SO_2	186	160	123	125

Source: Abbreviated from Zürn and Perscheid (1977).
Fermentation is with clarified must in volumes of 660 liters, each.
[1]1% inoculum.
[2]Inoculum, 10 g/hl.

Secondly, some strains of yeast produce appreciable amounts of SO_2 in fermenting must as has already been mentioned. This contributes to an undesirably high level of total SO_2 in wine. It is a point which has to be kept in mind when selecting a proper yeast strain (Kontek and Kontek 1977).

Another reason for careful selection of yeast strains is the ability of some strains to form H_2S by reduction of sulfate or sulfite compounds (Rankine 1963, 1964; Acree *et al.* 1972). This makes the wine unpalatable. There is a definite negative correlation between the organoleptic quality of the wine and the presence of volatile acids, ethyl acetate, acetaldehyde, and hydrogen sulfide.

Special *Saccharomyces* strains have to be selected for the production of sherry by surface growth of film forming yeasts or for the submerged culture process (Fornachon 1953; Rankine 1972). For the production of champagne one selects yeast strains which are well adapted to fermentations at low temperatures and which are highly flocculent. Yeasts which do not foam permit a better utilization of the fermentor volume (Eschenbruch and Rassell 1975; Dittrich and Wenzel 1976). For the production of sweet wines of the Tokay type, one selects strongly glucophilic yeasts in order to retain as much fructose as possible in the residual sugar fraction because of the greater sweetness of fructose (Minárik *et al.* 1978).

Finally there is considerable interest in yeasts which are capable of reducing the total acidity in fermenting must and wine. Many investigations have been carried out with strains of the genus *Schizosaccharomyces*. These yeasts ferment better at a high temperature. Therefore, a cool temperature reduces their numbers in competition with *Saccharomyces* yeasts. In general, such wines also suffer in their organoleptic quality. However, there are appreciable differences among different strains of this genus with regard to acid reduction, tolerance of ethanol or of cold temperatures, and aroma of the wine. This could become the basis for the selection of suitable strains (Benda 1974).

A further argument for fermentations with pure culture yeasts is the result of investigations of fermentations with mixed cultures of *S. cerevisiae* and *K. apiculata*. In such a mixed culture, alcohol yields were lower than with a pure culture of *S. cerevisiae* (Tarantola 1945–1946). Higher proportions of *K. apiculata* in mixture with *S. cerevisiae* inhibit the fermentation activity of the latter, while very small proportions have a positive effect (Schulle 1953).

The positive effect of pure culture fermentations with moldy grapes on the content of volatile acids of the wine can be seen from the data in Table 9.18.

TABLE 9.18.   CONCENTRATION OF VOLATILE ACIDS IN WINES AFTER INOCULATION OF MUSTS FROM VERY MOLDY GRAPES WITH PURE CULTURE YEASTS

Conduct of the Fermentation	Volatile Acids in g/liter
Spontaneous I	1.78
Spontaneous II	1.42
Liquid pure culture yeast, 3%	0.82
Active dry yeast I, 10 g/hl	0.36
Active dry yeast II, 10 g/hl	0.55
Active dry yeast III, 10 g/hl	0.10

Source: Benda (1975).
Fermentation is with clarified must in volumes of 1200 liters each.

In summary it can be concluded that the use of pure culture yeasts can often give a desirable support for the natural yeast flora, and that it is essential in some instances. Dittrich (1977) has listed such instances as follows:

(1) Fermentation of pasteurized musts, and musts made from re-diluted grape juice concentrates
(2) Fermentation of musts with very high sugar concentrations
(3) Fermentation of musts strongly infected with acetic acid bacteria, such as moldy grapes
(4) Fermentation of musts which may contain residual fungicides
(5) Fermentation of fruit pulp as distilling material
(6) Fermentation of fruit wines (other than grapes) and fruit dessert wines
(7) Addition of sugar to and secondary fermentation of wines too low in ethanol
(8) Secondary champagne fermentations

Pure Culture Yeast Processes.—The use of pure culture yeasts can be carried out with yeasts in several forms; as yeast suspensions (so-called "liquid pure culture yeasts"), as lyophilized yeasts, and in the form of active dry yeasts.

Liquid pure culture yeasts are obtained by inoculation of sterile must with a selected yeast strain. As soon as this pitching yeast is in the active growth phase, one adds 1 to 3% to the must for the large scale fermentation. This method is rather cumbersome since young, active pitching yeast must be available throughout the entire grape harvest. More rational methods for producing larger amounts of pitching yeast, including continuous processes, have been worked out (De Soto 1955; Fiechter 1962; Klemm 1975; Rankine *et al.* 1976; Schütz and Heinz 1978).

There has been a parallel development of other forms of yeast in the course of the past 20 years. In France lyophilized yeasts have been produced and used successfully in the production of wine (Brémond 1957). But such preparations have not led to extensive commercial applications. In contrast, the use of active, dry wine yeasts has found greater application (Reed and Peppler 1973). This development started with the production and use of compressed yeast (moist press cake) (Adams 1961; Thoukis *et al.* 1963), and led to the use of granulated active dry yeast which can maintain its activity for a 6 to 12 month period with cool, dry storage. Use of these preparations is simple and without problems since inoculation of the must can be done after a 10 to 15 min rehydration period in water or must. Until a few years ago the use of pure culture yeasts for wine production had been generally accepted in the United States, Australia, and South Africa, while it was used in Europe only by larger concerns and in special cases. The simplicity of the process and the conscious orientation for the production of high quality wines today indicates more general acceptance of this method by producers in all other wine producing areas.

In all cases it is decisive that the added pure culture yeast rapidly enters an active state. This is particularly important if the must has not been pasteurized and if it is desired to suppress the natural yeast flora. Apart from the activation of the yeast, it is important to consider the original cell count. With active dry yeast a high cell count can readily be obtained. A comparison of liquid pure culture yeast and active dry yeast for recommended cell counts shows that the lag time with active dry yeast is considerably shorter.

High initial cell counts rapidly form a reducing environment which lowers the required amount of SO_2, and alcohol yields are somewhat higher since less sugar is used for cell growth (Thoukis et al. 1963). However, Lafon-Lafourcade and Ribéreau-Gayon (1976) point out that the alcohol yield is again reduced at extremely high cell counts.

Analytical data and organoleptic evaluations of wines produced with active dry yeast show that this yeast is highly acceptable (Bauer and Kleinhenz 1978; Lafon-Lafourcade and Ribéreau-Gayon 1976; Weger 1976; Zürn and Perscheid 1977). The assurance that enologically important properties of the yeast strain are retained in spite of the drying process, that the preparations have low counts of bacteriological contaminants and foreign yeasts, and have very good stability are preconditions for commercial use of active dry yeasts.

Wine Defects Caused by Yeasts

Most of the defects due to yeasts are caused by after-fermentations, that is, fermentations which take place after the alcoholic fermentation and which are supported by residual, fermentable sugar. If the wine has already been bottled, it is unsalable because of clouding and formation of carbon dioxide. However, the wine can be made salable by an additional filtration which is cumbersome. In other cases yeasts may truly spoil the wine if the yeast metabolites result in the formation of an objectionable flavor or aroma. Therefore, some yeasts are conditional agents of spoilage, others are unconditional spoilage agents.

The most frequent defects are caused by species of *Saccharomyces*, such as *S. cerevisiae, S. bayanus,* and *S. bailii.* Of the latter the literature cites mostly the "oviformis" form of *S. bayanus* and the "acidifaciens" form of *S. bailii.* Spoilage yeasts of other genera are principally *Saccharomycodes ludwigii* and *Brettanomyces* species. To a lesser extent, film forming yeasts are among the spoilage organisms. These are strains of the genera *Hansenula, Pichia,* and *Candida.* Table 9.19, based on comprehensive study, tabulates yeast species isolated from cloudy wines which had undergone an after-fermentation.

***Saccharomyces* Species.**—*S. cerevisiae* occurs frequently in wines which have not been well filtered or which have been re-infected after filtration, which have an average alcohol concentration, and to which grape juice concentrate has been added. Wines with higher alcohol con-

TABLE 9.19.   YEASTS ISOLATED FROM WINES

| | France | | | South Africa | | |
Yeast	Red Wines	White Wines	Total	Red Wines	White Wines	Total
Number of wines examined	17	53	70	1	59	60
Number of yeast strains isolated	54	163	207	1	84	85
Saccharomyces cerevisiae						
"ellipsoideus"	17	17	34	—	9	9
"tetrasporus"	—	—	—	—	3	3
Saccharomyces bayanus						
"oviformis"	15	62	77	—	12	12
Saccharomyces bailii						
"acidifaciens"	5	49	54	—	11	11
"elegans"	4	2	6	—	1	1
Different *Saccharomyces* species	10	1	11	1	—	1
Saccharomycodes ludwigii	—	22	22	—	—	—
Brettanomyces spp.	3	0	3	1	29	30
Pichia membranaefaciens	—	—	—	—	8	8
Species of different genera	—	—	—	—	7	7

Source: Abbreviated from Domercq (1956) and Van der Walt and Van Kerken (1958).

centrations and with fermentable, residual sugars usually undergo after-fermentation by the "oviformis" form of *S. bayanus,* which shows higher resistance to ethanol. In France *S. cerevisiae* and *S. bayanus* ("oviformis") accounted for 16 and 26%, respectively, of the strains isolated from wines (Domercq 1956; Peynaud and Domercq 1959). These yeasts also occur frequently in South African wines (Van der Walt and Van Kerken 1958). In California Scheffer and Mrak (1951) could prove the development of these and of other yeast species in dry white wines containing merely some nonfermentable pentoses. *S. bayanus* ("oviformis") was still active in sweet sherry with an alcohol concentration of 18% by vol. (Rankine 1972). Such yeasts could also be isolated from sweet Tokay wines containing 19 to 20% ethanol by vol. (Minárik 1966). Sweet unfiltered Sauterne wines underwent an after-fermentation with *S. cerevisiae* unless the wines had been treated with at least 300 mg SO_2 per liter (Cruess 1943). Surface growth of "oviformis" strains can be observed in wines with high alcohol concentrations (Barret *et al.* 1955). The literature frequently refers to the occurrence of *S. bailii* ("acidifaciens") (Fig. 9.17). Domercq (1956) found this yeast in 30% of white wines which had undergone after-fermentation but rarely in red wines. The yeast also occurs in South African wines (Van Zyl and Du Plessis 1961) and French wines (Poulard 1978). It is highly resistant to SO_2, but also to other fungicides, pyrocarbonic acid diethyl ester, and sorbic acid (Eschenbruch 1973; Rankine and Pilone 1973).

Occasionally after-fermentations are caused by *S. chevalieri* (Scheffer and Mrak 1951; Domercq 1956).

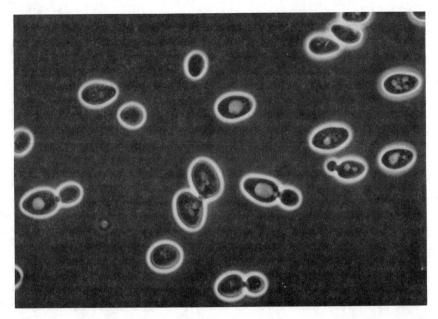

FIG. 9.17. *SACCHAROMYCES BAILII* VAR. *BAILII* (STRAIN "ACIDIFACIENS")

Saccharomycodes ludwigii.—This yeast is a true spoilage organism (see Fig. 9.18), and infections generally lead to complete spoilage of the wine. The yeast produces high concentrations of ethyl acetate (Ribéreau-Gayon and Peynaud 1960; Sponholz and Dittrich 1974). The French authors also show that high concentrations of esters are formed under anaerobic conditions. The fermentation by-products formed by this yeast give the wine an undesirable note. The yeast is particularly dangerous because of its great resistance to SO_2. It readily tolerates 80 to 120 mg SO_2 per liter (Domercq 1956) and up to 600 mg per liter (Sandu-Ville 1977). Therefore, it is often found in wines with exceptionally high sulfite concentrations. As shown in Table 9.19 it could frequently be isolated from French wines, 14% in vats and 27% in bottles (Domercq 1956). In Rumania 20% of the wines which contained residual sugar and which had become cloudy were infected by *Saccharomycodes ludwigii* (Sandu-Ville 1977). Its occasional isolation has also been reported by Minárik and Navara (1977).

Brettanomyces intermedius.—The older literature describes this species under the synonyms *B. vini* and *B. schanderlii*. Van der Walt (1970A) included them in the species *B. intermedius*. It is a true wine spoilage organism, and spoilage can occur in several ways. First, the yeast forms fine, powdery precipitates which do not sediment well. Secondly, under aerobic conditions it forms high concentrations of acetic acid and ethyl acetate by oxidation of ethanol. After a 12 month's growth period, 7.2 g acetic acid per liter could be found with several strains (Schanderl 1959).

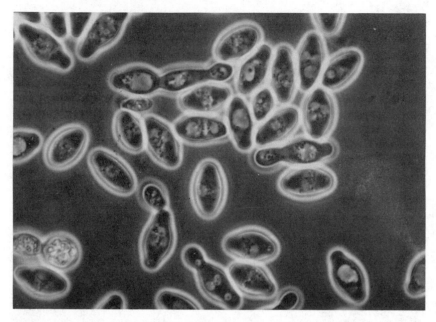

FIG. 9.18.   *SACCHAROMYCODES LUDWIGII*

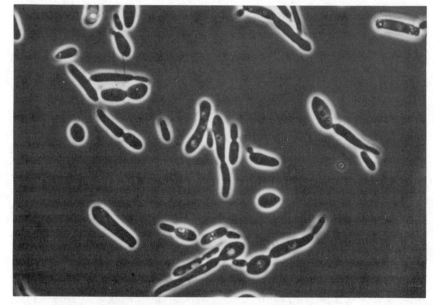

FIG. 9.19.   *BRETTANOMYCES INTERMEDIUS*

The yeast also forms large concentrations of isobutyric, caproic, and iso-valeric acid (Wenzel 1966). The quality of sherry can also be lowered appreciably by these fermentation by-products. The yeast can tolerate up to 15% ethanol by vol. (Van Zyl 1962), but its tolerance for sulfite is slight. *B. intermedius* is partially inhibited by 60 mg SO_2 per liter, and completely inhibited by 100 mg per liter.

It is an important spoilage organism in South Africa (Van der Walt and Van Kerken 1958; Van Zyl 1962) (see also Table 9.19). It occurred only occasionally in French and Italian wines (Domercq 1956; Castelli 1967), and more frequently in German champagne (Schanderl and Draczynski 1952). This species causes particular problems when the deposit of yeasts is loosened in bottles after the secondary fermentation. The quality of the wines is also lowered by the formation of acids. The occurrence of other *Brettanomyces* species has been described by Florenzano (1951).

Film Forming Yeasts.—Modern cellar practice has greatly reduced the importance of film forming yeasts as spoilage organisms. However, they are still important wherever wines of low alcohol concentrations are exposed to air because of their oxidative metabolism. Therefore, they are mainly found in vats which have not been completely filled. The yeasts form a film or skin on the surface of the wine. Species belonging to the film forming yeasts do not produce more than 1–4% by vol. of ethanol, and they can usually tolerate between 10 and 12% ethanol by vol. However, Scheffer and Mrak (1951) found strains in California which could tolerate up to 14% ethanol by vol. The resistance to SO_2 is high. Scheffer and Mrak found *Pichia* strains which tolerated up to 250 mg per liter of total SO_2. According to Schanderl (1959) the addition of 500 mg per liter of SO_2 does not guarantee complete inhibition.

The film forming yeasts cause a reduction of ethanol in wine and a decrease in "extract" substances, such as nonvolatile acids and glycerin (Van Zyl 1958). Acetaldehyde, acetic acid, and acetic acid esters are formed as the result of the oxidation of ethanol (Van Zyl 1958; Sponholz and Dittrich 1974).

The most frequently occurring film forming yeasts are *Pichia membranaefaciens, P. vini, P. fermentans, Hansenula anomala, Candida vini, C. valida, C. krusei, C. guilliermondii, C. zeylanoides,* and *C. mellini* (the two species *C. vini* and *C. valida* have been constituted from the yeast called *Candida mycoderma* in the older literature) (see Fig. 9.20–9.23).

The preceding yeasts occur frequently in South African wines (Van Zyl and Du Plessis 1961), where they account for 0.5 to 9.0% of the spoiled wines. *P. membranaefaciens* with 9% and *P. fermentans* with 4% have occurred most frequently. Czechoslovak wines contained most frequently *C. mycoderma, C. zeylanoides,* and *P. membranaefaciens* in young wines with high alcohol concentrations. White wines are more likely to spoil by the action of film forming yeasts than red wines (Minárik 1966).

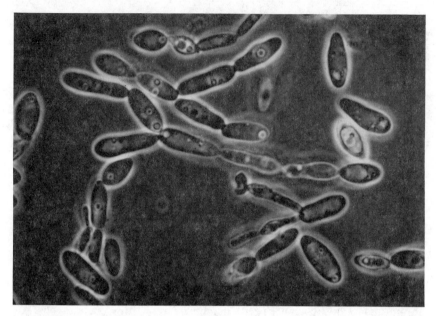

FIG. 9.20.   *PICHIA MEMBRANAEFACIENS*

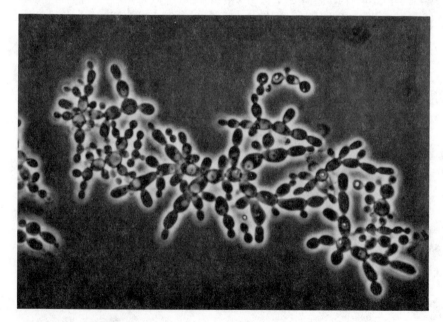

FIG. 9.21.   *PICHIA VINI*

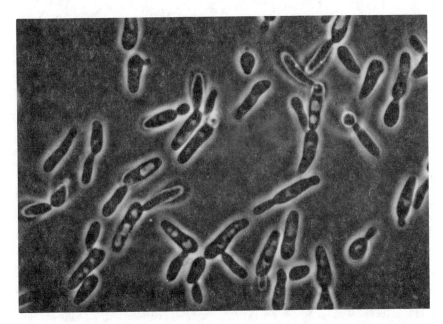

FIG. 9.22.   *HANSENULA ANOMALA*

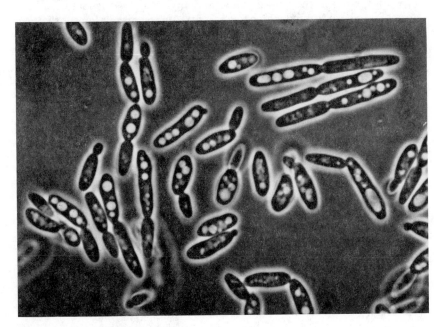

FIG. 9.23.   *CANDIDA VINI*

BACTERIAL PROCESSES DURING WINEMAKING

Must and wine are selective media for the development of bacteria since the low pH values permit only specialized groups of bacteria to grow and to multiply. Lactic acid bacteria and acetic acid bacteria belong to these groups, but the latter group loses its ability to survive as soon as reducing conditions prevail in the surrounding medium. In most instances bacteria occur as spoilage organisms, but lactic acid bacteria responsible for the metabolic degradation of malic acid are an exception, although a qualified one.

Lactic Acid Bacteria

A classification of the lactic acid bacteria of wines has been quite difficult since bacteria of highly acid media with low pH values have not been covered in the systematic classification of Breed *et al.* (1957). Therefore, additional work and an enlargement of the system of classification appeared necessary, and this was accomplished in the new edition, *Shorter Bergey's Manual of Determinative Bacteriology* (Holt 1977).

Classification and Physiology.—A French group of scientists under the leadership of Peynaud (among others) has contributed to the classification of the lactic acid bacteria that occur in wines. Therefore, the reader is first referred to the publications of Peynaud and Domercq (1967A,B, 1968, 1970) as well as to the review by Ribéreau-Gayon *et al.* (1975B). In this chapter only a few data on the classification and physiology of the lactic acid bacteria of wine are treated which are essential to the discussion that follows. For a more detailed review of the lactic acid bacteria of wine the reader is referred to Bidan (1967), Kunkee (1967A), Peynaud (1967), and Radler (1966).

The lactic acid bacteria of wines are facultative anaerobes or microphilic organisms. They are partly homofermentative, partly heterofermentative and belong to the genera *Pediococcus, Leuconostoc,* and *Lactobacillus* (Table 9.20). The homofermentative bacteria (all *Pediococcus* species and

TABLE 9.20.   CLASSIFICATION OF THE LACTIC ACID BACTERIA OF WINES

Characteristics	Species
Cocci	
homofermentative	*Pediococcus cerevisiae*
	Pediococcus pentosaceus
heterofermentative	*Leuconostoc gracile*
	Leuconostoc oenos
Rods	
homofermentative	*Lactobacillus plantarum*
	Lactobacillus casei
	Streptobacterium sp.
heterofermentative	*Lactobacillus fructivorans*
	Lactobacillus desidiosus
	Lactobacillus hilgardii
	Lactobacillus brevis

Source: Abbreviated from Ribéreau-Gayon *et al.* (1975B).

some *Lactobacillus* species) ferment glucose and fructose almost completely to lactic acid, that is, to about 95–100%. The heterofermentative bacteria (all *Leuconostoc* species and some *Lactobacillus* species) ferment glucose to lactic acid, ethanol, and CO_2, depending on the strain and the medium. The reaction may proceed further to lactic acid, acetic acid, and CO_2. The fermentation of fructose can lead to lesser or greater amounts of mannitol since a portion of the fructose can serve as a hydrogen acceptor. In addition, glycerin is formed as a side reaction during the fermentation of hexoses (Table 9.21). The lactic acid resulting from the fermentation of glucose may be the stereoisomeric form D(−) or a mixture of D(−)- and L(+)-lactic acid. The homofermentative species *Lactobacillus casei* is the exception and forms L(+)-lactic acid as the only isomer.

TABLE 9.21. FERMENTATION END PRODUCTS OBTAINED BY FERMENTATION OF 200 MILLIMOLES OF GLUCOSE OR FRUCTOSE WITH HOMOFERMENTATIVE AND HETEROFERMENTATIVE BACTERIA
End Products Are in Millimoles.

Fermentation End Product	Homofermentative Bacteria Glucose or Fructose	Heterofermentative Bacteria	
		Glucose	Fructose
Lactic acid	360	60	40
Carbon dioxide	0	100	50
Ethanol	0	100	0
Acetic acid	traces	120	100
Succinic acid	0	10	10
Mannitol	0	0	50
Butane-2,3-diol	0.3	10	10
Glycerin	10	80	40

Source: Ribéreau-Gayon *et al.* (1975B).

The ability to metabolize pentoses varies greatly among the lactic acid bacteria. Wine contains about 0.26 to 1.65 g/liter of arabinose and 0 to 0.44 g/liter of xylose (Melamed 1962). The ability to metabolize such pentoses is stable and can be used as a criterion for the identification of species (Nakagawa and Kitahara 1959; Peynaud and Domercq 1967A,1968; Vaughn *et al.* 1949). Melamed could show that with *L. arabinosus* the metabolism of arabinose is increased in the presence of hexoses. Xylose is metabolized to a minor extent. The metabolism of pentoses leads to the formation of considerable amounts of acetic acid.

The amino acids of the wine are largely responsible for the supply of nitrogen for the lactic acid bacteria (Du Plessis 1963; Peynaud and Domercq 1961; Radler 1958B). According to Radler the bacteria require arginine, cystine, cysteine, glutamic acid, histidine, leucine, phenylalanine, serine, tryptophan, tyrosine, and valine for growth. However, there are considerable differences in the requirements of different strains (see also Garvie 1967A,B). Gendron (1967) mentions the strong activation of growth of lactic acid bacteria and of the malo-lactic fermentation by arginine. Weiller and Radler (1976) investigated the extent to which amino acids are used for the synthesis of cell biomass, and also their degradation. Several strains of *Leuconostoc oenos* and *Lactobacillus brevis* decomposed arginine completely, some strains of *L. brevis* decomposed glutamic acid, one strain of *Pediococcus cerevisiae* degraded histidine, and one strain of *L. brevis* degraded tyrosine. Other amino acids were only partly degraded (Table 9.22). The end products of the decomposition of arginine, glutamic acid, and histidine are ornithine, 4-aminobutyric acid, and histamine, respectively, as well as NH_3 and CO_2. Arginine is degraded to ornithine by strains of *L. oenos* but not by *Pediococcus* species. Some yeast species are inhibited by ornithine, for instance *S. cerevisiae, S. carlsbergensis, Saccharomyces elegans, Hansenula* species, and *Debaryomyces* species (Boidron 1969). Künsch *et al.* (1974) welcomed the inhibiting effect of ornithine as a stabilizing factor against *Hansenula minuta.*

Malic and tartaric acids which occur in wine in higher concentrations as well as citric acid which occurs in lesser concentrations are metabolized by various lactic acid bacteria. The metabolism of malic acid will be discussed in greater detail following.

The ability to metabolize tartaric acid is not species specific and is only rarely present (Peynaud 1967). This author isolated 700 strains from wine and found only 6 strains which degraded tartaric acid completely, and 20 strains which degraded tartaric acid partially. These strains belong to the genera *Streptobacterium, Lactobacillus,* and *Leuconostoc.* The metabolism of lactic acid has also been treated by Barre (1969) and Krumperman and Vaughn (1966).

Radler and Yannissis (1972) demonstrated two pathways for the metabolism of tartaric acid by lactic acid bacteria. The homofermentative *Lactobacillus plantarum* metabolizes tartaric acid via oxalacetic acid and pyruvic acid to lactic acid, acetic acid, and CO_2. The heterofermentative *Lactobacillus brevis* metabolizes the intermediary compound, oxalacetic acid, to succinic acid, acetic acid, and CO_2. The metabolism of citric acid in wine also leads to an increase in acetic acid concentration. The pathway leads first to oxalacetic acid and acetic acid, and then from oxalacetic acid via pyruvic acid to lactic acid or to acetic acid and CO_2 (Charpentié *et al.* 1951; Charpentié 1954). In addition, acetoin and butanediol-2,3 may be formed by decarboxylation of pyruvic acid. Vaughn (1955) investigated bacterial processes in California wines and showed that a preference for the fermentation of either malic acid or citric acid by bacteria varies from strain to strain.

Fumaric acid in concentrations of 600 ppm inhibits lactic acid bacteria completely. This has been shown by investigations with *L. oenos* (Pilone *et al.* 1974; Tschelistcheff *et al.* 1971). In lesser concentrations fumaric acid can be decomposed by the enzyme, fumarase, which is present in lactic acid bacteria, and in that case fumaric acid activates growth. This is of eno-

TABLE 9.22. METABOLISM OF AMINO ACIDS BY VARIOUS LACTIC ACID BACTERIA

Bacterial Strain	Metabolism of Amino Acids in %										Formation Of			
	His	Try	Arg	Glu	Tyr	Ile	Thr	Asp	Phe	Ala	Orn	4-Ab	NH₃	x
Pediococcus cerevisiae B 16	100	–	–	.	.	.	.	.	.	.	.	.	–	.
Pediococcus cerevisiae B 56	–	50	–	.	.	.	.	.	.	.	.	.	–	.
Pediococcus cerevisiae B 89	–	20	–	.	.	.	.	.	.	.	.	.	–	.
Leuconostoc oenos B 65	–	–	100	20	–	–	–	30	–	–	+	(+)	+	–
Leuconostoc oenos B 70	–	–	100	–	–	40	–	30	–	–	+	–	+	+
Leuconostoc oenos B 66	–	–	100	–	–	–	–	–	–	–	+	(–)	+	–
Leuconostoc oenos B 121	–	–	20	.	.	.	.	.	–	.	+	–	+	–
Lactobacillus brevis B 26	–	–	100	–	60	–	–	–	–	–	+	–	+	+
Lactobacillus brevis B 31	–	–	100	100	–	50	50	40	–	–	+	+	+	–
Lactobacillus brevis B 37	–	–	100	20	100	–	–	–	–	–	+	(+)	+	+
Lactobacillus brevis B 41	–	–	100	100	20	50	40	40	20	–	+	+	+	+
Lactobacillus brevis B 54	–	–	100	100	30	40	–	40	20	20	+	(+)	+	+
Lactobacillus brevis B 63	–	–	100	20	20	40	30	30	–	–	+	+	+	+
Lactobacillus brevis B 73	–	–	100	100	–	30	20	–	–	–	+	+	+	+
Lactobacillus buchneri B 137	–	–	100	100	50	–	–	20	20	–	+	+	+	–

Source: Weiller and Radler (1976).

– = No metabolism or no formation, respectively.
. = Not investigated.
His = Histidine.
Arg = Arginine.
Tyr = Tyrosine.
Thr = Threonine.
Phe = Phenylalanine.
Orn = Ornithine.

+ = Formation.
x = Unknown compound reacting with ninhydrin.
Try = Tryptophan.
Glu = Glutamic acid.
Ile = Isoleucine.
Asp = Asparagine and aspartic acid.
Ala = Alanine.
4-Ab = 4-Aminobutyric acid.

logical interest since fumaric acid is sometimes used for the acidification of wines low in acid concentration (Pilone *et al.* 1973).

The sulfite binding substances acetaldehyde, pyruvic acid, and 2-keto-glutaric acid which are excreted by yeast into the medium during alcoholic fermentation can be metabolized by the lactic acid bacteria of wines, for instance by *Lactobacillus hilgardii* and *Leuconostoc oenos* (Boidron 1969; Fornachon 1963; Mayer *et al.* 1976A).

The presence of histamine and other biologically formed amines has recently been investigated more thoroughly because of their potential toxicity (Jakob 1978; Marquardt and Werringloer 1966; Ough 1971; Puputti and Suomalainen 1969; Quevauviller and Mazière 1969). Investigations in the author's laboratory have shown that metabolic processes of the lactic acid bacteria are responsible for the formation of these substances. Mayer and Pause (1973) found a close correlation between the presence of *Pediococcus cerevisiae* and the histamine content of the wines they had investigated. On the other hand, Weiller and Radler (1976), who tested 105 strains of bacteria, could find only a single strain of *P. cerevisiae* which formed histamine. This strain converted histidine to histamine in a stoichiometric ratio. The current assumption that simple decarboxylation leads to the formation of histamine has been questioned by Lafon-Lafourcade (1975). Concentrations for histamine reported in the literature vary between 1 and 20 mg per liter. Red wines generally show the higher values (see also Schneyder 1972). Lafon-Lafourcade (1975) grew some bacterial strains in must which had been fortified with histidine. Only little histamine was formed (1.6 mg/liter). Higher values were obtained with red wines. After seeding with heterofermentative rods and cocci, histamine values exceeded 6 mg per liter.

The following environmental factors of wines influence the development and the metabolism of lactic acid bacteria: temperature, pH, ethanol content, and the concentration of SO_2. The physiological dependence of the bacteria on the pH is of considerable enological significance. The pH optimum for growth is between pH 4.3 and 4.8, that is, higher than the normal pH of wine (Ingraham *et al.* 1960; Radler 1958B). But various strains can still grow up to a pH of 3.0. The lower pH limit for the metabolism of sugar and malic acid is generally in the pH region of 3.0 to 4.0 and varies from species to species (Peynaud and Domercq 1967A,B; 1968, 1970). The limits of sugar and malic acid metabolism can differ more or less among individual strains (Table 9.23). The difference in the limiting pH value between various strains for sugar and malic acid metabolism can be up to 0.8 pH. Sugar metabolism by lactic acid bacteria can result in the formation of volatile acidity. Therefore, these authors recommend a malo-lactic fermentation at pH values which limit sugar metabolism in order to obtain a "clean" fermentation. Mayer and Vetsch (1973) recommend for Swiss conditions a malo-lactic fermentation at pH 3.3 to 3.4. At this pH value *Leuconostoc oenos* ferments malic acid, but not the undesirable *Pediococcus cerevisiae*. A malo-lactic fermentation carried out by *P. cerevisiae* is often accompanied by the formation of diacetyl and histamine, and leads besides to the browning of red wines.

TABLE 9.23.   LOWER pH LIMIT FOR THE FERMENTATION OF MALIC ACID AND SUGAR BY BACTERIA IN WINE
Percentage of Bacterial Strains, Heterofermentative Cocci, Fermenting the Substrates at the Given pH Value.

Lower pH Limit	Malic Acid Fermentation, % of Bacteria	Sugar Fermentation, % of Bacteria
3.0	36	7
3.2	31	12
3.4	21	31
3.6	6	29
3.8	4	14
4.0	1	4
4.2	1	3

Source: Peynaud and Domercq (1968).

The lactic acid bacteria of wine are mesophilic organisms with a temperature optimum in the neighborhood of 25°C. Differences in temperature affect the growth rate and the rate of the metabolic reactions of the microorganisms (Peynaud 1967). The effect of the temperature also depends on the pH and the alcohol content whose negative effect is greater at higher temperatures. The rate of the malo-lactic fermentation is optimal at 20°–25°C. At 15°C and lower it is still possible though at a much slower rate.

The inhibiting effect of alcohol has been investigated several times (Flesch and Jerchel 1960; Peynaud 1967; Radler 1958A). Peynaud showed that an ethanol content of 6% by vol. in wine inhibited growth of several bacterial strains by 0 to 22% and the malo-lactic fermentation by 6 to 11%. At an alcohol concentration of 10% by vol. the respective values were 18 to 40% and 13 to 30% (Table 9.24). The effect of pH values between 3.0 and 5.5 and that of ethanol concentrations between between 5 and 11% by vol. is cumulative (Flesch and Jerchel 1960). Yet even at high alcohol concentrations (higher than 20% by vol.), growth of lactic acid bacteria is still possible (Bidan 1956; Vaughn 1955).

TABLE 9.24.   EFFECT OF ETHANOL ON THE GROWTH OF LACTIC ACID BACTERIA AND ON THE MALO-LACTIC FERMENTATION
Decrease in Growth Rate and Decrease of the Malo-lactic Fermentation in %.

| | 6% Ethanol by Vol. | | 10% Ethanol by Vol. | |
	Growth Rate	Malo-lactic Fermentation	Growth Rate	Malo-lactic Fermentation
Heterofermentative lactobacilli	22	10	40	21
Homofermentative lactobacilli	9	6	25	13
Heterofermentative cocci	0	11	18	30

Source: Abbreviated from Peynaud (1967).

The effect of SO_2 on the lactic acid bacteria of wine is of great enological importance. On the one hand, wines low in acidity require the use of this microbicide to prevent the danger of microbial spoilage. On the other hand, the development of a malo-lactic fermentation is desirable in wines high in acidity; and here the use of too high concentrations of SO_2 is contraindicated. Lactic acid bacteria are quite sensitive to free SO_2 and—in contrast to yeast—to bound SO_2 (Fornachon 1963; Lafon-Lafourcade and Peynaud 1974). The effect of "bound sulfurous acid" cannot be explained by the liberation of free SO_2 by bacterial decomposition of the acetaldehyde. The inhibiting effect depends on the pH of the medium (Lafon-Lafourcade and Peynaud 1974; Vetsch 1973) and all species of bacteria are affected both in the latent phase and during active growth. However, cocci are inhibited more than rods and growing cells moreso than cells in the latent phase. Sugar and malic acid metabolism are also inhibited. Free SO_2 has a stronger bactericidal effect than SO_2 bound to pyruvate or acetaldehyde (Lafon-Lafourcade and Peynaud 1974). Nevertheless, in a red wine with a pH of 3.5, 10 mg/liter of SO_2 bound to aldehyde showed inhibition, and at a concentration of 30 mg/liter, bacterial growth was arrested. It can further be shown that at 80 mg/liter of bound SO_2 the malo-lactic fermentation of the wine can be prevented. These findings which are important for the practice of wine making were confirmed by Mayer et al. (1975). These authors reported cases in which a concentration of 0 to 8 mg/liter of free SO_2 or 95 to 135 mg/liter of bound SO_2 inhibited the malo-lactic fermentation completely or almost completely. Mayer et al. (1976B) believe that it may be difficult to achieve a malo-lactic fermentation if yeast strains which can produce more than 50 mg SO_2 per liter develop spontaneously.

Leuconostoc species require CO_2 for their development (Mayer 1974). There was no growth in air free from CO_2 or in pure nitrogen gas, while strong growth occurred in an atmosphere of nitrogen plus 10% CO_2.

The Malo-lactic Fermentation

General Considerations.—The bacterial metabolism of malic acid to lactic acid and CO_2 is sometimes called a secondary fermentation by enologists. This process must be considered separately because of its great importance for particular wine growing regions although its effect may be quite different in nature. In some northern wine growing regions, the bacterial reduction of acids may be highly desirable, for instance in Switzerland and sometimes in France. In other areas in which the concentration of acids is low, it is important to prevent a further loss in acidity. This may be true in areas like California, South Africa, or Australia. But even here the malo-lactic fermentation prior to bottling may be of value in order to prevent later spoilage of wines with a high pH value. Grapes growing in a hot, dry climate have a higher proportion of tartaric acid. In cool and moist climates the ratio of the two principal acids is changed in favor of malic acid. For instance, in California the ratio of tartaric to malic acid is between 0.75 and 6.10 (Amerine et al. 1980), and in Germany between 0.65 and 2.30 (Vogt et al.

1974). Chemical methods are available for an adjustment of acids, that is, for a reduction of the total acidity of musts. But proponents of the malo-lactic fermentation suggest a favorable effect on wine quality (Eggenberger 1978; Ribéreau-Gayon *et al.* 1975B).

Chemistry of the Malo-lactic Fermentation.—The bacterial decomposition of malic acid leads quantitatively to lactic acid and CO_2. Different pathways have been found for lactic acid bacteria isolated from wines (Flesch 1969; Flesch and Holbach 1965; Lonvaud *et al.* 1977; Peynaud *et al.* 1968; Schütz and Radler 1973, 1974). The following pathways are possible:

(1)
$$\text{COOH-CH}_2\text{-CHOH-COOH} + \text{NAD}^+ \xleftrightarrow{\text{malate dehydrogenase}} \text{COOH-CH}_2\text{-CO-COOH} + \text{NADH} + \text{H}^+$$
malic acid $\qquad\qquad\qquad\qquad\qquad\qquad\qquad$ oxalacetic acid

$$\text{COOH-CH}_2\text{-CO-COOH} \xleftrightarrow{\text{oxalacetate decarboxylase}} \text{CH}_3\text{-CO-COOH} + \text{CO}_2$$
oxalacetic acid $\qquad\qquad\qquad\qquad\qquad\qquad$ pyruvic acid

$$\text{CH}_3\text{-CO-COOH} + \text{NADH} + \text{H}^+ \xleftrightarrow{\text{lactate dehydrogenase}} \text{CH}_3\text{-CHOH-COOH} + \text{NAD}^+$$
pyruvic acid $\qquad\qquad\qquad\qquad\qquad\qquad$ lactic acid

(2)
$$\text{COOH-CH}_2\text{-CHOH-COOH} + \text{NAD}^+ \xleftrightarrow{\text{malate enzyme}} \text{CH}_3\text{-CO-COOH} + \text{CO}_2 + \text{NADH} + \text{H}^+$$
malic acid $\qquad\qquad\qquad\qquad\qquad\qquad\qquad$ pyruvic acid

$$\text{CH}_3\text{-CO-COOH} + \text{NADH} + \text{H}^+ \xleftrightarrow{\text{lactate dehydrogenase}} \text{CH}_3\text{-CHOH-COOH} + \text{NAD}^+$$
pyruvic acid $\qquad\qquad\qquad\qquad\qquad\qquad$ lactic acid

(3)
$$\text{COOH-CH}_2\text{-CHOH-COOH} \xleftrightarrow[\text{NAD}]{\text{malo-lactate enzyme}} \text{CH}_3\text{-CHOH-COOH} + \text{CO}_2$$
malic acid $\qquad\qquad\qquad\qquad\qquad\qquad$ lactic acid

It has not been definitely determined which of these pathways actually is followed during the secondary fermentation of wine. Investigation favors the third pathway, which appears to be a likely path because the resulting lactic acid is always the L(+)-isomer (Peynaud *et al.* 1967, 1968). Otherwise the intermediate substrate, pyruvic acid, which occurs in the medium in free form would result in D(−)-lactic acid or a mixture of D(−)- and L(+)-lactic acid depending on the particular lactate dehydrogenase of the bacteria. It is not yet known whether the decomposition of malic acid by the malo-lactate enzyme is a simple decarboxylation. Nor is it known whether the malo-lactate enzyme is a single enzyme or an enzyme complex. Alizade and Simon (1973) believe that the path from malic acid to oxalacetic acid to pyruvic acid to lactic acid is catalyzed by a multi-enzyme complex and that the intermediary substrates, oxalacetic acid and pyruvic acid, are not released from the enzyme complex. The reaction depends on the presence of

NAD, and the pH optimum of enzyme action is in the acid range. Besides malic acid, glucose is required for induction. Since the reaction is not exergonic there is no gain in energy. Kandler *et al.* (1973) found no quantitative or qualitative change in the fermentation of glucose if malate was fermented simultaneously by *Leuconostoc mesenteroides*. But Pilone and Kunkee (1972, 1976) report a stimulating effect of the malo-lactic fermentation on the growth of *Leuconostoc oenos*. This effect is not due to a change of pH of the medium.

Different types of malic acid converting enzymes could be found in the lactic acid bacteria of wines (Lafon-Lafourcade and Peynaud 1970). For the majority of the strains the enzyme is constitutive, that is, for more than 60% of the investigated homo- and heterofermentative rods, and for more than 80% of the homo- and heterofermentative cocci. Some strains do not ferment malic acid.

Ecology of Lactic Acid Bacteria.—Numerous publications deal with the occurrence of lactic acid bacteria in wine and with their origin (Du Plessis and Van Zyl 1963; Fornachon 1964; Ingraham *et al.* 1960; Peynaud and Domercq 1961, 1967A,B, 1970; Weiller and Radler 1970, 1972). Table 9.25 gives examples of lactic acid bacteria found in French and German wines. The predominance of heterofermentative organisms in French wines and the predominance of homofermentative organisms in German wines is remarkable. Cocci, that is, *Pediococcus* and *Leuconostoc* species, constitute the majority of organisms in both countries. Spoiled wines have not been considered in these statistics. Such wines show a large number of lactobacilli, *L. hilgardii* and *L. brevis*. This observation parallels that of Fornachon (1963) for Australian wines and of Ingraham *et al.* (1960) for Californian wines in which mainly *L. hilgardii* was isolated. An extensive study of the wines of Piedmont, however, showed the frequent occurrence of *Leuconostoc* species (Malan *et al.* 1965).

TABLE 9.25. LACTIC ACID BACTERIA ISOLATED FROM WINES WITH MALO-LACTIC FERMENTATION

Bacterial Species	France, % from 323 Isolates	Germany, % from 49 Isolates
Homofermentative lactic acid bacteria		
Lactobacillus casei	2	25
Lactobacillus plantarum	4	—
Pediococcus cerevisiae	4	51
Pediococcus sp.	—	6
Heterofermentative lactic acid bacteria		
Lactobacillus hilgardii	7	—
Lactobacillus fructivorans	6	—
Lactobacillus brevis	3	8
Lactobacillus sp.	1	—
Leuconostoc oenos	42	10
Leuconostoc gracile	31	—

Source: Compiled from Ribéreau-Gayon *et al.* (1975B) and Weiller and Radler (1970).

The origin of the lactic acid bacteria of wines was investigated by Peynaud and Domercq (see Peynaud 1967). These authors show that the bacteria are present on grapes although they are not as numerous as yeasts or acetic acid bacteria. It could also be shown that contamination of musts occurs through contact with the equipment of the wine cellar. Lactic acid bacteria which can ferment malic acid could also be found on grape leaves, but only a part of these is typical of lactic acid bacteria occurring in wine (Weiller and Radler 1970, 1972). During the wine fermentation the number of lactic acid bacteria is quickly reduced because of unfavorable conditions of the medium, and a selection occurs in favor of particular species.

Bacteria which carry out a "clean" secondary fermentation are desirable, that is, bacteria that ferment little else but malic acid. There are great differences between the lactic acid bacteria from species to species but also from strain to strain so that "clean" fermentations, wine spoilage, or any intermediate results are possible. Environmental conditions are also very important. As Peynaud has shown, homofermentative bacteria are not always ideal for malo-lactic fermentation. Glucose is almost completely fermented to lactic acid, but other compounds are also fermented, such as fructose and sometimes pentoses, which leads to the formation of volatile acidity.

In wines which have undergone a normal malo-lactic fermentation, lactobacilli are not present in large numbers. Weiller and Radler (1970, 1972) have found a larger number of homofermentative *Lactobacillus casei* in wines, but these are probably not very significant. All of these strains were isolated from wines of a limited wine growing area.

However, homo- and heterofermentative cocci are more widely distributed (*Pediococcus cerevisiae* as well as *Leuconostoc oenos* and *L. gracile*). In spite of the frequent occurrence of *P. cerevisiae* in wines, this organism is not considered to be a specific bacterium for the fermentation of malic acid, since malic acid is not fermented preferentially ahead of hexoses at normal, low pH values (Peynaud and Domercq 1967B). Neither pentoses nor tartaric nor citric acid is fermented by *P. cerevisiae* and, therefore, this species has been considered to be harmless in young wines. On the other hand, *P. cerevisiae* could frequently be isolated from old wines which had undergone undesirable changes. Mayer (1974) and Mayer and Vetsch (1973) report also that the occurrence of *P. cerevisiae* in Swiss red wines leads to undesirable flavors. Mayer stresses the fact that this species hardly grows at pH values below 3.5. At a pH of 3.3 the number of cells doubled in 41 days, but at a pH of 3.5 it doubled in 5.3 days. Therefore, maintenance of a low pH value is considered to be the best method for a satisfactory malo-lactic fermentation.

Leuconostoc species, however, have a special significance for the malo-lactic fermentation. They are the specific agents of this fermentation. This genus comprises the species *L. gracile* (Synonym: *Bacterium gracile*) that does not ferment pentoses and *L. oenos* that includes strains which ferment pentoses (arabinose and xylose or arabinose or xylose). X^+ and P^+ strains are rarely found in wines, but *L. oenos* A^+ and to a lesser extent *L. gracile*

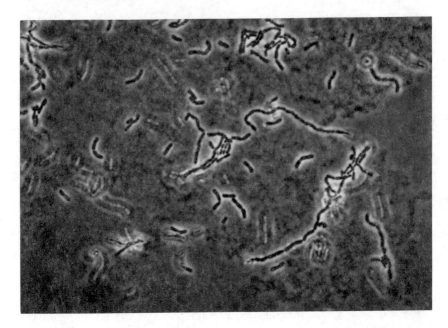

FIG. 9.24.   *LACTOBACILLUS CASEI*

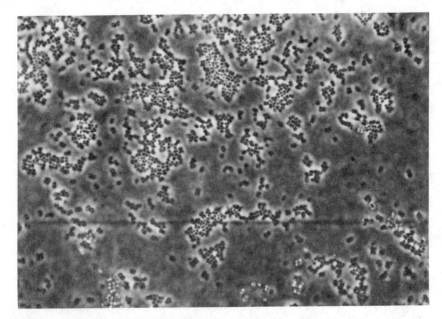

FIG. 9.25.   *PEDIOCOCCUS CEREVISIAE*

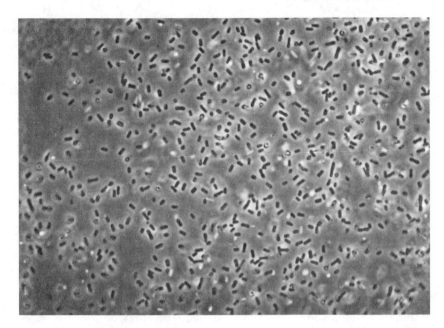

FIG. 9.26.   *LACTOBACILLUS BREVIS*

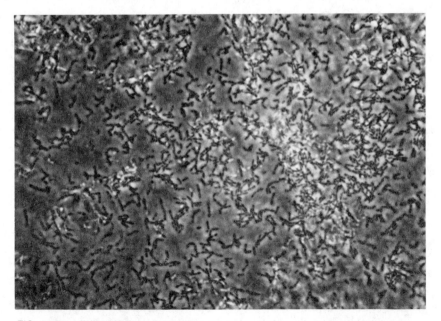

FIG. 9.27.   *LEUCONOSTOC OENOS*

occur frequently (Peynaud and Domercq 1968). *Leuconostoc* species are well adapted to the low pH values of wine in contrast to representatives of other genera. In general these species are preferred wherever a malo-lactic fermentation is desired.

It has often been attempted to start a malo-lactic fermentation in musts or wines by inoculation with pure cultures of *Leuconostoc* species (Cuinier 1973; Flesch and Jerchel 1961; Kunkee 1967B; Lafon-Lafourcade *et al.* 1968; Munyon and Nagel 1977; Vetsch 1973); but this method has not assumed much importance because it has been difficult to adapt these bacteria to wines. Besides, inoculation at an earlier stage of the fermentation (with residual sugar values above 2−2.5 g/liter is contraindicated because of the danger of undesirable bacterial fermentations.

Wine Defects Caused by Bacteria

Undesirable bacterial changes in wines extend over the whole scale from slight changes in flavor to complete spoilage. They are determined by the kind and concentration of the metabolic products of lactic acid and acetic acid bacteria.

Lactic Acid Bacteria as Spoilage Organisms.—The lactic acid bacteria of wines are biochemically quite heterogeneous as has already been discussed. If larger amounts of sugar in musts or wines are fermented by these organisms, variable amounts of CO_2, ethanol, volatile acids, and mannitol are formed depending on the particular species. Wines which have undergone undesired changes in flavor in this manner are said to have a "lactic acid flavor."

The wines have higher acidity because of the increase in lactic acid. The organoleptically perceptible "lactic acid flavor" is largely determined by the diketone, diacetyl, which occurs in such wines together with the closely related acetoin (Fornachon and Lloyd 1965; Radler 1962). Pilone *et al.* (1966) and Radler and Gerwarth (1971) investigated the formation of other fermentation by-products, but their importance for the aroma of the wine could not be clearly established. The analytical results of Pilone *et al.* (1966) show that not only lactic acid but also lactic acid ethyl ester are increased. Diacetyl is intensely aromatic. Wines with an off flavor had a diacetyl content of 0.9 mg/liter and up (Dittrich and Kerner 1964). The threshold value of perception in a light red wine is 1 ppm and in a fully aromatic red wine 4 ppm of diacetyl (Rankine *et al.* 1969). Radler (1972) investigated wines in which acids had been fermented by bacteria in comparison with those in which this fermentation had been absent. In the former case the concentration of diacetyl and acetoin was 0.9−34.9 mg/liter, and in the latter case 0.7−15.8 mg/liter. In Swiss wines in which the malo-lactic fermentation was due to *L. oenos*, the diacetyl content was low (0.78 mg/liter) but it was high in the presence of *P. cerevisiae* (3.9 mg/liter) (Mayer 1974). While high and exceptionally high concentrations of diacetyl have an undesirable effect on wine aroma, small concentrations probably have a positive effect. Diacetyl may be formed from acetoin which in turn may be formed from pyruvic acid via α-acetolactate (see also Radler 1962). The effect of diacetyl on the aroma of fermented beverages is also discussed in Chapter 10 on brewing.

An increase in volatile acids can also occur in wines which are dry and completely fermented with a normal concentration of 2 g of residual sugars per liter. These sugars are hexoses and pentoses which cannot be fermented by yeasts. The residual pentoses are mainly responsible for the subsequent increase in acetic acid concentration. Homofermentative as well as heterofermentative lactic acid bacteria which are capable of fermenting pentoses are responsible for this fermentation in resting wines. Such changes in dry wines occur frequently with red wines.

The fermentation of organic acids in wines can also lead to undesirable changes in aroma. Wines with low acidities are particularly susceptible to such changes. Several bacterial strains can ferment tartaric acid. This property is not tied to a particular species (Peynaud 1967). The lactic acid bacteria investigated by this author had only a small percentage of strains which could ferment tartaric acid completely (6 out of 700). Other strains fermented tartaric acid partially. Krumpermann and Vaughn (1966) isolated a large number of bacteria which ferment tartaric acid, which were homo- and heterofermentative species of *Lactobacillus*. The bacterial fermentation of tartaric acid leads to complete spoilage of the wine.

The amount of citric acid occurring in wine is relatively small. It is about 0–0.5 g per liter according to Amerine *et al.* (1980); but some countries permit the addition of citric acid to increase the total acidity of wine. The capability of fermenting citric acid is frequently found among lactic acid bacteria, mainly among the homofermentative cocci (Marques Gomes 1974). The fermentation of citric acid leads to an increase in acetic acid. Therefore, it is desirable to use only those bacteria for the malo-lactic fermentation which do not ferment pentoses or citric acid.

Various lactic acid bacteria can also ferment glycerin but this capability is rarely found in practice. Peynaud (1967) found in his investigation of French wines that only 4% of the isolated strains had that capability, which is not restricted to certain species. Both lactic and acetic acids are always produced during the fermentation of glycerin. According to Ribéreau-Gayon *et al.* (1975B) fermentation rarely degrades glycerin completely which would lead to the formation of acrolein. This substance is presumably responsible for the bitter flavor which variously occurs in wines (Rentschler and Tanner 1951). The formation of the bitter flavor substance results from a reaction of acrolein with polyphenols. Red wines are more likely to show this flavor defect because of their high concentration of polyphenols.

Even fortified wines with an ethanol concentration of 19–20% by vol. are subject to bacterial fermentations. In such wines, which generally have a high concentration of sugar and a low concentration of acid, the development of *Lactobacillus trichodes* could repeatedly be demonstrated. This heterofermentative microorganism does not ferment malic acid and has a high tolerance for ethanol (Douglas and McClung 1937; Fornachon *et al.*

1949). Wines which have been spoiled by *L. trichodes* show a strong, amorphous precipitate, a slight reduction in the concentration of sugar and an increase in the concentration of volatile and nonvolatile acids. Newer investigations have shown that lees wines infected by lactic acid bacteria contain higher concentrations of propanol-1 and butanol-2 (Bertrand and Suzuta 1976; Hieke and Vollbrecht 1974A). The experimental results indicate that butanediol-2,3, which is a by-product of the alcoholic fermentation, is reduced to butanol-2 by bacterial action. Therefore, the occurrence of higher concentrations of butanol-2 is an indication of bacterial spoilage. Particularly high concentrations of this alcohol were found in wines infected with *Lactobacillus brevis* strains (Hieke and Vollbrecht 1974B). The pathway of the formation of propanol-1 by lactic acid bacteria has not yet been elucidated. Extensive reviews of the isolation and description of lactic acid bacteria in spoiled wines have been published by Amerine *et al.* (1980), Cruess (1943), Du Plessis and Van Zyl (1963), Fornachon *et al.* (1949), and Vaughn (1955).

Acetic Acid Bacteria as Spoilage Organisms.—The role of acetic acid bacteria in wine making is not as important and not as complex as that of the lactic acid bacteria. They have become less important because of improvements in sanitation. Nevertheless, a lowering of the quality of wines by these bacteria is still possible, and in more serious cases may lead to complete spoilage.

In the enological literature the description of acetic acid bacteria which are found on grapes, in must, or in wine is usually based on the system of Frateur (1950). This system divides the genus *Acetobacter* into 4 biochemical subgroups (peroxydans, oxydans, mesoxydans, and suboxydans). Catalase activity, oxidation of acetate and lactate, and the formation of ketones and gluconic acid are used to differentiate the subgroups. On the other hand, the finding of peritrichous flagellation in some members of the genus *Acetobacter* leads to a division of the genus into 2 genera, *Acetobacter* with peritrichous flagellation and *Acetomonas* with polar flagellation (Leifson 1954). Besides, the subdivisions have been questioned because of the gradual change of characteristics from strain to strain which defies clear distinctions. For this reason Asai (1968) suggested a taxonomic division of acetic acid bacteria into the 2 genera, *Acetobacter* and *Gluconobacter* (identical with *Acetomonas*). Each genus contains only 1 species, namely, *Acetobacter aceti* and *Gluconobacter oxydans*; and all previously described forms should be considered varieties of these species. The reader is referred to Greenshields (1978) with regard to the foregoing discussion and with regard to the cited references. Table 9.26 gives a schematic view of the taxonomy of the acetic acid bacteria which should facilitate the orientation of the reader (Dittrich 1977). It takes account of the mentioned classification of acetic acid bacteria. (See also Chapter 19 on Vinegar.)

In contrast to the lactic acid bacteria, the acetic acid bacteria are aerobic organisms which do not grow in the absence of oxygen. Their importance as wine spoilage organisms rests not only on their ability to convert ethanol in

TABLE 9.26.   BIOCHEMISTRY AND PHYSIOLOGY OF THE ACETIC ACID BACTERIA,
ACETOBACTER

Group	Organism	Catalase	Acetate and Lactate Oxidation	Ketone Formation	Gluconic Acid Formation	Flagellation
Peroxydans	A. paradoxum	−	+	−	−	peritrichous
	A. peroxydans	−	+	−	−	peritrichous
Oxydans	A. ascendens	+	+	−	−	peritrichous
	A. rancens	+	+	(+)	+	peritrichous
	A. lovaniense	+	+	(+)	+	peritrichous
Mesoxydans	A. mesoxydans	+	+	+	+	peritrichous
	A. xylinum	+	+	+	+	peritrichous
	A. aceti	+	+	+	+	peritrichous
Suboxydans	A. suboxydans	+	−	+	+++	polar
= Acetomonas	A. melanogenum	+	−	+	+++	polar

Source: Abbreviated and modified from Frateur (1950) and Dittrich (1977).

wine to acetic acid, but an equally important danger is their growth on
grapes which have been mechanically damaged or which have become
moldy on the vines. The occurrence of acetic acid bacteria on grapes, in
must, and in wine has been investigated by Blackwood (1969A,B), Dupuy
(1957A), and Vaughn (1955). In wine, representatives of the peroxydans
and the oxydans group *(Acetobacter paradoxum, A. ascendens, A. rancens)*
predominate. Bacteria of the mesoxydans group *(A. mesoxydans, A. aceti)*
occur to a lesser extent in wine. Members of the suboxydans group occur
only rarely in wine, but such organisms *(A. suboxydans, A. melanogenum)*
always occur on grapes and in must. Mesoxydans and oxydans types can
also be isolated from grapes. These are generally species forming ketones.

Spoilage of wine can occur only if the wine is kept in vessels which are not
completely filled, since the acetic acid bacteria are aerobic organisms.
Depending on the species, growth can occur on the surface of the liquid as a
thin or a thick skin. Apart from the requirement for oxygen, growth of the
organisms is also affected by the temperature, the pH, and the ethanol
concentration of the wine (Dupuy 1957B; Vaughn 1942). The temperature
optimum for growth of these bacteria is between 30° and 35°C, the limit of
alcohol tolerance is between 12 and 15% by vol., and growth does not occur
at pH values lower than 3.2 to 3.0. The organisms are very sensitive to SO_2.

Acetic acid bacteria oxidize ethanol via acetaldehyde to acetic acid. The
esterification of acetic acid to ethyl acetate occurs regularly as a side
reaction, and the latter compound gives vinegary wines their characteristic
note (Peynaud 1975). There is no quantitative correlation between the
concentrations of acetic acid and ethyl acetate (Junge and Spadinger 1978).

Acetic acid bacteria are particularly harmful through their growth on
mechanically injured grapes or grapes infected with *Botrytis cinerea*. The
oxidation of ethanol to acetic acid and ethyl acetate can already take place
on the grapes. In addition, bacteria of the suboxydans and mesoxydans
group and some of the oxydans group can oxidize glucose to gluconic acid.

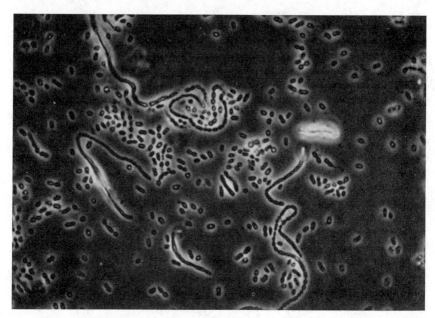

FIG. 9.28.   *ACETOBACTER RANCENS*

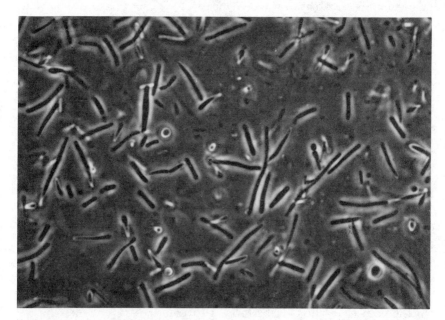

FIG. 9.29.   *ACETOBACTER XYLINUM*

Further oxidation results in the formation of keto acids (keto-2 and diketo-2,5-gluconic acids) (Peynaud and Sapis 1975). Also, fructose is oxidized to keto-5-fructose. This compound as well as diketo-2,5-gluconic acid are important because of their ability to bind SO_2.

The traditional production of red wines is another opportunity for the formation of undesirably high concentrations of acetic acid. The acetic acid bacteria can multiply readily in the cap which is not fully submerged in the liquid (see also Amerine et al. 1980). Also, in stuck fermentations in which the temperature has risen too high there may be a considerable increase in the number of thermophilic organisms. This is particularly true for very large vessels which have no provisions for cooling. For such cases a slow, controlled fermentation is indicated.

Wines containing higher concentrations of acetic acid must be considered spoiled. For this reason there are regulations governing the maximum concentrations of volatile acids in wine. Yeasts produce about 40 to 100 mg of acetic acid per 100 ml of wine. A total volatile acidity up to 300 mg per 100 ml can be tolerated. The threshold value for the detection of acetic acid in wine by expert wine tasters is about 70 mg per 100 ml.

TECHNOLOGICAL ASPECTS OF WINEMAKING

White Wine

The quality of wine depends greatly on the type of grape as well as on its maturity and health. The grapes are first separated from the stems and then crushed in order to facilitate pressing. De-stemming is important since it prevents the undesirable leaching of the woody stems if the grapes are fermented on the skins. This is even more important for red wines. Rapid processing of the grapes and the crushed grapes is important to inhibit the growth of undesirable microorganisms and especially that of acetic acid bacteria (see pages 358–361). This is still more important during warmer weather. In addition oxidative, enzyme-catalyzed processes occur on prolonged standing of the crushed grapes which may affect the color and organoleptic properties of the wine unfavorably. But a certain rest period is desirable because it increases the flow of free run juice and facilitates pressing which is mainly due to the action of pectic enzymes. However, it is important to treat the must immediately with SO_2 in order to inhibit oxidative enzymes and to prevent the growth of the mentioned microorganisms. In special cases the crushed grapes can be directly sulfited. The purpose of adding SO_2 and its effect on the microbiology of wines have already been discussed in some detail in the section on the alcoholic fermentation. Further, sulfiting is particularly important for the vinification of grapes with low concentrations of acid to prevent bacterial fermentation of the malic acid of the must (see pages 303–304).

In the next step of the process the pressed must or free run juice is freed from particulate matter by centrifugation, filtration, or the simple sedimentation which occurs on standing. This removal of the sediment has an important effect on the susceptibility of the must to oxidation and on the rate of fermentation and consequently on the color and aroma of the wine. Oxidative enzymes can also be inactivated by high temperature-short time heating of the must (to 85°C). This kills the microorganisms. In this case inoculation with a fermenting yeast is absolutely essential. But even if the must is not heated it is advantageous to inoculate it with a pure culture yeast, discussed on pages 331–336. The major problem of the fermentation is increase of the temperature by the exothermic reaction. This can lead to difficulties, particularly in very large fermentors. At temperatures above 38°C the fermentation can become "stuck" because of inactivation and loss of viability of the yeast. Besides, fermentations conducted at high temperatures lead to a loss of aroma and wine quality. Therefore, temperatures should not exceed 20°C. In California, temperatures between 10° and 15°C are preferred, but sometimes even lower temperatures are used.

After the end of the fermentation the wine is removed from the lees as soon as possible before the yeast begins to autolyze. However, autolysis of the yeast may aid in initiating the malo-lactic fermentation. After separation from the lees the wine is clarified and stabilized. The wine is bottled after a storage period which generally varies between 3 and 9 months, but which may be extended up to 2 years for high quality wines. During this bulk storage some chemical reactions, among them oxidations and esterifications, take place. The ethanol content of table wine is usually between 8 and 15% by vol.

Red Wine

The extraction of the color of the grape skins is very important for the production of red wines. For white wines the extraction of coloring matter and tannins is not desired. Therefore, the first stages of the production of red wines differ materially from those for white wines. The later stages of red wine production, namely the fermentation after pressing, are identical with those for white wine production. The various methods used in the production of red wines differ basically in the methods by which color is extracted.

The classical principle consists of the fermentation of the crushed grapes. The color is extracted by the alcohol which forms during the fermentation. The temperature of the fermentation is kept between 20° and 30°C so that the grape skins can be removed fairly soon in order to prevent excessive extraction of tannins. Fermentation can be carried out in open or closed vessels, although fermentations in open vessels are not common today. They require that the cap, which is lifted by evolving CO_2, be pushed under the surface of the liquid. This is done to increase color extraction and to prevent infection by spoilage organisms, such as acetic acid bacteria. This disadvantage can be met by fermenting in a closed vessel in which the cap is kept under a carbon dioxide atmosphere, or in which the particulate matter can be kept distributed throughout the vessel by other means. Continuous systems have been developed for the production of red wines in large fermenters. For such systems the continuous recirculation of the cap permits good extraction.

Color can also be extracted very well by short time-high temperature heating of the crushed grapes (to 85°C). The must obtained after pressing must be inoculated with a pure culture yeast. Red wines obtained by this method generally contain lower concentrations of tannins than red wines produced by fermentation of the crushed grapes.

Another interesting process for the production of red wines is the "macération carbonique." The uninjured grapes are kept for 8 to 10 days in a vessel under a layer of carbon dioxide gas, that is, under anaerobic conditions. During this time an intracellular fermentation occurs in the intact grapes which leads to the formation of 1.5 to 2.5% ethanol by vol., carbon dioxide, and by-products of the alcoholic fermentation. About 50% of the malic acid is fermented, and anthocyanins are brought into solution (Peynaud 1975). The intracellular fermentation is arrested as soon as the rising ethanol concentration leads to the death of the cells. The grapes are pressed when CO_2 evolution ceases, and the fermentation of the must is completed within about 48 hr. The following advantages have been claimed for this method: The aroma of the wines is more intense; the concentration of extracted polyphenols is low; and bacterial fermentation of malic acid is favored since there is also a decrease in acid concentration during the period of maceration.

Champagne Type Wines

The tradition of all of these wines is closely connected with that of champagne. Today the designation "champagne" is reserved for products produced in certain French districts and in determined amounts.[1] But the production of sparkling wines is also carried out in many other wine districts. In general, sparkling wines are those which foam readily because of the presence of high concentrations of dissolved CO_2. The CO_2 pressure is 4.05–5.06 Pa (4–5 atm) at 20°C. However, in the United States, wines containing a pressure of slightly more than 2.03 Pa (2 atm) may be called sparkling wines. The methods of production are the following:

(1) Champagne process (bottle fermentation, removal of yeast by disgorging)
(2) Transfer process (bottle fermentation, transfer to a tank, and removal of the yeast by filtration)
(3) Bulk fermentation
(4) Carbonation

Champagne Process.—In the classical bottle fermentation, a dry white wine (cuvé) undergoes a secondary fermentation after the addition of about 25 g per liter of sucrose and inoculation with a pure culture yeast. This secondary fermentation takes place in thick walled, tightly closed bottles at 9°–12°C. The fermentation requires several months. After that the wine

---

[1]In the United States this reservation is not observed, and sparkling wines made by natural fermentation may be called "champagne."

remains on the yeast for several months or years. During this rest period the yeast collects in the neck of the bottles, a process which is aided by shaking and by an increasing inclination of the bottles so that they approach a vertical position. Finally, the yeast deposit is frozen in the neck of the bottle and disgorges when the bottle is opened. The lost amount is then replaced by adding a solution of sucrose in wine. The sucrose concentration depends on the desired end product, and the bottles are tightly closed after addition of the dosage.

Transfer Process.—In this process the bottle fermentation is carried out as above, but the yeast is removed by transferring the wine from the bottles to a tank in a closed system and under nitrogen pressure. After addition of the dosage, the wine is filtered in a closed system and with nitrogen or carbon dioxide counterpressure and filled into bottles. This method permits a retention of the carbon dioxide in the wine.

Bulk Fermentation.—This process is suitable for mass production of sparkling wines and results in wines of somewhat lesser quality. The secondary fermentation is carried out in a pressurized vessel. A certain concentration of unfermented, residual sugar is retained in the wine so that there is no need for the addition of a dosage. After filtration the wine can be filled into bottles. In this process the carbon dioxide evolved during the secondary fermentation is also retained.

Carbonation.—In contrast to the preceding processes the sparkling character of the wine is obtained by impregnating the base wine with carbon dioxide. That means that there is no secondary fermentation. This process is suitable for the production of less expensive wines, and its quality is largely determined by the quality of the base wine. Also in contrast to the preceding processes, which involve a secondary fermentation, the carbon dioxide is only weakly bound and escapes more quickly after the bottles are opened.

The choice of a yeast is highly important for bottle fermentations since this fermentation is carried out under more demanding conditions. The alcohol concentration of the cuvé (about 11% by vol.), the low temperature, and the slowly increasing pressure of CO_2 are all inhibitory for the yeast. It is also important that the yeast be fairly flocculent and form a compact deposit. Strains of *S. cerevisiae* and *S. bayanus* are used in commercial practice.

Sherry

The traditional production of sherry originated in Spain. Today it is still carried out in a limited production area in Jerez de la Frontera. The original method, designated as the solera process, is also carried out in other countries, for instance in France, Australia, the United States (California) and the U.S.S.R. But other systems of sherry production, such as the submerged fermentation or the baking process, have also attained great importance.

Sherry is a dessert wine whose ethanol concentration of 19 to 21% by vol. is obtained by fortification. According to the desired type it contains various concentrations of sucrose. The wines have a characteristic, fine aroma.

Solera Process.—The bases of the process are the so-called flor yeasts, which were originally designated with the collective term *Mycoderma vini*. Later they were assigned to the genus *Saccharomyces* and described in the literature as *S. beticus*, *S. cheriensis*, or *S. rouxii*, among others. In today's nomenclature they are generally designated as *S. bayanus* (see also page 315). These yeasts are characterized by their ability to form a film on the surface of a base wine with 14–15% ethanol by vol. Their metabolism is adapted to the prevailing aerobic conditions. They respire available wine constituents, principally ethanol. The resulting metabolites are mainly acetaldehyde, but also other aldehydes, acetals, esters, and higher alcohols. These give sherry its characteristic aroma. There is also a considerable decrease in the concentrations of glycerin and malic acid, and there may be a decrease in volatile acidity.

For the production of sherry the harvested grapes are first dried in the sun to obtain a further concentration of the juice. The crushed grapes are "plastered" by addition of $CaSO_4$ in the form of finely ground gypsum in order to increase the acidity and lower the pH. The calcium sulfate reacts with the potassium acid tartrate with the formation of free tartaric acid. This reduces the susceptibility of the must to the growth of lactic acid bacteria which must be prevented in the production of sherry.

The must is fermented by *Saccharomyces* yeasts which are also responsible for the later aerobic stage. The alcohol content of the base wine is between 14.5 and 15.5% by vol. The base wine is then filled into barrels with a volume of about 500 liters. The barrels are put into a horizontal position, and they are only partially filled to allow maximum surface contact between the base wine and air. These barrels are arranged in a solera system which consists of 3 to 6 ranks of barrels. The young base wine is filled into the first (uppermost) rank of barrels. The formation of the flor begins already after about 2 weeks. The matured wine is withdrawn from the last (lowest) rank of barrels without disturbing the flor. This last rank is then refilled with wine from the next higher rank, and so forth through all the ranks of barrels. In a system with 5 ranks the wine withdrawn from the last rank has an average age of 12 years.

Submerged Fermentation.—The sherry process can be accelerated if the flor yeasts develop in submerged culture with strong aeration. This ensures an aerobic metabolism and leads to the rapid formation of acetaldehyde. The rate of acetaldehyde formation depends on the number of yeast cells. This process gives wines of good quality. However, they do not approach the quality of wines produced with the solera process. The submerged fermentation process is suitable for the mass production of less expensive sherries. The quality can be improved if the wines are stored in oak barrels after the submerged fermentation is completed.

Baking Process.—In California most of the sherry is produced by "baking." In this process the characteristic sherry flavor and aroma are obtained by treatment with elevated temperatures, and not by the action of yeasts. Before baking the base wine is fortified to an ethanol content of 17 to 21% by vol., and the sugar concentration is adjusted to concentrations between 10 and 100 g per liter depending on the desired degree of sweetness for dry, semi-dry and cream sherries. During fortification the yeast flocculates. The yeast sediment should be removed at this point of the process since its retention favors the development of *Lactobacillus trichodes* and leads to spoilage. The heat treatment is carried out at temperatures between 50° and 60°C for a period of 10 to 20 weeks. Caramelization and to some extent oxidative processes occur at this time, which later affect the aroma of the sherry. Following heat treatment the wines may be stored for up to 3 years, but in most cases the storage period is less than 6 months.

A modification of this process consists of bubbling air or oxygen through the wine during baking (Tressler process). The extent to which oxidative processes during this treatment affect the quality of the end product has not been fully clarified. The Tressler method is favored for wines made from *Vitis labrusca* grapes since it neutralizes the aroma of these grapes somewhat.

The processes discussed in this section on the "Technological Aspects of Winemaking" have been reviewed by Amerine *et al.* (1980), Peynaud (1975), Ribéreau-Gayon and Peynaud (1960), and Troost (1972).

AROMA COMPONENTS OF WINES

The aroma of the wine is determined by 400 to 600 volatile components which belong to diverse classes of compounds, such as alcohols, acids, esters, aldehydes, ketones, terpenes, lactones, and others. These compounds are either the primary aromatic substances of the grape which appear unchanged in the wine; or they are compounds which are chemically changed during vinification, or which are the result of microbial action during the fermentation and maturing of the wine. The total concentration of aroma substances is about 0.8 to 1.0 g/liter. About one-half of this amount is accounted for by higher alcohols formed during the fermentation, isobutanol, isoamyl alcohol, optically active amyl alcohol, as well as 2-phenylethanol (Rapp *et al.* 1978). All other aroma components are present in concentrations between 10^{-4} and 10^{-9} g/liter.

Primary Aroma Substances

Differences between grape varieties are sometimes considerable, but they are quantitative rather than qualitative (Drawert and Rapp 1966; Rapp and Hastrich 1976; Schreier *et al.* 1976B; Terrier and Boidron 1972). Rapp and Hastrich (1976) as well as Schreier *et al.* (1976A) have tried to use constant relationships between significant components for a characterization of grape varieties. On the other hand, aroma components with specific structures may be used to characterize the typical aroma of various varieties. Derivatives of terpene, furan, and pyran compounds can be used for this purpose. The following compounds have been found in somewhat higher concentrations in the more aromatic varieties of *Vitis vinifera:* Terpene alcohols, such as linalool, ho-trienol, geraniol, nerol, and α-terpineol; and the so-called "terpene oxides," such as neroloxide, the isomeric rose oxides, as well as trans-geranic acid (Schreier *et al.* 1976A,B; see also Ribéreau-Gayon *et al.* 1975A; Stevens *et al.* 1966; Webb *et al.* 1966). These substances can be detected unchanged in wine (Schreier *et al.* 1976B). For instance, 2-methoxy-3-isobutylpyrazene is such a typical compound for Cabernet-Sauvignon grapes (Bayonove *et al.* 1975). *Vitis rotundifolia* grapes contain higher concentrations of 2-phenylethanol (Kepner and Webb 1956). *Vitis labrusca* varieties, such as the Concord, Ives, and Niagara, contain rather high concentrations of methylanthranylate. This compound gives the grapes and wines a characteristic "foxy" flavor whose impact on wine quality has been a matter of controversy (Rice 1974; Stern *et al.* 1967). A steam distillation of this compound from grape must prior to the fermentation can reduce its concentration and improve the quality of the wine made from these varieties (Moyer *et al.* 1977).

The aroma components are not uniformly distributed in the grapes. They are usually concentrated in the skins and seeds. For instance, the terpene alcohols, nerol and geraniol, are mainly concentrated in the skin of the Muscat of Alexandria variety (Bayonove *et al.* 1974). This is important for the processing of these grapes (Bayonove *et al.* 1976).

Secondary Aroma Substances

The original aroma of the grapes is changed by the elimination of some compounds and by reaction of other compounds during the processing of the grapes, during the fermentation, and during the storage of the wine. But it is also greatly enriched by the formation of new compounds. Table 9.27 shows the volatile aroma compounds which have been identified (see Rapp *et al.* 1973). It also shows the increase in the number of components during the fermentation of the must to wine, and a further increase in the distillate.

The detailed review of Webb and Muller (1972) permits a good understanding of the volatile aroma substances of wine, and gives extensive references for individual components. A classification of aroma compounds identified in wine can also be found in the publications of Drawert *et al.* (1976) and Schreier *et al.* (1977). The aroma compounds of sherry have been investigated in detail by Webb *et al.* (1967) and Kepner *et al.* (1968).

Aroma Formation by Yeasts.—The most significant enrichment of the aroma of wine is due to the presence of the many and varied products of yeast metabolism. Suomalainen *et al.* (1974) have compared the composition of fermentation products of various industries. The central role of yeast metabolism is reflected in the great similarity of the chromatograms of the aroma components. This role of yeast is further emphasized by the fact that

TABLE 9.27. VOLATILE AROMA SUBSTANCES IN GRAPE MUSTS (1), WINES (2), AND BRANDY (3)[1]

Compound	1	2	3
Acids			
formic acid	+	+	
acetic acid	+	+	+
propionic acid	+	+	+
n-butyric acid	+	+	+
i-butyric acid			+
n-valeric acid	+		+
i-valeric acid		+	+
caproic acid	+	+	+
enanthic acid	+	+	+
caprylic acid	+	+	+
pelargonic acid	+	+	+
capric acid	+	+	+
undecanoic acid			+
lauric acid	+	+	+
myristic acid		+	
Esters			
ethyl formate	+	+	
methyl acetate	+	+	+
ethyl acetate	+	+	+
n-propylacetate	+		
i-propylacetate		+	
n-butylacetate	+	+	+
i-butylacetate	+	+	+
n-amylacetate	+	+	+
i-amylacetate	+	+	+
n-hexylacetate	+	+	+
n-heptylacetate	+	+	+
n-octylacetate	+		+
n-nonylacetate	+		
n-decylacetate			+
n-undecylacetate			+
benzylacetate			+
phenylacetate			+
β-phenylethylacetate		+	+
ethylpropionate	+	+	
n-propylpropionate	+	+	
i-butylpropionate		+	
n-amylpropionate	+		
i-amylpropionate		+	+
n-hexylpropionate			+
ethyl-i-butyrate		+	
i-amyl-i-butyrate			+
n-heptyl-i-butyrate			+
n-decyl-i-butyrate			+
n-undecyl-i-butyrate			+
ethyl-n-butyrate	+	+	+
n-propyl-n-butyrate			+
i-propyl-n-butyrate		+	
n-butyl-n-butyrate	+		
n-amyl-n-butyrate	+		
i-amyl-n-butyrate		+	+
n-heptyl-n-butyrate			+
n-octyl-n-butyrate			+
n-nonyl-n-butyrate			+
ethyl-i-valerate		+	+
n-butyl-i-valerate			+

TABLE 9.27. *(Continued)*

Compound	1	2	3
i-amyl-i-valerate			+
n-octyl-i-valerate			+
n-nonyl-i-valerate			+
n-decyl-i-valerate			+
ethyl-n-valerate	+	+	+
n-butyl-n-valerate			+
i-butyl-n-valerate			+
n-amyl-n-valerate			+
n-hexyl-n-valerate			+
n-heptyl-n-valerate			+
n-nonyl-n-valerate			+
ethyl-n-caproate		+	+
n-butyl-n-caproate			+
i-butyl-n-caproate		+	+
n-amyl-n-caproate			+
i-amyl-n-caproate		+	+
n-heptyl-n-caproate			+
methylenanthate			+
ethylenanthate		+	+
n-butylenanthate			+
n-amylenanthate			+
i-amylenanthate			+
n-hexylenanthate	+		
methylcaprylate			+
ethylcaprylate	+	+	+
i-butylcaprylate		+	+
n-amylcaprylate			+
i-amylcaprylate		+	+
n-hexylcaprylate	+		
methylpelargonate			+
ethylpelargonate		+	+
n-butylpelargonate			+
i-butylpelargonate			+
i-amylpelargonate			+
n-hexylpelargonate			+
ethylcaprate	+	+	+
n-propylcaprate			+
i-butylcaprate			+
n-amylcaprate			+
i-amylcaprate		+	+
ethylundecanate		+	+
i-butylundecanate			+
methyllaurate			+
ethyllaurate		+	+
i-amyllaurate		+	
ethylmyristate		+	+
ethyl-D-lactate		+	+
butyl-D-lactate			+
amyl-D-lactate			+
diethylsuccinate		+	+
ethylbenzoate		+	+
ethylcinnamate		+	
ethylsalicylate		+	
methylanthranilate	+	+	
Alcohols			
methanol	+	+	+
ethanol	+	+	+
1-propanol	+	+	+
2-propanol		+	

TABLE 9.27.   *(Continued)*

Compound	1	2	3
2-methyl-1-propanol	+	+	+
1-butanol	+	+	+
2-butanol		+	+
2-methyl-1-butanol	+	+	+
3-methyl-1-butanol	+	+	+
1-pentanol	+	+	+
2-pentanol		+	+
1-hexanol	+	+	+
2-hexanol		+	+
cis-3-hexanol-1	+		
1-heptanol	+	+	+
2-heptanol		+	+
1-octanol	+	+	+
2-octanol		+	+
3-octanol			+
1-nonanol		+	+
2-nonanol		+	+
1-decanol	+	+	+
2-decanol		+	+
3-decanol			+
1-undecanol	+	+	+
2-undecanol			+
2-dodecanol		+	+
benzyl alcohol			+
β-phenylethanol	+	+	+
geraniol	+	+	
α-terpineol	+	+	
linalool	+	+	
Aldehydes			
acetaldehyde	+	+	+
propionaldehyde	+	+	
n-butyraldehyde	+	+	
2-methyl-l-propanal		+	
n-valeraldehyde	+		+
2-methyl-l-butanal		+	
3-methyl-l-butanal		+	
n-hexanal	+		+
i-hexanal			+
enanthaldehyde	+	+	+
octanal		+	+
pelargonaldehyde		+	+
decanal	+	+	+
undecanal			+
1-dodecanal	+	+	
hexene-2-al-1	+		
benzaldehyde		+	+
cinnamaldehyde		+	
vanillin		+	
2-furfuraldehyde		+	+
citral	+	+	
limonene	+	+	
Ketones			
acetone	+	+	
2-butanone	+	+	
2-pentanone	+	+	
2-hexanone		+	

TABLE 9.27.   (Continued)

Compound	1	2	3
2-heptanone		+	
2-nonanone		+	
acetoin		+	+
diacetyl		+	+
α-ionone	+	+	
β-ionone	+		

Source: Rapp et al. (1973).
[1]Additional aroma components may have been identified since publication of these data.

fermented sugar solutions (free from added nitrogenous compounds) have chromatograms which resemble in many respects those of the alcoholic beverage products (Suomalainen and Nykänen 1966A).

Higher Alcohols.—Higher alcohols account for the major portion of the products of yeast metabolism. For California products Guymon and Heitz (1952) determined an average content of 250 mg/liter for white wines (120 samples) and 287 mg/liter for red wines (130 samples). But the effect of higher alcohols on the aroma of wine is smaller than one would expect on the basis of their quantitative dominance. In relatively small concentrations they probably contribute to the aroma of wine in a positive way. Higher concentrations lower the quality of the wine (Wagener and Wagener 1968). But the complexity of the substrates rarely permits clear-cut conclusions (Guymon and Heitz 1952).

The important higher alcohols in wines are n-propanol, 2-methyl-1 propanol (isobutanol), 2-methyl-1-butanol (optically active isoamyl alcohol), and 3-methyl-1-butanol (isoamyl alcohol). The aromatic alcohol, 2-phenylethanol, occurs in higher concentrations in spontaneously fermented wines (Wucherpfennig and Bretthauer 1970). Comprehensive reviews of the pathway leading to the production of higher alcohols have been published by Webb and Ingraham (1963), Rainbow (1970), and Suomalainen et al. (1968), including references to the original literature.

The corresponding α-keto acids, which can be formed by transamination of amino acids or from the sugars of the must, have a central role in the synthesis of higher alcohols. The α-keto acids are decarboxylated and then reduced to the corresponding alcohol. In a substrate containing amino acids the higher alcohols are formed both from the amino acids and from the sugars of the must (Äyräpää 1971). Usseglio-Tomasset (1971) found an increase in the concentration of n-propanol and an equally significant decrease in the concentration of isoamyl alcohol when he added ammonium or urea to must. But in agreement with Äyräpää he found that yeasts produce higher concentrations of isobutyl and isoamyl alcohol when they are forced to metabolize amino acids because of the lack of other nitrogenous substrates. The metabolism of amino acids to higher alcohols is restricted to the exponential growth phase (Vollbrecht and Radler 1973).

The formation of higher alcohols during the alcoholic fermentation depends very much on the yeast species (Rankine 1967B) and within a given species on the strain (Webb and Kepner 1961). There is a strong positive correlation between the activity of alcohol dehydrogenases specific for certain alcohols and the formation of these alcohols (Singh and Kunkee 1976). This would permit a quantitative prediction of the ability of various yeast strains for the formation of higher alcohols. Within the genus *Saccharomyces* large amounts of higher alcohols are produced by *S. ellipsoideus, S. bayanus* ("oviformis"), *S. chevalieri,* and *S. bailii* ("elegans"). Small amounts of these alcohols are produced by representatives of the genera *Torulopsis, Kloeckera,* and *Brettanomyces* (Peynaud and Guimberteau 1962), as well as by *Schizosaccharomyces malidevorans* (Rankine 1967B). Yeast strains have a strong effect on the production of 2-phenylethanol (Rankine and Pocock 1969; Usseglio-Tomasset 1967A). *Kloeckera apiculata, Metschnikowia pulcherrima,* and *Hansenula anomala* produce larger amounts of this alcohol. This may account for the effect of "wild yeasts" during spontaneous fermentations on the organoleptic properties of wines (Sponholz and Dittrich 1974). The same is also true for sherry in which significantly higher amounts of 2-phenylethanol were found by Webb and Kepner (1962) and Suomalainen and Nykänen (1966B). For dry wines the taste threshold for higher alcohols was as follows: 2-phenylethanol, 30 to more than 200 ppm; isoamyl alcohol, 100 to 900 ppm (average 300 ppm); isobutanol and n-propanol, 500 ppm (Rankine and Pocock 1969).

The infection of grapes with *Botrytis cinerea* (noble rot) leads to a lack of thiamin. During the fermentation of such musts the yeasts produce higher concentrations of keto acids. This in turn leads to an increased production of higher alcohols in such musts (Dittrich and Sponholz 1975). For instance, the concentration of isobutanol was twice that of normal musts.

The production of higher alcohols is also pH and temperature dependent (Peynaud and Guimberteau 1962). An increase in the temperature of the fermentation from 15° to 25°C increased the concentration of isoamyl alcohol by 24% and that of isobutanol by 39% but decreased the concentration of n-propanol by 17% (Rankine 1967B). And an increase in the pH from 3.0 to 4.2 led to average increases in the concentration of higher alcohols as follows: amyl alcohols by 28%, isobutanol by 85%, and n-propanol by 11%.

Esters.—Wine, as well as other fermented beverages, is rich in esters (Cordonnier 1971; Rapp et al. 1973; Schreier and Drawert 1974A; Webb 1967). They are very important for the total aroma of the wine because of their intense aroma at relatively small concentrations. The possibility of using esters for a characterization of grape varieties is rather limited, just as it is for other fermentation by-products (Schreier et al. 1976D). A total of more than 50 various esters have been identified in wines. Ethyl esters account for a major fraction of the total ester fraction (Table 9.27).

Nordström (1964) made detailed studies of the formation of esters by brewers' yeasts. These studies showed that for beverages with relatively low ethanol concentrations, such as beer and wine, enzymatic reactions are responsible for the formation of esters, while strictly chemical reactions do not proceed. Acyl coenzyme A has a key position in the biosynthesis of esters through its activation of a fatty acid or a keto acid. The final reaction in this chain is the alcoholysis of the acyl group to the corresponding ester. A detailed review of the work of Nordström (1964) was undertaken by Suomalainen et al. (1968) and Lehtonen and Suomalainen (1977). It is important to note that on the basis of these findings yeasts determine the composition and the amounts of the formed esters. Experimentally it has been found that, for instance, the formation of acetic acid esters in wine depends on the presence of various enzymes in individual yeast strains (Daudt and Ough 1973).

Wines made from Tokay grapes (Trockenbeerenauslese) which had been infected with *Botrytis cinerea* contain some esters which cannot be found in wine made from noninfected grapes (Schreier *et al.* 1976C). These are mainly ethyl esters of keto-, hydroxy-, and dicarboxylic acids. The formation of these esters must be due to changes in the composition of the must by *Botrytis cinerea* as well as by the acid metabolism of the mold (Dittrich *et al.* 1974; Hofmann 1968). The effect of esters on wine quality is difficult to quantify. There is a tendency toward higher quality ratings for wines with higher ester concentrations (Wagener and Wagener 1968).

The aroma of sherry is characterized by large concentrations of esters, among them ethyl lactate, diethylsuccinate, ethyl acetate (Suomalainen and Nykänen 1966B), as well as ethyl-4-hydroxybutyrate. Webb *et al.* (1967) consider the latter compound singularly characteristic of sherry.

Fatty Acids.—Wine always contains some volatile free acids, mainly aliphatic fatty acids (see Table 9.27). They occur in higher concentrations in spoiled wines because of the activity of film forming yeasts, and particularly of acetic acid bacteria and lactic acid bacteria. But they are also formed in normal wines during fermentation by *Saccharomyces* yeasts and during spontaneous fermentations because of the presence of *Kloeckera* species. Acetic acid constitutes regularly the largest amount of the volatile acids but its concentration varies greatly from sample to sample (Junge and Spadinger 1978). Acids with 4 to 10 carbon atoms occur only in small concentrations but are of importance because of their relatively strong aroma (Webb 1967). Significant amounts of such acids could be found in Cabernet wines, principally capric, caprylic, and caproic acids (Kepner *et al.* 1969; Suomalainen 1971). Among the volatile acids contributing to the aroma of wine, Drawert *et al.* (1974) have found relatively large concentrations of C_6, C_8, and C_{10} carbon acids. Fatty acids containing 14 carbon atoms could not be found in wine. Apparently the long chain fatty acids with 14 to 18 carbon atoms are eliminated during vinification. The aroma of sherry is also characterized by the presence of relatively larger concentrations of volatile fatty acids (Kepner *et al.* 1968).

According to Lynen the biosynthesis of long chain, even numbered fatty acids begins with acetyl coenzyme A which is formed by oxidative decarboxylation of pyruvate. The synthesis proceeeds to malonyl coenzyme A and to a condensation of acetyl coenzyme A with malonyl coenzyme A. The synthesis of odd numbered fatty acids does not start with acetyl coenzyme A but with propionyl coenzyme A. The unsaturated fatty acids are presumably derived from the corresponding saturated fatty acids or directly from palmityl- or stearyl-coenzyme A. A detailed review of the pathway of this synthesis has been published by Suomalainen et al. (1968) and by Lehtonen and Suomalainen (1977).

The manner in which yeasts affect the production of various fatty acids is not entirely clear. There are appreciable differences in the concentrations of various fatty acids in bakers' yeasts and brewers' yeasts. It is assumed that fatty acids are excreted by yeast cells during fermentation since they can be detected in the medium (Suomalainen 1971). Individual volatile acids are produced in widely varying concentrations by different yeast species. As an example, S. elegans produced relatively large concentrations of propionic acid, while S. ellipsoideus produced larger concentrations of caprylic acid (Usseglio-Tomasset 1967B).

Carbonyl Compounds.—Carbonyl compounds are always present as part of the aroma of wine (Rapp et al. 1973; Schreier et al. 1976C; Webb 1967). These are mainly aldehydes which are formed during the fermentation as intermediates in the formation of alcohol from sugars and amino acids; but some ketones also occur (Lehtonen and Suomalainen 1977). The diketone, diacetyl, which is a well known flavorant in many foods, occurs in wine where it is produced by yeasts during the alcoholic fermentation. Suomalainen and Ronkainen (1968) could show that the diacetyl produced during the fermentation is immediately metabolized by the yeasts. Its content in fermenting must is, therefore, zero. It rises slightly in the wine after the yeast is removed. Lactic acid bacteria produce higher concentrations of diacetyl, and under certain circumstances these may lower the quality of the wine.

Sherry has unusually high concentrations of aldehydes, yet there seems to be no correlation between their concentration and the quality of the wine (Amerine et al. 1980).

Other Compounds.—Various thio-ether compounds have been detected in wine (Schreier et al. 1974). Among these compounds the occurrence of 3-(methyl-thio)-1-propanol is particularly interesting since there is an obvious correlation between the methionine content of the must and the thio-ether content of the wine. This has been shown with Riesling grapes.

Gamma- and delta-lactones occur in sherry and in Cabernet wine and are probably produced enzymatically (Webb and Muller 1972; Muller et al. 1973). A few of these compounds have been detected and the presence of others is assumed. Gamma- and delta-lactones, up to 16 carbon atoms per molecule, are important aroma compounds of wines. A few lactones have also been isolated from German wines (Schreier and Drawert 1974B).

Finally, secondary amides may affect the aroma of wines (Schreier *et al.* 1975). Comparative analyses suggest that they are also formed during the alcoholic fermentation.

The flavor compounds of wine which are produced as the result of yeast metabolism are not useful for the characterization of grape varieties, in contrast to the components of the must. The enzymes of various yeasts are so numerous that by-products may be produced during the alcoholic fermentation from the most diverse must constituents. Some of these by-products are quite independent of the nitrogen compounds of the must.

Aroma Formation by Bacteria.—Lactic acid bacteria produce a number of compounds which may affect the aroma of wines. In general the volatile components of bacterial origin lower the quality of the wine. Volatile acids, ethyl acetate, and diacetyl are the most important of these compounds. But these compounds (just as other aroma components) have limiting concentrations at which a positive contribution to wine aroma turns into a negative contribution. For instance, the presence of 2–4 ppm of diacetyl in red wines serves to round off the flavor, while higher concentrations have a negative effect (Rankine *et al.* 1969). The most important compounds of bacterial origin have already been discussed in connection with wine spoilage. Some additional compounds will be mentioned but without the possibility of saying much about their organoleptic properties.

The formation of volatile by-products of the fermentation by lactic acid bacteria has been investigated by Radler and Gerwarth (1971). Through gas chromatographic analysis of the head space of fermentations by homofermentative lactic acid bacteria *(Lactobacillus plantarum, Pediococcus cerevisiae, Pediococcus pentosaceus)*, they found acetaldehyde, acetoin, and diacetyl, and traces of 2- or 3-methyl-1-butanol and isobutanol. For fermentations by heterofermentative lactic acid bacteria *(Lactobacillus brevis, Leuconostoc oenos)*, they found in addition traces of the following compounds: n-propanol, i-propanol, ethyl acetate, n-hexanol, 2,3-butanediol, n-octanol, and n-nonanol (or phenylacetaldehyde). Some authors have also reported the presence of several ethyl esters: ethyl lactate (Boidron 1966; Webb 1967), ethyl propionate, and ethyl butyrate (Bertrand 1968). The latter compounds were isolated from wines with a "lactic acid flavor."

The compounds which have been mentioned as well as other metabolites are often produced in such small amounts that aroma compounds derived from the must or from yeast metabolism make their recognition difficult or impossible (Pilone *et al.* 1966). On the other hand, experienced tasters are capable of distinguishing wines in which the malo-lactic fermentation has been carried on by different strains of lactic acid bacteria (Kunkee 1974).

BRANDY

Brandy is produced by distillation of wine. The starting material must be a fermented grape must. Distillates made from other raw materials, such as lees, pomace, or other fruits, must carry a different designation. Brandy is produced in all countries which produce wine. The famous French brandies, Cognac and Armagnac, are produced in two limited French wine growing districts. But in other French districts as well as in Spain, Italy, Greece, Germany, the U.S.S.R., and the United States, brandy production is well established. In the United States with a total production of about 34 million liters (9 million gal.) in 1972, brandy is mostly produced in California. Brandy production in the world and per capita consumption are shown in Tables 9.28 and 9.29. At that time French production of cognac was estimated at 150,000 hl and that of armagnac at 15,000 hl (pure alcohol).

TABLE 9.28.   BRANDY PRODUCTION IN THE WORLD, 1973[1]
In 1000 hl Pure Alcohol.

Country		Country		Country	
U.S.S.R.	2000	Brazil	48	Mexico	13
Spain	500	Australia	37	Cyprus	12
West Germany	328	Argentina	35	Norway	8
South Africa	242	Japan	24	Turkey	6
Yugoslavia	200	Greece	18	Iran	3.6
United States	187	Portugal	18	Rumania	3
Italy	175	Israel	17	Venezuela	2
Bulgaria	60	Canada	14	Uruguay	0.5
Holland	50				

Source: Anon. (1975).
[1] More recent data are not available.

TABLE 9.29.   BRANDY CONSUMPTION IN THE WORLD, 1972–1973[1]
Estimated per Capita Consumption in Liters per Year.

Country		Country		Country	
Cyprus	4.5	Australia	0.7	Canada	0.3
Spain	3	Rumania	0.7	Brazil	0.1
Yugoslavia	2	Norway	0.7	Colombia	0.1
U.S.S.R.	2	United Kingdom	0.6	Finland	0.06
Holland	1.5	Sweden	0.5	Japan	0.06
Bulgaria	1–1.5	Portugal	0.4	Mexico	0.05
South Africa	1.4	Greece	0.4	Turkey	0.05
West Germany	1.3	Argentina	0.3	Poland	very slight
Italy	0.7	United States	0.3	Uruguay	0.04

Source: Anon. (1975).
[1] More recent data are not available.

Technology of Brandy Production

The methods of brandy production vary from district to district but share basic features. Generally brandy is produced from white grapes. Brandy from red grapes is somewhat inferior to that from white grapes and contains larger concentrations of higher alcohols. This is true whether the fermentation is carried out with crushed grapes or with must (Guymon 1974).

The quality of the grapes is most important for the quality of the distilled product. Well selected, sound grapes must be used in order to minimize enzymatic oxidations and to inhibit the activity of undesirable microorganisms, such as acetic acid bacteria and lactic acid bacteria. For the production of high quality brandy, for instance for the production of Cognac, the de-stemmed berries are pressed only lightly.

It is desirable to avoid treatment of the must with SO_2 in order to prevent a binding of the aldehyde and to counteract its unfavorable effect on the aroma of the end product. The accumulation of aldehydes as well as the presence of SO_2 during the distillation is detrimental (Guymon 1974). The compound of acetaldehyde with sulfuric acid has the character of a sulfonic acid. It is a strong acid and corrodes distilling columns of copper. It is not clear whether the corrosion is due to the sulfonic acid or to dissociated free SO_2. The presence of sulfonic acids or bisulfite has the additional effect of favoring the reaction of ethanol and acetaldehyde to acetal which is catalyzed by H^+ ions (Guymon 1974).

The fermentation temperature should be kept rather low to prevent the production of larger concentrations of higher alcohols. According to Guymon it should be kept below 24°C. High concentrations of aldehydes, acetals, and fusel oils must be avoided.

Cognac.—The distillation takes place shortly after the fermentation ceases in the traditional copper pot stills heated with an open flame. The wine used for the distillation has not been filtered and contains a small amount of yeast. The first distillate contains about 28% ethanol by vol. During a second distillation the "heads" containing high concentrations of aldehydes and the "tails" containing larger concentrations of fusel oils are separated. The main fraction has an alcohol concentration of 70% by vol. The brandy is first stored in new barrels made of Limousin oak. Later it is stored in used barrels in order to prevent excessive extraction of tannins from the wood. Storage may last several years to develop a desirable bouquet. The matured cognac is sold with an ethanol concentration of 40% by vol.

Armagnac.—In contrast to cognac, armagnac is obtained in a single distillation from wine which does not contain any residual yeast. The continuous distillation is carried out in a distilling column with several bubble plates. The distillate has an alcohol concentration of 52–53% by vol.

Brandy.—Brandy production of other countries is based on the systems which have been described above. In California distillation is almost always carried out in a continuous fashion. The technology of brandy production has been described in detail by Amerine *et al.* (1980), Lafon *et al.* (1973), Peynaud (1975), and Suomalainen et al. (1968).

Aroma Components of Brandy

The aroma of brandy (as that of wine) is determined by aroma substances originating with the grapes and by compounds formed during the fermentation. Besides, the distillation process and storage conditions play an important role. Suomalainen (1975) investigated numerous alcoholic beverages, among them brandy. He found the most significant effect on the aroma to be that of the yeast and the conditions of fermentation, while the effect of the starting raw material was less important. For this reason there are great similarities in the volatile aroma components of these beverages. The differences can be found mainly in the concentrations of ethyl acetate, n-propanol, and n-hexanol (Drawert *et al.* 1967). Within the group of brandies, specific and quantitative differences between types can be found in the concentrations of ethyl acetate, isobutanol, and isoamyl alcohol (Drawert *et al.* 1967; Reinhard 1968). The effect of various chemical groups on the aroma of brandy has been determined by Litschew (1976) on the basis of intensive investigations of French and Bulgarian brandies. Accordingly, aromatic and higher aldehydes are responsible for the "vanilla" aroma and the less volatile esters for the "brandy" aroma. Terpene compounds give brandy a "flowery" aroma and higher alcohols determine principally the "fruity" aroma of freshly distilled brandy.

Higher Alcohols.—The higher alcohol content of brandy has been determined by Connell and Strauss 1974; Drawert *et al.* 1967; Guymon 1970, 1974; and Litschew 1976. The major alcohols are 3-methyl-1-butanol, 2-methyl-1-propanol, 2-methyl-1-butanol, and n-propanol, and in lesser amounts n-hexanol, n-butanol, and 2-butanol. The average concentration of higher alcohols in brandy is between 650 and 1000 mg per liter at 50% ethanol by vol. (Guymon 1974). It depends largely on the must, the yeast, and the conditions prevailing during the fermentation. The higher alcohols have a significant effect on the aroma of brandy. High concentrations are undesirable.

Fatty Acids and Esters.—Aliphatic acids and their esters are very important for the aroma of brandy. Their concentration is second only to that of the higher alcohols (Guymon 1974). The concentration of these compounds depends on the course of the fermentation, the distillation, and the aging process. The amount of yeast remaining in the wine at the time of distillation also affects the contribution of these compounds to the aroma of brandy. According to Guymon the high molecular weight fatty acids and esters are strongly adsorbed by the yeast cells while the low molecular weight compounds are excreted into the medium. Lauric, palmitic, palmitoleic, stearic, and oleic acids accumulate in the yeast cell and can be transferred to the distillate during the distillation (Suomalainen *et al.* 1974). Therefore, one generally finds in brandy the short chain fatty acids as well as fatty acids with up to 10 carbon atoms, and among them mostly caproic, caprylic, and capric acids. But if yeast has remained in the wine, longer chain fatty acids can be found, principally lauric acid (Suomalainen et al. 1974). Among the esters ethyl acetate has the highest concentration, but the values vary greatly from distillate to distillate. Generally, caprate, caprylate, and laurate esters occur in larger concentrations. During maturation of the brandy the high ethanol content causes esterifications and transesterifications (Suomalainen et al. 1968). The total concentration of esters increases greatly during storage. For instance, the content of total esters in an American brandy increased from 405 mg/liter to 713 mg/liter during a 4 year storage period (Valaer 1939).

The concentration of the more volatile esters, such as ethyl formate, ethyl acetate, and iso-pentyl acetate, decreases during the maturing of brandies. But the concentration of esters with higher boiling points, such as ethyl lactate, and the ethyl esters of fatty acids with 6, 8, 10, and 12 carbon atoms, increases. These esters are, therefore, an indication of a high quality, aged brandy (Litschew 1976). Connell and Strauss (1974) found a correlation between the occurrence of free alcohols and their esters in Australian brandies. Litschew (1976) states that the following esters which he identified in French and Bulgarian brandies are decisive for the aroma of the products: the ethyl-, hexyl- and iso-pentyl esters of myristic acid, the mono-ethyl ester of succinic acid, the capryl ester of enanthic acid, the ethyl ester of linoleic acid, and the phenyl ester of caproic acid.

Carbonyl Compounds.—During the distillation of brandy the first fraction of low boiling compounds contains mainly carbonyl compounds. They are largely separated from the major fraction of the distillate because of their intense odor (Suomalainen et al. 1968). Acetaldehyde, which is an important by-product of the fermentation, predominates, but smaller concentrations of other aldehydes are also present. Others are formed during storage. Ličev and Panajotov (1958) found that Bulgarian brandies contained a series of aldehydes from formaldehyde to valeraldehyde, as well as enanthaldehyde and furfuraldehyde. In addition they found some aromatic aldehydes which originated in the wood of the barrels, namely, syringic aldehyde, coniferaldehyde, and p-hydroxybenzaldehyde. Suomalainen attributes the increase in acetaldehyde concentration during the maturation of brandy to the oxidation of ethanol. Litschew (1977) points to the increase in furfuraldehyde during the aging process. Finally, it is interesting to consider the alcoholysis of lignin which occurs during aging in barrels. Together with oxidative processes this leads to the formation of aldehydes of the vanillin type, which are of great importance for the quality of brandy (Amerine et al. 1980).

All brandies contain diethylacetal which is formed by condensation of ethanol and acetaldehyde. The reaction is catalyzed by acids (Suomalainen et al. 1968). With time an equilibrium between aldehyde and acetal results which depends on the ethanol concentration of the brandy and which is important for its aroma. In this equilibrium the acetals account for about 10 to 20% of the total aldehydes (Suomalainen *et al.* 1968). Williams and Strauss (1975) found in the first fractions of wine distillates acetals which had been formed by reaction of diacetyl and acrolein, respectively, namely, 3,3-diethoxybutane-2-one and 1,1,3-triethoxypropane.

Other Compounds.—The effect of terpene compounds on the aroma of brandy has been investigated by Litschew (1976). He concludes that the relatively high concentrations of *cis*- and *trans*-farnesol, α- and β-ionone and the ethyl ester of linoleic acid determine the quality of Bulgarian brandies.

The maturing of brandy in wooden casks or barrels causes the appearance of several lactones. They are certainly of interest for the aroma of brandy and have been found in sherry by Muller *et al.* (1973) and by Salo *et al.* (1972) in whiskey. Several *cis*- and *trans*-β-methyl-γ-octolactones occur in brandy (Litschew 1976). These compounds could also be detected in extracts of oak wood and in brandy (Guymon and Crowell 1972). The *trans* form is of greater significance for the quality of the brandy than the *cis* form.

The development of the aroma of brandy during the maturation is a slow process. Litschew (1977) investigated old brandies over several decades both chemically and organoleptically. He concluded that the development of the aroma takes place in 3 stages. The aroma is formed during the first 10 to 15 years. Ripening occurs during the next 30 years. This is followed by a stage in which the quality of the brandy aroma declines. These 3 stages are affected by physical factors, such as the temperature, the relative humidity, and the size of the cask, as well as by the chemical composition of the distillate.

REFERENCES

ACREE, T.E., SONOFF, E.P. and SPLITTSTOESSER, D.F.   1972.   Effect of yeast strain on hydrogen sulfide production. Am. J. Enol. Viticult. *23*, 6–9.

ADAMS, A.M.   1961.   Commercial use of frozen stored wine starter. Rep. Ont. Hort. Exp. Stn. Prod. Lab., 102–103.

AINSWORTH, G.C.   1973.   Introduction and keys to higher taxa. *In* The Fungi, Vol. 4A. G.C. Ainsworth, F.K. Sparrow and A.S. Sussman (Editors). Academic Press, New York.

ALIZADE, M.A. and SIMON, H.   1973.   Mechanism and compartmentalization of the formation of L- and D-lactate from L-malate or D-glucose in *Leuconostoc mesenteroides*. Hoppe-Seyler's Z. Physiol. Chem. *354*, 163–168.

AMATI, A. and MINGUZZI, A.   1975.   Biological de-acidification of wine using re-fermentation by *Schizosaccharomyces pombe*. Vigne Vini (Bologna) *2*, 35–38.

AMERINE, M.A. and OUGH, C.S.   1957.   Studies on controlled fermentations. Am. J. Enol. Viticult. *8*, 18–30.

AMERINE, M.A. and THOUKIS, G.   1958.   The glucose-fructose ratio in California grapes. Vitis *1*, 224–229.

AMERINE, M.A. *et al.* 1980. The Technology of Wine Making, 4th Edition. AVI Publishing Co., Westport, Conn.

ANON. 1975. The brandy market of the world. Rep. Cent. Fr. Commer. Exter., AGREX, Div. 1, August.

ANON. 1980. Gross wine production by states. Wines Vines, May, 46.

ASAI, T. 1968. Acetic Acid Bacteria. Classification and Biochemical Activities. Univ. Tokyo Press, Tokyo.

ÄYRÄPÄÄ, T. 1971. Biosynthetic formation of higher alcohols by yeasts. Dependence on the nitrogenous nutrient level of the medium. J. Inst. Brew. London *77*, 266–276.

BAMBALOW, G. 1970. Changes in the glucose/fructose ratio in grape concentrate by *Zygosaccharomyces* yeasts. Weinberg Keller *17*, 87–90.

BARNETT, J.A. *et al.* 1972. Yeast counts on grapes of Bordeaux wines. Arch. Mikrobiol. *83*, 52–55.

BARRE, P. 1969. Numerical taxonomy of lactobacilli isolated from wine. Arch. Mikrobiol. *68*, 74–86.

BARRET, A., BIDAN, P. and ANDRÉ, L. 1955. Report on some failures of vinification due to flor yeasts. C.R. Acad. Agric. *41*, 426–430.

BAUER, H. and KLEINHENZ, J. 1978. Technological specifications for dry yeasts. Wein-Wiss. *33*, 188–199.

BAYONOVE, C., CORDONNIER, R. and DUBOIS, P. 1975. Study of an aroma fraction characteristic of the Cabernet-Sauvignon grape variety: Detection of 2-methoxy-3-isobutyl-pyrazine. C.R. Acad. Sci. *281*, 75–78.

BAYONOVE, C., CORDONNIER, R. and TARIER, R. 1974. Localization of the aroma of grapes: Muscat of Alexandria and Cabernet-Sauvignon grapes. Acad. Agric. Fr. P.V. Séances, 1321–1328.

BAYONOVE, C. *et al.* 1976. Extraction of the aroma fraction of Muscat grapes prior to vinification. C.R. Acad. Agric. *10*, 734–750.

BEECH, F.W. and CARR, J.G. 1977. Cider and perry. *In* Alcoholic Beverages. A.H. Rose (Editor). Academic Press, New York.

BELIN, J.-M. 1972. Study of the distribution of yeasts on the surface of grapes. Vitis *11*, 135–145.

BELIN, J.-M. 1977. Inframicroscopic morphology of *Vitis vinifera L.* Its effect on the distribution of yeasts. Connaiss. Vigne Vin *11*, 295–311.

BENDA, I. 1960. Unpublished data. Würzburg, W. Germany.

BENDA, I. 1962A. Ecological investigation of the yeast flora in the grape district of Franconia. Bayer. Landwirtsch. Jahrb. *39*, 595–614.

BENDA, I. 1962B. *Torulopsis burgeffiana nov. spec.*, a new yeast species isolated from grape berries. Antonie van Leeuwenhoek J. Microbiol. Serol. *28*, 208–214.

BENDA, I. 1964. The yeast flora of the wine district of Franconia. Weinberg Keller *11*, 67–80.

BENDA, I. 1970. Natural and controlled microbial processes in grape must and in young wine. Bayer. Landwirtsch. Jahrb. *47*, 19–29.

BENDA, I.   1974.   *Schizosaccharomyces* yeasts and their effect of acid reduction during vinification. Vignes Vins, No. Spec. Intern. Symp. Oenol., Arc Senans, 1973, 31–36.

BENDA, I.   1975.   Unpublished data. Würzburg, W. Germany.

BENDA, I. and SCHMITT, A.   1966.   Enological investigation of the biological decomposition of acids in must by *Schizosaccharomyces pombe*. Weinberg Keller *13*, 239–254.

BENDA, I. and SCHMITT, A.   1969.   Acid reduction in must by various strains of the genus *Schizosaccharomyces*. Weinberg Keller *16*, 71–83.

BENDA, I. and WOLF, E.   1965.   On the differentiation of strains of *Saccharomyces cerevisiae var. ellipsoideus*. Mitt. Rebe Wein Ser. A (Klosterneuburg) *15*, 300–316.

BERTRAND, A.   1968.   Application of liquid-gas chromatography to the determination of volatile constituents of wine. Connaiss. Vigne Vin *2*, 179–270.

BERTRAND, A. and SUZUTA, K.   1976.   Formation of butanol-2 by lactic acid bacteria isolated from wine. Connaiss. Vigne Vin *10*, 409–426.

BIDAN, P.   1956.   Several bacteria isolated from wines undergoing malolactic fermentation. Ann. Technol. Agric. *5*, 597–617.

BIDAN, P.   1967.   Growth factors for lactic acid bacteria in wine. Ferment. Vinification *1*, 195–213.

BIDAN, P., MEYER, J.-P. and SCHAEFFER, A.   1974.   The *Schizosaccharomyces* in enology. Bull. OIV *47*, 682–706.

BLACKWOOD, A.C.   1969A.   Acetic acid bacteria isolated from grapes and wine. I. Identification of isolated strains. Connaiss. Vigne Vin *3*, 227–241.

BLACKWOOD, A.C.   1969B.   Acetic acid bacteria isolated from grapes and wine. II. Formation of keto compounds. Connaiss. Vigne Vin *3*, 243–250.

BOIDRON, A.-M.   1969.   Two causes of the inhibition of yeasts by lactic acid bacteria. C.R. Acad. Sci. Paris *269*, 922–925.

BOIDRON, J.N.   1966.   Analytical identification of the aroma compounds of *Vitis vinifera* wine. 3rd Cycle Thesis *391*. Fac. Sciences, Univ. Bordeaux.

BONISH, P. and ESCHENBRUCH, R.   1976.   Sulfite reductase and ATP sulfurylase in low- and high-sulfite forming wine yeasts. Relationship to sulfite accumulation during fermentation. Arch. Mikrobiol. *109*, 85–88.

BRÉCHOT, P., CROSON, M. and MATSURA, S.   1969.   Fermentation and respiration of yeast in presence of sulfur dioxide. Antonie van Leeuwenhoek J. Microbiol. Serol. *35*, Supplement, 21–22.

BREED, R.S., MURRAY, E.G.D. and SMITH, N.R.   1957.   Bergey's Manual of Determinative Bacteriology, 7th Edition. Williams & Wilkins, Baltimore.

BRÉMOND, E.   1957.   Modern Processes of Vinification and Wine Preservation in Hot Countries. Librairie de la Maison Rustique, Paris.

BROCKMANN, M.C. and STIER, T.J.B.   1948.   Influence of temperature on the production of glycerol during alcoholic fermentation. J. Am. Chem. Soc. *70*, 413–414.

BRUGIRARD, A. and ROQUES, J. 1972. The *Schizosaccharomyces*. Utilization of the acid reducing power of these yeasts. Rev. Fr. Oenologie *13*, 26–36.

BURROUGHS, L.F. and SPARKS, A.H. 1973. Sulfite-binding power of wines and ciders. III. Determination of carbonyl compounds in a wine and calculation of its sulfite binding power. J. Sci. Food Agric. *24*, 204–217.

CANTARELLI, C. 1966. Influence of temperature of fermentation and of preservation of wine and of special wines on their chemical, microbiological and organoleptic properties. Bull. OIV *428*, 1191–1205.

CAPRIOTTI, A. 1954. Investigation of wine fermenting yeasts in Italy. Antonie van Leeuwenhoek J. Microbiol. Serol. *20*, 374–384.

CAPRIOTTI, A. 1955. Yeasts in some Netherland soils. Antonie van Leeuwenhoek J. Microbiol. Serol. *21*, 143–154.

CAPRIOTTI, A. 1965. The wine fermentation yeasts in various musts of Sardinia. Studi Sassar. *13*, 287–322.

CASTELLI, T. 1942. Should one always use cultures of *Saccharomyces cerevisiae* for vinification with selected yeasts? Ann. Microbiol. *20* (4) 131–134.

CASTELLI, T. 1957. Climate and agents of wine fermentation. Am. J. Enol. Viticult. *8*, 149–156.

CASTELLI, T. 1967. Ecology and systematic of wine yeasts. Vinification *1*, 89–105.

CASTELLI, T. and DEL GIUDICE, E. 1955. The agents of wine fermentation in the region of Etna. Riv. Vitic. Enol. Conegliano *8*, 127–141, 167–173.

CASTELLI, T. and HAZNEDARI, S. 1971. Consideration of the malo-alcoholic fermentation. Results of the four year period 1967–1970. Atti Acad. Ital. Vite Vino *23*, 3–28.

CASTELLI, T. and IÑIGO LEAL, B. 1957. The agents of wine fermentation in Rioja. Ann. Fac. Agrar. Perugia *13*, 1–20.

CASTOR, J.G.B. 1954. Fermentation products and flavor profiles of yeasts. Wines Vines *35* (8) 29–31.

CHARPENTIÉ, Y. 1954. Biochemical study of the factors causing acidity in wine. Ann. Technol. Agric. *3*, 89–167.

CHARPENTIÉ, Y., RIBÉREAU-GAYON, J. and PEYNAUD, E. 1951. On the citric acid fermentation by malo-lactic bacteria. Bull. Soc. Chim. Biol. *33*, 1369–1378.

CONNELL, D.W. and STRAUSS, C.R. 1974. Major constituents of fusel oils distilled from Australian grape wines. J. Sci. Food Agric. *25*, 31–44.

COOKE, W.B. 1962. A taxonomic study in the "black yeasts." Mycopath. Mycol. Appl. *17*, 1–43.

CORDONNIER, R. 1971. The aroma of wine and brandy. Bull. OIV *490*, 1128–1148.

CRUESS, W.V. 1943. The role of microorganisms and enzymes in wine making. Adv. Enzymol. *3*, 349–386.

CUINIER, C. 1973. The malolactic fermentation of wines. Vignes Vins *221*, 33–36.

DAUDT, C.E. and OUGH, C.S. 1973. Variations in some volatile acetate esters formed during grape juice fermentation. Effects of fermentation temperature, SO_2, yeast strain and grape variety. Am. J. Enol. Viticult. *24*, 130–135.

DE SOTO, R.T. 1955. Integrating yeast propagation with winery operation. Am. J. Enol. Viticult. *6*, 26–30.

DI MENNA, M.E. 1955. A quantitative study of airborne fungus spores in Dunedin, New Zealand. Trans. Br. Mycol. Soc. *38*, 119–129.

DI MENNA, M.E. 1959. Some physiological characters of yeasts from soils and allied habitats. J. Gen. Microbiol. *20*, 13–23.

DITTRICH, H.H. 1963A. The mechanism of the chemical decomposition of malic acid with a yeast of the genus *Schizosaccharomyces*. Wein-Wiss. *18*, 406–410.

DITTRICH, H.H. 1963B. Experiments in the decomposition of malic acid by a yeast of the genus *Schizosaccharomyces*. Wein-Wiss. *18*, 392–405.

DITTRICH, H.H. 1964. The alcoholic fermentation of L-malic acid by *Schizosaccharomyces pombe var. acidodevoratus*. Zentralbl. Bakteriol. Parasitenkd. Infektionskr. Hyg. Abt. 2 *118*, 406–421.

DITTRICH, H.H. 1977. Microbiology of Wines. Verlag Ulmer, Stuttgart.

DITTRICH, H.H. and KERNER, E. 1964. Diacetyl as an off flavor of wine. Cause and removal of the "lactic acid aroma." Wein-Wiss. *19*, 528–535.

DITTRICH, H.H. and SPONHOLZ, W.R. 1975. Decrease in amino acids in berries infected with *Botrytis* and the formation of higher alcohols in such musts during fermentation. Wein-Wiss. *30*, 188–210.

DITTRICH, H.H., SPONHOLZ, W.R. and KAST, W. 1974. Comparative investigation of musts and wines from healthy grape berries and berries infected with *Botrytis*. I. Vitis *13*, 36–49.

DITTRICH, H.H. and STAUDENMAYER, T. 1968. SO_2 formation, formation of hydrogen sulfide odor and its removal. Dtsch. Wein-Ztg. *24*, 707–709.

DITTRICH, H.H., STAUDENMAYER, T. and SPONHOLZ, W.R. 1973. The formation of yeast metabolic products which bind SO_2 during the fermentation and maturation of wine. Wein-Wiss. *28*, 84–93.

DITTRICH, H.H. and WENZEL, K. 1976. Dependence of the formation of foam during fermentation on the yeast and the treatment of the must. Wein-Wiss. *31*, 263–274.

DOBZHANSKY, T. *et al.* 1956. Studies on the ecology of *Drosophila* in the Yosemite region of California. IV. Ecology *37*, 544–550.

DOMERCQ, S. 1956. The classification of wine yeasts in the Gironde. Ingénieur-Docteur Thesis. Univ. Bordeaux.

DOTT, W., HEINZEL, M. and TRÜPER, H.G. 1976. Sulfite formation by wine yeasts. I. Relationship between growth, fermentation and sulfite formation. Arch. Mikrobiol. *107*, 289–292.

DOTT, W., HEINZEL, M. and TRÜPER, H.G. 1977. Sulfite formation by wine yeasts. IV. Active uptake of sulfate by "low" and "high" sulfite producing wine yeasts. Arch. Mikrobiol. *112*, 283–285.

DOUGLAS, H.C. and McCLUNG, L.S. 1937. Characteristics of an organism causing spoilage in fortified sweet wines. Food Res. *2*, 471–475.

DRAWERT, F., HEIMANN, W. and TSANTALIS, G. 1967. Gaschromatographic comparison of various brandies. Fresenius Z. Anal. Chem. *228*, 170–180.

DRAWERT, F. and RAPP, A. 1966. The compounds in must and wine. VII. Gas chromatographic investigation of aroma compounds in wine and their biogenesis. Vitis *5*, 351–376.

DRAWERT, F., SCHREIER, P. and SCHERER, W. 1974. Gaschromatographic-mass spectrometric investigation of volatile components of wine. III. Acids of the wine aroma. Z. Lebensm.-Unters.-Forsch. *155*, 342–346.

DRAWERT, F. *et al.* 1976. Gaschromatographic-mass spectrometric investigation of volatile components of wine. VII. Aroma compounds of Tokay wines b) organic acids. Z. Lebensm.-Unters.-Forsch. *162*, 11–20.

DUBOURG, N. 1896. *Cited by* Ribéreau-Gayon, J. and Peynaud, E. 1960. Treatise on Oenology, Vol. 1. Ripening of Grapes, Alcoholic Fermentation, Vinification. Librairie Polytechnique Ch. Béranger, Paris.

Du PLESSIS, L. DE W. 1963. The microbiology of South African winemaking. V. Vitamin and amino acid requirements of lactic acid bacteria from dry wines. S. Afr. J. Agric. Sci. *6*, 485–494.

Du PLESSIS, L. DE W. and VAN ZYL, J.A. 1963. The microbiology of South African winemaking. IV. The taxonomy and incidence of lactic acid bacteria from dry wines. S. Afr. J. Agric. Sci. *6*, 261–273.

DUPUY, P. 1957A. The *Acetobacter* of wine. Identification of several strains. Ann. Technol. Agric. *6*, 217–233.

DUPUY, P. 1957B. Factors causing formation of acetic acid in wine. Ann. Technol. Agric. *6*, 391–407.

EGGENBERGER, W. 1978. New facts about the problem of malolactic retrogradation. *In* 5th Intern. Oenological Symp., Auckland, 1978. E. Lemperle and J. Frank (Editors). Intern. Assoc. Mod. Winery Technol. Manage., Breisach.

ESCHENBRUCH, R. 1973. Contamination of wine by *Saccharomyces acidifaciens*, a yeast highly resistant to Baycovin, sulfur dioxide and sorbic acid. Wynboer (Stellenbosch) *496*, 23–24.

ESCHENBRUCH, R. 1974. Sulfite and sulfide formation during winemaking—a review. Am. J. Enol. Viticult. *25*, 157–161.

ESCHENBRUCH, R. and BONISH, P. 1976A. The influence of pH on sulfite formation by yeasts. Arch. Mikrobiol. *107*, 229–231.

ESCHENBRUCH, R. and BONISH, P. 1976B. Production of sulfite and sulfide by low and high-sulfite forming wine yeasts. Arch. Mikrobiol. *107*, 299–302.

ESCHENBRUCH, R., HAASBROEK, F.J. and DE VILLIERS, J.F. 1973. On the metabolism of sulfate and sulfite during the fermentation of grape must by *Saccharomyces cerevisiae*. Arch. Mikrobiol. *93*, 259–266.

ESCHENBRUCH, R. and RASSELL, J.M. 1975. The development of non-foaming yeast strains for wine making. Vitis *14*, 43–47.

FIECHTER, A.   1962.   Small scale equipment for the continuous cultivation of microorganisms. Chem.-Ing.-Tech. *34*, 692–696.

FLESCH, P.   1969.   On the malate dehydrogenase and the lactate dehydrogenase of L-malic acid decomposing bacteria. Arch. Mikrobiol. *68*, 259–277.

FLESCH, P. and HOLBACH, B.   1965.   L-Malic acid metabolism by lactic acid bacteria. I. On malic acid splitting enzymes of *Bacterium "L"* with special consideration of oxalacetic acid decarboxylase. Arch. Mikrobiol. *51*, 401–413.

FLESCH, P. and HOLBACH, B.   1970.   L-Malic acid splitting enzymes of *Schizosaccharomyces acidodevoratus*. Arch. Mikrobiol. *74*, 213–222.

FLESCH, P. and JERCHEL, D.   1960.   The cultivation of *Bacterium gracile* in natural media containing L-malic acid. Mitt. Rebe Wein Ser. A (Klosterneuburg) *10*, 1–13.

FLESCH, P. and JERCHEL, D.   1961.   The freeze dehydration of *Bacterium gracile*. Mitt. Rebe Wein Ser. A (Klosterneuburg) *11*, 12–16.

FLORENZANO, G.   1951.   Occurrence and enological significance of the yeast *Brettanomyces*. Atti Acad. Ital. Vite Vino *3*, 236–248.

FLORENZANO, G., BALLONI, W. and MATERASSI, R.   1977.   Contribution to the ecology of *Schizosaccharomyces* yeasts on grapes. Vitis *16*, 38–44.

FORNACHON, J.C.M.   1953.   Studies on the Sherry Flor. Australian Wine Board, Adelaide.

FORNACHON, J.C.M.   1963.   Inhibition of certain lactic acid bacteria by free and bound sulfur dioxide. J. Sci. Food Agric. *14*, 857–862.

FORNACHON, J.C.M.   1964.   A *Leuconostoc* causing malo-lactic fermentation in Australian wines. Am. J. Enol. Viticult. *15*, 184–186.

FORNACHON, J.C.M., DOUGLAS, H.C. and VAUGHN, R.H.   1949.   *Lactobacillus trichodes nov. spec.*, a bacterium causing spoilage in appetizer and dessert wines. Hilgardia *19*, 129–132.

FORNACHON, J.C.M. and LLOYD, B.   1965.   Bacterial production of diacetyl and acetoin in wine. J. Sci. Food Agric. *16*, 710–716.

FRATEUR, J.   1950.   Study of the systematic of *Acetobacter*. Cellule *53*, 287–389.

GALLANDER, J.F.   1977.   Deacidification of Eastern table wines with *Schizosaccharomyces pombe*. Am. J. Enol. Viticult. *28*, 65–68.

GARVIE, E.I.   1967A.   *Leuconostoc oenos sp. nov.* J. Gen. Microbiol. *48*, 431–438.

GARVIE, E.I.   1967B.   The growth factor and amino acid requirements of species of the genus *Leuconostoc*, including *Leuconostoc paramesenteroides* (spec. nov.) and *Leuconostoc oenos*. J. Gen. Microbiol. *48*, 439–447.

GEISS, W.   1952.   Pressure Fermentation. Sigurd Horn-Verlag, Frankfurt.

GENDRON, C.   1967.   Factors activating the development of malo-lactic acid bacteria in wines. Ferment. Vinification *1*, 215–218.

GOTTSCHALK, A.   1946.   The mechanism of selective fermentation of D-fructose from invert sugar by Sauternes yeast. Biochem. J. *40*, 621–626.

GREENSHIELDS, R.N.   1978.   Acetic Acid; Vinegar. Primary Products of Metabolism. A.H. Rose (Editor). Academic Press, London.

GUYMON, J.F.   1970.   Composition of California commercial brandy distillates. Am. J. Enol. Viticult. *21*, 61–69.

GUYMON, J.F.   1974.   Chemical aspects of distilling wines into brandy. *In* Chemistry of Winemaking. A.D. Webb (Editor). Adv. Chem. Ser. *137*. Am. Chem. Soc., Washington, D.C.

GUYMON, J.F. and CROWELL, E.A.   1972.   GC-separated brandy components derived from French and American oaks. Am. J. Enol. Viticult. *23*, 114–120.

GUYMON, J.F. and HEITZ, J.E.   1952.   The fusel oil content of California wines. Food Technol. *6*, 359–362.

HAUBS, H., MÜLLER-SPÄTH, H. and LOESCHER, T.   1974.   The effect of carbon dioxide on wine. Dtsch. Weinbau *26*, 930–934.

HAZNEDARI, S.   1976.   Considerations on the process of refermentation with *Schizosaccharomyces pombe Linder*. Vini Ital. *18*, 110–114.

HEINZEL, M. and TRÜPER, H.G.   1976.   Sulfite formation by wine yeasts. II. Properties of ATP-sulfurylase. Arch. Mikrobiol. *107*, 293–297.

HIEKE, E. and VOLLBRECHT, D.   1974A.   Volatile constituents of wine and of other alcoholic beverages. IV. Effect of yeast and lactic acid bacteria on the content of volatile components in lees wine. Chem. Mikrobiol. Technol. Lebensm. *3*, 65–68.

HIEKE, E. and VOLLBRECHT, D.   1974B.   The formation of butanol-2 by lactic acid bacteria and yeasts. Arch. Mikrobiol. *99*, 345–351.

HOFMANN, G.   1968.   Biochemical changes caused by *Botrytis cinerea* and *Rhizopus nigricans* in grape must. S. Afr. J. Agric. Sci. *11*, 335–348.

HOLT, G.J.   1977.   Shorter Bergey's Manual of Determinative Bacteriology. Williams & Wilkins, Baltimore.

INGRAHAM, J.L., VAUGHN, R.H. and COOKE, G.M.   1960.   Studies in the malolactic organisms isolated from California wines. Am. J. Enol. Viticult. *11*, 1–4.

INGRAM, M.   1948.   The germicidal effects of free and combined sulfur dioxide. J. Soc. Chem. Ind. *67*, 18–21.

INGRAM, M.   1958.   The antimicrobial effect of SO_2. Rev. Ferment. Ind. Aliment. *13*, 179–188.

IÑIGO LEAL, B. and ARROYO VARELA, V.   1964.   Metabolism of the film forming and non-film forming yeasts of the district of Montilla y Los Moriles. Rev. Cienc. Apl. *99*, 305–314.

IÑIGO LEAL, B., VÁZQUES MARTÍNEZ, D. and ARROYO VARELA, V.   1963.   The agents of wine fermentation in the Jerez district. Rev. Cienc. Apl.

17, 296–305.

JAKOB, L.   1978.   Some complementary statements on the histamine content of wine. Weinwirtsch. *114*, 126–127.

JUNGE, C. and SPADINGER, C.   1978.   Volatile acids in wine: Specific determination of acetic acid. *In* 5th Intern. Oenological Symp., Auckland, 1978. E. Lemperle and J. Frank (Editors). Intern. Assoc. Mod. Winery Technol. Manage., Breisach.

KANDLER, O., WINTER, J. and STETTER, K.O.   1973.   The influence of L-malate on the glucose fermentation by *Leuconostoc mesenteroides*. Arch. Mikrobiol. *90*, 65–75.

KEPNER, R.E. and WEBB, A.D.   1956.   Volatile aroma constituents of *Vitis rotundifolia* grapes. Am. J. Enol. Viticult. *7*, 8–18.

KEPNER, R.E., WEBB, A.D. and MAGGIORA, L.   1968.   Sherry aroma. VII. Some volatile components of flor sherry of Spanish origin. Acidic compounds. Am. J. Enol. Viticult. *19*, 116–120.

KEPNER, R.E., WEBB, A.D. and MAGGIORA, L.   1969.   Some volatile components of wines of *Vitis vinifera* varieties Cabernet-Sauvignon and Ruby Cabernet. II. Acidic compounds. Am. J. Enol. Viticult. *20*, 25–31.

KIELHÖFER, E.   1958.   Binding of sulfurous acid to wine constituents. Weinberg Keller *5*, 461–467.

KIELHÖFER, E. and WÜRDIG, G.   1960A.   Sulfurous acid bound to aldehyde. I. Acetal formation by enzymatic and non-enzymatic alcohol oxidation. Weinberg Keller *7*, 16–22.

KIELHÖFER, E. and WÜRDIG, G.   1960B.   Sulfurous acid bound to aldehyde in wine. II. Acetal formation during fermentation. Weinberg Keller *7*, 50–61.

KIELHÖFER, E. and WÜRDIG, G.   1960C.   The sulfurous acid bound to unknown wine constituents (rest SO_2) and its significance for the wine. Weinberg Keller *7*, 313–328.

KLEMM, K.   1975.   Continuous propagation and selection of wine yeasts for the fermentation in large vats. *In* 4th Intern. Oenological Symp., Valencia, 1975. E. Lemperle and J. Frank (Editors). Intern. Assoc. Mod. Winery Technol. Manage., Breisach.

KLIEWER, W.M.   1967.   The glucose-fructose ratio of *Vitis vinifera* grapes. Am. J. Enol. Viticult. *18*, 33–41.

KOCH, J. *et al.*   1953.   Formation of lactic acid in sweet must during storage in tanks under CO_2 pressure. Z. Lebensm.-Unters.-Forsch. *97*, 17–24.

KONTEK, A. and KONTEK, A.   1977.   Data on the selection of several yeast strains used for obtaining wines with low SO_2 content. An. Inst. Cercet. Vitic. Vinif. *8*, 419–425.

KROEMER, K. and KRUMBHOLZ, G.   1931.   Investigation of osmophilic yeasts. I. The fermentation of selected dry berries and the agents of that fermentation. Arch. Mikrobiol. *2*, 352–410.

KRUMPERMAN, P.H. and VAUGHN, R.H.   1966.   Some lactobacilli associated with the decomposition of tartaric acid in wine. Am. J. Enol. Viticult.

17, 185–190.

KUNKEE, R.E.   1967A.   Malo-lactic fermentation. Adv. Appl. Microbiol. *9*, 235–279.

KUNKEE, R.E.   1967B.   Control of malo-lactic fermentation induced by *Leuconostoc citrovorum*. Am. J. Enol. Viticult. *18*, 71–77.

KUNKEE, R.E.   1974.   Malo-lactic fermentation and winemaking. *In* Chemistry of Winemaking.   A.D. Webb (Editor). Adv. Chem. Ser. *137*. Am. Chem. Soc., Washington, D.C.

KUNKEE, R.E. and GOSWELL, R.W.   1977.   Table wines. *In* Alcoholic Beverages. A.H. Rose (Editor). Academic Press, New York.

KUNKEE, R.E. and OUGH, C.S.   1966.   Multiplication and fermentation of *Saccharomyces cerevisiae* under carbon dioxide pressure in wine. Appl. Microbiol. *14*, 643–648.

KÜNSCH, U., TEMPERLI, A. and MAYER, K.   1974.   Conversion of arginine to ornithine during malo-lactic fermentation in red Swiss wine. Am. J. Enol. Viticult. *25*, 191–193.

LAFON, M.   1955.   The formation of secondary products of the alcoholic fermentation. Sciences Thesis. Univ. Bordeaux.

LAFON, J., COUILLAUD, P. and GAY-BELLILE, F.   1973.   Cognac; Its Distillation. Baillière, Paris.

LAFON-LAFOURCADE, S.   1975.   Histamine in wine. Connaiss. Vigne Vin *9*, 103–115.

LAFON-LAFOURCADE, S., DOMERCQ, S. and PEYNAUD, E.   1968.   The inoculation of wines with bacteria causing malo-lactic fermentation. Connaiss. Vigne Vin *2*, 83–97.

LAFON-LAFOURCADE, S. and PEYNAUD, E.   1966.   The concentration of keto acids formed in the course of alcoholic fermentation. Ann. Inst. Pasteur *110*, 766–778.

LAFON-LAFOURCADE, S. and PEYNAUD, E.   1970.   Nature of the malic enzyme of lactic acid bacteria isolated from wine. C.R. Acad. Sci. *270*, 228–229.

LAFON-LAFOURCADE, S. and PEYNAUD, E.   1974.   The antibacterial action of sulfur dioxide in its free and its combined form.   Connaiss. Vigne Vin *8*, 187–203.

LAFON-LAFOURCADE, S. and RIBÉREAU-GAYON, P.   1976.   First observations on the use of dry yeasts in white wine fermentations. Connaiss. Vigne Vin *10*, 277–292.

LEHTONEN, M. and SUOMALAINEN, H.   1977.   Rum. *In* Economic Microbiology, Vol. 1.   Alcoholic Beverages. A.H. Rose (Editor).   Academic Press, New York.

LEIFSON, E.   1954.   The flagellation and taxonomy of species of *Acetobacter*. Antonie van Leeuwenhoek J. Microbiol. Serol. *20*, 102–110.

LE ROUX, G., ESCHENBRUCH, R. and DE BRUIN, S.J.   1973.   The microbiology of South African wine making. VIII. The microflora of healthy and *Botrytis cineria* infected grapes. Phytophylactica (Pretoria) *5*, 51–54.

LIČEV, V. and PANAJOTOV, I.M.   1958.   Cited by Suomalainen et al. 1968. Brandy. In Handbook of Food Chemistry, Vol. 7. Alcoholic Beverages. W. Diemair (Editor). Verlag Springer, Berlin-New York.

LITSCHEW, W.I.   1976.   The formation of bouquet in brandy. Mitt. Rebe Wein Ser. A (Klosterneuburg) 26, 139–148.

LITSCHEW, W.I.   1977.   The formation of bouquet in brandy. Mitt. Rebe Wein Ser. A (Klosterneuburg) 27, 11–13.

LODDER, J.   1970.   The Yeasts: A Taxonomic Study, 2nd Edition. North-Holland Publishing Co., Amsterdam.

LODDER, J. and KREGER van RIJ, N.J.W.   1952.   The Yeasts: A Taxonomic Study. North-Holland Publishing Co., Amsterdam.

LONVAUD, M., LONVAUD-FUNEL, A. and RIBÉREAU-GAYON, P. 1977. The mechanism of the malo-lactic fermentation of wines. Connaiss. Vigne Vin 11, 73–91.

MACRIS, B.J. and MARKAKIS, P.   1974.   Transport and toxicity of sulfur dioxide in Saccharomyces cerevisiae var. ellipsoideus. J. Sci. Food Agric. 25, 21–29.

MALAN, C.E.   1951, 1953, 1954, 1955.   The yeasts of the wine fermentation in Piedmont. Atti Acad. Ital. Vite Vino 3, 338–360; 5, 500–510; 6, 477–510; 7, 443–461.

MALAN, C.E.   1956.   Taste properties of sterile (grape) musts fermented in pure culture. Riv. Vitic. Enol. Conegliano 1, 11–22.

MALAN, C.E., OZINO, O.I. and GANDINI, A.   1965.   The Schizomycetes of the malo-lactic fermentation of various wines of the Piedmont: Blastomycetes, predominating in the course of their action. Atti Acad. Ital. Vite Vino 17, 235–240.

MALAN, C.E. and TARANTOLA, C.   1956.   Enological and biochemical characteristics of Saccharomyces uvarum Beijerinck. Atti Acad. Ital. Vite Vino 8, 559–579.

MARQUARDT, P. and WERRINGLOER, J.   1966.   Formation of amines in wine. Wein-Wiss. 21, 533–540.

MARQUES GOMES, J.V.   1974.   Ecology and taxonomy of the lactic acid bacteria. Vignes Vins, No. Spec. Intern. Symp. Arc Senans, 1973, 56–69.

MARQUES GOMES, J.V. and DE CASTRO REIS, A.M.L.   1961–1962. On the species Saccharomyces steineri in the fermentation of wines of Porto. Anais Inst. Vinho Porto, 6–67.

MAURON, P.   1979.   The status of viticulture in the world. Bull. OIV 52 (584) 827–862.

MAYER, K.   1974.   Microbiological and enological findings regarding the biological decomposition of acids. Schweiz. Z. Obst-Weinbau 110, 291–297.

MAYER, K. and DUFOUR, A.   1973.   Formation of sulfurous acid during fermentation. Schweiz. Z. Obst-Weinbau 109, 370–372.

MAYER, K. and PAUSE, G.   1973.   Non-volatile biogene amines in wine.

Mitt. Geb. Lebensmittelunters. Hyg. *64*, 171–179.

MAYER, K., PAUSE, G. and VETSCH, U. 1976A. Concentration of SO_2 binding substances in wine: Effect of fermentation and biological decomposition of acids. Schweiz. Z. Obst-Weinbau *112*, 309–313.

MAYER, K. and TEMPERLI, A. 1963. The metabolism of L-malate and other compounds by *Schizosaccharomyces pombe*. Arch. Mikrobiol. *46*, 321–328.

MAYER, K. and VETSCH, U. 1973. pH and biological decomposition of acids in wine. Schweiz. Z. Obst-Weinbau *109*, 635–639.

MAYER, K., VETSCH, U. and PAUSE, G. 1975. Inhibition of the biological decomposition of acids by bound SO_2. Schweiz. Z. Obst-Weinbau *111*, 590–596.

MAYER, K., VETSCH, U. and PAUSE, G. 1976B. Inhibition of the biological decomposition of acids by SO_2 forming yeasts. Schweiz. Z. Obst-Weinbau *112*, 89–92.

McKENZIE, J.A. 1974. The distribution of vineyard populations of *Drosophila melanogaster* and *Drosophila simulans* during vintage and non-vintage periods. Oecologia (Berlin) *15*, 1–16.

MELAMED, N. 1962. Determination of residual sugars in wine, and its relation to the malo-lactic fermentation. Ann. Technol. Agric. *11*, 5–32, 107–119.

MELAS-JOANNIDIS, Z. *et al.* 1958. Microbiological investigation of the fermentation of grape musts in the Peloponese. Ann. Microbiol. Enzymol. *8*, 118–137.

MILLER, M.W. and VAN UDEN, N. 1970. *Metschnikowia kamienski. In* The Yeasts: A Taxonomic Study. J. Lodder (Editor). North-Holland Publishing Co., Amsterdam.

MINÁRIK, E. 1963. The microflora of "Auslese" wines. Mitt. Rebe Wein Ser. A (Klosterneuburg) *13*, 185–188.

MINÁRIK, E. 1964. The yeast flora of fermenting musts in Czechoslovakia. Mitt. Rebe Wein Ser. A (Klosterneuburg) *14*, 75–82.

MINÁRIK, E. 1966. Ecology of Natural Wine-yeast Species in Czechoslovakia. J. Kolek (Editor). Biol. Work, Edn. Sci. Comm. Gen. Spec. Biol. Slovak Acad. Sci. XII/4, 1–107. (Czechoslovakian)

MINÁRIK, E. 1975. Reduction of sulfate to sulfite and its taxonomic significance for the classification of yeasts. Prog. Rech. Viti-Vinic. (Bratislava) 7, 279–298.

MINÁRIK, E., JUNGOVÁ, O. and EMERIAUD, M. 1978. Fructophilic yeasts and their effect on naturally sweet wines. Wein-Wiss. *33*, 42–47.

MINÁRIK, E. and NAVARA, A. 1967. Biological decomposition of malic acid in fermenting musts by various species of *Schizosaccharomyces*. Wein-Wiss. *22*, 385–395.

MINÁRIK, E. and NAVARA, A. 1977. Occurrence of *Saccharomycodes ludwigii Hansen* in sulfited wines low in alcohol content. Mitt. Rebe Wein Ser. A (Klosterneuburg) *27*, 1–3.

MINÁRIK, E. and RÁGALA, P. 1975. The selective action of fungicides on the microflora of grape berries. Mitt. Rebe Wein Ser. A (Klosterneuburg) 25, 187–204.

MOOSER, J. 1958. The occurrence of yeasts in bees, bumble bees and wasps. Zentralbl. Bakteriol. Parasitenkd. Infektionskr. Hyg. Abt. 2 111, 101–115.

MOSSEL, D.A. 1951. Investigation of a case of fermentation in fruit products rich in sugars. Antonie van Leeuwenhoek J. Microbiol. Serol. 17, 146–152.

MOYER, J.C., MILLER, R.C. and MATTICK, L.R. 1977. The effect of steam stripping of grape juice prior to fermentation. Am. J. Enol. Vitic. 28, 231–234.

MRAK, E.M. and McCLUNG, L.S. 1940. Yeasts occurring on grapes and in grape products in California. J. Bacteriol. 40, 395–407.

MULLER, C.J., KEPNER, R.E. and WEBB, A.D. 1973. Lactones in wines, a review. Am. J. Enol. Viticult. 24, 5–9.

MUNITIS, M.T., CABRERA, E. and RODRIGUEZ-NAVARRO, A. 1976. An obligate osmophilic yeast from honey. Appl. Microbiol. 32, 320–323.

MUNYON, J.R. and NAGEL, C.W. 1977. Comparison of methods of deacidification of musts and wines. Am. J. Enol. Viticult. 28, 79–87.

NAKAGAWA, A. and KITAHARA, K. 1959. Taxonomic studies on the genus Pediococcus. J. Gen. Appl. Microbiol. 5, 95–126.

NEISH, A.C. and BLACKWOOD, A.C. 1951. Dissimilation of glucose by yeast at poised hydrogen ion concentrations. Can. J. Technol. 29, 123–129.

NORDSTRÖM, K. 1964. Studies on the formation of volatile esters in fermentation with brewer's yeast. Sven. Kem. Tidskr. 76 (9) 1–34.

OHARA, Y., NONOMURA, H. and YUNOME, H. 1959. Dynamic aspect of yeast flora during vinous fermentation. Bull. Res. Inst. Ferm., Yamanashi Univ. 6, 77–88.

OSTERWALDER, A. 1934. Cold fermentations and cold fermentation yeasts. Zentralbl. Bakteriol. Parasitenkd. Infektionskr. Hyg. Abt. 2 90, 226–249.

OUGH, C.S. 1966A. Fermentation rates of grape juice. II. Effect of initial °Brix, pH and fermentation temperature. Am. J. Enol. Viticult. 17, 20–26.

OUGH, C.S. 1966B. Fermentation rates of grape juice. III. Effects of initial ethyl alcohol, pH and fermentation temperature. Am. J. Enol. Viticult. 17, 74–81.

OUGH, C.S. 1971. Measurement of histamine in California wines. J. Agric. Food Chem. 19, 241–244.

OUGH, C.S. and AMERINE, M.A. 1963. Use of grape concentrate to produce sweet table wines. Am. J. Enol. Viticult. 14, 194–204.

OUGH, C.S. and AMERINE, M.A. 1966. Effects of temperature on wine making. Calif. Agric. Exp. Stn. Bull. 827, 2–36.

OUGH, C.S., GUYMON, J.F. and CROWELL, E.A. 1966. Formation of higher alcohols during grape juice fermentation at various temperatures. J. Food Sci. *31*, 620–625.

PARFAIT, A. and JOURET, C. 1975. Formation of higher alcohols in rum. Ann. Technol. Agric. *24*, 421–435.

PARLE, J.N. and Di MENNA, M.E. 1966. The source of yeasts in New Zealand wines. N.Z. J. Agric. Res. *9*, 98–107.

PEYNAUD, E. 1967. Recent studies on the lactic acid bacteria of wine. Ferment. Vinification *1*, 219–256.

PEYNAUD, E. 1975. Knowledge and Study of Wine. Dunod, Paris.

PEYNAUD, E. and DOMERCQ, S. 1953. The yeasts of the Gironde. Ann. Inst. Natl. Rech. Agron. *4*, 265–300.

PEYNAUD, E. and DOMERCQ, S. 1955A. Yeast species selectively fermenting fructose. Ann. Inst. Pasteur *89*, 346–351.

PEYNAUD, E. and DOMERCQ, S. 1955B. Study of the microflora of the musts and wines of Bordeaux. C.R. Acad. Agric. *41*, 103–106.

PEYNAUD, E. and DOMERCQ, S. 1959. A review of microbiological problems in winemaking in France. Am. J. Enol. Viticult. *10*, 69–77.

PEYNAUD, E. and DOMERCQ, S. 1961. The lactic acid bacteria of wine. Ann. Technol. Agric. *10*, 43–60.

PEYNAUD, E. and DOMERCQ, S. 1967A. Study of some homofermentative lactic rods isolated from wines. Arch. Mikrobiol. *57*, 255–270.

PEYNAUD, E. and DOMERCQ, S. 1967B. Study of some homofermentative lactic acid cocci isolated from wines. Rev. Ferment. Ind. Aliment. Bruxelles *22*, 133–140.

PEYNAUD, E. and DOMERCQ, S. 1968. Study of 400 strains of heterofermentative lactic acid cocci isolated from wines. Ann. Inst. Pasteur Lille *19*, 159–170.

PEYNAUD, E. and DOMERCQ, S. 1970. A study of 250 strains of lactic acid bacteria. Arch. Mikrobiol. *70*, 348–360.

PEYNAUD, E. and GUIMBERTEAU, G. 1962. The formation of higher alcohols by wine yeasts. Ann. Technol. Agric. *11*, 85–105.

PEYNAUD, E., LAFON-LAFOURCADE, S. and GUIMBERTEAU, G. 1967. On the nature of the acids formed by the lactic acid bacteria of wines. Rev. Ferment. Ind. Aliment. *22*, 61–66.

PEYNAUD, E., LAFON-LAFOURCADE, S. and GUIMBERTEAU, G. 1968. On the mechanism of the malo-lactic fermentation. Mitt. Rebe Wein Ser. A (Klosterneuburg) *18*, 343–348.

PEYNAUD, E. and SAPIS, J.C. 1975. New findings on the binding of SO_2 and measures for SO_2 reduction. *In* 4th Intern. Oenological Symp., Valencia, 1975. E. Lemperle and J. Frank (Editors). Intern. Assoc. Mod. Winery Technol. Manage., Breisach.

PEYNAUD, E. and SUDRAUD, P. 1964. Utilization of the de-acidifying effect of *Schizosaccharomyces* in the vinification of grapes. Ann. Technol. Agric. *13*, 309–328.

PEYNAUD, E. *et al.* 1964. *Schizosaccharomyces* yeasts metabolizing L-malic acid. Arch. Mikrobiol. *48*, 150–165.

PHAFF, H.J. 1970. *Hanseniaspora zikes. In* The Yeasts: A Taxonomic Study. J. Lodder (Editor). North-Holland Publishing Co., Amsterdam.

PILONE, D.A., PILONE, G.J. and RANKINE, B.C. 1973. Influence of yeast strain, pH and temperature on degradation of fumaric acid in grape juice fermentation. Am. J. Enol. Viticult. *24*, 97–102.

PILONE, G.J. and KUNKEE, R.E. 1972. Characterization and energetics of *Leuconostoc oenos* ML 34. Am. J. Enol. Viticult. *23*, 61–70.

PILONE, G.J. and KUNKEE, R.E. 1976. Stimulatory effect of malolactic fermentation on the growth rate of *Leuconostoc oenos*. Appl. Microbiol. *32*, 405–408.

PILONE, G.J., KUNKEE, R.E. and WEBB, A.D. 1966. Chemical characterization of wines fermented with various malo-lactic bacteria. Appl. Microbiol. *14*, 608–615.

PILONE, G.J., RANKINE, B.C. and PILONE, D.A. 1974. Inhibiting malo-lactic fermentations in Australian dry red wines by adding fumaric acid. Am. J. Enol. Viticult. *25*, 99–107.

PITT, J.I. and MILLER, M.W. 1968. Sporulation in *Candida pulcherrima, Candida reukaufii* and *Chlamydozyma* species: Their relationships with *Metschnikowia*. Mycologia *60*, 663–685.

PONADE, N. *Cited by* Radler, F. 1973. Significance and possibility of using pure yeast cultures for wine fermentations. Weinberg Keller *20*, 339–350.

POULARD, A. 1978. Microbiological contamination during bottling of white wines of the Nantes region. Vignes Vins *267*, 25–28.

PUPUTTI, E. and SUOMALAINEN, H. 1969. Biologically formed amines of wines. Mitt. Rebe Wein Ser. A (Klosterneuburg) *19*, 184–192.

QUEVAUVILLER, A. and MAZIÈRE, M.A. 1969. Study of histamine in wines and its biological determination. Ann. Pharm. Franç. *27*, 411–414.

RADLER, F. 1958A. The biological decomposition of acids in wine. Isolation and characterization of malic acid decomposing bacteria. Arch. Mikrobiol. *30*, 64–72.

RADLER, F. 1958B. The biological decomposition of acids in wine. II. The nutrient and growth stimulant requirements of malic acid decomposing bacteria. Arch. Mikrobiol. *32*, 1–15.

RADLER, F. 1962. Formation of acetoin and diacetyl by bacteria causing biological decomposition of acids. Vitis *3*, 136–143.

RADLER, F. 1966. Microbiological basis of acid decomposition in wine. Zentralbl. Bakteriol. Parasitenkd. Infektionskr. Hyg. Abt. 2 *120*, 237–287.

RADLER, F. 1972. Problems of the bacterial decomposition of acids. Weinberg Keller *19*, 357–370.

RADLER, F. 1973. Significance and possibility of using pure yeast cultures for wine fermentations. Weinberg Keller *20*, 339–350.

RADLER, F. and GERWARTH, B. 1971. Formation of volatile by-products of the fermentation by lactic acid bacteria. Arch. Mikrobiol. *76*, 299–307.

RADLER, F. and YANNISSIS, C. 1972. Tartaric acid decomposition by lactic acid bacteria. Arch. Mikrobiol. *82*, 219–238.

RAINBOW, C. 1970. Brewer's yeast. *In* The Yeasts. A.H. Rose and J.S. Harrison (Editors). Academic Press, London.

RANKINE, B.C. 1953. Quantitative difference in products of fermentation by different strains of wine yeast. J. Appl. Sci. *4*, 590–602.

RANKINE, B.C. 1955. Yeast cultures in Australian winemaking. Am. J. Enol. Viticult. *6*, 11–15.

RANKINE, B.C. 1963. Nature, origin and prevention of hydrogen sulfide aroma in wines. J. Sci. Food Agric. *14*, 79–91.

RANKINE, B.C. 1964. Hydrogen sulfide production by yeasts. J. Sci. Food Agric. *15*, 872–877.

RANKINE, B.C. 1966. Decomposition of L-malic acid by wine yeasts. J. Sci. Food Agric. *17*, 312–316.

RANKINE, B.C. 1967A. Influence of yeast strain and pH on pyruvic acid content of wines. J. Sci. Food Agric. *18*, 41–44.

RANKINE, B.C. 1967B. Formation of higher alcohols by wine yeasts and relationship to taste thresholds. J. Sci. Food Agric. *18*, 583–589.

RANKINE, B.C. 1968A. The importance of yeast in determining the composition and quality of wines. Vitis *7*, 22–49.

RANKINE, B.C. 1968B. Formation of α-ketoglutaric acid by wine yeasts and its enological significance. J. Sci. Food Agric. *19*, 624–627.

RANKINE, B.C. 1972. Influence of yeast strain and malo-lactic fermentation on composition and quality of table wines. Am. J. Enol. Viticult. *23*, 152–158.

RANKINE, B.C. 1977. Modern developments in selection and use of pure yeast cultures for wine making. Aust. Wine Brew. Spirit Rev., June *27*, 31–33.

RANKINE, B.C. and BRIDSON, D.A. 1971. Glycerol in Australian wines and factors influencing its formation. Am. J. Enol. Viticult. *22*, 6–12.

RANKINE, B.C., FORNACHON, J.C.M. and BRIDSON, D.A. 1969. Diacetyl in Australian dry red wines and its significance in wine quality. Vitis *8*, 129–134.

RANKINE, B.C. and PILONE, D.A. 1973. *Saccharomyces bailii*, a resistant yeast causing serious spoilage of bottled table wine. Am. J. Enol. Viticult. *24*, 55–58.

RANKINE, B.C. and PILONE, D.A. 1974. Yeast spoilage of bottled table wine and its prevention. Aust. Wine Brew. Spirit Rev. *92*, 36–40.

RANKINE, B.C. and POCOCK, K.F. 1969. β-Phenethanol and n-hexanol in wines: Influence of yeast strain, grape variety and other factors and taste thresholds. Vitis *8*, 23–37.

RANKINE, B.C. *et al.* 1976. Practical use of pure yeast generators in wine-making. Aust. Grapegrower Winemaker *148*, 48−54.

RAPP, A. and HASTRICH, H. 1976. Gaschromatographic investigation of the aroma substances of grape berries. II. Possibility of characterization of varieties. Vitis *15*, 183−192.

RAPP, A., HASTRICH, H. and ENGEL, L. 1978. Possibilities of character-izing wine quality and vine varieties by means of gas chromatography. *In* 5th Intern. Oenol. Symp., Auckland, 1978; E. Lemperle and J. Frank (Editors). Intern. Assoc. Mod. Wine Technol. Manage., Breisach.

RAPP, A. *et al.* 1973. Gaschromatographic investigation of aroma sub-stances of musts, wines, and brandy. Chem.-Ztg. *97*, 29−36.

REED, G. and PEPPLER, H.J. 1973. Yeast Technology. AVI Publishing Co., Westport, Conn.

REHM, H.-J., WALLNÖFER, P. and KESKIN, H. 1965. Contribution to the understanding of the antimicrobial effect of sulfurous acid. IV. Dissociation and antimicrobial effect of sulfurous acid. Z. Lebensm.-Unters.-Forsch. *127*, 72−85.

REINHARD, C. 1968. Gas chromatographic investigation of wine, brandy, and distilled beverages. Wein-Wiss. *23*, 475−486.

RELAN, S. and VYAS, S.R. 1971. Nature and occurrence of yeasts in Hara-yana grapes and wines. Vitis *10*, 131−135.

RENTSCHLER, H. and TANNER, H. 1951. Development of bitterness in red wines. Occurrence of acrolein in beverages and its relation to the develop-ment of bitterness in wine. Mitt. Geb. Lebensmittelunters. Hyg. *42*, 463−475.

RIBÉREAU-GAYON, P. 1973. Understanding of the nature of combina-tions of SO_2 in wines. Bull. OIV *46*, 406−416.

RIBÉREAU-GAYON, P., BOIDRON, J.N. and TERRIER, A. 1975A. Aroma of Muscat grape varieties. J. Agric. Food Chem. *23*, 1042−1047.

RIBÉREAU-GAYON, P. and PEYNAUD, E. 1960. Treatise on Oenology, Vol. 1. Ripening of Grapes, Alcoholic Fermentation, Vinification. Librairie Polytechnique Ch. Béranger, Paris.

RIBÉREAU-GAYON, P. *et al.* 1975B. Treatise on Oenology, Vol. 2. Charac-ter of Wines, Ripening of Grapes, Yeasts and Bacteria. Dunod, Paris.

RICE, A.C. 1974. Chemistry of winemaking from native American grape varieties. *In* Chemistry of Winemaking. A.D. Webb (Editor). Adv. Chem. Ser. *137*. Am. Chem. Soc., Washington, D.C.

SALLER, W. 1955. Improvement in the Quality of Wines and Sweet Musts by Cooling. Sigurd Horn-Verlag, Frankfurt.

SALO, P., NYKÄNEN, L. and SUOMALAINEN, H. 1972. Odor thresholds and relative intensities of volatile aroma components in an artificial beverage imitating whisky. J. Food Sci. *37*, 394−398.

SANDU-VILLE, G. 1977. *Saccharomycodes ludwigii Hansen*, a pathogenic species of the microflora of wine. Cercet. Agron. Moldova (Iasi) *3*, 92−96.

SAPIS-DOMERCQ, S. 1969. Reactions of apiculated yeasts during vinifica-tion. Connaiss. Vigne Vin *4*, 379−392.

SAPIS-DOMERCQ, S. and GUITTARD, A.   1976.   Study of the yeast micro-flora of Roussillon. Connaiss. Vigne Vin *10*, 1–21.

SCHANDERL, H.   1959.   Microbiology of Must and Wine. Eugen Ulmer Verlag, Stuttgart.

SCHANDERL, H.   1962.   *Saccharomyces acidifaciens,* another dangerous "sulfur yeast" which has recently appeared in German white wines. Weinblatt *56*, 769–770.

SCHANDERL, H. and DRACZYNSKI, M.   1952.   *Brettanomyces,* an undesirable yeast genus in bottle fermented sparkling wine. Wein Rebe *20*, 462–463.

SCHEFFER, W.R. and MRAK, E.M.   1951.   Characteristics of yeasts causing clouding of dry white wines. Mycopath. Mycol. Appl. *5*, 236–249.

SCHMITTHENNER, F.   1949.   Effect of carbon dioxide on yeasts and bacteria. Dtsch. Weinbau (Wissensch. Beihefte) *3*, 147–187.

SCHNEYDER, J.   1972.   Comprehensive summary of our knowledge of histamine and similar compounds in wine. Mitt. Rebe Wein Ser. A (Klosterneuburg) *22*, 313–322.

SCHREIER, P. and DRAWERT, F.   1974A.   Gaschromatographic-mass-spectrometric investigation of volatile components of wine. I. Non-polar compounds of the wine aroma. Z. Lebensm.-Unters.-Forsch. *154*, 273–287.

SCHREIER, P. and DRAWERT, F.   1974B.   Gaschromatographic-mass-spectrometric investigation of volatile components of wine. V. Alcohol, hydroxyesters, lactones and other polar compounds of the aroma fraction of wine. Chem. Mikrobiol. Technol. Lebensm. *3*, 154–160.

SCHREIER, P., DRAWERT, F. and JUNKER, A.   1974.   Gaschromatographic-mass-spectrometric investigation of volatile components of wine. II. Thioether compounds of the aroma fraction of wine. Z. Lebensm.-Unters.-Forsch. *154*, 279–284.

SCHREIER, P., DRAWERT, F. and JUNKER, A.   1975.   Gaschromatographic-mass-spectrometric investigation of volatile components of wine. IV. Detection of secondary amines in wine. Z. Lebensm.-Unters.-Forsch. *157*, 34–37.

SCHREIER, P., DRAWERT, F. and JUNKER, A.   1976A.   Gaschromatographic-mass-spectrometric differentiation of the aroma components of various varieties of grapes, *Vitis vinifera.* Chem. Mikrobiol. Technol. Lebensm. *4*, 154–157.

SCHREIER, P., DRAWERT, F. and JUNKER, A.   1976B.   Identification of volatile constituents from grapes. J. Agric. Food Chem. *24*, 331–336.

SCHREIER, P., DRAWERT, F. and JUNKER, A.   1977.   Gaschromatographic analysis of aroma components of fermented beverages. X. Quantitative analysis of the aroma components of wine in the mg per liter range. Chem. Mikrobiol. Technol. Lebensm. *5*, 45–52.

SCHREIER, P. *et al.*   1976C.   Gaschromatographic-mass-spectrometric investigation of volatile components of wine. VI. Aroma components of Tokaj Aszu-wines. a) Neutral components. Z. Lebensm.-Unters.-Forsch. *161*, 249–258.

SCHREIER, P. *et al.* 1976D. Use of multiple differential analysis for the identification of grape varieties by means of the quantitative distribution of volatile wine components. Mitt. Rebe Wein Ser. A (Klosterneuburg) 26, 225–234.

SCHRÖDER, W. 1960. Comparative investigations of several osmotolerant yeasts in concentrated solutions of saccharose and invert sugar. Diss. Fac. Agric., Tech. Univ. Berlin.

SCHULLE, H. 1953. The significance of the apiculate yeasts for the fermenting activity of the true wine yeasts in high sugar musts. Arch. Mikrobiol. 18, 342–348.

SCHÜTZ, M. and HEINZ, W. 1978. Use of a "metabolistat" for the continuous production of pure culture yeast. Weinwirtsch. 114, 11–15.

SCHÜTZ, M. and RADLER, F. 1973. The "malate enzyme" of *Lactobacillus plantarum* and *Leuconostoc mesenteroides*. Arch. Mikrobiol. 91, 183–202.

SCHÜTZ, M. and RADLER, F. 1974. The occurrence of "malate enzyme" and "malo-lactate enzyme" in various lactic acid bacteria. Arch. Mikrobiol. 96, 329–339.

SINGH, R. and KUNKEE, R.E. 1976. Alcohol dehydrogenase activities of wine yeasts in relation to higher alcohol formation. Appl. Microbiol. 32, 666–670.

SLOOF, W.C. 1970. *Schizosaccharomyces Linder. In* The Yeasts: A Taxonomic Study. J. Lodder (Editor). North-Holland Publishing Co., Amsterdam.

SNOW, P.G. and GALLANDER, J.F. 1979. Deacidification of white table wines through partial fermentation with *Schizosaccharomyces pombe*. Am. J. Enol. Vitic. 30, 45–48.

SOLS, A. 1956. Selective fermentation and phosphorylation of sugars by Sauternes yeast. Biochim. Biophys. Acta 20, 62–68.

SOMAVILLA, J.F. *et al.* 1975. Osmophilic yeasts on grapes and dried fruits. Rev. Agroquim. Tecnol. Aliment. 15, 573–580.

SPONHOLZ, W.R. and DITTRICH, H.H. 1974. The formation of fermentation by-products which bind SO_2 of higher alcohols and esters by several pure culture yeasts and by enologically important "wild" yeasts. Wein-Wiss. 29, 301–314.

STERN, D.J. *et al.* 1967. Volatiles from grapes. Identification of volatiles from Concord essence. J. Agric. Food Chem. 15, 1100–1103.

STEVENS, K.L. *et al.* 1966. Volatiles from grapes. Muscat of Alexandria. J. Agric. Food Chem. 14, 249–252.

STEVIĆ, B. 1962. The importance of bees (*Apis* sp.) and wasps (*Vespa* sp.) as carriers of yeasts for the microflora of grapes and the quality of wines. Archiv Poljoprivredne Nauke, Beograd 15, 80–91.

SUOMALAINEN, H. 1971. Yeast and its effect on the flavor of alcoholic beverages. J. Inst. Brew. 77, 164–177.

SUOMALAINEN, H. 1975. Some general aspects of the composition of the aroma of alcoholic beverages. Ann. Technol. Agric. 24, 453–467.

SUOMALAINEN, H. and NYKÄNEN, L.   1966A.   The aroma components produced by yeast in nitrogen-free sugar solution. J. Inst. Brew. 72, 469–474.

SUOMALAINEN, H. and NYKÄNEN, L.   1966B.   The aroma compounds produced by sherry yeast in grape and berry wines. Suom. Kemistil. B 39, 252–256.

SUOMALAINEN, H., NYKÄNEN, L. and ERIKSSON, K.   1974.   Composition and consumption of alcoholic beverages—a review. Am. J. Enol. Viticult. 25, 179–187.

SUOMALAINEN, H. and RONKAINEN, P.   1968.   Mechanism of diacetyl formation in yeast fermentation. Nature 220, 792–793.

SUOMALAINEN, H. et al.   1968.   Brandy.   In Handbook of Food Chemistry, Vol. 7. Alcoholic Beverages. W. Diemair (Editor).   Springer Verlag, Berlin-New York.

ŠVEJCAR, V.   1969.   Yeast flora of wine grapes and that of vine growing soil in the vineyards of the Academy of Agriculture in Lednice, Czechoslovakia. Publ. Univ. Horticult. 33, 207–211.

TARANTOLA, C.   1945–1946.   New contribution to the study of apiculate yeasts. Ann. Acad. Agric. Torino 88, 115–133.

TARANTOLA, C.   1955.   Cited by Ribéreau-Gayon, J. and Peynaud, E. 1960. Treatise on Oenology, Vol. 1. Librairie Polytechnique Ch. Béranger, Paris.

TCHELISTCHEFF, A.   1948.   Comments on cold fermentation. Univ. Calif. Wine Technol. Conf., Aug. 1948, Univ. Calif., Davis.

TCHELISTCHEFF, A., PETERSON, R.G. and VAN GELDEREN, M.   1971. Control of malo-lactic fermentation in wine. Am. J. Enol. Viticult. 22, 1–5.

TEMPERLI, A. et al.   1965.   Purification and properties of the decarboxylating malate dehydrogenase from yeast. Biochim. Biophys. Acta 110, 630–632.

TERRIER, A. and BOIDRON, J.N.   1972.   Identification of terpene derivatives in the grapes of certain varieties of Vitis vinifera. Connaiss. Vigne Vin 6, 147–160.

THOUKIS, G., REED, G. and BOUTHILET, J.R.   1963.   Production and use of compressed yeast for winery fermentation. Am. J. Enol. Viticult. 14, 148–154.

TOLEDO, O. and TEIXERA, C.G.   1955.   Advantages of the association of yeasts in wine fermentations: Reduction of the volatile acidity in wine. Agric. Ital. 55, 155–164.

TRAUTWEIN, K. and WASSERMANN, J.   1931.   The pH sensitivity of respiring and fermenting beer yeasts. Switch from fermentation to respiration. Biochem. Z. 236, 35–53.

TROOST, G.   1972.   Wine Technology. Eugen Ulmer Verlag, Stuttgart.

USSEGLIO-TOMASSET, L.   1967A.   β-Phenyl ethanol in wine. Riv. Vitic. Enol. 20, 10–13.

USSEGLIO-TOMASSET, L.   1967B.   The volatile acids (homologues of acetic acid) in fermentations with various yeast species. Atti Acad. Ital. Vite Vino, Siena 19, 165–183.

USSEGLIO-TOMASSET, L.   1971.   Ethyl acetate and the higher alcohols in wine. Riv. Vitic. Enol. *24*, 236, 276, 303.

VALAER, P.   1939.   Brandy. Ind. Eng. Chem. *31*, 339–353.

VAN DER WALT, J.P.   1970A.   *Brettanomyces Kufferath et van Laer. In* The Yeasts: A Taxonomic Study, 2nd Edition.   J. Lodder (Editor).   North-Holland Publishing Co., Amsterdam.

VAN DER WALT, J.P.   1970B.   *Saccharomyces Meyen emend. Reess. In* The Yeasts: A Taxonomic Study, 2nd Edition.   J. Lodder (Editor).   North-Holland Publishing Co., Amsterdam.

VAN DER WALT, J.P. and VAN KERKEN, A.E.   1958.   The wine yeasts of the Cape. Part I. A taxonomical survey of the yeasts causing turbidity in South African table wines. Antonie van Leeuwenhoek J. Microbiol. Serol. *24*, 239–252.

VAN UDEN, N. and VIDAL-LEIRIA, M.   1970.   *Torulopsis berlese. In* The Yeasts: A Taxonomic Study, 2nd Edition. J. Lodder (Editor). North-Holland Publishing Co., Amsterdam.

VAN ZYL, J.A.   1958.   Contribution to the biology and metabolism of Jerez yeast and film forming yeasts. Zentralbl. Bakteriol. Parasitenkd. Infektionskr. Hyg. Abt. 2 *111*, 33–79.

VAN ZYL, J.A.   1962.   Turbidity in South African dry wines caused by the development of *Brettanomyces* yeast. Sci. Bull. Dep. Agric. Technol. Serv. Pretoria *381*.

VAN ZYL, J.A. and DU PLESSIS, L. DE W.   1961.   The microbiology of South African winemaking. I. The yeasts occurring in vineyards, musts and wines. S. Afr. J. Agric. Sci. *4*, 393–403.

VAUGHN, R.H.   1942.   The acetic acid bacteria. Wallerstein Lab. Commun. *5*, 5–26.

VAUGHN, R.H.   1955.   Bacterial spoilage of wines with special reference to California conditions. Adv. Food Res. *6*, 67–108.

VAUGHN, R.H., DOUGLAS, H.C. and FORNACHON, J.C.M.   1949.   The taxonomy of *Lactobacillus hilgardii* and related heterofermentative Lactobacilli. Hilgardia *19*, 133–139.

VERONA, O.   1951.   Note on the microbiology of the wines of Sardinia. Ann. Fac. Agrar. Univ. Pisa 12, 123–145.

VETSCH, U.   1973.   Growth of *Bacterium gracile (Leuconostoc oenos)* during the biological decomposition of acids in wine. Schweiz. Z. Obst-Weinbau *109*, 468–479.

VOGT, E. *et al.*   1974.   Wine. Eugen Ulmer-Verlag, Stuttgart.

VOLLBRECHT, D. and RADLER, F.   1973.   Formation of higher alcohols by amino acid deficient mutants of *Saccharomyces cerevisiae*. I. The decomposition of amino acids to higher alcohols. Arch. Mikrobiol. *94*, 351–358.

VON ARX, J.A.   1968.   Fungi. Verlag Cramer, Lehrte, W. Germany.

WAGENER, W.W.D. and WAGENER, G.W.W.   1968.   The influence of ester and fusel alcohol content upon the quality of dry white wine. S. Afr. J. Agric. Sci. *11*, 469–476.

WEBB, A.D.   1967.   Some aroma compounds produced by vinous fermentation. Biotechnol. Bioeng. *9*, 305–319.

WEBB, A.D. and INGRAHAM, J.L. 1963. Fusel oil. Adv. Appl. Microbiol. *5*, 317–353.

WEBB, A.D. and KEPNER, R.E.   1961.   Fusel oil analysis by means of gas liquid partition chromatography. Am. J. Enol. Viticult. *12*, 51–59.

WEBB, A.D. and KEPNER, R.E.   1962.   The aroma of flor sherry. Am. J. Enol. Viticult. *13*, 1–14.

WEBB, A.D., KEPNER, R.E. and MAGGIORA, L.   1966.   Gas chromatographic comparison of volatile aroma materials extracted from eight different muscat-flavored varieties of *Vitis vinifera*. Am. J. Enol. Viticult. *17*, 247–254.

WEBB, A.D., KEPNER, R.E. and MAGGIORA, L.   1967.   Sherry aroma. VI. Some volatile components of flor sherry of Spanish origin. Neutral substances. Am. J. Enol. Viticult. *18*, 190–199.

WEBB, A.D. and MULLER, C.J.   1972.   Volatile aroma components of wines and other fermented beverages. Adv. Appl. Microbiol. *15*, 75–146.

WEGER, B.   1976.   Experiments on the application of dry yeasts and new pectolytic enzymes. Wein-Wiss. *31*, 197–201.

WEILLER, H.G. and RADLER, F.   1970.   Lactic acid bacteria from wine and grape leaves. Zentralbl. Bakteriol. Parasitenkd. Infektionskr. Hyg. Abt. 2 *124*, 707–732.

WEILLER, H.G. and RADLER, F.   1972.   Vitamin and amino acid requirements of lactic acid bacteria from wine and from grape leaves. Mitt. Rebe Wein Ser. A (Klosterneuburg) *22*, 4–18.

WEILLER, H.G. and RADLER, F.   1976.   On the amino acid metabolism of lactic acid bacteria isolated from wine. Z. Lebensm.-Unters.-Forsch. *161*, 259–266.

WENZEL, K.W.O.   1966.   Gas chromatographic analysis of fermentation byproducts in wine. M.Sc. Thesis (Chem.). Univ. Stellenbosch, S. Afr.

WHITE, J. and MUNNS, D.J.J.   1951.   Influence of temperature on yeast growth and fermentation. J. Inst. Brew. *57*, 280–284.

WILLIAMS, P.J. and STRAUSS, C.R.   1975.   3,3-Diethoxybutane-2-one and 1,1,3-triethoxypropane: Acetals in spirits distilled from *Vitis vinifera* grape wines. J. Sci. Food Agric. *26*, 1127–1136.

WINDISCH, S.   1969.   Studies on osmotolerant yeasts. Proc. 2nd Symp. Yeasts, Bratislava,Czechoslovakia, 1966, Inst. Slovak Acad. Sci., 127–131.

WOLF, E. and BENDA, I.   1965.   Quality and resistance. III. Feed selectivity of *Drosophila melanogaster* for wine yeast species and strains. Biol. Zentralbl. *84*, 1–8.

WOLF, E. and BENDA, I.   1967.   Differentiation of yeast strains by *Drosophila melanogaster* with regard to representatives of the genus *Schizosaccharomyces*. Weinberg Keller *14*, 163–166.

WUCHERPFENNIG, K. and BRETTHAUER, G.   1970.   Formation of volatile aroma components in grape wine as a function of the pre-treatment of the must and of the used yeast strain. Mitt. Rebe Wein Ser. A (Klosterneuburg)

20, 36–46.

WÜRDIG, G. and SCHLOTTER, H.-A.   1971.   The occurrence of SO$_2$ forming yeasts in the natural yeast flora of grape must. Dtsch. Lebensm.-Rundsch. *67*, 86–91.

YANG, H.Y.   1973A.   Effect of pH on the activity of *Schizosaccharomyces pombe*. J. Food Sci. *38*, 1156–1157.

YANG, H.Y.   1973B.   Deacidification of grape musts with *Schizosaccharomyces pombe*. Am. J. Enol. Viticult. *24*, 1–4.

YANG, H.Y.   1975.   Effect of sulfur dioxide on the activity of *Schizosaccharomyces pombe*. Am. J. Enol. Viticult. *26*, 1–4.

YOKOTSUKA, I.   1954.   Studies on the Japanese wine yeasts. Bull. Res. Inst. Ferment. Yamanashi Univ., Korfu, *1*.

ZÜRN, F.   1976.   Influence of various cellar treatments on the sulfur requirement of wines. Wein-Wiss. *31*, 145–159.

ZÜRN, F. and PERSCHEID, M.   1977.   Use of various yeast strains during fermentation of grape must. Dtsch. Weinbau *32*, 1198–1201.

10

Beer

J. Raymond Helbert

Historical Notes

Brewing Process.—A knowledge of brewing is so ancient that its origins are lost in prehistory. Archaeological evidence clearly indicates that the Egyptians had a well-established process for producing a fermented beverage from barley and other cereals at least 60 centuries ago. Similar evidence for the Sumerians and Babylonians, though more fragmentary, is even older. Of all the fermented foods known to man, beer and bread are among the oldest.

Ancient brewing and breadmaking were in fact closely related. Barley or other grains were sprouted, then ground into a meal which was made into a dough and dried or partially baked. The loaves so prepared were broken into pieces, mixed with water, and allowed to ferment spontaneously. A sweet beer was prepared by straining the dough fragments from the mixture after 1 or 2 days of fermentation. A more alcoholic and acidic product was prepared by allowing fermentation to continue for a longer period of time. Neither of these products contained hops and would not have tasted much like modern beer.

While the ancients added various spices or herbs to wine and beer for special occasions or purposes, the use of hops in beer cannot be documented until medieval times—during the 8th century in Bavaria. From there, the practice spread gradually to the rest of Europe where the use of hops in beer and ale was quite general by the 17th century.

Bavaria was also the area where lager brewing developed. For many years, lager beer was made exclusively by Bavarian brewers since neither the process nor the yeast used in it could be exported legally. Elsewhere, malt beverages were ales. However, in 1842, both the lager process and its yeast were smuggled into Czechoslovakia. Within a few years, lager brewing had spread to northern Germany, Denmark, and the Lowlands, and also to the United States via immigrant brewmasters. Lager brewing gradually displaced ale brewing everywhere except in England and in a few of her colonies. The process has persisted with minor modifications to the present day.

Brewing Microbiology.—Although the brewing process is very ancient, the microbiology of brewing is modern. An account of its development begins logically with Antonie van Leeuwenhoek (1632–1723) who is credited with inventing the simple microscope about 1660. With it, he discovered the microbial world. He explored this new-found world for the remainder of his life, studying almost every object that could be examined microscopically. His "little animals" included all of the major types of microorganisms known today—protozoa, molds, yeasts, algae, and bacteria—and he described many of them with such accuracy that their species can sometimes be inferred from his account.

Simple microscopes capable of high magnification were difficult to construct and inconvenient to use. Although van Leeuwenhoek was able to develop their construction and use to a high art, obtaining magnifications of 50–300 diameters, few if any of his contemporaries could match his skill. Most of them employed compound microscopes which were easier to use but were optically inferior to simple microscopes. It was not until the 1830s and onward that compound microscopes were produced with optical qualities approaching those of today. Consequently, the microscopic explorations begun by van Leeuwenhoek were not continued by his immediate successors but were delayed for more than 100 years until improved compound microscopes became available.

Spontaneous Generation.—In the meantime, microbiology was developing largely in terms of brewing and other alcoholic fermentations, along lines which eventually clarified the role of microorganisms in these processes. The initial impetus for this development came from the long-standing controversy over spontaneous generation.

The concept of spontaneous generation, namely that living organisms can in some sense arise spontaneously from nonliving matter, has been widely held throughout history. In fact, it was revived recently by Haldane and Operin to explain the origin of the first primitive forms of life on earth. However, their conception of the process differed radically from that held in the 17th and 18th centuries. At that time, the process was regarded as capable of giving rise to rather highly developed forms of life in a relatively short period of time, and it was regarded as an ongoing process. Most of what was considered evidence for this view was simply a misinterpretation of biological phenomena, such as the appearance of maggots in decomposing organic matter. Such misconceptions had gradually been corrected by the work of a series of naturalists beginning during the Renaissance so that by the 1660s the hypothesis was regarded as untenable by many philosophers and scientists. However, the discovery of microorganisms at about that time, particularly those observed during the fermentation of beer and wine, which were both spontaneous processes, revived the waning hypothesis. It was in this context that the following microbiological developments occurred.

Exploratory Phase.—During the 1750s, Lazzaro Spallanzani (1729–1799) discovered that by heating hermetically sealed infusions of organic matter in boiling water, he could prevent indefinitely the appearance of "little animals"; also, neither fermentation nor putrefaction ever occurred. However, as soon as he made a small crack in the glass, admitting new air to the flask, the "little animals" would reappear along with fermentation or putrefaction. From these findings, Spallanzani concluded that the "little animals" did not arise spontaneously but, being airborne, were introduced into the sterile flasks with the entering air.

For several reasons, Spallanzani's experiments did not seem as convincing to his contemporaries as they do to us. The main reason was that some investigators were unable to confirm his results because of inadequate heating. Another reason was that the chemists of that day adduced quite a different explanation than Spallanzani did. Most of them ascribed his results to the presence or absence of oxygen.

The science of chemistry was developing side by side with microbiology—also in terms of brewing and other alcoholic fermentations. Gay-Lussac elaborated the stoichiometry of alcoholic fermentations

$$C_6H_{12}O_6 \rightarrow 2C_2H_5OH + 2CO_2$$

which appeared to be a simple chemical reaction. Since many chemical reactions observed in the laboratory were attended by the evolution of gas and the formation of a precipitate of one kind or another, chemists tended to see alcoholic fermentation simply as a chemical process. Moreover, the amorphous sediment which formed during fermentation, even if viewed microscopically, did not provide compelling evidence for a living organism. The prevalent view of chemists therefore was that yeast was a chemical compound, albeit a very complex one.

Meanwhile, developments along microbiological lines had continued, mainly with a view to testing the concept of spontaneous generation. C. Cagniard-Latour observed that although yeast was nonmotile, it reproduced by budding; he assigned it to the vegetable kingdom. Theodor Schwann (1810–1882) observed spore formation and likewise concluded that yeast was a living organism; he called it "sugar fungus," a term which survives as the genus name, *Saccharomyces.* Yet another investigator, F. Kützing, also concluded from microscopic studies that yeast was a vegetable organism. Although he and Schwann and Cagniard-Latour all worked independently, they reported their results at about the same time, concluding that fermentation did not spontaneously generate yeast but rather that yeast was the causative agent of fermentation. Kützing (1837) added, "It is obvious that chemists must now strike yeast from the list of chemical compounds, since it is not a compound but an organized body, an organism."

These 3 reports were received at first with dismay and then with scorn and sarcasm—not from the proponents of spontaneous generation but from chemists who regarded such views as "vitalistic." Thus, a sharp dichotomy over the nature of fermentation developed between chemists and microbiologists, both of whom opposed the doctrine of spontaneous generation. Notable among these chemists were Berzelius (1779–1848), Wöhler (1800–1882), and Liebig (1803–1873). Much of the emotionalism in this matter stemmed from the fact that chemists regarded these ideas as retrogressive, a return to the vital-force concept from which organic chemistry had been rescued by Wöhler's synthesis of urea in 1828.

Berzelius regarded fermentation as a catalytic process and yeast cells simply as complex catalysts. Liebig's views were somewhat different. He explained fermentation as the interaction of the oxygen in air with the nitrogenous matter of fermentable plant juices to produce a ferment. This ferment was unstable and transmitted its instability to the sugar which then decomposed into ethanol and carbon dioxide. All of the sugar which disappeared was accounted for by the ethanol and carbon dioxide.

Liebig's objections to the views of Schwann et al. were numerous. If fermentations required a causative organism, how could one explain the lactic fermentation of milk for which no causative organism was known? Furthermore, isolated ferments such as diastase which converted starch to sugar produced chemical effects without an amorphous precipitate which could be regarded as an organism. This view was supported by the finding that diastase, after being heated, irreversibly lost its capacity to convert starch into sugar.

At about this time, Pasteur (1822–1895), who had been educated as a chemist, became interested in fermentation because of the optical isomerism of some of its products, viz., isoamyl alcohol and tartaric acid. He published his first paper on fermentation in 1857, describing an organism associated with a lactic fermentation. One of Liebig's objections was thus removed. Pasteur repeated Spallanzani-type experiments which differed from those of his predecessors in that no kind of filtration or other treatment of air re-entering the flask was employed. The contents of his flasks communicated directly with the outside atmosphere through a narrow-bore, swan-neck tube. Such an arrangement, because of gravitational and adsorptive effects, prevented the entry of particles, even those as small as bacteria. His sterile flasks never fermented spontaneously even in the presence of air and oxygen. Moreover, he established the presence of microorganisms in air by drawing large quantities of it through a guncotton filter, then dissolving the latter in an alcohol-ether mixture. The residue was examined microscopically and also was used to inoculate a sterile broth; both methods indicated microorganisms.

Pasteur then examined Liebig's ideas about the interaction of oxygen in air with the nitrogenous organic matter of fermentable liquids as the means by which fermentation was initiated. He simply eliminated nitrogenous organic matter from the fermentable medium, preparing a simple, chemically defined medium with sugar, ammonium tartrate, and mineral phosphate, to which he added yeast; the result was not only an alcoholic fermentation but a crop of yeast which exceeded the inoculum. From these results, Pasteur (1860) concluded: "The chemical act of fermentation is essentially a phenomenon correlative with a vital act, commencing and ceasing with it. I am of the opinion that alcoholic fermentation never occurs without simultaneous organization, development, multiplication of cells, or the continued life of cells already formed." Alluding to a hoax published anonymously by Wöhler and Liebig (Anon. 1839), Pasteur (1860) concluded: "Should we say that yeast feeds on sugar and excretes alcohol and carbonic acid? Or should we say that yeast during its development produces some albuminous substance which acts upon the sugar and then disappears, since no such substance is found in fermented liquids? I have nothing to reply to these hypotheses. I neither admit them nor reject them, and wish only to restrain myself from exceeding the facts. And the facts tell me simply that all true fermentations are correlative with physiological phenomena."

Liebig's reply was a long time coming. It appeared in 1870 and consisted in part of (1) an accusation that Pasteur's explanation begged the question, his "physiological phenomena" being the very thing that required explanation, and (2) a reiteration that fermentation accompanied the breakdown, not the buildup, of cell constituents. Pasteur's reply was brief. He volunteered to grow as much yeast as desired in a simple, synthetic medium, the components of which were to be provided by Liebig. The challenge was never accepted.

From this time on, less and less was heard from the proponents of spontaneous generation, and the polemic between chemists and microbiologists also subsided. In fact, the chemical and microbiological views of fermentation began to converge. As early as 1858, Moritz Traube proposed the theory that all fermentations produced by microorganisms are caused by ferments which are definite chemical substances produced within the organism. He considered ferments to be closely related to proteins and thought that their function was to transfer the oxygen and hydrogen of water to different parts of the fermentable molecule. In this way he explained the apparent intramolecular oxidation-reduction effected by fermentation. None of these views, however, was supported by much experimental evidence. Consequently, Pasteur's view, that without life there could be no fermentation, prevailed—and did so until 1897.

In that year, Eduard Buchner (1860–1917) succeeded in preparing a cell-free yeast extract which was capable of converting sugar into ethanol and carbon dioxide. Efforts to make such a preparation had begun as early as 1845. A large number of investigators, including Liebig and Pasteur, had tried to make such preparations but failed. If their preparations fermented sugar, they invariably contained some intact yeast cells which rendered the results inconclusive. On the other hand, if the process of maceration lasted long enough to ensure that no intact yeast cells remained, the preparation did not ferment sugar. The latter situation is now easily understood in terms of enzyme denaturation. Buchner succeeded by grinding yeast cells with sand and then filtering the crude preparation through kieselguhr. The filtrate, prepared for animal nutrition studies, was unstable; and because other preservatives were unsatisfactory, sucrose was added for this purpose. Buchner's genius enabled him to recognize the importance of the serendipitous result.

The discovery of zymase, as Buchner's crude preparation was called, established the enzyme theory of fermentation and brought the controversy between chemists and microbiologists to an end. Clearly, neither group had been entirely right nor entirely wrong.

Initially, zymase was thought to be a single enzyme rather than a system of enzymes. The efforts of chemists to unravel the complexities of zymase eventually led to an understanding of the metabolism of carbohydrates and to the science of biochemistry. As biochemistry and microbiology continued to develop, the two disciplines became less and less exclusively concerned with alcoholic fermentation: biochemists became interested in the intermediary metabolism of muscle as well as of yeast; microbiologists extended their purview to include the diseases of animals and man, besides those of wine and beer.

Descriptive Phase.—The first applications of microbiology to brewing pertained to spoilage problems and consisted of finding and describing the various microbes associated with the "diseases" of beer. Notable among those so engaged were Pasteur in France, Horace T. Brown in England, E.C. Hansen in Denmark, and J. Balcke in Germany. Pasteur in his *Studies on Beer* (1876) described simply as Ferment No. 7 what was subsequently described by Hansen in 1879 as *Sarcina* and by Balcke in 1884 as *Pediococcus*, the generic name now accepted. By 1871, Horace Brown, who quickly followed Pasteur's lead, was also involved in spoilage problems. In England at that time, because refrigeration had not yet been developed, ale was brewed only from October to May. Stock ales which were produced to be stored and subsequently sold during the warm months often spoiled, and on many occasions the spoilage was characterized by a "silky" turbidity followed by the formation of lactic acid. Brown observed but did not name the organism, and Pasteur, again in *Studies on Beer*, merely called it "bacilles des bières tournées" (bacilli of sour beer). In 1892, van Laer named it *Saccharobacillus pastorianus*, but it was not until much later in the 1930s and 1940s that this heterofermentative rod was finally classified as a member of Beijerinck's genus *Lactobacillus*.

In *Studies on Wine* (1866) and in *Studies on Vinegar* (1868), Pasteur described the acetic acid bacteria, recognizing two subgroups which he called *Mycoderma aceti* and *Mycoderma vini*. Both of these were capable of oxidizing ethanol to acetic acid. The latter, *Mycoderma vini*, could further oxidize acetic acid to carbon dioxide and water. *Mycoderma aceti* was subsequently changed to *Acetomonas* and then to *Gluconobacter*, while *Mycoderma vini* was changed to *Acetobacter*.

It should be noted that in the period prior to about 1900, many of the names applied to microorganisms were simply descriptive, without the specialized taxonomic significance associated with them today. The term, *Mycoderma*, for example, was used by Pasteur and his contemporaries to signify any film-forming organism, yeast or bacteria. Likewise, the terms *Bacterium* or *Bacillus*, were often vaguely applied to almost any rod-shaped microbe. It is easy to understand how the multiplicity of names which accumulated for the same or closely related microorganisms eventually led to much confusion and to the need for systematic methods of classification. It should likewise be noted that all of the early work by Pasteur and others was done with naturally occurring mixed cultures. The development of pure culture techniques was pioneered by O. Brefeld who worked with fungi. In 1870, he introduced the use of single cell isolates to start pure cultures. However, since micromanipulators were not yet developed, his method could not readily be applied to yeast and bacteria. However, only 8 years later, Joseph Lister (1827–1912) developed the dilution technique and isolated Pasteur's "lactic ferment"—the first bacterial pure culture.

Emil C. Hansen, working in the laboratory of the Old Carlsberg Brewery in Copenhagen, applied these techniques and prepared the first pure brewers' yeast. His work, like Pasteur's and Brown's, was prompted by a beer spoilage problem which the Danes and Germans called "yeast turbidity" and the British called "frets." From spoiled beer of this kind, Hansen isolated pure cultures of 3 different yeasts which he called *Saccharomyces cerevisiae, S. pastorianus,* and *S. ellipsoideus.* He seeded sterile wort with these cultures, both separately and as mixtures. Only *S. cerevisiae* produced a good quality beer without "yeast turbidity."

Hansen not only developed laboratory methods for isolating and preserving pure cultures of brewers' yeast but also introduced, in 1883, special equipment for propagating such pure cultures on a scale suitable for use in the brewery. Although he contributed significantly to yeast taxonomy, the practical management of pure yeast was probably his most important achievement.

Although the use of pure yeast cultures in brewing spread quickly and widely throughout the Continent and North America, its acceptance in England was slow; in fact, it is not universally accepted there even today. G.H. Morris, who worked with Horace T. Brown at Worthington's Brewery, tried repeatedly to brew English stock ales with pure yeast cultures but without success. Somewhat later in 1904–1905, N.H. Claussen, also from the Carlsberg Brewery, studied this problem and found that the secondary fermentation of English stock ales was due to a different yeast than the one which produced the primary fermentation. He isolated the yeast and, because of its peculiar importance to British brewing, named it *Brettanomyces.* Although Claussen and subsequently some others considered using mixtures of pure cultures, the idea has generally been regarded as impractical and has never been seriously developed.

Microbial Metabolism and Nutrition.—By the 1930s, the exploratory and descriptive phases of brewing microbiology were substantially completed, and subsequent developments took on a distinct biochemical flavor. The important role of brewers' yeast in the elucidation of glycolysis has been alluded to already.

Ehrlich, between 1906 and 1912, explained the formation of fusel alcohols during fermentation as by-products of the nitrogen metabolism of brewers' yeast. The process was described as the deamination and decarboxylation of amino acids to yield ammonia and an aldehyde; the ammonia was utilized by the cell, but the aldehyde was reduced to an alcohol and excreted. In the light of subsequent work, mainly that of SentheShanmuganathan (1958, 1960A,B), we now understand the reaction to be transamination with the formation of glutamate, rather than deamination with the formation of ammonia. Otherwise, Ehrlich's description remains unchanged. The importance of Ehrlich's reaction for yeast nutrition became apparent when R.S.W. Thorne, in the 1940s, learned that about 40% of the nitrogen requirements of fermenting yeast were obtained from the free amino acids in wort.

Work on the nutritional requirements of brewers' yeast had, of course, started much earlier. Wildiers in 1901 found that he could not grow yeast on a defined medium without adding small amounts of extract from malt or yeast. From observations of this kind, he developed the concept of "bios" by which he meant substances needed only in trace amounts for yeast growth. Further work indicated that "bios" was not simple but complex, and interest in it escalated with the realization that such growth factors were essential for human and animal nutrition as well as for microbial. Subsequently, Bios I was identified as meso-inositol, Bios IIA as pantothenic acid, and Bios IIB as biotin. On the basis of work published by C. Rainbow (1939), P.R. Burkholder (1943), A.S. Schultz, L. Atkin (1949), L. Atkin et al. (1949), and S.R. Green (1955) in the 1940s and 1950s, we now know that most brewers' yeasts require biotin, and some strains also require pantothenate and/or inositol. Few brewers' yeasts have an absolute requirement for any other vitamins although many are stimulated by thiamin and pyridoxine.

The intermediary metabolism and nutritional requirements of beer spoilage microorganisms, both bacteria and "wild" yeasts, have likewise been studied. In general, the lactic acid beer spoilers characteristically were found to have numerous nutritional requirements, including most amino acids, several components of the B vitamin complex, and one or more of the purine and pyrimidine bases. By comparison, the acetic group had few requirements. "Wild" yeasts, including those of the Saccharomyces genus, were found usually to metabolize dextrins and oligosaccharides larger than maltotriose and also to utilize lysine as the sole source of nitrogen. These characteristics distinguish spoilage yeasts from the culture yeasts, S. cerevisiae and S. uvarum. Until 1970 (Lodder 1970), S. uvarum was called S. carlsbergensis.

Yeast Genetics.—Lindegren (1949) credits Kruis and Satava with discovering the sexual cycle of yeasts in 1918. However, it was not until Ö. Winge in 1935 described the life cycle of Saccharomyces ellipsoideus in terms of haploid endospores and diploid vegetative cells that sustained effort was devoted to yeast genetics. Winge soon showed (Winge and Lausten 1939) that another yeast species, Saccharomycodes ludwigii, exhibited sexuality and Mendelian segregation. Between 1945 and 1960, Carl C. and Gertrude Lindegren made a number of fundamental contributions to this field.

Although a few hybrid yeasts have been produced and the genetic control of some yeast characteristics, like flocculence, has been demonstrated, not much practical application has resulted. A number of reasons for this situation have been proposed, including: (1) the conservative nature of the brewing industry; (2) the practical difficulties of yeast hybridization; and (3) the inability of anyone to specify an ideal brewing yeast. While each of these possibilities has some substance, it seems that the answer may lie elsewhere, viz., the brewing industry, having no need to change its materials or process, did not need what yeast genetics could provide.

That situation is changing rapidly. The brewing industry is moving away from some of its less efficient, traditional practices; and yeast genetics is in the midst of a renaissance. This combination may produce some very interesting and useful results.

Magnitude of the Brewing Industry

The magnitude of the brewing industry will be indicated in terms of the volume of malt beverages produced. Table 10.1 gives a breakdown by continents of world production for 1976, viz., 825.7 million hl. Europe outproduced everyone, including all of the Americas combined. In fact, European production was slightly more than one-half of world production.

TABLE 10.1. MALT BEVERAGE PRODUCTION BY CONTINENTS FOR 1976

Rank	Continent	Quantity in Millions hl	bbl[1]
1	Europe	433.1	369.0
2	North and South America	286.2	243.8
3	Asia	55.6	47.4
4	Africa	26.9	22.9
5	Australia/Oceania	23.9	20.4
	World	825.7	703.5

Source: Joh. Barth & Sohn, Nürnberg, W. Germany.
[1]U.S. beer barrel = 31.0 U.S. gal. = 117.335 liters.

Malt beverage production for 1976 is shown in Table 10.2. Here we see that the United States produced about twice as much as West Germany, which ranked second. However, if we consider the populations of these countries (see Table 10.3) and assume that most of the malt beverages produced in a country were also consumed there, we note that in terms of per capita *consumption*, West Germany ranked first while the United States ranked thirteenth. Ten of the top 14 countries in terms of per capita consumption were European.

There are a number of interesting differences between the brewing industries of the two top producing countries, viz., the United States and Germany. In the United States during 1974, there were 118 breweries; the 10 largest produced 75% of that country's output. During the same year in West Germany, there were 1636 breweries (1122 of them in Bavaria alone), almost all of which were small to medium enterprises. Also, the larger

TABLE 10.2.   MALT BEVERAGE PRODUCTION BY COUNTRY FOR 1976

Rank	Country	Quantity in Millions	
		hl	bbl[2]
1	United States[1]	192.183	163.739
2	West Germany	95.675	81.515
3	U.S.S.R.[3]	72.000	61.344
4	United Kingdom	64.029	54.553
5	Japan	36.393	31.007
6	France	23.869	20.336
7	Czechoslovakia	22.500	19.170
8	East Germany	21.000	17.892
9	Canada	20.227	17.233
10	Australia	19.717	16.799
11	Mexico	18.941	16.138
12	Brazil	18.000	15.336
13	Spain	17.127	14.592
14	Belgium	14.630	12.465
15	Netherlands	13.862	11.810

Source: Joh. Barth & Sohn, Nürnberg, W. Germany.
[1]Includes taxable and nontaxable quantities.
[2]U.S. beer barrel = 31.0 U.S. gal. = 117.335 liters.
[3]Estimated.

TABLE 10.3.   PER CAPITA CONSUMPTION OF MALT BEVERAGES BY COUNTRY FOR 1975

Rank	Country	Quantity	
		liters	U.S. gal.
1	West Germany	147.6	39.0
2	Czechoslovakia	142.3	37.6
3	Australia	141.9	37.5
4	Belgium	140.0	37.0
5	New Zealand	133.2	35.2
6	Luxembourg	129.1	34.1
7	East Germany	117.7	31.1
8	United Kingdom	117.7	31.1
9	Denmark	117.3	31.0
10	Austria	103.7	27.4
11	Canada	85.9	22.7
12	Ireland	84.4	22.3
13	United States	81.8	21.6
14	Netherlands	78.7	20.8

Source: Katz (1979).

brewers in the United States market their products on a national or regional basis; but in West Germany, most breweries sell on a local basis only. Yet another interesting comparison between the United States and West Germany is shown in Table 10.4. Here the 8 most popular beverages in both countries have been ranked in terms of per capita consumption per annum. The top 4 beverages in the United States were also the top 4 beverages in West Germany, but they ranked quite differently: while coffee ranked first in the United States, beer ranked first in West Germany; and while beer ranked fourth in the United States, soft drinks ranked fourth in West Germany. Milk ranked third in both countries.

Table 10.5 is a historical summary of beer production and the number of operating breweries in the United States. In 1975, there were fewer than

TABLE 10.4. PER CAPITA CONSUMPTION OF VARIOUS BEVERAGES IN THE UNITED STATES AND WEST GERMANY FOR 1973–1974

Beverage	Rank	United States liters	U.S. qt	Rank	West Germany liters	U.S. qt
Coffee	1	135.0	142.7	2	136.1	143.8
Soft drinks	2	115.1	121.6	4	76.0	80.3
Milk	3	96.0	101.5	3	93.0	98.3
Beer	4	76.0	80.3	1	147.1	155.4
Tea	5	27.0	28.5	5	29.0	30.7
Fruit juices	6	19.0	20.1	7	12.0	12.7
Distilled spirits	7	7.0	7.4	8	7.0	7.4
Wine	8	6.0	6.3	6	17.0	18.0

Source: Piendl (1977).

TABLE 10.5. HISTORICAL SUMMARY OF UNITED STATES BEER PRODUCTION

Year[1]	Quantity Produced in Millions hl	bbl[3]	No. of Breweries[2]
1940	64.4	54.9	611
1945	101.6	86.6	468
1950	104.1	88.8	407
1955	105.3	89.8	292
1960	110.8	94.5	229
1965	126.7	108.0	197
1970	158.0	134.7	154
1975	185.2	157.9	117

Source: U.S. Brew. Assoc. (1976).
[1]Fiscal year ended June 30.
[2]Operating for any part of the year.
[3]U.S. beer barrel = 31.0 U.S. gal. = 117.335 liters.

one-fifth as many breweries as in 1940, but they produced almost 3 times as much beer. Some other interesting changes have occurred during this same period. Today, only about one-third of all beer sold is consumed outside the home (i.e., in a tavern or restaurant); 30 years ago, two-thirds was consumed there. Today, about 40% of beer is sold in food stores; but then, none was. Consistent with these changes in consumer habits, less than one-fourth of the beer consumed today is sold on-draft; but in 1940, three-fourths of it was dispensed in this manner.

Types of Beer

There are two major types of beer which are distinguished mainly on the basis of fermentation differences. Bottom-fermented beers are named from the manner in which the yeast behaves during fermentation. As the fermentation subsides, bottom yeast *(Saccharomyces uvarum)* tends to flocculate and settle. Top-fermented beers are named analogously. During fermentation, top yeast *(S. cerevisiae)* rises to the surface where it is recovered by skimming. Some of the yeast during a top fermentation does settle out, but this portion is discarded.

Lager Beers.—Bottom-fermented beers are generally called lagers. This term derives from the German verb, lagern, to lay down or store. Historically, such beers were brewed to be stored for later use, like British stock ales. This significance no longer exists although modern lager beers are stored cold for a period of time as part of the maturation or conditioning process. There are two kinds of lager beer, light and dark. These terms have traditionally referred to the color and to the body of the product, and this is still the case. Recently, however, low-calorie lager beers, called diabetic beers in Germany, are coming to be called light beers in the United States.

In Europe, beers traditionally have been named after the town or city in which they were produced. Since the character of such beers was often unique, being determined by the barley, hops, water, and microflora endemic to the area, their names have come to be used in a generic sense. Pilsner and Dortmunder, for example, originally referred simply to the beer produced in Pilsen and Dortmund, but now refer to any beer with the characteristics peculiar to beers produced in those areas. Pilsner is a very pale-colored beer, fully attenuated (i.e., fermented as completely as possible), with a high level of hops. This beer is light bodied with a dry, crisp bitterness. Dortmunder is also pale in color and fully attenuated but with a medium level of hops. It too is light bodied with a dry but mellow bitterness. Münchener is a dark-colored beer, not fully attenuated, and with a relatively low level of hops—qualities characteristic of the beer produced in Munich. This beer is full bodied, much sweeter than Pilsner or Dortmunder, and quite aromatic. Bock beer is also a dark-colored beer which is rich and heavy bodied. The name may be a shortened form of Einbeck, a German town famous for this kind of beer. It is usually sold in the early spring and therefore is sometimes referred to as March beer.

Ales.—Most top-fermented beers are called ales. These, like lagers, are made in several varieties distinguished on the basis of color and body. Pale ale is pale in color with a very high level of hop bitterness and a dry palate. Mild ale has an intermediate level of hop bitterness and is sometimes darker in color and sweeter. The darker, sweeter mild ales are occasionally called brown ales. Stout is a very dark-colored ale, full bodied, with a relatively low level of hop bitterness. This kind of ale is sometimes called porter, but the term is becoming archaic. Dark lagers and ales derive their color from roasted malt or caramel or both.

A top-fermented beer, called Weissbier, is brewed in Germany, particularly in the Berlin area. This beer is prepared from malted wheat rather than barley and is light colored and light bodied. Weissbier is sold with yeast present in the bottle. Another top-fermented beer produced on the Continent is called Lambic beer which is peculiar to the Brussels area. The mash is prepared from grist containing 60% malted barley and 40% raw wheat. The fermentation is not seeded but spontaneous, and a significant part of the flavor derives from the microflora indigenous to the Brussels area. The microorganisms present during fermentation are *Brettanomyces* plus other yeasts, lactic acid bacteria, and other bacteria. The fermentation may take several years. The product is very aromatic and very acidic (a pH of 3.2–3.5). Lambic beer is often blended with lager beer.

Scope of Material Covered

This chapter will deal primarily with the microbiological aspects of lager brewing in North America, particularly in the United States. However, the salient differences between ale and lager brewing in Europe and North America will be described at appropriate points in the text. The microbiological aspects of beer making will be emphasized while the brewing process itself will be presented mainly as a frame of reference.

ELEMENTS OF THE BREWING PROCESS

Ingredients

The ingredients used to produce beer are malt, adjuncts, water, hops, yeast, and a few miscellaneous additives.

Malt.—Malt is prepared from barley which is sprouted under controlled conditions and then dried in a kiln. It is prepared from 2 species of barley, 2-rowed *(Hordeum distichum)* and 6-rowed *(Hordeum hexastichum)*. These names describe the arrangement of the barley corns on the spike. The 2-rowed variety is the principal type grown and used in Europe; and the 6-rowed variety is the principal type grown and used in North America. Regular malts are kilned enough to arrest growth but not enough to impart much color or to impair enzymatic activity. Caramelized or roasted malts, on the other hand, are kilned at higher temperatures for a longer time and always have a reduced enzymatic activity. Such malts impart a pronounced, characteristic flavor and color; they are rarely used in excess of 10% of the mash bill.

Malt provides carbohydrates and proteins and also the carbohydrases and proteases which break them down. The simple sugars and amino acids produced serve as nutrients for yeast growth and metabolism.

Adjuncts.—Brewing adjuncts are prepared from unmalted cereals—mainly corn (maize) or rice—and are relatively pure carbohydrates without significant amounts of lipids, enzymes, or other proteins. Brewers use adjuncts to impart desirable qualities to their products. For example, the 6-rowed barley malts used in North America would yield dark-colored, heavy-bodied beers if adjuncts were not used. In this case, adjuncts dilute the nitrogenous components of malt and enable American brewers to produce light-colored, light-bodied beers from this type of barley malt. Adjuncts are also frequently less expensive than malt, but this is not invariably the case.

Adjuncts are of 2 types: grits or flakes which are added during mashing; and syrups which are added during kettle boil. Grits and flakes are prepared by milling decorticated and degermed corn or rice. Syrups are prepared from cornstarch by acid hydrolysis, enzymatic degradation, or a combination of the two. Enzymatic methods now permit the preparation of brewers' syrups, i.e., high-maltose syrups with relatively low levels of nonfermentable carbohydrates.

Water.—Good quality water is a paramount requirement for brewing beer. Water quality is most usefully defined in terms of purity and mineral composition. Pure water is free of perceptible turbidity, color, odor, and taste. It is free of organic matter, including microbial contamination, and also free of iron, heavy metals, sulfides, and nitrites. Some of the more important aspects of mineral composition are summarized in terms of hardness.

Total hardness is a measure of the calcium and magnesium content expressed as calcium carbonate. The amount of hardness equivalent to the carbonate and bicarbonate alkalinity, expressed as calcium carbonate, is called carbonate hardness. It is also often referred to as temporary hardness because it can be eliminated by boiling. The amount of total hardness in excess of carbonate hardness is called permanent hardness. This type of hardness represents the amount of calcium and magnesium associated with anions other than carbonate and bicarbonate, principally sulfate and chloride ions. The hardness associated with brines is a pseudohardness due to their high sodium content which prevents the solution of sodium soaps by a common ion effect.

The types and concentrations of dissolved salts have a profound influence on such things as enzyme activity and stability during mashing, hop extraction, the precipitation of proteins and tannins, and the growth and metabolism of yeast. Therefore, the mineral composition of brewing water—particularly the proportion of temporary to permanent hardness—considerably influences the type as well as the quality of beer produced. The mineral characteristics of water associated with some well known types of beer are summarized in Table 10.6.

TABLE 10.6.   HARDNESS OF WATER FROM VARIOUS BREWING AREAS

Type of Hardness	mg/liter as $CaCO_3$				
	Pilsen	Dortmund	Burton-on-Trent	Munich	Milwaukee
Carbonate	15	450	235	275	133
Noncarbonate	10	295	695	5	4
Total	25	745	930	280	137

Historically, water was used without modification and therefore dictated, in part, the type of beer which could be most successfully produced. The water in Milwaukee, a famous American brewing center, resembles that in Munich. If unmodified, its characteristics suggest Münchener as the best type for the area. However, nowadays, any water supply can be readily modified to suit the requirements for producing any type of beer. The type of beer produced in any area is now dictated by consumer preference rather than by the water supply.

Hops.—The hop plant, *Humulus lupulus*, is a climbing, herbaceous perennial which is dioecious, that is, its staminate and pistillate flowers are borne on different plants. Only the pistillate flowers, which are cone-like in appearance, are used in brewing and are what is commonly referred to as hops. In most hop-growing areas in America and on the Continent, the stamen-bearing plants are systematically eradicated so that hops are seedless. This is not the case in Great Britain where hops may be 25% seeds by weight. For some varieties of hop plants, the seedless form is richer in essential oil and α-resin which contribute most of the brewing value of hops.

The essential oil from hops is a mixture of numerous compounds. More than half of them are hydrocarbons, mainly terpenes, such as myrcene, farnesene, humulene, and caryophyllene (Fig. 10.1). The remainder are oxygenated, i.e., esters, alcohols, and carbonyl compounds. Hops are only about 1% essential oil, but this portion produces virtually all of the characteristic aroma.

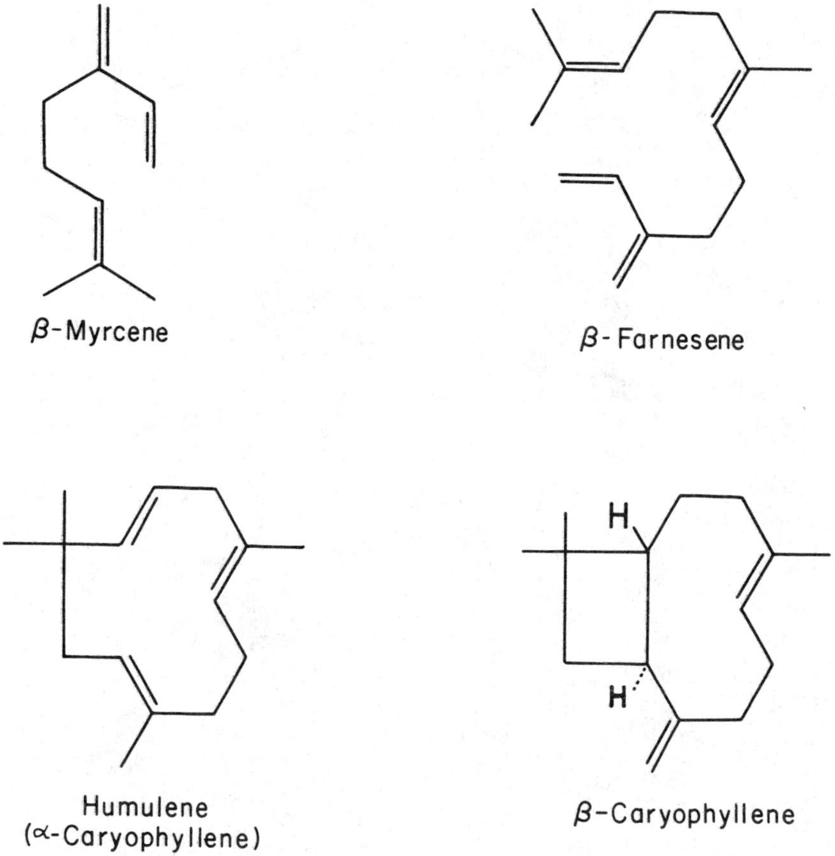

β-Myrcene

β-Farnesene

Humulene
(α-Caryophyllene)

β-Caryophyllene

FIG. 10.1.   SOME TERPENES FOUND IN THE ESSENTIAL OIL OF HOPS

Resins, which are about 15% by weight of hops, are responsible for the pleasant bitterness which hops impart to beer. The total resins can be fractionated on the basis of their solubility in hexane into two portions, hard resins (insoluble) and soft resins (soluble). Hard resins have no brewing value. Soft resins are further subdivided into α-resins and β-resins. Alpha-resins consist principally of 3 α-acids: humulone, cohumulone, and adhumulone. Most of the flavor impact of hops in beer is due to the α-acids. Similarly, the β-resins consist mainly of the β-acids, lupulone, colupulone, and adlupulone. The chemical formulas for the α- and β-acids are given in Table 10.7. The acidity of these compounds derives from the π-electron system associated with the triketo-methane configuration.

TABLE 10.7.   THE α- AND β-ACIDS OF THE SOFT RESINS OF HOPS

α-ACIDS   β-ACIDS

Side Chain (-R)		Name	pK_a	Name
$(CH_3)_2CHCH_2-$	isobutyl	Humulone	5.5	Lupulone
$(CH_3)_2CH-$	isopropyl	Cohumulone	4.7	Colupulone
$C_2H_5(CH_3)CH-$	sec-butyl	Adhumulone	5.7	Adlupulone

There are numerous varieties of hops grown in different parts of the world. A few examples are Hallertauer and Spalter in Germany, Fuggles and Northern Brewer in Great Britain, and Bullion and Cascade in the United States. The characteristic bitterness produced by different varieties is due mainly to different proportions of the various α- and β-acids. For example, lupulone is the predominant β-acid in Continental varieties of hops while colupulone is predominant in English and American varieties. While the α-acid, adhumulone, is found in most varieties at the fairly constant level of 10–15%, the proportion of humulone and cohumulone varies considerably. Whereas Hallertauer has 65–70% humulone and about 20% cohumulone, Bullion has 40–45% humulone and about 45% cohumulone.

Because of the reactive nature of many hop components, measures must be taken to avoid their oxidation during storage. The β-acids are particularly susceptible to oxidation, yielding hulupones, while the α-acids are gradually converted to hard resins. In general, these oxidation products have less desirable brewing qualities than the α- and β-acids. To minimize such deterioration, hops are dried quickly after harvesting to a moisture content of 10−12%, baled under considerable pressure to press the flowers closed and exclude air, and stored at cold temperatures.

Some brewers use whole hops. Others use hops ground to a powder and formed into pellets. Still others use hop extracts. Both hop pellets and hop extracts are usually packaged in hermetically sealed containers which exclude air to help avert oxidative changes. These forms are more economical to ship and to refrigerate and more convenient to use.

Culture Yeast.—Since spontaneous fermentation is rarely practiced in modern breweries anywhere, culture yeast is an essential ingredient. However, because this topic will be treated in much more detail than other ingredients, a discussion of it is deferred to a more appropriate place later in this chapter.

Miscellaneous Additives.—The use of minor ingredients varies widely from one brewer to another. Although such additives are usually used during conditioning or finishing, some are used earlier in the process. Some brewers use yeast foods to assure a vigorous fermentation. These additives provide such nutrilites for yeast as trace metals which may be present at suboptimal levels in wort. Chillproofing agents, like proteolytic enzymes or tannic acid, are rather generally used, especially in the United States where beer is served very cold. The proteolytic enzymes most commonly used for this purpose are papain, bromelin, and ficin. These agents are often added just prior to aging. The action of proteolytic enzymes is slow because of the low storage temperatures, and they are partially inactivated by heat denaturation during pasteurization. Reducing agents, such as potassium metabisulfite, generally referred to as KMS, are used at relatively low levels (50 mg per liter or less) and function primarily as oxygen scavengers. Clarifying agents, e.g., isinglas, Nylon-66, or polyvinylpyrrolidone, are used to facilitate removal of colloidally suspended material. Alginates are sometimes added as foam stabilizers or enhancers.

Priming sugars or syrups are generally added during conditioning or finishing. These additives are used in some ales to improve the balance between sweetness and bitterness. At lower levels, often below the taste threshold, priming sugars improve the body and mouth-feel of lager beers.

Process

Malting.—In the past, malting was usually done as part of the brewhouse operations. At present, breweries often obtain all or part of their malt from independent malting companies. While larger breweries may have their own malting facilities, these generally are not operated as part of the brewery.

When barley is first received for malting, it is cleaned of dust and foreign material and then stored. Storage is necessary because newly harvested barley germinates unevenly. The duration of storage depends upon the barley variety and such conditions as temperature and humidity at harvesting. Following storage, broken barley grains and any residual foreign material are removed. The barley is then seived into 3 sizes. The large- and medium-sized corns are malted separately while the small-sized ones are sold for animal feed. All of these preliminary steps are aimed at obtaining uniform germination which is essential for the preparation of good quality malt. Different barley varieties are always malted separately for the same reason.

Malting begins with steeping of the barley in water at 12°–15°C for 2 or 3 days. During this period, the steep water is changed 3 to 6 times, partly to remove the microbial flora and partly to replenish the supply of dissolved oxygen. The microflora initially present on the husks flourish in this environment and if not removed would impart undesirable characteristics to the malt. Also, their rapid growth reduces the dissolved oxygen available for the grain. Steep water is sometimes aerated in order to maintain an adequate concentration of dissolved oxygen.

After steeping, germination begins. The embryo is activated, and cytases from it dissolve the wall of the endosperm; other hydrolytic action on the endosperm moves progressively outward from the embryo. This process is called modification. A properly modified malt is readily friable and can be milled with little fragmentation of the husks.

Germination is arrested after 7 or 8 days by drying at or below 50°C. The germinating barley is dried from about 45% moisture to 5% or less. The dried malt is cured by heating at 80°C, or even higher for dark malts. The color and aroma of a malt are determined at this stage. The final step in malting is screening, a process which removes the rootlets (culms) which are undesirable in brewing.

Brewing.—Modern brewhouse operations consist of milling, mashing, kettle boil, and hop addition. The malt is ground to a meal or coarse flour, generally by a dry milling process although wet milling is also used. An advantage claimed for wet milling is that husks are fragmented less, with the result that run-off in the lauter tub is more rapid, and fewer tannins find their way into the wort and beer.

Mashing.—Mashing begins with doughing-in which is simply the mixing of ground malt with brewing water. This mash is then heated. Basically, there are two methods of heating mash, infusion and decoction; but there are numerous variations and combinations of the two. Infusion mashing consists of step-wise heating to several successively higher temperatures below the boiling point. Decoction mashing on the other hand is accomplished by withdrawing a portion of the mash, heating it to boiling in a mash kettle, and pumping it back to the mash tun, thereby raising the temperature of the entire mash. This sequence is usually repeated several times. Some form of the infusion method of mashing is generally used in preparing worts for top fermentations and is used extensively in Great Britain. A form of decoction mashing is often used to prepare worts for bottom fermentations; this is invariably the case in central Europe.

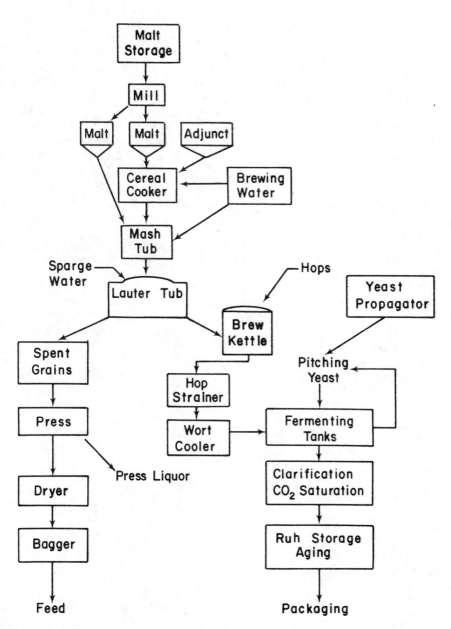

FIG. 10.2. SALIENT FEATURES OF THE BEER-MAKING PROCESS

In North America, especially in the United States, where adjuncts are used extensively, a double-mash system with some of the features of both the infusion and decoction methods is common. In this system, an all-malt mash is heated at 45°–60°C, the optimum temperature range for malt proteases. Simultaneously, in a separate vessel, the adjunct of corn or rice is heated at gradually increasing temperatures to gelatinize the starch and is finally brought to a boil. The adjunct is then added to the malt with the result that the temperature of the mixture is brought to 60°–65°C, which is optimum for malt amylases. At this temperature, the starch is liquefied mainly by the action of α-amylase on amylose and amylopectin. Liquefaction is followed by saccharification, a process due to the action of α- and β-amylases and which yields predominantly maltose with some monoses, oligosaccharides, and dextrins. The mixture is "mashed off" by heating to about 75°C for a brief period. The final steps are filtration and sparging to yield sweet wort which is pumped to the brew kettles.

Kettle Boil and Hop Addition.—Sweet wort is brought to a rolling boil in large brew kettles, usually called coppers in Europe, which refers to the metal from which they have traditionally been made. Stainless steel is also used today.

During kettle boil, a number of essential changes occur. The wort is sterilized. The enzymes which were active during mashing are denatured. Hops which are added during kettle boil are extracted, and the extracted resins undergo changes which improve their solubility. During kettle boil, a flocculant precipitate (or "hot break") forms which contains proteins, polyphenols (tannins), hop resins, and metal ions. Many of these substances, particularly the polyphenols and proteins, would impair the physical and sensory qualities of the beer if not removed. The color of the wort tends to become darker during kettle boil because of the caramelization of sugars, the formation of melanoidins, and the oxidation of tannins from hops. The degree of color change, however, can be controlled. If a very dark beer is desired, caramel may be added at this point in the process. Also, the reducing power of wort increases during boiling, partly due to the formation of reductones from carbohydrates and partly due to the fact that dissolved oxygen is driven off. Depending upon the duration of kettle boil which is generally 1 hr or more, concentration of the wort may be an important and desired effect.

The dissolved matter in wort is called extract, and its concentration is usually expressed in degrees Plato. The Plato scale is a slightly corrected version of the older Balling scale; both scales are related to specific gravity. A 12° Plato wort has the same specific gravity as a 12% (w/w) sucrose solution.

One of the more important changes which occurs during kettle boil is isomerization of the humulones to isohumulones and allo-isohumulones, as shown in Fig. 10.3. To some extent, the isohumulones are hydrolyzed to the corresponding humulinic acids which have no bitter taste. It is now known that the lupulones, like the humulones, can be isomerized. However, lupu-

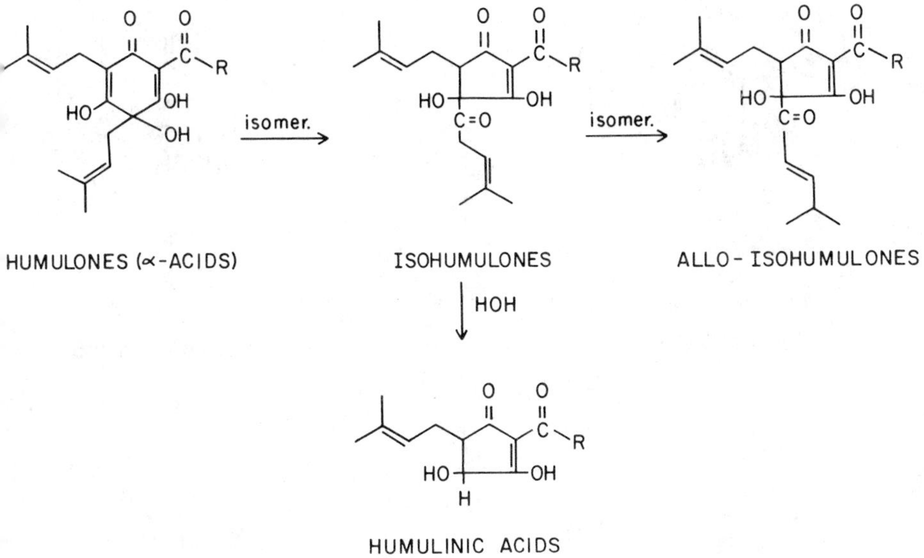

FIG. 10.3.   ISOMERIZATION AND HYDROLYSIS OF HUMULONES
-R has the same significance as in Table 10.7.

lones are much less important both quantitatively and qualitatively. Hop extracts are sometimes isomerized to preform the isohumulones; such preparations are generally not used as kettle additives but rather as post-kettle additives during conditioning or finishing.

Upon exposure to sunlight, beer tends to become "sunstruck" or "skunky." The unpleasant aroma which develops is thought to be due to the formation of isopentenyl mercaptan by the postulated photochemical reaction shown in Fig. 10.4. This undesirable sequence of events can be substantially eliminated by reducing the isohumulones with sodium borohydride, a reaction also shown in Fig. 10.4. The ρ-isohumulones thus produced do not enter into the photochemical reaction. However, they are reported (Hough et al. 1971) to be less bitter than the unreduced isohumulones.

At the completion of kettle boil, the hot wort is pumped through a strainer or hop back if whole hops were used. This device removes the hop plant residue and with it some of the "hot break." However, if hop pellets or extracts were used, the hot wort is pumped directly to hot wort tanks where trub settles out, and the wort begins to cool. Microbial contamination is not a problem during this part of the process because the temperature is $60°-70°C$ or higher. Below $60°C$, another precipitate called "cold break" forms and tends to settle out, becoming part of the trub. Besides removal by straining or sedimentation, trub is sometimes removed by filtration or centrifugation. The amount of trub produced is about $30-60$ g/hl. Trub, when first precipitated, has a large surface area and adsorbs iron, copper, zinc, and other heavy metals.

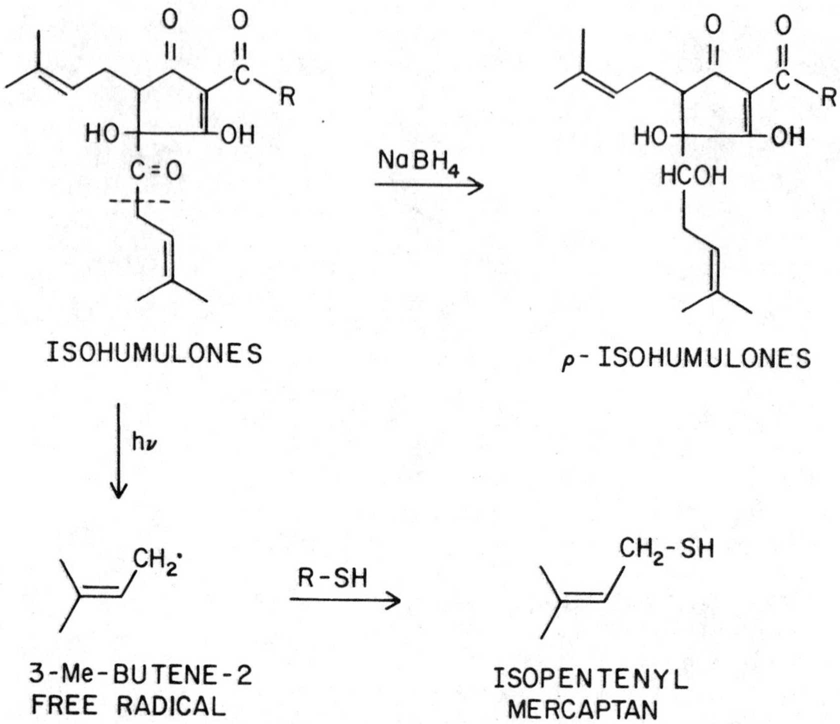

FIG. 10.4.   BOROHYDRIDE REDUCTION AND HYPOTHETICAL PHOTOREACTION OF ISOHUMULONES
-R has the same significance as in Table 10.7.

Before inoculating or pitching with yeast, the clarified wort is cooled rapidly to $10°-15°C$ to minimize microbial contamination. The potentially most dangerous range is $20°-40°C$. Rapid cooling is usually effected in an enclosed plate heat exchanger, while at the same time the wort is saturated with oxygen $(8-10$ mg/liter) by exposure to sterile air.

Fermentation.—The clarified and aerated wort is inoculated or pitched with yeast to give an initial concentration of about 10 million cells per ml. The yeast is a suitable strain which has been grown in a pure culture propagator or recovered from an earlier fermentation.

Primary Fermentation.—Twenty-four to 48 hr after pitching, clumps of foam called kräusen appear on the surface due to evolution of carbon dioxide; they gradually increase in size until the entire surface is covered to a depth of 1 m (1.1 yd). This period of vigorous fermentation is called high kräusen and corresponds to the logarithmic growth phase of the yeast. Yeast growth during a lager fermentation amounts to 4 or 5 times the inoculum, while growth during an ale fermentation will be about double that amount. As the fermentation proceeds, the specific gravity decreases due to utilization by yeast of carbohydrates, amino acids, etc., and also due to the production of ethanol and other compounds with densities less than that of water. This process is referred to as attenuation.

The carbon dioxide evolved during fermentation is recovered, purified, and stored as a liquid for adjusting the level of carbonation during finishing. Since most brewery fermentations are carried out nowadays in closed vessels, the recovery of carbon dioxide is easily and efficiently accomplished. Older breweries may have some open fermentors, but these are disappearing not only because carbon dioxide is less efficiently recovered from them, but also because they are more difficult to clean and maintain. Such vessels, usually made of wood or concrete and lined with some inert material, are being replaced by stainless steel vessels.

Much heat is liberated during fermentation. It is necessary, therefore, to have cooling coils in fermentors to control the temperature. At the end of primary fermentation, the fermentor is cooled to about 0°C. During this period, most of the yeast in suspension settles out. At this time, the young beer, or "green" beer as it is often referred to, is removed from the settled yeast and pumped to a storage tank at $0°-2°C$—a process called fassing, from the German word *Fass*, meaning barrel or cask. Fassing is entirely analogous to the process of racking in wine making.

Fassing initiates the next step in beer making which is variously referred to as lagering, conditioning, ruh storage, or aging. However, before discussing this phase of the process, we should note that while brewing is most commonly a batch process, it may be a continuous process.

Continuous Fermentation.—The idea of continuous fermentation is not very new. The continuous fermentation of beer was suggested as early as 1892 (Delbrück), and since the beginning of this century there has been a small but steady stream of patents describing processes and equipment for reducing this concept to practice. The underlying principles as well as the classification of continuous fermentation systems have been described (Aiba *et al.* 1973; Hough *et al.* 1971).

It will suffice for our present purpose simply to enumerate the salient advantages and disadvantages of continuous fermentation relative to conventional batch fermentation. Continuous fermentors and ancillary equipment are much more efficient and much smaller in size. Consequently, the equipment costs per barrel are lower. Also, the manpower requirements per barrel are less because the process is automatically controlled, and less cleaning is required. Continuous runs in excess of 1 year have been made. On the other hand, the equipment required for continuous fermentation is relatively complex and a high level of technical skill is required to operate it. Although the fermentation time per barrel is greatly reduced by continuous fermentation, the process is relatively inflexible in the sense that changes in production rate are not readily made, and changes from one type of beer to another are difficult. Moreover, while the character of continuously fermented beer may be quite acceptable, it is usually different in character from batch fermented beer. Hence, a brewer would find it difficult to produce the same beer in two different plants, if one were equipped for continuous fermentation and the other for batch fermentation. Finally, the conditions in continuous fermentors are often conducive to bacterial development, particularly the lactic acid bacteria. Preventive as well as remedial measures can be quite expensive.

Aging and Finishing.—Aging may or may not involve a secondary fermentation. If it does not, the beer is fully attenuated during primary fermentation, and the finishing process includes clarification, carbonation, and the use of various additives.

Dry hops, usually in the form of pellets, are added to ales at the beginning of conditioning in order to enhance the hop bouquet of the finished product. Purified hop extracts are sometimes added to lager beers at about the same point in the process to adjust the level of hop bitterness and thus obtain a more uniform product. In both cases, the hop character of the final product tends to be improved because desirable hop components are retained with this method of addition which are lost in kettle addition due to volatilization or to adsorption. Other additives, like caramel as a colorant, KMS as an antioxidant, and proteolytic enzymes or tannic acid as chillproof agents, may also be used.

The purpose of clarification is not only to remove residual yeast and other relatively large particulate matter but also to remove the larger molecular aggregates, such as the protein-tannin complexes, which render the beer colloidally unstable at low temperatures (e.g., 5°C). Consequently, beer is aged near 0°C to foster the precipitation of such colloidal complexes. Also, material of this kind may be adsorbed on finings, such as isinglass, nylon, or polyvinylpyrrolidone, or may be solubilized by proteolytic enzymes. Suspended material may likewise be removed by centrifugation and filtration. The taste of beer becomes more mellow with the removal of tannins, which tend to be harsh and astringent. The two principal effects of aging and finishing are clarity and improved flavor.

Turbidity during aging sometimes arises from microbial contaminants. Most microorganisms can be quite effectively removed by filtration prior to packaging. Residual microorganisms plus those which may enter during the packaging operation are usually controlled by pasteurization.

When a secondary fermentation is desired as a part of the aging process, beer is fassed before it is fully attenuated. Secondary fermentation during aging is simply an extension of primary fermentation—but at a much reduced rate due to the lower temperature and the lesser concentrations of both yeast and fermentable carbohydrates. In order to revive the yeast and accelerate the secondary fermentation, priming sugars or syrups may be added. Such is often the practice in brewing ales. A similar result is achieved in brewing lagers by adding some fermenting beer from a primary fermentation at high kräusen. Kräusening, as this technique is called, may be an integral part of the process, or it may be used only as a corrective measure when the beer has been fully attenuated prior to fassing. Secondary fermentation is carried out in closed tanks so that the beer becomes fully carbonated during the process. Other aspects are much the same as described previously for aging without secondary fermentation.

After final filtration, the beer is packaged—mostly in cans and bottles in North America. In Europe, the major portion is packaged in barrels. In some areas, beer is also distributed in bulk, e.g., in tank cars or trucks. This practice is perhaps most prevalent in Great Britain.

PURE CULTURE YEAST

Quality

The quality of a brewing yeast is judged ultimately by the quality of the beer it produces. Clearly then, the quality of brewing yeast depends not only upon its intrinsic characteristics, but also upon the environment provided by the brewer—particularly the type of wort and other fermentation conditions. Consequently, the number of parameters potentially useful for selecting brewing yeast is very large, and their relationships are complex. In spite of these difficulties, progress has been made in reducing and systemizing the criteria for typing brewing strains.

Thorne (1972) has developed a scheme by which brewing yeasts can be adequately and usefully described in terms of the following characteristics: rate and extent of growth, rate and extent of fermentation, flocculence, and flavor and aroma of the beer produced. Obviously, uniform and predictable behavior are also desirable qualities. Adequate growth at an adequate rate is needed for a normal fermentation, for acceptable flavor, for an adequate supply of pitching yeast, and for minimizing microbial contamination. The growth rate is closely related to the fermentation rate which in turn is important for scheduling production in a modern brewery. Flocculence, the clumping or agglomeration of yeast cells during the latter part of primary fermentation, often plays an important role in separating yeast from beer. In some breweries, where this separation is effected with centrifuges, this phenomenon may be unimportant or even undesirable. The flavor and aroma of beers can be characterized adequately enough to discriminate among yeast strains in terms of the following sensory characteristics: aromatic and estery, cleanness and freshness, sulfury off-flavor, and diacetyl off-flavor. Finally, uniform and predictable behavior are largely a consequence of genetic purity and stability.

It is interesting to note that these yeast characteristics, which are so useful for typing brewing strains, are useless to the taxonomist because they are affected too much by the environment. A taxonomist looks for intrinsic differences which tend to persist in spite of ordinary environmental changes. On the other hand, stable characteristics which are taxonomically useful are often unimportant to the brewer. For example, the capacity to ferment melibiose distinguishes *S. uvarum* from *S. cerevisiae*, but melibiose is an unimportant part of wort. This is not to imply that taxonomy is unimportant to the brewer—quite the contrary. The brewer, however, is interested in other more subtle differences as well.

Sources of Pure Culture

Most larger breweries isolate, select, and maintain pure cultures of yeast strains for their own use. Others obtain this service from commercial laboratories. Such laboratories often maintain a large collection of brewing cultures which can be useful to any brewer seeking a new strain of yeast.

A new pure culture can be started readily by subculturing single cell isolates from recognized brewing strains. When a brewer needs to replace his yeast culture, the best and most readily available source is his own yeast propagator system. By making single cell isolates from this source, he eliminates contaminants such as bacteria, wild yeast, and undesirable mutants. Strains with desirable properties are invariably present. Other sources from which isolates may be made are cultures from another brewery, pure cultures from a type culture collection, or pure cultures from a commercial laboratory.

Besides isolating and selecting from existing strains, a brewer can, at least in principle, develop new strains by induced mutation, hybridization, transformation, or transduction. Induced mutation is severely limited as a method for developing useful brewing strains. Induced mutants are the result of what tends to be a deletion process which has rarely produced anything that a brewer could regard as an improvement. Moreover, even when a desirable property appears, it is often accompanied by other undesirable changes. Although potentially more effective than mutation for producing new brewing strains, hybridization has not been extensively used. One reason is that brewing yeasts, especially $S.$ $uvarum$, do not sporulate readily and when they do, many of the spores are not viable. These and other genetic complications are due to the fact that brewing yeasts are frequently polyploid or aneuploid. Nonetheless, a number of hybrids of brewing yeasts have been reported; in fact, one of them has been patented (Windisch et al. 1967). Although transformation was first described in 1944 and transduction in 1952, little has been reported on the genetic transformation of yeasts which is of interest to the brewer and, at this writing, no report on the transduction of yeasts is known to the author. Of what value these techniques may ultimately be to the brewing-yeast geneticist remains to be seen.

Maintenance of Pure Culture

Stock Cultures.—After isolation and selection are completed, a stock culture or master culture is prepared in order to preserve the new strain unchanged. The media for this culture may be liquid or solid, or the culture may be preserved by lyophilization.

One of the oldest media for this purpose is a solution of 10 g of sucrose or lactose in 100 ml of water. The level of inoculum used is very low so that the culture appears to be clear. The yeast remains dormant, with growth so minimal that the mutation rate is virtually nil. Such cultures are revived in wort about every 2 years before reculturing in sucrose solution. Another liquid medium, more common today, is a broth of malt extract, yeast extract, glucose, and peptone, abbreviated MYGP (Wickerham 1951). Cultures in this medium are initially incubated without shaking for 2−4 days at a temperature near 28°C, then stored. Subculturing is repeated every 2−3 months. In general, liquid media are stored at 2°−4°C.

Stock cultures are more commonly preserved on solid media. Both MYGP Agar and Universal Beer Agar, UBA (Kozulis and Page 1968), are used in North America. Agar slants or slopes are inoculated, incubated at about 28°C for 2−4 days, then stored at about 4°C. Subculturing is usually repeated every 6 months. In a frequently used modification of this method, sterile paraffin oil is layered over the culture after the initial incubation, prior to storage. Screw-capped culture tubes are generally used. Subculturing is repeated annually. Some advantages adduced for this modification are (1) the loss of water from the medium is greatly reduced and with it the increased salt concentration and osmotic pressure which contribute much to the demise of older cultures, (2) there is a slow diffusion of oxygen through the oil so that microaerophilic organisms can be preserved in this way, and (3) partial transfers of a culture can be made without loss of the remaining part.

Since Wickerham and Andreasen (1942) proposed lyophilization for the long-term preservation of yeast cultures, its use has gradually increased. The method is attractive not only because it significantly reduces the likelihood of contamination, but also because it saves time and money. Although many of the major culture collections now maintain most of their yeasts in the lyophilized state, the brewing industry has been reluctant to accept the method. This situation is understandable since, as already noted, brewers are concerned with more subtle and variable yeast characteristics than taxonomists are. Moreover, some early reports (Atkin *et al.* 1962; Wynants 1962) indicated that mutational changes occurred during storage in the lyophilized state, contrary to claims made for the method. Other more recent reports, however, have tended to confirm Wickerham's claims (Kirsop 1974B; Richards 1975; Barney and Helbert 1976). Another deterrent to acceptance of this method is the high dead-cell counts observed—as high as 90% or more. Recent findings (Hall and Webb 1975) indicate that dead-cell counts can be significantly reduced by properly selecting the cryoprotective agent and the method of resuscitation. Moreover, even in cases where the dead-cell count was high, the residual viable cells were still representative of the original culture (Barney and Helbert 1976). Besides these considerations, the successful use of this method depends upon (1) rapid freezing at an optimum temperature (−30°C has been reported for *S. uvarum*), (2) complete dehydration, and (3) maintenance of a high vacuum or inert atmosphere during preparation and storage. Recently, a few breweries have begun to use this method of preservation to supplement more traditional methods.

Pure Culture Propagation.—Starting with a few tenths of a gram of pure culture, it is necessary to propagate 70−100 kg or more of pure yeast in order to pitch a commercial fermentor. This increase in yeast quantity of 100,000-fold or greater is achieved in a step-wise manner. About a 10-fold increase is achieved at each step. The first 2 or 3 steps are made in the laboratory, with the remaining steps made in a small yeast plant ancillary to the brewery, generally referred to as a pure culture propagator system.

Laboratory.—Stock culture is aseptically transferred to 100 ml of sterile wort in the laboratory. When the fermentation reaches high kräusen in 2–4 days, it is transferred *in toto* to about 1 liter of sterile wort. In turn, when this fermentation reaches high kräusen, it is used either to inoculate another larger laboratory fermentation or to inoculate the first plant propagator, depending upon the size of that vessel.

Another technique is to start the fermentation with a small volume of wort in a relatively large vessel, and at high kräusen to double the volume

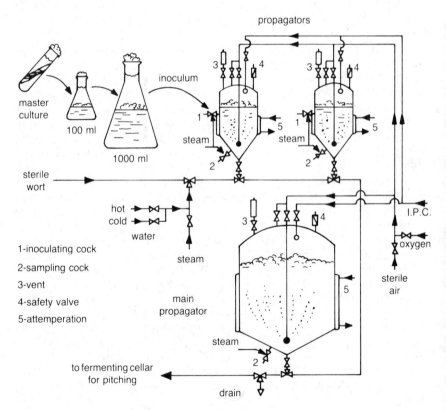

From Piesley and Lom (1977)

FIG. 10.5.   PURE YEAST CULTURE PROPAGATOR SYSTEM

by adding fresh wort. This topping-off technique is also used in the brewery where it often effects economies of time, space, and energy.

Brewery.—In North America, propagation of pure yeast in the brewery is generally modeled on the concepts of C. Hansen; the equipment used is functionally similar to that devised in the 1880s by Hansen and Kühle. This process is fermentative in nature so that yeast is produced under conditions similar to those in the brewery. The process differs from a brewery fermentation in that it is operated aseptically and aerated with sterile air. Although aeration is used, the yeast metabolism is not respiratory. It remains fermentative due to the relatively high concentration of carbohydrates present throughout the process (Crabtree effect). When the fermentation is complete and the yeast has settled, the supernatant beer is decanted. The residual yeast is roused or resuspended by agitation and pumped by sterile air pressure to the next larger propagator where it serves as inoculum. A residuum of yeast is left in the first propagator as inoculum for the next cycle. Consequently, this vessel is inoculated only occasionally with fresh culture from the laboratory, e.g., if microbial contamination occurs due to mechanical failure or human error, or when mutational changes accumulate to an undesirable level.

A number of variations on this process are practiced. Some brewers do not retain a residuum of yeast in the first propagator to reinitiate the cycle, but rather start each cycle with pure culture from the laboratory. In this way, mutational changes are minimized, and a more uniform yeast is obtained. Some brewers aerate intermittently rather than continuously in order to reduce foam and permit the use of smaller vessels. While the original Hansen system was operated several degrees above brewery fermentations, many brewers today, particularly those who ferment at elevated temperatures, operate their propagation systems at the same temperature as the brewery. Numerous other more minor variations exist.

A fundamentally different propagator system was devised by Curtis and Clark (1957, 1960). This system operates on the same principle as a bakers' yeast plant. The carbohydrate source is maintained at a low level so that the Crabtree effect is eliminated and the yeast metabolizes via the respiratory pathway. The much greater energy efficiency of this pathway significantly enhances yeast growth so that yeast crops greater than 100-fold can be achieved at each propagation step. Although this system is sound in both principle and practice, it appears to be little used outside Great Britain. This may be because the system was initially developed for ale yeasts, and no one has yet evaluated it for lager yeasts.

Evaluation of Yeast Produced.—The quality of yeast produced by a propagator system must be evaluated at regular intervals. This evaluation is accomplished by the same battery of tests used initially to select a pure yeast strain. Since laboratory propagations can so readily be done aseptically, the testing may be less extensive if done at this stage. Some of the tests commonly used to detect microbial infection and undesirable levels of mutants are discussed in the following sections.

Microscopic Examination.—The cellular morphology of yeast varies with the composition of wort or media, with the time and temperature of fermentation or incubation, and with other environmental parameters. Nevertheless, if test conditions are carefully standardized, cellular morphology can be very useful as a presumptive indication of cellular change. A skilled observer can readily detect differences in granulation and vacuolation, significant changes in cell size and shape, and differences in the degree and type of budding. Although such observations are largely subjective, this kind of information can often alert a careful and skillful microbiologist to incipient problems.

Cycloheximide in UBA.—Cycloheximide (actidione) is added to Universal Beer Agar at a concentration of $2-4$ μg/ml. This medium enables one to detect bacteria and some wild yeasts in the presence of culture yeast. *Saccharomyces* spp., particularly *S. uvarum*, are sensitive to the antibiotic cycloheximide and fail to grow at the concentration specified. Beer-spoilage bacteria and wild yeasts from genera other than *Saccharomyces* grow well (Harris and Watson 1968; Green 1955).

Because cycloheximide is temperature sensitive, the medium is best prepared by aseptically adding the filter-sterilized antibiotic to autoclaved UBA. Inoculated plates may be incubated aerobically at 28°C for 3 days and also anaerobically in a CO_2 atmosphere at 28°C for 7 days. The aerobic conditions facilitate detection of obligate aerobes (seldom present) and facultative organisms such as wild yeasts from genera other than *Saccharomyces*. The anaerobic conditions facilitate detection of obligate anaerobes, such as *Zymomonas anaerobia*, and microaerophilic organisms such as the lactic acid group.

Lysine, Crystal Violet, Fuchsin-sulfite Media.—These 3 media are used to detect wild yeasts in the presence of culture yeast. The basis for differential growth in each case is as follows. Brewing strains of *S. cerevisiae* and *S. uvarum* cannot utilize lysine as the sole source of nitrogen in an otherwise complete growth medium; wild yeasts can, except those of the genus *Saccharomyces* (Walters and Thiselton 1953). The growth of brewers' culture yeasts is inhibited by crystal violet at a concentration of 18 μg/ml in wort agar; but the growth of wild yeasts, including those of the genus *Saccharomyces*, is not (Kato 1967). Most brewing strains of *S. cerevisiae* and *S. uvarum* are inhibited by 0.35% fuchsin-sulfite mixture in a complete growth medium; wild yeasts are not (Brenner *et al*. 1970). Because brewers' yeasts may differ in their response to these media, it is imperative that each brewer verify the concentrations effective with his own strain. It may be necessary to modify the medium.

Serological Tests.—During the past $10-15$ years, serological methods have been applied to the problem of detecting relatively low levels of wild yeasts in culture yeast (Campbell 1972; Richards 1969; Chilver *et al*. 1978).

Antibodies to wild yeasts are developed in rabbits and the antiserum is cross-adsorbed with culture yeast in order to remove antibodies to common antigens. A specimen of culture yeast treated with the cross-adsorbed antiserum will not bind the antibodies present, but wild yeasts will. Wild yeast-antibody complexes are detected by a second antibody, usually derived from sheep or goats and tagged with a fluorochrome, e.g., fluorescein isothiocyanate, or with radioactive iodine. The second antibody agglutinates with the first, thereby tagging the wild yeast with either a fluorochrome or radioactive iodine. The first is an immunofluorescent assay (IFA); the second is a radioimmunoassay (RIA). Both methods are very sensitive. An IFA is reported capable of detecting 10 wild yeasts per 1 million culture yeasts in 3 hr (Chilver *et al.* 1978).

In spite of their sensitivity, speed, etc., serological methods are not routinely used in the brewing industry, except perhaps in England. This situation probably derives from the fact that *S. cerevisiae* is more amenable to this technique than *S. uvarum.*

Triphenyltetrazolium Chloride.—The presence of respiratory deficient mutants such as petites can be detected by layering a film of triphenyltetrazolium agar over colonies grown on a nutrient medium. Colonies of normal cells can reduce the tetrazolium ion to a formazan which is red (Fig. 10.6). Colonies of petites are unable to effect this chemical change and therefore remain unchanged in color (Ogur *et al.* 1957).

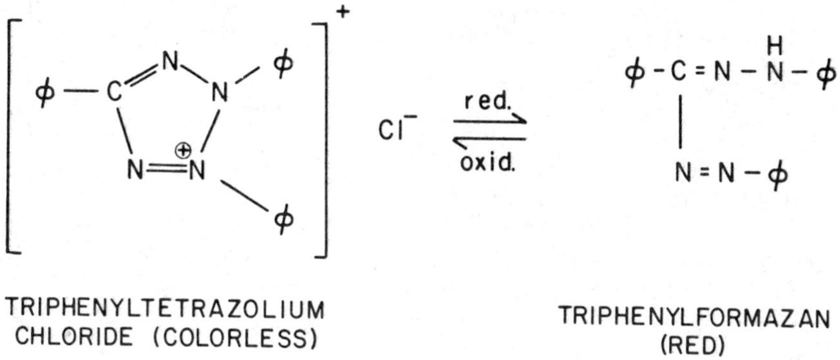

TRIPHENYLTETRAZOLIUM
CHLORIDE (COLORLESS)

TRIPHENYLFORMAZAN
(RED)

FIG. 10.6.   CHEMICAL BASIS FOR THE TTC TEST

Wallerstein Laboratories Nutrient (WLN) Agar.—Another less well-defined class of mutants are called verdants from the coloration of their colonies when grown on WLN agar. Verdants represent a diverse class of mutants. Some of them, like petites, are respiratory deficient, i.e., they lack some cytochromes in the electron transport chain and, therefore, are incapable of respiratory metabolism. Others are like normal cells and have functional electron transport systems (Kot and Helbert 1977).

Verdants, unlike petites, do not always grow more slowly than normal cells under brewery conditions; they may grow as fast and occasionally even faster than culture yeast and, thus, are potentially more dangerous to the brewer than petites. Some verdants produce beers with a high level of diacetyl (Czarnecki and Van Engel 1959). Whether these cells produce unusually large quantities of alpha-hydroxyketo acids (precursors of the vicinal diketones) or lack the metabolic capacity to reassimilate them is unknown.

Verdants may represent a mutational state intermediate between normal cells and petites, but this too is unknown. In any case, verdants, like petites, are undesirable in brewing yeasts because, if present in sufficiently high concentrations, they have always yielded beers with unsatisfactory flavor.

Test Fermentations.—Laboratory test fermentations in glass cylinders are pitched with propagator yeast and the following parameters measured: initial and final pH, initial and final degrees Plato, and flocculence [estimated by a test such as Helm's modification of the Burns' test (Helm *et al.* 1953)]. These data are then compared with those obtained earlier in order to verify that the yeast crop is comparable to the original strain.

Recycled Yeast.—Most of the yeast crop from a well-run brewery is usable for pitching subsequent fermentations. Consequently, it is general practice to use such yeast for this purpose, thereby reducing significantly the magnitude and expense of the pure culture propagator system. The main steps in pitching with plant yeast, which of course is not pure culture, are enumerated below with emphasis on the microbiological aspects.

Separation and Collection.—In the conventional lager process, the yeast separates spontaneously by flocculation and is collected for pitching after the beer has been decanted. Methods of handling yeast crops to be used for pitching vary considerably from brewery to brewery. In some breweries, only yeast free of microbial contamination is used. In others, microbial contaminants are removed by washing or other treatments prior to pitching; only contaminated yeast is washed. In still others, all yeast recycled for pitching is washed without prior testing.

In open fermentors, a preliminary selection of yeast is often made during the collection process by using only the middle layer of the settled yeast. This layer is selected because it tends to contain a minimal amount of dead cells, trub, and other debris. In the more modern closed tanks, both horizontal and vertical, such a selection process is not practical. In these cases, all of the settled yeast is collected, and undesirable elements, such as trub, are removed by vibrating mechanical screens. Dead cells, bacteria, and some wild yeasts may be removed by a washing operation.

Many modern breweries do not depend upon flocculation, but separate yeast by centrifugation. In such cases, the fermentors are invariably of the closed type, and yeast is harvested for pitching as just described.

Storage.—Pitching yeast slurries should, in general, be stored for as brief a time as possible at a temperature of $0°-2°C$. Lager yeast is generally used within a matter of hours after harvesting. Ale yeast, because it is collected much earlier in the fermentation, must be stored somewhat longer; however, the slow fermentation of the residual carbohydrates helps maintain the yeast in good condition. In fact, when pitching yeast slurries must be stored for longer periods of time, one of the best ways to do so is to inoculate regular wort with them, ferment slowly at low temperatures (2°–4°C), and store under the beer. In this way, pitching yeast has been preserved in good condition for 5–6 weeks. An even better way is to store the yeast as pressed cake for several weeks at 0°–2°C in sealed containers.

Microbial Contaminants.—The microorganisms which may contaminate plant yeast are obviously those which can grow successfully in fermenting wort. They may be bacteria or wild yeasts. The bacteria most commonly encountered are the lactic acid group (*Lactobacillus* spp. and *Pediococcus* spp.), *Hafnia* spp., and *Zymomonas anaerobia*. The most common wild yeasts are *Saccharomyces* spp. These organisms will be discussed in more detail later in this chapter. Methods for detecting and enumerating them have been discussed already.

Yeast Washing.—If plant yeast intended for pitching is infected with bacteria, the situation can often be remedied by one of several procedures. In the first of these, the contaminated yeast is suspended in sterile water and stirred thoroughly, after which the yeast settles out. The supernatant water containing bacteria, some wild yeasts, degenerate and dead culture yeast cells, etc., is decanted. This process may be repeated several times. Obviously, the process can only be used effectively with flocculent culture yeast.

In a second method, the contaminated culture yeast is steeped in a dilute solution of an acid such as sulfuric, phosphoric, or tartaric. The use of tartaric acid for this purpose was introduced by Pasteur. To effectively eliminate the lactic and acetic acid groups of bacteria by this treatment, the pH must be about 2.4, or even as low as 2.2. The recommended exposure time is 2–4 hr. At the acid concentrations represented by pH's of 2.2–2.4, yeast cells are also adversely affected to some degree. Consequently, if excessively high kills of culture yeast are to be avoided, the recommended exposure time and storage temperature (5°–10°C) must be carefully observed. Also, in order to avoid inordinately high local concentrations of hydrogen ions and heat, strong acids such as sulfuric and phosphoric must be prediluted and added slowly with continuous stirring. Although often called acid washing, this process is more accurately referred to as acid treatment since the entire mixture of yeast and acid is used to pitch a fermentor.

A modification of the foregoing procedure involves the use of sodium or ammonium persulfate at a concentration of 0.75% of the yeast slurry (Bruch *et al.* 1964; Brenner 1965). It is claimed that this modified acid treatment will effectively eliminate bacterial contaminants at pH's which are less detrimental to yeast. However, there is some question about its effectiveness against lactic acid bacteria.

Although washing with sterile water may be useful for removing some wild yeasts which are powdery rather than flocculent, acid treatment will be of little value in eliminating this kind of contamination. At higher levels of wild yeast contamination, there is no effective remedy except to discard the yeast and sanitize the equipment involved.

FERMENTATION

Fermentation is central to the beer-making process; its principal ingredients are wort and yeast.

Wort

Wort is the raw material from which beer is made by the metabolic action of brewers' yeast; it is important not only as a nutritional substrate for yeast, but also as the underlying basis for the sensory properties of beer. Since wort is derived mainly from agricultural products, viz., barley malt and hops, its chemical composition is complex. The class of compounds present in greatest abundance is carbohydrates. There are many less abundant compounds, viz., nitrogenous compounds, inorganic ions, vitamins, polyphenols, hop compounds, lipids, and oxygen. In spite of their relatively low concentrations, many of these have a profound influence on yeast metabolism and growth—and, therefore, also have a profound influence on the characteristics and quality of the beer produced.

Carbohydrates.—It is useful to classify wort carbohydrates as fermentable or nonfermentable on the basis of whether or not they can be metabolized by brewers' yeast. The more important fermentable carbohydrates are maltose, maltotriose, glucose, fructose, and sucrose (Table 10.8). Fructose and sucrose come directly from malt; glucose, maltose, and maltotriose are formed during the mashing operation and may also be added as components of liquid adjunct (syrup). Maltose is the most abundant, comprising about 50% of the total carbohydrates in wort. There are also many minor carbohydrates, such as the following: galactose, maltulose [O-α-D-glucopyranosyl-(1 → 4)-D-fructofuranose], isomaltose [O-α-D-glucopyranosyl-(1 → 6)-D-glucopyranose], melibiose [O-α-D-galactopyranosyl-(1 → 6)-D-glucopyranose], panose [O-α-D-glucopyranosyl-(1 → 6)-O-α-D-glucopyranosyl-(1 → 4)-D-glucopyranose], and raffinose [O-α-D-galactopyranosyl-(1 → 6)-α-D-glucopyranosyl-β-D-fructofuranoside].

Since the capacity to ferment various carbohydrates is species or strain dependent, not all of the carbohydrates listed can be fermented by every brewers' yeast. Melibiose and raffinose, for example, can be completely fermented by *S. uvarum* but not by *S. cerevisiae*. While there are a few strains which cannot ferment maltotriose, most brewers' yeasts can; on the other hand, there are a few superattenuating strains which can partially assimilate and metabolize maltotetraose (e.g., *S. cerevisiae*, NCYC-422). The fermentation of isomaltose or panose is also strain dependent. Excepting experimentally induced mutants, all strains of brewers' yeast can ferment glucose, fructose, galactose, maltose, and sucrose.

TABLE 10.8.   CHEMICAL COMPOSITION OF TOTAL SOLIDS (EXTRACT) OF HOPPED WORTS[1]

Component	Composition	
Carbohydrates		90−92%[2]
Maltose	43−46%[2]	
Maltotriose	10−13	
Glucose	4−8	
Sucrose (saccharose)	1−3	
Fructose	1−2	
Nonfermentable	22−25	
Nitrogen Components (N × 6.25)		3−6%
Free amino acids	1.0−1.5%	
Peptides and proteins	1.5−3.0	
Nucleic acid derivatives, etc.	0.5−1.0	
Inorganic Ions		1.5−2.0%
Calcium	40−200 mg/liter	
Sodium	10−100	
Potassium	200−500	
Magnesium	30−150	
Manganese	0.05−0.25	
Zinc	0.10−0.30	
Copper	0.02−0.15	
Iron	0.01−0.50	
Phosphate	400−800	
Chloride	100−400	
Sulfate	100−600	
Miscellaneous Components		0.5−1.5%
Vitamins	2.5−3.5 mg/liter	
Polyphenols	0.1−0.2%	
Hop compounds	0.05−0.1%	
Lipids	10−50 mg/liter	
Oxygen	8−12 mg/liter	

[1]North American lager worts ranging from all-malt to 40% adjunct.
[2]All percentages are w/w in terms of total wort solids, dry basis.

Nonfermentable carbohydrates account for about 25% of total wort carbohydrates. Most of these are α-glucans with DP (degree of polymerization) > 3 or 4; the higher polymeric forms are commonly referred to as dextrins. Some of the less abundant nonfermentable carbohydrates in wort are the following: β-glucans, arabinose, ribose, xylose, and pentosans. Nonfermentable carbohydrates survive fermentation and may contribute to the sweetness of beer. Dextrins, along with some peptides and proteins, are thought to contribute to the "body" or "palate fullness" of beer.

Nitrogen Compounds.—These compounds, estimated as N × 6.25, are 3−6% of total wort solids (Table 10.8). All-malt worts are at the upper end of this range, 5−6%, while worts made with adjuncts are at the lower end. This class of compounds consists principally of free amino acids, peptides, proteins, and nucleic acid derivatives. About 30% of wort nitrogen is in the form of free amino acids. In an all-malt wort, this is a concentration of about 1200 mg/liter; if a portion of the malt is replaced by adjuncts, the concentration of free amino acids will be correspondingly lower. Besides the 20 amino acids commonly found in proteins, wort usually contains aminobutyric acid.

The concentration of proline in wort is relatively high, being about 20% of total free amino acids. Proline is not assimilated to any significant degree by brewers' yeast; consequently, its concentration in beer is approximately the same as that in wort.

Free amino acids are important not only for yeast nutrition but also for beer flavor. Free amino acids are the main, if not the sole, source of nitrogen for yeast during most of the fermentation. About one-half of the free amino acids in wort are utilized by the yeast. Those which remain in the beer contribute to its flavor.

Peptides and proteins account for about 50% of the wort nitrogen. An acidic protein fraction (<2% of wort nitrogen) combines with tannin to form a complex with a molecular weight greater than 60,000 daltons; the portion of this complex which finds its way into beer is responsible for haze instability. A neutral glycoprotein fraction (2–4% of wort nitrogen), with a molecular weight somewhat greater than 12,000 daltons, combines with hop compounds to form a complex which contributes to stable foam in both wort and beer. An ill-defined group of polypeptides and proteins, with molecular weights less than 12,000 daltons, are thought to have some flavor impact, including body or palate-fullness. Inasmuch as *Saccharomyces* species secrete proteases, some peptides (particularly those with molecular weights less than 1200 daltons) can be utilized by brewers' yeast.

The nucleic acids in malt are degraded during mashing to yield the nucleotides, nucleosides, purines, and pyrimidines which occur in wort and account for about 20% of wort nitrogen. Only free bases are readily assimilated by brewers' yeast during fermentation. During ruh storage or maturation, on the other hand, brewers' yeast *excretes* nucleotides. Although the concentrations of nucleotides in beer are low, they may have some influence on flavor.

A number of compounds important for beer flavor and color are produced by the reaction of amines and reducing sugars. Some of these compounds are formed during the kilning of malt, but many others are formed during the boiling of wort. Primary amines (most amino acids) react with reducing sugars, such as glucose, fructose, and maltose, to produce complex compounds called melanoidins; these contribute to the characteristic color of beer. Compounds with a luscious caramel flavor are also produced by this reaction. The reaction is known as the Maillard or nonenzymatic browning reaction (Hough *et al.* 1971).

Secondary amines (e.g., proline) react with reducing sugars to produce another group of compounds with caramel flavor, viz., maltol [3-hydroxy-2-methyl-4H-pyran-4-one], isomaltol [1-(3-hydroxy-2-furanyl)-ethanone], and their derivatives. Secondary amines also react with hexoses, such as glucose, fructose, and galactose, to produce aminohexose reductones. These compounds are active reducing agents which readily combine with oxygen; they also react with amino acids to produce complex brown pigments.

Inorganic Ions.—Some of the more abundant and more important inorganic ions found in wort are listed in Table 10.8. The calcium, sodium, chloride, and sulfate ions are mainly derived from brewing water. Their importance for water hardness and for the resulting organoleptic properties of beer has already been discussed. Calcium ions are also essential for the proper flocculation and sedimentation of yeast. Potassium and phosphate ions are very abundant, arising mainly from the malt. Magnesium, manganese, copper, iron, and zinc are all required as cofactors for various enzymes. Kinases, which catalyze the transfer of phosphate groups, commonly require magnesium ions. Biotin-coupled carboxylases require manganese or zinc ions. Some dehydrogenases, e.g., ethanol dehydrogenase, also require zinc ions. Hemoproteins which contain iron and non-heme, iron-containing proteins, such as ferredoxins, catalyze oxidation-reduction reactions. Other metalloproteins, which catalyze oxidation-reduction reactions, contain copper or manganese.

Requirements for zinc, copper, and iron have been reported for brewers' yeast in the following ranges: zinc, 0.2–0.5 mg/liter; copper, 0.010–0.015 mg/liter; and iron, 0.070–0.080 mg/liter. The ion most commonly deficient in wort is zinc. Reduced yeast growth, retarded fermentations, sulfurous odors, and other organoleptic defects have all been correlated with zinc deficiencies.

Miscellaneous Components.—All other components of wort which have been identified and which are known to have some importance in the beer-making process are listed as miscellaneous in Table 10.8. For convenience, these compounds have been divided into subgroups; vitamins, polyphenols, etc.

Vitamins.—The vitamins or growth factors most abundant in wort are those of the B-complex. Some characteristic values are as follows: thiamin (B_1), 200–600 μg/liter; riboflavin (B_2), 100–500 μg/liter; niacin, 1.5–2.5 mg/liter; pantothenate, 200–500 μg/liter; biotin, 5–10 μg/liter; pyridoxine (B_6), 200–500 μg/liter; *meso*-inositol, 50–150 mg/liter; and *p*-aminobenzoate, 20–30 μg/liter. These concentrations exceed brewers' yeast requirements severalfold, so that vitamins almost never limit brewery fermentations.

Polyphenols.—Polyphenols or wort tannins are derived from both malt and hops, but mainly from malt. Among the components of wort tannins are relatively simple, unpolymerized compounds which are derivatives of hydroxybenzoic acid, hydroxycinnamic acid, flavanols, anthocyanogens, or catechins. There are also polymerized compounds which are subdivided into hydrolyzable and nonhydrolyzable tannins. The hydrolyzable tannins are polymers of gallic acid and are true tannins in the sense that they are the predominant components of commercial tanning materials; they are often referred to as gallotannins. Gallotannins occur in hops but have not been detected in malt. Malt, however, contains the nonhydrolyzable or condensed tannins. Most of the wort tannins are precipitated with the hotbreak, the coldbreak, or the chill haze formed during conditioning; the trace amounts which remain pass into the beer substantially unchanged.

Although many polyphenols have been identified in wort and beer, there are doubtless a great many more which have not. The diverse character of this class of compounds gives rise to diverse effects. For example, some polyphenols contribute to chill haze and to astringent flavors, which are undesirable properties; others retard flavor deterioration by scavenging free radicals or oxygen, which is a desirable property. Both the composition and the influence on beer quality of this group of compounds are ill defined.

Hop Compounds.—In addition to what has been said already about hops, it should be noted that hop compounds are also polyphenols. They therefore react with proteins and other nitrogenous compounds of wort and probably also copolymerize with some of the other polyphenols just mentioned. Although the concentration of hop compounds in wort is low, the concentration in beer is lower. The reduced pH of beer (4.0–4.2) and the reduced temperature during conditioning (0°C) significantly depress the solubility of these compounds. The concentration of dissolved humulone in beer, for example, is only about 5 mg/liter. The iso-α-acids, however, are somewhat more soluble.

Lipids.—Less than 2% of the lipids in malt appear in the wort. Prominent among the components of this heterogeneous group of compounds are free fatty acids, glycerides (mainly triglycerides), sterols, and phospholipids. The majority of these compounds arise from malt, but a trace of lipids and waxes may be derived from hops. Most of the fatty acids present in wort are in the C-12 to C-18 range.

Although their concentrations are slight, lipids are important components of wort. They are related to yeast viability, ester formation, gushing, foam stability, and flavor deterioration. Brewers' yeast requires a small amount of ergosterol or unsaturated lipids for budding and for cell viability during storage. Unsaturated fatty acids suppress gushing (the spontaneous ebullition of beer from a can or bottle just opened) while saturated fatty acids have the opposite effect. Some wort lipids, particularly triglycerides, have an antifoam effect; this is especially true when they are present at concentrations higher than normal. Enzymatic oxidation of linoleic acid is thought to yield precursors of the unsaturated aldehydes reported to be partially responsible for the unpleasant aroma and taste of overaged or thermally abused beer. Of course, lipids are only one of numerous factors which influence flavor stability; some others are heat, light, air content, and the concentrations of metal ions and reductones. For reasons not well understood, brewing with adjuncts or with high-gravity worts generally enhances flavor stability.

Oxygen.—In brewery fermentations, yeast growth and fermentation rate depend on the initial presence of some dissolved oxygen (Kirsop 1974A). Since dissolved oxygen is effectively expelled from wort during kettle boil, it must be replaced. This is done most frequently by saturating cold wort with sterile air. Air saturation at 100% is equivalent to oxygen saturation at about 20%. At this initial concentration of oxygen, Markham (1969) has shown that both the yeast crop (Fig. 10.7) and the fermentation rate (Fig. 10.8) are near maximal levels. Below 20% oxygen saturation, both parameters are very dependent upon the level of oxygen.

Although oxygen is initially present and stimulates yeast growth, brewery fermentations (including the initial part) do not involve respiration to any significant extent (Markham 1969; Swanson and Clifton 1948). Molecular oxygen seems to be negligibly involved as an electron acceptor, functioning instead as an essential nutrilite or growth factor (Andreasen and Stier 1953, 1954). The latter view is corroborated by the fact that the level of dissolved oxygen normally drops from the initial concentration to nil in the first $10-15$ hr—before most of the metabolism or growth has occurred (Fig. 10.9).

Yeast

Growth.—The growth of yeast cells and their vegetative multiplication are a normal and necessary part of brewery fermentations. Yeast cells, like most other unicellular microorganisms in a liquid medium, exhibit 4 growth phases: a lag phase, an exponential or logarithmic phase, a stationary phase, and a decline or death phase. Generalized growth curves depicting these phases are available elsewhere, e.g., Fig. 9.5 in Rose (1976) or Fig. 9.2 in Stanier *et al.* (1976).

Growth rate is constant during the exponential phase; this is the only phase of the growth cycle in which a metabolically steady state (balanced growth) can reasonably be assumed. It is consequently the only phase of growth which can readily be described mathematically. The specific or exponential growth rate is defined as

$$\mu = (1/N)dN/dt = d(\ln N)/dt$$

where N is the concentration of yeast cells at time t and ln N is the natural logarithm of N. Rearranging, one obtains

$$d(\ln N) = \mu \, dt$$

A plot of ln N vs. t will yield a straight line with the slope equal to μ and the intercept equal to the initial cell concentration, N_0. Also from this equation, one can derive an expression for estimating the mean doubling time (t_d) or the mean generation time (g).

$$g = t_d = 0.693/\mu$$

Derivations of these expressions are available elsewhere (Rose 1976; Stanier *et al.* 1976; Aiba *et al.* 1973).

From chemical kinetics it will be apparent that yeast growth mimics first-order, autocatalytic chemical reactions. This seems remarkable since yeast growth evidently involves a network of hundreds, perhaps thousands,

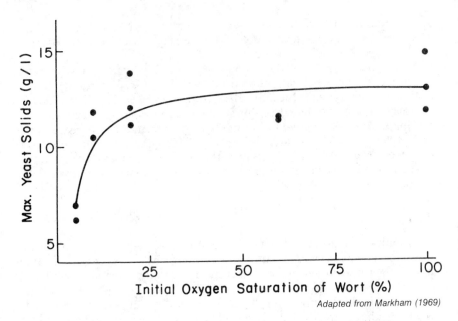

Adapted from Markham (1969)

FIG. 10.7.   INFLUENCE OF INITIAL OXYGEN LEVELS ON YEAST GROWTH

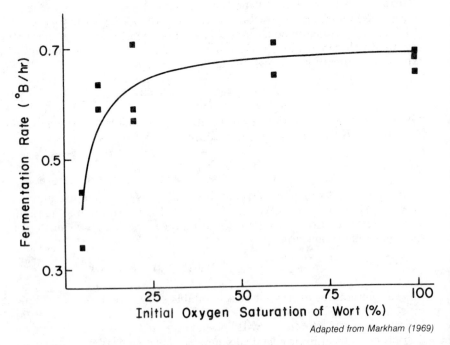

Adapted from Markham (1969)

FIG. 10.8.   INFLUENCE OF INITIAL OXYGEN LEVELS ON FERMENTATION RATE

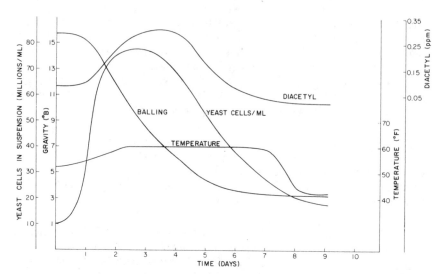

FIG. 10.9. SOME CHARACTERISTIC PARAMETERS OF A TYPICAL HIGH-GRAVITY LAGER FERMENTATION

of chemical reactions. Moreover, it seems improbable that there should be but a single rate-determining step in such a network (Monod 1949). Consequently, the quantitative interpretation of exponential growth-rate data is severely restricted. However, such data can provide general information which is very useful. For example, Monod (1949) derived some relatively simple and useful empirical relationships between the exponential growth rate and the concentrations of essential nutrients.

Yeast growth during a brewery fermentation is represented in Fig. 10.9, along with several other relevant parameters. Note that the yeast concentration is expressed as suspended cells rather than as total cells. There are two reasons for this: (1) for most of the fermentation, total cell count is experimentally inaccessible in a brewery because some cells settle out; and (2) settled cells do not contribute much to the beer-making process. However, the suspended cell count and the total cell count are the same during the exponential phase due to the vigorous metabolic activity. Therefore, for this phase the equations given previously can be applied to the data in Fig. 10.9 with the following results: the exponential growth rate, μ, is about 0.045 hr^{-1}, and the mean generation time, g, is about 15 hr. Such values are approximate and lack general validity for several reasons. First, note that the temperature changed continuously throughout the exponential phase of growth. Furthermore, μ and g are constant only if fermentation conditions are constant. If the composition of wort or the degree of agitation (Rice and Helbert 1974) is altered significantly, so are μ and g. However, a brewer using essentially the same yeast, the same wort, and the same operating conditions, could use the mean generation time to monitor his fermentation process. Note, finally, that the decline in yeast concentration following what appears to be a stationary phase (Fig. 10.9) is not due to a death phase, but to the flocculation and sedimentation of yeast cells. In lager fermentations, most of the settled yeast is viable and remains so throughout the process. Only the lag and exponential phases of growth in Fig. 10.9 have the same meaning as those in generalized growth curves.

Transport Processes.—Before the metabolic activities which support growth and produce beer can occur, the nutrients in wort must be selectively transported into the yeast cell and the by-products of metabolism removed. The two organelles primarily involved in these processes are the cell wall and the plasma membrane.

Cell Wall.—The cell wall of yeasts is very thick—about 70 nm, which is 10 times the thickness of the plasma membrane. This heavy wall not only serves as an exoskeleton, protecting and supporting the rather fragile plasma membrane, but also is metabolically important. It is now thought that the wall consists of an inner layer of glucan (probably the structural component) and an outer layer of mannan, with proteins embedded between the two. The proteins are mainly glycoprotein enzymes, such as invertase, melibiase, acid phosphatase, glucanases, aryl β-glucosidase, phospholipases, and proteases. These enzymes appear to be covalently linked to both the glucan and the mannan (perhaps more extensively to the latter) and to function anabolically as well as catabolically. A significant part of the extracellular enzyme activity of yeasts is localized in the cell wall and in periplasmic spaces (spaces between the cell wall and the plasma membrane). Although the foregoing conception of the cell wall was derived from *S. cerevisiae*, which has been much more extensively studied than other yeasts, the structure and function of the cell wall of *S. uvarum* is probably very similar.

Plasma Membrane.—The plasma membrane is about 7–8 nm thick and lies immediately inside the cell wall. Chemically, this membrane is approximately half lipid and half protein. It is now often described as a continuous bilayer of phospholipid molecules with globular proteins and some sterols embedded in it. This model, known as the fluid mosaic model (Lehninger 1975; Rose 1976), now tends to supplant the older triple-layered model. However, there is not yet a consensus on this matter.

The plasma membrane has several functions. Two of them have already been mentioned, viz., to admit nutrients into the cell and to eliminate by-products of metabolism. In addition, this organelle ultimately defines the yeast cell: yeast cells can exist with the cell wall partially or wholly removed; but if the plasma membrane is also removed, the cells cease to exist. All of these functions depend upon the fact that the plasma membrane is selectively permeable. The selectivity is often unidirectional and can function in opposition to a concentration gradient. Some of the transport mechanisms which generate this selectivity are discussed in the following paragraphs.

Simple diffusion in response to a concentration or electrochemical gradient rarely occurs across the plasma membrane of brewers' yeast. Water molecules may be the only ones which can enter or leave the yeast cell by *passive diffusion* or *unmediated transport* as this mechanism is sometimes called. During passive diffusion, solute molecules are not modified chemically nor are they associated with another molecule.

Mediated transport on the other hand does involve association of the solute with another molecule and may involve chemical change. All types of mediated transport involve specific carrier proteins or transport systems. When the specific carrier protein is a membrane-bound enzyme, the solute is chemically changed during the transport process at the expense of metabolic energy. This process is called *group translocation.*

There are other carrier proteins, also membrane-bound, which are similar to enzymes in that they selectively bind specific solutes, but are unlike enzymes in that they do not catalyze chemical change. These proteins are often referred to as permeases, a term deplored by some, because it incorrectly implies that these molecules are enzymes. This type of mediated transport is called *facilitated* or *mediated diffusion* if the driving force is a concentration gradient, or *active transport* if the process is coupled to a source of metabolic energy. The precise manner in which a carrier protein physically moves a bound solute molecule from one side of the plasma membrane to the other has not yet been elucidated.

Most of the molecular movement across the plasma membrane of brewers' yeast is by facilitated diffusion or active transport. For example, maltose, maltotriose, and phosphate ions are taken up by active transport, but most other ions, e.g., NH_4^+, K^+, Mg^{+2}, Fe^{+2}, and SO_4^{-2}, move across the membrane by mediated diffusion. Glucose, fructose, sucrose, and perhaps amino acids are taken up both by mediated diffusion and by active transport. In fact, it has been shown that some active transport systems, if uncoupled from their sources of metabolic energy, can function by mediated diffusion.

The carrier proteins involved in mediated transport, like enzymes, may be either constitutive or inducible. In brewers' yeast, for example, the carrier proteins for glucose appear to be constitutive, while those for maltose are inducible. The rate at which molecules like glucose and maltose are transported across the cell membrane may significantly influence the rate at which such molecules are metabolized.

Metabolism.—The compounds transported into the cell can serve as nutrients by furnishing either energy or chemical building blocks for new cell substance, or both. The sequences of chemical rearrangements whereby nutrient compounds are disassembled and then reassembled in the form of new cell substance are generally referred to as intermediary metabolism. When such rearrangements of organic compounds occur without utilizing oxygen, they are called fermentations. In most fermentations, the starting material is a carbohydrate; such is the case in brewery fermentations where the starting materials are mainly glucose, fructose, maltose, and maltotriose.

Carbohydrates.—The Embden-Meyerhof-Parnas (EMP) or glycolytic pathway is one of the more important and primitive pathways (in an evolutionary sense) for metabolizing carbohydrates. It occurs not only in a large array of aerobic, anaerobic, and facultative microorganisms but also in mammalian and other muscle. This pathway is discussed in detail in most general biochemistry texts and is shown in an abbreviated form in Fig. 10.10.

The EMP pathway is the principal route by which carbohydrates are utilized during brewery fermentations. Note that 2 ATPs are required for phosphorylating glucose and that subsequently 2 ATPs are recovered between glyceraldehyde-3-phosphate and pyruvate. Since dihydroxyacetone phosphate and glyceraldehyde-3-phosphate are readily interconvertible, this pathway can produce a total of 4 ATPs from each glucose—with a net gain of 2 ATPs—and is the main source of cellular energy during alcoholic fermentations. Other carbohydrates, such as galactose, maltose, and maltotriose, enter this pathway after being converted to glucose. Brewers' yeast does not excrete pyruvate but decarboxylates it to yield acetaldehyde which is then reduced to ethanol. Oxidation-reduction is thereby balanced: the NADH produced during the oxidation of glyceraldehyde-3-phosphate to 1,3-diphosphoglyceric acid is utilized in the reduction of acetaldehyde to ethanol; the NAD^+ regenerated in this way is available for further oxidation of glyceraldehyde-3-phosphate. Consequently, CO_2 and ethanol are the principal by-products of this metabolic pathway.

In a similar way, some of the NADH produced during the oxidation of glyceraldehyde-3-phosphate can be used to reduce dihydroxyacetone phosphate to glycerol phosphate and finally to glycerol. The extent of these reactions is relatively minor.

Two other metabolic pathways of interest to the brewer are the homolactic and heterolactic fermentations (Fig. 10.10). Brewers' yeast does not utilize these pathways, but beer spoilage bacteria of the lactic acid group do. Homolactic fermenters use the EMP pathway with one additional step, viz., the reduction of pyruvate to lactate. This reduction is effected with the NADH produced during the oxidation of glyceraldehyde-3-phosphate. The lactate excreted causes the unpleasant flavor which characterizes beer spoiled by these microorganisms.

Heterolactic fermenters produce not only lactate but ethanol and CO_2 as well. The metabolic scheme involved is more complex and is sometimes referred to as the phosphoketolase (PK) pathway. The initial part of this pathway, from glucose-6-phosphate to xylulose-5-phosphate, produces the CO_2 which evolves. Xylulose-5-phosphate is cleaved with phosphorylation by a phosphoketolase to form glyceraldehyde-3-phosphate and acetyl phosphate. The former yields lactate via pyruvate in a manner analogous to the homolactic fermentation. The acetyl phosphate, however, is dephosphoryl-

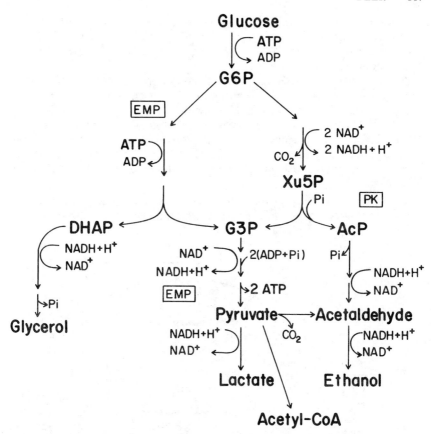

FIG. 10.10.   AN ABBREVIATED REPRESENTATION OF SOME IMPORTANT FERMEN-
TATIONS (→→INDICATES THE OMISSION OF ONE OR MORE INTERMEDIATES)

Alcoholic Fermentation: EMP→Acetaldehyde→Ethanol
Homolactic Fermentation: EMP→Lactate

Heterolactic Fermentation: Glucose→G6P→→Xu5P →

G3P→Pyruvate→Lactate
+
AcP→Acetaldehyde→Ethanol

ABBREVIATIONS

EMP = Embden-Meyerhof-Parnas or Glycolytic Pathway:
        Glucose→G6P→→DHAP + G3P→Pyruvate.
PK = Phosphoketolase Pathway: *Cf.* heterolactic fermentation.
G6P = Glucose-6-phosphate.
DHAP = Dihydroxyacetone phosphate.
G3P = Glyceraldehyde-3-phosphate.
NAD$^+$ and NADH = Oxidized and reduced forms of nicotinamide adenine dinucleotide.
Xu5P = Xylulose-5-phosphate.
AcP = Acetyl phosphate.
ADP and ATP = Adenosine diphosphate and triphosphate.
P$_i$ = Inorganic phosphate.

ated and reduced to acetaldehyde which in turn is reduced to ethanol. The NADH produced in the initial part of the PK pathway is utilized in these two reductions, and the NAD$^+$ thus regenerated becomes available for more oxidations.

The initial part of the PK pathway—glucose-6-phosphate →→ xylulose-5-phosphate → glyceraldehyde-3-phosphate—is analogous to a portion of the phosphogluconate pathway (Fig. 10.11) which, however, requires the coenzyme NADP$^+$ (nicotinamide adenine dinucleotide phosphate) rather than NAD$^+$. The phosphogluconate pathway is also known as the pentose phosphate pathway or the hexose monophosphate shunt. The last term derives from a consideration of this pathway an an alternate loop, or shunt, to the EMP pathway.

Less than 20% of the glucose utilized by brewers' yeast during fermentation is metabolized by way of the phosphogluconate pathway. Although this

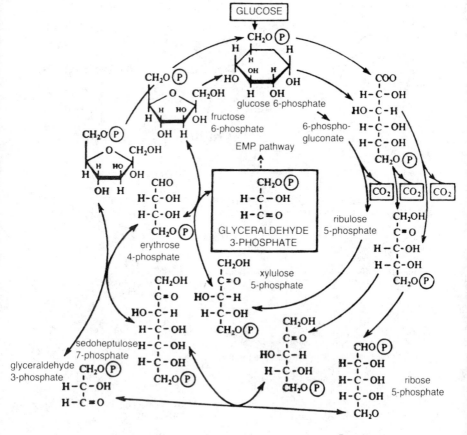

From Hough et al. (1971)

FIG. 10.11.   THE PHOSPHOGLUCONATE PATHWAY

pathway can yield as many as 35 ATPs for each glucose molecule degraded under *aerobic* conditions, under *anaerobic* conditions it may yield as few as 1 ATP for each glucose. Consequently, because of the anaerobic conditions that exist in a brewery fermentation, the phosphogluconate pathway is not important as a source of energy. It is important, however, as a source of pentoses needed for the synthesis of RNA and DNA because brewers' yeast cannot utilize exogenous pentoses. It is equally important as a source of reducing power (NADPH) needed to carry on the biosynthesis of sterols and fatty acids.

Pyruvate is a pivotal compound leading to a number of important metabolites. A few of these are shown in Fig. 10.10. A very important one is

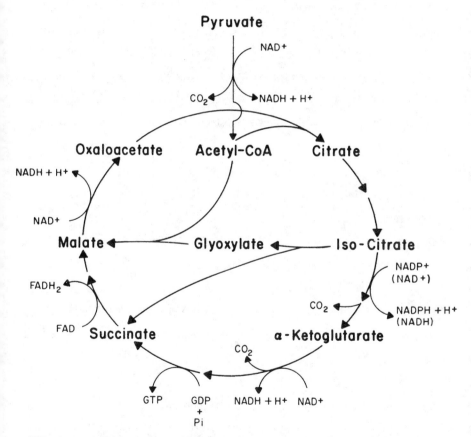

FIG. 10.12.   ABBREVIATED TRICARBOXYLIC ACID (TCA) AND GLYOXYLATE CYCLES

ABBREVIATIONS (For those not listed, see Fig. 10.10)

$NADP^+$ and NADPH = Oxidized and reduced nicotinamide adenine dinucleotide phosphate.
GDP and GTP = Guanosine diphosphate and triphosphate.
FAD and $FADH_2$ = Oxidized and reduced flavin adenine dinucleotide.

acetyl-CoA which leads directly to the TCA and glyoxylate cycles (Fig. 10.12) and to lipid synthesis.

The TCA cycle, like the phosphogluconate pathway, is not important as a source of energy for brewers' yeast. Because of the anaerobic conditions during brewery fermentations, the electron transport chain and the oxidative phosphorylation associated with it are inoperative. Consequently, the 36 additional ATPs which can arise from the *aerobic* breakdown of glucose via the EMP and TCA pathways cannot be realized. However, the TCA cycle has important metabolic functions other than the production of energy—and can operate without being coupled to electron transport and oxidative phosphorylation. It is an amphibolic pathway essential for the intermediary metabolism of many substances besides carbohydrates; it is important in the catabolism of amino acids (Fig. 10.13) and also as a source of precursors for the biosynthesis of α-keto-acids, amino acids, and higher aldehydes and alcohols (Fig. 10.14). These pathways and compounds are important for both yeast metabolism and beer flavor.

It now seems clear from the work of Schatz and his coworkers (Criddle and Schatz 1969; Paltauf and Schatz 1969; Plattner and Schatz 1969) and Watson *et al.* (1970) that *anaerobically* grown *Saccharomyces* retain mitochondrial structures which contain enzymes of the TCA cycle. Schatz

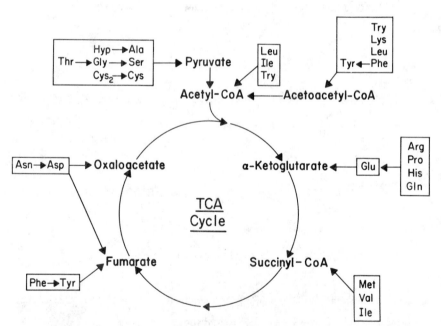

FIG. 10.13.   CATABOLISM OF AMINO ACIDS RELATIVE TO THE METABOLISM OF CARBOHYDRATES
Abbreviations are those commonly accepted for amino acids; numerous intermediates have been omitted.

named these structures "promitochondria." They are morphologically and functionally simpler than mitochondria from *aerobically* grown yeast; presumably, in facultative organisms like *Saccharomyces*, they are convertible to mitochondria in the presence of oxygen. Promitochondria have few if any cristae and (as already indicated) lack functional electron transport and oxidative phosphorylation systems.

At glucose concentrations of about 1% or more, even the normal mitochondrial structures of respiring *Saccharomyces* tend to degenerate with a loss of cristae and other morphological features. Correlative with these changes are reductions in activity of the enzymes involved in respiration, including those of the TCA and glyoxylate cycles. The overall effect is an inhibition of respiration by fermentation which is frequently called the "Crabtree effect." It is also known as the "reverse Pasteur effect." The Pasteur effect itself, namely, the inhibition of fermentation by respiration, does not occur in brewing; this is so either because of the overriding influence of the Crabtree effect or simply because of the absence of oxygen.

In brewery fermentations, TCA cycle intermediates are diverted into anabolic pathways and must be replaced if the cycle is to continue to function. Replacement can be achieved by the operation of the glyoxylate cycle (a truncated version of the TCA cycle, Fig. 10.12), by amino acid catabolism (Fig. 10.13), and by other noncyclic pathways. The glyoxylate cycle diverges from the TCA cycle at isocitrate which is cleaved to form succinate and glyoxylate. The latter combines with acetyl-CoA to form malate (Fig. 10.12). In bypassing both of the CO_2-evolving steps of the TCA cycle, the glyoxylate pathway conserves carbon and augments the concentrations of the 4-carbon intermediates.

Amino Acids, Keto-acids, Etc.—Both 4-carbon and 5-carbon intermediates of the TCA cycle are augmented by the catabolism of amino acids. How this occurs is indicated in Fig. 10.13. For the purpose of clarity, many details are omitted; these can be found in treatises on biochemistry (Lehninger 1975; McGilvery 1975). Only the most characteristic reaction of amino acid catabolism, namely, the removal of the α-amino group, will be briefly discussed here. Of the enzymatic mechanisms for such removal, transamination is most important in brewery fermentations. Transamination is a reaction in which the amino group from an α-amino acid is moved to an α-keto-acid. Examples of transamination are:

$$\text{L-Asp} + \alpha\text{-ketoglutarate} \rightleftarrows \text{oxaloacetate} + \text{L-Glu}$$
$$\text{L-Ala} + \alpha\text{-ketoglutarate} \rightleftarrows \text{pyruvate} + \text{L-Glu}$$

All such reactions are freely reversible and occur in the cytosol as well as in the mitochondria of eukaryotic cells.

There are numerous transaminases (aminotransferases) which facilitate these reactions. Most transaminases require α-ketoglutarate as one of the amino-group acceptors. Although these enzymes are conventionally named for the amino-group *donor* (e.g., aspartate transaminase and alanine trans-

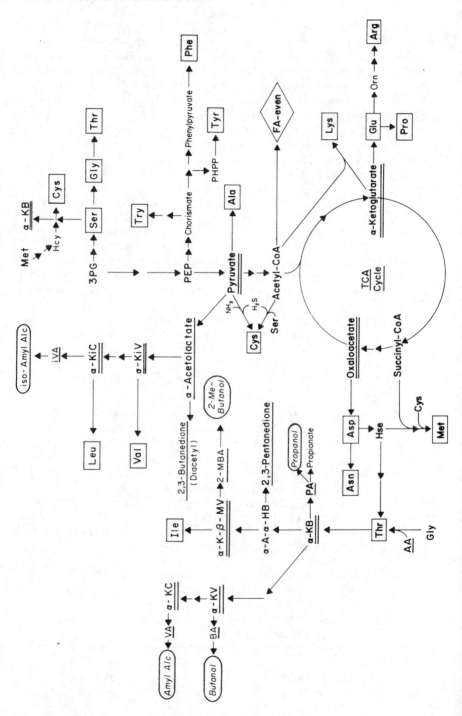

FIG. 10.14. SOME BIOSYNTHETIC PATHWAYS IMPORTANT IN BREWERY FERMENTATIONS

ABBREVIATIONS AND SYMBOLS

AA = Acetaldehyde.
α-A-αHB = α-Aceto-α-hydroxybutyrate.
α-KB = α-Ketobutyrate.
α-K-β-MV = α-Keto-β-methylvalerate.
α-KC = α-Ketocaproate.
α-KV = α-Ketovalerate.
α-KiC = α-Ketoisocaproate.
α-KiV = α-Ketoisovalerate.
BA = n-Butyraldehyde.
Hcy = Homocysteine.
Hse = Homoserine.
iVA = iso-Valeraldehyde.
2-MBA = 2-Methylbutyraldehyde.
Orn = Ornithine.
PA = Propionaldehyde.
PEP = Phosphoenolpyruvate.
3PG = 3-Phosphoglycerate.
PHPP = p-Hydroxyphenylpyruvate.
VA = n-Valeraldehyde.

NOTES

Aldehydes and ketones are underscored singly.
Ketoacids are underscored doubly.
Biosynthesized amino acids are enclosed in a rectangle.
Fusel alcohols are enclosed in an oval.
Fatty acids with an even number of carbons are enclosed in a rhombus.

aminase for the previous reactions), their specificity for the amino acid is often much less than that for α-ketoglutarate. All transaminases have pyridoxal phosphate as the prosthetic group, and their reaction mechanisms appear to be the same. The following 12 amino acids are known to participate in transamination reactions: alanine, arginine, asparagine, aspartic acid, cysteine, isoleucine, leucine, lysine, phenylalanine, tryptophan, tyrosine, and valine.

Simplified biosynthetic pathways for the formation of amino acids and some other compounds important in brewery fermentations are shown in Fig. 10.14. This figure and Fig. 10.13 indicate those intermediates in the EMP and TCA pathways which are common to both the anabolism and catabolism of amino acids. Since anabolic pathways are never simple reversals of corresponding catabolic pathways, these intermediates provide a connection between the two. These pathways do, however, often contain common steps, e.g., transamination in the case of amino acids. Both pathways can operate simultaneously, generally by occurring in different compartments of eukaryotic cells. This operation makes complete interconversion of amino acids possible. Consequently, for brewers' yeast, there are no nutritionally indispensable or essential amino acids. Auxotrophs are obviously excepted.

Some of the α-keto-acids formed during the metabolism of amino acids and carbohydrates are decarboxylated to produce aldehydes (Fig. 10.14). These, in turn can be reduced to alcohols which are collectively referred to as fusel alcohols, viz., propanol, butanol, 2-methylbutanol (optically active amyl alcohol), amyl alcohol, iso-amyl alcohol, and β-phenylethanol. Fusel alcohols are secreted into the beer to a total concentration of 75–125 mg/liter. Related aldehydes and α-keto-acids, like many other metabolic intermediates, are lost from yeast cells in significantly lesser amounts and mainly as a consequence of autolysis.

Other compounds related to amino acid metabolism are the vicinal diketones, 2,3-butanedione (diacetyl), and 2,3-pentanedione. One pathway for their biosynthesis (shown in Fig. 10.14) is by nonenzymatic decomposition of the appropriate α-aceto-α-hydroxyacid: α-acetolactate yields diacetyl, and α-aceto-α-hydroxybutyrate yields 2,3-pentanedione. The formation of diacetyl by brewers' yeast is also known to occur by direct condensation of acetyl-CoA with "active" acetaldehyde, i.e., 2(α-hydroxyethyl)thiamin pyrophosphate. It is definitely not formed from acetoin (acetylmethyl-carbinol). Nor is acetoin formed from diacetyl, but rather by condensation of acetaldehyde with "active" acetaldehyde. None of the latter reactions is shown in Fig. 10.14. Acetoin, diacetyl, and 2,3-pentanedione are directly or indirectly secreted by brewers' yeast during fermentation and all are important flavor components. Their approximate relative proportions in beer are 150:3:1.

The decarboxylation of α-keto-acids to yield aldehydes can be followed by oxidation to fatty acids rather than by reduction to alcohols which was described previously. The acid produced in greatest abundance by this pathway is acetic acid from acetaldehyde; lager beers contain this acid at concentrations of 50–150 mg/liter, which is about twice the concentration found in ales. Trace amounts of as many as 80 other organic acids have been detected in beer. Of course, not all of them are biosynthesized by this pathway. Some fatty acids, especially those with more than 12 carbon atoms, are derived ultimately from acetyl-CoA (Fig. 10.14). Biosynthesis of palmitic acid from acetyl-CoA is catalyzed by a cluster of 7 enzymes called the fatty-acid synthetase complex. For yeast, this complex cannot be dissociated without loss of activity. The biosynthesis begins with a single molecule of acetyl-CoA to which 7 2-carbon units are added successively via malonyl-CoA with the loss of 7 molecules of CO_2. NADPH and ATP are required. The malonyl-CoA is derived from acetyl-CoA in a preliminary reaction catalyzed by acetyl-CoA carboxylase which is allosterically modulated by citrate. The entire process occurs in the cytosol and is mediated by completely different enzymes from those used for fatty acid oxidation, which occurs in the mitochondria.

Palmitic acid is the precursor of longer chain fatty acids, both saturated and unsaturated. Stearic acid formed from palmitic can be desaturated to form oleic. The formation in this way of monoenoic fatty acids by brewers' yeast requires molecular oxygen (Bulder and Reinink 1974; Rattray et al. 1975). Furthermore, all known eukaryotes (except Schizosaccharomyces japonicus) lack an anaerobic pathway for synthesizing unsaturated fatty acids. The biosynthesis of sterols (by quite another pathway) also requires molecular oxygen at several points: first, for the cyclization of squalene to form lanosterol; and then, for the demethylation and desaturation steps to convert lanosterol to ergosterol (Aries and Kirsop 1978). These facts, together with the fact that a significant reduction in yeast sterols and unsaturated fatty acids leads to functionally abnormal membranes, may explain in part why yeast cannot be grown anaerobically for more than a few generations.

MICROBIAL CONTAMINATION

Although the number of microorganisms capable of spoiling beer are relatively few, the number which can indirectly have an adverse effect on it are more numerous. In this section, microorganisms other than pitching yeast will be discussed, whether they directly or indirectly influence the process or product. Beer-making will be considered step-by-step in terms of microbial contamination.

Barley and Malt

The microflora naturally associated with barley are mainly fungi, some bacteria, and perhaps a few yeasts. These microorganisms are also found in air, in soil, and on vegetation in general. Their type and abundance on barley prior to harvest depend mainly on weather conditions, but also on varietal susceptibility. Damp harvest and storage conditions are very conducive to the development of such microflora.

Fungi.—The following genera of field fungi have been reported to occur abundantly on barley: *Alternaria, Helminthosporium, Fusarium, Epicoccum, Cladosporium, Stemphylium, Penicillium, Aspergillus, Rhizopus,* and *Nigrospora.* The same field fungi have been reported from such diverse places as California, North Dakota, Scotland, Denmark, and Japan. Field fungi gradually lose viability during storage and are replaced by storage fungi, most of which are species of *Aspergillus* and *Penicillium.* When present at low levels on barley, these fungi produce no noticeable adverse effects. However, when damp harvest and storage conditions lead to high levels, a number of undesirable consequences can occur.

Fungi are responsible for the initial heating of moist stored grain. Fungi such as *Aspergillus glaucus, Aspergillus candidus,* and *Aspergillus flavus* can raise the temperature of stored grain to 55°C and keep it there for several weeks. The temperature can then be raised as high as 75°C by thermophilic bacteria. Beyond this point, chemical processes not associated with microorganisms can heat the grain to the point of combustion. Heavy infections of these fungi can also damage the germ of the barley kernel, thus impairing the germinative capacity of the barley and the diastatic power of the malt. *Helminthosporium sativum* and *Fusarium roseum* have been consistently associated with these effects. *Fusarium moniliforme* can interfere with the malting process by restricting the oxygen available to the germinating barley. Fungi, both field and storage types, are responsible for gushing of the beer produced from heavily infected grain. Most of the genera listed above have been indirectly implicated in this problem. However, three of them—*Rhizopus, Stemphylium,* and *Fusarium*—have been directly implicated.

Amaha *et al.* (1974) isolated and characterized chemical entities produced by the growing mycelium of *Rhizopus, Stemphylium,* or *Fusarium* on germinating barley; these chemical entities were capable of inducing vigorous gushing at very low concentrations, viz., 0.05 mg/liter. *Fusarium graminearum* (synonym: *Gibberella zeae*) produced an especially active gushing factor. All of the gushing factors isolated thus far have been relatively small, stable molecules, viz., oligopeptides or peptidoglycans. These molecules, formed during malting, can survive the entire brewing process to become a component of the beer; the organisms which produce them, however, scarcely survive malting and certainly cannot survive the brewing process.

Under suitable conditions of growth and storage, some of the fungi which grow on barley produce mycotoxins. Those species encountered most frequently, viz., *Aspergillus ochraceus, Penicillium viridicatum,* and *Penicillium citrinum,* produce ochratoxins and citrinin, neither of which is carcinogenic (Hesseltine 1974). Moreover, it has been reported that more than 80% of ochratoxin-A and citrinin are lost during malting and brewing (Krogh *et al.* 1974; Nip *et al.* 1975). Mycotoxins have never been reported present in commercially produced beer.

Bacteria.—Those bacteria most commonly associated with barley in the field are divided into 3 groups—lactic acid bacteria, enterobacteria (coliforms), and *Pseudomonas*. One species, *Pseudomonas syringae* (synonym, *Pseudomonas atrofaciens*), can cause the barley germ to deteriorate. Many of these 3 bacterial groups survive the malting process; and although they cannot survive mashing and brewing, some become airborne in the brewery and can infect the wort. Therefore, they will be discussed in conjunction with sweet and hopped worts. [*Note*: the 8th Edition of *Bergey's Manual* (Buchanan and Gibbons 1975) has been used as the standard for bacterial nomenclature. Accepted names appear in italics without parentheses. Other names, likely to be encountered in the brewing literature and considered synonymous with an accepted name, follow it in italics within parentheses.]

Mash and Sweet Wort

Because of the elevated temperatures during mashing and lautering, it is not surprising that microbial infections are uncommon at this stage. However, a few thermophilic lactic acid bacteria can grow during some portions of the process. For example, *Lactobacillus delbrückii* is a Gram-positive, homofermentative, thermophilic rod with an optimum growth temperature of 45°C, but it can grow at a temperature as high as 54°C. If the temperature of mash tun worts should fall into this range, this organism will grow rapidly, resulting in spoilage of the product.

Similarly, *Pediococcus acidilactici* (synonym: *P. lindneri*) is a Gram-positive, homofermentative, thermophilic coccus with an optimum temperature for growth of 40°C and a maximum temperature for growth of 52°C. At temperatures in this range, *P. acidilactici* can also spoil mash tun worts. Both of these organisms are sensitive to hops and cannot grow in hopped wort or beer.

Unfermented Hopped Wort

Hopped wort will, nevertheless, support the growth of some bacteria and wild yeasts. It becomes microbiologically vulnerable when cooled for pitching. Enterobacteria, *Zymomonas, Acinetobacter, Pseudomonas,* and the acetic acid bacteria can all grow in hopped wort. These are the Gram-negative genera shown in Table 10.9. Of these, the strict aerobes are probably less common than they once were because, in a modern brewery, the exposure of cooled wort to nonsterile air is minimal. In American breweries today, the enterobacteria and *Zymomonas* are the most common.

Those genera of the Enterobacteriaceae family which have been found in breweries are *Citrobacter, Klebsiella, Enterobacter, Hafnia, Serratia,* and *Erwinia* (Priest *et al.* 1974; van Vuuren *et al.* 1978). They are Gram-negative rods which are facultative anaerobes. All are motile except *Klebsiella.* Although some enterobacteria are important pathogens of both plants and animals, not all of them are intestinal parasites and many are

TABLE 10.9.   GENERA OF BREWERY BACTERIA

Gram-negative, strict aerobes
  Pseudomonas
  Gluconobacter
  Acetobacter
  Acinetobacter

Gram-negative, facultative anaerobes
  Citrobacter
  Klebsiella
  Enterobacter
  Hafnia
  Serratia
  Erwinia
  Zymomonas

Gram-positive, facultative anaerobes/microaerophiles
  Lactobacillus
  Pediococcus

not pathogenic. They are widely distributed in nature, occurring in soil and water and in association with plants and plant products.

As knowledge of these and other bacteria has increased over the years, their nomenclature has changed many times. As an aid for the reader in perusing the microbiological literature of brewing, the currently accepted names of relevant Gram-negative and Gram-positive bacteria with some of their synonyms have been tabulated (see Table 10.10). Only those synonyms germane to or derived from the brewing literature have been included; the list is not definitive. Note that *Aerobacter* has been abandoned as a generic name; the nonmotile forms of *Aerobacter aerogenes* have been assigned to *Klebsiella pneumoniae*, while the motile forms have been assigned to *Enterobacter aerogenes*. In 1936, Shimwell discovered and named *Flavobacterium proteus* which in 1963 he renamed *Obesumbacterium proteus*. More recently, Priest *et al.* (1974) renamed this organism *Hafnia protea*, including it among the enterobacteria.

Although wild yeasts can grow in hopped wort, they generally are unable to multiply appreciably in the interval between wort cooling and the onset of fermentation by culture yeast. However, they can always in some degree compete with culture yeast and can survive fermentation to appear in pitching yeast and beer. They will be discussed in more detail.

Fermenting Wort

Enterobacterial organisms which find their way into wort prior to fermentation continue to grow and produce unsavory metabolic products until the pH falls below 4.4 and the ethanol concentration rises significantly above 2 g/100 ml. Since many of the enterobacteria produce hydrogen sulfide and other malodorous metabolic products, they can cause a distinctly substandard product even though they do not survive fermentation. *Hafnia protea* is reported to survive brewery fermentations somewhat better than other enterobacteria and is consequently a common contaminant in pitching yeast.

TABLE   10.10.   CURRENTLY   ACCEPTED   NAMES   OF   GRAM-NEGATIVE   AND
GRAM-POSITIVE BREWERY BACTERIA WITH SYNONYMS

Currently Accepted Names	Synonyms
Gluconobacter oxydans	*Acetomonas oxydans*
G. oxydans subsp. *industrius*	*Acetobacter capsulatum* [sic]
	Acetobacter viscosum [sic]
	Gluconobacter industrius
	Gluconobacter capsulatus
Acetobacter aceti	
A. aceti subsp. *xylinum*	*Acetobacter xylinoides*
A. pasteurianus	*A. vini-acetati*
	A. agglutinans
	A. alcoholophilus
Citrobacter freundii	*Escherichia freundii*
C. intermedius	*E. intermedia*
	Paracolobactrum intermedium
Klebsiella pneumoniae	*Klebsiella aerogenes*
	Aerobacter aerogenes (non-motile forms)
Enterobacter cloacae	*Aerobacter cloacae*
E. aerogenes	*A. aerogenes* (motile forms)
Hafnia alvei	
Hafnia protea	*Obesumbacterium proteus*
	Flavobacterium proteus
Serratia spp.	
Erwinia herbicola	*Enterobacter agglomerans*
Zymomonas mobilis	*Termobacterium mobile*
	Pseudomonas lindneri
	Saccharomonas lindneri
Z. anaerobia	*Achromobacter anaerobium* [sic]
	Saccharomonas anaerobia
Z. anaerobia var. *immobilis*	*Saccharomonas anaerobia* var. *immobilis*
Z. anaerobia var. *pomaceae*	
Lactobacillus brevis	*Lactobacillus lindneri* (Henneberg)
	L. plantarum
	L. pastorianus (Van Laer)
	L. pastorianus var. *brownii*
	L. brownii
	L. pastorianus var. *diastaticus*
	L. diastaticus
L. buchneri	*L. parvus*
	L. frigidus
Pediococcus cerevisiae	*Pediococcus damnosus* (Mees)
	P. damnosus var. *diastaticus*
	P. mevalovorus

Although the optimal growth temperature of *Zymomonas* is 30°C, it can grow at 4°–10°C. It also tolerates ethanol at the concentrations that occur in fermenting wort and can grow over a wide pH range, viz., 2.5–8.0. Although this organism does not ferment maltose, it does ferment glucose and other simple sugars via the Entner-Doudoroff pathway to yield ethanol and carbon dioxide. Other metabolic products include acetaldehyde and hydrogen sulfide which together produce an odor reminiscent of rotten apples. This organism clearly has the potential to survive brewery fermentations with very undesirable consequences.

The lactic acid bacteria, viz., lactobacilli and pediococci, are also well adapted to survive brewery fermentations. These organisms are Grampositive, facultative anaerobes which tolerate hops and ethanol very well. Moreover, they grow well at the pH's normally encountered during fermentation. These organisms will be discussed further in the next section.

Beer

Beer can be spoiled either by bacteria or by wild yeasts. In American breweries, the most common spoilage bacteria are those of the lactic acid group, while the most common spoilage yeasts are those of the *Saccharomyces* genus.

Bacteria.—The bacteria generally associated with beer-spoilage problems are lactic acid bacteria, acetic acid bacteria, and *Zymomonas anaerobia*. Of the 3 groups, the lactic acid bacteria are most prevalent both in beer during processing and in packaged beer. The two most prominent species are *Lactobacillus pastorianus* and *L. diastaticus*. Morphologically, culturally, and physiologically, the two are quite similar; they differ in that *L. pastorianus* is able to hydrolyze dextrins and starch. Taxonomists regard both of these names as synonyms for *Lactobacillus brevis*. As already noted, what is important for brewers may not be important for taxonomists, and vice versa. The lactobacilli of importance to brewers are all heterofermentative, except the thermophilic species, *L. delbrückii*, already discussed. When these microorganisms spoil beer, they do so by producing a "silky" haze due to microbial growth and by producing excessive acidity due to the formation of lactic acid. The cocci of the lactic acid group are now assigned to the genus *Pediococcus*; at one time, they were considered part of the genus *Streptococcus* and before that they were called *Sarcina*. The most common name among brewers for these organisms is *Pediococcus damnosus*. Taxonomists consider this name a synonym for *Pediococcus cerevisiae*. These microorganisms are also Gram-positive, facultative anaerobes (microaerophiles). Unlike most brewery lactobacilli, the pediococci are all homofermentative. Furthermore, *P. cerevisiae* requires carbon dioxide for growth and generally produces diacetyl which is one of the characteristics of "Sarcina sickness" produced by this organism.

The acetic acid bacteria are Gram-negative, strict aerobes. Consequently, they are able to spoil beer only if defective containers fail to exclude air. There are two genera in this group, *Gluconobacter* and *Acetobacter*; in the presence of air, both are able to oxidize ethanol to acetic acid. This is the principal effect when these bacteria spoil beer.

Because *Zymomonas anaerobia* is so readily killed by heat (5 min at 60°C), this organism rarely occurs in bottles or cans of beer which have been pasteurized. However, keg beers, especially those which have been primed with glucose, are very susceptible to this microorganism. In the United States, where most beer is sold in bottles or cans, *Z. anaerobia* is not much of a problem for brewers. In England, however, where most beers and ales are sold on draught, this organism can be very destructive.

Zymomonas is very unusual in that it anaerobically catabolizes glucose via the Entner-Doudoroff pathway, which occurs mainly in strictly aerobic bacteria. In addition, *Zymomonas* has an incomplete TCA cycle and polar flagella. For these reasons, as well as others, *Zymomonas* resembles the acetic acid bacteria, especially *Gluconobacter*. In fact, Swings and DeLey (1977) have suggested that *Zymomonas, Gluconobacter*, and *Acetobacter* have a common phylogenetic origin.

Wild Yeasts.—Wild, or nonculture, yeasts can spoil beer at almost any point in the process, including the packaged product. A great many such yeasts have been reported (see Table 10.11). These organisms may cause one or more of the following defects: overattenuation, excessive acidity, off-odors and flavors, and haze due to suspended cells.

The most common and the most pernicious wild yeasts are those from the genus *Saccharomyces*, especially *S. diastaticus*. Although not listed in Table 10.11, *Saccharomyces uvarum* (synonym: *S. carlsbergensis*) is re-

TABLE 10.11.   SOME NONCULTURE YEASTS WHICH CAN SPOIL BEER

Saccharomyces bayanus (synonym: *S. pastorianus*[1])
S. diastaticus
S. inusitatus

Pichia membranaefaciens

Candida utilis
C. lambica

Torulopsis colliculosa
T. inconspicua

Debaryomyces spp.

Hansenula anomala

Rhodotorula spp.

Kloeckera apiculata

Hanseniaspora spp.

Dekkera bruxellensis
D. intermedia

Brettanomyces bruxellensis
B. lambicus
B. anomalus (synonym: *B. dublinensis*[1])
B. clausenii
B. intermedius

Saccharomycodes ludwigii

[1]See Lodder (1970).

garded as a wild yeast in top fermentations; analogously, *S. cerevisiae* is regarded as a wild yeast in lager fermentations. The detection and elimination of wild yeasts, especially those from the genus *Saccharomyces,* are difficult tasks. As discussed earlier, wild yeasts can be detected by the use of 3 selective media: (1) a liquid or solid medium containing cycloheximide, (2) a liquid or solid medium containing lysine as sole source of nitrogen, and (3) a liquid or solid medium containing crystal violet. These 3 media are not equivalent; each detects a somewhat different spectrum of wild yeasts. Consequently, all 3 should be used routinely for best results. While it is true that wild yeasts can survive in numerous places in a brewery, it is generally agreed that pitching yeast is the principal reservoir of such infection.

REFERENCES

AIBA, S., HUMPHREY, A.E. and MILLIS, N.F. 1973. Biochemical Engineering, 2nd Edition. Academic Press, New York, London.

AMAHA, M. *et al.* 1974. Gushing inducers produced by some mould strains. Eur. Brew. Conv., 14th Proc. Cong., Salzburg, 1973. Elsevier Scientific Publishing Co., Amsterdam, London, New York.

ANDREASEN, A.A. and STIER, T.J.B. 1953. Anaerobic nutrition of *Saccharomyces cerevisiae*. I. Ergosterol requirement for growth in a defined medium. J. Cell. Comp. Physiol. *41*, 23–36.

ANDREASEN, A.A. and STIER, T.J.B. 1954. Anaerobic nutrition of *Saccharomyces cerevisiae*. II. Unsaturated fatty acid requirement for growth in a defined medium. J. Cell. Comp. Physiol. *43*, 271–281.

ANON. 1839. The mystery of alcoholic fermentation is solved! (Liebig's) Ann. Chem. Leipzig *29*, 100–104. (German)

ARIES, V. and KIRSOP, B.H. 1978. Sterol biosynthesis by strains of *Saccharomyces cerevisiae* in the presence and absence of dissolved oxygen. J. Inst. Brew. *84*, 118–122.

ATKIN, L. 1949. Yeast growth factors. Wallerstein Lab. Commun. *12*, 141–151.

ATKIN, L., GRAY, P.P., MOSES, W. and FEINSTEIN, M. 1949. Growth and fermentation factors for different brewery yeasts. Wallerstein Lab. Commun. *12*, 153–170.

ATKIN, L., MOSES, W. and GRAY, P.P. 1962. The preservation of yeast cultures by lyophilization. Wallerstein Lab. Commun. *25*, 153–155.

BARNEY, M.C. and HELBERT, J.R. 1976. Use of lyophilization for long-term storage of brewers yeast. J. Am. Soc. Brew. Chem. *34*, 61–64.

BARTH, J. AND SON. 1977. Hops. Joh. Barth and Sohn, Nürnberg.

BRENNER, M.W. 1965. Disinfection of brewing yeast with acidified ammonium persulfate. J. Inst. Brew. *71*, 290.

BRENNER, M.W., KARPISCAK, M., STERN, H. and HSU, V.P. 1970. A differential medium for detection of wild yeast in the brewery. Am. Soc. Brew. Chem. Proc., 79–88.

BRUCH, C.W., HOFFMAN, A., GOSINE, R.M. and BRENNER, M.W.   1964. Disinfection of brewing yeast with acidified ammonium persulfate.   J. Inst. Brew. 70, 242–246.

BUCHANAN, R.E. and GIBBONS, N.E.   1975.   Bergey's Manual of Determinative Bacteriology, 8th Edition.   Williams & Wilkins, Baltimore.

BULDER, C.J.E.A. and REININK, M.   1974.   Unsaturated fatty acid composition of wild type and respiratory deficient yeasts after aerobic and anaerobic growth.   Antonie van Leeuwenhoek J. Microbiol. Serol. 40, 445–455.

BURKHOLDER, P.R.   1943.   Vitamin deficiencies in yeast.   Am. J. Bot. 30, 206–211.

CAMPBELL, I.   1972.   Simplified identification of yeasts by a serological technique.   J. Inst. Brew. 78, 225–229.

CHILVER, M.J., HARRISON, J. and WEBB, T.J.B.   1978.   Use of immunofluorescence and viability stains in quality control.   J. Am. Soc. Brew. Chem. 36, 13–18.

CRIDDLE, R.S. and SCHATZ, G.   1969.   Promitochondria of anaerobically grown yeast. I. Isolation and biochemical properties.   Biochem. 8, 322–334.

CURTIS, N.S. and CLARK, A.G.   1957.   Experiments on growing culture yeast for the brewery.   Eur. Brew. Conv., Proc. Cong., Copenhagen, 1957. Elsevier Publishing Co., Amsterdam, London, New York, Princeton, N.J.

CURTIS, N.S. and CLARK, A.G.   1960.   New yeast culture plant.   J. Inst. Brew. 66, 287–292.

CZARNECKI, H.T. and VAN ENGEL, E.L.   1959.   The isolation and identification of diacetyl producing brewers yeast.   Brew. Dig. 34 (3) 52.

DELBRÜCK, M.   1892.   Accelerated fermentation with controlled yeast. Wochenschr. Brau. 9, 695. (German)

GREEN, S.R.   1955.   A review of differential techniques in brewing microbiology.   Wallerstein Lab. Commun. 18, 239–251.

HALL, J.F. and WEBB, T.J.B.   1975.   Factors affecting the survival of lyophilized brewery yeast strains.   J. Inst. Brew. 81, 471–475.

HARDEN, A.   1932.   Alcoholic Fermentation.   Longmans, Green, and Co., London.

HARRIS, J.O. and WATSON, W.   1968.   The use of controlled levels of actidione for brewing and nonbrewing yeast strain differentiation.   J. Inst. Brew. 74, 286–290.

HELM, E., NOHR, B. and THORNE, R.S.W.   1953.   The measurement of yeast flocculence and its significance in brewing.   Wallerstein Lab. Commun. 16, 315–325.

HESSELTINE, C.W.   1974.   Natural occurrence of mycotoxins in cereals. Mycopathol. Mycol. Appl. 53, 141–153.

HOUGH, J.S., BRIGGS, D.E. and STEVENS, R.   1971.   Malting and Brewing Science.   Chapman and Hall, London.

KATO, S.   1967.   A new measurement of infectious wild yeasts in beer by means of crystal violet medium.   Bull. Brew. Sci. 13, 19–24.

KATZ, P.C.   1979.   National patterns of consumption and production of beer. *In* Fermented Food Beverages in Nutrition. C.F. Gastineau, W.J. Darby and T.B. Turner (Editors).   Academic Press, New York.

KIRSOP, B.   1974A.   The stability of biochemical, morphological, and brewing properties of yeast cultures maintained by subculturing and freeze-drying.   J. Inst. Brew. *80*, 565–570.

KIRSOP, B.H.   1974B.   Oxygen in brewery fermentation.   J. Inst. Brew. *80*, 252–259.

KOT, E.J. and HELBERT, J.R.   1977.   Two verdant types of *Saccharomyces uvarum*.   J. Am. Soc. Brew. Chem. *35*, 25–28.

KOZULIS, J.A. and PAGE, H.E.   1968.   A new universal beer agar medium for the enumeration of wort and beer microorganisms.   Am. Soc. Brew. Chem. Proc., 52–58.

KROGH, P., HALD, B., GJERTSEN, P. and MYKEN, F.   1974.   Fate of ochratoxin A and citrinin during malting and brewing experiments.   Appl. Microbiol. *28*, 31–34.

KÜTZING, F.   1837.   Microscopic investigations of yeast and mother of vinegar together with other associated vegetable forms. (Abbrev. Lect. Given Meet. Soc. Nat. Hist. Harz, Alexisbad, July 26, 1837.)   J. Prakt. Chem. *11*, 385–409. (German)

LEHNINGER, A.L.   1975.   Biochemistry, 2nd Edition.   Worth Publishers, New York.

LINDEGREN, C.C.   1949.   The Yeast Cell, Its Genetics and Cytology.   Educational Publishers, St. Louis.

LODDER, J.   1970.   The Yeasts, A Taxonomic Study, 2nd Edition.   North-Holland Publishing Company, Amsterdam.

MARKHAM, E.   1969.   The role of oxygen in brewery fermentations.   Wallerstein Lab. Commun. *32*, 5–12.

McGILVERY, R.W.   1975.   Biochemical Concepts.   W.B. Saunders Co., Philadelphia.

MONOD, J.   1949.   The growth of bacterial cultures.   Annu. Rev. Microbiol. *3*, 371–394.

NIP, W.K., CHANG, F.C., CHU, F.S. and PRENTICE, N.   1975.   Fate of ochratoxin A in brewing.   Appl. Microbiol. *30*, 1048–1049.

OGUR, M., ST. JOHN, R. and NAGAI, S.   1957.   Tetrazolium overlay technique for population studies of respiratory deficiency in yeast.   Science *125*, 928–929.

PALTAUF, F. and SCHATZ, G.   1969.   Promitochondria of anaerobically grown yeast. II. Lipid composition.   Biochemistry *8*, 335–339.

PASTEUR, L.   1860.   *Cited by* Harden, A. 1932. Alcoholic Fermentation. Longmans, Green, and Co., London.

PASTEUR, L.   1866.   Studies on wine. Imprimerie Imperiale, Paris. (French)

PASTEUR, L.   1868.   Studies on vinegar.   Gauthier-Villars, Paris. (French)

PASTEUR, L.   1876.   Studies on beer.   Gauthier-Villars, Paris. (French)

PIENDL, A.   1977.   Bavarian vs. U.S. beers—an analytical comparison. Brew. Dig. *52* (4) 58–62.

PIESLEY, J.G. and LOM, T.   1977.   Yeast—strains and handling techniques. *In* The Practical Brewer, 2nd Edition. H.M. Broderick (Editor).   Master Brewers Assoc. of the Americas, Madison, Wis.

PLATTNER, H. and SCHATZ, G.   1969.   Promitochondria of anaerobically grown yeast. III.   Morphology. Biochemistry *8*, 339–343.

PRIEST, F.G., COWBURNE, M.A. and HOUGH, J.S.   1974.   Wort enterobacteria—a review.   J. Inst. Brew. *80*, 342–356.

RAINBOW, C.   1939.   The Bios requirements of various strains of *Saccharomyces cerevisiae*.   J. Inst. Brew. *45*, 533–545.

RAINBOW, C.   1948.   *p*-Aminobenzoic acid, a growth factor for certain brewer's yeasts.   Nature (London) *162*, 572–573.

RATTRAY, B.M., SCHIBECI, A. and KIDBY, D.K.   1975.   Lipids of yeasts. Bacteriol. Rev. *39*, 197–231.

RICE, J.F. and HELBERT, J.R.   1974.   The quantitative influence of agitation on yeast growth during fermentation.   Am. Soc. Brew. Chem. Proc. *34* (2) 94–96.

RICHARDS, M.   1969.   The rapid detection of brewing contaminants belonging to the genus *Saccharomyces*—examination of lager yeasts.   J. Inst. Brew. *75*, 476–480.

RICHARDS, M.   1975.   Preservation of brewing yeast characteristics by freeze-drying.   Am. Soc. Brew. Chem. Proc. *33*, 1–4.

ROSE, A.H.   1976.   Chemical Microbiology, 3rd Edition.   Plenum Press, New York.

SENTHESHANMUGANATHAN, S.   1960A.   Biochem. J. *74*, 568–576.

SENTHESHANMUGANATHAN, S.   1960B.   Biochem. J. 77, 619–625.

SENTHESHANMUGANATHAN, S. and ELSDEN, S.R.   1958.   Biochem. J. *69*, 210–218.

STANIER, R.Y., ADELBERG, E.A. and INGRAHAM, J.L.   1976.   The Microbial World, 4th Edition.   Prentice-Hall, Englewood Cliffs, N.J.

SWANSON, W.H. and CLIFTON, C.E.   1948.   Growth and assimilation in cultures of *Saccharomyces cerevisiae*.   J. Bacteriol. *56*, 115–124.

SWINGS, J. and DeLEY, J.   1977.   The biology of *Zymomonas*.   Bacteriol. Rev. *41*, 1–46.

THORNE, R.S.W.   1972.   The problem of typing brewery yeast.   Master Brew. Assoc. Am. Tech. Q. *9*, 173–182.

U.S. BREW. ASSOC.   1976.   Brewers Almanac. U.S. Brewers Assoc., Washington, D.C.

Van VUUREN, H.J.J. *et al.*   1978.   *Enterobacter agglomerans*—a new bacterial contaminant isolated from lager beer breweries.   J. Inst. Brew. *84*, 315–317.

WALTERS, L.S. and THISELTON, M.R. 1953. Utilization of lysine by yeasts. J. Inst. Brew. *59*, 401–404.

WATSON, K., HASLAM, J.M. and LINNANE, A.W. 1970. J. Cell Biol. *46*, 88–96.

WICKERHAM, L.J. 1951. Taxonomy of yeasts. Techniques of classification. U.S. Dep. Agric. Tech. Bull. *1029*.

WICKERHAM, L.J. and ANDREASEN, A.A. 1942. The lyophil process: its use in the preservation of yeasts. Wallerstein Lab. Commun. *5*, 165–169.

WINDISCH, S., OESER, H., STECHOWSKI, U. and EMEIS, C.C. 1967. Process for the production of superattenuated beer. Ger. Pat. 1,792,747. July 6. (German)

WINGE, Ö. 1935. Haplophase and diplophase in some *Saccharomycetes*. C.R. Trav. Lab. Carlsberg *21*, 77.

WINGE, Ö. and LAUSTEN, O. 1939. *Saccharomycodes ludwigii* Hansen: A balanced heterozygote. C.R. Trav. Lab. Carlsberg *22*, 357.

WYNANTS, J. 1962. Preservation of yeast cultures by lyophilization. J. Inst. Brew. *68*, 350–354.

SOME USEFUL MONOGRAPHS, SYMPOSIA, REVIEWS, AND ARTICLES
(In Chronological Order)

HANSEN, A. 1948. Jorgensen's Micro-organisms and Fermentation. Charles Griffin and Co., London.

DeCLERCK, J. 1958. A Textbook of Brewing, Vol. 1 and 2. (Translated by K. Barton-Wright). Chapman and Hall, London.

SHENEMAN, J.M. and HOLLENBECK, C.M. 1960. Microbial patterns in malt. Lactic acid bacteria. Am. Soc. Brew. Chem. Proc., 22–27.

SHENEMAN, J.M. and HOLLENBECK, C.M. 1961. Microbial patterns in malt. II. Aerogenes-type bacteria. Am. Soc. Brew. Chem. Proc., 93–97.

AULT, R.G. 1965. Spoilage bacteria in brewing—A review. J. Inst. Brew. *71*, 376–391.

ROSE, A.H. and HARRISON, J.S. 1969–1971. The Yeasts, Vol. 1, 2, and 3. Academic Press, London, New York.

LODDER, J. 1970. The Yeasts, A Taxonomic Study, 2nd Edition. North-Holland Publishing Co., Amsterdam.

AULT, R.G. and NEWTON, R. 1971. Spoilage organisms in brewing; brewing hygiene and biological stability of beers. *In* Modern Brewing Technology. W.P.K. Findlay (Editor). Macmillan Press, London.

FINDLAY, W.P.K. 1971. Modern Brewing Technology. Macmillan Press, London.

HOUGH, J.S., BRIGGS, D.E. and STEVENS, R. 1971. Malting and Brewing Science. Chapman and Hall, London.

KLEYN, J. and HOUGH, J.   1971.   The microbiology of brewing.   Annu. Rev. Microbiol. *25*, 583.

REED, G. and PEPPLER, H.J.   1973.   Yeast Technology.   AVI Publishing Co., Westport, Conn.

EUR. BREW. CONV.   1974.   E.B.C. Wort Symp., Monogr. 1, Nov. 1974. Zeist, Netherlands.

PRIEST, F.G., COWBURNE, M.A. and HOUGH, J.S.   1974.   Wort enterobacteria—A review.   J. Inst. Brew. *80*, 342–356.

BUCHANAN, R.E. and GIBBONS, N.E.   1975.   Bergey's Manual of Determinative Bacteriology, 8th Edition.   Williams & Wilkins, Baltimore.

LEHNINGER, A.L.   1975.   Biochemistry, 2nd Edition.   Worth Publishers, New York.

McGILVERY, R.W.   1975.   Biochemical Concepts.   W.B. Saunders, Philadelphia.

MEILGAARD, M.C.   1976.   Wort composition: With special reference to the use of adjuncts.   Master Brew. Assoc. Am. Tech. Q. *13*, 78–90.

ROSE, A.H.   1976.   Chemical Microbiology, 3rd Edition.   Plenum Press, New York.

STANIER, R.Y., ADELBERG, E.A. and INGRAHAM, J.L.   1976.   The Microbial World, 4th Edition.   Prentice-Hall, Englewood Cliffs, N.J.

BRODERICK, H.M.   1977.   The Practical Brewer, A Manual for the Brewing Industry, 2nd Edition.   Master Brewers Assoc. of the Americas, Madison, Wis.

ETCHEVERS, G.C., BANASIK, O.J. and WATSON, C.A.   1977.   Microflora of barley and its effects on malt and beer properties: A review.   Brew. Dig. *52* (1) 46–50, 53.

ROSE, A.H.   1977.   Alcoholic Beverages.   Academic Press, London, New York.

SWINGS, J. and DeLEY, J.   1977.   The biology of Zymomonas.   Bacteriol. Rev. *41*, 1–46.

11

Distilled Beverage Alcohol

D.A. Brandt

Fermentation alcohol production in the United States generated nearly $12,000,000,000 for public treasuries during 1976. One of every 50 Americans was employed in the alcohol industries and well over $12,000,000,000 was spent in acquiring materials and services from supporting sources. In addition, the estimated total beverage alcohol product was in excess of $34,000,000,000 for 1976. For fiscal 1977 total production of distilled beverage spirits was 159 million tax gallons with the following distribution (in millions of tax gallons): whiskey 82, brandy 19, rum 1, gin 26, and vodka 33 (Anon. 1978, 1979).

The total quantity of distilled spirits produced during the 12 months preceding June 30, 1976, was 668,000,000 tax gallons. A tax gallon (also called proof gallon) is a standard United States gallon (3.8 liters) containing 50% alcohol by volume. These gallons were manufactured in the following categories—whiskeys, brandies, rum, gin, vodka, alcohol, and spirits.

Beverage alcohols are probably as old as mankind. Legendary evidence has been obtained from 800 B.C. when descriptions of crude distillations were made about observations of condensation of warmed wine on the ceilings of caves. Master Salernus, an Italian physician, reported on his fractionation of alcohol in the 12th century. Some years later Albertus Magnus added significant improvements in the separation of alcohol from aqueous solution.

Early uses of alcohols were primarily pharmaceutical, and probably provided the pattern for the prescription in more modern times for mild tranquilizing applications.

What was originally a cottage industry has steadily expanded to distilleries capable of producing thousands of tax gallons per day. Similarly, the product line has been broadened; and with the adoption of consumer testing techniques by a steadily increasing portion of the industry, it is quite likely expansion will continue into the future.

The triumvirate of quality, service, and cost has led to a diminishing number of producing units. Pennsylvania and Maryland during one period after the repeal of Prohibition (1933) were major manufacturing states. Currently, Kentucky, Indiana, Illinois, and Tennessee distill a major portion of the total.

The chemical composition of the distilled beverages discussed is shown in Table 11.1.

WHISKEY

Definition

Whiskey as a category is defined by the United States government as an alcoholic distillate from a fermented mash of grain received in a cistern tank at less than 190° proof; has the organoleptic characteristics normally attributed to whiskey; is aged in oak barrels (exception: corn need not be stored in wood); and is bottled at a minimum of 80° proof. Bourbon, rye, wheat, or malt whiskey is further described as being distilled at less than 160° proof; made from a mash bill of at least 51% corn, rye, wheat, or malted barley, respectively, and matured in charred new oak containers at not more than 125° proof. Corn whiskey is distilled at less than 160° proof; made from a mash of 80% corn or more; and ordinarily aged in used charred barrels.

If the above described whiskeys are stored for a minimum of 2 years, they may be designated further as straight (bourbon) whiskey or straight (corn) whiskey.

In January 1968, a new type of whiskey production was authorized and identified as "light." The definition specifies distillation at more than 160° proof and maturation under usual procedures in used charred oak barrels.

Raw Materials

The corn selected for beverage alcohols is typically USDA #1, #2, or #2 distillers' grade and is probably grown in Indiana, Illinois, Iowa, Minnesota, Nebraska, or Kentucky. The rye is USDA #1 or #2 plump and is raised in Minnesota or North Dakota. Similarly, the malt is made from barley grown in these 2 states.

Upon receipt at the distillery the grains are sampled, graded, and checked for possible presence of undesirable contaminants such as aflatoxin in corn or unusually high bacterial counts in the malted barley.

Airveyors or bucket elevators may be used to convey the grain over magnetic metal scavengers, through cleaners, and into storage or milling. Ecological and safety requirements demand that suitable dust collecting devices be utilized to minimize air pollution and explosion danger.

Mashing and Conversion

Dry milling may be accomplished with attrition, hammer, or roller mills. A distiller's aim is to achieve maximum starch exposure with minimum flouring. Attrition and hammer mills work satisfactorily for atmospheric

TABLE 11.1.   COMPONENTS OF WHISKEY

(Typical values reported as g/100 liters/100° proof.)

Components	Bourbon		Rye Whiskey		Corn Whiskey		Light Whiskey		Tennessee Whiskey		Scotch Whiskey		Irish Whiskey
	Unaged	Aged	Unaged	Aged	Unaged	Aged	Unaged	Aged	Unaged	Aged	Unaged	Aged	Aged
Solids		127.4		164.7				145.6		168.8	3.7	120.5	100.3
Total acids, as acetic		44.7		71.2				27.9		69.9	5.3	37.5	14.9
Volatile acids, as acetic		39.4		61.4				19.6		57.6	5.3	27.8	10.4
Tannins		53.4		54.9		28.3		17.4		67.0	0.0	27.3	15.7
pH		3.9		3.9				3.9		4.2	5.5	4.0	4.0
Color (yellow-green filter)		52.0		51.0				65		48	—	65.5	65
Ethyl acetate	4.9	27.8	5.0	63.7	4.8	6.3	0.3	11.2	5.0	35.3	25.1	20.3	5.9
Ethanol	0.7	4.1	0.3	10.8	1.5	6.7	0.2	2.0	0.3	3.2	4.6	2.5	1.3
Methanol	11.4	9.8	2.1	8.3	3.6	3.0	5.5	3.1	12.8	16.3	2.3	2.4	2.3
1-Propanol	13.5	16.2	16.2	13.9	8.3	10.7	10.4	6.7	18.4	22.4	24.8	14.7	29.5
2-Methylpropanol	30.5	28.2	76.8	69.4	28.3	41.4	0.6	9.0	41.1	46.2	41.0	70.8	33.4
2-Methylbutanol	34.1	34.5	48.5	60.0		51.5	0.0	10.8	29.4	41.3	18.8	28.9	14.6
3-Methylbutanol	95.8	98.7	135.0	126.5	158.3	91.0	0.0	25.0	100.5	124.2	58.9	84.2	28.2
Total fusel oils	173.9	177.6	276.5	269.9		194.6	11.0	51.5	189.4	234.1	143.5	198.6	105.7
Ethyl-2-hydroxypropanoate	4.1	8.1	1.0	1.0	0.23	0.6	0.6	0.6	30.1	9.3	2.8	0.0	0.3
Ethyl octanoate	0.43	0.55	0.5	1.2	0.66	1.1	0.1	0.1	0.5	0.7	1.1	2.1	0.4
Furfural	0.00	0.83	0.0	1.5	0.01	0.2	0.4	0.4	trace	1.6	trace	0.8	0.1
Ethyl decanoate	1.00	0.91	0.7	1.5	1.1	1.5	0.3	0.3	0.7	1.2	3.5	2.5	0.5
Ethyl dodecanoate	0.90	0.39	0.2	0.5	0.37	0.5	0.2	0.2	0.8	0.3	3.2	1.1	0.1
2-Phenylethanol	3.13	2.5	0.9	2.1	3.4	2.5	0.6	0.6	4.6	3.5	2.0	2.2	0.3
Ethyl tetradecanoate	0.14	0.16	0.3	0.4	0.1	0.1	0.1	0.1	0.3	0.1	0.6	0.1	0.2
Ethyl hexadecanoate	0.83	0.84	1.8	0.4	0.13	0.2	1.0	1.0	trace	0.2	1.9	0.5	0.4

Source: Schenley Distill. (1979).

mashing. Roller mills may be preferred for pressure cooking due to relatively low incidence of flouring when operated properly. Willkie and Prochaska (1943) compare productive capacity of these 3 types of mills when 25% of the ground corn remains on a #12 screen as follows:

Attrition—with 40.6 cm (16 in.) grinding plates—3520 liters (100 bu)/hr

Hammer—with 60.9 cm (24 in.) wide grinding surface—10,560 liters (300 bu)/hr

Roller with 22.8 cm × 76.2 cm (9 in. × 30 in.) rolls, 3 break—3520 liters (100 bu)/hr

Meal should be held for only short periods of time due to susceptibility to hygroscopic action and to infection. Ideally meal will be slurried almost immediately after grinding. Scale hoppers are most generally used for weighing meal into batch cookers while continuous feeding belt devices may be used for continuous cooking.

Wet milling may also be used in the manufacturing of distilled spirits. A sketch of this process would indicate a steeping of the corn in water for approximately 48 hr. The hulls and germ are then separated by milling from the starch and the gluten. The starch and the gluten are slurried once again and separated. In order to fractionate the starch an acid treatment is introduced. One of the streams then is pumped to the distillery for conversion, pH adjustment, and fermentation.

In the most typical situation in a beverage distilled spirits plant, the meal is slurried with water, and in some applications stillage or spent beer. In a further move toward starch exposure, the corn slurry may then be mashed at atmospheric temperatures or cooked in batch or continuous vessels under elevated pressures.

Volume of liquid introduced for mashing or cooking may vary from 76 to 114 liters (20 to 30 gal.) per bushel. Temperatures will range from 66° to 177°C (150° to 350°F) and holding times at selected temperature from 30 min to 1 min. Whatever technique is used should produce a mash of pH 5.2–5.5 and good starch exposure. Barley malt at 0.5% added to the mash during or immediately prior to heating will aid materially in reducing the viscosity of the mash. Cooling may be accomplished by flashing to atmospheric pressure and applying vacuum to reduce to malting temperature of 63°C (145°F).

The barley malt meal is slurried separately at approximately 46°–49°C (115°–120°F) and is added when the mash is cooled to 63°C (145°F). Conversion of the starch to fermentable sugar begins immediately, with the α-amylase attacking the long chain sugar molecules. If the cooling to fermentation temperature is not completed in the heating vessel, transfer of the converting mash to double pipe coolers or plate heat exchangers may begin within 5 to 10 min after introducing the malt. Cooling should continue rapidly to fermentation set temperature.

Fermentation

Yeast propagation for whiskey making may be accomplished in various ways and will range from pure culture methods exclusively through combinations of pure cultures and backstocking to the use of outside purchase of commercially prepared dried yeast. A typical procedure will maintain a pure culture on slants which will provide the inoculation for plant scale liquid culture in regular short term cycles. Multi-plant companies may provide starter cultures from a central location where the yeast will be transferred under sterile conditions from the slant to a small liquid medium. At 18–20 hr intervals the volume of culture media will be increased until desired plant dona volume has been achieved. A dona is a starter culture.

In the distillery there may be 2 additional stages of volume acclimatization prior to the inoculation of a plant size fermentor. The plant yeast mash may be prepared from regular fermentor mash by fortification with approximately 15% malt. The yeast mash when received in a sterile yeast tank is inoculated with a lactic acid producing bacterial culture, and maintained at 53°C (128°F) until a pH of 3.9 is reached. The soured mash may then be heated to 63°C (145°F) and held until needed or pasteurized at 82°C (180°F) for 30 min, cooled to 23°–24°C (73°–75°F) and inoculated from the dona (starter culture). Temperatures during culturing should not exceed 28°–29°C (83°–84°F). The yeast should be used when approximately half of initial balling is reached or cooled rapidly to 13°C (55°F) to hold until required.

The barley malt in the mash serves 2 purposes: conversion of starch into fermentable sugar and contribution of flavor to the whiskey. Especially if "hi-con" malt (high diastatic power malt) is used, as little as 5% malt might be sufficient for conversion. This low quantity may sacrifice some flavor. If flavor is not considered important, as little as 2% "hi-con" malt might be used in conjunction with fungal enzymes, as pointed out by Brandt (1975). At least in distillers' mash, pH control is critical when fungal enzymes are used. At pH 5.5 best activity is achieved, but the range should not exceed pH 5.2–5.5. Optimum conversion takes place when the enzymes are added to the fermentor at 21°–24°C (70°–75°F).

A quick start of fermentation is very desirable as the "race" soon begins with the viable biological systems in the mash. Either the goal of producing ethanol is achieved or the unwanted development of too high concentrations of acetaldehyde, ethyl acetate, acetic or butyric acids, and other undesirable fermentation components occurs. While total time is important from this point of view, high temperature which might reduce the favorable condition for rapid fermentation must be avoided so that environment for bacteria or wild yeast will be as unaccommodating as possible.

A small grain, rye, is typically used to add flavor to the bourbon mash. After the repeal of Prohibition, mash bills were composed of 25–30% rye, but today significantly lower quantities are used by many distillers. As rye mash may become very viscous, a choice has to be made between combining with corn meal to cook under pressure or mashing separately at 74°–77°C (165°–170°F) for 30 min. If mashed alone, the rye is added to the cooling corn at about 66°–68°C (150°–155°F). Formation of dough balls can be prevented by carefully watching temperatures and by adding the rye mash to the corn at a slow rate.

Sanitary conditions must be very carefully controlled in preparing the malt slurry for conversion. Beginning with the unloading and continuing on through storage bins, conveyors, mills, and slurry vessels, cleanliness must be the watchword. The malt meal is fed into warm water (46°–49°C or 115°–120°F) and heating is continued to 60°–63°C (140°–145°F). When the corn-rye mash is cooled to about 66°C (150°F), the malt is added. Conversion begins immediately. Within 5 min liquefaction should have proceeded far enough to start pumping through mash coolers to the set temperature selected.

Fermentors may be constructed from concrete, wood, steel, or stainless steel. Good practice would dictate immediate rinsing, washing with suitable detergents, and steaming prior to acceptance of new mash. As soon as the first converted cooled mash is received in the fermentor, a 2.5 to 3% by volume yeast inoculum should be introduced. Beers of 2, 3, 4, or even 5 days may be produced dependent upon the distiller's desire, the availability of temperature control, and the maximum temperature deemed acceptable. Possibly, ideal conditions might dictate a set temperature of 22°C (72°F) and a maximum of 29°C (85°F). Practical limitation might extend from 19° to 32°C (67° to 90°F).

Good biological control will demand a vigorous continuous effort to maintain good housekeeping practices throughout the entire operation.

At the set of the fermentor, representative samples of the filled mash vessel should be obtained for laboratory analysis. Balling, pH, titratable acidity, and temperature should be routinely recorded. Progress samples should be taken thereafter at suitable intervals but at least every 8 hr. Data should be obtained for balling, pH, and temperature.

Operating schedule should be so designed that when laboratory indicates fermentor is "worked out," the beer should be put into a beer "well" for immediate distillation. The laboratory analyses at finish of fermentation might profitably include the alcohol in addition to the "set" items. This will provide means to compare expected yield with the actual volume from the still.

Distillation

Distillation units in use range from kettle-column batch types to continuous multicolumn vacuum stills. Over the past several years final proof of distillation has tended to increase. A typical product from a beer still could be drawn off at 110°–115° proof. A further step—"doubling"—could produce a whiskey from 120° to 145° proof dependent upon the distiller's desire for flavor level considered "best." Some distillers produce beverage spirits with only the beer still. In such instances the proof would in most cases be higher than would be expected when further distillation was planned. The future might bring an expansion in multicolumn whiskey distillation units which are capable of making uniform, light-bodied bourbons which mature more rapidly than the heavier flavored whiskeys. In addition to reducing maturation time, these lighter bourbons might counteract present slowly declining market trends as consumer surveys show sharp interest for bourbons with fewer congenerics. Equally clear is the fact that a strong potential continues to exist for bourbon and it is unlikely this uniquely American product will ever be less than a very strong contender in the beverage alcohol market.

Bourbon Whiskey

The most popular category in the family of American whiskeys continues to be bourbon. Flavor thresholds as recognized organoleptically and defined analytically by appropriate wet and instrumental chemical techniques, have shown a steady trend toward more mild beverages. Contrasting trends as illustrated by the growth of Canadian and Scotch whiskeys make equally clear a demand for flavor even though expressed as significantly less intense.

Bourbons by law must contain at least 51% corn, and in previous years a mash bill of 60% corn, 30% rye, and 10% barley malt was frequently used. If in addition the distiller utilized a particular strain of yeast or employed prolonged or continued backstocking for inoculum, the whiskey produced might well be very high in congenerics. Distillation methods for several years following repeal did not utilize fractionation. The whiskey entered into the barrel for aging contained practically everything produced during the biological chemical process of fermentation. The only component removed was a portion of the water. The level of precursors present led naturally to high fusel oil, abundant esterification, and more than ample quantities of acetic, propionic, butyric, lactic, and other organic acids.

Current bourbon production methods may specify a mash bill of 80–85% corn, 8–10% rye, and 8–12% malt. In yeasting pure culture techniques may be employed exclusively or backstocking may be strictly limited. Distillation may also specify at least minimum removals of "heads and tails."

Bourbon consumption is principally as an "independent" flavor although tremendous volumes are used as components in blended whiskeys, in cordials and liqueurs, and in cocktails.

Strong efforts have been made initially by the Bourbon Institute and more recently by the Distilled Spirits Council of the United States to get worldwide agreement on the specification that if a whiskey is defined as bourbon, it must be produced in the United States.

Rye Whiskey

Rye whiskey by law must be made from a mash containing at least 51% rye grain. Composition in the past ranged from the minimum rye plus 34% corn, and 15% barley malt to 88% rye, and 12% rye malt. Flavors of rye whiskeys may range from a mild, pleasant, pleasing sensation to very harsh, hardly palatable characteristics. Previous distillation methods included the use of "charge" stills as one of the types of equipment to produce rye. A charge still is a series of chambers in a vertical stack arrangement. The vapors rise through the still but the liquids are manually transferred from top to bottom. The mash is exposed to high temperatures for extended periods of time, and as a result protein decomposition is induced with correspondingly heavy, unusual, and less desirable flavor development.

Pleasingly "sweet" rye whiskey may be produced, however, in a rye excess (51%) formula, with clean fermentation and careful minimum burn distillation. This category of whiskey is not nearly as popular at this time as it has been in the past.

Ryes produced currently are usually in the "excess" category and are primarily utilized as desirable components in blended whiskey. Rye distilleries which were fairly numerous in both Maryland and Pennsylvania have been almost completely abandoned. This type of whiskey may have lost most of its popularity due to a rise in preference by the American consumer for blandness in beverages and foods. The decline of rye as a product by itself may illustrate the necessity for the American distiller not only to be aware of market trends but also to try to counteract unfavorable situations through new product development.

Corn Whiskey

Corn whiskey by law must be distilled from a mash containing at least 80% corn. Little, if any, rye is used because of relatively lower potential yield. A typical mash bill would be 92% corn and 8% barley malt. A good corn whiskey will have a pleasingly mild, slightly sweet, clean taste. As a consequence corn blends well with other whiskeys. High usage levels add only small amounts of flavor which do not dominate the characteristics of the completed blend.

Light Whiskey

Light whiskey is a new category authorized by the BATF[1] in 1968. The distillation proof specified ranges from 160° to less than 190°. Permission to distill this beverage alcohol was given so that American producers could more favorably compete with Canadian and Scotch whiskey makers. Lightness in flavor, provision for maturing in previously used barrels, and the

---

[1]The BATF is the Bureau of Alcohol, Tobacco and Firearms of the U.S. Treasury.

opportunity for the production of a new blended 100% whiskey beverage were created with the establishment of this classification.

No requirements for particular mash bills were set for light whiskey and some distillates appeared to have been made from mashes containing small grains. Significant quantities were produced with corn content in excess of 90%. Proof of distillation ranged from 160° to 189° +. As in other types of whiskeys, flavors covered the spectrum from "heavy" to extremely "light." Fractionation accordingly extended from "token" draws of congenerics to significant removal of "heads" and "tails." Very light, pleasant, smooth flavors may be developed after proper maturation in charred white oak.

Tennessee Whiskey

Tennessee is another category of American whiskey which is not defined as bourbon even though similar mash bills are used, nor is it designated as "straight" whiskey even though it is matured in new, charred white oak for more than 2 years. Through the mashing or cooking stages Tennessee whiskey is similar to the others. Following starch exposure, the hot mash is pumped into a "souring" vessel. The mash undergoes natural souring at this stage, that is, lactobacilli produce lactic acid by fermentation. When the proper level of souring has been achieved, the process is stopped. The mash is then cooled to conversion temperature, 63°C (145°F), and is converted with slurried barley malt. After cooling to approximately 20°C (68°F), the mash is inoculated with pure culture yeast. Fermentation proceeds over several days with temperatures not exceeding 32.2°C (90°F). The fermented mash—distillers' beer—is then pumped into a beer still for the initial separation of water from the alcohol. The next stage may be varied if the distiller desires. The options are: (1) leach and double or (2) double and leach.

Leaching is a process which consists of a battery of vessels which have been packed with char produced by burning maple. The charcoal has tremendous surface area which adsorbs some of the larger molecules of selected esters. The charcoal filtration process in fact so materially changes the organoleptic properties of the product that "straight" is not included in the label statement.

In addition to the leaching operation, Tennessee whiskeys undergo a second distillation or doubling. The second stage distillation similar to the typical bourbon method is not truly a fractionation step. While "heads" and "tails" are separated to some degree, recycling materially lessens congeneric removal.

Maturing

Bourbons, ryes, and Tennessee whiskeys are aged in new charred white oak barrels. In former years, the price of cooperage was not as big an economic factor as it is now. A standard barrel will contain ±189 liters (±50 gal.) Current costs are approximately $50.00. The cost of the aging vessel therefore is about $0.26 per liter ($1.00 per gal.). From a practical point of view a barrel is a semipermeable membrane. Over an 8 year period of maturation, the whiskey volume in a barrel may shrink from 189 liters (50 gal.) to 113.5 + liters (30 + gal.) dependent upon the soundness of the container originally and the atmospheric conditions existing in the maturing warehouse. Unfortunately, present law prohibits the reuse of the barrel for maturing any of these types of whiskey.

Corn and light whiskeys may be finished in used barrels. Corn may be designated as "straight" after 2 years in appropriate cooperage but light whiskey may not, as proof of distillation exceeds 160°.

All whiskeys except light may not be entered into barrels for aging in excess of 125° proof. Usually the proof declines for the first several months in the warehouse. The extent of the decrease and the length of recovery time are influenced by the atmospheric conditions maintained in the maturing facility. If water evaporates more readily than ethanol, the proof will increase over a period of time. On the other hand if ethanol is lost at a faster rate, the proof will go down.

Without exception, unaged distillates are mellowed by entering in wood even for brief periods. While vodka and gin may not ordinarily be softened in barrels, the smoothness achieved is easy to perceive. Much remains to be learned about the aging process of distilled spirits, but various analyses reveal changes in solids, acids, color, esters, and tannins. To be kept in mind is the fact that the greater the concentration of fusel oil and certain other congenerics the more vigorous the maturation cycle has to be. From a net volume point of view, the shorter the aging period the more liters (gallons) available for sale. The proper balance must also be maintained in wood flavor acquired by the beverage.

If the whiskey is very light at time of distillation, for example, the aging period must be carefully watched so that the whiskey does not acquire a dominant woody flavor. A clean fermentation, controlled fractionation, and good maturation should produce a palatable product in as few as 2 years. This type of beverage probably should not be entered for more than 4 years as its desirable attributes may be lost soon after 48 months in wood.

To be considered, too, is the adverse result of excessive aging even on heavy-bodied whiskeys. After many years in wood—12 to 15—whiskeys will tend to develop a more common flavor and be less distinguishable from one distillation to another. This may be the result of the emergence of a definite wood taste which tends to dominate other usually characterizing flavors.

The maturation cycle should be established by considering proposed marketing ages; type of beverage produced equating congeneric levels with projected sales aims; and the investment allocated for inventory. Other aspects of achieving the goal may be accomplished readily by providing suitable equipment, knowledgeable personnel, and good quality control.

As the whiskeys will be developing over a period of years, standards will be required as references. Sampling plans will have to be established to determine if desirable goals are being achieved. One suitable way is to randomly sample, rotating each month's production on an annual basis as long as the whiskey is in the maturing cycle. Each individual sample should be examined organoleptically for abnormalities by a group of at least 3 experienced tasters. After screening out the abnormal samples, a proportionate composite should be made of the acceptable items. Objective analyses should be obtained for sensory evaluation by using procedure E-339-67 of the American Society for Testing and Materials. The physical-chemical analyses should be performed using appropriate methods established by the Association of Official Analytical Chemists, current edition. Nonconforming annual samples should be characterized for specific faults and prescription for correction should be made by the experienced taste panel.

Canadian Whiskey

Canadian whiskeys are distilled only in Canada from mash bills of corn, rye, and barley malt. Milling, cooking, and distilling are very similar to United States industry practices. Possibly 90% of the whiskey produced in Canada is distilled at high proofs similar to the United States light whiskey category. The remainder are distilled in procedures related to U.S. bourbons or ryes and these contribute more substantially to the flavor level of the finished blend. The bulk of the whiskeys are matured in used barrels of approximately 189 liter (50 gal.) capacity. Some of the "flavor" whiskeys may be aged in a combination of new and used wood. In order to be designated Canadian whiskey, the spirits must be aged for a minimum of 3 years. Both Canadian and Scotch whiskeys at this time enjoy a label statement advantage, when sold outside the United States, as 3 years in wood enables the bottling without age declaration. In the United States whiskey must be in wood for a minimum of 4 years before bottling is permitted without an age statement.

An appealing characteristic of Canadian whiskey is lightness of flavor, and the increasing popularity of these products may be related to a presently marked trend toward lightness or blandness in many foods and beverages.

Possibly due to severe climatic conditions during winter months, the most prevalent construction is of more substantial buildings with adequate means for heating and ventilating. Brick and concrete warehouses are typical for aging structures. By design, as stated, Canadian whiskeys are lighter or less flavorful than many other distilled spirits and under usual circumstances mature in shorter periods of time. Nonetheless, some special premium whiskeys are matured for long time intervals and are marketed as superior products with very smooth, light, clean organoleptic properties.

Distilleries are located from the eastern provinces across Canada to British Columbia. At the present time, the industry is concentrated primarily in Quebec and Ontario. As in the United States, distilled spirits provide a significant contribution to the Canadian government, with exportation revenue particularly desirable.

Important contributions to the industry have been made by Canadian distillers in applications in the fields of enzymes, yeasts, distilling, and aging of spirits.

Scotch Whiskey

Scotch whiskeys by law and by gentlemen's agreement are made only in Scotland and in the Scottish Isles (Purvis 1977). There are 2 basic types—malts and grains. Rarely are the 2 types produced by the same distiller in one location.

Malt is considered a key component of Scotch whiskey, so important, in fact, that many malt distillers until fairly recent years made their own malt. The process of malting the barley begins by cleaning the grain and is followed by steeping the kernels for 2 to 3 days in warm water. The water is drained from the grain and the soaked kernels in the "floor" method are spread over the floor to an even depth for germination. Periodically during the 10 to 12 day sprouting cycle the grain is turned. In the "box" method mechanical turning devices are built in so that the maltster can closely regulate the frequency of turning.

When the desired level of germination has been achieved, the malt is heated over burning peat to stop further development. At this stage the option exists to produce a malt with more or less flavor. The object of malting or germination is the production of the α- and β-amylases which provide the enzymatic power to convert starch in the grains into sugars which the yeast can ferment. In addition the desired level of "peatiness" is achieved.

The mashing process begins by preparation of the "grist," that is, grinding of the malt. Hot water is drawn into the mash tun and the grist is added. Rakes revolve in the tun which is made with a slotted bottom. Usually 4 hot water rinses are used to produce the wort which is collected in the "backs" (fermentors) for fermentation. After the fermentable sugars have been completely extracted from the tun, the remaining "draff" is removed from the distillery for use as an animal feed.

While the "backs" have traditionally been made from wood, steel is gradually being introduced as the replacement material. Fermentation temperatures range from 22° to the 27° to 32°C range (72°F to the 80°s). The inoculation of the yeast, *Saccharomyces cerevisiae*, may originate from the distiller's own pure culture; may be supplied in dry form by commercial yeast producers; may be "washed" brewers' yeast; or may be combinations of sources. After fermentation of the wort is completed, the "wash" is deposited in a pot still for the distillation of "low wines."

After condensation, the low wines are pumped into another pot still for redistillation. From a mash of 8–9% alcohol, the finished whiskey is drawn off the second still between 115° and 120° U.S. proof. Portions of distillate known as "feints" are also drawn from the stills to be recycled in future distillations. The malt whiskey may then be reduced to approximately 110° U.S. proof and entered into various size cooperage for aging. Capacities range from "barrels" [approximately 189 liters (50 U.S. gal.)] through "hogsheads" [208–360 liters (55–95 U.S. gal.)] to "puncheons" [378.5–416 liters (100–110 U.S. gal.)]. While smaller cooperage than the "barrel" is still used to some extent, the trend appears to be more toward standardization on the "barrel" size. Availability of the desirable wood, white oak, is probably the main reason for selection of the 189 liter (50 gal.) container. Most of these barrels come from the whiskey industry in the United States. Formerly these barrels were "shooked"—disassembled, bundled and shipped. A large Scotch distillery needed a considerable number of skilled coopers to reassemble and make the barrels tight again prior to filling with Scotch. With the advent of containerized shipping facilities, barrels can be safely transported "standing up," and the added expense on both the shipping and receiving points can be saved.

Malts are identified by the regions in which they are produced—Highlands, Lowlands, Islay, or Campbeltown.

Grain whiskeys are made from maize (corn) and barley malt. The maize is grown in South Africa, France, or North America.

The maize is ground, probably cooked under pressure, cooled, converted, yeasted, and fermented. In the early 1800s a Scot named Aeneas Coffey built a 2-column continuous still which was designed not only to be more economical to operate than a pot still but was also constructed to permit the fractionation of the spirit into various components. By the removal of "heads" and "tails" a relatively light flavored spirit is produced. The importance of this type of beverage alcohol cannot be underestimated as the blending of malts and grains has contributed enormously to the successful distribution of Scotch whiskey throughout the world. Grain whiskeys also have important economic advantages to the Scots as the comparatively low level of congenerics enables a shorter maturation cycle. A clean fermentation plus proper distillation and good maturation may result in a pleasant, blendable whiskey in 3 to 4 years. In contrast, many of the heavier-bodied malt whiskeys require aging periods extending over 3 to 5 times the number of years for grain.

Cooperage may be stored on ricks—wood strips—2 or 3 levels high or in wood or steel racks 8 to 10 barrels high. A new trend may be developing by storing on pallets.

Atmospheric conditions generally are cool temperatures with comparatively high relative humidities. Such environment is conducive to a slow rate of aging. As previously stated, if "short" maturing cycles are the distiller's aim, the whiskey should be "light," the cooperage fairly new, and the warehousing more aggressive.

After the whiskeys have reached the projected level of maturity, the blending operation consists of selecting the malts and grains to be utilized. "Reciprocals" are available to most blenders. These are whiskeys produced by other distillers and exchanged either at the time of distillation or after various stages of maturity. Continuity of finished product taste uniformity is sought by the blender, and he may strive to use the same component whiskeys over a considerable number of years. Some blenders may incorporate 15 to 20 different malts and 4 to 5 different grains.

When the mixing of these various whiskeys has been completed, the proof is reduced with water. Some distillers use untreated "soft" water while others use demineralized water.

At blending time some distillers may elect to age further in vats—marry—or some may draw off the whiskey into small wood for marrying. Still others may complete the operation by finishing the product in a continuous sequence.

Many distillers are considering, and some are utilizing, an additional clarifying, stabilizing stage prior to bottling. The completed whiskey is chilled to some suitable temperature and tight filtered to produce a bright, sparkling liquid. With proper temperature—around $-4°C$ (mid-20's F)—good shelf stability can be achieved if proof has been reduced with low mineral content water.

Irish Whiskey

Irish whiskey has gradually evolved from a highly flavored beverage alcohol to a lighter, smoother, more pleasant spirit. In the mid-1960s a typical mash bill would be composed of 60% barley malt, 25% barley, 5% rye, and the balance from wheat and/or oats. Pot stills were used to separate the alcohol from the fermented mash. Currently, much Irish whiskey is a blend of malt and grain produced from different mash bills and distilled with other techniques.

For malt whiskeys the grain is ground usually on a 3-break roller mill. The mash bill is composed of 100% malted barley. If the malt is to be heavily peated, only peat is used in the drying operation. Lightly peated malt though is dried with coal to which a small amount of peat has been added. The wort is prepared in a cast iron mash tun known as a kiev. This vessel is equipped with an open rake for agitation. To the hot water is added the meal and the desired temperature is maintained by means of "coppers." After approximately 3 hr at $63°C$ ($145°F$), the mash is allowed to settle for 2 hr. The wort is then drained through the slotted bottom to the "under-back." This is a large copper vessel used for pasteurizing the wort at approximately $65°C$ ($149°F$). The wort is then cooled through a plate heat exchanger to fermentation temperature and is pumped into a "wash-back" vessel. The entire mashing operation must be carefully controlled from a sanitation point of view, and sufficiently high temperatures must be maintained to avoid possible contamination by thermophilic bacteria.

The wash-back is set at $20°C$ ($68°F$) with a distillers' yeast such as DCL"M"[1]. The culture is acquired in a semisolid form and is acclimated to distillery conditions by growing the inoculum in sterile wort to which yeast foods such as phosphates have been added. In the 24 hr growth phase, the temperature is kept below $29.4°C$ ($85°F$).

Fermentation in the wash-back is completed in 72−96 hr at temperatures not exceeding $26.7°C$ ($80°F$). The completed wash has an alcohol range of 8.75−9.0%.

The spent solids in the mash tun are dug out and sold wet for animal feed.

---

[1]Distiller's Co., Ltd., London.

The distillation system for malts follows this sequence: pot still, which is direct fired, and then spirit still. The product is collected after a small initial "fore shot" and consists of 22,700–29,500 liters (6000–7000 gal.) of malt whiskey from a charge of 60,500 liters (16,000 gal.). The next "cut" taken is identified as "strong feints," which are composed of at least 25% ethanol. The remainder of the charge is "weak feints." Through a series of recyclings, various products are fed back into the system for redistillation. The proof at completion of distillation is approximately 144° U.S. The malt whiskey is drawn into barrels for aging at 125°–126° U.S. proof. Irish distillers' policy is to mature this type of beverage spirit for at least 9 years.

Current procedures for distilling grain whiskeys call for a mash bill of 30% malt and 70% barley. Mashing is the same as for malt. Similarly, fermentation practices follow the same plan. Distillation is different in that a single column continuous still is utilized. The product is drawn from the still at 174° U.S. proof. For aging, the grain whiskeys are reduced to 142° U.S. and mature over a 4 year period.

Processing

To minimize problems associated with undesirable metals, stainless steel for all surfaces which contact whiskey is highly recommended.

Processing the matured whiskeys prior to packaging involves removal of whiskey from the warehouse, emptying the barrel, proof reduction, special treatment if necessary, filtration, laboratory analyses, bottling, and final laboratory check.

Filtration is a very important step in the processing procedure. There are several different types of equipment available. Each has advantages and disadvantages. In common use are tank types, plate and frame units, screens, and to a lesser extent membrane models. Concentration levels of congenerics and proof may be profitably considered when evaluating filtration equipment requirements. Small pore diameter filter media alone may not be sufficient to ensure a stable, haze, cloud and sediment-free whiskey on the shelf.

Clarity and stability of straight whiskeys, for example, bourbon and blends of straights, may be achieved by chilling to the desired temperature and filtering through tight media at the reduced temperature. Principally the fatty acid esters are precipitated through this technique but calcium and magnesium levels may also be lowered. Various instrumental means may be used to measure the clarity of the spirits. Among these are turbidimeters, particle counters, and nephelometers.

Filter aids in use are pads, papers, cellulose granules, activated carbon, and diatomaceous earth. Specifications for these materials should insist on low metal content—particularly for calcium, iron, and copper.

Pads and papers might safely be limited to 900 ppm calcium, 100 ppm iron, and 5 ppm copper. Precoat and body feed materials should contain no more than 2100 ppm calcium, 100 ppm iron, and 5 ppm copper. Statistical sampling plans should be used on incoming supplies of filtration materials and analytical data should be obtained to evaluate acceptability prior to use in processing.

More efficient use of the filtration step may be achieved through a 2-stage effort—coarse and polish. The first pass could remove the macro particles while the polishing stage would be designed to remove the smaller solids. In any event if screen filters are used, it is essential that a second filtration follow through a pad.

Until recently combination asbestos-cellulose pads were in broad, general use. With the advent of concern for ingestion of asbestos fibers, cellulose pads have been steadily replacing asbestos. Experience to date with the 100% cellulose pads shows that whiskeys can still be well filtered, but that the sparkling, brilliant finish obtained when asbestos is properly used cannot be attained.

Filters should be very carefully prepared. When precoating, the most satisfactory approach is to slurry the material in the product to be filtered. Recirculation of the slurry through the filter back to the tank until the liquid clears should ensure a good precoat. In order to extend the life of the filtration media, body feed may be added sparingly to the tank while filtration is in progress. For best results pressure should be at relatively low levels, and the pressure drop across the filter should not exceed 10 lb.

The entire "clean" lot may be lost by careless blowdown of the filter at the end of the cycle. Best practice is to blow the last few liters (gallons) of filtrate into a separate tank for further filtration.

Blended whiskeys prepared from combinations of straight whiskeys, plus light whiskey or grain spirits, need not necessarily be chill filtered. Clarity and brilliance will be better if the procedure described previously is followed. A satisfactory clarification can be achieved, however, by using a tight pad and suitable filter aids. This whiskey type will be stable under usual conditions without chilling due to the fact that troublesome components are present in lower concentrations than are found in straight whiskeys.

Excellent clarity, long term stability, and product integrity are achievable with good processing techniques, well designed properly operated equipment, and high quality inert packaging supplies.

Grain Neutral Spirits

In the United States grain neutral spirits may be produced from any combination of cereal grains. Dependent upon cost, milo or corn is the usual raw material. Mashing or cooking procedures may be the same as for U.S. whiskeys, but wet milling techniques are used by some large U.S. producers. Modern equipment in service is multicolumn continuous distillation in design, and very clean products are the result.

Conversion of starch into fermentable sugars may be accomplished with 5% malted barley; with 0.5% malt plus amyloglucosidase; or with fungal enzymes exclusively. Fermentation may be carried out with proprietary pure culture methods through various combinations to commercial dried or compressed yeast.

If compressed or dried yeast is used, some may elect to rehydrate with 43°C (110°F) water prior to seeding the commercial fermentor. Other applications have demonstrated that the yeast may be added directly to the fermentor as described earlier.

Volume requirements for grain neutral spirits have exhibited a steady increase for the past several years. The chief uses for beverage spirits are for gin, vodka, blended whiskey, and cordials.

Through the development process extending over the past several years, the true neutrality as evaluated organoleptically has improved to the extent that neutral today may be very close to odorless and tasteless. While the law specifies 190° proof or more, existing beverage spirit equipment is installed which is capable of routine production at high commercial rates of 194° + proof. The typical gas chromatographic analysis of such a spirit will result in almost a straight line with the exception of the ethanol peak.

Vodka

As the grain neutral spirits may be very clean as distilled, the BATF has ruled that such spirits may be redesignated as vodka grade without activated charcoal treatment. Formerly, to qualify as vodka the spirits had to be subjected to activated charcoal treatment. If a charcoal filtered treatment is claimed on the label, however, the spirits must be exposed to charcoal.

Gin

Grain neutral spirits may also be used for the production of gin. Domestic as well as imported gins are identified as distilled gin or merely gin. If the label calls for distilled gin, the product must be made by subjecting various botanicals such as juniper berries, coriander seeds, dried orange peel, angelica, cardamom, and many others to neutral spirits in a gin still. The steam may be supplied directly or through coils in the kettle. Heads and tails are removed and the product is withdrawn from the center cut of the run. A wide range of proofs may be selected, but the most common probably is from 140° to 180° U.S. proof.

The other designation—gin—is claimed when typical botanical flavors are extracted from the herbs. These flavors may be placed in a tank of neutral spirits at an appropriate proof, mixed, filtered, and bottled.

Distilled gin may also be entered in wood for usually short periods of time. Even though brief, the aging aids in achieving smoothness without losing the desirable characteristics. Color development may be controlled by the type of barrel used, the length of time in wood, and decolorizing charcoal.

Gin sales in the past were concentrated chiefly in the warm months. While this is still true to some extent, this type of beverage now enjoys sales throughout the year.

By-product Recovery

By-product recovery is a very important aspect of the operation of a modern distillery. After the alcohol has been distilled from the beer, a residual rich in protein, vitamins, minerals, organic acids, fibers, and fats is further processed into a valuable animal feed component. For many years the Distillers Feed Research Council has sponsored research projects at various universities. These investigations not only have provided the identification of valuable ingredients, but also have outlined specific uses for this material from data obtained from animal feeding studies.

The spent beer or stillage for recovery may be split into at least 4 principal fractions: light grains, dark grains, solubles, and/or syrup.

The solids content of the spent beer from the still ranges from 5 to 10% based on the preceding processing operation.

One other obvious disposition for spent beer is a wet feeding cattle operation where the entire by-product is taken from the base of the still to a feedlot. For smaller distilleries this may not be too difficult as far as transportation of the liquid is concerned, but for larger plants such as a 704 kl (20,000 bu) distillery, huge volumes of liquid must be continuously removed from the base of the still. In order to establish a better energy balance, the wet feeding proposition is attractive as more than half of the total energy used in a distillery is expended in the drying operation. Experimental projects have investigated the collection of all the waste from cattle fed in slotted concrete floored pens. Fermentation of the wastes produces methane which in turn is consumed in the generation of steam for manufacturing the principal product.

A typical by-product recovery scheme currently in wide use in the distilling industry begins with separation of the solids from the liquids. Newest techniques employ continuous centrifuges but most commonly utilized procedures follow some form of screening and pressing. Vibratory or "travel and paddle" inclined screens may be the most commonly used. Screening is followed by pressing. At this point moisture content of the solids portion is about 60–65%. Possibly this product could become useful in a feedlot within a relatively short distance of the distillery. Significant water removal has been achieved and the transportation of the semiwet feed would be more economical. Nutritive value is good. Disposition of the liquid would remain a problem.

Feed from the presses may be conveyed into horizontal double tubed rotary dryers mounted on a declining angle. Grain tumbles from the higher to the lower end and is discharged at 8–10% moisture. Dried product is identified as distillers' light or dark grains dependent upon whether the syrup produced from the liquid portion of the spent grains has been added at the feed end. Care must be taken in the operation of the rotary dryers to make sure that moisture levels are controlled to prevent caking of grains on the steam tubes. Excessive accumulation of cake on tubes will prevent adequate heat transfer. The dryer will eventually fail to perform and the system will become clogged by wet grain.

More recent installations have included direct fired or flash dryers in which the wet grain is blown at high velocity through hot gases. This type of dryer accomplishes the same purpose and produces the same finished grains as the rotary type. Possibly the flash dryer offers more potential for good control in that conditions of operation may be altered and will be measurable in a shorter time span. Danger from fire, however, might be greater in the flash dryer.

The liquid portion of the spent beer may be concentrated from less than 10% solids to 28–50% in evaporators. Different types of equipment may be used but principally long or shorter tubes, forced or natural circulation, pressure, or vacuum units are in typical use. They may be operated as forward, backward, or mixed feed systems. Careful operation will prevent fouling of tubes and loss of heat transfer. Short intervals—10 to 14 day cycles—between complete boil-outs should contribute to efficient operation and uniformity of product. The 28—50% solids syrup may be sold as a finished item, drum dried or spray dried to 8–10% moisture, and distributed as distillers' solubles. Another common alternative is to mix the syrup with the light grains produced in the rotary dryer to manufacture a feed identified as distillers' dark grains. The moisture of either light or dark grains should be about 10%.

If distillers' by-products are to be stored for extended periods of time, provision should be made for rotation of inventory due to the hygroscopic nature of the grains.

Dependent upon various factors, a recovery of at least 8 kg (18 lb) of dried grains per 25 kg (56 lb) bu mashed should be attained. The feeds should contain about 28% protein, 8% fiber, and 10% fat. The acid content is relatively high—approximately 1%.

Current research may result in finding applications suitable for human utilization of distillers' by-products. Various amino acids might provide valuable human supplements. Another possibly interesting application might be the utilization of some of the organic acids as flavors or enhancers of flavors in baked goods.

Rum

Rum is defined by law as an alcoholic distillate from the fermented juice of sugar cane syrup, sugar cane molasses, or other sugar cane by-products, produced at less than 190° proof . . . and bottled at not less than 80° proof . . . .

Rum is one of the oldest distilled beverages in the world. Fermentation and distillation plants are distributed widely—probably on every continent. In the United States in the past, rum was produced in New England and in Kentucky. These rums were used for both beverage and flavoring purposes.

The rums best known to North Americans are produced in Puerto Rico, St. Croix, Virgin Islands, Martinique, Jamaica, Barbados, and Cuba. Some distillers use the pure cane juice, but more typical is the use of blackstrap molasses obtained from the refining of cane in the sugar making process.

While general opinion supports the premise that further conversion to fermentable sugars is not required, there is some evidence that fungal amylase might enhance the yield. Investigative work is in progress to determine whether or not yields may be improved.

The popularity of rum is increasing rapidly and may well be due to the light, clean characteristics of many of the beverages now being offered to the consumer.

The molasses is diluted with water to approximately 20° Balling. Many distillers use pure culture yeast but some West Indies producers continue to use spontaneous fermentations. A very short cycle—about 36 hr—is all that is required to utilize the available sugar. It is desirable to hold maximum temperatures at around 32°–33°C (in the low 90°s F).

Most large rum distilleries use continuous multicolumn stills. Good instrumentation facilitates uniform operating conditions which in turn produce the distillation of similar lots on a day-to-day basis.

Very light-bodied, medium, and highly flavored products provide good options for aging and blending. Rums are improved by exposure to white oak barrels as are other spirits. The light rums mellow in short periods while the most flavorful require longer periods of time.

Processing and bottling methods are similar to other beverage alcohol products.

Tequila

Tequila is an old beverage alcohol in terms of the time since its introduction into the world but is relatively new to widespread distribution in the United States. The source for the fermentable material is a variety of cactus identified as agave or mezcal. Plants of various ages are grown in the same field. After selection of the matured plant, harvesting is accomplished by cutting the individual plant with a machete. Under ordinary circumstances the plants are about 7 years old when cut. The body of the plant is cut away and the "head" or "pina" portion is transported to the distillery for processing.

At the tequila factory the "heads" are split into quarters and loaded into an oven. The heads are heated for 24 hr at nearly 93°C (200°F). Molasses is drained from the bottom of the oven during heating and the flow continues while the oven cools over the next several hours. The heads are then shoveled into a shredder. The strips are fed into a roller mill which presses out the juice. The solids are separated out by screening and washed to recover the remaining sugar.

Usually the mash is set at 9.5° Balling. A short fermentation time, about 38 hr, under ordinary circumstances is sufficient to produce about 4.5% ethanol.

Mexican law permits the addition of "piloncilla" brown sugar to the agave liquor to increase the alcohol production potential during fermentation. Yeasting practices range from pure culture through continuous backstocking. Yeasting and fermentation methods closely follow long standing traditional plans.

Tequila is distilled typically in pot stills and the product is drawn from 76°–110° U.S. proof. The classification system is divided into 3 categories: (1) Tequila White—no aging in wood; (2) Tequila Reposado—aged in barrels to which caramel coloring may be added; (3) Tequila Anejo—aged in barrels for a minimum of 1 year.

Tequila traditionally was consumed straight with a pretaste of salt—possibly the rim of the glass was rolled in salt. Tequila is increasing in the United States as the spirit portion of mixed cocktails such as "Margarita" or "Sunrise."

Cordials and Liqueurs

Cordials and liqueurs are defined in Title 27 CFR, Chapter 1, Paragr. 5.22 (b) (Anon. 1977) as being produced from distilled spirits and flavors, and containing not less than 2.5% sugar by weight of the finished product. This category is one of the most rapidly developing in the beverage alcohol industry. The appeal of fruity flavors, pleasantly sweetened, and highlighted by an appropriate distilled spirit has been expanding at such a rapid rate that few fruits or combinations of fruits commonly known are not readily available. Many countries throughout the world have developed their own specialties, such as the United Kingdom, France, Spain, Italy, and Greece.

As an expansion of this general concept the premixed cocktail potential market is being developed. Many of the multi-ingredient cocktails—particularly those which require special processing techniques—are being offered on a very broad basis.

The expansion of packaged cocktails has also introduced some new problems to the industry. Generally biological contamination was not of great concern due to the presence of adequate alcohol concentrations. With a continuing trend toward less and less alcohol, more and more effort has to be expended to ensure biological as well as organoleptic and physical stability. In some facilities it has become necessary to install clean-in-place techniques and equipment. Sanitary tanks, pumps, lines, and package filling machines, as well as pasteurizers and aseptic bulk handling facilities, have become an integral part of this phase of the industry. Processing equipment such as membrane filters, homogenizers, and special mills have also found applications in these installations.

Not only have new sanitary facilities and methods been employed, but also a greatly expanded function of the biological laboratories has been necessary. Sampling procedures, broad selection of suitable culture media, and microbiologists have been added to the increased activity of the biological control services.

Atmospheric conditions surrounding a distillery usually have an abundant supply of molds and yeasts. Care must be exercised to ensure that these specimens do not create hazardous or unusual contamination problems.

Water

Water is a very important component in distillery operation. Abundant supplies are required. Freedom from excessive quantities of iron, calcium, sulfur, undesirable odors and tastes—as well as cold (13°C or 56°F)—water aid materially in the production of good distilled spirits. Chemicals used for the treatment of water for steam generation must be carefully selected as steam becomes a part of the product in slurrying, cooking or mashing, and distilling.

For reducing proof prior to entry of whiskey into the barrel for aging and for the future reduction of the matured product prior to bottling, demineralized or distilled water is required. This water should not have more than a total solids content of 20 ppm. The iron content should not exceed 0.5 ppm, copper should not be more than 2.0 ppm, and the acceptable pH range is from 4.5 to 7.0. As water is perishable, careful estimates are to be made for requirements—ideally for the next 24 hr only. Water storage tanks must be cleaned on a regular short-term basis to avoid microbial contamination.

At scheduled intervals all water sources for processing should be sampled for biological analyses. Good sanitary control should provide a safe, reliable water supply.

Regular monitoring of demineralizing and charcoal filtering equipment must be established to prevent foci of infection from developing. Special vigilance is necessary on water sources due to the insidious nature of deterioration of water. Not only should water be protected by a clean environment but also short time limits for acceptability should be established after water has been drawn into a tank for use.

Congeners

The components of whiskey from various raw materials are shown in Table 11.1. The analytical data shown are typical or representative values. The components can be grouped into acids (and volatile acids), the fusel oil fraction consisting largely of higher boiling alcohols, esters, and tannin. The latter compound is derived from the wood barrels in which the whiskey is stored.

The analysis of whiskey and other distilled alcoholic beverages is difficult. At present, gas chromatographic methods are preferred (Batiz and Rosada 1978). Many of the recent papers on analytical methods contain useful data on the actual composition of distilled beverages (Reinhard 1977; Postel and Adam 1977; Lehtonen and Suomalainen 1979). Analytical techniques may be used to distinguish various types of alcoholic beverages (Smedt and Liddle 1975). The effect of maturing of whiskey on its composition can be seen in Table 11.1. Tracing the fate of individual compounds requires more sophisticated techniques. For instance, the conversion of ethanol to acetic acid, acetaldehyde, ethyl acetate, and ethanol lignin can be followed by using radioactive ethanol (Reazin et al. 1976).

However, in many instances gas chromatographic methods are not sufficient to detect odor compounds in distilled alcoholic beverages. Therefore organoleptic methods are commonly used in the industry. The descriptive sensory analysis of whiskey flavors has been worked out by Piggott and Jardine (1979).

General descriptions of the production of distilled beverages have been provided by the following authors: Willkie and Prochaska (1943); Harrison and Graham (1970); Reed and Peppler (1973); Brandt (1975); Rose (1977); and Maisch *et al.* (1979).

REFERENCES

ANON. 1964. Laboratory Practice, London. United Trade Press, London.

ANON. 1974. Annual Book of ASTM Standards of Identity, Vol. 46. Am. Soc. for Testing and Materials, Philadelphia.

ANON. 1975. Methods of Analysis AOAC, 12th Edition. Assoc. Official Analytical Chemists, Washington, D.C.

ANON. 1976. DISCUS Facts Book. Distilled Spirits Council U.S., Washington, D.C.

ANON. 1977. Code of Federal Regulations, Title 27, Paragr. 5.21. Standards of Identity. Off. Fed. Reg., Gen. Serv. Admin., Washington, D.C.

ANON. 1978. Distilled Spirits Industry; Annual Statistical Review. Distilled Spirits Council U.S., Washington, D.C.

ANON. 1979. Fiscal Year 1977; Alcohol, Tobacco, and Firearms, Summary Statistics. Bureau of Alcohol, Tobacco and Firearms, Washington, D.C.

BATIZ, H. and ROSADA, E. 1978. A single gas chromatographic method for the direct trace analysis of high boiling components in rum. J. Agric. Univ. P.R. *62* (4) 330−342.

BRANDT, D.A. 1975. Distilled alcoholic beverages. *In* Enzymes in Food Processing, 2nd Edition. G. Reed (Editor). Academic Press, New York.

HARRISON, J.S. and GRAHAM, C.J. 1970. Yeasts in distillery practice. *In* The Yeasts, Vol. 3. A.H. Rose and J.S. Harrison (Editors). Academic Press, London.

LEHTONEN, M. and SUOMALAINEN, H. 1979. The analytical profile of some whisky brands. Process Biochem. *14* (2) 5−6, 8−9, 26.

MAISCH, W.F., SOBOLOV, M. and PETRICOLA, A.J. 1979. Distilled beverages. *In* Microbial Technology, 2nd Edition. H.J. Peppler and D. Perlman (Editors). Academic Press, New York.

PIGGOTT, J.R. and JARDINE, S.P. 1979. Descriptive sensory analysis of whisky flavour. J. Inst. Brew. London *85*, 82−85.

POSTEL, W. and ADAM, L. 1977. Gas chromatographic characterization of whisky. Branntweinwirtschaft *117* (12) 229−234.

PURVIS, C. 1977. Malt whisky—the spirit of Scotland. Chem. Ind. London *24*, 975−977.

REAZIN, G.H. *et al.*   1976.   Determination of the congeners produced from ethanol during whiskey maturation. J. Assoc. Off. Anal. Chem. *59* (4) 770–776.

REED, G. and PEPPLER, H.J.   1973.   Yeast Technology.   AVI Publishing Co., Westport, Conn.

REINHARD, C.   1977.   Analysis and evaluation of whisky. Dtsch. Lebensm. Rundsch. *73* (4) 124–129.

ROSE, A.H.   1977.   Economic Microbiology, Vol. 1. Alcoholic Beverages and Potable Spirits. Academic Press, New York.

SCHENLEY DISTILL.   1979.   Central Analytical Services. Schenley Distillers, Cincinnati.

SMEDT, P. DE and LIDDLE, P.   1975.   Differentiation between rums and other spirits. Ann. Technol. Agric. *24* (3/4) 269–286.

WILLKIE, H. and PROCHASKA, J.   1943.   Fundamentals of Distillery Practice. Joseph E. Seagram & Sons, Louisville, Ky.

12

Oriental Fermented Foods

Hwa L. Wang[1]
and C.W. Hesseltine[1]

Fermented foods are essential elements of diets in all parts of the world. We know the great contribution that cheeses, other fermented dairy products, pickles, sauerkraut, fermented sausages, wines, beer, and baked goods have made to the diet of the Western world. However, very few realize the popularity and importance of fermented foods in the Orient, because it is generally believed that the development of fermented foods depends on rather sophisticated microbiology and food technology. It was, indeed, a remarkable achievement in the early history of the Orient.

We do not know how or when fermented foods were developed in the Orient. We can speculate that fermentation was one of the early forms of food preservation, even though man had no idea what happened except that animal and plant materials under certain conditions could be kept for long periods. The discovery of fermentation was purely by chance.

Because of their long history, the Chinese have made major contributions to the development of fermented foods. The Buddhist religion, in which meat was excluded from the diet, undoubtedly played a significant part in the use of salt and fermentation in the preparation of plant materials, striving for flavor in a bland vegetable diet. It is commonly reported that development of the rather sophisticated soybean products was the result of Buddhist monks working in their monasteries. Bush (1959) stated that Buddhism was well established in China and Korea by the 4th century and was introduced into Japan between 500 and 600 A.D. It may be that the cultivation of soybeans (Brandemuhl 1963), as well as their use in food, including fermented foods, was introduced then in Japan.

---

[1]Northern Regional Research Center, Agricultural Research Service, U.S. Department of Agriculture, Peoria, Illinois 61604.
The mention of firm names or trade products does not imply that they are endorsed or recommended by the U.S. Department of Agriculture over other firms or similar products not mentioned.

Oriental food fermentations are often spoken of as "traditional fermentations," but one must not leave the impression that Oriental fermentation processes are primitive operations carried out in the traditional manner in the home, with the art passed on from one generation to the next. Certainly, this does occur, and the preparation of fermented food is a household art. However, the ancient methods of making fermented foods are changing rapidly through modern microbiology and technology. Today, some of the most modern and sophisticated fermentations in the world are found in the Oriental fermented food industry; Japan has become a leader in the field of industrial microbiology.

A great many fermented foods are known in the Orient. In this chapter, only those Oriental food fermentations which have been specifically studied will be discussed in detail.

SOY SAUCE

Of the many Oriental fermented products, soy sauce is the one most widely consumed and the only one that has become well known in the cookery of Western countries. Soy sauce is a dark brown liquid with a salty taste and a distinct pleasant aroma suggestive of meat extracts. It is a seasoning agent used as substitute for salt in preparation of food as well as a table condiment. It enhances the flavor and adds to the color of meats, seafoods, vegetables, and other foods. The product is known as chiang-yu in China; kecap in Indonesia; shoyu in Japan; kanjang in Korea; toyo in the Philippines; and see-iew in Thailand. In the Western world, the name soy sauce has been adopted.

Soy sauce is made by fermentation of a combination of soybeans and cereal, usually wheat, and salt. Some historical accounts seem to indicate that the fermentation of soy sauce may have been patterned after a fermented fish product used along the coast of southeast Asia. The fermentation was said to have originated in China during the Chou dynasty (1121–220 B.C.), and the Chinese made it in ancient times as a household industry. Today, soy sauce fermentation is one of the most advanced, sophisticated, and studied food industries. Nevertheless, in some parts of the Orient, soy sauce fermentation has remained as a family art.

According to Ebine (1976), 253,000 MT of whole soybeans and 178,000 MT of defatted soybean meal were used in Japan in 1974 to produce shoyu, miso, and natto. His data are summarized in Table 12.1. The total production of shoyu in Japan exceeds 1 million kl a year. There are more than 4000 shoyu producers in Japan; however, the biggest 4 or 5 companies produce about 50% of the total production. The annual per capita consumption is estimated at about 10.2 liters. Similar information is not available from other nations of the Orient, but the per capita consumption figure may well apply to other Oriental populations, especially to the Chinese. There seems to be little doubt that Japan leads the soy sauce industry in the world. Japan not only has the largest fermentation plant but also employs the most

TABLE 12.1.   PRODUCTION OF FERMENTED SOYBEAN FOODS IN JAPAN IN 1974

Name of Food	Raw Material	Amount (Metric Tons)
Miso		587,228
	Soybeans	191,621
	Defatted soybean meal	2,200
	Rice	102,104
	Barley	22,280
	Salt	80,265
Shoyu		1,213,350
	Soybeans	14,278
	Defatted soybean meal	176,138
	Wheat	176,319
	Salt	209,674
Natto		90,000
	Soybeans	47,000

Source: Ebine (1976).

advanced technology developed through comprehensive research on all aspects of soy sauce fermentation.

In addition to the fermentation process, soy sauce is also made by a chemical method (Hesseltine and Wang 1978) in which acid hydrolyzes the proteins and carbohydrates. Acid hydrolysis usually results in a more complete breakdown of the substrates than enzyme hydrolysis; however, acid hydrolysis cannot perform many of the other specific reactions or interreactions of hydrolyzed products as carried out by the multiple enzyme systems produced by molds, yeasts, and bacteria. Chemical soy sauce, therefore, does not possess the flavor and odor of fermented soy sauce.

The technology of soy sauce fermentation, as well as of other traditional fermentations, was formerly a closely guarded family art. Now, the major steps are no longer a secret, but the important fine points are still confidential information. Many improvements have been made since the early development of the fermentation, but the basic method of manufacture is almost unchanged (Yokotsuka 1960; Hesseltine and Wang 1978; Yong and Wood 1974) and is illustrated in Fig. 12.1. In preparation for fermentation, soybeans are usually soaked overnight, drained, and steamed for several hours. The effect of cooking conditions on the enzymatic digestion of soybean proteins has been studied; cooking at higher temperatures and for shorter times seems to be the current trend (Yokotsuka 1971). Tateno and Umeda (1955) and the Noda Soy Sauce Co. (1955) claimed that the soybean protein was best utilized when the beans were soaked in water for 10–12 hr at room temperature and then autoclaved at 69–90 kPa (10–13 psi) for about 1 hr. A batch-type cooker was then developed so that the temperature could be raised rapidly to over 120°C and lowered quickly immediately after cooking. Because of economic advantages, over 90% of shoyu in Japan is now made from defatted meal or flakes. According to Umeda et al. (1969), defatted soybean products first are moistened by spraying with water amounting to about 130% of soybean weight and then are steamed at 90 kPa

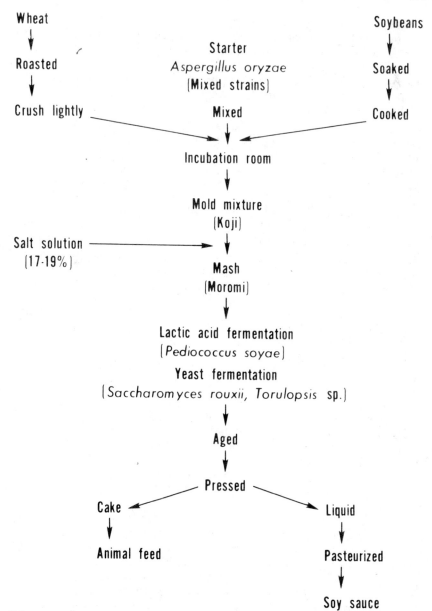

FIG. 12.1.   FLOW SHEET FOR MANUFACTURE OF SOY SAUCE

(13 psi) for 45 min. The cooked soybeans are mixed with wheat that has been roasted and cracked or coarsely ground. The proportion of soybeans to wheat may vary from one manufacturer to another. The best soy sauce is generally believed to be made from a soybean to wheat ratio of 50:50 by

weight, or 52:48 by volume (Yokotsuka 1960). The use of wheat decreases the total nitrogen content of soy sauce, but it contributes aroma and flavor.

The soybean-wheat mixture is next inoculated with a starter which is known as tane koji in Japan and as chung chu (seed mold) in China. A good starter culture must give characteristic aroma and flavor to the soy sauce, have high proteolytic and amylolytic activities, and must be easy to culture. Molds used in soy sauce fermentation have been extensively investigated. The commercial starter or tane koji is made under strictly controlled conditions with tested strains of *Aspergillus oryzae* and *A. soyae*, such as *A. oryzae* NRRL 1988 and 1989. The pure cultures are grown separately on cooked rice for 3–5 days at 30°C until the rice is covered with spores, which are then harvested. Tane koji is usually a mixture of pure spores consisting of several strains in appropriate proportions. A detailed account of the preparation of tane koji—whether for shoyu or other fermentation—can be found in a review of Shibasaki and Hesseltine (1962). The Japanese manufacturers recommend inoculating the soybean-wheat mixture with tane koji at a rate of 0.1 to 0.2%. The inoculated mixture is distributed in shallow wooden boxes to about 5 cm in depth and then incubated at 30°C. After 24 hr of incubation, the mixture is covered with a thin white growth of mold. As mold growth continues, the temperature of the mixture could rise above room temperature to 40°C or higher. Therefore, the mixture should be turned or stirred periodically to maintain uniform temperature, moisture, and aeration. The mixture should also be free of lumps to minimize bacterial propagation. As the incubation time increases, the molds continue to grow and their growth turns yellow and dark green. The moisture of the mixture gradually decreases. After about 72 hr of incubation, the molded mixture, which is called shoyu koji in Japan and chiang chu in China, is ready for brine fermentation. In recent years, automatic koji-making processes have been developed to replace the traditional way of making koji, which involved wooden trays and hand mixing. The new equipment includes automatic inoculator, automatic mixer, large perforated shallow vats in closed chambers equipped with forced filtered air devices, temperature controls, and mechanical devices for turning the substrates during incubation.

Shoyu koji of superior quality has a dark green color, pleasant aroma, sweet but bitter taste and, most important of all, high activities of proteases and amylases. The relationship between enzyme yield and culturing conditions varies with the organisms. *A. oryzae* (Maxwell 1952) and *A. soyae* (Yamamoto 1957) produce greater amounts of proteases at temperatures lower than their optimum growth temperature. Similar findings with other fungi were reported by Wang *et al.* (1974). The effect of temperature on enzyme production observed in these studies emphasizes the importance of a common practice in koji making, i.e., frequent turning of the growth mass or use of thin layers of solid substrates. Otherwise, the heating which results from active growth will increase incubation temperature and affect enzyme production. Harada (1951) found it necessary to keep moisture content of koji at 27 to 37% in order to maintain high enzyme activity.

So moisture and temperature are the two more important factors in making koji of superior quality. Current industry practice is to increase moisture content of the soybean-wheat mixture and to lower incubation temperature.

The second step in preparation of soy sauce is brine fermentation. Traditionally, the shoyu koji is transferred to a deep vessel in which an equal volume of salt solution (17 – 19%) is added to make a mash or moromi as it is called in Japan, but recently the volume of salt solution has been increased to 1.1 to 1.2 times that of the koji. It was found (Yokotsuka 1960) that moromi made with a ratio of salt solution to koji greater than one had a better utilization of the total nitrogen of the raw material; also the lower the concentration of salt in the salt solution the better the utilization of nitrogen. On the other hand, when the concentration of the salt solution is less than 16%, putrefaction may occur.

Since koji making is not done under aseptic conditions, one would expect the presence of yeasts and bacteria in the moromi. However, pure cultures of yeasts and bacteria are sometimes added to the mash to accelerate the fermentation and to improve the flavor of the final product. Strains of *Saccharomyces rouxii*, *Torulopsis* yeasts, and *Pediococcus soyae* were found to be important flavor producers. Representative strains maintained in the Agricultural Research Culture Collection (NRRL) have been designated as *S. rouxii* NRRL Y-6681, *T. etchellisii* NRRL Y-7583, *T. versatilis* NRRL Y-7584, and *P. halophilus* NRRL B-4243 and NRRL B-4244. The initial pH 6.5 – 7.0 of the mash gradually decreases as the lactic acid fermentation advances, and at a pH of around 5.5, yeast fermentation takes place.

The moromi is stirred occasionally during the early stage of the process to provide enough aeration for good growth of yeast, to prevent the growth of undesirable anaerobic microorganisms, to maintain uniform temperature, and to facilitate the removal of carbon dioxide. However, too much stirring hinders the fermentation. The change of temperature is said to be important for the normal progress of fermentation. Therefore, shoyu fermentation in Japan usually starts in April and takes a year to complete. In general, low temperature fermentation gives better results; because the rate of enzyme inactivation is slow, the enzymes remain active longer (Komatsu 1968). Watanabe (1969) indicated that good quality shoyu can be obtained by 6 month fermentation when the temperature of moromi is controlled as follows: starting at 15°C for 1 month, followed by 28°C for 4 months, and finishing the fermentation at 15°C for 1 month.

The matured mash is then pressed and the liquid is pasteurized at 70° – 80°C to stop the microbial and enzymatic reactions, filtered to remove precipitates, and bottled for market. In Japan, either benzoic acid or butyl-*p*-hydroxybenzoate is added as a preservative.

Average composition of soy sauce made from whole soybeans and defatted soybean meal by 11 factories in Japan is given in Table 12.2 (Umeda 1963; Umeda *et al.* 1969). A good soy sauce has a salt content of about 18%. Its pH is between 4.6 and 4.8; below that the product is considered too acid, suggesting that acid has been produced by undesirable bacteria. It is also generally recognized in Japan that the quality and price of shoyu are

TABLE 12.2.   AVERAGE COMPOSITION OF SOY SAUCE MADE FROM WHOLE SOYBEANS AND DEFATTED SOYBEAN MEAL

Conditions	Raw Material	
	Whole	Defatted Meal
Baumé (°)	22.7	23.4
NaCl (%)	18.5	18.0
Total nitrogen (%)	1.6	1.5
Amino nitrogen (%)	0.7	0.9
Reducing sugar (%)	1.9	4.4
Alcohol (%)	2.1	1.5
Acidity I [1]	10.1	14.0
Acidity II [2]	9.8	13.6
pH	4.8	4.6
Glutamic acid (%)	1.3	1.2
Nitrogen yield (%)	75.7	73.7

Source: Umeda (1963); Umeda et al. (1969).
[1] Ml of 0.1 N NaOH required to neutralize 10 ml of soy sauce to pH 7.3.
[2] Ml of 0.1 N NaOH required to bring the pH of 10 ml of soy sauce from pH 7.3 to 8.3.

determined by nitrogen yield, total soluble nitrogen, and the ratio of amino nitrogen to total soluble nitrogen. The nitrogen yield is the percentage of nitrogen of raw material converted to soluble nitrogen, which shows the efficiency of enzymic conversion. The total soluble nitrogen is a measure of the concentration of nitrogenous material in the shoyu and indicates a standard of quality. A ratio of greater than 50% of amino nitrogen to total soluble nitrogen is also evidence of quality. These results can be affected by many factors such as raw materials, steaming conditions, tane koji, koji making, and brine fermentation. Technology to improve these values is constantly being sought. During the last 20–30 years, a shorter cooking time at higher temperatures has been adopted. Koji-making technology has been improved so as to increase enzymatic activity. New strains of koji molds, yeasts, and bacteria have been developed by induced mutations or by diploid formation. Temperatures for moromi fermentation have been controlled. As a result of these improvements, nitrogen recovery has increased from 60% in 1945 to about 90% in 1975 (Ebine 1976), the fermentation period has been reduced significantly, and flavor has been greatly improved.

The chemical changes in the production of shoyu and in its flavor are complicated. Yokotsuka (1960) has written a complete review on these changes. He states that of the total nitrogen, about 40–50% is amino acids, 40–50% peptides and peptones, 10–15% ammonia, and less than 1% protein. Seventeen common amino acids are present, with glutamic acid and its salts being the principal flavoring constituents. The organic bases, believed to be hydrolyzed products of nucleic acids, are adenine, hypoxanthine, xanthine, guanine, cytosine, and uracil. Sugars present are glucose, arabinose, xylose, maltose, and galactose; also 2 sugar alcohols, glyercol and mannitol. Organic acids reported in shoyu are lactic, succinic, acetic, and pyroglutamic. The color of soy sauce is generally recognized to be the result of a nonenzymatic browning reaction. More than 100 compounds have been reported as flavor components of soy sauce; the compounds that decisively characterize soy sauce, however, remain unknown. The guaiacyl compounds seem to have an important effect upon the overall flavor of Japanese shoyu.

MISO

Miso is the Japanese name given to paste-like products made by fermenting cereal, soybeans, and salt with molds, yeasts, and bacteria. Similar products are also made and consumed in other parts of the Orient. Each nation has its own name for the product: chiang in China; tauco in Indonesia; doenjang in Korea; and tao-chieo in Thailand. Literally, they all mean bean paste. Bean paste has the consistency of peanut butter, some smooth and some chunky, and its color varies from light yellow to reddish-brown. It has a distinctive pleasant aroma resembling that of soy sauce, and it is typically salty, although the degree of saltiness may vary; some may even have a sweet taste. Like soy sauce, bean paste is used as a flavoring agent in cooking as well as a table condiment. It is used in place of soy sauce and salt to prepare special flavored foods. These products blend well with varieties of foods including fish, meats, and vegetables.

Variations of bean paste are innumerable. In Japan alone, there are as many types of miso as there are different varieties of cheese in the United States. They are made by varying the cereal used, the ratios of beans to cereal, salt content, length of fermentation, and addition of other ingredients such as hot pepper, which is very popular in China and Korea. Although the manufacturing method may differ from country to country, or from variety to variety, the principle is believed to be the same. This account will be general in nature and follows the procedures for making Japanese miso, because much of the literature on the subject has resulted from studies on Japanese miso. Laboratory methods of making miso have been investigated in the United States by Hesseltine and his coworkers (Hesseltine 1965).

Miso, the most popular fermented food in Japan, is typically used in soup served hot every morning in almost every Japanese home. The soup usually consists of seasonal vegetables with the addition of a spoonful of miso as flavoring agent. About 590,000 MT of commercial miso (Table 12.1) and 150,000 MT of homemade miso are produced in Japan annually (Ebine 1976). The per capita yearly consumption is about 6.7 kg.

According to Ebine (1971), Japanese miso is categorized into 3 major types based on the raw materials used: rice miso made from rice, soybeans, and salt; barley miso from barley, soybeans, and salt; and soybean miso from soybeans and salt. These 3 types are further classified on the basis of taste into sweet, medium salty, and salty groups; each of these groups is again divided by color into white, light yellow, and red. Approximately 80% of the total industrial production is rice miso. The production process, as shown in Fig. 12.2, generally consists of cooking soybeans, preparing koji, mixing cooked soybeans with rice koji and salt, fermenting and ripening in

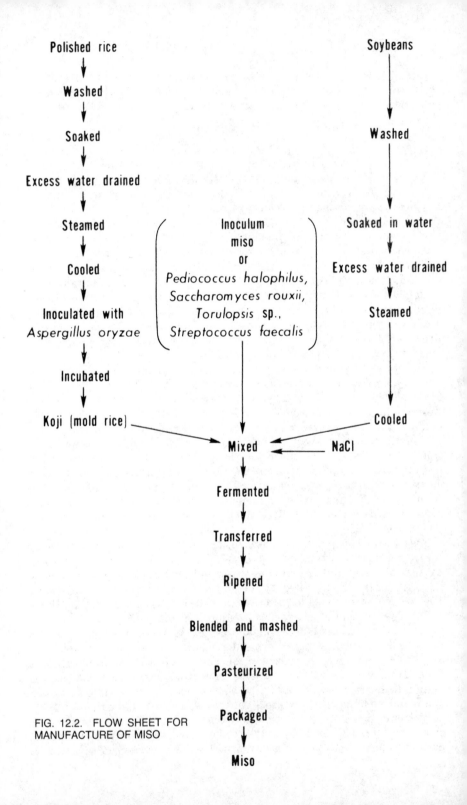

FIG. 12.2. FLOW SHEET FOR MANUFACTURE OF MISO

a tank, blending, pasteurizing, and then packing. The characteristics of 3 types of rice miso in relation to the fermentation conditions are presented in Table 12.3 (Ebine 1967). With respect to microorganisms and fermentation principles, miso and soy sauce fermentations are similar.

Whole soybeans are generally used for making miso. Dehulled soybeans or full-fat soybean grits are sometimes used for making white or light yellow rice miso. A patent by Smith *et al.* (1961) covers the use of grits. Unlike soy sauce fermentation, in which over 90% of shoyu produced in Japan is made from defatted soybean products, defatted soybean products are not suitable for making miso of good quality.

Soybean quality is of utmost importance in miso fermentations. The soybean variety was found to significantly affect the quality and organoleptic scores of the final product. In general, the miso industry prefers to use soybeans of a large size (more than 170 g/1000 seeds) because the ratio of hull to cotyledon is lower in the large beans. Soybeans that have a pale yellow hull and hilum are especially acceptable for making white or light yellow miso. The soybean should have high water-absorbing capacity, and when cooked under the described conditions, the beans should be homogeneously soft with fine texture and bright color. A bright color of the cooked beans is extremely important when white miso or light yellow miso is made. The Japanese miso makers found the domestic soybeans to be the most suitable, followed by Chinese and then United States soybeans. The major difference between Japanese and U.S. soybeans is in the amounts of oil and carbohydrates (total sugar); the Japanese varieties have a higher carbohydrate and a lower oil content than United States varieties. Tests have shown no correlation between the capacity to absorb water and the protein and oil content, but there is a high correlation between carbohydrate content and water-absorbing capacity. Among the United States varieties tested, Kanrich, Mandarin, and Comet are the most promising ones.

TABLE 12.3.   CHARACTERISTICS OF RICE MISO IN RELATION TO FERMENTATION PROCESS

Item	White Miso	Light Yellow Salty Miso	Red Salty Miso
Fermentation process			
Soybean:rice:salt	100:200:35	100:60:45	100:50:48
Fermentation time	2–4days	30 days	60 days
Fermentation temperature	50°C	30°–35°C	30°–35°C
Characteristics			
Color	Bright light yellow	Light yellow	Yellow-red
Taste	Very sweet	Salty	Salty
NaCl (%)	5	12–13	12.5–13.5
Moisture (%)	43	48	50
Protein (%)	8	10	12
Sugar (%)	20	13	11
Shelf life	Short	Fairly long	Long

Source: Ebine (1967).

To prepare whole soybeans for fermentation (Fig. 12.2), they are washed, soaked in water for about 20 hr at 16°C, and drained. The soaked beans are then cooked in water (white miso) or steamed at a temperature of 115°C for about 20 min in a closed cooker. The new batch-type cooker used in the shoyu fermentation, in which the temperature can be raised rapidly to 120°C and lowered quickly immediately after cooking, has also been introduced to the miso industry.

For koji preparation, polished rice is used since it is essential that the mold mycelium quickly penetrate the rice kernels. Polished rice is soaked in water (15°C) overnight or until the moisture content is about 35%. Excess water is drained off, and the soaked rice is steamed at atmospheric pressure for 40 min. A continuous cooker was developed and is now widely used for cooking rice and barley (Ebine 1976). The belt conveyor in this type of cooker allows substantial savings in time and labor. After the steamed rice is cooled to 35°C, tane koji is sprayed over the rice and mixed well. One gram of tane koji (koji inoculum) containing 10^9 or more viable spores is recommended for the inoculation of 1 kg of raw rice. Tane koji for miso fermentation is a blended mixture of several different strains of *A. oryzae* (e.g., *A. oryzae* NRRL 3484, 3485, and 3488) prepared as described under "Soy Sauce." The inoculated rice is then incubated in a temperature- and humidity-controlled room. Koji fermentors of various types (Ebine 1971) are now used instead of koji rooms; a rotary fermentor is the most popular. The temperature and humidity of the air in the fermentor can be regulated so as to promote growth and enzyme production of the mold. In about 40–48 hr at 30°–35°C, the rice is completely covered with white mycelium of the inoculated *A. oryzae* strains. Harvesting is done while the koji is white and before any sporulation has occurred. At this time, the koji has a pleasant smell, lacks any musty or moldy odors, and is quite sweet in taste. The koji is removed from the fermentor and mixed well with salt to stop any further development of the mold.

The next fermentation is carried out under anaerobic conditions by yeast and bacteria. Cooked soybeans are slightly mashed and mixed with the salted koji, and then inoculated with a starter containing pure cultures of yeasts and bacteria. The well-blended mixture, now known as green miso, is tightly packed into a vat or tank for fermentation at around 25°–30°C. Fermentation time varies widely, depending on the type of miso desired; for example, white miso takes 1 week, salty miso 1–3 months, and soybean miso over 1 year. During the fermentation period, the green miso is transferred from one vat to another at least twice to improve fermentation conditions. At the end of the fermentation, the mass is kept at room temperature for about 2 weeks to ripen. The aged product is then blended, mashed, pasteurized, and packaged.

In the past, miso from a previous batch was used as the inoculum of bacteria and yeast; however, the present pure culture starters speed up the fermentation and reduce the influence of weed yeasts and bacteria. Strains of *Saccharomyces rouxii*, *Torulopsis*, *Pediococcus halophilus*, and *Streptococcus faecalis* are the most important yeasts and bacteria in the miso fermentation. Although the industrial strains are not readily available, S. *rouxii* NRRL Y-2547 maintained in the ARS Culture Collection was isolated from miso.

The yield of miso is determined by the length of fermentation and ripening, type, and moisture content. Generally, 3300 kg of light yellow salty miso with a moisture content of 48% is made from 1000 kg of soybeans, 600 kg of rice, and 430 kg of salt.

Since different types of miso are made by varying the ratios of the raw materials, the composition of miso varies with the type. Table 12.4 presents the standard chemical composition of several types of miso as reported by Ebine (1971). Miso has a fairly narrow moisture range of 44–50%. Protein content ranges from 8 to 19% and fat from 2 to 10%, reflecting the increase of soybean used in the fermenting mixture. Except for the white and sweet types, miso contains more than 10% salt. Because of the high salt concentration, miso may be kept for considerable periods without refrigeration; but, for the same reason, the use of miso as a protein food is limited.

To reduce the salt concentration of miso products, enzyme-digested soybean mash has been investigated (Ito *et al.* 1965). Defatted soybean meals are sprayed with an equal weight of water and then steamed under pressure (0.7 kg/cm^2) for 40 min. The enzyme preparation, Takadiastase SS, is added to the cooked material at the level of 0.2%, and the digestion is carried on for 3–4 hr. The enzyme-digested soybean mash is then mixed with miso and allowed to ripen for several days. The new type of miso has about 6.3% salt and 17.6% protein.

Miso has a strong buffer activity due to the presence of protein, peptides, amino acids, phosphoric acid, and various organic acids produced during the fermentation. This property plays an important role as a seasoning in a variety of foods to which miso renders a palatable flavor.

The strong characteristic flavor and the salty taste of miso are not familiar to Westerners, and it may even be considered offensive to some. However, when miso is used in small amounts as a flavoring agent to enhance or to alter the flavor of a food, the taste is distinct and pleasant. Recently we have taken a look at Western cookery and have found (Hesseltine 1976) that miso can substitute or replace part of tomato paste, mayonnaise, and

TABLE 12.4. COMPOSITION OF VARIOUS TYPES OF MISO

Variety	Moisture (%)	Protein (%)	Reducing Sugar (%)	Fat (%)	Sodium Chloride (%)
White miso	44	8	33	2	5
Edo sweet miso	46	10	20	4	6
Salty light yellow miso	49	11	13	5	12
Salty red miso	50	12	14	6	13
Salty barley miso	48	12	11	5	12
Soybean miso	47	19	2	10	10

Source: Ebine (1971).

salt in recipes for salad dressings, barbeque sauces, snack dips, and spreads. The resulting miso-containing products can be considered new items, because they have a distinctly different flavor consumers can easily detect that is mild and nonoffensive. These miso-containing items are similar in appearance and texture to existing products in the same category, and they are used in the same way. The future potential for miso in the United States is good. Perhaps the research on the miso fermentation should be focused on using defatted soybean products as substrate and on reducing the salt concentration of the final product.

TEMPEH

Tempeh is the only Oriental fermented product that has been extensively investigated by the scientists in the West. It originated in Indonesia and is widely consumed in the region of Malaysia and Indonesia, but it was not known elsewhere. Tempeh, or tempe kedelee as it is called in Indonesia, is made by fermenting dehulled soybeans with a mold, *Rhizopus*; the mycelia bind the soybean cotyledons together in a cake-like product. The fresh tempeh has a clean, yeasty odor, but does not have the beany flavor some people find unpleasant in whole soybeans. When fried in oil, it has a pleasant flavor, aroma, and texture that are familiar and highly acceptable even to people of the Western world. Unlike most of the other fermented soybean foods, which are usually used as flavoring agents or relishes, tempeh is used as a main dish and meat substitute in Indonesia. Because of its high protein content and universally acceptable taste and texture, tempeh is a potential source of low-cost protein. The tempeh fermentation is rather simple and quick. Traditionally, soybeans are soaked in tap water overnight until the hulls can be easily removed by hand. Some prefer to boil the soybeans for a few minutes to loosen the hulls and then to soak the beans overnight. After dehulling, the beans are boiled with excess water, drained, and spread for surface drying. Small pieces of tempeh from a previous fermentation, or ragi tempeh (commercial starter), are mixed with the soybeans which are then wrapped in banana leaves and allowed to ferment at room temperature for 24 to 48 hr until the beans are covered with white mycelium and bound together as a cake (Hesseltine *et al.* 1963B). The cake is either sliced thin, dipped in a salt solution, and deep fat-fried in coconut oil, or cut into pieces and used in soups.

Biochemical and microbiological studies are necessary in order to understand the fermentation and to develop uniform, high quality products and well-defined, economically feasible processes to manufacture them. In the late 1950s, scientists at the New York Agricultural Experiment Station, Geneva, N.Y., and at the Northern Regional Research Center, Peoria, Ill., began to study this century-old fermentation. As a result, a pure culture fermentation method was developed on a laboratory scale. Changes in soybeans during fermentation and the nutritional value of tempeh were studied in detail. The physiology and biochemistry of the tempeh mold were also studied. More recently, a freeze-dried tempeh starter was developed so that the tempeh fermentation can be carried out at home as easily as bread making or yogurt fermentation.

The mold used for tempeh fermentation was reported earlier to be *Rhizopus oryzae* (Stahel 1946; Van Veen and Schaefer 1950; Steinkraus *et al.* 1960). We have received cultures isolated from different lots of tempeh in Indonesia, and we found that only *Rhizopus* could make tempeh in pure culture fermentation. Of the 40 strains of *Rhizopus* received, 25 are *R.oligosporus* Saito; others are *R. stolonifer* (Ehren) Vuill, *R. arrhizus* Fischer, *R. oryzae* Went and Geerligs, *R. formosaensis* Nakazawa, and *R. achlamydosporus* Takeda. Apparently, *R. oligosporus* is the principal species used in Indonesia for tempeh fermentation; a strain identified as *R. oligosporus* Saito NRRL 2710 is one of the better producers of a good product (Hesseltine *et al.* 1963B). This strain is characterized by sporangiospores showing no striations and being very irregular in shape under any condition of growth. The sporangiophores are short, unbranched, and arise opposite rhizoids that are very reduced in length and branching. All isolates show large numbers of chlamydospores (Hesseltine 1965).

The utilization of various carbon and nitrogen compounds by *R. oligosporus* Saito NRRL 2710 was investigated by Sorenson and Hesseltine (1966). They found that the principal carbohydrates of soybeans, i.e., stachyose, raffinose, and sucrose, are not utilized as sole sources of carbon, whereas common sugars such as glucose, fructose, galactose, and maltose supported excellent growth, as does xylose. Various vegetable oils can be substituted for sugars as sources of carbon with excellent growth. Since the soybean sugars are not utilized by *R. oligosporus,* and since strong lipase activity has been reported for *Rhizopus* cultures used in tempeh fermentation (Wagenknecht *et al.* 1961; Wang and Hesseltine 1966), it is likely that lipid materials, and particularly fatty acids, are the primary sources of energy for the tempeh fermentation. Ammonium salts and amino acids such as proline, glycine, aspartic acid, and leucine are excellent sources of nitrogen. Other amino acids are less suitable, and tryptophan supports no growth at all. However, the fungus does not depend upon the presence of any specific amino acid in the medium for growth. Sodium nitrate is not utilized as the sole source of nitrogen.

R. oligosporus is highly proteolytic, which is important in tempeh fermentation because of the high protein content of the substrate. Two proteolytic enzyme systems were observed (Wang and Hesseltine 1965); one has an optimum pH at 3.0 and the other at 5.5. Both enzyme systems have maximum activities at $50°-55°C$ and are fairly stable at pH 3.0–6.0, but rapidly denatured at pH below 2 or above 7. The proteolytic enzyme having optimum pH at 3.0 has been purified and separated into 5 active fractions; crystalline enzymes were obtained from the 2 major fractions (Wang and Hesseltine 1970B). In addition to high protease activity, the mold possesses strong lipase activity (Wagenknecht *et al.* 1961; Wang and Hesseltine 1966) but low amylase activity and no detectable pectinase (Hesseltine *et al.* 1963B).

Pure culture methods of making tempeh with *R. oligosporus* were then developed. In many respects, the procedures are similar to those used in Indonesia. Hesseltine *et al.* (1963B) have carried out pure culture fermentations in petri dishes, a testing procedure which proved to be very satisfactory. The preparation of soybeans for fermentation is the same as the traditional manner. Later, Martinelli and Hesseltine (1964) introduced full-fat soybean grits for tempeh fermentation (Fig. 12.3). Soybean cotyledons are mechanically cracked into 4 to 5 pieces. Since soybean grits absorb water

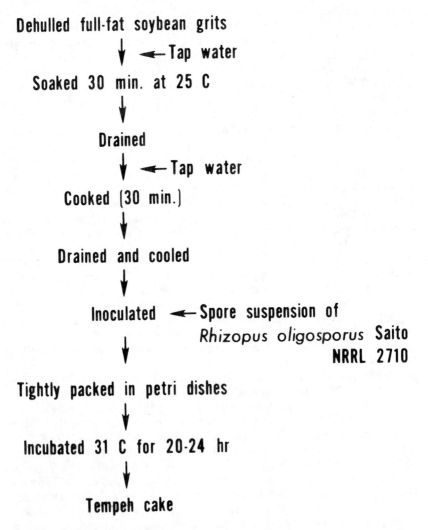

FIG. 12.3.   FLOW SHEET FOR TEMPEH FERMENTATION

easily, the soaking time can be reduced from more than 20 hr to 30 min. Furthermore, since the hulls are removed mechanically in producing grits, much labor can be saved. The beans are boiled for 30 min, drained, cooled, and inoculated with spores of *R. oligosporus*, which have been grown on potato-dextro-agar slants at 28°C for 5–7 days. The spore suspension is prepared by adding a few milliliters of sterilized distilled water to the slant. The inoculated beans are mixed, packed tightly into petri dishes, and placed in an incubator at 30°–31°C for about 20 hr. *R. oligosporus* does not require much aeration as do many other molds; as a matter of fact, too much aeration may cause spore formation. It is, therefore, important to pack the petri dishes tightly; even so, some sporulation may still occur at the edge of the dish, but it will not affect the product. This procedure can also be adopted for making tempeh either in shallow wooden or metal trays with perforated bottoms and covers or in perforated plastic bags and tubes (Martinelli and Hesseltine 1964).

Steinkraus and his coworkers (1960) suggested the use of 0.85% lactic acid as soaking water. The dehulled, soaked beans are also cooked in the acid solution. This treatment would bring the pH of the beans to a range of 4.0–5.0. At this pH range, the growth of contaminating bacteria will be inhibited, but not that of the tempeh mold. However, we have not encountered bacterial growth in our process. Because *R. oligosporus* produces an antibacterial agent (Wang *et al*. 1969), and because this organism also has the unique characteristic of fast growth there is little chance for bacteria to gain ground before the tempeh fermentation is complete. Ko (1970) has further investigated this matter. He purposely inoculated with different amounts of *Escherichia coli, B. mycoides, Pseudomonas pyocyanea, Proteus* sp. or *P. cocovenenaus* along with *R. oligosporus* in making tempeh as described by Hesseltine *et al*. (1963B). His results indicated that the fermentation is not interfered with by the presence of inoculated bacteria. Ko commented that prefermentation during soaking or addition of acid to the soaking water may not be very important in the process of tempeh fermentation.

The dehulling, soaking, washing, cooking, and fermenting steps employed in the preparation of tempeh all contribute to loss of soybean constituents. Average values for these losses obtained from a number of tempeh preparations are given in Table 12.5. The total losses of solids range from 24.5–48.3%, depending on the variety and type of soybeans as well as the processes used. The more significant differences in solid losses are in dehulling and cooking. Steinkraus *et al*. (1960) used an abrasive vegetable peeler to loosen the hull of the hydrated beans and found a loss of 17.1%, of which only 9.6% was due directly to the removal of hulls; whereas Smith *et al*. (1964) removed the hulls by hand and observed a loss of only 7.9%. On the other hand, Smith *et al*. (1964) reported a much greater loss during cooking than Steinkraus *et al*. (1960). It can be explained, in part, by the fact that Steinkraus *et al*. soaked and cooked the beans in an acid solution of pH 5, which is close to the isoelectric point of soybean proteins and may have reduced the leaching effect. Although mechanically dehulled soybean grits

TABLE 12.5.   SOLIDS LOSSES FROM TEMPEH PROCESSING OF VARIETIES OF WHOLE SOYBEANS AND SOYBEAN GRITS

Material and Procedure	Harosoy[1] (%)	Variety of Soybeans Hawkeye[1] (%)	Seneca[2] (%)
Whole soybeans			
Soaking	4.7	1.6	4.9
Dehulling	7.9	7.9	17.1
Cooking	11.0	14.0	1.6
Fermentation	3.4	1.0	3.9
Total loss	27.0	24.5	27.5
Soybean grits			
Mechanical dehulling	9.5	9.5	
Soaking	7.2	5.1	
Cooking	27.1	13.9	
Fermentation	4.5	4.0	
Total loss	48.3	32.5[3]	

[1]Source: Smith et al. (1964).
[2]Source: Steinkraus et al. (1960).
[3]Source: Wang and Swain (1977).

are preferred for tempeh making, one disadvantage appears to be the greater solids losses during soaking. Further studies to determine the varietal effect of soybeans on the total solids losses in tempeh processing are warranted.

To prevent the loss of water soluble substances during preparation and cooking, soybeans were treated in a minimum amount of water, just enough to soak the beans thoroughly, or were sprayed with a certain volume of water before autoclaving. But Smith et al. (1964) found that when this procedure was followed, the tempeh showed less mold development and much sporulation. The product also had an unpleasant odor and poor flavor. The presence of a water-soluble and heat-stable mold inhibitor in soybeans was suggested by Hesseltine et al. (1963A). Later, Wang and Hesseltine (1965) found that the water-soluble and heat-stable fraction of soybeans also inhibited the formation of proteolytic enzymes by R. oligosporus. Therefore, soaking and cooking of soybeans in excess water which is later discarded are essential in making tempeh.

As mentioned earlier, the tempeh fermentation is characterized by its simplicity and rapidity. The lack of a suitable inoculum, however, could be a hindrance, since it is essential that the inoculum be pure and that spores have the ability to germinate immediately. Without these conditions, the fermentation process is almost inoperable. Traditionally, small pieces of tempeh from a previous fermentation serve as inoculum. The fungus is then propagated mainly by means of fast-growing mycelia. This practice can lead to contamination by undesirable microorganisms, and the inability of mycelia to survive adverse temperatures and dehydration makes mycelia unsuitable for long-term preservation of their viability.

In the laboratory, preparing agar media for mass production of spores is expensive and time-consuming, not at all adapted either for industrial processes in advanced countries or for use in less industrially advanced countries. Steinkraus *et al.* (1965) used a powdered lyophilized tempeh mold to inoculate soybeans for a pilot-plant process for producing dehydrated tempeh. They used 3 g of the lyophilized mold culture to each kilogram of precooked soybeans. The inoculum was grown as pure culture on sterilized, hydrated soybeans in 3 liter Fernbach flasks with 500 g of soaked beans per flask. The flasks, after autoclaving and cooking, were inoculated and incubated for 4 days at 37°C. The sporulated culture was freeze-dried and pulverized in a sterilized laboratory burr mill.

Wang *et al.* (1975) developed a tempeh inoculum having a high viable spore count that would maintain its viability for a long time with minimal attention. The spores of *R. oligosporus* Saito NRRL 2710 were made by fermenting rice at a 40% moisture level for 4–5 days at 32°C. The fermenting mass was made into a slurry by blending with sterilized water and then was freeze-dried. On a dry basis, the viable spore count per g of preparation was about 1×10^9 before freeze-drying and 1×10^8 after freeze-drying. When the freeze-dried preparation was kept in a closed plastic bag at 4°C up to 6 months, the spore counts showed typical experimental variations and were comparable to their original counts. At room temperature, a significant decrease in viability was noted after 2 months (from 2×10^7 to 1×10^6); thereafter, no further decrease was observed. The bacterial count of the preparation was minimal; therefore, bacterial contamination was not found to be a problem either during the process of fermentation or in storage. Wheat bran is also a good substrate for sporulation of *R. oligosporus*. When soybeans were used to prepare spores, an unpleasant odor often resulted after 4–5 days of fermentation, perhaps due to their high protein content. Among the substrates tested for spore production by *R. oligosporus*, wheat was the poorest. The stickiness of wheat substrate might have created an anaerobic condition unfavorable for sporulation. We also found that poor growth and sporulation occurred when the ratio of water to substrate was below 4:10, but growth increased as the ratio increased. When the ratio of water to substrate was raised above 8:10, sporulation was significantly less, even though growth was abundant. Therefore, the moisture content of the substrate is of utmost importance in solid fermentations.

The amount of inoculum required to make satisfactory tempeh is significant because fermentation time becomes too critical if the amount of inoculum is too large. On the other hand, too small an amount of inoculum provides a chance for contaminating bacteria to grow. We recommend using 1×10^6 *R. oligosporus* spores per 100 g of cooked soybeans.

In Indonesia, copra (pressed coconut cake) is sometimes used in tempeh fermentation; the product is then known as tempeh bongkrek. We have developed (Hesseltine *et al.* 1967) new tempeh-like products by fermenting cereal grains, such as wheat, oats, barley, rice, or mixtures of cereals and soybeans with *Rhizopus*. Good tempeh (Gandjar and Slamet 1972; Wang *et al.* 1975) can also be made from the water-insoluble fraction of soybeans which is the residue of making soybean milk and tofu, the two main food products derived from the water extraction of soybeans (Wang 1967A). The moisture content of this water-insoluble fraction, however, must be reduced to less than 80% so that the texture appears crumbly before this fraction is suitable for fermentation. On a dry basis, this water-insoluble fraction contains 32% protein, which has the highest quality among several soybean fractions studied by Hackler *et al.* (1963); full-fat soybean flour, water-extract of soybeans, acid-precipitated curd, and whey protein.

Tempeh is perishable and usually is consumed the day it is made because the release of ammonia by enzymatic action causes the product to become obnoxious. Its shelf life, however, can be prolonged by various methods. In Indonesia, they cut the tempeh into slices which are then dried under the sun. We found that the most satisfactory way to keep tempeh is first to blanch the sliced tempeh to inactivate the mold and enzymes, and then to freeze it. Steinkraus *et al.* (1965) developed a pilot-plant process to dehydrate tempeh by a hot air dryer at 93°C for 90–120 min. However, hot air drying causes a reduction in soluble solids, including soluble nitrogen (Table 12.6) as reported by Steinkraus *et al.* (1960). Whether or not the reduction in solubles represents a serious loss in nutritional value has yet to be determined. Lyophilization was found to have less effect on soluble components. Iljas (1969) evaluated the acceptability and stability of tempeh preserved in a sealed can for 10 weeks. There was no significant change in acceptability of the tempeh when the can was sealed and immediately stored at −29°C or when the can was filled with water, steam-vacuum sealed, heat-processed at 115°C for 20 min, and stored at room temperature. However, when tempeh was first air dried at 60°C for 10 hr and then sealed in a can which was stored at room temperature, acceptability of the tempeh tended to decrease as storage progressed. Another way to prolong the shelf life of tempeh is to defer the fermentation (Martinelli and Hesseltine 1964; Wang *et al.* 1975). Preinoculated beans are packaged, stored in the freezer, and allowed to ferment when needed.

TABLE 12.6. COMPARATIVE EFFECT OF FREEZE-DRYING VERSUS HOT AIR DEHYDRATION (69°C) ON TEMPEH

Tempeh	pH	Reducing Substances (%)	Soluble Solids (%)	Soluble Nitrogen (%)
Fresh	6.3	0.71	17.6	2.31
Lyophilized	6.2	0.41	19.5	1.19
Hot air dried	5.3	0.28	13.8	0.61

Source: Steinkraus *et al.* (1960).

The effects of *R. oligosporus* on soybeans have been studied by several investigators and reviewed by Hesseltine and Wang (1978) and Iljas *et al.* (1973). Steinkraus *et al.* (1960) found that the temperature of fermenting beans rises to above that of the incubators as fermentation progresses, but that it falls as the growth of mold subsides. The pH increases steadily, presumably because of the protein breakdown. After 69 hr of incubation, soluble solids rise from 13 to 28%; soluble nitrogen also increases from 0.5 to

2.0%, whereas total nitrogen remains fairly constant, and reducing substances slightly decrease, probably due to utilization by the mold. Similar changes were observed when wheat was fermented by *R. oligosporus* (Wang and Hesseltine 1966). A decrease in ether-extractable substances of soybeans after fermentation was reported by Murata *et al.* (1967) and Wang *et al.* (1968), indicating that the mold uses the soybean oil as its energy source (Sorensen and Hesseltine 1966). Wagenknecht *et al.* (1961) reported that one-third of the total ether-extractable soybean lipid is hydrolyzed by the mold after 69 hr of incubation, and among all the fatty acids, 40% of the linoleic acid is utilized by the mold.

Although the total nitrogen remains fairly constant during the fermentation, free amino acids in tempeh increase. The amino acid composition of soybeans, on the other hand, is not significantly changed by fermentation (Smith *et al.* 1964; Stillings and Hackler 1965). Perhaps the amount of mycelial protein present in tempeh is not high enough to alter greatly the amino acid composition of the soybeans, nor does the mold depend upon any specific amino acid for growth as suggested by Sorenson and Hesseltine (1966).

Niacin, riboflavin, pantothenic acid, and vitamin B_6 contents of soybeans increase after fermentation, whereas thiamin does not change significantly (Roelofsen and Talens 1964; Murata *et al.* 1967). Wang and Hesseltine (1966) also noticed in fermenting wheat with *R. oligosporus* that the amount of niacin and riboflavin of the wheat tempeh greatly exceeds that of unfermented wheat, while thiamin appears to be less. Apparently, *R. oligosporus* has a great synthetic capacity for niacin, riboflavin, pantothenic acid, and vitamin B_6, but not for thiamin.

As mentioned earlier, *R. oligosporus* produces very little amylase. Since starch is seldom found in mature soybeans, it is not particularly important that this species produce amylase during tempeh fermentation. Lipase is produced by the mold to hydrolyze soybean lipids. Proteases are, perhaps, much more important enzymes in tempeh fermentation. The ability of *Rhizopus* to produce proteolytic enzymes varies greatly between different strains of the same species as well as between species (Wang and Hesseltine 1965). The proteolytic enzyme systems have optimal pH at 3.0 and 5.5, with the pH 3.0 type predominating in submerged cultivation and pH 5.5 type predominating in tempeh fermentation.

Lipids in tempeh were found to be more resistant to autoxidation than those in control soybeans. The peroxide value of tempeh was 1.1, whereas that of control soybeans was 18.3 to 201.9 (Iljas 1969). Ikehata *et al.* (1968) found that the peroxide value of lyophilized tempeh stored at 37°C for 5 months increased from 6 to 12 compared with 6 to 426 of unfermented soybeans. The autoxidant activity of tempeh was further substantiated by Packett *et al.* (1971), who reported that corn oil containing 50% tempeh showed higher antioxidant potential than oil containing 25% tempeh, 0.01% α-tocopherol, or 0.03% α-tocopherol. In 1964, György *et al.* isolated a new isoflavone from tempeh designated as "Factor 2," which was then identified as 6,7,4'-trihydroxyisoflavone. 6,7,4'-Trihydroxyisoflavone was later chemically synthesized and proved to be a potent antioxidant for Vitamin A and for linoleate in aqueous solution at pH 7.4 (Ikehata *et al.* 1968). However, when the isoflavone was mixed with soybean powder or soybean oil, it did not prevent their autoxidation. These authors speculated that the insolubility of the isoflavone in the oil and the difficulty of dispersion into soybean powder may be some of the reasons for its failure to prevent autoxidation. Therefore, the compound responsible for the antioxidant activity of tempeh has not yet been determined.

In the production of tempeh, soybeans are only partially cooked and they remain nearly as firm as the soaked beans. After fermentation the beans are soft and similar in texture to completely cooked soybeans. An earlier cytological study (Steinkraus *et al.* 1960) showed only slight penetration of the mycelia into the underlying tissue of the bean, suggesting that the digestion was mainly enzymatic. However, a recent study (Jurus and Sundberg 1976) revealed hyphae infiltration to a depth of 742 μm or about 25% of the average width of a soybean cotyledon. These authors speculated that the extreme depth of mycelial infiltration partially explains the rapid physical and chemical changes occurring during tempeh fermentation. The hyphae may mechanically push the bean cells apart prior to, or in conjunction with, enzymatic digestion; thus, the beans become soft. Likewise, the penetration of enzymatic activity could also be enhanced, since the distance over which diffusion of enzymes must occur is greatly reduced.

Indonesians consider tempeh to be a nourishing and easily digestible food. Van Veen and Schaefer (1950) observed beneficial effects of tempeh on patients with dysentery in the prison camps of World War II, and they suggested that tempeh was much easier to digest than soybeans. However, animal feeding experiments have not substantiated this conclusion (Hackler *et al.* 1964; Smith *et al.* 1964; Murata *et al.* 1967), even though more than half of the soybean protein, fat, and N-free extract could be solubilized by 72 hr fermentation (Van Buren *et al.* 1972). The protein efficiency ratio (PER) of tempeh is also not significantly different from that of the unfermented soybeans (Hackler *et al.* 1964; Smith *et al.* 1964; Wang *et al.* 1968; Murata *et al.* 1971). The cooking procedures, however, affect the nutritional value of tempeh; the PER value of tempeh significantly declined after more than 3 min of frying in oil; on the other hand, steaming up to 2 hr had no effect (Hackler *et al.* 1964). The quality of tempeh protein can be improved by making tempeh from mixtures of cereals and soybeans (Hesseltine *et al.* 1967). For example, the PER value of wheat-soybean (1:1) tempeh was comparable to that of casein (Wang *et al.* 1968).

The superior nutritive value of tempeh over unfermented soybeans has been noted by György (1961) on animals fed low-protein diets. His results resemble those obtained with animals fed antibiotics added to their protein source. We found that *R. oligosporus* indeed produces an antibacterial agent during tempeh fermentation as well as in submerged culture (Wang *et al.* 1969). The compound is especially active against some Gram-positive bacteria, including both microaerophilic and anaerobic bacteria, e.g., *Streptococcus cremoris, Bacillus subtilis, Staphylococcus aureus, Clostridium perfringens,* and *C. sporogenes.* The compound contains polypeptides hav-ing a high carbohydrate content. Its activity is not affected by pepsin or *R. oligosporus* proteases, is slightly decreased by trypsin and peptidase, but is rapidly inactivated by pronase. It is well established that antibiotics, in addition to minimizing infections, elicit growth-stimulating effects in ani-mals, especially those whose diets are deficient in any one of several vita-mins, proteins, or other growth factors. Oriental people are constantly exposed to overwhelming sources of infection and their diets are frequently inadequate. Therefore, the finding of antibacterial agents produced by *R. oligosporus* may offer a clearer understanding of the value of tempeh in the diet of Indonesians, and, perhaps, of fermented foods in the diets of all Orientals.

Although stachyose and raffinose, known as flatulence factors, are not utilized as the sole source of carbon (Sorenson and Hesseltine 1966), stach-yose in soybeans was found to decrease as fermentation progressed (Shallen-berger *et al.* 1966). Calloway *et al.* (1971) found that tempeh did not increase gas production over baseline values of healthy young men and caused a significant delay in the time of gas forming, suggesting temporary suppres-sion of intestinal bacteria. The delay could well be due to the presence of the antibiotic substance produced by the mold *Rhizopus.*

As mentioned earlier, tempeh is used as a main dish in Indonesia. Total production data on tempeh are not available, but in the Province of Central Java alone, 35,100 MT of tempeh were made in 1972 (Winarno 1976).

Because of the high acceptability of taste and texture, lack of beany flavor, nutritional advantages, and its simple, low-cost processing tech-niques, tempeh appears to be a good candidate for any country searching for a low-cost and high-protein food. With the recently increasing interest of vegetarians in foods of vegetable origin, tempeh consumption has been on an upsurge in the United States. In addition to tempeh-making as a home project, several commercial tempeh producers have been established. Tem-peh may soon be a regular item in the United States market.

ONTJOM

Ontjom, or oncom, a product closely related to tempeh, is another tradi-tional Indonesian food, especially popular in West Java. Like tempeh, ontjom is a solid cake-like product and is commonly served deep fat-fried or cooked in other native dishes.

Peanut press cake is generally used as substrate in ontjom fermentation, although coconut press cake, cassava press cake, and residues from making soybean milk and tofu are sometimes used wholly or in part, either to improve the texture or to reduce the cost of the product. The fermentation is carried out by strains of *Neurospora* or *Rhizopus; Neurospora* fermentation results in a pink or orange cake and *Rhizopus* in a cake of ash-grey color.

The preparation of ontjom in Indonesia as described by various investiga-tors (Van Veen and Graham 1968; Winarno 1979; Saono *et al.* 1974; Ho 1976) is summarized in Fig. 12.4. The peanut press cakes are broken into

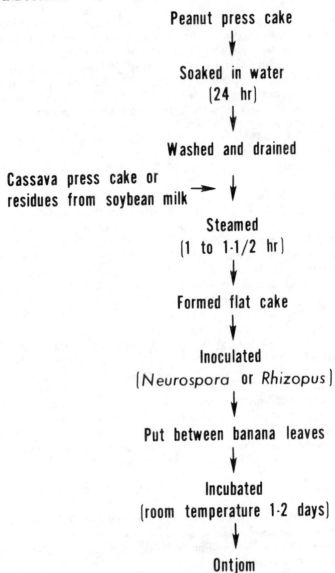

FIG. 12.4.   FLOW SHEET FOR ONTJOM FERMENTATION

pieces and soaked in water for 24 hr or until soft. They are washed, drained, and gently pressed to remove the excess water. Some fatty materials that float over the surface during soaking should also be removed. The soaked peanut press cakes are mixed thoroughly with cassava press cakes which have been crumbled and screened through a coarse sieve. The resulting mixture is steamed for 1 to 1-1/2 hr and then transferred to a mold to form a flat cake about 2−3 cm in thickness. After cooling, the cakes are inoculated with powdered ontjom from an earlier preparation and placed in bamboo trays, which are then kept at 25°−30°C for 1−2 days. The cakes are actually exposed to the air during fermentation, even though some manufacturers cover the cake with banana leaves. This practice favors sporulation, so that the fermented cake is covered with the orange spores of *Neurospora* to form an orange cake, or covered with the black spores of *Rhizopus* to form an ash-grey cake. In tempeh fermentation sporulation is undesirable and is largely prevented by restricting aeration.

Typical ontjom cultures maintained in the AR Culture Collection are designated as *N. sitophila* NRRL 2884 and *R. oligosporus* NRRL 2710. However, Ho (1976) recently made an investigation of the ontjom culture and reclassified the *Neurospora* as *N. intermedia* based on meiotic sterility.

Van Veen and Graham (1968) used pure cultures of *Neurospora* isolated from ontjom samples from Indonesia to study the ontjom fermentation. They found that the mold grew well on substrate with a pH lower than 6.0 and that its growth was stimulated by the addition of tapioca. Incubation temperatures of 25° or 30°C were equally suitable for fermentation; higher temperatures, however, were not advantageous. Also, the quality of the final product was greatly affected by the heat treatment of the substrate. They concluded that extraction of the peanut press cake with hot water, addition of 1% tapioca followed by pasteurization (30 min at 65°C), and adjustment of pH to 4.5 would produce the best product judging from mold growth, sporulation, color, and flavor development.

According to Winarno (1979), fresh ontjom has a moisture content of 57%; protein, 13%; fat, 6%; and carbohydrates, 22%. As in the tempeh, the proximate composition of ontjom resembles that of the unfermented substrate, but its constituents are probably hydrolyzed in various degrees by the enzymes produced by the mold (Beuchat and Worthington 1974; Worthington and Beuchat 1974; Beuchat *et al.* 1975; Beuchat and Basha 1976).

The protein efficiency ratio of the peanut protein as determined by the rat assay is not changed by the fermentation (Van Veen and Graham 1968), and there is also no effect on the apparent digestibility after fermentation. Nevertheless, rats receiving the ontjom diet had a higher food intake than the rats receiving the diet containing unfermented peanut. The increased food intake resulted in a concurrent increase in weight gain. Therefore, the fermentation process may have increased the palatability of the peanut press cake.

HAMANATTO

Hamanatto is a Japanese name for a salty fermented whole-soybean product that is quite popular in many Oriental countries. The product is known as "tou-shih" in China; "tao-si" in the Philippines; and "tao-tjo" in the East Indies. It has a pleasant flavor, resembling that of a soy sauce, and is nearly black. The fermented beans are used as a side dish to be consumed with bland foods, such as rice gruel, or they can be cooked with vegetables, meats, and seafoods as a flavoring agent.

Preparation methods may vary from country to country, but the essential features are similar. Soybeans are soaked and steamed until soft, drained, cooled, mixed with parched wheat flour, and then inoculated with *Aspergillus oryzae*. After incubation for 1–2 days, the beans, which are now covered with green mold, are dried in the sun to about 10–12% moisture. The dry beans are packed in a container with the desired amount of salt, spices, wine, sugar, and water, tightly covered, and aged for several weeks or months. After fermentation, the beans are dried again. The composition of the brine solution varies with each country; thus, the final product differs somewhat in taste and appearance. Japanese hamanatto is rather soft, having a high moisture content. Chinese tou-shih is not as soft and has a lower moisture than that of hamanatto. Tao-tjo tends to have a sweet taste because sugar is often added to the brine.

Investigations on the hamanatto fermentation are inadequate; thus, methods of producing the product have not been modernized. Recently, Kon and Ito (1975) found that the main microorganisms considered responsible for the hamanatto are strains of *A. oryzae*, *Streptococcus*, and *Pediococcus*. The *A. oryzae* strain is dark olive-green and produces strong proteolytic, but not amylolytic, activity.

SUFU

In the Western world, sufu has been referred to either as Chinese cheese or bean cake. Both names give a good description of the product. Sufu is a soft cheese-type product made from cakes of soybean curd (tofu) by the action of a mold. It is widely consumed by Chinese, but is not generally known by other Orientals. Because of the numerous dialects used in China, the product is also known as fu-ju or tou-fu-ju by many Chinese. Chao of Vietnam, tahuri of the Philippines, taokoan of Indonesia, and tao-hu-yi of Thailand are the same as sufu, and are usually consumed by Chinese immigrants.

The process of making sufu was considered a natural phenomenon. Not until 1929 was a microorganism believed to be responsible for sufu fermentation isolated and described by Wai (1929). He identified the microorganism to be an undescribed species of *Mucor* and proposed the name *Mucor sufu*. Wai also thought that this fungus originated on rice straw, because rice straw was always used to cover the tofu cubes for fermentation in the traditional way. Almost 40 years later, Wai (1968) reinvestigated the microorganism in sufu fermentation; as a result, a pure culture fermentation for making sufu was developed. His report was summarized by Wang and Hesseltine (1970A). Three steps are normally involved in making sufu: preparing tofu, molding, and brining (Fig. 12.5).

To make tofu, soybeans are first washed, soaked overnight, and then ground with water; a water to dry bean ratio of 10:1 is commonly used. After boiling or steaming the ground mass for about 20 min, the hot mass is

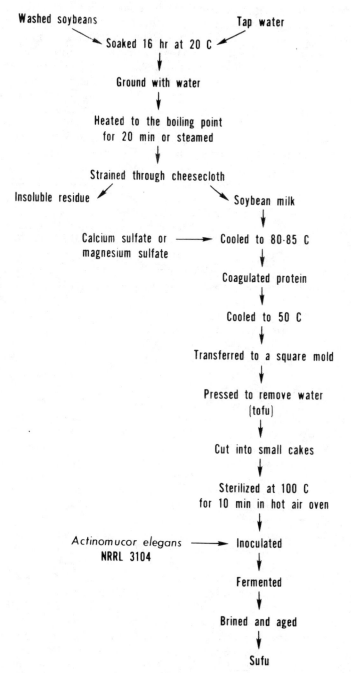

FIG. 12.5.   FLOW SHEET FOR THE PREPARATION OF SUFU

strained through a cloth bag, double-layered cheesecloth, or a fine metal screen to separate the soybean milk from the insoluble residue. Curdling is achieved by the addition of a coagulant, such as calcium salts, magnesium salts, and acid. The curd is then transferred into a cloth-lined wooden box and pressed with weight on top to remove whey. A soft but firm cake-like curd (tofu) forms. So, tofu is to soybean milk what cottage cheese is to cows' milk, except that the curdling of soybean milk is traditionally brought about by a calcium salt and only occasionally by acid. Tofu has a bland taste and a high water content (about 90%). It can be consumed directly and is so eaten extensively throughout the Far East. But the water content of tofu for making sufu is lower than that of tofu consumed directly; otherwise, it is likely to be spoiled by bacterial growth. Typically, tofu used for sufu fermentation has a water content of 83%; protein, 10%; and lipids, 4%.

To prepare tofu for fermentation, the cubes (2.5 × 3 × 3 cm) are first soaked in a solution containing 6% NaCl and 2.5% citric acid for 1 hr, and then subjected to hot air treatment at 100°C for 15 min. This treatment prevents growth of contaminating bacteria but does not affect the growth of fungi needed in making sufu. Tofu cubes should be separated from one another and placed in a tray with pinholes in the bottom and top to aid air circulation because mycelia must develop on all sides of the cubes. After cooling, the cubes are then inoculated over their surfaces by rubbing with a pure culture of an appropriate fungus grown on filter paper impregnated with a culture solution. The inoculated cubes are incubated at 20°C or lower for 3–7 days depending on the culture. The freshly molded cubes, known as pehtze, have a luxurious growth of white mycelium and no disagreeable odor. The pehtze has a water content of 74%; protein, 12.2%; and lipid, 4.3% as reported by Wai (1968).

In order to obtain a fermented product of good quality, the mold used in this fermentation has to have certain qualities. The organism probably utilizes the carbon in lipids as an energy source, because proper types of carbohydrates are not readily available in the substrate. The organism must develop enzyme systems having high proteolytic and lipolytic activity, since the mold grows on a protein- and lipid-rich medium. The mold must have white or light yellowish-white mycelium to ensure that the final product has an attractive appearance. The texture of the mycelial mat should also be dense and thick, so that a firm film will be formed over the surface of the fermented tofu cubes to prevent any distortion in their shape. Of course, it is also important that the mold growth not develop any disagreeable odor, astringent taste, or mycotoxin. Dr. Wai and his coworkers confirmed that *Actinomucor elegans, Mucor hiemalis, Mucor silvaticus,* and *Mucor subtilissimus* possess all these characteristics and can be used to make good quality sufu. Among them, *A. elegans* seems to be the best organism and is the one used commercially in the Orient.

The last step in making sufu is brining and aging. The molded cubes can be placed in various types of brine solutions, according to the flavor desired. A basic and most common brine suitable to Chinese taste is one containing 12% NaCl and rice wine containing about 10% ethyl alcohol. The immersed cubes are allowed to age for about 40–60 days. The product is then bottled with the brine, sterilized, and marketed as sufu.

Freshly molded tofu is bland in taste. The flavor and aroma of sufu develop during the brining and aging process. Other additives, either to give color or flavor, are frequently incorporated into the brine. Red rice and soy mash are added to the brine, imparting a red color to the product; when these ingredients are added, the final product is known as red sufu or hon-fang. Fermented rice mash or a large amount of wine can be added to the brine, so that the product has more of the alcoholic bouquet. This product is known as tsui-fang or tsue-fan, which means drunk sufu. The addition of hot pepper to the brine would make hot sufu. Rose sufu can be made by aging in a brine containing rose essence. Therefore, the taste and aroma of sufu, in addition to its own characteristically mild one, can be easily enhanced or modified by the ingredients of the brine solution.

During the brining and aging process, the enzymes elaborated by the mold act upon their respective substrates, yielding various hydrolytic products. Since sufu fermentation has a rather simple substrate—55% protein and 30% lipids on a dry basis—it is likely that the hydrolytic products of protein and lipids provide the principal constituents of the mild characteristic flavor of sufu. The added alcohol reacts with the fatty acids chemically or enzymatically to form esters providing the pleasant odor of the product. Ethanol also prevents the growth of microorganisms.

The added salt imparts a salty taste to the product, as well as retarding mold growth and growth of contaminated microorganisms. Most important, the salt solution releases the mycelia-bound proteases. In sufu fermentation, the mold growth is limited to the surface of the cubes, and the mycelium does not penetrate into the tofu cubes. The enzymes produced by the mold, on the other hand, are not extracellular (Wang 1967B). They are loosely bound to the mycelium, possibly by ionic linkage. However, the enzymes can be easily eluted by NaCl or other ionic salt solutions, but not by water or nonionizable salt. Therefore, the NaCl brine solution also serves to elute the enzyme from mycelia which, in turn, penetrate into the molded cubes and act upon the substrate protein.

Changes which occur during the aging process have also been studied by Wai (1968). After 30 days of aging at room temperature, total soluble nitrogen increased from 1.00 to 2.74% and total insoluble nitrogen decreased from 7.89 to 6.05%, while total nitrogen changed slightly. The soluble nitrogenous compounds were reported to consist of soluble proteins, peptides, and amino acids, including aspartic acid, serine, alanine, leucine/isoleucine, and glutamic acid. Lipids in pehtze were also partially hydrolyzed through the aging period. Free fatty acids increased from 12.8 to 37.1% and total lipids remained unchanged.

Traditionally, sufu is consumed directly as a relish or is cooked with vegetables or meats. Either way, sufu adds a zest to the bland taste of a rice-vegetable diet. Because sufu has the texture of cream cheese, it would be suitable to use in Western countries as a cracker spread or as an ingredient for dips and dressings. The traditional product, however, has too high a salt content to suit Westerners' taste; nonetheless, a less salty product can be obtained by reducing the salt concentration of the brine solution from 12% to not less than 3%, which is the minimal salt concentration to release the mycelium-bound proteases (Wang 1967B).

NATTO

In Oriental soybean fermentations, molds usually dominate, but the natto fermentation is an exception in which bacteria predominate. The bacterium *Bacillus natto*, identified as *B. subtilis*, is the organism responsible for this fermentation. Thus, natto possesses the characteristic odor and perisistent musty flavor of this organism, and is also covered with the viscous, sticky polymers that this organism produces. Perhaps because of its characteristic odor, flavor, and slimy appearance, natto, even though it is well known in Japan, especially in Northern Japan, is not so popular nor so widely consumed as compared with other fermented soybean foods. However, there seems to be an upswing in consumption; the production of natto in Japan was 30,000 MT in 1959 and 90,000 MT in 1974. The yearly per capita consumption in 1974 was about 760 g as reported by Ebine (1976). Other than Japan, natto is also produced and consumed in Korea.

In Japan, natto is seasoned with soy sauce, salt, or sometimes mustard, and served with rice. Making natto is a simple operation and can be easily done at home. After the beans are soaked, they are boiled, drained, cooled, wrapped in rice straw, and kept in a warm place for 1–2 days. The quality of the product is then ascertained by the stickiness of the beans and their flavor. Rice straw is credited not only with supplying the fermenting organism, but also with providing the aroma of straw, which many consumers are fond of, and with absorbing the unpleasant odor of ammonia from natto.

In industrial practice (Ohta *et al.* 1967), the beans are first soaked in water for 16–20 hr at 15°C or 12–16 hr at 20°C, and then steamed under pressure for 30–40 min. The cooked beans are inoculated with natto starter, which is *B. natto* grown in peptone-glucose media at 37°C for 24 hr. The inoculated beans are wrapped in a paper-thin sheet of pine wood or packed in plastic packages weighing about 80–120 g. The fermentation lasts for 15–20 hr at 40°–43°C in the package in which they are sold.

Many papers have been published concerning the microorganisms in natto fermentation; however, it is now well established that bacilli are the most important ones. In 1960, Sakurai reconfirmed that *B. natto* is an aerobic, Gram-positive rod, and classified it as a related strain of *B. subtilis*. There are two types of *B. natto* in the laboratory of the Food Research Institute, Ministry of Agriculture and Forestry, Tokyo, Japan. One has optimum temperature from 30° to 45°C and the other, from 35° to 45°C. He recommended the culture known as *B. natto* SB-3010, with an optimum temperature of 30°–45°C, as the one more suitable for making natto. In our AR Culture Collection, we have several strains of *B. subtilis* suitable for natto fermentation, namely NRRL B-3383, B-3384, B-3385, B-3386, B-3387, and B-4008.

Recently, Ohta *et al.* (1976) investigated the effects of fermentation time on the organoleptic quality of natto; they found that natto made with 8 hr fermentation had the highest overall scores, whereas natto made with the. traditional 16 hr fermentation received a poor rating. In general, when the ammonium production exceeds a level of 1%, the product becomes obnoxious. Perhaps short-period fermentation would result in a product of mild flavor that would gain acceptance from more people.

Natto has a short shelf life, partly because it has a moisture content of 60% (Ohta *et al.* 1967) and partly because it is usually prepared in small-scale plants with poor quality control. Cold storage has been used to extend its shelf life.

There are many reports concerning the changes which occur during natto fermentation. Hayashi (1959) made one of the most comprehensive studies. His data indicated that there was no change in fat and fiber contents of soybeans during a 24 hr period of fermentation, but that carbohydrate almost totally disappeared. A great increase in water-soluble nitrogen and ammonia nitrogen was noted during fermentation as well as during storage. The amino acid composition remained the same. Boiling markedly decreased the thiamin level of soybeans; but fermentation by *B. natto* enhanced the thiamin content of natto to approximately the same level as soybeans. Riboflavin in natto greatly exceeded that in soybeans. Vitamin B_{12} in natto was found by Sano (1961) to be higher than in soybeans. Conflicting results on the nutritive value of natto have been published by several investigators (Arimoto 1961; Hayashi 1959; Sano 1961). They disagreed on the nutritive value of natto protein as being superior to that of boiled soybeans. But they agreed that rats fed a diet containing natto and rice grew as well as rats receiving a complete laboratory diet; whereas, rats fed a diet of boiled soybeans and rice did not grow as well as the controls. However, Hayashi found that addition of thiamin to the diet of boiled soybeans and rice corrected the deficiency; the growth of rats receiving a thiamin-enriched diet of boiled soybeans and rice was comparable to that of rats fed natto and rice.

IDLI

A great variety of fermented foods is consumed daily in India and her neighboring countries (Batra and Millner 1974, 1976), but idli is the only one that has been studied in much detail.

Idli is a breakfast dish in most parts of India, especially popular in southern India. It is usually prepared by steaming a fermented batter consisting of rice and black gram *(Phaseolus mungo)*—a legume, and is served like pancakes with butter and honey, or with jams and other sauces. The steamed pancakes are soft and spongy and have a desirable sour taste and flavor.

Like most of the fermented foods consumed in the Asiatic countries, idli is made by natural fermentation. Generally (Desikachar *et al.* 1960), decorticated black gram known as black gram dahl and rice are soaked separately in water for 3–10 hr before grinding into a paste. Parboiled rice semolina is frequently used, but dry black gram flour is not suitable. The two pastes are mixed with some salt and enough water to make a batter of desirable consistency and then allowed to ferment at 25°–30°C for 14 to 16 hr.

While the proportions of rice to black gram have varied from 4:1 to 1:4 in various studies, the best result was obtained when the proportion of rice to black gram was 1:2 (Radhakrishnamurty et al. 1961). When the black gram was less than 25%, the steamed idli was hard and organoleptically unacceptable; when it was more than 50%, the product was too sticky to be acceptable.

The amount of water required to prepare batters of desirable and uniform consistency depends upon the proportions of rice to black gram dahl and varies from 1.5 to 2.2 times the dry weight of the ingredients.

Both rice and black gram act as substrates as well as providing microorganisms, but the major component contributing to the fermentation appears to be the black gram. Presoaking was found to be an important step of the fermentation; it provides extra fermentation time and microorganisms essential for the fermentation.

Mukherjee et al. (1965) made a thorough study of the microorganisms responsible for the leavening action of idli. They first washed the black gram with tap water and then distilled water, presumably to remove some of the surface microorganisms. The washed beans were soaked for 8 hr in distilled water, then ground and mixed with rice semolina and salt to prepare the batter. The batter was then allowed to ferment at 30°C. Soaking water, as well as fermenting batter, was sampled at 4 hr intervals for plate count of microorganisms by using a tryptone, glucose, and yeast extract broth. The microorganisms responsible for the fermentation were then isolated and identified. The authors found that the rate of growth of microbial populations was exceedingly high during soaking and the early stage of fermentation. Of these isolates, about 95% were Leuconostoc mesenteroides. In the later stages of fermentation, growth of Streptococcus faecalis and, still later, of Pediococcus cerevisiae became significant. Thus, the authors found that the microorganism responsible for souring, as well as for gas production, was L. mesenteroides. Rajalakshmi and Vanaja (1967) also identified the microorganisms responsible for idli fermentation as L. mesenteroides which was present in black gram. However, Batra and Millner (1974) consistently isolated Torulopsis candida (Saito) Lodder and Trichosporon pullulans (Lindner) Diddens and Lodder from samples of idli batter collected from the major center of production—Madras, India. They prepared idli batter with these two yeasts, either singly or in combination, and found that idli having the characteristic consistency, appearance, and flavor can only be made by the combined action of the two yeasts; both of the yeasts imparted the characteristic acidity, and T. candida also produced gas. Desikachar et al. (1960) have shown that both yeasts and bacteria participate in the idli fermentation. However, acid and gas production have been found to be mostly dependent on the bacterial growth.

Some investigators (Desikachar *et al.* 1960; Joseph *et al.* 1961) have added yeasts and buttermilk as inocula, but Steinkraus *et al.* (1967) found that idli prepared with added dry yeast and/or sour buttermilk inoculum was similar to that prepared without inocula. However, they experienced no fermentation failures with added inocula, while occasional failures were encountered without the inocula. A variation often used in India is to mix the ingredients with water at 80°C (Khandwala *et al.* 1962). Steinkraus *et al.* (1967) reported that the normal bacterial flora was markedly decreased from 15,000 to about 600/g, even if the batter was maintained at 80°C for only a matter of minutes. This practice generally resulted in a slower fermentation and an idli with an altered aroma and flavor.

The two more significant changes during idli fermentation are leavening and acidification of the batter; these have been used as the criteria for judging the progress of fermentation. During fermentation, the volume of idli batter has been reported to increase, ranging from 1.6 to 3.1 times its original volume. The pH falls from 6.0 to 4.3–5.3, and acidity increases from 3.2 to 19.0 ml of 0.1 N lactic acid per 25 g idli batter (Desikachar *et al.* 1960; Steinkraus *et al.* 1967).

Soluble solids generally showed a slight increase during fermentation, whereas soluble nitrogen showed a decrease (Steinkraus *et al.* 1967). This is a contrast to some fungal fermentations, such as tempeh, which usually show considerable increases both in soluble solids and nitrogen during fermentation.

Idli is generally considered a nourishing and easily digestible food. However, results obtained from the growth and nitrogen balance experiments with rats do not show that the nutritive value and digestibility of proteins of the rice and black gram mixture are improved by the fermentation process (Ananthachar and Desikachar 1962; Khandwala *et al.* 1962; Van Veen *et al.* 1967). On the other hand, Radhakrishna Rao (1961) disclosed that feeding of fermented idli reduces the liver fat content of rats fed a high fat and low protein diet. The results were further substantiated by his findings that there was an increase in choline, methionine, and folic acid content of the rice and black gram mixture after fermentation. Rajalakshmi and Vanaja (1967) also found idli superior to an unfermented product. Varied results on the effect of fermentation on methionine and cystine content of the substrate mixture, showing an increase in some and a decrease in others, were observed by Steinkraus *et al.* (1967). They found an apparent increase in both methionine and cystine in idli prepared from a 1:1 proportion of rice to black gram. Obviously, more research is needed on the synthesis of methionine in idli fermentation which would be a most valuable contribution to Oriental fermented foods, because these foods are usually made from plant materials that are generally low in methionine.

ANG-KAK

Chinese red rice, or ang-kak, originated in China; it is a product made by fermenting rice with strains of *Monascus purpureus*. Only those strains

(Church 1920) that produce a dark red growth throughout the rice kernels, at low enough moisture levels to allow the individual grains to remain separate from one another, are suitable for the fermentation. Because of its color, ang-kak is used for coloring various foods, and as a color additive for manufacturing fermented food such as sufu, red wine, fish sauce, and fish paste in the Orient.

M. purpureus NRRL 2897 maintained in the AR Culture Collection was isolated from an ang-kak sample bought in the Philippines market and has been demonstrated to carry out the fermentation successfully. Palo *et al.* (1960) studied various conditions of the fermentation and found that the optimum temperature for pigment formation is about 27°C. Growth will occur as low as 20°C and as high as 37°C, but at these extremes poor pigmentation results. The mold will produce the pigment over a wide range of pH values (3 to 7.5). All varieties of rice are suitable except the glutinous ones, which are unsatisfactory because the rice becomes gluey and the grains stick together.

The procedure for preparing ang-kak on a laboratory scale as developed by Palo *et al.* (1960) is to first wash the rice, soak it in water 24 hr, and then drain it thoroughly. The rice is placed in a beaker or other suitable container that is large enough to have considerable air space above the rice. It is then covered with a milk filter disc, autoclaved for 30 min at 121°C, cooled, and inoculated with a sterile water suspension of ascospores removed from 25-day-old cultures of *M. purpureus* grown on Sabouraud's agar. This medium is excellent for growth and pigment production. The culture should show much red color on this medium. After the suspension is thoroughly mixed with the rice, it should be incubated at from 25° to 32°C. At the time of inoculation the rice will seem rather dry, but this state is one of the secrets to the production of good ang-kak. The rice should never be wet or mushy. In 3 days, the rice should begin to redden. At this time, the material should be stirred and shaken to redistribute the rice in the center and bottom of the fermentor to retain an even moisture at the surface. We have found that an occasional vigorous shaking of the fermentation jar is quite adequate. Addition of some sterile water may be required from time to time to replace lost moisture. Care must be taken only to moisten and not to soak the rice. In 3 weeks, the rice should be a deep purplish-red with each rice kernel unattached to its neighbors. The material is then dried in an oven at 40°C. When the kernels are broken, the pigment should have infiltrated completely through the rice. Each grain should crumble easily between one's fingers.

Lin (1973) isolated a strain of *Monascus* sp. F-2 from kaoliang koji and found that it produced large amounts of pigment by submerged culture with rice as a sole carbon source. N-Methyl-N'-nitro-N-nitrosoguanidine treatment and successive isolation (Lin and Suen 1973) greatly improved the yield of the pigment production by *Monascus* sp. F-2. Two hyperpigment-producing strains, R-1 and R-2, were isolated which produced 5 times more red pigment than the original parent strain F-2 did.

The ang-kak studied by Nishikawa (1932) consisted of 2 pigments, principally monascorubrin, a red coloring material with a formula of $C_{22}H_{24}O_5$, and a small amount of monascoflavin, a yellow pigment ($C_{17}H_{22}O_4$). The pigments tend to accumulate in the microorganisms because of their poor solubility. Recently Yamaguchi *et al.* (1973) reported that the pigment can be solubilized by reacting it with a water-soluble protein, a water-soluble peptide, an amino acid, or a mixture of these. The reaction is carried out at pH from 5 to 8.5 and requires from several minutes to several hours. As the reaction progresses, the color of the pigment changes from a brownish-red to a deep scarlet and the pigment becomes water soluble. The solubility of the pigment in water varies with the type of protein, peptide, or amino acid used. The same authors also indicated that the water-soluble pigment can be produced directly by the mold grown in a medium containing soluble protein, peptide, or amino acid at a pH of 7–9 under aerobic conditions and at a temperature of about 27°C for 48 to 72 hr. The mechanism of this reaction is not clear; the authors speculated that the water-soluble pigment is a complex resulting from binding between the pigment and the chosen amino group. The water-soluble pigment is suitable for coloring foodstuffs, particularly sour-type drinks, candy, milk products, and meat products.

FERMENTED FISH PRODUCTS

Parallel with fermented soybean products, soy sauce, and miso (soybean paste) are two fermented fish products—fish sauce and fish paste. These products have a long tradition as a food condiment in the Orient. Both fish and soybean fermentations involve the use of large amounts of salt to select certain organisms and to inhibit food-poisoning ones; both are carried out under semianaerobic if not actual anaerobic conditions; both involve enzymes that break protein down into amino acids; and both are highly flavored and complement a bland rice diet. However, soy sauce and miso fermentations have become highly technological operations, whereas the fish fermentation remains a traditional one.

Many different types of fish sauce and fish paste are obtained through lengthy but simple fermentation processes (Van Veen 1953; Saisithi *et al.* 1966; Subba Rao 1967; Arcega 1971). Whole small fish with viscera or shrimp are heavily salted, packed into a sealed container, and allowed to undergo natural fermentation for periods varying from a few days to more than a year. Therefore, most of these products are the results of enzymatic digestion ensuing from tissue, especially viscera, or from microorganisms. In fish paste fermentation, other materials such as roasted rice powder, red rice, and pineapple are frequently added, and the whole fermented mass is ground into a fine paste. Fish sauce, on the other hand, is the liquid fraction of the fermented mass. It is usually a clear yellow to amber liquid rich in salt and soluble nitrogen compounds. Both fish paste and sauce have a distinctive odor and flavor, generally not appreciated by Westerners. The products are described as having a characteristic salty cheese flavor and a slight fish odor.

The fermented fish products are especially popular in Southeast Asia and are consumed by almost everyone. To the rich, these products add zest to the food; to others, they make the bland rice and vegetable diet more palatable. Fish sauce and fish paste made in various countries are known under different names, such as ngam-pya-ye and ngapi for fish sauce and fish paste, respectively, in Burma; nuoc-mam and mams in Cambodia and Vietnam; patis and bagoong in the Philippines; and nampla and kapi in Thailand. Although the basic principles of making these products are the same, a large number of variations develop. By following more or less constant techniques over hundreds of years, each country develops a fairly standardized product having the characteristic texture, appearance, and flavor that appeals to its own people. However, scientific studies on these products are scanty.

Nuoc-mam, a Vietnam fish sauce, has been studied by a number of investigators. Because many of the findings may be applicable to other fermented fish products in Southeast Asia, an account of nuoc-mam is given.

The production of nuoc-mam is a very active industry in Vietnam. It is produced in large plants as well as in so-called cottage industries. Total production was estimated at 40 million liters per year in the early 1970s, and daily per capita consumption ranged from 15 to 60 ml depending on one's income, with greater consumption for low-income people.

In the most primitive way of manufacturing nuoc-mam, small fish are kneaded and pressed by hand. They are salted and tightly packed into earthenware pots which are sealed and buried in the ground for several months. When the pots are opened, the supernatant liquid or the pickle that has been formed is decanted. This liquid is the nuoc-mam.

In the commercial production of nuoc-mam (Van Veen 1953) cylindrical vats made with local wood and encircled with twisted bamboo are used. The vats are equipped with taps near the bottom. The inside end of the tap is buried beneath shells, sometimes mixed with rice husks, and the inside opening is plugged with hairs in order to improve the filtering power.

The fresh, uncleaned fish are mixed with a small amount of salt and packed in the vat in alternate layers with additional salt. A final layer of salt is placed on the top. After 3 days, the collected liquid, known as nuoc-boi, is drained off and reserved for later use. Meanwhile, the fish have settled below the top of the vat, and the salt has almost disappeared. The fish are now packed down thoroughly, and the surface is smoothed. The contents are covered with a layer of coconut leaves and then with bamboo trays on top of which weights are placed. The drained nuoc-boi is poured back into the vat so that a layer of liquid, about 10 cm, covers the fish. The fish are now left to ferment.

Fermentation time varies with the kind of fish used, ranging from a few months for small fish to 1 year or 18 months for large fish. The amount of salt also varies with the species of fish used, usually 1 part of salt for every 1–1.5 parts of fish by volume. Generally, the higher the concentration of salt, and the more the proteins are hydrolyzed to products of low molecular weight, the better will be the keeping quality of the nuoc-mam.

After maturing, the liquid is run off through the tap at a rate of 300–400 liters per day; this is first quality nuoc-mam. A product of less quality is then obtained by extracting the residual mass with fresh brine solution. The undissolved residue is usually sold as manure. In some countries, the residue is not repeatedly extracted but is used as fish paste.

One part of fish gives from 2 to 6 parts of nuoc-mam depending upon the nitrogen content of the end product. Because nuoc-mam is an important item in the diet of the people of Vietnam, its manufacturing, trade, and quality are strictly controlled by governmental legislation. The quality of nuoc-mam depends upon the quantity of fish, the degree of fermentation, and the nutritive value. The quantity of fish is assessed by the total nitrogen in nuoc-mam, not less than 16 g/liter being the first class; the degree of fermentation is based on the formol titratable nitrogen, not less than 50% of total nitrogen; nutritive value is evaluated by the amount of ammonia nitrogen, not more than 50% of formol nitrogen (Subba Rao 1967). The average composition of nuoc-mam is given in Table 12.7 (Subba Rao 1967).

TABLE 12.7.   COMPOSITION OF NUOC-MAM

Chemical Composition	First Quality Nuoc-mam, g/liter	Ordinary Quality Nuoc-mam, g/liter
Acidity	3	2.5
NaCl	275	280
Total nitrogen	22	11.0
Organic nitrogen	15	7.5
Formol titrable nitrogen	16	8
Ammonia nitrogen	7	3.5
Amino acid nitrogen	9	4.5

Source: Subba Rao (1967).

Nuoc-mam has a high content of methyl ketones (Van Veen 1953) which probably accounts for the cheese-like odor. The same author found that nuoc-mam did not contain significant amounts of thiamin, riboflavin, or pyridoxine. The product, however, was found to contain 25–200 μg of vitamin B_{12} per liter (Subba Rao 1967).

The hydrolysis of fish protein appears to be due mainly to the enzymes in the fish. Glycerol extracts of the fish internal organs were found especially active, even in the presence of salt. Baens-Arcega et al. (1969) found that the addition of proteolytic enzymes from Aspergillus oryzae to the fermentation mass could greatly reduce fermentation time and increase yield. Salt does retard the rate of enzymatic action, but it also reduces the growth of putrefactive bacteria which are responsible for the off-flavor of the fermented product. However, the product made aseptically lacked the typical flavor of nuoc-mam. Therefore, some microorganisms might be responsible for the characteristic flavor.

Boez and Guillerm (1930A,B) have isolated anaerobic, spore-forming bacteria belonging to the *Clostridium* group said to produce the typical flavor of nuoc-mam. Saisithi *et al.* (1966) investigated the fish sauce made in Thailand and found that the total viable counts decreased as the fermentation advanced. Approximately 70% of the isolates from a 9-month-old fish sauce were halophiles of bacillus types. These isolates were found to produce volatile acid; the largest amount of acid, however, was produced by *Staphylococcus* strain 109. They extracted the compounds responsible for the aroma from the fish sauce with diethylether-alcohol mixture and separated them with ion exchange chromatography. The typical aroma fraction was eluted from the ion exchange resin with 4 N formic acid, suggesting that it is probably an acid. Silica gel thin layer chromatography further revealed the presence of amino acids as well as amines, glucosamine, histamine, glutamine, and trimethylamine. The aroma present in the fish sauce, therefore, could be a blend of acids and amines; the bacteria obviously play a role in the forming of these compounds.

ABSENCE OF MYCOTOXIN IN FERMENTED FOODS

Since molds are known to produce a number of mycotoxins (literally, toxins produced by fungi), one might question whether these foods that are based on mold fermentation offer a health hazard. The mycotoxin in which most interest has developed is the carcinogen, aflatoxin, produced by *Aspergillus flavus* and *A. parasiticus*. These species are in the same group as *A. oryzae*, a species widely used in Oriental food fermentations. For this reason, the Japanese have extensively looked at the aflatoxin problem in fermented products.

In probably the most important paper on the problem of whether aflatoxin occurs in shoyu and other mold-fermented foods, Yokotsuka *et al.* (1967A,B) investigated 73 industrial strains of *Aspergillus* (27 for shoyu, 16 for miso, 28 for alcoholic beverages, and 2 wild strains) for production of fluorescent compounds after cultivation on a Czapek Dox medium containing zinc. These strains were mostly *A. oryzae* and *A. sojae*, but included *A. niger* and *A. usami*. No aflatoxin was found, but about 30% of these strains produced compounds that were very similar to those of aflatoxins in fluorescence spectra and R_f values in thin layer chromatograms. Materials were also tested in animals with negative results. A wild strain which gave a high production of aspergillic acid on Czapek Dox medium failed to produce aspergillic acid on a mixture of soybeans and wheat or soybeans and rice. This compound could not be found in any fermented foods; and when mold-fermented foods were fed to mice, no toxicity occurred. Of 68 aspergilli strains that are widely used in Japanese food industries, 26 strains produced aspergillic acid in modified Mayer's medium at 30°C by surface culture after more than 10 days, and on solid soy sauce koji fermentation after 4 to 10 days (Yokotsuka 1971). However, the koji is usually harvested within 2 to 3 days. The same author also reported that a small amount of β-nitropropionic acid was produced by one of the tested strains in the beginning of the culturing, but it rapidly disappeared. Kojic, oxalic, or formic acids were formed by some molds during koji making, but only in trace amounts. Considering the low toxicity of these compounds, they create no problems.

Kinosita *et al.* (1968) obtained 24 fermented foods from several regions of Japan known to have a high death rate caused by cardiovascular diseases, stomach cancer, and hepatoma. These were homemade products or commercial products from local plants. From these 24 food samples, 37 strains of fungi were isolated and 21 of these produced toxic culture filtrates. None, however, produced aflatoxin although many formed kojic acid and β-nitropropionic acid.

Murakami *et al.* (1967, 1968A,B, 1969) likewise studied aflatoxin in industrial strains of *Aspergillus* and found all of them to be negative. Thirteen strains were selected which produced fluorescent substances, but when grown on rice they proved to produce substances other than aflatoxin. Some of the strains appeared to be similar to aflatoxin-producing strains but were found to be negative.

Earlier, our laboratory (Hesseltine *et al.* 1966) examined samples of soy sauce, miso, tou-shih, and soybean, rice, and wheat tempehs, but found no aflatoxin.

Matsuura *et al.* (1970) and Matsuura (1970) investigated 238 koji mold strains used for the manufacturing of miso and soy sauce obtained from the All Japan Koji Starter Association. Each strain was grown on both solid and liquid media and analyzed for aflatoxin. None was found. Since rice is an important ingredient in the fermentation, it was examined and in 46 domestic and 11 imported samples, no aflatoxin was detected. As a further investigation of the safety of these fermented foods, 28 kinds of rice koji were collected from various miso factories; 108 samples of miso, 21 samples of homemade miso, and 6 representative brands of soy sauce were examined and found negative for aflatoxin. In this survey, a number of chloroform-soluble fluorescent compounds were detected in 52 strains of *A. oryzae* and *A. sojae*, but 161 strains produced no fluorescent materials. However, with various chemical and physical means, the fluorescent compounds proved to be nontoxic (Manabe *et al.* 1968, 1972; Manabe and Matsuura 1972A,B). In one of these studies, aflatoxin was added to miso before fermentation. Results indicated that about 50% of aflatoxin B_1 and G_1 remained in the miso after 1 month of fermentation.

The only incidence of aflatoxin in soy sauce is a report by Shank *et al.* (1972B), who collected foods in Thailand and Hong Kong. In Hong Kong, 22 out of 878 samples of foods contained aflatoxin. Among these was at least 1 sample of soy sauce, but the level was low. However, the same authors (Shank *et al.* 1972A) did not mention the detection of any aflatoxin-producing strains of *A. oryzae* from their Southeast Asia survey.

A small amount of aflatoxin, 0.001–0.01 ppm, was reported in the mycelia of *A. oryzae* in Taiwan by Tung and Ling (1968), but no further details were given.

Recently, El-Hag and Morse (1976) have claimed that a strain of *Aspergillus oryzae* NRRL 1988 produced aflatoxin but this report was apparently based upon contamination of the culture in their laboratory (Fennell 1976; Stoloff *et al.* 1977; Morgan-Jones 1977). Maing *et al.* (1973) prepared soy sauce with a strain of *A. oryzae* NRRL 1988 and with an aflatoxin-producing strain of *A. parasiticus* NRRL 2999. No aflatoxin was produced in the soy sauce prepared with *A. oryzae*, but large amounts of aflatoxin were formed by *A. parasiticus* when it was grown with *A. oryzae*. After brining for 6 weeks, large amounts of aflatoxin remained in the fermentation in which *A. parasiticus* was grown. Obviously, if an aflatoxin-producing strain of *A. flavus* or *A. parasiticus* contaminated the culture which was used to make koji for soy sauce, the soy sauce would contain toxin.

SUMMARY

Traditional fermented protein-rich foods are highly acceptable to millions of people. These fermented foods are easily made and are generally more attractive to the consumers than the cooked original materials (substrates) from which they are made. In general, the organoleptic characteristics of the substrate are improved by the fermentation process. Many fermented foods are used as flavoring agents; the undesirable flavor and odor of soybeans are sometimes either masked or destroyed by the fermentation process. Fermentation may also increase the nutritional value of the substrate. The amounts of many vitamins in these fermented foods are significantly greater than in the unfermented materials, and the digestibility of the fermented products is also improved. However, the protein quality is usually not changed by fermentation. Fermentation often reduces cooking time and increases keeping quality. If properly fermented, these foods are not hazardous to health since the microorganisms responsible for these processes are not toxin producers. Nevertheless, further studies are needed to develop uniform high-quality products and well-defined economical processes to manufacture them. If the wholesomeness of these foods can be improved and their use can be expanded, they will become more acceptable to those people who are now unfamiliar with them. With interest increasing in soybeans as human food, the Western world may find these foods quite desirable.

REFERENCES

ANANTHACHAR, T.K. and DESIKACHAR, H.S.R. 1962. Effect of fermentation on the nutritive value of idli. J. Sci. Ind. Res. Sect. C *21*, 191–192.

ARCEGA, L.B. 1971. Oriental fermented sauces. *In* Global Impacts of Applied Microbiology (GIAM), December 7–12, 1969. Y.M. Freitas and F. Fernandes (Editors). University of Bombay, India.

ARIMOTO, K. 1961. Nutritional research on fermented soybean products. *In* Meeting the Protein Needs of Infants and Preschool Children. N.A.S.-N.R.C. Publ. *843*.

BAENS-ARCEGA, L. *et al.* 1969. New and quick process of manufacturing patis and bagoong. Philipp. Pat. 4113. Jan. 7.

BATRA, L.R. and MILLNER, P.D. 1974. Some Asian fermented foods and beverages, and associated fungi. Mycologia *66*, 942–950.

BATRA, L.R. and MILLNER, P.D. 1976. Asian fermented foods and beverages. Dev. Ind. Microbiol. *17*, 117–128.

BEUCHAT, L.R. 1978. Food and Beverage Mycology. AVI Publishing Co., Westport, Conn.

BEUCHAT, L.R. and BASHA, S.M.M. 1976. Protease production by the ontjom fungus, *Neurospora sitophila*. Eur. J. Appl. Microbiol. *2*, 195–203.

BEUCHAT, L.R. and WORTHINGTON, R.E. 1974. Changes in the lipid content of fermented peanuts. J. Agric. Food Chem. *22*, 509–512.

BEUCHAT, L.R., YOUNG, C.T. and CHERRY, J.P. 1975. Electrophoretic patterns and free amino acid composition of peanut meal fermented with fungi. Can. Inst. Food Sci. Technol. J. *8*, 40–45.

BOEZ, L. and GUILLERM, J. 1930A. Microorganism in the making of an Indochinese sauce (nuoc-mam). C.R. Acad. Sci. (Paris) *190*, 534–535. (French)

BOEZ, L. and GUILLERM, J. 1930B. Microorganism in the making of nuoc-mam. Arch. Inst. Pasteur Indochine *3*, 17–21. (French)

BRANDEMUHL, W. 1963. Soybean History. Aspects of Buddhist Influence. Anthropology Dep., Univ. of Wisconsin, Madison.

BUSH, L. 1959. Land of the Dragonfly. Robert Hale, London.

CALLOWAY, D.H., HICKEY, C.A. and MURPHY, E.L. 1971. Reduction of intestinal gas-forming properties of legumes by traditional and experimental food processing methods. J. Food Sci. *36*, 251–255.

CHURCH, M.B. 1920. Laboratory experiments on the manufacture of Chinese ang-khak in the United States. J. Ind. Eng. Chem. *12*, 45–46.

DESIKACHAR, H.S.R. *et al.* 1960. Studies on idli fermentation: Part 1—Some accompanying changes in the batter. J. Sci. Ind. Res. Sect. C *19*, 168–172.

EBINE, H. 1967. Evaluation of dehulled soybean grits from the United States varieties for making miso. U.S. Dep. Agric. Final Tech. Rep., P.L. 480 Proj. *UR-A11-(40)-2.*

EBINE, H. 1971. Miso. Proc. 5th Intern. Symp. Convers. Manuf. Foodst. Microorg., Jpn. Inst. Food Technol., Kyoto, Dec. 5–9, 1971, 127–132.

EBINE, H. 1976. Fermented soybean foods. Proc. Conf. Expanding Use of Soybeans for Asia and Oceania. Chiang Mai, Thailand, Feb. 1976; Intern. Soybean Program (INTSOY) Ser. *10*, Univ. Illinois, Urbana.

EL-HAG, N. and MORSE, R.E. 1976. Aflatoxin production by a variant of *Aspergillus oryzae* (NRRL strain 1988) on cowpeas *(Vigna sinensis)*. Science *192*, 1345–1346.

FENNELL, D.I. 1976. *Aspergillus oryzae* (NRRL strain 1988): A clarification. Science *194*, 1188.

GANDJAR, I. and SLAMET, D.S.   1972.   Tempe made from residue of making tofu. Penelitian Gizi dan Makanan Jilia 2, 70–79. (Indonesian)

GYÖRGY, P.   1961.   The nutritive value of tempeh. In Meeting the Protein Needs of Infants and Preschool Children. N.A.S.-N.R.C. Publ. 843.

GYÖRGY, P., MURATA, K. and IKEHATA, H.   1964.   Antioxidants isolated from fermented soybeans (tempeh). Nature 203, 870–872.

HACKLER, L.R., HAND, D.B., STEINKRAUS, K.H. and VAN BUREN, J.P. 1963.   A comparison of the nutritional value of protein from several soybean fractions. J. Nutr. 80, 205–210.

HACKLER, L.R., STEINKRAUS, K.H., VAN BUREN, J.P. and HAND, D.B. 1964.   Studies on the utilization of tempeh protein by weanling rats. J. Nutr. 82, 452–456.

HARADA, Y.   1951.   Effect of temperature change of material in koji culture. Rep. Tatsuno Inst. Soy Sauce 2, 51–55.

HAYASHI, U.   1959.   Experimental studies of the nutritive value of "natto" (a fermented soybean). Jpn. J. Nation's Health 28, 568–596.

HESSELTINE, C.W.   1965.   A millennium of fungi, food and fermentation. Mycologia 57, 149–197.

HESSELTINE, C.W.   1976.   New uses for miso. Sci. Technol. Miso 263, 5–7.

HESSELTINE, C.W., DECAMARGO, R. and RACKIS, J.J.   1963A.   A mould inhibitor in soybeans. Nature 200, 1226–1227.

HESSELTINE, C.W., SHOTWELL, O.L., ELLIS, J.J. and STUBBLEFIELD, R.D.   1966.   Aflatoxin formation by Aspergillus flavus. Bacteriol. Rev. 30, 795–805.

HESSELTINE, C.W., SMITH, M., BRADLE, B. and DJIEN, K.S.   1963B. Investigations of tempeh, an Indonesian food. Dev. Ind. Microbiol. 4, 275–287.

HESSELTINE, C.W., SMITH, M. and WANG, H.L.   1967.   New fermented cereal products. Dev. Ind. Microbiol. 8, 179–186.

HESSELTINE, C.W. and WANG, H.L.   1978.   Fermented soybean food products. In Soybeans: Chemistry and Technology, Vol. 1, Revised Edition. A.K. Smith and S.J. Circle (Editors).   AVI Publishing Co., Westport, Conn.

HO, C.C.   1976.   Microbiology of ontjom, an Indonesian fermented food. 5th Intern. Ferment. Symp., Institute for Fermentation Technology and Biotechnology, 1976, Berlin. (abstract)

IKEHATA, H., WAKAIZUMI, M. and MURATA, K.   1968.   Antioxidant and antihemolytic activity of a new isoflavone, "factor 2" isolated from tempeh. Agric. Biol. Chem. 32, 740–746.

ILJAS, N.   1969.   Preservation and shelf-life studies of tempeh. M.S. Thesis. Ohio State Univ.

ILJAS, N., PENG, A.C. and GOULD, W.A.   1973.   Tempeh: An Indonesian fermented soybean food. Dep. Hortic., Ohio Agric. Res. Dev. Center, Wooster, Hortic. Ser. 394.

ITO, H., EBINE, H. and NAKANO, M.   1965.   Miso manufacturing of high protein and low salt. Rep. Food Res. Inst. (Tokyo) 19, 127–135.

JOSEPH, K. *et al.* 1961. Studies on the nutritive value of idli fortified with Indian multipurpose food. J. Sci. Ind. Res. Sect. C *20*, 269–272.

JURUS, A.M. and SUNDBERG, W.J. 1976. Penetration of *Rhizopus oligosporus* into soybeans in tempeh. Appl. Environ. Microbiol. *32*, 284–287.

KHANDWALA, P.K., AMBEGAOKAR, S.D., PATEL, S.M. and RADHA-KRISHNA RAO, M.V. 1962. Studies in fermented foods: Part 1—Nutritive value of idli. J. Sci. Ind. Res. Sect. C *21*, 275–278.

KINOSITA, R. *et al.* 1968. Mycotoxins in fermented foods. Cancer Res. *28*, 2296–2311.

KO, S.D. 1970. Personal communication. Albardoweg 31, Wageningen, Netherlands.

KOMATSU, Y. 1968. Changes of some enzyme activities in shoyu brewing. 1. Changes of the constituents and enzyme activities in shoyu fermentation after low-temperature mashing. Seas. Sci. *15* (2) 10–20. (Japanese)

KON, M. and ITO, H. 1975. Studies on hamanatto. Rep. Natl. Food Res. Inst. (Tokyo) *30*, 232–237.

LIN, C.F. 1973. Isolation and cultural conditions of *Monascus* sp. for the production of pigment in a submerged culture. J. Ferment. Technol. *51*, 407–414.

LIN, C.F. and SUE, S.J-T. 1973. Isolation of hyperpigment-productive mutants of *Monascus* sp. F-2. J. Ferment. Technol. *51*, 757–759.

MAING, I.V., AYERS, A.C. and KOEHLER, P.F. 1973. Persistence of aflatoxin during the fermentation of soy sauce. Appl. Microbiol. *25*, 1015–1017.

MANABE, M. and MATSUURA, J. 1972A. Studies on the fluorescent compounds in fermented foods. Part III. Test on aflatoxin contamination and its possibility in rice. J. Food Sci. Technol. (Tokyo) *19*, 268–274.

MANABE, M. and MATSUURA, J. 1972B. Studies on the fluorescent compounds in fermented foods. Part IV. Degradation of added aflatoxin during miso fermentation. J. Food Sci. Technol. (Tokyo) *19*, 275–279.

MANABE, M., MATSUURA, S. and NAKANO, M. 1968. Studies on the fluorescent compounds in fermented foods. Part I: Chloroform-soluble fluorescent compounds produced by koji-molds. J. Food Sci. Technol. (Tokyo) *15*, 341–346.

MANABE, M., OHNUMA, S. and MATSUURA, J. 1972. Studies on the fluorescent compounds in fermented foods. Part II. Test on aflatoxin contamination of miso and miso-koji in Japan. J. Food Sci. Technol. (Tokyo) *19*, 76–80.

MARTINELLI, A.F. and HESSELTINE, C.W. 1964. Tempeh fermentation: package and tray fermentations. Food Technol. *18*, 167–171.

MATSUURA, S. 1970. Aflatoxins and fermented foods in Japan. Jpn. Agric. Res. Q. *5*, 46–51.

MATSUURA, S., MANABE, M. and SATO, T. 1970. Surveillance for aflatoxins of rice and fermented rice products in Japan. *In* Proc. 1st U.S.-Jpn. Conf. Toxic Micro-Organisms, U.S.-Jpn. Nat. Resources Program, Honolulu, Oct. 7–10, 1968.

MAXWELL, M.E. 1952. Enzymes of *Aspergillus oryzae*. Aust. J. Sci. Res. Ser. B. *5*, 43–55.

MORGAN-JONES, G. 1977. *Aspergillus oryzae* (NRRL strain 1988). Science *196*, 1354.

MUKHERJEE, S.K. *et al.* 1965. Role of *Leuconostoc mesenteroides* in leavening the batter of idli, a fermented food of India. Appl. Microbiol. *13*, 227–231.

MURAKAMI, H., HARA, S., TAHAHASHI, T. and RYU, G. 1969. Taxonomic studies on the Japanese industrial strains of *Aspergillus*. Part 12. Pink coloration of conidia by anisic acid and the related compounds, reexamination on the production of kojic acid and aflatoxin and the determination of decomposition of hydrogen peroxide. Rep. Res. Inst. Brew. *142*, 13–22.

MURAKAMI, H., SAGAWA, H. and TAKASE, S. 1968A. Nonproductivity of aflatoxin by Japanese industrial strains of *Aspergillus*. Part 3. Common characteristics of the aflatoxin-producing strains. J. Gen. Appl. Microbiol. *14*, 251–262; Res. Inst. Brew. Jpn. *141*, 208–220.

MURAKAMI, H., TAKASE, S. and ISHII, T. 1967. Nonproductivity of aflatoxin by Japanese industrial strains of *Aspergillus*. Part 1. Production of fluorescent substances in agar slant and shaking cultures. J. Gen. Appl. Microbiol. *13*, 323–334; Res. Inst. Brew. Jpn. *141*, 1801–1892.

MURAKAMI, H., TAKASE, S. and KUWABARA, K. 1968B. Nonproductivity of aflatoxin by Japanese industrial strains of *Aspergillus*. Part 2. Production of fluorescent substances in rice koji and their identification by absorption spectrum. J. Gen. Appl. Microbiol. *14*, 97–110.

MURATA, K., IKEHATA, H., EDANI, Y. and KOYANAGI, K. 1971. Studies on the nutritional value of tempeh. Agric. Biol. Chem. *35*, 233–241.

MURATA, K., IKEHATA, H. and MIYAMOTO, T. 1967. Studies on the nutritional value of tempeh. J. Food Sci. *32*, 580–588.

NISHIKAWA, H. 1932. Biochemistry of filamentous fungi. I. Coloring matters of *Monascus purpureus* Went. J. Agric. Chem. Soc. Jpn. *8*, 1007–1015.

NODA SOY SAUCE CO. 1955. The Treatment of Soy Sauce Fermentation. Noda-shi, Chiba-ken, Japan.

OHTA, T., NAKANO, M., KOBAYASHI, Y. and MUTO, H. 1967. The report of the 1st All Japan Natto Exhibition. Rep. Natl. Food Res. Inst. (Tokyo) *22*, 68–91.

OHTA, T. *et al.* 1976. Manufacturing new-type fermented soybean food products employing *Bacillus natto*. Rep. Natl. Food Res. Inst. (Tokyo) *31*, 52–59.

PACKETT, L.V., CHEN, L.H. and LIU, J.T. 1971. Antioxidant potential of tempeh as compared to tocopherol. J. Food Sci. *36*, 798–799.

PALO, M.A., VIDAL-ADEVA, L. and MACEDA, L.M. 1960. A study on ang-kak and its production. Philipp. J. Sci. *89*, 1–22.

RADHAKRISHNA RAO, M.V. 1961. Some observations on fermented foods. Meeting the protein needs of infants and children. N.A.S.-N.R.C. Publ. *843*.

RADHAKRISHNAMURTY, R., DESIKACHAR, H.S.R., SRINVASAN, M. and SUBRAHMANYAN, V. 1961. Studies on idli fermentation: Part II—Relative participation of black gram flour and rice semolina in the fermentation. J. Sci. Ind. Res. Sect. C 20, 342–345.

RAJALAKSHMI, R. and VANAJA, K. 1967. Chemical and biological evaluation of the effects of fermentation on the nutritive value of foods prepared from rice and grams. Br. J. Nutr. 21, 467–473.

ROELOFSEN, P.A. and TALENS, A. 1964. Changes in some B-vitamins during molding of soybeans by Rhizopus oryzae in the production of tempeh kedelee. J. Food Sci. 29, 224–226.

SAISITHI, P., KASEMSARN, B., LISTON, J. and DOLLAR, A.M. 1966. Microbiology and chemistry of fermented fish. J. Food Sci. 31, 105–110.

SAKURAI, Y. 1960. Report of the researches on the production of high-protein food from fermented soybean products. Food Res. Inst. Min. Agric. Forestry, Fukagawa, Tokyo. UNICEF Res. Grant Rep. 1958–1959.

SANO, T. 1961. Feeding studies with fermented soy products (natto and miso). In Meeting the Protein Needs of Infants and Preschool Children. N.A.S.-N.R.C. Publ. 843.

SAONO, S., GANDJAR, I., BASUKI, T. and KARSONO, H. 1974. Mycoflora of ragi and some other traditional fermented foods of Indonesia. Ann. Bogor. 5 (4) 187–204.

SHALLENBERGER, R.S., HAND, D.B. and STEINKRAUS, K.H. 1966. Changes in sucrose, raffinose, and stachyose during tempeh fermentation. Rep. 8th Dry Bean Res. Conf., Bellaire, Mich., Aug. 11–13, 1966. U.S. Dep. Agric., Agric. Res. Serv.

SHANK, R.C., WOGAN, G.N. and GIBSON, J.B. 1972A. Dietary aflatoxins and human liver cancer. I. Toxigenic moulds in foods and foodstuffs of Tropical Southeast Asia. Food Cosmet. Toxicol. 10, 51–60.

SHANK, R.C., WOGAN, G.N., GIBSON, J.B. and NONDASUTA, A. 1972B. Dietary aflatoxins and human liver cancer. II. Aflatoxins in market foods and foodstuffs of Thailand and Hong Kong. Food Cosmet. Toxicol. 10, 61–69.

SHIBASAKI, K. and HESSELTINE, C.W. 1962. Miso fermentation. Econ. Bot. 16, 180–195.

SMITH, A.K., HESSELTINE, C.W. and SHIBASAKI, K. 1961. Preparation of miso. U.S. Pat. 2,967,108. Jan. 3.

SMITH, A.K. et al. 1964. Tempeh: Nutritive value in relation to processing. Cereal Chem. 41, 173–181.

SORENSON, W.G. and HESSELTINE, C.W. 1966. Carbon and nitrogen utilization by Rhizopus oligosporus. Mycologia 58, 681–689.

STAHEL, G. 1946. Foods from fermented soybeans as prepared in the Netherlands Indies. J. N.Y. Bot. Gard. 47, 261–267.

STEINKRAUS, K. et al. 1960. Studies on tempeh, an Indonesian fermented soybean food. Food Res. 25, 777–788.

STEINKRAUS, K.H., VAN BUREN, J.P., HACKLER, L.R. and HAND, D.B.   1965.   A pilot-plant process for the production of dehydrated tempeh. Food Technol. *19*, 63–98.

STEINKRAUS, K.H., VAN VEEN, A.G. and THIEBEAU, D.B.   1967. Studies on idli—an India fermented black gram-rice food. Food Technol. *21*, 916–919.

STILLINGS, B.R. and HACKLER, L.R.   1965.   Amino acid studies on the effect of fermentation time and heat processing of tempeh. J. Food Sci. *30*, 1043–1049.

STOLOFF, L., MISLIVEC, P. and SCHINDLER, A.F.   1977.   *Aspergillus oryzae* (NRRL strain 1988). Science *196*, 1353–1354.

SUBBA RAO, G.N.   1967.   Fish processing in the Indo-Pacific area. Indo-Pac. Fish. Counc., Reg. Studies *4*. Food Agric. Organ. Reg. Ofc. Asia Far East, Bangkok.

TATENO, M. and UMEDA, I.   1955.   Cooking method of soybean and soybean-cake, in the soy sauce manufacturing. Jpn. Pat. 204,858. Jan. 13.

TUNG, T.C. and LING, C.H.   1968.   Study on aflatoxin of foodstuffs in Taiwan. J. Vitaminol. *14*, 48–52.

UMEDA, I.   1963.   Comparison of United States and Japanese soybeans for making shoyu. U.S. Dep. Agric. Final Tech. Rep., P.L. 480 Proj. *UR-A11-(40)-(C)*. Nat. Agric. Library, Beltsville, Md.

UMEDA, I., NAKAMURA, K., YAMATO, M. and NAKAMURA, Y.   1969. Investigations of comparative production of shoyu (soy-sauce) from defatted soybean meals obtained from United States and Japanese soybeans and processed by United States and Japanese methods. U.S. Dep. Agric. Final Tech. Rep., P.L. 480 Proj. *UR-A11-(40)-21*. Nat. Agric. Library, Beltsville, Md.

VAN BUREN, J.P., HACKLER, L.R. and STEINKRAUS, K.H.   1972. Solubilization of soybean tempeh constituents. Cereal Chem. *49*, 208–211.

VAN VEEN, A.G.   1953.   Fish preservation in Southeast Asia. Adv. Food Res. *4*, 209–231.

VAN VEEN, A.G. and GRAHAM, D.C.W.   1968.   Fermented peanut press cake. Cereal Sci. Today *13*, 96–99.

VAN VEEN, A.G., HACKLER, L.R., STEINKRAUS, K.H. and MUKHER-JEE, S.K.   1967.   Nutritive quality of idli, a fermented food of India. J. Food Sci. *32*, 339–341.

VAN VEEN, A.G. and SCHAEFER, G.   1950.   The influence of the tempeh fungus on the soya bean. Trop. Geogr. Med. *2*, 270–281.

WAGENKNECHT, A.C. *et al.*   1961.   Changes in soybean lipids during tempeh fermentation. J. Food Sci. *26*, 373–376.

WAI, N.   1929.   A new species of *Mono-Mucor, Mucor sufu*, on Chinese soybean cheese. Science *70*, 307–308.

WAI, N.   1968.   Investigation of the various processes used in preparing Chinese cheese by the fermentation of soybean curd with mucor and other fungi. U.S. Dep. Agric. Final Tech. Rep., P.L. 480 Proj. *UR-A6-(40)-1*. Nat. Agric. Library, Beltsville, Md.

WANG, H.L. 1967A. Products from soybeans. Food Technol. *21*, 115–116.

WANG, H.L. 1967B. Release of proteinase from mycelium of *Mucor hiemalis*. J. Bacteriol. *93*, 1794–1799.

WANG, H.L. and HESSELTINE, C.W. 1965. Studies on the extracellular proteolytic enzymes of *Rhizopus oligosporus*. Can. J. Microbiol. *11*, 727–732.

WANG, H.L. and HESSELTINE, C.W. 1966. Wheat tempeh. Cereal Chem. *43*, 563–570.

WANG, H.L. and HESSELTINE, C.W. 1970A. Sufu and lao-chao. J. Agric. Food Chem. *18*, 572–575.

WANG, H.L. and HESSELTINE, C.W. 1970B. Multiple forms of *Rhizopus oligosporus* protease. Arch. Biochem. Biophys. *140*, 459–463.

WANG, H.L., RUTTLE, D.I. and HESSELTINE, C.W. 1968. Protein quality of wheat and soybeans after *Rhizopus oligosporus* fermentation. J. Nutr. *96*, 109–114.

WANG, H.L., RUTTLE, D.I. and HESSELTINE, C.W. 1969. Antibacterial compound from a soybean product fermented by *Rhizopus oligosporus*. Proc. Soc. Exp. Biol. Med. *131*, 579–583.

WANG, H.L. and SWAIN, E.W. 1977. Personal communication. Northern Regional Res. Center, Peoria, Ill.

WANG, H.L., SWAIN, E.W. and HESSELTINE, C.W. 1975. Mass production of *Rhizopus oligosporus* spores and their application in tempeh fermentation. J. Food Sci. *40*, 168–170.

WANG, H.L., VESPA, J.B. and HESSELTINE, C.W. 1974. Acid protease production by fungi used in soybean food fermentation. Appl. Microbiol. *27*, 906–911.

WATANABE, T. 1969. Industrial production of soybean foods in Japan. UNIDO Expert Group Meet. Soybean Process. Use, U.S. Dep. Agric., Peoria, Ill., Nov. 17–21, 1969.

WINARNO, F.G. 1976. Progress report of phase-1, National Soybean Survey. Bogor Agricultural Univ., Bogor, Indonesia.

WINARNO, F.G. 1979. Fermented vegetable protein and related foods of Southeast Asia with special reference to Indonesia. J. Am. Oil Chem. Soc. *56*, 363–366.

WORTHINGTON, R.E. and BEUCHAT, L.R. 1974. α-Galactosidase activity of fungi on intestinal gas-forming peanut oligosaccharides. J. Agric. Food Chem. *22*, 1063–1066.

YAMAGUCHI, Y. *et al.* 1973. Water soluble Monascus pigment. U.S. Pat. 3,765,905. Oct. 16.

YAMAMOTO, K. 1957. Koji: Effects of cultural temperature on the production of mold protease. Bull. Agric. Chem. Soc. Jpn. *21*, 319–324.

YOKOTSUKA, T. 1960. Aroma and flavor of Japanese soy sauce. Adv. Food Res. *10*, 75–134.

YOKOTSUKA, T.   1971.   Shoyu. Proc. 5th Intern. Symp. Convers. Manuf. Foodst. Microorg., Jpn. Inst. Food Technol., Kyoto, Dec. 5–9, 1971, 117–125.

YOKOTSUKA, T. *et al.* 1967A. Production of fluorescent compounds other than aflatoxins by Japanese industrial molds. *In* Biochemistry of Some Foodborne Microbial Toxins.   M.I.T. Press, Cambridge, Mass.

YOKOTSUKA, T. *et al.* 1967B. Studies on the compounds produced by molds. Part I. Fluorescent compounds produced by Japanese industrial molds. J. Agric. Chem. Soc. Jpn. *41*, 32–38.

YONG, F.M. and WOOD, B.J.B.   1974.   Microbiology and biochemistry of soysauce fermentation. Adv. Appl. Microbiol. *17*, 157–194.

Section III

Food Related Fermentations

13

Microbial Biomass, Single Cell Protein, and Other Microbial Products

Gerald Reed

This chapter deals with the production of microbial biomass and single cell protein as well as some derived products which are not treated more extensively in other chapters, such as yeast autolysates and 5′-nucleotides. Biochemicals which find their major application in the fields of medicine or analysis or which are research chemicals will not be discussed.

With some notable exceptions man has always relied on macrobial life as a source of food. However, microbes have contributed traditionally to the supply of food through fermentation processes. In many instances these have been carried out since prehistoric times. Such fermentations include the alcoholic drinks (beer, wine, sake), cheese, vinegar, bread, miso, tempeh, and soy sauce, to mention just a few. In almost all of these fermentations the nutritional contribution of microbial protein is difficult to assess because the number, the weight, and the composition of the microbes is not known. In modern white bread where this can be accurately determined, yeast protein contributes about 5% to the total protein content, a contribution which is very important because of the high lysine content of yeast protein.

The incidental consumption of microbes by humans in fermented foods and that of distillers' and brewers' spent grains by domestic animals is quite old. But the conscious attempt to grow microbes for the human diet started in Germany with the drying of spent brewers' yeast (about 1910). By 1914 about 9000 MT (10,000 tons) of dried yeast became available for use in feeds (Butschek 1962). The acute shortage of protein in Europe during the Second World War stimulated the erection of industrial plants for the production of yeasts, usually *Candida utilis*. Wood hydrolysates and sulfite waste liquor were the commonly used substrates. Several plants which had been built during the war years were dismantled after the end of hostilities, for instance, an installation which produced feed yeast from cane molasses in Jamaica. But at the same time new plants were built after the war, notably in the United States for the production of *C. utilis* from waste sulfite liquor and in Eastern European countries for the production of feed yeast on wood hydrolysates.

Since the 1960s studies in this field have multiplied, and commercial interest has covered a wider range of substrates and microorganisms. This is due to 3 events which all occurred at about the same time, that is, sometime between 1960 and 1970. The first was the recognition of acute shortages of protein both for human consumption and for animal feeding. The second was the pioneering work of A. Champagnat at the Societé Française de Petroles BP who studied the microbial oxidation of petroleum fractions by microorganisms beginning about 1957. Finally, the need to combat pollution by industrial wastes of high biological oxygen demand was recognized in the developed countries of the world. It is difficult to say which of these factors was the most important one. Most likely the exciting work with gas oils and n-paraffins catalyzed the whole field of inquiry and led academic institutions and leading companies in private industry into an ever widening inquiry into methods of producing microbial cell mass for feeds and foods. Following the lead of research workers at the Massachusetts Institute of Technology the products which are discussed in this chapter are usually called "single cell protein" or "SCP." This designation blurs the distinction between microbial cells as such which may contain from 40 to 65% protein and the proteins or protein concentrates which may be extracted from them. In this chapter this distinction will be retained by calling products consisting of whole cells "biomass" or "microbial biomass" and calling proteins, protein extracts, or protein concentrates "single cell protein" or "SCP." It is important to retain this distinction because a considerable fraction of the nitrogen of the microbial biomass is not protein nitrogen, a fact which is generally recognized by experts in the field but often overlooked in practice.

In yeast cells the nonprotein nitrogen fraction consists largely of purine and pyrimidine compounds of the nucleotides and may account for 10–15% of the total nitrogen. A factor of 6.25 is commonly used to derive the protein content of microbial biomass from the nitrogen value. This factor is indeed correct if it is applied to isolated yeast protein, as one can readily calculate from the composition of its amino acids. It is not correct if it is applied to microbial biomass and leads to an overestimate of its protein content by 10–20%. In animal feeding studies the protein content of microbial biomass in the ration is usually determined by use of the factor 6.25. This results in lower values for the PER or the biological value of microbial protein because less protein is actually fed than the control protein (casein). Most of the research on the production of microbial biomass and all industrial efforts place the emphasis almost entirely on the formation of protein because of the well recognized protein shortage. Microbial biomass also contains nutritionally important vitamins and minerals, but most of these can be obtained rather inexpensively from other sources. While this is true of those vitamins which are routinely used for the fortification of foods or feeds it is not true for some nutritional factors which have only been recognized within the last 10 to 20 years, and it may not be true for some yet unidentified nutritional factors. This will be mentioned further later.

The traditional substrates for the production of biomass have been carbohydrates either in the form of assimilable sugars or in the form of starchy materials which can be readily hydrolyzed by enzymes. These sources are molasses, starchy grains, wood sugars from spent sulfite liquor or from wood hydrolysates, and to a lesser extent whey. The microbes grown are *Saccharomyces cerevisiae, S. uvarum, Candida utilis*, and to a minor extent *Kluyveromyces fragilis*. Since the early 1960s many other carbon sources have been investigated, most prominently some hydrocarbons such as gas oil, and more specifically its n-paraffin fraction, methane, methanol, and ethanol. Others are whey, cellulose, and large numbers of agricultural and food processing waste materials. These raw materials require the use of microorganisms other than those mentioned above. They are yeasts of the genera *Candida, Endomycopsis, Kluyveromyces*, molds of the genera *Paecilomyces* and *Trichoderma*, and *Pseudomonas (Methylococcus)* bacteria, as well as several algae and photosynthetic and lithotropic bacteria. Only a few of these processes have led to commercial or semicommercial installations. Some of these are listed in Table 13.1.

The use of microbes other than the commonly accepted food yeasts (*Saccharomyces* sp., *Candida utilis*, and *Kluyveromyces fragilis*) has created problems of proving the safety of the microbial biomass for use in foods or feeds. In some instances unreasonable demands for extensive and exhaustive toxicological examinations have delayed or prevented the erection or use of production facilities, notably in Japan and Italy (O'Sullivan 1978). The nutritional properties and the toxicology of various products will be discussed below. At this point it must merely be added that in Western countries objections to the use of microbial biomass in feeds or foods have a political or ideological rather than a factual basis.

TABLE 13.1.   UNITED STATES PRODUCTION FACILITIES FOR MICROBIAL BIOMASS

Company	Location	Capacity[1]	Substrate	Organism
Amber Div., Milbrew, Inc.	Juneau, WI	20	whey beer wort	*K. fragilis* *S. uvarum*
Amoco	Hutchinson, MN	15	ethanol	*C. utilis*
Anheuser-Busch	Jacksonville, FL	10	beer wort	*S. uvarum*
Boise-Cascade	Salem, OR	10	sulfite liquor	*C. utilis*
Lake States Div., St. Regis Paper Co.	Rhinelander, WI	10	sulfite liquor	*C. utilis*
Yeast Products, Inc.	Clifton, NJ	—	beer wort	*S. uvarum*

[1]Capacity = Estimated capacity in million lb (1 lb = 0.45 kg) of dried biomass annually.

ECONOMICS

It is difficult to obtain a clear picture of the economics of producing microbial biomass. Cost estimates given in the literature are usually supplied by scientists working in academia or for consulting firms, not by those employed by the manufacturers of biomass. Nevertheless a general picture emerges from the numerous articles which deal with the economics of biomass production. The process is capital intensive; it has a relatively high cost for raw materials and energy, and a relatively low cost for labor. For various commercial processes Moo-Young (1977) estimates the following breakdown: Depreciation 6–10%; raw materials 45–75%; energy 12–37%; and labor 5–11%. Table 13.2 shows these values as they are assigned to 5 commercial processes. The major energy costs for fermentation are for oxygen transfer and for heat removal. Table 13.3 gives examples of such costs for 8 substrates (Abbott and Clamen 1973).

A particular element of uncertainty exists for processes which utilize waste materials such as waste sulfite liquor, whey, or other agricultural or food processing waste. In such cases the cost of the raw material may be assumed to be zero or it may be assigned a negative cost because of the cost incurred in disposing it through municipal sewage systems or open field disposal. In some cases the cost of the raw material may be attractive for the production of microbial biomass but (as in the case of whey) it may not be available in sufficient quantity at a central location, and hence its transportation costs may be prohibitive.

The following publications deal with the economic analysis of processes for the production of microbial biomass: Abbott and Clamen (1973); Micholt et al. (1975); Mateles (1975); Pace and Goldstein (1975); Peschard-Mariscal and Viniegra-Gonzales (1977); Litchfield (1977B); and Moo-Young (1977).

SUBSTRATES AND NUTRIENT REQUIREMENTS

The nutritional requirements of various microorganisms may differ appreciably but all of them require carbon, nitrogen, and phosphorus sources, as well as other minerals and frequently vitamins. Sources of nitrogen (ammonia, ammonium salts, nitrates) and of phosphorus are well known. These as well as sources for other minerals such as calcium, magnesium, potassium and trace elements are infrequently mentioned in the technical literature because of the overriding importance of the carbon source for the technical and economic feasibility of the biomass process. In some instances, vitamins such as biotin, thiamin, or others have to be supplied to the growth medium. For bench scale experiments this is usually done in the form of a yeast extract. For commercial processes the requirements for vitamins and trace elements have to be determined accurately; and they have to be added in the form of soluble mineral compounds and synthetic vitamins, usually with a yeast or yeast extract supplement. A specific example of the nutrient requirements of *Candida utilis* grown on ethanol is shown in Table 13.4 (Ridgeway et al. 1975). Nutritional requirements have been reviewed by Suomalainen and Oura (1971) and fermentation substrates by Ratledge (1977).

TABLE 13.2. COMPARATIVE PRODUCTION COSTS OF BIOMASS PRODUCTION

Production Item	Yeast Paraffin (%) (Italy)	Bacteria Methanol (%) (England)	Yeast Ethanol (%) (Czechoslovakia)	Fungi Sulfite Liquor (%) (Finland)	Bacteria Bagasse (%) (United States)
Depreciation	9.3	5.8	5.8	9.1	11.5
Raw materials	58.5	73.8	77.1	55.1	43.6
Substrate	29.4	47.4	63.9	17.0	25.7
Phosphoric acid	11.1	11.8	3.2	16.2	5.7
Ammonia	9.9	12.0	4.8	13.3	3.6 (?)
Mineral salts	2.9	2.6	1.9	4.2	4.2
Miscellaneous	5.2	—	3.3	4.4	4.4
Utilities	23.8	14.2	12.0	24.8	36.6
Labor, etc.	8.4	6.2	5.1	11.0	8.3
Total	100%	100%	100%	100%	100%

Source: Moo-Young (1977).
Note: Substrate cost assumption: paraffin = 1.3 methanol;   ethanol = 2.6 methanol.

TABLE 13.3.   EFFECT OF SUBSTRATE AND YIELD COEFFICIENTS ON OPERATING COSTS OF FERMENTATION

All costs expressed in terms of 1973 dollar values.

Substrate	Substrate Cost ¢/lb Substrate	¢/lb Cells	Oxygen Transfer Cost ¢/lb Cells	Heat Removal Cost ¢/lb Cells	Combined Cost ¢/lb Cells
Malate	0	0	0.46	0.75	1.2
Glucose[1]	2.0	3.9	0.23	0.54	4.7
Paraffins	4.0	4.0	0.97	1.4	6.4
Methanol	2.0	5.0	1.2	1.9	8.1
Methane	1.0	1.6	3.3	3.7	8.6
Ethanol	6.0	8.8	0.75	1.3	11.0
Isopropanol	5.0	11.6	2.7	3.1	17.4
Acetate	6.0	16.7	0.62	1.1	18.4

Source: Abbott and Clamen (1973).
[1] Glucose equivalents of molasses.

TABLE 13.4.   INORGANIC NUTRIENT REQUIREMENTS OF C. UTILIS GROWN ON ETHANOL

Nutrient Element	Typical Compound	Nutrient Input per 100 g Cells Produced
Macro-nutrients		
Phosphorus	H_3PO_4	2–4 g
Potassium	KCl	2–3 g
Magnesium	$MgCl_2 \cdot 6H_2O$; $MgSO_4$	0.3–0.6 g
Calcium	$CaCl_2$	0.001–0.2 g
Sodium	$Na_2CO_3 \cdot H_2O$; NaCl	0.01–0.2 g
Micro-nutrients		
Iron	$Fe(C_3H_4(OH)(COOH)_3)$	6–13 mg
Manganese	$MnSo_4 \cdot H_2O$	4–8 mg
Zinc	$ZnSO_4 \cdot 7H_2O$	2–6 mg
Molybdenum	$Na_2MoO_4, 2H_2O$	1–2 mg
Iodine	KI	1–3 mg
Copper	$CuSO_4 \cdot 5H_2O$	0.5–1 mg

Source: Ridgeway et al. (1975).

Carbon Sources

The traditional carbon sources for cell mass production are carbohydrates in the form of fermentable sugars. Beet or cane sugar molasses containing about 50% fermentable sugars and sulfite liquors containing 2 to 3% of hexoses and pentoses are the most common substrates. These substrates require a minimum of preparation, generally stripping of SO_2 and clarification for the removal of insoluble solids. Whey containing about 5% lactose, demineralized whey, or ultrafiltered whey also requires little processing. Starches from grains require enzymatic hydrolysis to fermentable sugars, except that some molds, bacteria, and yeasts themselves have sufficient extracellular amylase activity. The use of malt amylases has often been replaced by the use of fungal and bacterial amylases. Cellulose or cellulose containing wastes require fairly extensive pretreatment by mechanical subdivision (milling) or treatment with alkali or acids. Ethanol (either synthetic or fermentation ethanol) and methanol require no pretreatment.

Methane and n-alkanes are hydrocarbon compounds that may be used for the production of biomass. Methane is a major constituent of natural gas. The n-alkanes occur in gas oils which are commonly marketed as kerosene, diesel fuel, or heating oil. Gas oils with a boiling point between 200° and 380°C may contain from 8 to 20% of n-alkanes. Gas oil can be used directly as a substrate. Its use requires solvent extraction of the produced biomass.

N-alkanes isolated by molecular sieve adsorption methods and with chain lengths from 10 to 23 carbons are most suitable. With shorter chain length compounds, microbial growth rates are too slow, and with longer chain length compounds the viscosity of the substrate is too high (Neumann 1975; Litchfield 1977A). Figure 13.1 shows a schematic diagram of the range of substrates, microorganisms and end products. Extensive cost figures for fermentation substrates have been provided by Ratledge (1977). Specific uses of these substrates as well as experimental or unusual substrates will be mentioned.

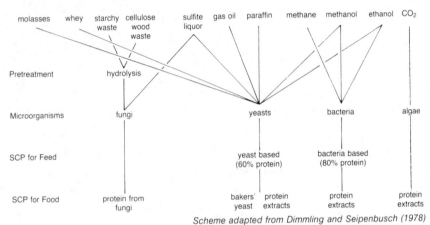

Scheme adapted from Dimmling and Seipenbusch (1978)

FIG. 13.1.   SUBSTRATES FOR PRODUCTION OF SCP FROM CLASSES OF MICROORGANISMS

Table 13.5 shows a comparison of the yield of biomass, the oxygen consumed, and the calories evolved for bacterial growth on some of the previously mentioned substrates (Abbott and Clamen 1973). It is apparent that hydrocarbon substrates which have a higher carbon content than carbohydrates give higher cell yields, require more oxygen per gram of cells grown, and evolve more heat.

Fermentation

In contrast to the production of antibiotics and enzymes many processes for the production of microbial biomass do not require conditions of absolute sterility. However, the fast growth rates of microorganisms in biomass production impose stringent requirements for efficiency in the supply of oxygen and for the removal of heat evolved in the aerobic process. Faust (1975) has calculated the cost of producing biomass on paraffin (n-alkanes) in a large plant and finds the following distribution of costs (as a percentage of total costs): Raw material 49%; energy 17%; labor and overhead 8%; and depreciation 26%. Of the energy costs almost one half was required for

TABLE 13.5. COMPARISON OF YIELD COEFFICIENTS FOR BACTERIA GROWN ON VARIOUS CARBON SOURCES

Substrate	Y_{sub}	Y_O	Y_{kcal}
Malate	0.34[1]	1.02	0.30
Acetate	0.36[1]	0.70[1]	0.21
Glucose equivalents (molasses, starch, cellulose)	0.51[1]	1.47	0.42
Methanol	0.40[1]	0.44	0.12
Ethanol	0.68[1]	0.61[1]	0.18
Isopropanol	0.43[1,2]	0.23[1,2]	0.074
n-Paraffins	1.03[1]	0.50	0.16
Methane	0.62[1]	0.20[1]	0.061

Source: Abbott and Clamen (1973).
[1]Indicates experimental data.
[2]Indicates yields measured with n-propanol.
Y_{sub} = g cell solids per g substrate.
Y_O = g cell solids per g oxygen consumed.
Y_{kcal} = g cell solids per kcal evolved.
(For a comparison of yeasts and bacteria see Humphrey 1970.)

cooling. The increasing cost of energy has caused a re-evaluation of the classical, agitated fermentor with its relatively high energy requirement for agitation and the concomitant need for additional heat removal.

The simplest form of a fermentor consists of a cylindrical vessel equipped with air sparger tubes at the bottom of the fermentor and internal or external cooling coils. One can induce directional flow of the fermentor liquid by the use of draft tubes. One such design is shown schematically in Fig. 13.2. Such air lift fermentors (also called bubble column fermentors or loop fermentors) have been described by Lehman et al. (1975), Cooper et al. (1975), and Hatch (1975). Figure 13.3 shows schematically the salient features of some commercially available fermentors using turbines for mixing and improved oxygen transfer. Some of the details of the effect of turbines on the rate of mixing and oxygen transfer by impellers have been worked out by McManamey et al. (1973). Moser and Lafferty (1975) have dealt with the theoretical problems of fermentor design, particularly with regard to the optimization of the oxygen transfer rate.

Nutritional Value of Microbial Biomass and SCP

The principal value of microbial biomass as a component of feeds or foods is its contribution to protein nutrition. Table 13.6 show the gross chemical composition of various classes of microorganisms. The protein content is highest in bacteria and lowest in the filamentous fungi. It is intermediate in yeasts and algae. For use in feeds the high nucleic acid (NA) content of

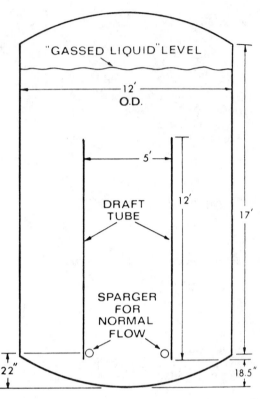

FIG. 13.2. WASCO AIR LIFT FERMENTOR WITH 5 FT DRAFT TUBE (AIR SPARGED, NO AGITA-TOR)

From Cooper et al. (1975)

microorganisms causes no problems since mammals, with the exception of man and some primates, convert uric acid to allantoin, which is very soluble and readily excreted in urine. Man does not possess urate oxidase (uricase), and the ingestion of purine compounds leads to increased plasma levels of uric acid and may lead to metabolic disturbances, specifically gout. A figure of 2 g of NA per day per adult is used as the acceptable upper limit. This would permit the consumption of up to 10 g of yeast or bacterial protein (in the form of cell biomass) per day.

The quality of various proteins may be evaluated by feeding tests with experimental animals of which the rat is the most widely used species. The determination of the coefficient of digestibility, the biological value, the net protein utilization (NPU), nitrogen balance studies, and the protein efficiency ratio (PER) are widely used. Each of these methods has advantages and disadvantages which have been discussed in the context of microbial biomass evaluations by Young and Scrimshaw (1975).

In the United States and Canada the PER method is accepted by regulatory agencies as the method of choice. It is relatively simple and gives fairly reproducible results if a reference protein, such as casein, is used as a standard of comparison. In general the method reflects the deficiency of the most limiting amino acid of a protein. Unfortunately, the PER method is the least appropriate method for determining the contribution of a protein to a diet containing several animal and vegetable proteins. But this is exactly the proposed use of microbial protein in the human diet, and it is also its most likely practical application (see also Bodwell 1977).

Experiments with mixed protein diets are costly and time consuming. One such noteworthy experiment is illustrated in Fig. 13.4. The quality of the protein is expressed as the minimum amount required for the maintenance of nitrogen balance in humans. Mixtures of egg protein with wheat, corn, potato, or algal proteins were used. A mixture of 60% egg protein and 40% algal protein was optimal and nitrogen balance was maintained at levels of less than 0.5 g N per kg body weight in humans (Kofranyi and Jekat 1967).

The potential nutritional value of a protein can be estimated from the percentages of essential amino acids. This is indeed no more than an estimate because digestibility and specific amino acid imbalances affect the nutritional value. Table 13.7 shows the amino acid composition of commercially available biomass preparations. Extensive data on the amino acid composition of microcial biomass are available in the following reviews:

(F = fungi; Y = yeasts; B = bacteria; A = algae)

Dabbah (1970)	F,A,Y,B	Frahm and Lembke (1975)	A,Y,B
Snyder (1970)	A,Y,B	Young and Scrimshaw (1975)	F,A,Y,B
Kihlberg (1972)	A,Y,B	Anderson et al. (1975)	F
Reed and Peppler (1973)	Y	Cooney and Levine (1975)	Y
Schulz and Oslage (1975)	A,Y	Rehm (1976)	A,Y,B

Such compilations of individual data are not shown here because the amino acid composition of biomass preparations varies with the strain and the conditions of growth, and data obtained with a single species often show greater variation than between species. For instance, Wolf et al. (1975) emphasize the differences in amino acid composition of C. utilis grown by them on sulfite liquor and values given in the literature for the same yeast grown on the same substrate.

In general microbial proteins are rich in lysine and relatively poor in sulfur-containing amino acids. In this respect they resemble soybean proteins. But each preparation of microbial biomass or SCP must be analyzed separately to determine its amino acid composition. The danger of generalizing from the literature is shown by the work of Erdman et al. (1977A), who determined the amino acid concentrations in the proteins of Lacto-

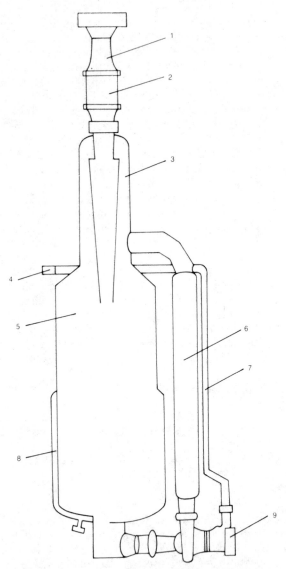

FIG. 13.3.   DEEP JET AERATION FERMENTOR
1—Air inlet duct.
2—Air filter.
3—Overflow shaft.
4—Safety relief valve.
5—Vessel body.
6—Pump discharge pipe.
7—Degassifier pipe.
8—Cooling jacket.
9—Circulation pump.

TABLE 13.6. GROSS CHEMICAL COMPOSITION OF MICROBIAL BIOMASS FOR CLASSES OF MICROORGANISMS

Chemical Composition %	Filamentous Fungi	Algae	Yeasts	Bacteria
Nitrogen	5–8	7.5–10	7.5–9	11.5–13.3
N × 6.25	31–50	47–63	47–56	72–83
Nucleic acids	9.2[1]	3–8	6–12	8–16
Ash	9–14	8–10	5–9.5	3–7
Lipids	2–8	7–20	2–6	1.5–3

Source: Modified from Kihlberg (1972).
[1]From Anderson *et al.* (1975).

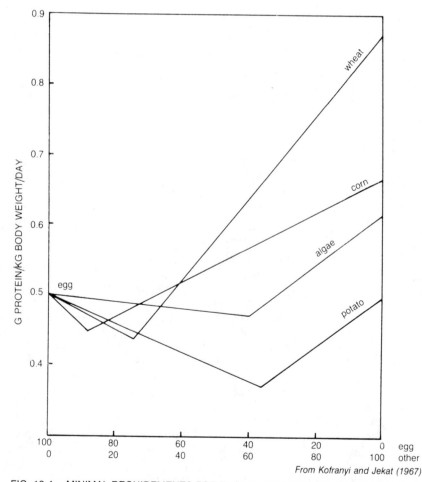

From Kofranyi and Jekat (1967)

FIG. 13.4. MINIMAL REQUIREMENTS FOR THE MAINTENANCE OF NITROGEN EQUILIBRIUM IN HUMANS FOR MIXTURES OF EGG PROTEIN WITH WHEAT, CORN, POTATO, AND ALGAL PROTEIN

TABLE 13.7.   AMINO ACID COMPOSITION OF COMMERCIAL BIOMASS PRODUCTS

(In g per 16 g nitrogen)

Amino Acid	C. lipolytica[1]	P. methylotropha[2]	C. utilis[3]	S. cerevisiae[4]	P. varioti[5]
Lysine	7.4	7.32	7.17	8.43	6.4
Valine	5.9	6.51	5.33	6.67	5.1
Leucine	7.4	8.43	6.97	8.13	6.9
Isoleucine	5.1	5.35	4.30	5.19	4.3
Threonine	4.9	5.71	4.71	5.20	4.6
Methionine	1.8	3.00	1.02	1.63	1.5
Cystine	1.1	0.76	0.61	1.6	1.1
Phenylalanine	4.3	4.27	3.69	4.50	3.7
Tryptophan	1.4	1.11	0.04	1.1	1.2
Histidine	2.1	2.29	2.05	3.14	—
Tyrosine	3.6	3.82	3.28	4.87	—
Arginine	5.1	5.56	7.17	5.35	—

[1]Source: Allison (1975).
[2]Source: Gow et al. (1975).
[3]Source: Ridgeway et al. (1975).
[4]Source: Universal Foods (1977).
[5]Source: Romantschuk (1975).

bacillus acidophilus, L. bulgaricus, L. casei, L. fermenti, L. plantarum, and L. thermophilus. They found 3-fold variations in isoleucine content and a 2-fold variation in lysine content between the highest and the lowest percentages of these amino acids. In assessing the nutritional value of proteins one must be aware of the distinction between whole cells and isolated SCP which differ somewhat in their amino acid composition. For SCP the method of extraction and recovery of the protein influences amino acid composition as Vananuvat and Kinsella (1975A,B) have demonstrated for Kluyveromyces fragilis. Charatyan and Wolnowa (1975) have shown the differences in biological value between the biomass and the isolated SCP of various yeasts with Tetrahymena pyriformis as the test organism. The yeasts were Candida guilliermondii grown on hydrocarbon, C. scottii grown on wood hydrolysate, Saccharomyces cerevisiae grown on a carbohydrate, and Mycoderma vini grown on alcohol.

Methionine is the limiting amino acid in most preparations of microbial biomass. Synthetic D,L-methionine is rather inexpensive and may be used to improve the quality of the protein. For instance, the PER of primary grown bakers' yeast was 2.02. It could be increased to 2.27 with the addition of 0.16% of D,L-methionine and to 2.77 with the addition of 0.50% of

D,L-methionine based on the weight of the yeast solids. Yeast protein was added to the diet of rats based on true protein, not on "crude" protein.[1] Seely (1975) reports a PER of 2.2 for a protein fraction from bakers' yeast essentially free from nucleic acids.

Inactive dried brewers' or bakers' yeasts are widely used as dietary supplements. They contribute some protein and vitamins (except for vitamin C, vitamin B_{12}, and fat soluble vitamins) as well as trace minerals. There are frequent allusions to unidentified nutritional factors of yeast in the trade literature. These should not be dismissed lightly because some of these factors have actually been identified. For instance, the significance of selenium for animal and human nutrition was first discovered in experiments with brewers' yeast and C. utilis yeast (Schwarz and Foltz 1957). Also, yeast is currently the best source of the glucose tolerance factor, a chromium-containing organic compound, which mediates the action of insulin. The factor is essential for the aged who have lost the ability to synthesize the glucose tolerance factor from inorganic chromium of the diet (Mertz 1975). Robbins and Seeley (1977) showed that the isolated, comminuted cell walls of bakers' yeast had a strong effect in reducing cholesterol levels in rats fed a hypercholesterolemic diet. The effect appeared to be larger than that of other nondigestible polymeric carbohydrates which can serve as sources of fiber.

Reduction of the Nucleic Acid Content of Microbial Biomass

There is no need for the removal of nucleic acid (NA) from biomass for use in feeds. However, for use in food and with the purpose of supplying a major portion of the required protein from biomass, the presence of NA presents a serious obstacle. This is expressed well in a blunt statement by Viikari and Linko (1977): "As long as safe processes for the removal of nucleic acids continue to be economically prohibitive, the nucleic acid content constitutes a universal and major limitation to the use of SCP as human food." This accounts for the voluminous literature which deals with this important question. It should be kept in mind that the high costs to which these authors refer are due to the poor yield of isolated protein and rarely to the specific treatment for NA removal. As a matter of fact all methods for isolating protein from microbial biomass suffer from the same high costs due to poor yields whether they are coupled with a specific attempt to reduce NA or not. [A recently announced process (Anon. 1978A,B) for treating the dried cell mass with anhydrous NH_3 and methanol may well present a more promising approach.]

The NA content of microbial biomass shows considerable variation, not only in its relation to total cell mass but also in relation to its protein

---

[1]Universal Foods Corp., unpublished data.

content. For instance, Sinskey and Tannenbaum (1975) demonstrated the great effect of growth rate on NA concentration. *Aerobacter aerogenes* grown at a dilution rate of 0.25/hr had a protein to NA ratio of about 5 to 7, while at a dilution rate of 0.7/hr, the ratio of protein to NA was about 3. Apart from such variations the problems caused by high NA concentrations are similar for fungi, algae, yeasts, and bacteria. Bacteria, which have the highest protein content, also have the highest concentration of NA. In general about 15 to 20% of the total nitrogen of biomass preparations is NA nitrogen. The content of other nitrogenous fractions is usually quite small although there are notable exceptions. For instance, *Fusarium graminearum* contains about 73% of its nitrogen in the protein and amino acid fraction, 15% in the NA fraction, and 10% as n-acetyl glucosamine (Anderson *et al.* 1975). The nitrogen content of the free amino acid pool accounted for 7% of the total nitrogen. It must be added that almost all authors include free amino acid nitrogen in the protein fraction when dealing with microbial biomass. For SCP preparations which have undergone precipitation this is, of course, not the case.

Several processes have been developed to reduce the NA content of biomass preparations. These can be divided into processes which aim at the production of biomass with a low NA content and those which aim at the production of isolated SCP. Sinskey and Tannenbaum (1975), Hedenskog and Mogren (1973), Kihlberg (1972), and others have reviewed the various methods which have been proposed. For the treatment of whole cells, heat shock or alkaline extraction or a combination of both methods seems to be the simplest procedure. For instance, Viikari and Linko (1977) achieved a reduction of NA from 9 to 3% by a 30 min treatment of cells of *Paecilomyces varioti* with 0.125 N NaOH at 50°C. The use of heat shock is described in a recent patent (Imp. Chem. Ind. 1977) in which cells of *Pseudomonas methylotropha* are heat shocked at a pH below 5 and at a temperature exceeding 60°C followed by raising of the pH to between 6 and 10. Similar results have been reported for *Fusarium graminearum* (Rank Hovis McDougal 1977). Trevelyan (1976) obtained almost complete removal of NA by extraction with 5% NaCl solution at 120°C. The methanol/anhydrous NH_3 process has already been mentioned (Anon. 1978A,B).

Toxicity and Other Safety Aspects

The most important aspect of the safety of microbial biomass preparations for humans is the presence of NA, which has already been discussed. It limits the daily consumption to about 10 g of microbial protein (in the form of whole cell preparations). A reduction of the NA content is technically feasible but costly, and isolated microbial proteins (SCP) have not appeared on the market in commercial quantities.

A major part of the presently available microbial biomass is produced with *S. cerevisiae* grown on ale wort or molasses, *S. uvarum* grown on beer wort, *Candida utilis* grown on sulfite liquor or molasses, and *Kluyveromyces* sp. grown on whey. These yeasts have been generally recognized as safe (GRAS) for human consumption.

On the other hand, processes for microbial biomass developed during the past 15 years have encountered vigorous opposition by portions of the public and by regulatory agencies in Western countries, an opposition which does not appear to be justified on the basis of the available evidence. This opposition has led to the abandonment of planned installations, for instance in Italy and Japan, and in the closing of some facilities. At present the prospect for acceptance of petroleum grown yeast for human consumption is not good even if all the conditions of the guidelines of the Protein Advisory Group of the United Nations were to be fulfilled (Anon. 1972 A,B,C). However, extensive research on novel processes continues, and it is likely that such processes will result in the commercial production of microbial biomass for feeds.

Nucleic Acids.—High uric acid levels in human plasma cause gout. Sources of NA in diets are principally organ meats and secondarily beer and possibly other fermented beverages. Careful studies by Edozien *et al.* (1970) and Waslien *et al.* (1970) have established the relationship between the feeding of yeast and algae and the elevation of the plasma uric acid level in humans. On the basis of their work an upper limit of 2 g of NA per day has been generally accepted. Some caution is indicated for persons whose normal diet contains a considerable amount of organ meats (liver, lungs, brain, spleen).

Polycyclic Aromatic Hydrocarbons.—The carcinogenic activity of some polycyclic hydrocarbons is well established. Typical carcinogens are benzo (a) pyrene and benzo (a) anthracene. Grimmer (1974) has determined the presence of 13 polycyclic hydrocarbons including benzo (a) pyrene and benzo (a) anthracene in bakers' yeast from France, Britain, Germany, and the U.S.S.R., in various European food yeasts, and in yeasts grown on gas oil. The yeasts grown on gas oil generally showed lower levels of polycyclic hydrocarbons than the yeasts currently used in the production of bread or as food supplements. Wolf *et al.* (1975) have reported on the concentration of various benzpyren compounds in food yeasts and algae in comparison with that in common vegetables.

Lysinoalanine.—Some proposed processes for the isolation of microbial protein make use of alkaline extraction as the first step (Lindblom 1974). A nephrotoxic factor has been found in alkali treated soybean protein and identified as lysinoalanine (Woodard and Short 1973). It appears that treatment of a protein at a pH above 10.5 at 25°C or treatment at a pH above 8 at 100°C leads to the formation of some lysinoalanine, but small quantities of lysinoalanine can be found in most heat treated protein foods. If the problem has any practical significance it is of a general nature and not specifically a problem associated with the production of SCP, but it is obvious that some attention has to be paid in any process which employs alkaline extraction at a high temperature (Anon. 1976A).

Other Factors.—Yeast autolysates may contain small concentrations of histamine and tyramine which result from the decarboxylation of the corresponding amino acids. Between 0.1 and 1.6 mg of tyramine and between 0.2 and 2.8 mg of histamine have been reported per g of yeast autolysate (Blackwell *et al.* 1969). The use of yeast autolysates as condiments limits the actual amount of histidine and tyramine consumed.

Microorganisms assimilate some heavy metals and other toxic materials from nutrient media. Payer *et al.* (1975) investigated the contamination of algae with toxicants derived from their environment. They concluded that the level of contaminants is smaller or within the range of that of other food products.

Human Feeding Studies

Several reports in the literature indicate that feeding of microbial biomass to humans results in discomfort to some individuals and occasionally in stronger gastrointestinal disturbances, desquamation of the skin, and other symptoms, particularly at high levels of feeding. This literature has been reviewed by Scrimshaw (1975) and Calloway and Waslien (1971). Some of the reactions appear to have been the result of sensitization of individuals and suggest allergic reactions. Feldheim (1975) reviewed the results of several reports of mass feeding trials with humans (with species of algae) and found generally good acceptance. It is important to note that reactions to microbial biomass from various species of organisms may vary as much as reactions to various species of edible plants. Hence, biomass preparations have to be judged on the basis of specific tests and on the basis of practical experience by inclusion in the human diet.

PRODUCTION OF MICROBIAL BIOMASS

Traditional processes for the production of biomass such as the production of brewers' yeast and *Candida utilis* yeast will be treated only in summary form. Newer processes will be described with somewhat more detail; and some proposed processes will be mentioned if they are of sufficient economic importance or academic interest.

Brewers' Yeast

During the fermentation of brewers' wort there is a 3- to 8-fold multiplication of yeast cells. This yeast can be readily recovered by centrifuging, and a small portion is often used for inoculation of the next batch of unfermented wort. Of the excess yeast the largest portion is dried together with brewers' spent grains and sold as feed. In the United States a considerable portion, about 11.25 million kg (25 million lb) of yeast solids, is dried separately or used in the production of brewers' yeast autolysate. This represents approximately 20% of the total available amount of brewers' yeast.

Feed grade yeast is dried directly from the centrifuged yeast cream which contains about 10–15% yeast solids. Food grade yeast is debittered by an alkaline wash at pH 8 or above followed by washing with water. It is then dried by spray drying or occasionally by atmospheric drum drying. It is used as an ingredient in fabricated foods or sold as a dietary supplement.

Processes for the production of brewers' yeast have been described by Quittenton (1966) and Hunt (1969). Brewers' yeast in liquid form which does not meet a minimum content of 40% of crude protein may be stored at 4°C until endogenous metabolism of reserve carbohydrates leads to a higher protein level in the yeast. While the total amount of protein does not change during this storage, its relative percentage of cell solids increases as shown in Fig. 13.5 (Ingledew et al. 1977).

Distillers' and Wine Yeast

In neither of these industries is it feasible to separate the considerable amounts of yeast which have grown during the fermentation. In distilleries, the yeast is recovered in the spent distillers' grains and sold to the feed industry. In the wine industry, the yeast is discarded with the lees.

Bakers' Yeast

Inactive dried bakers' yeast is too expensive for use in the feed industry. However, it is used as a nutritional supplement. Since this is a primary grown yeast, i.e., a yeast grown specifically for sale as such, it is possible to increase the crude protein content to 50–55% and to induce high levels of thiamin and nicotinic acid during the growth process. (See also the chapter on the production of bakers' yeast.)

Other Carbohydrate Substrates

Sulfite Liquor.—A widely available carbohydrate substrate is spent sulfite liquor which is a by-product of paper pulp mill operations. Liquor from hardwoods contains about 2–3% of fermentable sugars of which about 80% are pentoses and 20% hexoses. Liquor from softwoods contains about 80% of the sugars in the form of hexoses and 20% as pentoses. Romantschuck (1975) gives the following breakdown for the various sugars in spruce

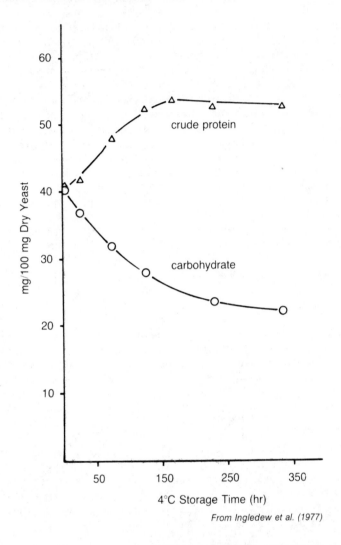

FIG. 13.5.   CHANGES IN PERCENTAGE PROTEIN AND CARBO-
HYDRATE IN SPENT BREWERS' YEAST SLURRY

sulfite liquor: mannose 50%, glucose 12%, galactose 12%, xylose 20%, arabinose 4%. In addition, sulfite liquor contains appreciable amounts of acetic,
galacturonic and formic acids.

Candida utilis, which assimilates pentoses, hexoses, and many organic
acids, has been produced commercially for several decades. It is surprising
that this fermentation has not been treated more extensively in the litera-

ture. The basic papers describing current operations have been published by Inskeep *et al.* (1961) and Butschek and Krause (1962), and the process has been reviewed by Reed and Peppler (1973). The aerobic process is carried out continuously with a dilution rate of 0.27 to 0.3 (Peppler 1970), at a pH of 4.5 and a temperature of 32°C. A nitrogen source (ammonia), phosphate, and potassium have to be supplied. Addition of biotin is not required. The effect of dilution rate on sugar utilization and yield has been investigated by Lorenz *et al.* (1967).

C. utilis is recovered by centrifuging. Minimal washing results in a feed yeast which still contains about 10% lignosulfonic acids. Additional washing results in a food grade yeast free from lignosulfonic acids. In the United States a major portion of the *C. utilis* produced on sulfite liquor is used in fabricated foods. One of the more recent plant installations has been described by Anderson *et al.* (1974).

C. utilis for use in feeds is also grown extensively in Eastern Europe as well as in Cuba and Taiwan. In the latter two countries molasses is the substrate. The production of *C. utilis* on ethanol is mentioned later.

In Finland, sulfite liquor is used in the production of a microfungus (probably *Paecilomyces varioti*) for use as a feed ingredient. The gross chemical composition of the dried mycelium, its RNA content, and its amino acid composition are fairly similar to that of *C. utilis*. The process is carried out on a continuous basis with a retention time of 4.5 to 5 hr. The process has an advantage in that the fungal mass can be recovered with a rotary filter instead of a centrifuge. The process is known as the "Pekilo" process (Romantschuk 1975). Additional process details have been described by Romantschuk and Lehtomaki (1978), and the reduction of the NA content of the "Pekilo protein" by Viikari and Linko (1977). Two 360 m^3 fermentors (agitated) are in operation. The yield of biomass is 55% based on the consumption of reducing materials; and the productivity is 2.7–2.8 g of fungal solids per liter per hr.

Whey.—Cheese whey is an excellent substrate for the production of biomass. It contains about 5% lactose, 0.8% protein, 0.7% mineral matter, and from 0.2 to 0.6% lactic acid. Lactose is assimilated by "dairy" yeasts. *Kluyveromyces fragilis* and *K. lactis* are the species used. The whey proteins may be separated by ultrafiltration prior to the fermentation; or they may be partially removed by acid/heat coagulation.

The fermentation may be carried out with incremental feeding of the whey or whey concentrate or as a continuous process. A yeast inoculum of 1×10^9 cells per ml, a pH of 4.5, and a temperature of 30°C are quite satisfactory. For the continuous fermentation a dilution rate of 0.125/hr can be maintained. Almost all authors describing the growth of dairy yeasts on whey supplement the medium with yeast extract. The principal purpose seems to be supplementation with biotin (Harju *et al.* 1976). The fermentation may be carried out aerobically for the production of yeast biomass or with minimum aeration for the production of ethanol. A schematic representation of the process is shown in Fig. 13.6 (Bernstein *et al.* 1977).

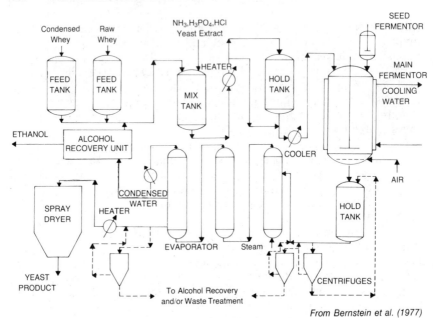

From Bernstein et al. (1977)

FIG. 13.6.   SCHEMATIC FOR THE PRODUCTION OF ALCOHOL AND YEAST FROM WHEY

If the entire content of the fermentor is dried, the end product will be a feed grade material which contains yeast, residual whey proteins, a fairly high ash content, and some lactic acid. If the yeast cells are separated by centrifuging and washed, the dried end product is a food grade material with a crude protein content of 45–55% and a PER of 2.26. The process gives a yield of about 45–55% of yeast solids based on lactose consumed. Addition of whey can be regulated on the basis of an automatic assay of lactose in the fermentor liquid (Bernstein and Plantz 1977). An economic analysis has been prepared by Pace and Goldstein (1975) based on an earlier process by Wasserman et al. (1961).

A 2-step fermentation process for the production of biomass has been described by German workers. Starting with whey or whey ultrafiltrates, they converted lactose to lactic acid by an anaerobic fermentation. This was followed by an aerobic fermentation with a lactic acid assimilating, thermophilic yeast (Moebus et al. 1977). A similar approach has been used by Skupin et al. (1977), who fermented whey anaerobically with Propionibacterium shermanii or two other species of propionibacteria, followed by an aerobic fermentation with K. fragilis. The whey medium was fortified with 5,6-dimethylbenzimidazole, a precursor of vitamin B_{12}. This results in the production of propionibacteria with high levels of vitamin B_{12}. These bacteria also contribute a higher level of sulfur-containing amino acids to the biomass.

A feed product may be produced from whey by fermentation with lacto-bacilli and neutralization with ammonia which results in the formation of ammonium lactate. The concentrated fermentor liquor is referred to as fermented, ammoniated condensed whey. In this product 70% of the nitro-gen is derived from ammonium lactate, 17% from whey proteins, and 8% from the cells of *L. bulgaricus* (Reddy *et al.* 1976; Erdman *et al.* 1977A,B; Gerhardt and Reddy 1977).

Petrochemical Substrates

The petrochemical substrates that have been considered or used for the production of biomass are n-alkanes, alkane-containing gas oils, methane, methanol, and ethanol. Commercial work has progressed furthest with alkane fractions, although some pilot plant or demonstration plants have been closed, and production start-up in other plants has been blocked by legal procedures.

N-Alkanes and Gas Oil.—The microorganisms most suitable for growth on n-alkanes are yeasts of the genus *Candida*. Generally bacilli do not grow on alkanes or they grow only weakly, although some are able to oxidize alkanes in the presence of other carbon sources (Kachholz and Rehm 1977). Use of alkane fractions with $10-23$ carbon atoms, and boiling points from $175°$ to $280°C$ have already been mentioned.

The use of water-insoluble alkanes as substrate adds an extra phase to the fermentor content; that is, there are now 4 phases: the liquid alkane phase, the aqueous phase, the gas phase, and the cells themselves. Aside from the carbon source, the nutrients added are those well known from the carbohydrate fermentation, namely, ammonia, phosphoric acid, K, Mg, Mn, Zn, and Fe (Litchfield 1977A). Operation is usually on a continuous basis at a pH at or below 4 and at a temperature of $30°-32°C$. The process developed by British Petroleum and practiced at its Grangemouth plant (3600 MT or 4000 tons/year) was operated aseptically and oxygen transfer was achieved with the aid of mechanical agitation. Other companies have developed processes operating nonaseptically and with the use of simple air lift fer-mentors (Gulf, Kanegafuchi, Liquichimica).

The alkane based processes have been described by Cooper *et al.* (1975), Kanazawa (1975), Lainé and Chaffaut (1975), Knecht *et al.* (1977), and Allison (1975), among others. There are some fundamental differences between fermentations based on carbohydrates and those based on hydro-carbons. For hydrocarbons the size of the hydrocarbon droplets is critical since growth of the yeast cells takes place through direct contact of the cell wall and the surface of the alkane droplets (for tridecane and higher al-kanes). Oxygen requirements are higher, and approximately 2 g of O_2 are required for each g of yeast solids grown on alkanes. The heat evolved during the fermentation is, therefore, proportionately high, approaching

0.11 kcal per mMol O_2 consumed (Cooney *et al.* 1969). Yields are high in comparison with carbohydrate substrates. For n-paraffins they are between 0.95 and 1.15 g of cell solids per g of substrate. Productivities of up to 4 g/liter/hr can be achieved but it is not economical to operate at that rate since energy requirements increase disproportionately at productivities exceeding 2.5 g/liter/hr (Knecht *et al.* 1977).

Alkane grown yeasts can be recovered by simple centrifuging. After washing, the cell mass can be spray dried resulting in a light tan powder of about 6% moisture. The gross chemical composition is 9.5% N, 1.6% P, 6% ash, and 8–10% lipids (after acid hydrolysis). The species most widely used is *C. lipolytica*. Knecht *et al.* (1977) more properly refer to this yeast which has a sexual cycle as *Endomycopsis lipolytica*.

Yeast grown on gas oil has to be extracted with solvents after recovery to remove adhering hydrocarbons. This results also in the extraction of some lipids from the cells and accounts for the relatively higher protein content of cells grown on gas oil. The earlier literature on the production of yeasts from hydrocarbons has been reviewed by Lainé *et al.* (1976), Johnson (1967), Humphrey (1970), Fiechter (1967), and Gounelle de Pontanel (1972).

Methane.—The use of methane gas as substrate poses some extraordinary problems. Methane oxidizing organisms such as *Methylococcus capsulatus* are readily inhibited by small concentrations of extracellular products of metabolism. For instance, methanol inhibits such organisms strongly. This requires the removal of methanol from the fermentation medium as soon as it is formed. While this can be achieved by continuous dialysis it is not a practical solution. An alternate solution is the use of mixed cultures of methane oxidizing organisms with organisms assimilating methanol, a procedure which is designed to remove the inhibiting material as fast as it is formed (Hamer *et al.* 1975; Naguib 1975; and Shell Oil Co. 1977).

Methanol.—Both methanol and ethanol have some advantages in comparison with paraffins and methane. They are water soluble and they are commercially available in relatively pure form. Methanol is assimilated by several genera of yeasts and bacteria. Sahm and Wagner (1975) list the methanol-using organisms as follows:

Bacteria.

Obligate methylotrophic.   (CH_4, CH_3OH, CH_3-O-CH_3)

Methylobacter, Methylococcus, Methylomonas, Methylocystis, Methylosinus.

Facultative methylotrophic.   (CH_3OH, carbohydrates, organic acids)

Arthrobacter, Bacillus, Hyphomicrobium, Protaminobacter, Pseudomonas, Rhodopseudomonas, Streptomyces, Vibrio.

Yeasts.

Facultative methylotrophic.   (CH_3OH, carbohydrates, organic acids)

Candida, Hansenula, Kloeckera, Pichia, Torulopsis, Trichoderma.

Foo (1978) dealt with the pathways of methanol formation by obligate and facultative methylotrophic organisms. The substrates for obligate methylotrophs include methane, methanol, formic acid, formamide, carbon monoxide, dimethyl ether, dimethyl amine, and trimethyl amine. The methanol metabolism of bacteria and yeasts has been reviewed by Sahm and Wagner (1975) and Schlanderer *et al.* (1975) and will not be further discussed. Präve and Sukatsch (1975) grew a strain of *Pseudomonas* at 37°C and at a pH of 6.65 and obtained a specific growth rate of 0.2/hr with a cell concentration of 18.9 g/liter and a yield of 41.5% based on methanol. Dostalek and Molin (1975) used a strain of *Methylomonas methanolica* and achieved a maximum specific growth rate of 0.53/hr and a yield of 48%. They investigated the interrelationship among several variables (specific growth rate, temperature, yield, and productivity). Figure 13.7 shows the effect of temperature on productivity.

The process developed by Imperial Chemical Industries which is nearing the commercial stage has been described by Littlehailes (1975) and Gow *et al.* (1975). The organism, *Pseudomonas methylotropha*, is a Gram-negative rod, catalase positive, and shows oxidative metabolism. Its optimum growth temperature is between 34° and 37°C and its optimum pH range is from 6.5 to 6.9. It could be grown at a specific growth rate of 0.5 per hr. A carbon conversion of 62% could be achieved and a cell concentration of 30 g/liter. The continuous aerobic fermentation is carried out under aseptic conditions in an air lift fermentor which is pressurized and which has an external loop (see also Fig. 13.2). The crude protein content of the microbial mass is 85% and the true protein content 64%. A demonstration plant producing a strain of *Methylomonas clara* was placed in operation by Hoechst Germany in 1978 (Anon. 1978B).

The isolation of methanol utilizing strains of yeasts has been described by Oki *et al.* (1972). Yeasts generally give lower yields than bacteria, namely 29 to 45% based on the weight of methanol. Specific growth rates for yeasts are also lower. Cooney and Makiguchi (1977) and Cooney and Levine (1975) have reported on their extensive work with *Hansenula polymorpha*. The authors have addressed the problem of selecting the most economical conditions for biomass production in an exemplary manner. Figure 13.8 shows the effect of dilution rate and the resulting methanol concentration of the medium on dry cell weight and productivity. The authors conclude that "it is essential to give a strong bias to the effect of dilution rate on cell yield in

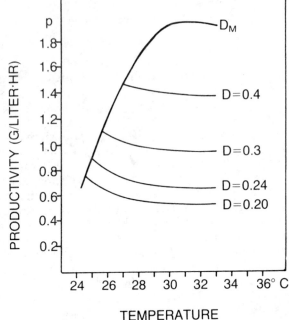

FIG. 13.7. INFLUENCE OF TEMPERATURE ON PRODUCTION OF BIO-MASS (G/LITER·HR) AT DIFFERENT DILUTION RATES, D/HR
Curve D_M—Maximum productivity at individual temperatures.

From Dostalek and Molin (1975)

the selection of an optimum dilution rate. Thus the selection of an economically optimum dilution rate is comprised of several factors but, predominantly, it is the effect of dilution rate on cell yield which determines the selection."

Ethanol.—The following genera of yeasts and bacteria have been considered for the production of biomass from ethanol.

Yeasts: *Candida, Debaryomyces, Endomycopsis, Hansenula, Mycoderma, Pichia, Rhodotorula, Saccharomyces*

Bacteria: *Acetobacter, Acinetobacter, Arthrobacter, Bacillus, Brevibacterium, Corynebacterium, Hyphomicrobium, Nocardia, Pseudomonas*

In addition some edible mushrooms have been grown on ethanol, such as *Lentinas, Pleurotus,* and *Schizophillum* (Laskin 1977). In the United States, *C. utilis* is produced commercially on ethanol. The process uses a continuous fermentation in an agitated, aerated vessel of special construc-

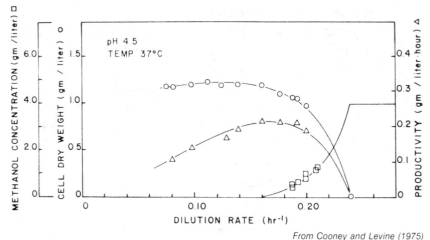

From Cooney and Levine (1975)

FIG. 13.8. PRODUCTIVITY AS A FUNCTION OF DILUTION RATE

tion (Standard Oil Co. 1977). Details have been described by Ridgeway *et al.* (1975).

Demonstration plants have been built in Japan for the production of *C. ethanothermophilum* and in Czechoslovakia for the production of an unspecified yeast.

The assimilation of ethanol by bakers' yeast has been known for a long time. Ethanol produced during oxygen-limited fermentations by *S. cerevisiae* is assimilated during later stages of yeast growth. However, ethanol assimilation is rather inefficient and it is not practical to grow bakers' yeast on ethanol.

Japanese workers have also emphasized the need for a substrate which avoids the alleged toxicity problems with hydrocarbon media. Masuda (1974) and Masuda *et al.* (1976) have utilized the thermophilic yeast, *C. ethanothermophilum*. The yeast is grown at a temperature of 40°C, which reduces the cost of cooling considerably, at a pH of 3.5, and with a specific growth rate of 0.45/hr. The fermentation is carried out in an air lift type fermentor at a cell solids concentration of 20 g/liter.

A bacterial culture, *Acinetobacter calcoaceticus*, has been used by Laskin (1977A) for biomass production. The optimum pH for growth was between 6.5 and 7.5 and the optimum temperature between 32° and 35°C. The fermentation has to be carried out under ethanol-limiting conditions and high growth rates in order to prevent the accumulation of acetic acid and aldehyde which inhibit the fermentation. This is also desirable to maximize the yield coefficient. Figure 13.9 shows the protein yield coefficient as a function of specific growth rate. The variation of the yield coefficient with growth rate indicates a fairly high requirement for energy for maintenance metabolism. This was calculated as 0.11 g of ethanol per g of biomass per hr.

Cellulose and Agricultural Waste

The term "waste" in the above subtitle cannot be used with precision. A substrate used for the purpose of producing biomass ceases to be "waste." In that sense molasses or whey cannot be considered waste when it is used for an industrial fermentation. Another difficulty arises from the fact that it is not always simple to distinguish between fermentation processes designed for the disposal of waste and those designed to yield a salable biomass. Only those processes will be discussed which are clearly intended for the production of microbial biomass and the commercial sale of that biomass.

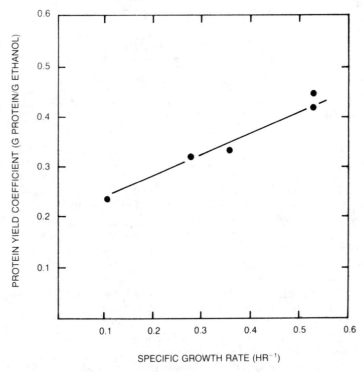

From Laskin (1977B)

FIG. 13.9.   EFFECT OF SPECIFIC GROWTH RATE ON THE PROTEIN YIELD COEFFICIENT OF A. CALCOACETICUS GROWING UNDER ETHANOL LIMITATION IN A CHEMOSTAT CULTURE

The materials which are potentially suitable for the production of biomass can be classified into timber, wood residues, and wood pulp (Stone 1977), agricultural residues including feedlot waste (Sloneker 1977), and solid waste from food processing operations (Cooper 1977). Almost all of these materials contain cellulose as the principal carbon source. Price, availability, and transportation problems differ greatly for the various substrates. However, one problem is common to all of them. This problem is the difficulty of converting cellulose or other polymeric carbohydrates such as xylan to fermentable sugars. Cellulose is well digested by some microorganisms, particularly by some wood-rotting fungi but the "low growth rates of cultures on untreated cellulosics have made economically feasible biomass production from cellulose difficult to envision" (Dunlap 1975). Cellulolytic enzymes isolated from wood-rotting fungi have a low rate of activity, and there is little correlation between the efficiency of fungal attack on cellulose fiber and that of a cell-free extract of the fungi (Kulp 1975).

It is generally assumed that the principal obstacle to efficient enzymatic hydrolysis of cellulosic materials is the degree of crystallinity of cellulose and lignification (Cowling and Kirk 1976). Bellamy (1975) stated emphatically that the "effective and rapid utilization of lignin and the cellulose lignin complex is the major obstacle to economic recycling of cellulosic waste." The amount of cellulose in different raw materials varies greatly from 90% in cotton fibers to 15−20% in leaves. Hardwood stems contain 40−50% cellulose, 20−40% hemicellulose, and 18−25% lignin; softwood stems contain 45−50% cellulose, 25−35% hemicellulose, and 25−35% lignin. Grasses such as bamboo, palms, wheat, rice, sugar cane, etc., contain 25−40% cellulose, 25−50% hemicellulose, and 10−30% lignin. Newsprint generally contains the same percentages as the woods from which it was made. Animal manure contains relatively more lignin than other sources, presumably because of the digestion of cellulosic material in the rumen. Susceptibility of the various raw materials to enzymatic attack differs greatly. This has led to a broad search for pre-treatment methods of the substrate and for suitable microorganisms. Only some of these can be mentioned in this chapter.

There are basically two approaches which can be used for the production of biomass on cellulosic materials. The first is the hydrolysis of cellulose by acid or enzymes followed by the growth of microorganisms on the formed glucose. The second is the growth of fungi which hydrolyze cellulose and assimilate the sugars directly.

Most of the work on acid hydrolysis of cellulose has been done in Eastern Europe and the literature on the process is not extensive. Earlier processes of acid hydrolysis had the inherent disadvantages of poor yields due to the formation of sugar reversion products and the need to dispose of sizeable quantities of salts derived from the added hydrochloric or sulfuric acid and alkali. Recently the reaction of dry sawdust with 33% HCl and dry HCl has been suggested with a predicted yield of 95% glucose after a 90 min treatment. Grethlein (1978) found acid hydrolysis of newsprint more economical than enzymatic hydrolysis. Using a plug flow reactor for hydrolysis with 1% sulfuric acid, he found optimum conditions for a residence time of 0.19 min at a temperature of 230°C. One part of glucan yielded 0.55 parts of glucose, 0.24 parts of decomposed glucose, and 0.21 parts of unreacted glucan.

The use of concentrated H_2SO_4 as a solvent for cellulose has been emphasized recently. This can later be diluted and used as a catalyst for cellulose hydrolysis (Tsao 1978A,B).

Enzymatic conversion of cellulose has been treated extensively in a recent symposium (Gaden 1976). Major obstacles to fast and efficient enzymatic hydrolysis have been the inaccessibility of cellulose due to poor penetration of the enzyme and the inhibition of cellulase activity by cellobiose. Yields of wood sugars can be greatly increased by pre-treatment of the cellulose: chemically by treatment with alkali, amines, or ammonia or mechanically by comminution, preferably by ball milling (Millet *et al.* 1976). Working with newsprint as raw material, Wilke *et al.* (1976) have arrived at the most attractive cost projections. These were based on 50% conversion of cellulose with the cellulase of *Trichoderma viride* for 40 hr at 45°C with recycling of the sugar enzyme solution to agitated, concrete digesters (see also Cysewski and Wilke 1976).

The inhibition of cellulase by cellobiose may be overcome by diffusion of the formed cellobiose through membranes during hydrolysis. Alternately, the sugars formed during enzymatic hydrolysis may be removed through concurrent growth of a microorganism or through concurrent production of ethanol by fermenting microorganisms. Meyers (1978) added *C. utilis* during the saccharification of cellulose by *T. viride* cellulase. In this coupled saccharification-fermentation process the rate of saccharification was about the same as in the absence of the yeast but it proceeded for a longer time. Ultimately, 20% more saccharification was achieved as shown in Fig. 13.10. The process can also be carried out with *S. cerevisiae* (Savarese and

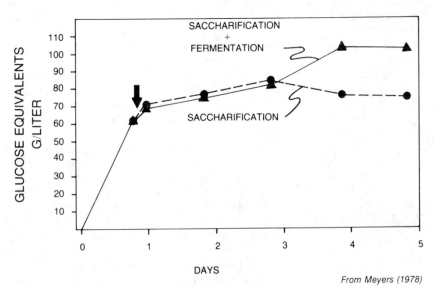

From Meyers (1978)

FIG. 13.10.   EFFECT OF FERMENTATION DURING THE SIMULTANEOUS SACCHARIFICATION AND FERMENTATION OF CELLULOSE

Young 1978). The principle is the same as that traditionally used in the production of distilled beverages where starch hydrolysis by malt enzymes proceeds simultaneously with the fermentation of the formed maltose by distillers' yeast.

Direct cultivation of microorganisms on cellulose seems to promise the best results with thermophilic organisms. Since direct growth of microorganisms on cellulose is slow, the loss of substrate due to the maintenance metabolism of the organism is rather high. Use of thermophiles partially overcomes this obstacle. Humphrey *et al.* (1977) worked with pure cellulose substrates (Avicel) and obtained yields of up to 45 g cell solids per 100 g of cellulose utilized and growth rates of 0.45/hr with a *Thermoactinomyces* species. Some work with other thermophiles on mixed agricultural wastes is listed below. Dunlap (1975) used mesophilic organisms, a *Cellulomonas* species, and *Alcaligenes faecalis* with bagasse as substrate. The waste material required pre-treatment by swelling in dilute alkali. *Cellulomonas* has also been used in the process developed by Louisiana State University (La. State Univ. 1976).

A great amount of work has been done with agricultural waste and food processing waste in various industries. Only a few examples can be given. Gregory *et al.* (1976) grew *Aspergillus fumigatus* on cassava at a pH of 3.5 and a temperature of 45°–50°C, obtaining yields of cell solids of 50%. Macris and Kokke (1977) used the extract of carob bean pods to grow *Fusarium moniliforme*. Ladisch *et al.* (1977) and Church *et al.* (1977) worked with residues of the corn milling industry using fungi such as *T. viride* and *Gliocladium deliquescens* for conversion to biomass. Chahal *et al.* (1977) grew *Chaetomium cellulolyticum* on wheat straw. Effluents from coffee plantations are good media for the growth of a *Verticillium* species at 33°C and at a low pH of 3.5 (Espinosa *et al.* 1977). Sánchez-Marroquín (1977) used mixed cultures of yeast and bacteria for the production of biomass on agave juice. Tanner and Hussain (1977) hydrolyzed kudzu starch with α-amylase and grew strains of *S. cerevisiae* on the hydrolysate. By-products of the malting and brewing industry have been considered as substrates by Pomeranz (1976) and Shannon and Stevenson (1975). A general review of earlier work with microfungi growing on carbohydrate wastes was published by Anderson *et al.* (1975).

It should be noted that the easily hydrolyzed starchy substrates can be readily fermented by yeasts of the *Saccharomyces* and *Candida* genera. Carbohydrates more resistant to enzymatic hydrolysis and containing pentosans, cellulose, or more complex polymers appear to be more suited for direct attack by fungi. Secondly, it appears that much of the work reported aims at the reduction of the BOD of effluents rather than at the production of biomass per se. For instance, Fig. 13.11 shows the growth of *C. utilis* on sauerkraut brine and the concomitant reduction in BOD (Huang 1974).

Feedlot waste consisting largely of cattle manure contains much nonprotein nitrogen, but a considerable part of its nitrogenous constituents is protein of microbial origin resulting from the growth of rumen organisms. In view of the variability of the substrate it is not surprising that various

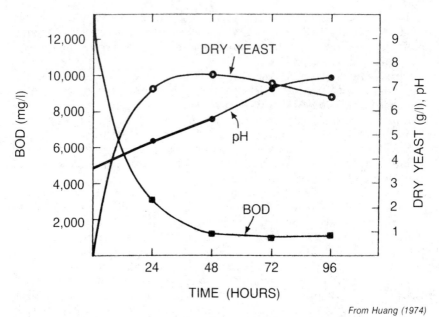

From Huang (1974)

FIG. 13.11.   REDUCTION OF THE BOD OF ACID BRINE BY YEAST FERMENTATION

authors report widely differing values for this fraction. Morrison *et al.* (1977) found that 45% of the protein of sieve-fractionated manure is of microbial origin. They describe a procedure for recovery of a protein fraction by extraction with 0.1 N NaOH followed by ammonium sulfate precipitation.

The use of cattle manure as a substrate has been considered by several research workers. It may be less attractive than other agricultural wastes because it is already partly digested by microbes, and transportation is an obvious problem. The latter problem can be minimized by building fermentation facilities on the site of the feedlots. A process for the anaerobic fermentation of feedlot waste has recently been put into operation. It generates methane as the principal end product, and the residual fermentation sludge is used as cattle feed.

Reddy and Erdman (1977) studied the liquid extract of feedlot waste as a potential substrate for fermentation. The waste, including urine, straw, wasted feed particles, etc., was slurried and filtered. The filtrate fraction contained 3.0 to 5.2% nitrogen, 35% ash, 3% cellulose, 7.0% lignin, and 6.0% noncellulose carbohydrate on a total solids basis. It was not practical to ferment this filtrate without the addition of other agricultural wastes high in carbohydrate content. But the additives actually used in the experimental work, such as whey, molasses, cornstarch, or potato starch, are probably not available for such supplementation in a practical manner. The fermentation of the liquid extract of feedlot waste with some of the mentioned additives proceeded best with the natural flora and less efficiently with added *L. bulgaricus* cultures or cultures of rumen organisms.

The preceding paragraphs deal mostly with descriptions of laboratory or pilot plant processes for various agricultural wastes. It is not surprising that the few plants in commercial operation are based on food processing wastes containing principally starch. These are wastes from potato processing plants or from the wet milling of corn. One of these processes is based on the hydrolysis of starch by *Endomycopsis fibuliger* followed by growth of *C. utilis*. This "Symba" process is practiced in the Scandinavian countries. Growth of the two species of yeasts indicates a symbiotic relationship. Because of the faster growth of *C. utilis*, the percentage of *E. fibuliger* cell solids is only 4% of the harvested biomass. The process has been described by Jarl (1969) and Skogman (1976). Figure 13.12 shows a picture of the fermenter installation.

In the United States a system for treating waste water from potato processing operations is designed mainly for the reduction of BOD but biomass becomes available as a by-product for use as a feed (Anon. 1977).

Courtesy of PEC, Process Engineering Co., Männedorf, Switzerland

FIG. 13.12. OVERALL VIEW OF A SYMBA PLANT
AT LEFT—Seed and substrate preparation. AT CENTER—Harvesting equipment. AT RIGHT—300 m^3 fermentor.

Details of this process have not been published. But a process for growing lactobacilli on acid hydrolyzed starch from potato processing waste has been described (Forney and Reddy 1977). The principal end product is ammonium lactate resulting from the neutralization of lactic acid with NH_3. It is an excellent rumen feed. Ammonium lactate formation from whey has already been mentioned (Erdman et al. 1977A,B).

Photosynthetic Bacteria

Bacterial photosynthetic processes use molecular hydrogen, reduced sulfur compounds, such as H_2S, or some organic compounds as electron donors. Water is not used as electron donor which means that oxygen is not liberated as the result of bacterial photosynthesis. The process is basically anaerobic. Table 13.8 shows the photosynthetic reactions of plants and bacteria in a schematic way. Using one species of non-sulfur purple bacteria, Shipman et al. (1977) grew Rhodopseudomonas gelatinosa on an infusion of wheat bran with suitable sunlight or incandescent illumination at a pH of 7 to 7.2 and a temperature of 37°C. Retention time in the fermentor was about 24 hr.

Chemolithotrophic Microbes

Lafferty et al. (1975) experimented with aerobic organisms which can use molecular hydrogen as electron donor. These belong to the genera Hydrogenomas and Alcaligenes. Their rate of growth surpasses that of the photo-

TABLE 13.8.   BASIC PHOTOSYNTHETIC REACTIONS OF PLANTS AND MICRO-ORGANISMS

(1) Higher plants and unicellular algae

$$CO_2 + H_2O \xrightarrow[\text{chlorophyll}]{\text{solar energy}} \underset{\text{biomass}}{(CH_2O)} + O_2$$

(2) Certain algae and bacteria

$$CO_2 + 2H_2 \xrightarrow[\substack{\text{chlorophyll or} \\ \text{bacteriochlorophyll}}]{\text{solar energy}} (CH_2O) + H_2O$$

(3) Photosynthetic sulfur bacteria

$$CO_2 + H_2S \xrightarrow[\text{bacteriochlorophyll}]{\text{solar energy}} (CH_2O) + S + H_2O$$

(4) Purple photosynthetic bacteria

$$CO_2 + \underset{\substack{\text{organic} \\ \text{matter}}}{CHO} + H_2O \xrightarrow[\text{bacteriochlorophyll}]{\text{solar energy}} (CH_2O) + H_2 + H_2O$$

Source: Adapted by Shipman et al. (1977) from Stanier et al. (1970).

trophic organisms. These mesophilic organisms grow well between 28° and 35°C and at a pH between 6.8 and 7.2. Hydrogen, oxygen, and carbon dioxide (as the carbon source) have to be supplied as well as nitrogen in the form of ammonia, nitrates, or urea. The need to supply the organisms with an explosive gas mixture, and the formation of poly-β-hydroxybutyric acid as an energy depot of the cell are disadvantages. The formation of poly-β-hydroxybutyric acid can be blocked by genetic manipulation.

Algae

Among phototrophic organisms, algae have been investigated widely as sources of biomass for food or feed. It is interesting to note that information on processes is meager in contrast to fairly extensive work on the feeding of algae to humans and animals. The use of algae as part of the human diet by the Aztecs of Central America and the natives of the Lake Chad area in Africa has been well established, but quantitative data on dietary intakes are lacking.

Algae use carbon dioxide as the carbon source. They derive energy from sunlight or artificial light in the range of 700 nm by means of the chlorophyll of the cells (see also Table 13.8). Since algae grow abundantly on stagnant ponds, attempts to use these microbes as food or feed are a borderline case between traditional agriculture and modern microbial biomass production. From a practical point of view algal cultures should be carried out between the latitudes of 35°N and 35°S from the equator. Soeder and Mohn (1975) distinguish between the so-called clean processes with well-defined media and the waste water processes.

Clément (1975) described the cultivation of *Spirulina* in a semi-natural culture basin of Sosa Texcoco, South America, in cooperation with the Institut Français du Pétrole. The culture medium contains mineral salts and a certain amount of sodium carbonate and bicarbonate. Optimum growth rates are obtained at a pH between 8.5 and 11 and at a temperature of 32°C. The mass transfer coefficient (K_L) for CO_2 absorption by suspensions of *Scenedesmus quadricauda* has been investigated by Livansky *et al.* (1973). They found it to be independent of the chemical reactivity of CO_2 in the algal suspension.

The algae which have been considered for biomass production belong principally to the genera *Chlorella*, *Spirulina*, and *Scenedesmus*. The cell walls of algae interfere with full utilization of the protein, and, therefore, the results of feeding trials depend greatly on the treatment of the biomass after harvest. For untreated *Scenedesmus* or *Chlorella* digestibility was only 25%. Spray drying, air drying, vacuum drying, and freeze drying improved the digestibility, but the best results, 70–80% digestibility, were obtained with drum drying or simple boiling of the biomass for 6–8 min (Pabst 1975). The feeding of algae to humans has been recently reviewed by Pabst (1975), Feldheim (1975), Frahm and Lembke (1975), and Pokrovsky (1975).

Secondary Processing of Microbial Biomass

Secondary processing includes the recovery of biomass from the fermentor, concentration, and drying, as well as other processes designed to sterilize the material or to fractionate it. Recovery of yeasts is generally done by centrifuging of the fermentor liquid resulting in a product with 10–20% solids content. This material may be washed with water once or several times followed by recentrifuging. In the case of paraffin grown yeast, a surfactant is added to remove traces of paraffins; however, for yeasts grown on gas oil a solvent extraction step must be included. Bacteria are more difficult to harvest because of their smaller size and lower density. The rate of centrifugal separation is a function of the difference in density between the cells and the liquid medium, and it varies also with the square of the diameter of the cell. For algae growing on the surface of ponds, simple skimming is the most economical method of harvest (Labuza 1975).

Drying has been done traditionally on atmospheric drum dryers of the single or double drum type. Recent installations favor spray drying. An intermediate evaporation step is rarely required since centrifuging to a 15 to 20% solids concentration of cell solids can be carried out. Besides, yeast slurries such as bakers' yeast or brewers' yeast cannot be pumped well once the solids content exceeds 22–23%. Labuza et al. (1972A,B) found that below 10% solids the slurry showed Newtonian characteristics with viscosities below 4–5 centipoises.

For both drum drying and spray drying it is desirable to heat the yeast slurries prior to the drying step in order to kill contaminants and to reduce the number of viable cells of the biomass. For spray drying Labuza et al. (1972B) reported that in the range from 21° to 46°C for outlet air temperature, cell death ranged from 3 to 7 log cycles of kill. It must be kept in mind that pathogenic contaminants are rarely present in the biomass at the end of the fermentation period, but problems may arise during secondary treatment steps by recontamination. A preliminary heating step is also useful to induce the formation of "leached" yeast solids which aid in drying.

Products obtained by spray drying are generally lighter in color, milder in flavor, and more powdery. Drum dried products are often darker with a slightly toasted flavor which may be desirable in some uses.

The isolation of protein concentrates from biomass has been studied extensively. The first step usually consists of the mechanical disintegration of the cells to permit separation of the protoplasmic cell contents from the cell wall material. High pressure homogenizers, high speed ball mills, colloid mills, and sonic probes have been used for that purpose (Dunnil and Lilly 1975). Most investigators have found homogenization to be most practical. Pressures of 48,300 to 69,000 kPa (7000 to 10,000 psi) are used, and several passes through the homogenizer are usually required to obtain disintegration of more than 90% of the cells.

Disintegration of the cells is usually followed by alkaline extraction of the protein. For instance, Hedenskog et al. (1970) extracted algae and yeasts at pH 11 to 11.5, recovering 80–85% of total cell nitrogen in the extract. This was followed by precipitation of the protein at pH 4 yielding a concentrate with 70% crude protein. Alternately, Hedenskog and Mogren (1973) precipitated the protein at an alkaline pH by heating, which permitted the recovery of 60% of the original cell nitrogen in the protein precipitate.

Cunningham *et al.* (1975) also used alkaline extraction followed by separation from cell wall debris and precipitation at pH 3.8. They obtained a concentrate containing 70% crude protein and recovered 41.5% of the original cell nitrogen in the precipitate. All available reports from the literature led to the conclusion that one can not hope to recover much more than 50% of the cell protein by such methods on a commercial scale. The previously mentioned process of treating the biomass with anhydrous ammonia and methanol may be an exception (Anon. 1978A).

A fraction of the cell protein is tied to the cell wall, and a large fraction of the extracted protein cannot be precipitated by the methods which have been mentioned. Recovery of only one-half of the total cell protein doubles the raw material cost per kg (lb) of recovered protein. This has been a major obstacle to the use of microbial biomass in food applications which require both a concentration of the protein fraction and a reduction of the NA content. A considerable improvement can be obtained by an additional processing step. If the alkaline cell extract is adjusted to a mildly acidic pH and incubated, the yeast nucleases hydrolyze nucleic acids so that they are not precipitated and carried into the protein precipitate. Such a process has been described by Seeley (1977) for bakers' yeast. It produces 3 fractions from the whole cell. The first fraction consists of the cell wall residue, called "glycan fraction," which contains only 11% crude protein and 0.3% NA. This fraction has been recommended as a thickener for salad dressings because its functional properties resemble those of vegetable gums. The second fraction contains 75% protein and only 0.7% NA, which would permit its use in various food applications without concern for the effect of NA. The third fraction contains the remaining soluble solids and has been called a "flavor fraction." The process is probably applicable to microbial biomass other than bakers' yeast.

USE OF MICROBIAL BIOMASS IN FOODS AND FEEDS

For use in the feed industry microbial biomass is used as such by blending with other dry feed ingredients or by wet processing of grain mashes with active yeasts. For use in foods additional processing methods, such as the isolation of SCP, have been abundantly described in the literature, but at the present time such processes are not in commercial use.

SCP can be spun by well known methods, for instance by solubilization of the protein at an alkaline pH followed by extrusion into an acid dope. The processes reported so far have resulted in protein fibers of relatively low tensile strength (Daly and Ruiz 1974; Rha 1975). Addition of some hydrocolloids improves the tensile strength of the fiber (Rha 1975).

While the major fraction of isolated yeast protein has a molecular weight of approximately 100,000, the concentration of this fraction is rapidly reduced at elevated temperatures and at higher pH levels (Lindblom 1974). Texturization of SCP as described by Tannenbaum (1977) and Akin (1973) is probably a more practical way of improving the functional properties than spinning. The whipping properties of the protein fractions, their gelling properties, and their water binding capacity have also been investigated. The water binding capacity is probably due to the cell wall material whose isolation and use in foods has already been mentioned (Seeley 1977; Robbins 1976). The most comprehensive review of the use of biomass and SCP in foods is that of Chen and Peppler (1977).

Use in Foods

Inactive dry yeasts (*C. utilis, S. cerevisiae, S. uvarum, K. fragilis*) are often included in fabricated foods such as baked goods, baby foods, geriatric foods, soups, gravies, meat extenders, etc. Concentrations vary from a few tenths of a percent to about 2%. Apart from the nutritional value of the microbial biomass, it contributes a desirable flavor if it is added in small concentrations. In bread baking, inactive dry yeasts make doughs more extensible due to the reducing action of -SH groups (Reed and Peppler 1973). Most of the applications which have been mentioned have never been described in the scientific literature, but widespread use of yeasts is attested to by inclusion in the ingredient statements of many prepared foods. Inactive dry yeast preparations are also sold in the health food industry in the form of tablets or dry powders. In this application yeasts are often fortified with water soluble vitamins. A recent review of food uses of microbial biomass has been prepared by Kharatyan (1978).

Use in Feeds

The use of yeasts as an ingredient in feeds and feed concentrates is well established. Current uses have been described by Peppler and Stone (1976). Table 13.9 from this publication lists the specifications for yeasts used in the United States. A major fraction of the brewers' yeast and *C. utilis* produced in the United States is used in feeds. A special application is the use of active dry yeasts or compressed yeasts in the preparation of "yeast culture." This feed supplement is prepared by seeding a cereal grain mash with yeast cells, incubation, and then drying of the fermented mash in such a manner that live yeast cells, enzymes, and other heat sensitive nutritional factors are preserved.

The development of biomass processes in Europe, Japan, Russia, and Eastern Europe is designed to relieve an acute shortage of protein for the feeding of monogastric animals and ruminants. The production of *C. utilis* yeast grown on molasses or on sulfite liquor has already been mentioned. The Pekilo process (*P. varioti*) and the Symba process using *C. utilis* and *E. fibuliger* have also been mentioned. Finally the production of bacterial biomass (*P. methylotropha*) is scheduled for commercial production in England to supply the feed market.

The pet food industry is a major outlet for microbial biomass. Again this application has not been well documented, but dog foods, cat foods, and to a lesser extent bird foods and fish foods contain small concentrations of yeast. In the case of dog and cat foods the addition of yeast or yeast extracts makes these pet foods more palatable to the animals.

TABLE 13.9. AAFCO SPECIFICATIONS FOR FEED YEASTS

Product	Fermentative	Genus	Protein[1]	Comment
96.1 Primary dried yeast	no	*Saccharomyces*	40%	[2]
96.2 Active dry yeast	yes	NS[3]	NS[3]	[4]
96.3 Irradiated dried yeast	no	NS	NS	[5]
96.4 Brewers' dried yeast	no	*Saccharomyces*	40%	[6]
96.5 Grain distillers' dried yeast	no	*Saccharomyces*	40%	[2]
96.6 Molasses distillers' dried yeast	no	*Saccharomyces*	40%	[2]
96.7 Torula dried yeast	no	*Torulopsis*[7]	40%	[2]
96.8 Yeast culture	yes	NS	NS	[8]

Source: Peppler and Stone (1976).
[1]Crude protein (nitrogen × 6.25).
[2]Separated from the mash or medium.
[3]NS = not specified; however, it is *Saccharomyces* with more than 45% protein.
[4]15 × 10^9 live cell count per g; no added cereal or filler permitted.
[5]Source of vitamin D_2 (calciferol).
[6]Nonextracted.
[7]Now in genus *Candida*.
[8]Yeast plus medium.

YEAST AUTOLYSATES

Yeast autolysates (self digests) are produced by the action of intracellular enzymes, principally proteases, on polymeric proteins and other polymers of the yeast cell. The process is valuable because it results in the formation of protein split products which have a pronounced meat-like flavor. In general, proteolysis is carried on to the point where 25–50% of the nitrogen of the cell is present as α-amino nitrogen. This does not really give a good picture of the distribution of the various protein split products among free amino acids, low molecular weight peptides, and polypeptides; and it is surprising that such data can not be found in the literature. Autolysis is not restricted to protein but carbohydrases and nucleases hydrolyze their respective substrates as well.

The technical process of yeast autolysis has been described by Hough and Maddox (1970), Peppler (1967), and Reiff *et al.* (1962). The process is carried out at temperatures which kill the yeast cells but do not inactivate the hydrolases, that is, at temperatures between 40° and 55°C. It is often advantageous to initiate the autolytic process by the addition of plasmolyzing agents which may also serve as antiseptics to suppress the growth of thermophiles. Such antiseptics are chloroform, toluene, thymol, phenol, ether, ethyl acetate, and formaldehyde.

Autolysis is carried out at a slightly acid pH. It may proceed from 12 to 36 hr depending on the degree of hydrolysis desired. The autolyzed mixture is then pasteurized at from 80° to 90°C, cooled, and filtered with diatomaceous earth. It can then be concentrated in a vacuum concentrator to a paste of about 80% solids or spray dried. Frequently salt is added to the autolysate before drying since the material is generally used as a condiment. A typical composition of such a paste is: total solids 80%; salt 15%, ash (other than added salt) 6%, total nitrogen 6.5 to 7.5%, α-amino nitrogen 2 to 4%, and pH 5 to 6 (Reed and Peppler 1973).

Yeast autolysates may be prepared from brewers' yeast (either debittered or undebittered) or from primary grown yeasts such as bakers' yeast. The composition of commercially available autolysates varies greatly with the source, the processing conditions, and with the addition of salt, sodium glutamate, or 5'-nucleotides.

Yeast autolysates are also called "yeast extracts." The majority of the commercial products are indeed extracts since autolysis has been followed by filtration to remove the insoluble cell wall debris. The extracts are completely water soluble and form clear water solutions whose color varies between amber and dark brown. Some yeast autolysates (not extracts) are also available in which the cell wall material has been concentrated and dried with the rest of the cell material.

Yeast autolysates have a pleasant, meat-like flavor and aroma which is the reason for their use in soups, gravies, meat dishes, pet foods, and generally as condiments. In Australia and New Zealand and to some extent in Great Britain, autolysates in paste form are used as bread spreads.

Yeast, yeast autolysates, and yeast extracts and dialysates also find use in the food and fermentation industries as a fermentation nutrient.

5'-NUCLEOTIDES

The presence of a potent flavor enhancer was discovered in bonito more than 50 years ago. Since then the compound has been identified as 5'-inosinic acid; and both 5'-inosinic acid (IMP) and 5'-guanylic acid (GMP) have been identified in beef, pork, chicken, whale, and fish (Shimazono 1964). IMP, GMP, and 5'-xanthylic acid are flavor enhancers, but only GMP and IMP are commercially available.

IMP and GMP are produced from nucleosides by a phosphohydrolase which splits the linkage between the 3-carbon of the ribose moiety and the phosphate link. This leaves the phosphate attached to the 5-carbon of the next ribose unit. The starting material for this enzymatic process is RNA isolated from C. utilis biomass, usually by extraction with hot NaCl solutions followed by acid precipitation (Reed and Peppler 1973).

IMP and GMP are used as flavor enhancers in meats, meat broths, soups, and fabricated foods. They have a synergistic effect with glutamate, yeast extracts, and hydrolyzed vegetable proteins.

For general reviews of the field of biomass production the reader is referred to the following references which include books, book chapters, and more extensive review articles: Davis (1974); Gaunelle de Pontanel (1972); Humphrey et al. (1977); Kharatyan (1978); Laskin (1977B); Litchfield

(1977B,C); Mateles and Tannenbaum (1968); Peppler (1970); Perlman (1977); Reed and Peppler (1973); Rockwell (1976); Tannenbaum (1977); Tannenbaum and Wang (1975); Wagner (1975).

REFERENCES

ABBOTT, B.J. and CLAMEN, A. 1973. The relationship of substrate, growth rate, and maintenance coefficient to single cell protein production. Biotechnol. Bioeng. *15*, 117–127.

AKIN, C. 1973. Process for texturizing microbial cells by alkali-acid treatment. U.S. Pat. 3,781,204. Dec. 11.

ALLISON, K. 1975. Protein from oil: The BP alkane process. *In* Symposium Microbial Production of Protein 1975. F. Wagner (Editor). Verlag Chemie, Weinheim, W. Germany. (German)

ANDERSON, A., WEISBAUM, R.B. and ROBE, K. 1974. Converts waste to 50% protein Torula. Food Process. *35* (7) 58–59.

ANDERSON, C. *et al.* 1975. The growth of microfungi on carbohydrates. *In* Single Cell Protein II. S.R. Tannenbaum and D.I.C. Wang (Editors). MIT Press, Cambridge, Mass.

ANON. 1972A. Preclinical testing of novel sources of protein. Protein Advisory Group Guideline 6. United Nations, New York.

ANON. 1972B. Human testing of supplementary food mixtures. Protein Advisory Group Guideline 7. United Nations, New York.

ANON. 1972C. Production of single cell protein for human consumption. Protein Advisory Group Guideline 12. United Nations, New York.

ANON. 1976A. Processed protein foods and lysinoalanine. Nutr. Rev. *34* (4) 120–122.

ANON. 1976B. ICI to scale up single cell protein process. Chem. Eng. News *54* (42) 25–26.

ANON. 1977. System treats waste water, grows SCP. Canner/Packer (2) 82.

ANON. 1978A. Hoechst starts up SCP pilot plant. Chem. Eng. News *56* (17) 18.

ANON. 1978B. Research on chemicals from biomass aired. Chem. Eng. News *56* (14) 31.

BELLAMY, W.D. 1975. Conversion of insoluble agricultural wastes to SCP by thermophilic microorganisms. *In* Single Cell Protein II. S.R. Tannenbaum and D.I.C. Wang (Editors). MIT Press, Cambridge, Mass.

BERNSTEIN, S. and PLANTZ, P.E. 1977. Ferments whey into yeast. Food Eng. *49* (11) 74–75.

BERNSTEIN, S., TZENG, C.H. and SISSON, D. 1977. The commercial fermentation of cheese whey for the production of protein and/or alcohol. *In* Single Cell Protein from Renewable and Nonrenewable Resources. A.E. Humphrey and E.L. Gaden, Jr. (Editors). Biotechnol. Bioeng. Symp. 7. John Wiley & Sons, New York.

BLACKWELL, B., MABBIT, L.A. and MARLEY, W. 1969. Histamine and tyramine content of yeast products. J. Food Sci. *34*, 47–51.

BODWELL, C.E. 1977. Evaluation of Proteins for Humans. AVI Publishing Co., Westport, Conn.

BUTSCHEK, G. 1962. Food and feed yeasts. *In* The Yeasts, Vol. 2. F. Reiff *et al.* (Editors). Verlag Hans Carl, Nurnberg, W. Germany. (German)

BUTSCHEK, G. and KRAUSE, G. 1962. Sulfite waste liquor. *In* The Yeasts, Vol. 2. F. Reiff *et al.* (Editors). Verlag Hans Carl, Nurnberg, W. Germany. (German)

CALLOWAY, D.H. and WASLIEN, C.I. 1971. Bioregeneration of food. Environ. Biol. Med. *1*, 229–241.

CHAHAL, D.S., SWAN, J.E. and MOO-YOUNG, M. 1977. Protein and cellulase production by *Chaetomium cellulolyticum* grown on wheat straw. Dev. Ind. Microbiol. *18*, 433–442.

CHARATYAN, S.G. and WOLNOWA, A.I. 1975. Biological value of the biomass and protein isolates from different yeasts. Nahrung *19* (10) 885–890.

CHEN, S.L. and PEPPLER, H.J. 1977. Single-cell proteins in food applications. Dev. Ind. Microbiol. *19*, 79–94.

CHURCH, B.D., WIDMER, C.M. and BOON, W.C. 1977. Water reuse and animal feed production from continuous fungal bioconversion of agro-industrial and food processing waste. Proc. 2nd Intern. Biochem. Symp., 1977, Toronto.

CLÉMENT, G. 1975. Producing spirulina with CO_2. *In* Single Cell Protein II. S.R. Tannenbaum and D.I.C. Wang (Editors). MIT Press, Cambridge, Mass.

COONEY, C.L. and LEVINE, D.W. 1975. SCP production from methanol by yeast. *In* Single Cell Protein II. S.R. Tannebaum and D.I.C. Wang (Editors). MIT Press, Cambridge, Mass.

COONEY, C.L. and MAKIGUCHI, N. 1977. An assessment of single cell protein from methanol-grown yeast. *In* Single Cell Protein from Renewable and Nonrenewable Resources. A.E. Humphrey and E.L. Gaden, Jr. (Editors). Biotechnol. Bioeng. Symp. 7. John Wiley & Sons, New York.

COONEY, C.L., WANG, D.I.C. and MATELES, R.I. 1969. Measurement of heat evolution and correlation with oxygen consumption during microbial growth. Biotechnol. Bioeng. *11*, 269–281.

COOPER, J.L. 1977. The potential of food processing solid wastes as a source of cellulase for enzymatic conversion. *In* Bioeng. Biotechnol. Symp. *6*. E.L. Gaden, Jr. (Editor). John Wiley & Sons, New York.

COOPER, R.G., SILVER, R.S. and BOYLE, J.P. 1975. Semi-commercial studies of a petroprotein process based on n-paraffins. *In* Single Cell Protein II. S.R. Tannenbaum and D.I.C. Wang (Editors). MIT Press, Cambridge, Mass.

COWLING, E.B. and KIRK, T.K. 1976. Properties of cellulose and lignocellulosic materials as substrates for enzymatic conversion processes. *In* Enzymatic Conversion of Cellulosic Materials. Biotechnol. Bioeng. Symp. *6*. E.L. Gaden, Jr. (Editor). John Wiley & Sons, New York.

CUNNINGHAM, S.D., CATER, C.M., MITTIL, K.F. and VANDERZANT, C. 1975.   Rupture and protein extraction of petroleum grown yeast. J. Food Sci. 40, 732–735.

CYSEWSKI, G.R. and WILKE, C.R.   1976.   Utilization of cellulosic materials through enzymic hydrolysis. I. Fermentation of hydrolysate to ethanol and single cell protein. Biotechnol. Bioeng. 18, 1297–1313.

DABBAH, R.   1970.   Protein from microorganisms.   Food Technol. 24, 659–665.

DALY, W.H. and RUIZ, L.P.   1974.   Reduction of RNA in single cell protein in conjunction with fiber formation. Biotechnol. Bioeng. 16, 285–287.

DAVIS, J.P.   1974.   Single Cell Protein.   Academic Press, New York.

DIMMLING, W. and SEIPENBUSCH, R.   1978.   Raw materials for the production of SCP. Process Biochem. 13 (3) 9–15, 34.

DOSTALEK, M. and MOLIN, N.   1975.   Studies of biomass production of methanol oxidizing bacteria. In Single Cell Protein II. S.R. Tannenbaum and D.I.C. Wang (Editors).   MIT Press, Cambridge, Mass.

DUNLAP, C.E.   1975.   Production of single cell protein from insoluble agricultural wastes by mesophiles. In Single Cell Protein II. S.R. Tannenbaum and D.I.C. Wang (Editors).   MIT Press, Cambridge, Mass.

DUNNIL, P. and LILLY, M.D.   1975.   Protein extraction and recovery from microbial cells. In Single Cell Protein II. S.R. Tannenbaum and D.I.C. Wang (Editors).   MIT Press, Cambridge, Mass.

EDOZIEN, J.C., UDO, U.U., YOUNG, V.R. and SCRIMSHAW, N.S. 1970.   Effect of high levels of yeast feeding on uric acid metabolism of young men. Nature 228, 180.

ERDMAN, M.D., BERGEN, W.G. and REDDY, C.A.   1977A.   Amino acid profiles and presumptive nutritional assessment of single-cell protein from certain lactobacilli. Appl. Environ. Microbiol. 33, 901–905.

ERDMAN, M.D., REDDY, C.A. and BERGEN, W.G.   1977B.   Nutritional and chemical evaluation of the protein fraction from fermented ammoniated whey. J. Dairy Sci. 60, 1509–1514.

ESPINOSA, R., MACDONALDO, O., MENCHU, J.F. and ROLZ, C. 1977. Aerobic non-aseptic growth of Verticillium on coffee waste waters and cane blackstrap molasses on a pilot plant scale. In Single Cell Protein from Renewable and Nonrenewable Resources. A.E. Humphrey and E.L. Gaden, Jr. (Editors). Biotechnol. Bioeng. Symp. 7.   John Wiley & Sons, New York.

FAUST, U.   1975.   Practical experience with unconventional fermenters. In Symposium Microbial Production of Protein 1975. F. Wagner (Editor).   Verlag Chemie, Weinheim, W. Germany. (German)

FELDHEIM, W.   1975.   Feeding trials with man and animals for the evaluation of single cell protein. In Symposium Microbial Production of Protein 1975. F. Wagner (Editor).   Verlag Chemie, Weinheim, W. Germany. (German)

FIECHTER, A.   1967.   Production of microbial protein from hydrocarbons. Chimia 21, 501–508.

FOO, E.L.   1978.   Microbial production of methanol. Process Biochem. *13* (3) 23–28.

FORNEY, L.J. and REDDY, C.A.   1977.   Fermentative conversion of potato processing waste into a crude protein feed supplement by Lactobacilli. Dev. Ind. Microbiol. *18*, 135–143.

FRAHM, H. and LEMBKE, A.   1975.   Nutritional-physiological and toxicological problems on feeding microbial proteins to man and animals. *In* Symposium Microbial Production of Protein 1975. F. Wagner (Editor). Verlag Chemie, Weinheim, W. Germany. (German)

GADEN, E.L., JR.   1976.   Enzymatic Conversion of Cellulosic Materials. Biotechnol. Bioeng. Symp. *6.* John Wiley & Sons, New York.

GERHARDT, P. and REDDY, C.A.   1977.   Conversion of agroindustrial wastes into ruminant feedstuffs by ammoniated organic acid fermentation: a brief review and preview. Dev. Ind. Microbiol. *19*, 71–78.

GOUNELLE DE PONTANEL, H.   1972.   Proteins from Hydrocarbons. Academic Press, New York.

GOW, J.S., LITTLEHAILES, J.D., SMITH, S.R.L. and WALTER, R.B. 1975. SCP production from methanol: Bacteria. *In* Single Cell Protein II. S.R. Tannenbaum and D.I.C. Wang (Editors).   MIT Press, Cambridge, Mass.

GREGORY, K.F. *et al.*   1976.   Conversion of carbohydrates to protein by high temperature fungi. Food Technol. *30* (3) 30, 32, 35.

GRETHLEIN, H.E.   1978.   Comparison of the economics of acid and enzyme hydrolysis of newsprint. Biotechnol. Bioeng. *20*, 503–525.

GRIMMER, G.   1974.   Detection and occurrence of polycyclic hydrocarbons in yeasts cultured on mineral oils. Dtsch. Lebensm. Rundsch. *70* (11) 394–397.

HAMER, G., HARRISON, D.E.F., HARWOOD, J.H. and TOPIWALA, H.H. 1975.   SCP production from methane. *In* Single Cell Protein II. S.R. Tannenbaum and D.I.C. Wang (Editors).   MIT Press, Cambridge, Mass.

HARJU, M., HEIKONEN, M. and KREULA, M.   1976.   Nutrient supplementation of Swiss cheese whey for the production of feed yeast. Milchwissenschaft *31* (9) 530–534.

HATCH, R.T.   1975.   Fermenter design. *In* Single Cell Protein II. S.R. Tannenbaum and D.I.C. Wang (Editors).   MIT Press, Cambridge, Mass.

HEDENSKOG, G. and MOGREN, H.   1973.   Some methods of processing for single cell protein. Biotechnol. Bioeng. *15*, 129–142.

HEDENSKOG, G., MOGREN, H. and ENEBO, L.   1970.   A method for obtaining protein concentrates from microorganisms. Biotechnol. Bioeng. *12*, 947–959.

HOUGH, J.S. and MADDOX, I.S.   1970.   Yeast autolysis. Process Biochem. *5* (5) 50–52.

HUANG, F. and RHA, C.   1978.   Formation of single-cell protein filament. J. Food Sci. *43*, 780–782.

HUANG, Y.D.   1974.   Production of food yeast from acid brine. Proc. 4th Intern. Symp. Yeasts, Vienna, 1974, Part I, 157–158, Intern. Comm. Yeasts (ICY).

HUMPHREY, A.E.   1970.   Microbial protein from petroleum. Process Biochem. *5* (6) 19–22.

HUMPHREY, E.A., MOREIRA, A., ARMIGER, W. and ZABRISKIE, D. 1977. Production of single cell protein from cellulose wastes. *In* Single Cell Protein from Renewable and Nonrenewable Resources. E.L. Gaden, Jr. (Editor). John Wiley & Sons, New York.

HUNT, L.A.   1969.   Brewers' grains and yeast: Products, not by-products. Brew. Dig. *44* (1) 42–43, 46–48.

IMP. CHEM. IND.   1977.   Treatment of unicellular proteins. Belg. Pat. 844,127. Jan. 14.

INGLEDEW, W.M. *et al.*   1977.   Spent brewers' yeast—Analysis, improvement, and heat processing considerations. Tech. Q. Master Brew. Assoc. Am. *14* (4) 231–237.

INSKEEP, G., WILEY, A.J., HOLDERBY, J.M. and HUGHES, L.P.   1951. Food yeast from sulfite liquor. Ind. Eng. Chem. *43*, 1702–1711.

JARL, K.   1969.   Symba yeast process. Food Technol. *23* (8) 23–26.

JOHNSON, M.J.   1967.   Growth of microbial cells on hydrocarbons. Science *155*, 1515–1519.

KACHHOLZ, T. and REHM, H.-J.   1977.   Degradation of long chain alkanes by bacilli. Europ. J. Appl. Microbiol. *4*, 101–110.

KANAZAWA, M.   1975.   The production of yeast from n-paraffins. *In* Single Cell Protein II. S.R. Tannenbaum and D.I.C. Wang (Editors). MIT Press, Cambridge, Mass.

KHARATYAN, S.G.   1978.   Microbes as foods for humans. Annu. Rev. Microbiol. *32*, 301–327.

KIHLBERG, R.   1972.   The microbe as a source of food. Annu. Rev. Microbiol. *26*, 427–466.

KNECHT, R., PRÄVE, P., SEIPENBUSCH, R. and SUKATSCH, D.A. 1977. Microbiology and biotechnology of SCP produced from n-paraffin. Process Biochem. *12* (4) 11–14.

KOFRANYI, E. and JEKAT, F.   1967.   Determination of the biological value of food proteins. XII. Mixture of egg with rice, corn, soy beans and algae. Hoppe Seyler's Z. Physiol. Chem. *348*, 84–88.

KULP, K.   1975.   Carbohydrases. *In* Enzymes in Food Processing. G. Reed (Editor).   AVI Publishing Co., Westport, Conn.

LABUZA, T.B.   1975.   Cell collection: Recovery and drying for SCP manufacture. *In* Single Cell Protein II. S.R. Tannenbaum and D.I.C. Wang (Editors). MIT Press, Cambridge, Mass.

LABUZA, T.P., SANTOS, D.B. and ROOP, R.N.   1972A.   Engineering factors in single cell protein production: Fluid properties and concentration of yeast by evaporation. Biotechnol. Bioeng. *12*, 123–134.

LABUZA, T.P. *et al.*   1972B.   Engineering factors in single cell protein production: Spray drying and cell viability. Biotechnol. Bioeng. *12*, 135–140.

LADISCH, M.R., GONG, C.S. and TSAO, G.T.   1977.   Corn crop residues as a potential source of single cell protein: Kinetics of *Trichoderma viride* cellobiase action. Dev. Ind. Microbiol. *18*, 157–168.

LAFFERTY, R.M., MOSER, A. and STEINER, W.   1975.   Production of single cell protein from chemo-lithotrophic bacteria. *In* Symposium Microbial Production of Protein 1975. F. Wagner (Editor). Verlag Chemie, Weinheim, W. Germany. (German)

LAINÉ, B.M., SNELL, R.C. and PEET, W.A.   1976.   Production of single cell protein from n-paraffins. Chem. Eng. (London) *310*, 440–443, 446.

LAINÉ, B.M. and CHAFFAUT, J. du.   1975.   Gas-oil as a substrate for single cell protein production. *In* Single Cell Protein II. S.R. Tannenbaum and D.I.C. Wang (Editors).   MIT Press, Cambridge, Mass.

LASKIN, A.I.   1977A.   Single cell protein. *In* Annual Reports on Fermentation Processes, Vol. 1. D. Perlman and G.T. Tsao (Editors).   Academic Press, New York.

LASKIN, A.I.   1977B.   Ethanol as substrate for single cell protein production. *In* Single Cell Protein from Renewable and Nonrenewable Resources. E.L. Gaden, Jr. (Editor). Biotechnol. Bioeng. Symp. *7*.   John Wiley & Sons, New York.

LA. STATE UNIV.   1976.   Edible protein from cellulose by lignin degradation and cellulose fermentation. U.S. Pat. 3,761,355. Nov. 24.

LEHMANN, J. *et al.*   1975.   Fluid dynamic, mass transport, and growth in bubble column reactors. *In* Symposium Microbial Production of Protein 1975. F. Wagner (Editor).   Verlag Chemie, Weinheim, W. Germany. (German)

LINDBLOM, M.   1974.   The influence of alkali and heat treatment on yeast protein. Biotechnol. Bioeng. *16* (11) 1495–1506.

LITCHFIELD, J.H.   1977A.   Use of hydrocarbon fractions for the production of SCP. *In* Single Cell Protein from Renewable and Nonrenewable Resources. E.L. Gaden, Jr. (Editor). Biotechnol. Bioeng. Symp. *7*.   John Wiley & Sons, New York.

LITCHFIELD, J.H.   1977B.   Comparative technical and economic aspects of single cell proteins. Adv. Appl. Microbiol. *22*, 267–305.

LITCHFIELD, J.H.   1977C.   Single cell protein. Food Technol. *31* (5) 175–179.

LITTLEHAILES, J.D.   1975.   The ICI single cell protein process. *In* Symposium Microbial Production of Protein 1975. F. Wagner (Editor).   Verlag Chemie, Weinheim, W. Germany. (German)

LIVANSKY, K. *et al.*   1973.   Some problems of CO_2 absorption by algae suspensions. *In* Biotechnol. Bioeng. Symp. *4*.   John Wiley & Sons, New York.

LORENZ, M. *et al.*   1967.   Synthetic efficiency of *C. utilis* as a function of temperature and continuous cultivation. Z. Allg. Mikrobiol. *7*, 363–371.

MACRIS, B.J. and KOKKE, R.   1977.   Kinetics of growth and chemical composition of *Fusarium moniliforma* cultivated on carob aqueous extract for microbial protein production. Eur. J. Appl. Microbiol. *4*, 93–99.

MASUDA, Y.   1974.   Production of SCP from ethanol. Chem. Econ. Eng. Rev. 6 (11) 54–60.

MASUDA, Y., NAKANISHI, M. and SAKAKURA, Y.   1976.   Make SCP from ethanol. Hydrocarbon Process. (Nov.) 113–116.

MATELES, R.I.   1975.   Production of SCP in Israel. *In* Single Cell Protein II. S.R. Tannenbaum and D.I.C. Wang (Editors).   MIT Press, Cambridge, Mass.

MATELES, R.I. and TANNENBAUM, S.R.   1968.   Single Cell Protein. MIT Press, Cambridge, Mass.

McMANAMEY, J.W., LOUCAIDES, R. and LEWIS, J.M.   1973.   Mixing and mass transfer for simulated fermentation systems in turbine agitated vessels. *In* Advances in Microbial Engineering, Part I. B. Sikyta *et al.* (Editors). John Wiley & Sons, New York.

MERTZ, W.   1975.   Effects of metabolism of glucose tolerance factor. Nutr. Rev. *33* (5) 129–135.

MEYERS, S.G.   1978.   Ethanolic fermentation during enzymatic hydrolysis of cellulose. Proc. 2nd Pac. Chem. Eng. Congr., 1977, Denver, Am. Inst. Chem. Eng.

MICHOLT, C., KANNAERTS, J. and NYNS, E.J.   1975.   Evaluation of chances of hydrocarbon grown single cell protein to gain a share of the world gross potential market. Rev. Ferment. Ind. Aliment. *30* (1) 3–16.

MILLET, M.A., BAKER, A.J. and SATTER, L.D.   1976.   Physical and chemical pretreatments for enhancing cellulose saccharification. *In* Enzymatic Conversion of Cellulosic Materials. E.L. Gaden, Jr. *et al.* (Editors). Biotechnol. Bioeng. Symp. *6*.   John Wiley & Sons, New York.

MOEBUS, O., KIESBYE, P. and TEUBER, M.   1977.   Improvement in the technology of single cell protein production from highly polluted dairy solutions. Studies on the technical production of *Saccharomyces cerevisiae* poor in ribonucleic acid. Kiel. Milchwirtsch. Forschungsber. *29* (2) 131–139.

MOO-YOUNG, M.   1977.   Economics of SCP production. Process Biochem. *12* (4) 6–10.

MORRISON, S.M., ELMUND, G.K., GRANT, D.W. and SMITH, V.J. 1977. Protein production from feed lot waste. Dev. Ind. Microbiol. *18*, 145–155.

MOSER, A. and LAFFERTY, R.M.   1975.   Newer results and biotechnological aspects of biomass production. *In* Symposium Microbial Production of Protein 1975. F. Wagner (Editor).   Verlag Chemie, Weinheim, W. Germany. (German)

NAGUIB, M.   1975.   Investigations of the physiology of growth of the obligatory methane oxidizing strain M 102. The effect of extracellular metabolites and a possible inhibitory mechanism. *In* Symposium Microbial Production of Protein 1975. F. Wagner (Editor).   Verlag Chemie, Weinheim, W. Germany. (German)

NEUMANN, H.J.   1975.   Petrochemical raw materials for microbial protein production: Manufacture, availability, and economic aspects. *In* Symposium Microbial Production of Protein 1975. F. Wagner (Editor). Verlag Chemie, Weinheim, W. Germany. (German)

OKI, T., KOUNO, K., KITAI, A. and OZAKI, A.   1972.   New yeasts capable of assimilating methanol. J. Gen. Appl. Microbiol. *18*, 295–305.

O'SULLIVAN, D.A.   1978.   BP contests petroprotein plant health issue. Chem. Eng. News *56* (12) 12.

PABST, W.   1975.   Feeding trials for the determination of the nutritional value of micro-algae. *In* Symposium Microbial Production of Protein 1975. F. Wagner (Editor). Verlag Chemie, Weinheim, W. Germany. (German)

PACE, G.W. and GOLDSTEIN, D.J.   1975.   Economic analysis of ultrafiltration—fermentation plants producing whey protein and SCP from cheese whey. *In* Single Cell Protein II. S.R. Tannenbaum and D.I.C. Wang (Editors). MIT Press, Cambridge, Mass.

PAYER, H.D. *et al.*   1975.   The contamination of micro-algae with some environmental toxins. *In* Symposium Microbial Production of Protein 1975. F. Wagner (Editor). Verlag Chemie, Weinheim, W. Germany. (German)

PEPPLER, H.J.   1967.   Microbial Technology.   Reinhold Publishing Corp., New York.

PEPPLER, H.J.   1970.   Food yeast. *In* The Yeasts. A.H. Rose and J.S. Harrison (Editors). Academic Press, New York.

PEPPLER, H.J. and STONE, C.W.   1976.   Feed yeast products. Feed Manage. *27* (8) 17–18.

PERLMAN, D.   1977.   Fermentation industry—quo vadis? Chem. Technol. *7* (7) 434–443.

PESCHARD-MARISCAL, E. and VINIEGRA-GONZALEZ, G.   1977.   Cost analysis of yeast protein and RNA production by aerobic fermentation of cane molasses. *In* Single Cell Protein from Renewable and Nonrenewable Resources. E.L. Gaden, Jr. (Editor). Biotechnol. Bioeng. Symp. *7*.   John Wiley & Sons, New York.

POKROVSKY, A.   1975.   Some results of SCP medico-biological investigations. *In* Single Cell Protein II. S.R. Tannenbaum and D.I.C. Wang (Editors). MIT Press, Cambridge, Mass.

POMERANZ, Y.   1976.   Single cell protein from by-products of malting and brewing. Brew. Dig. *51* (1) 49–55, 60.

PRÄVE, P. and SUKATSCH, D.A.   1975.   Production of biomass from methanol. *In* Symposium Microbial Production of Protein 1975. F. Wagner (Editor). Verlag Chemie, Weinheim, W. Germany. (German)

QUITTENTON, R.C.   1966.   An assessment of brewery by-products. Tech. Q. Master Brew. Assoc. Am. *3*, 174–177.

RANK HOVIS McDOUGAL.   1977.   Reducing nucleic acid content of edible proteins prepared by cultivating fungi imperfecti in suspension. U.S. Pat. 4,041,189. Sept 8.

RATLEDGE, C.   1977.   Fermentation substrates. *In* Annual Reports on Fermentation Processes, Vol. 1. D. Perlman and G.T. Tsao (Editors). Academic Press, New York.

REDDY, C.A. and ERDMAN, M.D.   1977.   Production of a ruminant protein supplement by anaerobic fermentation of feed lot waste filtrate. *In* Single Cell Protein from Renewable and Nonrenewable Resources. E.L. Gaden, Jr. (Editor).   John Wiley & Sons, New York.

REDDY, C.A., HENDERSON, H.E. and REDMAN, M.D.   1976.   Bacterial fermentation of cheese whey for production of a ruminant feed supplement rich in crude protein. Appl. Environ. Microbiol. *32*, 769–776.

REED, G. and PEPPLER, H.J.   1973.   Yeast Technology.   AVI Publishing Co., Westport, Conn.

REHM, H.-J.   1976.   Biotechnology of the production of single cell protein. Ernaehr. Umsch. *23* (10) 307–311.

REIFF, F., LINDEMANN, M. and HOLLE, H.   1962.   Yeast in foods. *In* The Yeasts. F. Reiff *et al.* (Editors).   Verlag Hans Carl, Nurnberg, W. Germany.

RHA, C.K.   1975.   Utilization of SCP in human food. *In* Single Cell Protein II. S.R. Tannenbaum and D.I.C. Wang (Editors).   MIT Press, Cambridge, Mass.

RIDGEWAY, J.A. *et al.*   1975.   Single cell protein materials from ethanol. U.S. Pat. 3,865,691. Feb. 11.

ROBBINS, E.A.   1976.   Manufacture of yeast protein isolate having a reduced nucleic acid content by a thermal process. U.S. Pat. 3,991,215. Nov. 9.

ROBBINS, E.A. and SEELEY, R.D.   1977.   Cholesterol lowering effect of dietary yeast and yeast fractions. J. Food Sci. *42* (3) 694–698.

ROCKWELL, P.J.   1976.   Single Cell Proteins from Cellulose and Hydrocarbons.   Noyes Data Corp., Park Ridge, N.J.

ROMANTSCHUK, H.   1975.   The Pekilo process: Protein from spent sulfite liquor. *In* Single Cell Protein II. S.R. Tannenbaum and D.I.C. Wang (Editors).   MIT Press, Cambridge, Mass.

ROMANTSCHUK, H. and LEHTOMAKI, M.   1978.   Operational experiences of first full scale Pekilo process SCP-mill application. Process Biochem. *13* (3) 16–17.

RUT, M. *et al.*   1976.   Mass balance in batch cultivation of *C. utilis* in synthetic alcohol. Kvasny Prum. *22* (5) 111–114.

SAHM, H. and WAGNER, F.   1975.   Methanol metabolism in microbes. *In* Symposium Microbial Production of Protein 1975. F. Wagner (Editor). Verlag Chemie, Weinheim, W. Germany. (German)

SÁNCHEZ-MARROQUÍN, A.   1977.   Mixed cultures in the production of single cell protein from agave juice. *In* Single Cell Protein from Renewable and Nonrenewable Resources. E.L. Gaden, Jr. (Editor). Biotechnol. Bioeng. Symp. *7*.   John Wiley & Sons, New York.

SAVARESE, J.J. and YOUNG, S.D.   1978.   Combined enzyme hydrolysis of cellulose and yeast fermentation. Biotechnol. Bioeng. *20*, 1291–1293.

SCHLANDERER, G. *et al.*   1975.   Methanol metabolism and its regulation in yeasts. *In* Symposium Microbial Production of Protein 1975. F. Wagner (Editor).   Verlag Chemie, Weinheim, W. Germany. (German)

SCHREIER, K.   1974.   Bioreactor: Stage of development and industrial application, especially with regard to systems for gas transfer. Proc. 4th

Intern. Symp. Yeasts, Vienna, 1974, Part I, 137–138. Intern. Comm. Yeasts (ICY).

SCHULZ, E. and OSLAGE, H.J. 1975. Analytical values and feeding trials for the nutritional and physiological evaluation of single cell protein. *In* Symposium Microbial Production of Protein 1975. F. Wagner (Editor). Verlag Chemie, Weinheim, W. Germany. (German)

SCHWARZ, K. and FOLTZ, C.M. 1957. Selenium as an integral part of factor 3 against dietary necrotic liver degeneration. J. Am. Chem. Soc. *79*, 3292–3293.

SCRIMSHAW, N.S. 1975. Single cell protein for human consumption— an overview. *In* Single Cell Protein II. S.R. Tannenbaum and D.I.C. Wang (Editors). MIT Press, Cambridge, Mass.

SEELEY, R.D. 1975. Functional aspects of single cell protein area to potential markets. Food Prod. Dev. *9* (7) 46, 51.

SEELEY, R.D. 1977. Fractionation and utilization of bakers' yeast. Tech. Q. Master Brew. Assoc. Am. *14* (1) 35–39.

SHANNON, L.J. and STEVENSON, K.E. 1975. Growth of fungi and BOD reduction in selected brewery wastes. J. Food Sci. *40*, 826–829.

SHELL OIL CO. 1977. Growing methane utilizing microorganisms in a liquid medium containing methanol using and non-methylotrophic microorganisms. U.S. Pat. 4,042,458. Aug. 16.

SHIMAZONO, H. 1964. The distribution of 5′ nucleotides in foods and their application to foods. Food Technol. *18*, 294–303.

SHIPMAN, R.H., FAN, L.T. and KAO, I.C. 1977. Single cell protein production by photosynthetic bacteria. Adv. Appl. Microbiol. *21*, 161–182.

SINSKEY, A.J. and TANNENBAUM, S.R. 1975. Removal of nucleic acid in SCP. *In* Single Cell Protein II. S.R. Tannenbaum and D.I.C. Wang (Editors). MIT Press, Cambridge, Mass.

SKOGMAN, H. 1976. Production of Symba yeast from potato wastes. *In* Food from Waste. Conf., 1976, Univ. Reading, England, 167–179.

SKUPIN, J. *et al.* 1977. Nutritive value of propionibacteria and lactose fermenting yeast grown in whey. J. Food Process. Preservation *1*, 207–216.

SLONEKER, J.H. 1977. Agricultural residues, including feed lot wastes. *In* Single Cell Protein from Renewable and Nonrenewable Resources. E.L. Gaden, Jr. (Editor). Biotechnol. Bioeng. Symp. *6*. John Wiley & Sons, New York.

SNYDER, H.E. 1970. Microbial sources of protein. Adv. Food Res. *18*, 85–140.

SOEDER, C.J. and MOHN, H. 1975. Technological aspects of the culture of micro-algae. *In* Symposium Microbial Production of Protein 1975. F. Wagner (Editor). Verlag Chemie, Weinheim, W. Germany. (German)

STANDARD OIL CO. 1977. Aerobic fermentation vessel for protein production. U.S. Pat. 4,019,962. Apr. 26.

STANIER, R.Y., DOUDOROFF, M. and ADELBERG, E.A. 1970. The Microbial World, 3rd Edition. Prentice-Hall, Englewood Cliffs, N.J.

STONE, R.N. 1977. Timber, wood residues, and wood pulp as sources of cellulose. *In* Single Cell Protein from Renewable and Nonrenewable Resources. E.L. Gaden, Jr. (Editor). Biotechnol. Bioeng. Symp. *6*. John Wiley & Sons, New York.

SUOMALAINEN, H. and OURA, E. 1971. Yeast nutrition and solute uptake. *In* The Yeasts, Vol. 2. A.H. Rose and J.S. Harrison (Editors). Academic Press, New York.

TANNENBAUM, S.R. 1977. Single cell protein. *In* Food Proteins. J.R. Whitaker and S.R. Tannenbaum (Editors). AVI Publishing Co., Westport, Conn.

TANNENBAUM, S.R. and WANG, D.I.C. 1975. Single Cell Protein II. MIT Press, Cambridge, Mass.

TANNER, R.D. and HUSSAIN, S.S. 1977. Kudzu (*Pueraria lobata*) root starch as a substrate for the lysine-enriched bakers' yeast and ethanol fermentation process. Pap. 174th Natl. Meet. Am. Chem. Soc., Div. Microb. Biochem. Technol., Chicago, Aug. 30, 1977.

TREVELYAN, W.E. 1976. Chemical methods for the reduction of the purine content of baker's yeast, a form of single cell protein. J. Sci. Food Agric. *27* (3) 225–230.

TSAO, G.T. 1978A. Cellulosic material as a renewable resource. Process Biochem. *13* (10) 12–14.

TSAO, G.T. 1978B. Personal communication. Purdue Univ., Lafayette, Ind.

UNIVERSAL FOODS. 1977. Unpublished data. Milwaukee, Wis.

VANANUVAT, P. and KINSELLA, J.E. 1975A. Protein production from crude lactose by *S. fragilis*. Continuous culture studies. J. Food Sci. *40*, 823–825.

VANANUVAT, P. and KINSELLA, J.E. 1975B. Amino acid composition of protein isolates from *Saccharomyces fragilis*. J. Agric. Food Chem. *23*, 595–597.

VIIKARI, L. and LINKO, M. 1977. Reduction of nucleic acid content of SCP. Process Biochem. *12* (4) 17–19, 35.

WAGNER, F. 1975. Symposium Microbial Production of Protein 1975. Verlag Chemie, Weinheim, W. Germany. (German)

WANG, D.I.C. 1968. Cell recovery. *In* Single Cell Protein. R.I. Mateles and S.R. Tannenbaum (Editors). MIT Press, Cambridge, Mass.

WASLIEN, C.I. 1975. Unusual sources of protein for man. Crit. Rev. Food Sci. Nutr. *6* (1) 77–151.

WASLIEN, C.I., CALLOWAY, D.H., MARGEN, S. and COSTA, F. 1970. Uric acid levels in men fed algae and yeast as protein sources. J. Food Sci. *35*, 294–298.

WASSERMAN, A.E., HANSON, J.E. and ALVARE, N.F. 1961. Large scale production of yeast in whey. J. Water Pollut. Control Fed. *33*, 1090–1094.

WILKE, C.R., YANG, R.D. and VON STOCKAR, U. 1976. Preliminary cost analysis for enzymatic hydrolysis of newsprint. *In* Enzymatic Conversion of Cellulosic Materials. E.L. Gaden, Jr. (Editor). Biotechnol. Bioeng. Symp. *6*. John Wiley & Sons, New York.

WOLF, A. *et al.* 1975. Evaluation of new protein sources from a health point of view. Nahrung *19* (8) 657–667.

WOODARD, J.C. and SHORT, D.D. 1973. Toxicity of alkali-treated soy protein in rats. J. Nutr. *103* (4) 569–574.

YOUNG, V.R. and SCRIMSHAW, N.S. 1975. Clinical studies on the nutritional value of single cell proteins. *In* Single Cell Protein II. S.R. Tannenbaum and D.I.C. Wang (Editors). MIT Press, Cambridge, Mass.

14

Production of Bakers' Yeast

Gerald Reed

HISTORICAL INTRODUCTION

The art of fermenting doughs from cereals was practiced before recorded history. This and the production of liquid, fermented mashes from cereals are closely related processes. It is likely that the liquid from a fermented mash was drunk as a slightly alcoholic beverage, while the semisolid mash was formed into a dough and baked. Even today yeast strains used in the production of ale and bread are those of a single species, *Saccharomyces cerevisiae*. Until well into the middle of the 19th century bakers obtained their yeast from breweries. At that time lager beer strains of *Saccharomyces uvarum* (synonym: *S. carlsbergensis*) were introduced into Central Europe and later the United States. These strains do not tolerate the high osmotic pressure in a dough, and bakers were forced to look for another source of yeast. Distillers' yeasts which are also strains of *S. cerevisiae* perform reasonably well in bread baking, but they are difficult to separate from the distillers' mash. This led to the establishment of a separate industry which produced bakers' yeast on a large commercial scale for sale to bakers and for home baking.

It is entirely possible to produce baked goods by keeping a portion of an actively fermenting bread sponge and by using it to inoculate a new sponge. This is how active fermentations were perpetuated on farms or in bakeries before bakers' yeast became available as a commodity. This method of inoculation from preceding fermentations is still used to a considerable extent in the wine industry, and it is common in the brewing industry (see also Chapters 8 and 9 on wine making and brewing).

There are two important reasons why this method of propagation is not practical for the commercial production of baked goods. Both reasons have to do with the plastic consistency of doughs. Yeast grows very slowly in doughs, and very long fermentation and proof times are required unless the doughs are inoculated with large numbers of yeast cells. Secondly, the semisolid, plastic nature of doughs makes it difficult to store and transfer them for subsequent use, and to mix them with more flour and water. Table

14.1 shows that the number of yeast cells per gram of dough exceeds that of a starting beer or wine fermentation by a factor of about 50. During fermentation, there is a 5- to 10-fold growth in the number of yeast cells in beer and wine, while there is scarcely any growth in dough during the short period of a bakers' fermentation. These differences explain the requirement of the baking industry for a large-scale commercial source of yeast.

The earlier bakers' yeast plants, well into the 20th century, produced a mixture of alcohol and bakers' yeast from grain mash fermentations. Both alcohol and yeast were sold. Beginning with the early decades of the century, yeast producers used higher aeration and, from about 1920, incremental feeding of their fermentations. Such highly aerobic, incrementally fed (fed batch) fermentations produce more yeast and very little alcohol so that ethanol is not recovered any longer. During the past 100 years production has shifted from a typical distillers' fermentation to a highly aerobic fermentation of molasses worts which is now characteristic of the production of bakers' yeast.

OUTLINE OF THE PRODUCTION OF BAKERS' YEAST

The principal carbon and energy source for the production of bakers' yeast is cane or beet molasses. The nitrogen sources are ammonia, ammonium salts, and urea, and the phosphorus source is ortho-phosphates or phosphoric acid. The fermentation medium is also supplemented with minerals (magnesium and trace minerals) and vitamins (biotin and thiamin).

The final trade fermentations are carried out under highly aerobic conditions, and with incremental feeding of the molasses wort. This fermentation is carried out at a pH between 4 and 6, at a temperature of 30°C, and for periods of 8 to 20 hr. The multiplication of yeast cells is 5- to 10-fold, and the concentration of yeast solids may reach 4 to 6% at the end of the fermentation period.

After the fermentation the yeast cells are concentrated by centrifuging to a yeast cream of 15 to 20% solids. This cream is cooled and it is either pressed in a filter press or filtered in a rotary, vacuum filter. The resulting yeast press cake is extruded in the form of semiplastic blocks and packaged in wax paper, or it is crumbled and sold in bulk. The slightly moist yeast cakes or the crumbled yeast is called "bakers' compressed yeast." For the production of active, dry yeast the press cake is extruded in the form of fine strands. These are dried in tunnel driers on endless steel mash belts, in rotary louver driers, or in air lift driers. Compressed yeast has a solids content of about 30%; active dry yeast contains 90 to 95% yeast solids.

YEAST STRAINS

Bakers' yeasts are strains of *S. cerevisiae*. They are propagated by pure culture methods in the laboratories of yeast producers. Suitable strains are also available from known public culture collections (see Chapter 3). In contrast to the brewing, distilling, and wine industry, the yeast strains used

TABLE 14.1. CHARACTERISTICS OF YEAST FERMENTATIONS

Fermentation Characteristic	Bread Dough	Lager Beer	Table Wine	Whiskey
Raw material	flour, sugar	wort	grape juice	cereal mash
Fermentation time	1–5 hr	8–10 days	5–10 days	3 days
pH, at start	5.5	5.5	3–4	5.0
at end	4.8	4.5	3–4	4.0
Yeast cell count[1]				
at start	275	6–10	5–10	5–10
at end	300	30–70	50–150	50–150
Final ethanol concentration, %	2–3	4	11–13	7–9

Source: Modified from Reed (1974).
[1]Cell counts in millions per ml or per g.

by bakers are not proprietary strains. That is, they are freely available to anyone wishing to make single cell isolates for his own culture collection. The only protection which a producer could obtain for a proprietary strain would be through appropriate patents. For almost all of the commercially available yeasts such patent protection has not been sought.

Within the last 10 years there have been some exceptions to this general rule. In 1969 Lodder *et al.* described the hybridization of bakers' yeast strains, and several patents have since been issued which specifically aim to protect such hybrids (Lodder 1968; Langejan and Khoudokormoff 1976). One of the major problems in improving yeast strains by hybridization is the difficulty of evaluating many hundreds of hybrids with regard to all of the qualities required by a bakers' yeast. Some of these qualities are related to the production of such yeasts: yield, growth rate, stability on storage of the compressed yeast, ability to withstand drying. Others are related to performance of such hybrids in the bakery: gas production rates in lean doughs, in doughs for normal white pan bread, in sweet doughs or in frozen doughs.

Attempts have also been made to hybridize *S. cerevisiae* with *Saccharomyces rosei* in order to confer better osmotolerance to the yeast. This is particularly important if such yeasts are to be used in cookie doughs which are now made with chemical leavening (Windisch and Schubert 1973). At this time such hybrids are not used commercially.

RAW MATERIALS

Carbon and Energy Sources

The most widely used carbon sources are cane and beet molasses with a fermentable sugar concentration between 50 and 55%, and a Brix of about 80°. Table 14.2 shows the constituents of molasses as a percentage of total solids. Since molasses is a by-product of the sugar industry there may be considerable variations in composition. Yeast manufacturers have no control over the composition of molasses and are rarely able to choose the raw material on the basis of known performance.

The pH of molasses is somewhere in the range of 6.5 to 8.5. Yeast ferments and assimilates sucrose as rapidly as invert sugar (glucose plus fructose) because of its high invertase activity. The fructose moiety of the trisaccharide, raffinose, is fermented by bakers' yeast. The residual melibiose moiety is fermented by some strains and not by others.

Other compounds in molasses, such as acetic acid, lactic acid, succinic acid, tartaric acid, and glycerol, can be assimilated in the presence of mono- and diglycerides (Kautzmann 1969A). Some amino acids can serve as both carbon and nitrogen sources.

Nitrogen Sources

The nitrogenous compounds in cane or beet molasses cannot be relied upon to serve as an adequate source since only some of them are assimi-

TABLE 14.2.   COMPOSITION OF MOLASSES AND DESUGARED MOLASSES
(As % of total solids.)

Composition	Cane Sugar			Beet Sugar		
	Molasses %	Alt. 1 & 2 %	Alt. 3 %	Molasses %	High Sugar %	Low Sugar %
Sugars	73.1	27.8	56.3	66.5	21.5	10.0
sucrose	45.5	10.2	9.7	63.5	18.0	6.0
raffinose	—	—	—	1.5	1.5	1.7
invert sugar	22.1	8.1	40.1	—	—	—
other	5.5	9.5	6.5	1.5	2.0	2.3
Organic	15.5	40.8	24.7	23.0	54.5	62.0
GA & PY[1]	2.4	6.4	3.9	4.0	9.4	10.8
other N	3.1	8.3	5.0	—	—	—
other amino acids	—	—	—	3.0	7.0	8.1
betaine	—	—	—	5.5	15.5	17.8
organic acids	7.0	18.8	11.4	5.5	13.8	14.9
pectin, etc.	2.7	7.3	4.4	5.0	9.6	10.4
Inorganic	11.7	31.4	19.0	10.5	24.0	28.0
K_2O	5.3	14.2	8.6	6.0	14.0	16.1
Na_2O	0.1	0.3	0.2	1.0	2.3	2.6
CaO	0.2[2]	0.5	0.3	0.2	0.6	0.7
MgO	1.0	2.7	1.6	0.2	0.4	0.5
$Al_3O_3 \cdot Fe_2O_3$	—	—	—	0.1	0.25	0.3
SiO_2	—	—	—	0.1	0.25	0.3
Cl	1.1	2.9	1.8	1.7	3.4	3.9
$SO_2 + SO_3$	2.3	6.2	3.7	0.5	1.2	1.4
P_2O_5	0.8	2.2	1.3	0.1	0.2	0.2
N_2O_5	—	—	—	0.4	0.9	1.0
others	0.9	2.4	1.5	0.2	0.5	1.0

Source: Hongisto and Laakso (1978).
[1]Means glutamic acid plus pyrrolidine carboxylic acid.
[2]Means after Ca precipitation.
Alt. means alternate process.

lated. This fraction varies with the different types of molasses. Ammonia nitrogen and that of many of the amino acids is assimilated. Glutamic and α-aminobutyric acid can serve as sources of assimilable nitrogen; aspartic acid, alanine, glycine, and lysine can serve as partial sources (Kautzmann 1969B).

In practice most of the required nitrogen is supplied by added ammonium salts, liquid ammonia, or urea. The nitrogen of nitrates is not assimilated.

Vitamins

Bakers' yeast requires biotin for growth, and compressed yeast contains about 0.75 to 2.5 ppm of this vitamin (dry weight basis). Cane molasses supplies ample amounts of biotin (0.5 to 0.8 ppm); beet molasses does not (0.01 to 0.02 ppm). Therefore, at least 20% of cane molasses has to be blended with beet molasses in the preparation of the feed wort, or the feed has to be supplemented with synthetic biotin. At current prices for biotin such additions are economically feasible. If urea is used as a source of nitrogen, higher amounts of biotin are required. L(+)-Aspartic acid can partially replace biotin, and L(+)-aspartic acid plus oleic acid can completely replace biotin in growth media for bakers' yeast (Suomalainen and Keranen 1963).

Bakers' yeast will adapt to the absence or a deficiency of pantothenate and inositol, but these vitamins are required for optimum growth. They are generally present in sufficient quantities in molasses.

For optimum growth it is also advisable to supplement the thiamin content of molasses with this vitamin. Thiamin is almost quantitatively taken up by bakers' yeast during growth. Sufficient thiamin is usually added to the medium to obtain a content of 50 to 10 μg per g of final yeast solids because it improves the activity of compressed yeast in dough systems. Other vitamins are present in molasses in sufficient quantities or they are not needed for yeast growth. Literature references on the requirements of bakers' yeast for various vitamins are usually based on the concentration of the vitamins in the growth media but fail to give the amount of yeast grown. Therefore, they are not particularly helpful for the practice of commercial fermentations.

Minerals

For growth and good performance in fermentations, bakers' yeast requires the addition of phosphates. The amounts added should give a final composition of the yeast of 2.5 to 3.5% P_2O_5 for yeasts containing 7 to 9.5% nitrogen (all based on dry weights). Phosphates are almost quantitatively taken up by the yeast during growth. The common sources of phosphorus are phosphoric acid, alkali phosphate salts, or ammonium phosphate. The latter can also serve as a source of nitrogen.

Molasses contains sufficient potassium to supply the requirements of yeast for this element. The same is generally true of calcium, but molasses has to be supplemented with a magnesium salt, generally magnesium sulfate. Molasses contains sufficient sources of sodium and sulfur to supply these elements. Yeast ash contains 0.4 to 0.5% sodium as NaO_2 and 0.2 to 0.25% sulfate as SO_3. If sodium chloride is added to yeast cream to aid in its filtration, the sodium concentrations may be somewhat higher.

Bakers' yeast also requires the presence of some elements in trace amounts. These are Fe, Zn, Cu, Mn, and Mo although the information in the scientific literature leaves some doubt as to whether these are the only trace elements required. As with vitamins, requirements are generally expressed in terms of nutrient concentration in the medium without reference to the amount of yeast grown. Therefore, quantitative interpretation is difficult. In general such trace elements are supplied in sufficient quantity by molasses with the possible exception of zinc. This metal may be added in the form of zinc sulfate. Table 13.4 in Chapter 13 shows the trace metal requirements of *Candida utilis* yeast grown on ethanol with a zinc requirement of 2 to 6 mg per 100 g of yeast solids.

Fermentation Activators and Inhibitors

Many products have been reported to be activators of yeast growth, such as flour milling waste, sludge from aerobic digesters, etc. It is probable that such reports are based on the stimulatory effects of these materials in growth media which have been deficient in one or another vitamin or trace element. Occasionally, well defined plant growth factors, such as indolyl acetic acid, have been reported as yeast growth stimulants (Jakubowska and Wlodarczyk 1969), but as far as is known these are not used on a commercial scale.

SO$_2$ inhibits yeast growth but concentrations up to 800 ppm in molasses can be well tolerated (Bergander 1969). *S. cerevisiae* adapts well to the presence of even higher concentrations of SO$_2$ as is known from the use of this species in the wine industry where fermentations are often carried out in the presence of 80 to 100 ppm of SO$_2$.

Molasses contains variable amounts of nitrate which can be reduced to nitrite by bacterial action during the production of yeast. A considerable loss in yield has been reported for concentrations of 0.004 to 0.001% nitrite (Notkina *et al.* 1975).

Other Carbon and Energy Sources

Any sugar-containing raw material or any starchy material that can be hydrolyzed to fermentable sugars may serve as a carbon and energy source for the production of bakers' yeast. These sugars are sucrose, maltose, glucose, fructose, and mannose. Lactose is not fermented by bakers' yeast, and galactose is fermented only very slowly. Such sugar-containing raw materials may be sugar cane juice or molasses, grape juice concentrates, date juice, wood hydrolysates, starch hydrolysates, or waste sulfite liquor. Up to the present time economics have dictated the use of molasses. Waste sulfite liquor is used to some extent in Finland. This liquor from paper pulp mills contains a mixture of hexoses and pentoses at very low concentrations. *S. cerevisiae* assimilates only the hexoses, and consequently very large volumes of liquor have to be passed through the fermentors (Piš 1970).

During the past 5 years the economic picture has changed so that the use of molasses is not as attractive as it has been in the past. There are two basic reasons for this change, both of a technical nature. The first is the improved recovery of sucrose from beet or cane juice which leaves a molasses with a lower concentration of fermentable sugars and with a higher concentration of compounds which are of no value to the yeast producer. In Europe molasses is now available with a fermentable sugar concentration of 40% as compared with a traditional concentration of 50–55%. A newly developed process is capable of removing a major portion of sucrose from molasses. This results in a substrate with very low concentrations of fermentable sugars and high concentrations of inorganic and organic (non-sugar) compounds (Hongisto and Laakso 1978). This is shown in the second, third, fifth, and sixth columns of Table 14.2 (see also Huncikova 1976).

The second problem is the high BOD of fermentor effluents if molasses is used as fermentation substrate. Obviously, this problem will be further aggravated if the fermentable sugar content of molasses is reduced and that of organic non-sugar compounds is increased. Together these problems have rekindled interest in the use of alternate carbohydrate sources for the production of bakers' yeast. At present the only available sources which are technically acceptable and economically attractive are starchy materials from cereal grains, principally from corn (maize).

Grain mashes were used traditionally for the production of bakers' yeast until the 1930s, and high quality bakers' yeast can be produced with them. Of course, starchy grains require conversion to fermentable sugars, a task which can be carried out with bacterial and fungal amylases more efficiently than in the early decades of the century. Alternately, corn syrups of high dextrose equivalent can be used as a suitable carbon source.

Ethanol is aerobically assimilated by bakers' yeast after a period of adaptation (Drews and Hessler 1967; Krajovan et al. 1971). However, it is not a satisfactory raw material for the production of bakers' yeast.

Losses of ethanol through aeration of the fermentor are negligible for bakers' yeast fermentations which produce less than 0.1% ethanol in the fermentation medium (Starik et al. 1975).

PRINCIPLES OF AEROBIC GROWTH

Growth Rate

For most anaerobic batch fermentations the increase in yeast cell mass follows a predictable pattern. At the beginning of the fermentation there is slow growth and fermentation may be barely perceptible. This early lag phase is followed by a period of exponential growth, that is, growth where cell divisions take place at identical intervals. This phase is often called "stormy" in the beer and wine industry. Finally, the exponential growth phase is followed by the latent phase in which growth declines or stops completely. Fermentation continues during the latent phase, although at a reduced rate, until the amount of substrate is exhausted. This pattern is well known from the fermentation of alcoholic beverages.

In contrast, aerobic, continuous fermentations show exponential growth throughout the length of the fermentation. For such fermentations the increase in yeast cell mass can be expressed as the generation time or the specific growth rate constant for exponential growth. Generation time is simply the time required for each doubling of the yeast population. The specific growth rate constant (μ) is defined by the following equation

$$\frac{dP}{dt} = \mu \times P$$

in which P is the mass of yeast and t is time. The equation can be written

$$\frac{dP}{P} = dt \times \mu$$

Upon integration, one obtains $\ln (P_t/P_0) = \mu \times t$; and for $t = 1$ hr, one obtains $\mu = \ln (P_t/P_0)$ or $2.31 \times \log_{10} (P_t/P_0)$. This is the specific growth rate for exponential growth, and for continuous fermentations its value is identical with that of the dilution rate. The relation of μ to generation time is expressed by the following equation

$$\text{Generation time (hr)} = \frac{2.303 \times \log_{10} 2}{\mu}$$

Table 14.3 shows some of the actual values spanning a practical range.

TABLE 14.3. RELATION OF THE GENERATION TIME TO THE SPECIFIC GROWTH RATE CONSTANT

Generation Time in hr	P_t/P_0	Specific Growth Rate Constant μ
1	2.00	0.693
2	1.41	0.345
3	1.26	0.230
4	1.19	0.173
5	1.15	0.139

P_t means mass of product at time $= 1$ hr.
P_0 means mass of product at time $= 0$.

Commercial yeast fermentations are so-called "fed batch" fermentations. They are carried out with incremental feeding of the substrate for growth. There is no simultaneous removal of the fermentor content, and the fermentation must be finished when the fermentor is full. The total time for such a fermentation is generally between 8 and 20 hr. In the course of the fermentation, the growth rate drops and the generation time increases from 3 to 5 hr and finally to 7 hr. This leads to an 8-fold multiplication of yeast cells in a 15 hr fermentation period.

Oxygen Requirement and Aeration

The maximum theoretical yield coefficient under anaerobic conditions is $Y_s = 0.075$, which means a yield of 7.5 kg of yeast solids per 100 kg of fermentable sugar. Under strictly aerobic conditions the highest possible yield is $Y_s = 0.54$. The purpose of a bakers' yeast plant is the production of cell mass, and, therefore, fermentations are carried out under aerobic conditions which maximize yield.

However, two additional conditions have to be met to obtain maximal yields. The growth rate, μ, must not exceed values of about 0.2, and the amount of substrate present at any one time must not exceed a given limiting value. Figure 14.1 shows the effect of dilution rate (or specific growth rate) on yeast yield (Dellweg *et al.* 1977). Strikingly similar results have been obtained by Meyenburg (1969) except that his yields did not drop until a value of $\mu = 0.23$ had been reached. At growth rates below 0.2 the respiratory quotient (RQ) $= Q_{CO_2}/Q_{O_2}$ is about 1. At higher growth rates carbon dioxide development is greatly accelerated, the RQ increases rapid-

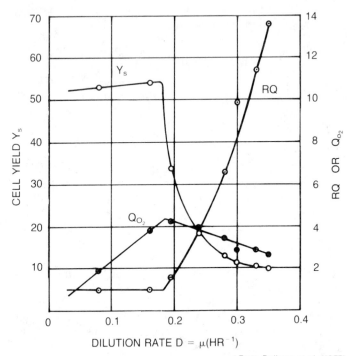

From Dellweg et al. (1977)

FIG. 14.1.   EFFECT OF DILUTION RATE ON YEAST METABOLISM IN
CONTINUOUS CULTURE
Y_s—Cell yield in g solids per 100 g of substrate.
Q_{O_2}—Respiration rate in moles per g yeast solids per hr.
RQ—Respiratory quotient.

ly, and ethanol is formed from the fermentable sugar. This state is called
"aerobic fermentation." For practical purposes it is, therefore, necessary to
keep growth rates below 0.20.

The amount of oxygen required for yeast growth is in the neighborhood of
1 g of O_2 per g of yeast solids (Harrison 1967; Mateles 1971). The elemental
composition of yeast has been determined by Harrison (1967) to correspond
to the formula $C_6H_{10}NO_3$. The elemental composition C = 45%, H = 6.8%, N
= 9.0%, O = 30.6% as determined by Wang *et al.* (1977) approximates the
above formula when translated into percentages by weight. The required
oxygen may be supplied by pure oxygen gas, or in the form of hydrogen
peroxide, but in practice it is always supplied by blowing air through the
liquid fermentor contents.

The capacity of an aeration system to transfer oxygen from air into the liquid phase is expressed in terms of its volumetric oxygen transfer coefficient, $K_L a$ (hr^{-1}), where K_L is the oxygen transfer coefficient (m/hr) and a is the interfacial area between air bubbles and liquid per unit volume of liquid (m^2/m^3 or 1/m). In actual practice, the rate of oxygen transfer in a fermentation system is often expressed in terms of millimoles O_2/liter-hr, which is equal to $K_L a \times C^*$, where C^* is the equilibrium concentration of dissolved oxygen in the liquid phase (millimoles O_2/liter). One can readily calculate that an oxygen transfer rate of 140 mM/liter-hr is required to produce 4.5 g of yeast solids per liter per hr.

The efficiency of aeration systems is quite variable. For fermentors without agitators the efficiency of oxygen utilization may not be higher than 20%; that is, for incoming air with an oxygen concentration of 22% the exit air would have an oxygen concentration of 17%. That means that the amount of air blown through the fermentor would have to be 5 times the amount that can be calculated from the required oxygen transfer rate. Better dispersion of the incoming air, and hence smaller bubbles and a larger bubble surface area, can be obtained by mechanical agitation at the point at which the air enters the fermentor. With such systems 40 to 50% of the available O_2 can be transferred to the liquid phase and utilized by the yeast. Such systems have been described by Aiba et al. (1965).

Oxygen transfer rates are often determined by the so-called sulfite oxidation method (Cooper et al. 1944). This measures the rate of oxidation of a solution of sulfite to sulfate in the presence of a metal catalyst. The method is very useful for determining the effect of variables (agitator speed, addition of surfactants, etc.) on oxygen transfer in a given fermentor. Its usefulness for predicting the performance of aerating systems for producing a given amount of yeast is somewhat limited because the rate of oxygen transfer is affected by the catalyst used (Co or Cu) as well as by other extraneous factors (Linek and Beneš 1978; Reith and Beek 1973). The addition of surface active agents affects the oxygen transfer rate and the effect on the oxygen transfer coefficient (K_L) may differ from that on the bubble surface area (a). This is another reason why actual aeration requirements cannot be accurately predicted from sulfite oxidation measurements. They have to be determined by varying the aeration rate during actual fermentations and determining the effect of these variations on yeast yield.

The level of dissolved oxygen during a fermentation can be determined in the fermentor by oxygen electrodes which are commonly available. They can be calibrated by aerating the fermentor medium in the absence of yeast and assuming an oxygen concentration of 7.7 ppm at 30°C under these conditions. During an actual fermentation these values may be considerably lower and often below 0.4 ppm (or about 5% that of the saturation level). Figure 14.2 shows some of these values as they vary throughout a fermentation for an air sparged, nonagitated fermentor with an oxygen transfer capability of 150 mM per liter per hr.

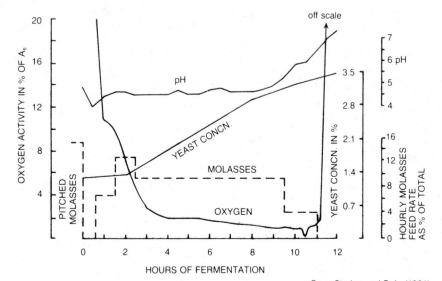

FIG. 14.2. SUPERIMPOSED PLOTS SHOW RELATION OF RAW MATERIAL FEED, YEAST CONCENTRATION, pH, AND DISSOLVED O_2 ACTIVITY IN PILOT PLANT SPECIAL *S. CEREVISIAE* PROPAGATION

If the fermentor is aerated with gas mixtures containing a very high percentage of oxygen, there is a disturbance of yeast metabolism. Glucose consumption per g of yeast produced increases, ethanol is formed, and the ability of the yeast to ferment and to respire glucose is impaired. Figure 14.3 shows such conditions for various percentages of oxygen in the air supply. The most favorable conditions for yeast growth are obtained at oxygen levels between 21 and 30% in the gas. At such levels the RQ approaches 1, ethanol production is negligible, and yield is maximized (Oura 1974). Dellweg *et al*. (1977) have suggested several reasons for the effect of hyperbaric oxygen pressures on yeast fermentations.

Up to this point it has been assumed that oxygen transfer from the gas to the fermentor liquid is the limiting step in oxygen transfer rates and in the rate of oxygen uptake by yeast cells. The effect of bulk mixing on microbial systems has rarely been studied. Einsele *et al*. (1978) worked with continuous, aerobic fermentations of bakers' yeast. Mixing times were determined as the response of intracellular NADH to a glucose step change during the continuous fermentation. The response time of this particular system was 4.4 sec. This indicates that bulk mixing times are only important if they are long in relation to the response time of the yeast cells.

Concentration of Fermentable Sugars

It is well known that the rate of glucose fermentation, that is, the production of ethanol and CO_2, is faster under anaerobic conditions than under

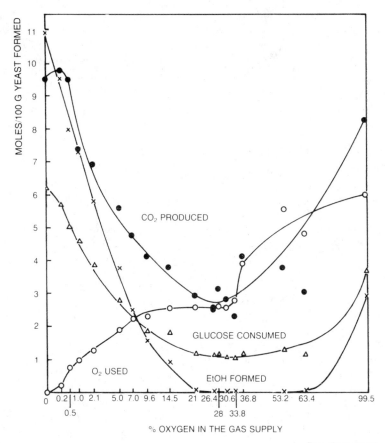

From Oura (1974)

FIG. 14.3. EFFECT OF THE INTENSITY OF AERATION ON THE USE OF OXYGEN (O—O), ON THE CONSUMPTION OF GLUCOSE (△—△), AND ON THE PRODUCTION OF ETHANOL (x—x) AND CARBON DIOXIDE (●—●) DURING CONTINUOUS CULTURE OF BAKERS' YEAST

aerobic conditions. This is called the Pasteur effect. The presence of glucose at higher concentrations inhibits respiration even in the presence of excess oxygen (Franz 1964; Görts 1967). This is also expressed in a decrease of the levels of enzymes of the electron transport system and of the citric acid cycle (Suomalainen 1969; Rogers and Stewart 1973). The respiratory coefficient is a sensitive tool for determining the degree of glucose repression of respiration. Wang *et al.* (1977) have calculated the percentages of glucose which are fermented or respired for different values of RQ. Dellweg *et al.* (1977) report that "a yeast culture growing at 1.1 mM glucose (0.2 g/liter) shows the highest respiration rate. It is, of course, lower at lower glucose concentrations, but above this concentration oxygen uptake is again diminished as the result of the so-called glucose effect" (see also Rikard and Hogan 1978). The carbon source, glucose or fructose, must therefore be provided in very small concentrations. This can be achieved by incremental feeding as a sugar or molasses solution. In practice maximal yields are obtained by limiting the growth rate to $\mu = 0.2$ or less, and by incremental feeding which gives an RQ of approximately 1 and which reduces ethanol formation to negligible levels.

pH and Temperature

Bakers' yeast can tolerate a fairly wide range of hydrogen ion concentrations. It can be grown at pH levels between 3.6 and 6 with optimum levels between 4.5 and 5. At lower pH levels contamination by bacteria is minimized. But the adsorption of coloring material from the molasses substrate at low pH values must be considered. Therefore, most commercial fermentations start at low pH levels. At the end of the fermentation the pH is raised by the addition of ammonia or alkali. Kautzmann (1969C) suggests a starting pH of 4.2 to 4.5 and a final pH of 4.8 to 5.0. The pH values of a commercial fermentation are also shown in Fig. 14.2. Eroshin *et al.* (1976) determined optimum yields with *S. cerevisiae* (not necessarily a bakers' strain). In continuous fermentations with a growth rate of $\mu = 0.1$, he found optimum yields at a pH of 4.1.

White (1954) determined the generation times for bakers' yeast growth at various temperatures and found the following values: 5 hr at 20°C, 3 hr at 24.5°C, 2.2 hr at 30°C, 2.1 hr at 36°C, 4 hr at 40°C, and approximately 8 hr at 43°C. Similar results have been obtained by Keszler (1967) with a top fermenting brewers' yeast strain of *S. cerevisiae*. But temperature optima for maximum yield are lower. Eroshin *et al.* (1976) obtained maximum yields at 28.5°C. In practice yields are optimized by constant fermentation temperatures between 28° and 30°C.

Yield and Development of Heat

Yield of bakers' yeast is best expressed as the yield coefficient, Y_s, that is, grams of yeast solids per gram of substrate used. For practical purposes this value is about 0.5. Several investigators have reported somewhat higher values. Chen (1959) and Chen and Gutmanis (1976) obtained a yield coefficient of 0.5; Oura (1974) reported a value of 0.52 and Dellweg *et al.* (1977) a value of 0.54.

A certain amount of substrate is used for the maintenance of the metabolism of the cells. For fermentations with incremental feeding Wang *et al.* (1977) found that 0.08 g of sugar was required by 1 g of yeast cell solids per hr for maintenance. Since the maintenance requirement is probably independent of growth rate, one can expect higher yields at higher growth rates, at least up to a growth rate of $\mu = 0.18$. At higher growth rates yields are reduced because aerobic fermentation interferes with yeast growth.

In the trade, yields of bakers' yeast are often expressed in terms of compressed yeast as a percentage of the molasses used. This makes sense for practical purposes of production costs since molasses is bought and compressed yeast is sold on a per kg (lb) basis. But it makes it difficult to judge the efficiency of the operation since the sugar content of molasses and the solids content of compressed yeast vary. In the United States compressed yeast generally has a solids content of 30%, while in Europe it is often assumed to be 27 or 28%.

Osmotic Pressure

The effect of high osmotic pressure in inhibiting yeast growth and fermentation is well known. Most investigations have dealt with the ability of various yeast species to ferment foods with high sugar concentrations such as jams and jellies. With regard to baked goods Windisch *et al.* (1976) have investigated the effect of very high osmotic pressures in cookie doughs on yeast fermentation. *S. cerevisiae* is not an osmophilic yeast, yet there are appreciable differences among strains of bakers' yeast. It is also known that the conditions of growth affect the osmotolerance of the yeast.

White (1954) found that bakers' yeast grown with incremental feeding is more osmotolerant than yeast grown in set fermentations. He also described the effect of growth rate on osmotolerance. The addition of salt decreases the specific growth coefficient of bakers' yeast (Watson 1970).

The conditions affecting osmotolerance in bakers' yeast have not been studied sufficiently. This is unfortunate because the osmotolerance of bakers' yeast has assumed greater importance with higher sugar concentrations in modern bread formulations and with the production of high sugar sweet goods.

Yeast Concentration in the Fermentor

In commercial practice yeast concentrations of 4 to 6% (yeast solids) are obtained. But growth rate decreases rapidly at concentrations exceeding 3 to 4%. Fries (1962) found a generation time of 4 hr at the beginning of a commercial fermentation and of 8 hr at the end. Many reasons have been advanced for this drastic reduction in growth rate, for instance, the exhaustion of growth factors, the inability of nutrients (sugar, oxygen) to reach the surface of the yeast cells due to increased viscosity or inadequate mixing, increasing concentrations of inhibitors which may be present in molasses, increased osmotic pressure, and others.

In particular it has been suggested that the increase in CO_2 concentration in the fermentor accounts for the decrease in growth rate (Karolus 1974). However, Chen and Gutmanis (1976) tested this hypothesis by growing bakers' yeast under conditions approximating those of a commercial fermentation. Aeration was carried out with gas streams containing 21% oxygen but varying concentrations of CO_2. Inhibition of yeast growth was negligible below 20% CO_2 in the gas mixture, and slight inhibition was found at the 40% level.

Actually, one can grow bakers' yeast to concentrations exceeding 10% yeast solids in experimental fermentations provided a sufficient transfer of oxygen can be achieved. Other factors may, of course, play a role at higher yeast concentrations. At a concentration of 10% yeast solids the cell volume may occupy 25% of the liquid fermentor volume, and at concentrations of 22 to 23% solids the mass cannot be pumped any more. That means that there is certainly a practical limit to the concentration of yeast solids that can be achieved in the fermentor.

Periodicity and Budding

During the propagation of bakers' yeast the supply of molasses is reduced toward the end of the fermentation period in order to "mature" the yeast. The maturing period results in a compressed yeast with somewhat lower fermentation activity but better storage stability, and in a yeast with reduced numbers of budding cells. A strict comparison of the glycolytic activity of bakers' yeast cells at various stages of the cell cycle has not been undertaken except by Meyenburg (1969).

Extensive work with *Schizosaccharomyces pombe* has shown that this yeast can be grown in synchronous culture, and that the rate of glycolysis increases in a linear fashion between the synchronous divisions and with a doubling of the rate of increase at each division (Kramhøft *et al.* 1978). It is unlikely that synchronous growth can be induced in *S. cerevisiae* as readily as in *S. pombe*, at least in a manner that would permit industrial application. Figure 14.4 shows the results that can be achieved if one starts with an inoculum of yeast cells which are at an identical stage of the cell cycle. After several cycles of growth, periodicity is less pronounced and finally abolished. The figure does indicate a pronounced change in the respiratory quotient at various stages of the cell cycle (Wiemken *et al.* 1970).

PRACTICE OF THE AEROBIC GROWTH OF BAKERS' YEAST

Preparation of Substrate

Molasses is received at about 80° Brix and with a fermentable sugar concentration of 50 to 55%. For use in fed batch fermentations it is diluted and clarified. Clarification is required to remove insoluble solids which would otherwise be carried into the compressed yeast, and in order to lighten the color of the final yeast. Molasses is generally diluted to a Brix of 40° and the pH is adjusted to about 5 with acids such as sulfuric acid. Insoluble solids are removed in a desludger centrifuge, that is, a solid bowl centrifuge with intermittent discharge of the sludge. Beet molasses may also be clarified by filtration but cane molasses is difficult to filter. The clarified molasses is then sterilized in a high-temperature short-time heat exchanger and cooled (Rosen 1977).

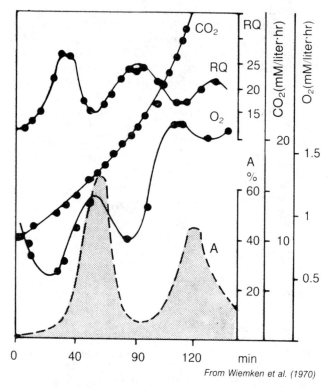

From Wiemken et al. (1970)

FIG. 14.4.   GAS EXCHANGE IN A SYNCHRONOUS CULTURE OF
SACCHAROMYCES CEREVISIAE (GENERATION TIME ABOUT 70
MIN). O_2 UPTAKE AND CO_2 RELEASE ARE EXPRESSED IN mM
GAS PER LITER OF CULTURE MEDIUM PER HR
A—Percentage of initial budding cells.

It is important that the molasses not be heated too long to prevent
darkening and the destruction of sugars. This is also true of the diluted
molasses wort which should not be stored at elevated temperatures before
use in the fermentation.

Other substrates do not require particular preparation and can be fed
from individual feed tanks to the fermentor.

Fermentation

Fermentation Tanks.—The construction material is generally stainless
steel and in modern factories wort storage tanks, tanks for storage of other
liquid nutrients, and all piping are also of stainless steel. The size of
commercial fermentors may vary greatly but fermentors for the final trade
fermentation with a volume of 200 m^3 or more are common. One of the

critical considerations in the design of a fermentor is the ratio of height to diameter. Oxygen transfer is greatly improved with greater height of the liquid column, but for very tall fermentors the capital investment for a given volume is higher. Above a liquid height of 3 to 5 m the air must be supplied by compressors to overcome the hydrostatic pressure of the liquid column. For squat fermentors air blowers or self-priming aerators may be used.

Bakers' yeast fermentations need not be carried out under conditions of absolute sterility because of the low pH. Therefore, the fermentation vessel need not be pressurized which lowers the cost of production in comparison with fermentors of equal size used in the production of enzymes or antibiotics.

Aeration.—It must be re-emphasized that the supply of oxygen is the most critical factor for the construction of the fermentor and for the conduct of the fermentation. It is relatively easy to add molasses, acids, ammonia, minerals, or vitamins to a fed batch fermentation, but it is difficult and costly to add oxygen, and this nutrient is usually the limiting factor in achieving high productivity. High oxygen transfer rates can be achieved with specially designed aeration equipment. Such fermentors are shown, for instance, in Fig. 13.2 and 13.3 of Chapter 13.

In the United States, simpler systems of supplying oxygen are generally used. Air is fed into the fermentor through perforated tubes at the bottom of the tank. A suitable arrangement is that shown in Fig. 14.5 (DeBecze and Liebmann 1944). Rosen (1977) describes such a stationary system which consists of a horizontal tube with 24 side tubes provided with 30,000 holes of a diameter of 1.5 mm. It is generally believed that holes of very small diameter are required to create air bubbles of small diameter and hence to improve oxygen transfer. This view has been questioned since the diameter of air bubbles in a turbulent liquid does not depend on the diameter of the air outlet orifice once the air bubble has reached a distance of about 10 cm from that orifice.

Distribution of air bubbles may be assisted by internal agitators and most United States fermentors now have such aeration systems. The use of agitators lowers the requirement for air volume but adds to the cost of the equipment and may have a slightly higher total energy requirement. Air requirements are often stated in terms of volume of air per fermentor volume per min (VVM). Such figures are not meaningful in comparing oxygen transfer capabilities of various fermentors since this depends also on bubble size and dwell time of the air bubbles. The critical factors which determine the oxygen transfer capability and hence productivity of the fermentor are the oxygen transfer rate and the density of the mixture of air and fermentor liquid. The latter is, of course, a measure of that portion of the fermentor volume occupied by liquid. Fermentors with simple air spargers have lower oxygen transfer rates but a higher usable fermentor volume than systems in which the air bubbles are finely dispersed and have a long dwell time (see also the discussion of aerobic fermentors in Chapter 13 and Hoehne (1975) for bakers' yeast fermentations).

AIR

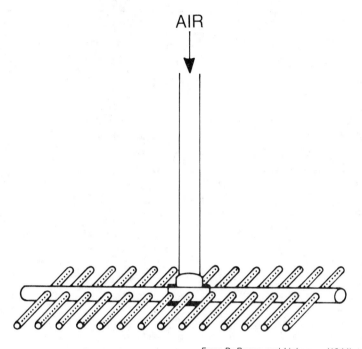

From DeBecze and Liebmann (1944)

FIG. 14.5.   NETWORK OF PERFORATED AIR TUBES

Cooling.—Cooling can be carried out with internal cooling coils or by means of external heat exchangers. Bakers' yeast fermentations are carried out at 28° to 30°C. The cooling requirements are great since about 3.5 kcal are generated per g of yeast solids produced (aerobically).

Defoaming.—There is considerable foaming in highly aerated bakers' yeast fermentations. This can be reduced by addition of suitable anti-foaming agents which may be silicones, fatty acid derivatives, or other edible surface active materials. Antifoam agents affect oxygen transfer in a complex manner. They tend to depress oxygen transfer coefficients but increase the total interfacial area between air bubbles and fermentor liquid (Finn 1969).

Feed Rates.—Bakers' yeast fermentations are not carried out under conditions of exponential growth. In a fed-batch fermentation a constant feed rate does not permit exponential growth but can provide only for a constantly diminishing growth rate. Figure 14.6 shows some experimental and some commercial molasses feed rates as a function of fermentation time. For comparison a truly exponential curve has been included. A diminished growth rate does not mean diminished productivity since a large yeast population at lower growth rates may have greater productivity than a

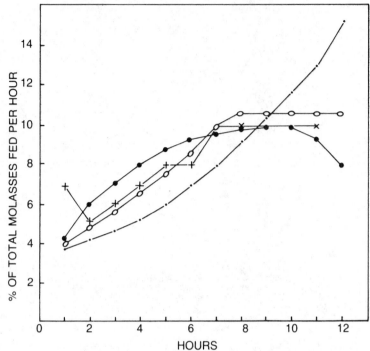

HOURS

From (1) Drews et al. (1962) and (2) Butschek and Kautzmann (1962)

FIG. 14.6.   MOLASSES FEED CURVES (EXPERIMENTAL AND COMMERCIAL)
Exponential, exptl (1)   ● —— ●
Exponential and constant, exptl (1)   X —— X
Commercial (1)   O —— O
Commercial (2)   · —— ·

small population with a high growth rate. Productivity is here meant to be "grams of yeast solids produced per liter fermentor volume per hour." This is the decisive factor which determines the capacity of a fermentor for the production of bakers' yeast. Productivity varies from a low value at the beginning of the fermentation to higher productivity in the latter stages. On the average one can expect a productivity of 3 g/liter-hr for a 15 hr fermentation in which yeast solids concentration increases from 0.6 to 5.1%.

The time of a bakers' yeast fermentation may vary from 10 to 13 hr for a 4-fold multiplication and from 16 to 20 hr for an 8-fold multiplication. Shorter fermentations can be started with higher initial yeast concentrations and hence improved productivity during the fermentation. However, the turnaround time of the fermentor (charging, discharging, and cleaning) becomes a higher percentage of the total available time.

Toward the end of the fermentation the feed rate is sharply reduced in order to permit "maturing" of the yeast cells. This maturity is expressed in a low percentage of budding yeast cells and greater stability of the compressed yeast on storage. Panek (1975) has provided a more convincing rationale for the physiological changes during this maturation period.

In the commercial production of bakers' yeast ethanol formation is minimized in order to maximize cell yield. Ethanol concentrations in the fermentor liquid should be kept below 0.1% and preferably below 0.05%. In that case the yield of bakers' yeast with 30% solids will be 80 to 100% of the weight of the molasses used. As mentioned earlier the yield of yeast solids based on fermentable sugar in molasses is 45 to 50%.

In some European countries it is customary to produce both yeast solids and ethanol in the course of production. This is not economical since low concentrations of yeast solids and low concentrations of ethanol raise the cost of recovery of both materials disproportionally. It may be justified on the basis of legal and economic considerations regarding the production of potable alcohol.

Sequence of Fermentations.—The bakers' yeast fermentations which have been discussed in the preceding sections have always been the final or trade fermentation stages. This final stage is, of course, decisive for yield and productivity of a plant, and it is most important for product quality. In actual operation the trade fermentation is preceded by a sequence of smaller fermentations in which the pitching yeast is grown. Table 14.4 shows a typical sequence of such fermentation stages (Suomalainen 1963). The first two stages are called R1 and R2. They are carried out in small scale equipment and under conditions of complete sterility. The R1 fermentor is inoculated from a laboratory grown pure yeast culture (for methods of handling pure cultures see Chapter 3). There is, of course, no incremental feeding during the earlier stages of the fermentation and yields are low. Fermentation times for the early stages may be from 8 to 16 hr.

Beginning with the third stage, R3, the fermentors are not pressurized and a low level of contaminants begins to appear. At the end of the F3 stage enough pitching yeast has been grown to divide the contents of the fermentor and to pitch 3 additional fermentations. The same is true at the end of the F4 stage. One may also separate the yeast from the F3 stage by centrifuging and storage of the yeast cream for later pitching of the F4 stage. This has two advantages. It increases the flexibility of the plant operation by permitting staggered operation of the final fermentation stages, and it reduces the number of bacterial contaminants. The fermentation stages designated F4 and F5 are highly aerobic and yields and productivities are those indicated previously for trade fermentations. The entire sequence shown in Table 14.4 involves 24 generations of yeast.

Harvesting of Yeast Cells

Yeast cells are separated from the fermentor liquor of the final fermentation by centrifuging in a vertical, nozzle type, continuous centrifuge which develops a G of 4000 to 5000. This is sufficient to affect complete separation of the cells which have a water content of 62 g per 100 g of cells and a density of 1.133 g/cm^3 (Sambuichi et al. 1971). During the first pass through such centrifuges the yeast concentration can be tripled and on additional passes (with or without washing with water) a concentration of 18 to 20% yeast solids can be obtained. This pumpable liquid which has a whitish appear-

TABLE 14.4. SUCCEEDING STAGES IN THE PROPAGATION OF BAKERS' YEAST
The amount of yeast grown in each stage is shown.

Pure Culture # 1 (R1)	Pure Culture # 2 (R2)	Open Tank (R3)	Open Tank (F1)	Open Tank (F2)	Open Tank (F3)	Open Tank (F4)	Open Tank (F5)
							1666–11,000 kg
						833–5000 kg	1666–11,000 kg
							1666–11,000 kg
							1666–11,000 kg
0.2–0.8 kg	0.8–3.5 kg	3.5–25 kg	25–120 kg	120–420 kg	420–2500 kg	833–5000 kg	1666–11,000 kg
							1666–11,000 kg
							1666–11,000 kg
						833–5000 kg	1666–11,000 kg
							1666–11,000 kg
0.8 kg	3.5 kg	25 kg	120 kg	420 kg	2500 kg	15,000 kg	100,000 kg

Source: After Suomalainen (1963).

ance is called yeast "cream." It may be stored for several days at temperatures between 1° and 4°C without detriment to yeast quality.

Yeast from the cream is further concentrated by filtration (or by pressing which accounts for the word "compressed yeast"). Filtration can be carried out with plate and frame filter presses without use of a filter aid. More commonly it is done with rotary, continuous vacuum filters. The filter surface of the rotary filter must be coated with an edible material since very small concentrations of this coating may be found in the press cake. The best material is potato starch or sago palm starch which has a sufficiently large granule size to permit efficient filtration. Such filters produce a press cake of about 27 to 28% solids. Higher solids levels can be achieved on this equipment if the cream is salted just before filtration. The osmotic effect of the salt treatment reduces the moisture content of the cells. The salt is then removed directly on the filter by water sprays (Kuestler and Rokitansky 1960). The resulting crumbly mass of yeast cells is called a press cake. A scientific investigation of the filtration and extrusion of bakers' yeast has been reported by Sambuichi et al. (1974).

The yeast press cake is now mixed in a blender with small amounts of emulsifiers and/or cutting oils. These additives, which may be of the order of 0.1 to 0.2%, facilitate extrusion and provide a better, lighter appearance of the yeast cake. The emulsifiers used are mono- or diglycerides, sorbitan esters, or lecithin (MacDonald and Geisler 1965).

The yeast press cake is now extruded through nozzles in the form of thick strands with a rectangular cross section. These are cut into appropriate lengths to form the well known shape of packaged bakers' yeast with weights of 1 lb or 500 g. The 500 g (1 lb) cakes are wrapped in wax paper and packaged 22.5 kg (50 lb) to a case. Today most of the compressed yeast sold in the United States to wholesale bakers is packaged in the form of a crumbled cake with irregularly shaped particles in 22.5 kg (50 lb) plastic lined bags. The free flow characteristics of such crumbled yeast can be improved by admixture of hydrophobic and hydrophilic additives (Luca et al. 1979).

Starting with a yeast cream with a temperature of 3° to 4°C, the mass warms up through the filtration, mixing, extrusion, and packaging operations. Therefore, immediate and rapid cooling is required after packaging. This is done in refrigerators with vigorous circulation of cold air and with stacking of the cases or bags in such a manner that the air can circulate freely around them. Generally a 24 to 48 hr cooling period is required. Compressed yeast is shipped to bakers in refrigerated trucks and stored in the bakery in the refrigerator until used (Schuldt and Seeley 1966).

It is entirely feasible to ship bakers' yeast in the form of a pumpable yeast cream. Some attempts to introduce this method of distribution have been made but have not been successful, probably because of the requirement for refrigerated storage tanks at the bakery. The shipment of refrigerated yeast creams in Russia has been described by Volkova and Roiter (1973) and Pasivkin (1973).

Stability of Compressed Yeast

The storage stability of compressed yeast at refrigerator temperatures of 5° to 8°C is quite good. To some extent stability depends on processing conditions, for instance, low nitrogen concentrations in the yeast and a low percentage of budding cells (less than 5 to 10%) favor storage stability. In general there is a loss of 3 to 5% of gassing activity over the period of 1 week. During storage there may be some loss of moisture. A small rate of respiration of endogenous carbohydrate leads to a slight decrease in this fraction and to a relative increase in the nitrogen content of stored yeast.

A liquid yeast cream could be kept satisfactorily at 4° to 6°C for 10 days and at 20°C for 1 day (Volkova et al. 1974). For storage at 23°C the activity of Finnish compressed yeast showed no drop in activity but after a 2 week storage period the drop was precipitous. For storage at 35°C about one-third of the gassing activity was lost on 2 days storage. The loss of activity is generally greater for maltose fermentations than for the fermentation of glucose, fructose, or sucrose (Hautera and Lovgren 1975).

Efforts have been made to correlate the degree of instability with changes in cell constituents. The best correlations could be obtained with the trehalose content of yeast. A high trehalose content indicates full "maturation" of the cells which occurs at reduced growth rates toward the end of the fermentation. The higher the trehalose content the better the storage stability of the yeast (Bocharova et al. 1976; Panek 1975).

Compressed yeasts with nitrogen concentrations above 8% (based on solids) have a shorter shelf-life than yeast with nitrogen levels of 7% or below. Therefore, yeasts with high nitrogen levels and high rates of gas production as they are used in the United States are always shipped under refrigeration. Such yeasts have now been introduced into Canada. In Europe where such yeasts are known as "Schnellhefen" (fast yeasts), refrigerated shipment and storage are also recommended. Yeasts which have to be distributed at ambient temperatures or which require an exceptionally long shelf-life (for instance, consumer yeasts) are produced at nitrogen levels between 6 and 7%.

The stability of compressed yeast can also be checked by determining the effect of storage on viability. The most reliable method is a yeast plate count which reveals the number of cells capable of reproduction. Staining techniques with methylene blue are commonly used but their accuracy is questionable. Parkhinen et al. (1976) found good correlation between live cell counts and staining methods if he used mixtures of live cells and cells which had been killed by heating. But if death occurred as the result of storage at 35°C the correlation was poor as shown in Fig. 14.7. The authors obtained better results with fluorochromes, such as primuline, than with methylene blue. Yeasts stored at 5°C lost only 1 to 2% in viability during a 16 day storage period. Storage at 20°C led to the death of 10% of the cells in 16 days.

Contaminants

Since bakers' yeast is not grown under pure culture conditions, it contains various microbial contaminants. The most numerous are lactic acid bacte-

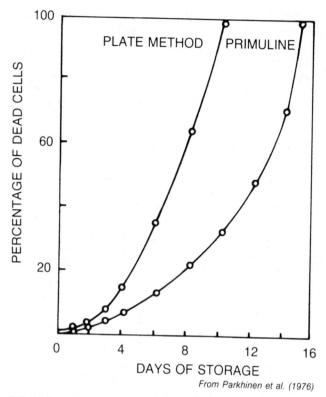

From Parkhinen et al. (1976)

FIG. 14.7. PERCENTAGE OF DEAD CELLS IN BAKERS' YEAST STORED AT 35°C
By the plate method and by the fluorescence method using primuline.

ria belonging generally to the genera *Lactobacillus* or *Leuconostoc*. The total bacterial count which generally reflects the presence of these bacteria is usually between 10^4 and 10^9 cells per g. Carlin (1958) has reported somewhat higher values. These particular contaminants are without influence on the production of bread in normal commercial operation. They are a good source of "sourdough" organisms. In liquid pre-ferments they are reduced to one-tenth their original number (Robinson *et al.* 1958). Some coliform organisms and sometimes *E. coli* can be found. The same bacteria also occur in active dry yeast. Most bacteria are more sensitive than yeast to drying and storage in dried form. Therefore, total bacterial counts in active dry yeast are reduced, and a further reduction takes place on prolonged storage.

Yeast is an excellent growth medium for molds, and mold growth often occurs on yeast cakes which have been stored for 3 to 4 weeks or longer in the refrigerator. On rare occasions massive infections with *Oidium lactis* (machine mold) or with other wild yeasts occurs. These yeasts generally belong to species with a faster growth rate than *S. cerevisiae*. The following species have been identified: *Saccharomyces paradoxus, Candida utilis, Torulopsis minor, Candida krusei,* and *Candida mycoderma* (Goncharova *et al.* 1965); and *C. krusei, C. mycoderma, C. tropicalis, Trichosporon cutaneum, Torulopsis candida,* and *Rhodotorula mucilaginosa* (Fowell 1965). The most common genera are *Candida* and *Torulopsis* (Podel'ko *et al.* 1975). Microbiological techniques and media for the detection and estimation of bacteria, yeasts, and molds in bakers' compressed yeast have been described by Fowell (1967).

Automatic Control of the Fermentation

Some type of automatic control has been practiced in the industry for some time. For instance, automatic control of the pH at a set value or at a pH which rises slowly during the fermentation is entirely practical. The principles of such control devices are similar whether one wishes to control the pH or some other variable. It requires a sensor, a recorder, and a device which activates a solenoid valve whenever a preset value has been reached or exceeded. The principles are basically the same as those used for thermoregulation of laboratory water baths. The problems with such systems are engineering and maintenance problems (Sher 1961).

Apart from pH control, which is relatively simple and reliable, there is an incentive to control the addition of molasses feed. This is important because underfeeding leads to a loss of productivity and overfeeding leads to the production of ethanol and a loss in yield. At the present time molasses wort additions are preset so that given amounts of the wort are fed to the fermentor in given time intervals. Such feed curves have been developed pragmatically. One such feed curve is shown in Fig. 14.2. In practice the ability of yeast to grow at a given rate depends among other things on the amount of pitching yeast originally present, on the amount of fermentable sugars in molasses, on the availability of essential nutrients in adequate amounts, on the availability of sufficient oxygen, and on the physiological state of the yeast. All of these factors show some variation in practice, and as a consequence yeast growth rates vary from batch to batch. It would indeed be highly desirable to feed molasses wort "on demand," that is, based on the concentration of yeast at any given moment and on the concentration of glucose in the fermentor liquid. But neither of these two variables is currently suitable for on-line monitoring. Automatic cell counting and optical density measurements are not precise enough for the measurement of cell solids, and the determination of glucose requires the removal of yeast (catalase) prior to the enzymatic assay.

Some other parameters can be used to obtain indirect measurements of the cell concentration at any given time. Some of these show good promise. The oldest method depends on the determination of ethanol in the effluent gas and throttling of the wort feed as soon as this concentration exceeds a pre-set value (Rungaldier and Braun 1961). New sensors for continuous monitoring of ethanol in the effluent gas are now available and permit full control of feed additions. These so-called hydrocarbon analyzers react with all oxidizable organic compounds, but it is assumed that ethanol is the major oxidizable constituent of the exit gas (Bach *et al.* 1978).

Feed control may also be based on the level of dissolved oxygen. This requires a reliable and sufficiently sensitive oxygen electrode. Such a device has been used by Miśkiewicz *et al.* (1975) in a bakers' yeast fermentation. The control device was set at a point corresponding to a 12% saturation of the fermentor liquid with oxygen. Molasses additions are automatically reduced if the level of oxygen falls below this pre-set level.

Finally, the RQ, that is the rate of carbon dioxide evolution divided by the rate of oxygen uptake, can be used to regulate molasses feed. For respiration of a carbohydrate the Q_{CO_2} should equal Q_{O_2}, that is, the RQ should be 1. Figure 14.8 shows the extent of ethanol production at RQ values higher than 1. For a cellular yield of 0.5 g per g of sugar used, the RQ should be regulated at 1.04 (Wang *et al.* 1977). For a well conducted bakers' yeast fermentation the uptake of oxygen in the fermentor and the evolution of carbon dioxide show a constant RQ between 0.95 and 1 as shown in Fig. 14.9. This figure also gives an indication of the precision which can be

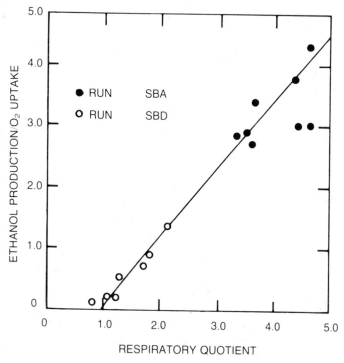

From Wang et al. (1977)

FIG. 14.8.   EXPERIMENTAL CORRELATION BETWEEN THE ETHANOL PRODUCTION RATE AND THE RQ
The slope of 1 is consistent with the theoretical analysis.

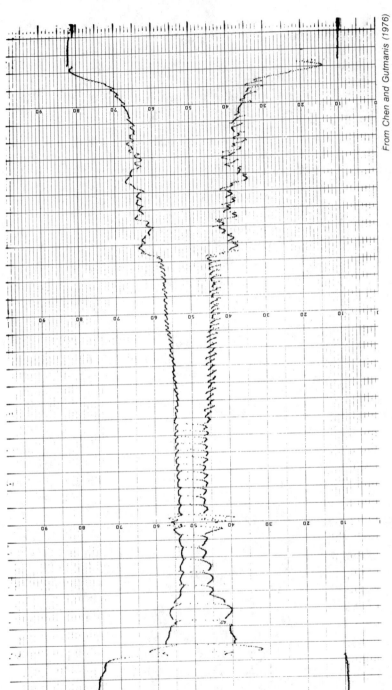

From Chen and Gutmanis (1976)

FIG. 14.9. MEASUREMENT OF CO₂ AND O₂ IN FERMENTOR EXIT GAS DURING BIOMASS PRODUCTION

Starting from the right hand side of the graph, the upper curve shows the increase in CO_2 content due to yeast metabolism. These two curves are mirror images of each other during the entire growth period, indicating a constant CO_2 O_2 ratio (each division represents 30 min, starting from the right hand side of the graph). The respiratory quotient for this uniform metabolic pattern lies between 0.95 and 1.0. The oscillations in these two curves are due to uneven wort additions inherent in the feeding device.

achieved with automatic measurement of the fermentor exit gas. The principle of RQ measurement for automatic feed control to the fermentor has also been used by Aiba *et al.* (1976) and Whaite *et al.* (1978). All authors have used computers for on-line analysis of the output of sensors.

There are still technical questions to be overcome before this type of sophisticated control can be applied to commercial fermentations. Such difficulties reside in oscillations of the system and in the requirement for accurate delivery of the feed by pumps. But the principle of computer control has been adequately demonstrated to be applicable to bakers' yeast fermentations. "Speed of computation, often a difficulty with rapidly responding systems, is not a real problem with microbiological cultures, since any metabolic delays are substantially greater than computational ones" (Whaite *et al.* 1978).

Continuous Fermentation

Bakers' yeast may be produced in continuous fermentation which would permit better utilization of the total available fermentor volume. Since a maturing step is required to obtain a stable compressed yeast, at least 2 vessels in series are required. One such continuous fermentation was carried out commercially in England during the 1960s and has been adequately described by Olson (1961), Sher (1961), and Burrows (1970). The system operated as a 5-stage, open, homogeneous, and continuous fermentation. The total operating time of a continuous run lasted from 5 to 7 days.

One can foresee several problems with continuous bakers' yeast fermentations, but it is not quite clear why this particular operation was discontinued. The growth of contaminants certainly presents greater problems in a continuous fermentation than in a fed batch culture involving several stages. Morphological changes have been observed after several days operation of continuous fermentations. These consist of cell elongations similar to those which have been observed in continuous brewing systems (Beran and Zemanova 1969).

Effects of Ultrasound.—The use of low frequency ultrasound on bakers' yeast cream, compressed bakers' yeast, and liquid pre-ferments has been investigated by a group of Russian workers. A noticeable increase in fermentation rate of bread doughs has been claimed resulting in a 25 to 30% decrease in fermentation time (Ermachenko *et al.* 1974; Rusanova *et al.* 1974, 1976). So far confirmatory reports from other groups of investigators are lacking.

Active Dry Yeast

Introduction.—The production of active dry yeast (ADY) is a technical triumph. Bakers' yeast is the only vegetative microorganism which is dried commercially without significant loss of viability of the cells. While there is rarely a significant loss of viability there is always some loss in bake activity, and this loss may be larger or smaller depending on processing conditions. Therefore, active dry yeast has not replaced compressed yeast in many wholesale bakery operations, at least not in areas with good means of refrigerated distribution. ADY has largely replaced compressed yeast in consumer and institutional markets where storage stability is decisive.

Drying Methods.—It is entirely possible to dry compressed yeast by crumbling it or by extruding it in the form of small strands, and by spreading it on paper in a dry room. The resulting dry product shows a considerable loss in baking activity but could be used to raise doughs. One can also mix, for instance, 1 kg of compressed yeast with 5 kg of low moisture starch or flour and obtain an ADY with a relatively low moisture content (Rupprecht and Popp 1970). Apart from these simple procedures the literature reveals numerous more sophisticated methods of dehydration beginning in the 1920s. The history of the development of ADY has been reviewed by Frey (1957).

None of the earlier methods calling for the addition of substantial amounts of dry, edible materials (flour, starch, calcium salts) has been commercialized. At present the only methods used on a commercial scale start with yeast press cake, which is subdivided before drying into thin strands or smaller particles. Drying is carried out with currents of air at temperatures which keep the temperature of the yeast itself below 40°C.

Continuous belt tunnel driers are widely used for the commercial production of ADY. In this method the strands of extruded press cake are deposited on the wire mesh screen of an endless belt. This belt carries the yeast through several drying chambers in which the airflow is directed through the yeast bed, alternately in an upward or downward direction. Drying times vary from 2 to 4 hr and air inlet temperatures from 28° to 42°C (Belokon 1962). Figure 14.10 shows a schematic drawing of such a drier. Tunnel drying may also be carried out as a batch process for smaller installations. In this case the yeast strands are layered onto the wire mesh of rectangular screens. These screens are mounted on racks which are rolled into the drying tunnel. The results of this type of drying are comparable to the continuous tunnel driers.

Recently air lift (fluidized bed) driers have been used for the commercial production of ADY. This can also be done on a continuous or batch basis. Figure 14.11 shows a schematic diagram of a batch type air lift fermentor which is used for the production of ADY. Its operation has been described by Simon (1976). Generally air lift driers can be operated with shorter drying periods because they permit the use of more finely granulated yeast press cake. Drying times of 1 to 2 hr are satisfactory but much shorter drying times (as short as 10 min) have been suggested in the patent literature (Langejan 1974). Using an airstream at 100° to 150°C at the beginning of the drying period and keeping the temperature of the yeast particles within the 25° to 40°C range, drying times of 10 to 30 min have been used (Langejan 1972). Temperatures of the yeast particle up to 50°C are not detrimental at the end of the drying period (Burrows 1976).

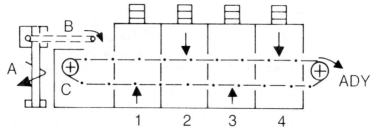

Courtesy of Proctor and Schwartz, Inc.

FIG. 14.10. TUNNEL DRIER (CONTINUOUS)
A—Oscillating feeder. B—Feeder belt. C—Endless wire mesh belt. 1, 2, 3, and 4—Drying chambers.
Arrows indicate direction of airflow.

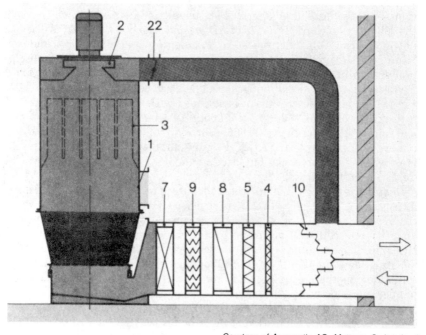

Courtesy of Aeromatic AG, Muttenz, Switzerland

FIG. 14.11. SCHEMATIC SETUP OF FLUID BED DRIER FOR ACTIVE DRY BAKERS' YEAST
1—Explosion relief flap. 2—Centrifugal fan. 3—Exhaust air filter. 4—Coarse dust filter. 5—Fine dust filter. 7—Air heater. 8—Cooler. 9—Droplet catcher. 10—By-pass flap. 22—Air flap.

Air lift driers are also available for continuous operation. The use of a vibrating screen may provide some advantages in obtaining a uniform suspension of yeast particles in the airstream. The drying of yeast press cake with a "vibrating fluidized bed method" has been described by Russian workers (Karavaev *et al.* 1974; Rysin *et al.* 1975).

Finally, roto-louver type driers are used for the production of ADY in the form of small, round pellets. Drying times are generally between 10 and 20 hr in a drum which carries the yeast press cake on the inside and which rotates at about 4 rpm. Hot air is blown through louvers in the surface of the drum and passes through the yeast bed. Other drum drying methods, including vacuum drum drying, have been described by Russian workers (Shishatskii and Bocharova 1973; Sysojeva and Gorochova 1965). Vacuum drying can also be carried out with tunnel driers in which liquid yeast cream is spread on endless, wide steel belts (Hartmeier 1977). Spray drying has not been particularly successful on a commercial scale. Freeze drying results in considerable losses in viability.

Traditionally ADY has been made without the use of any additives, but within the last 20 years some additives have been tried and used on a commercial scale. These additives are usually emulsifiers which improve the rehydration characteristics of ADY, particularly if the yeast is dried to low moisture levels. Sucrose esters (Pomper and Akerman 1969) and sorbitan esters (Mitchell and Enright 1957) have been used. In commercial practice sorbitan monostearate is used in concentrations between 0.5 and 2% based on the dry weight of yeast solids. Protection of low moisture ADY against oxidative deterioration is obtained with the addition of butylated hydroxyanisole (BHA) at levels of about 0.1% based on yeast solids (Chen and Cooper 1962; Chen et al. 1966). Such yeasts are available in the trade under the name "protected ADY." Storage stability of such yeasts is greatly improved and approaches but does not match the complete removal of oxygen from the package.

During the drying of bakers' yeast the rate of water removal is rapid until a moisture content of 15 to 20% is reached. It is assumed that the water removed is "free" water. At a moisture content of about 20%, changes in viability and cell wall permeability are minor. Harrison and Trevelyan (1963) find an optimum rehydration temperature of ADY of 21°C for a 23% moisture level but 42.5°C for a 5.8% moisture level. There is some loss of cell solids during the early stages of drying due to increased respiration at temperatures between 30° and 40°C.

As drying is continued below 15 to 20% moisture, the rate of drying decreases sharply. Respiration stops (Koga et al. 1966) and the yeast cell leaches constituent compounds into the surrounding water if it is rehydrated. When losses of gassing activity or viability occur, this will be apparent at moisture levels below 15%. There is undoubtedly injury to the yeast cell which is not restricted to the cell wall. Beker (1977), who has done the most extensive work in this area, documents such changes in morphology. For instance, invaginations in the cytoplasmic membrane of compressed yeast have a depth not exceeding 50 nm, but in ADY they extend to a depth of 200 nm. The deformation of the cell nucleus which occurs on dehydration is seen in Fig. 14.12.

There is a positive correlation between the concentration of trehalose in bakers' compressed yeast and its ability to withstand drying (Grba et al. 1976; Panek 1975). Krasnikov et al. (1975), who accelerated the rate of

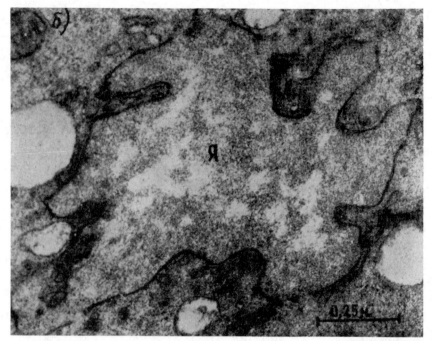

From Beker (1977)

FIG. 14.12.   DEFORMATION OF THE YEAST CELL NUCLEUS ON DEHYDRATION

drying by application of a high frequency field, specified a level of at least 11 to 12% of trehalose (based on solids). It is not clear whether the presence of trehalose is the cause of the stability of yeast during drying or whether it merely reflects conditions of aerobic growth or a low protein concentration of the yeast.

Generally ADY is dried to a moisture content of 7.5 to 8.5%. This level represents a compromise between the demand for good bake activity which is higher at higher moisture levels and good stability which is better at lower moisture levels. For use in baking ADY is rehydrated in water. If the yeast is rehydrated in cold water, about 20 to 25% of the cell solids leach out of the yeast cell. This results in a considerable loss in baking activity. Rehydration in warm water minimizes this leaching as shown in Fig. 14.13 (Chen *et al.* 1966). A temperature of 40°C is optimal. The time required for rehydration is generally less than 5 min.

Leaching of solids from the yeast can be entirely prevented by vapor rehydration, but this is not practical in bakeries. Leaching is also mini-mized if the ADY is added directly to the dry flour before mixing. Presuma-bly this reduces leaching because both the flour and the yeast compete for the small amount of available water. For direct addition to flour the ADY has to be finely ground or the particles of ADY must have a small diameter such as those obtained by air lift drying (Greup 1974). For such air lift dried

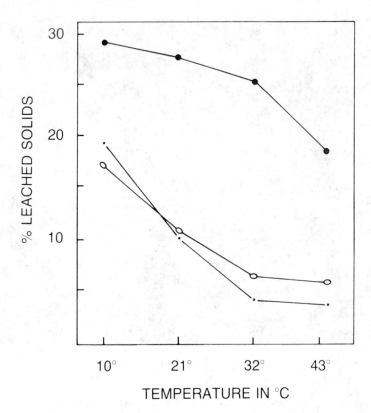

FIG. 14.13.   LEACHED SOLIDS OF ADY AS A FUNCTION OF REHY-
DRATION TEMPERATURE
· ADY—8.1% moisture.
● ADY—5.3% moisture.
O ADY—4.0% moisture, with 2% sorbitan monostearate.

ADY the bake activity is often better if it is added directly to the flour than if
it is separately rehydrated in water.

The leaching phenomenon has been investigated by Herrera *et al.* (1956)
and Peppler and Rudert (1953). The phenomenon is not well understood.
Beker (1977) has provided an excellent review and possible explanations
based on the behavior of biomembranes.

Stability of ADY, Form of Particles, and Packaging.—ADY from the
roto-louver process consists of small pellets with a diameter of about 1 to 2
mm. The surface of the pellets is smooth which tends to improve their
stability in air. ADY from tunnel drying methods takes the form of irregularly shaped, highly fissured strands of a diameter of 1 to 2 mm and of
varying length. This yeast can be ground to a powder which passes a No. 20
mesh sieve. This type of yeast is routinely sold to bakers without protective

packaging in polyethylene lined bags or drums. It loses about 7% of its bake activity per month at ambient temperatures and has a useful storage life of 1 to 2 months. If it is packaged in an inert atmosphere or under vacuum it loses only about 1% of its activity per month with an annual loss not exceeding 10%.

Air lift dried ADY usually has very small cylindrical particles with rounded ends and a diameter of less than 1 mm. It generally has a moisture content of 4 to 6% and is less stable in air, which makes protective packaging mandatory.

Protective packaging for ADY means packaging in an atmosphere with less than 2% oxygen and preferably less than 0.5% oxygen, or vacuum packaging. Yeast for the consumer trade is usually packaged in foil pouches in which the aluminum foil is laminated to a heat sealable plastic film such as Pliofilm or Saran. Such films are also suitable for larger packages in 500 g and 10 kg sizes. The inert gas for the smaller consumer packages (7 g) is commonly nitrogen. Larger foil bags are usually sealed under vacuum and so are 2 lb tins. Tin cans containing 10 to 15 kg ADY must be flushed with nitrogen before sealing. ADY adsorbs carbon dioxide readily (Amsz *et al.* 1956). If this gas is used with larger packages, a sufficient vacuum is created to give the appearance and performance of vacuum packaging. While the size and shape of the yeast particles affect their stability in air, they do not affect their stability in protective packages.

Chemical Composition of ADY.—The regular ADY of commerce has a moisture level of 7.5 to 8.5%. Air lift dried yeasts and protected ADYs usually have lower moisture levels (4 to 6%). The nitrogen level is normally about 7% (based on solids) except for some highly active yeasts with nitrogen levels up to 9.5% (Langejan and Khoudokormoff 1976). The concentration of phosphorus expressed as P_2O_5 is generally one-third that of the nitrogen.

GENERAL REFERENCES

White (1954) wrote a book on the production of bakers' yeast which, in spite of its age, is still the standard reference work for the industry. Another book dealing exclusively with bakers' yeast and its production has been published by Sato (1966). Extensive reviews in the form of book chapters have been published by Butschek and Kautzmann (1962), Harrison (1963), Burrows (1970), Peppler (1967), and Reed and Peppler (1973).

The determination of the activity of bakers' yeast and actual data on gas production rates are treated in Chapter 10.

REFERENCES

AIBA, S., HUMPHREY, A.E. and MILLIS, N.F. 1965. Biochemical Engineering. Academic Press, New York.

AIBA, S.A., NAGAI, S. and NISHIZAWA, Y. 1976. Fed batch culture of *S*.

cerevisiae: A prospective of computer control to enhance the productivity of baker's yeast cultivation. Biotechnol. Bioeng. *18*, 1001–1016.

AMSZ, J., DALE, R.F. and PEPPLER, H.J.   1956.   Carbon dioxide sorption by yeast. Science *123*, 463.

BACH, H.P., WOEHRER, W. and ROEHR, M.   1978.   Continuous determination of ethanol during aerobic cultivation of yeasts. Biotechnol. Bioeng. *20*, 799–807.

BEKER, M.E.   1977.   Biomembranes of microorganisms during the process of dehydration, rehydration, and reactivation. *In* Biomembranes. M.E. Beker and G.Ya. Dubura (Editors).   Publishing Co. "Science," Riga, U.S.S.R.

BELOKON, V.N.   1962.   Yeast drying on a belt drier. Spirt. Promst. *1*, 40–42.

BERAN, K. and ZAMANOVA, J.   1969.   Cell growth and population activity of *S. cerevisiae* in two-stage continuous cultivation. Biotechnol. Bioeng. *11*, 853–862.

BERGANDER, E.   1969.   The effect of several fermentation inhibitors in molasses. Lebensm. Ind. *16*, 219–221.

BEUCHAT, L.R.   1978.   Food and Beverage Mycology.   AVI Publishing Co., Westport, Conn.

BOCHAROVA, N.N., CHERNYSH, V.G. and OZEROVA, V.P.   1976.   Keeping characteristics of pressed yeast. Khlebopek. Konditer. Promst. (8) 37–38.

BURROWS, S.   1970.   Bakers' yeast. *In* The Yeasts, Vol. 3. A.H. Rose and J.S. Harrison (Editors).   Academic Press, New York.

BURROWS, S.   1976.   Process of drying yeast. U.S. Pat. 3,962,467. June 8.

BUTSCHEK, G. and KAUTZMANN, R.   1962.   Production of bakers' yeast. *In* The Yeasts, Vol. 2. F. Reiff *et al.* (Editors). Verlag Hans Carl, Nurnberg, W. Germany.

CARLIN, G.T.   1958.   The fundamental chemistry of bread making. II. Proc. Am. Soc. Bakery Eng., 56–63.

CHEN, S.L.   1959.   Carbohydrate assimilation in actively growing yeast, *S. cerevisiae*. I. Metabolic pathways for (1-^{14}C) glucose utilization by yeast during aerobic fermentation. Biochem. Biophys. Acta *32*, 470–479.

CHEN, S.L. and COOPER, E.J.   1962.   Production of active dry yeast. U.S. Pat. 3,041,249. June 26.

CHEN, S.L., COOPER, E.J. and GUTMANIS, F.   1966.   Active dry yeast. I. Protection against oxidative deterioration during storage. Food Technol. *20*, 1585–1589.

CHEN, S.L. and GUTMANIS, F.   1976.   Carbon dioxide inhibition of yeast growth in biomass production. Biotechnol. Bioeng. *18*, 1455–1462.

COONEY, C., WANG, D.I.C. and MATELES, R.I.   1969.   Measurement of heat evolution and correlation with oxygen consumption during microbial growth. Biotechnol. Bioeng. *11*, 269–281.

COOPER, C.M., FERNSTROM, G.A. and MILLER, S.A.   1944.   Performance of agitated gas liquid contactors. Ind. Eng. Chem. *36*, 504–509.

DeBECZE, G. and LIEBMANN, A.J.   1944.   Aeration in the production of compressed yeast. Ind. Eng. Chem. *36*, 882–890.

DELLWEG, H., BRONN, W.K. and HARTMEIER, W.   1977.   Respiration rates of growing and fermenting yeast. Kem. Kemi *4* (12) 611–615.

DREWS, B. and HESSLER, K.   1967.   Oxidative assimilation of ethanol by baker's yeast. Monatsschr. Brau. *20*, 224–241.

DREWS, B., SPECHT, H. and HERBST, A.M.   1962.   Growth of baker's yeast in concentrated molasses wort. Branntweinwirtschaft *102*, 245–247.

EINSELE, A., RISTROPH, D.L. and HUMPHREY, A.E.   1978.   Mixing times and glucose uptake measured with a fluorometer. Biotechnol. Bioeng. *20*, 1487–1492.

ERMACHENKO, V.A. *et al.*   1974.   Effect of ultrasound on the biological properties of baker's yeast. Prikl. Biokhim. Mikrobiol. *10* (3) 402–409.

EROSHIN, V.K. *et al.*   1976.   Influence of pH and temperature on the substrate yield coefficient of yeast growth in a chemostat. Biotechnol. Bioeng. *18*, 289–295.

FINN, R.K.   1969.   Energy costs of oxygen transfer. Process Biochem. *4* (6) 17, 22.

FOWELL, M.S.   1965.   The identification of wild yeast colonies on lysine agar. J. Appl. Bacteriol. *28*, 373–383.

FOWELL, M.S.   1967.   Infection control in yeast factories and breweries. Process Biochem. *2* (12) 11–15.

FRANZ, B. VON.   1964.   Investigation of the energy metabolism of *S. cerevisiae*, an interpretation of the Pasteur effect. Branntweinwirtschaft *104*, 468–470.

FREY, C.N.   1957.   History and development of active dry yeast. *In* Yeast, Its Characteristics, Growth and Function in Baked Products. Proc. Symp., U.S. Quartermaster Food Container Inst., Chicago, 1957, 7–33.

FRIES, H. VON.   1962.   Peculiarities of yeast growth in aerated fermentations. Branntweinwirtschaft *102*, 442–445.

GIST-BROCADES.   1978.   Procedure for the production of compressed yeast or active dry yeast. Ger. Pat. Appl. (Auslegeschr.) 21 17 901.

GONCHAROVA, L.A. *et al.*   1965.   Effect of wild yeasts on the yield and quality of baker's yeast. Microbiology (USSR) *34*, 157–162.

GÖRTS, C.P.M.   1967.   Effect of different carbon sources on the regulation of carbohydrate metabolism in *S. cerevisiae*. Antonie van Leeuwenhoek J. Microbiol. Serol. *33*, 451–463.

GRBA, S., OURA, E. and SUOMALAINEN, H.   1976.   On the formation of glycogen and trehalose in baker's yeast. Eur. J. Appl. Microbiol. *2* (1) 29–37.

GREUP, D.H.   1974.   Active dry baker's yeast. Getreide, Mehl Brot *28* (10) 259–263.

HARRISON, J.S.   1963.   Baker's yeast. *In* Biochemistry of Industrial Microorganisms. C. Rainbow and A.H. Rose (Editors).   Academic Press, New York.

HARRISON, J.S. 1967. Aspects of commercial yeast production. Process Biochem. 2 (3) 41–45.

HARRISON, J.S. and TREVELYAN, W.E. 1963. Phospholipid breakdown in baker's yeast during drying. Nature 200, 1189–1190.

HARTMEIER, W. 1977. Active dry yeast and method of its production. Ger. Pat. Appl. 25 15 029.

HAUTERA, P. and LOVGREN, T. 1975. The fermentation activity of baker's yeast. Its variation during storage. Baker's Dig. 49 (3) 36–37, 49.

HERRERA, T. et al. 1956. Loss of cell constituents on reconstitution of active dry yeast. Arch. Biochem. Biophys. 63, 131–143.

HOEHNE, R. 1975. Aeration system for high capacity fermenters. Branntweinwirtschaft 115 (22) 400–401.

HONGISTO, H.J. and LAAKSO, P. 1978. Application of the Finn-sugar-Pfeifer and Langen molasses desugaring process in a beet sugar factory. 20th Gen. Meet., Am. Soc. Sugar Beet Technol., San Diego, Aug. 6, 1978.

HUNCIKOVA, S. 1976. Evaluation of molasses quality from the point of view of the production of baker's yeast. Kvasny Prum. 22 (3) 58–62.

JAKUBOWSKA, J. and WLODARCZYK, M. 1969. Observations on yeast growth and metabolism influenced by beta-indolylacetic acid. Antonie van Leeuwenhoek J. Microbiol. Serol. 35 (Suppl. Yeast Symp.) G17.

KARAVAEV, M.N. et al. 1974. The AI-VGS machine for baker's yeast granulation and drying. Tr. Vses. Nauchno-Issled. Eksp. Konstr. Inst. Prodovol. Mashinostr. 38, 3–9.

KAROLUS, B. 1974. The effect of air exchange rate on baker's yeast production. Przem. Ferment. Rolny 18 (12) 8–10.

KAUTZMANN, R. 1969A. Utilization of non-sugar carbon sources by baker's yeast strains. Branntweinwirtschaft 109, 333–339.

KAUTZMANN, R. 1969B. Effect of amino acids on yield and quality of baker's yeast. Branntweinwirtschaft 109, 215–222.

KAUTZMANN, R. 1969C. Methods of aerobic yeast culture. Branntweinwirtschaft 109, 193–200.

KESZLER, H.J. 1967. Some factors determining growth of yeast. Zentralbl. Bakteriol. Parasitenkd. Infektionskr. Hyg. Abt. 2 121, 129–178.

KOGA, S., ECHIGO, A. and NUNOMURA, K. 1966. Physical properties of cell water in partially dried S. cerevisiae. Biophys. J. (Japan) 6, 665–674.

KRAJOVAN, V. et al. 1971. Studies on changes of physiological properties of baker's yeast during aerobic growth on ethanol as the only carbon source. Proc. 1st Special. Symp. Yeast, Intern. Comm. Yeasts, 1971, Smolenice, Czechoslovakia.

KRAMHOFT, B. et al. 1978. The cell cycle and glycolytic activity of Schizosaccharomyces pombe synchronized in defined medium. Carlsberg Res. Commun. 43, 227–239.

KRASNIKOV, V.V. et al. 1975. Intensification of baker's yeast drying. Khlebopek. Konditer. Promst. (7) 27–30.

KUESTLER, E. and ROKITANSKY, K.   1960.   Process of producing yeast of increased dry solid content and reduced plasticity. U.S. Pat. 2,947,668. Aug. 2.

KUNINORI, T. and SULLIVAN, B.   1968.   Disulfide-sulfhydryl interchange studies of wheat flour. II. Reaction of glutathione. Cereal Chem. 45, 486–495.

LANGEJAN, A.   1972.   A novel type of active dry baker's yeast. Ferment. Technol. Today, 669–671.

LANGEJAN, A.   1974.   Dried baker's yeast. U.S. Pat. 3,843,800. Oct. 22.

LANGEJAN, A. and KHOUDOKORMOFF, B.   1976.   High protein active dried baker's yeast. U.S. Pat. 3,993,783. Nov. 23.

LESAFFRE.   1978.   New culture of baker's yeasts, procedure to obtain them, and yeasts obtained from these cultures. Belg. Pat. 862,191. June 22.

LEVIT, KH.D. and GALASHOV, G.I.   1970.   Thermal effects of yeast cultivation. Khlebopek. Konditer. Promst. 14 (1) 35–37.

LINEK, V. and BENEŠ, P.   1978.   Enhancement of oxygen absorption into sodium sulfite solutions. Biotechnol. Bioeng. 20, 697–707.

LODDER, J.   1968.   Process for the manufacture of bread with the aid of yeast. U.S. Pat. 3,394,008. July 23.

LODDER, J., KHOUDOKORMOFF, B. and LANGEJAN, A.   1969.   Melibiose fermenting baker's yeast hybrids. Antonie van Leeuwenhoek J. Microbiol. Serol. 35 (Suppl. Yeast Symp.) F9.

LUCA, S., THOMMEL, J. and BRONN, W.K.   1979.   Pulverulent free flowing fresh baker's yeast preparation and process for its production. U.S. Pat. 4,160,040. July 3.

MacDONALD, I. and GEISLER, A.S.   1965.   Plasticized yeast. U.S. Pat. 3,167,433. Jan. 26.

MATELES, R.I.   1971.   Calculation of the oxygen required for cell production. Biotechnol. Bioeng. 13, 581–582.

MEYENBURG, H.K. VON.   1969.   Energetics of the budding cycle of S. cerevisiae during glucose-limited aerobic growth. Arch. Mikrobiol. 66, 289–303.

MIŚKIEWICZ, T., LESNIAK, W. and ZIOBROWSKI, J.   1975.   Control of nutrient supply in yeast propagation. Biotechnol. Bioeng. 17, 1829–1832.

MITCHELL, J.H. and ENRIGHT, J.J.   1957.   Effect of low moisture levels on the thermostability of active dry yeast. Food Technol. 11, 359–362.

NOTKINA, L.G., BALYBERDINA, L.M. and LAVRENCHUK, L.D.   1975.   The effect of nitrites in baker's yeast manufacture. Khlebopek. Konditer. Promst. (2) 28–31.

OLSON, A.J.C.   1961.   Manufacture of baker's yeast by continuous fermentation. I. Plant and process. Soc. Chem. Ind. (London) Monograph 12, 81–93.

OURA, E.   1974.   Effect of aeration intensity on the biochemical composition of baker's yeast. I. Factors affecting the type of metabolism. Biotechnol. Bioeng. 16, 1197–1212.

PANEK, A.D.   1975.   Trehalose synthesis during starvation of baker's yeast. Eur. J. Appl. Microbiol. 2 (1) 39–46.

PARKHINEN, E., OURA, E. and SUOMALAINEN, H. 1976. Comparison of methods for the determination of the viability of stored baker's yeast. J. Inst. Brew. *82* (5) 283–285.

PASIVKIN, A.I. 1973. Effect of yeast content in a suspension on the accuracy of dosing equipment. Khlebopek. Konditer. Promst. (6) 44–45.

PEPPLER, H.J. 1967. Microbial Technology. Reinhold Publishing Corp., New York.

PEPPLER, H.J. and RUDERT, F.J. 1953. Comparative evaluation of some methods for estimating the quality of active dry baker's yeast. Cereal Chem. *30*, 146–152.

PIŠ, E. 1970. Combined production of baker's yeast and feed yeast. Kvasny Prum. *14* (2) 41–42.

PODEL'KO, A.D. *et al.* 1975. The effect of ultrasound on the microflora in yeast manufacturing plants. Khlebopek. Konditer. Promst. (9) 23–24.

POMPER, S. and AKERMAN, E. 1969. Preparation of active dry yeast. U.S. Pat. 3,448,010. June 3.

REED, G. 1974. Comparison of the use of commercial yeasts. Process Biochem. *9* (9) 11–12, 32.

REED, G. and PEPPLER, H.J. 1973. Yeast Technology. AVI Publishing Co., Westport, Conn.

REITH, T. and BEEK, W.J. 1973. The oxidation of aqueous sodium sulfite solutions. Chem. Eng. Sci. *28*, 1331–1339.

RIKARD, P.A.D. and HOGAN, C.B.J. 1978. Effects of glucose on the activity and synthesis of fermentative and respiratory pathways of *Saccharomyces* sp. Biotechnol. Bioeng. *20*, 1105–1110.

ROBINSON, R.J. *et al.* 1958. The aerobic microbiological population of preferments and the use of selected bacteria for flavor production. Cereal Chem. *35*, 295–305.

ROGERS, P.J. and STEWART, P.R. 1973. Respiratory development in *Saccharomyces cerevisiae* grown at controlled oxygen tensions. J. Bacteriol. *115* (1) 88–97.

ROSEN, K. 1977. Production of baker's yeast. Process Biochem. *12* (3) 10–12.

RUNGALDIER, K. and BRAUN, E. 1961. Method and device for controlling and growth of microbial cultures. U.S. Pat. 3,002,894. Oct. 3.

RUPPRECHT, H. and POPP, L. 1970. Preparation of stable concentrate of baking flours containing yeast. U.S. Pat. 3,510,312. May 5.

RUSANOVA, T.V. *et al.* 1974. Application of ultrasonically activated yeast in bread production. Khlebopek. Konditer. Promst. (5) 28–30.

RUSANOVA, T.V. *et al.* 1976. Intensification of dough making by ultrasound treatment. Prikl. Biokhim. Mikrobiol. *12* (6) 914–921.

RYSIN, A.P. *et al.* 1975. Granulation and drying of baker's yeast using a vibrating fluidized bed method. Khlebopek. Konditer. Promst. (12) 24–27.

SAMBUICHI, M. *et al.* 1974. Filtration and extrusion characteristics of baker's yeast. I. Results of compression permeability test and constant pressure filtration. Hakko Kogaku Zasshi *49*, 880–885.

SATO, T. 1966. Baker's Yeast. Korin-Shoin Publisher, Tokyo.

SCHULDT, E.H. and SEELEY, R.D. 1966. Bulk yeast. Baker's Dig. *40* (2) 42–44.

SHER, H.N. 1961. Manufacture of baker's yeast by continuous fermentation. II. Instrumentation. Soc. Chem. Ind. (London) Monograph *12*, 94–115.

SHISHATSKII, Y.I. and BOCHAROVA, G.A. 1973. Vacuum drying of baker's yeast. Izv. Vyssh. Uchebn. Zaved. Pishch. Tekhnol. (5) 73–77.

SIMON, E.J. 1976. Drying of microorganisms in fluid bed driers under mild conditions. Chemie-Technik *7*, 277–280.

STARIK, E.V., ASHKINUZI, Z.K. and KOVALENKO, A.D. 1975. Losses of biomass and ethanol in baker's yeast production. Fermentn. Spirt. Promst. (3) 19–21.

STROHM, J.A. and DALE, R.F. 1961. Dissolved oxygen measurement in yeast propagation. Ind. Eng. Chem. *53*, 760–764.

SUOMALAINEN, H. 1963. Changes in the cell constitution of baker's yeast in changing growth conditions. Pure Appl. Chem. *7*, 639–651.

SUOMALAINEN, H. 1969. Trends in physiology and biochemistry of yeasts. Antonie van Leeuwenhoek J. Microbiol. Serol. *35* (Suppl. Yeast Symp.) 83–111.

SUOMALAINEN, H. and KERANEN, A.J.A. 1963. The effect of biotin deficiency on the synthesis of fatty acids by yeast. Biochim. Biophys. Acta *70*, 493–503.

SYSOJEWA, J.I. and GOROCHOVA, N.W. 1965. Production of active dry yeast. Khlebopek. Konditer. Promst. *9* (6) 37–41.

VOLKOVA, G.A., DROBOT, V.I. and ROITER, I.M. 1974. Use of liquid yeast concentrate in bread baking. Kharchova Promst. (6) 32–34.

VOLKOVA, G.A. and ROITER, I.M. 1973. Changes in the quality of yeast cream during storage. Khlebopek. Konditer. Promst. (11) 13–16.

WANG, H.Y., COONEY, C.L. and WANG, D.I.C. 1977. Computer aided baker's yeast fermentation. Biotechnol. Bioeng. *19*, 69–86.

WATSON, T.G. 1970. Effects of sodium chloride on steady-state growth and metabolism of *S. cerevisiae*. J. Gen. Microbiol. *64* (Pt. 1) 91–99.

WHAITE, P. *et al.* 1978. Microprocessor control of respiratory quotient. Biotechnol. Bioeng. *20*, 1459–1463.

WHITE, J. 1954. Yeast Technology. Chapman and Hall, London.

WIEMKEN, A., MATILE, P. and MOOR, H. 1970. Vacuolar dynamics in synchronously budding yeast. Arch. Mikrobiol. *70*, 89–103.

WINDISCH, S., KOWALSKI, S. and ZANDER, I. 1976. Dough-raising tests with hybrid yeasts. Eur. J. Appl. Microbiol. *3*, 213–221.

WINDISCH, S. and SCHUBERT, B.A. 1973. Breeding of new yeasts for biscuit doughs. Gordian *73* (7/8) 288–290, 292–293.

15

Enzyme Production

J.T.P. Böing

For centuries microbial enzyme reactions have been conducted in fermentation processes by using enzymes as an integral part of microbial cells. But the intercession of microorganisms may have a number of drawbacks:

(1) A high proportion of substrate may be converted to biomass
(2) Optimum conditions for growth and product formation may diverge
(3) Conversion rate to desired product may be poor
(4) Side reactions may occur
(5) Separation of desired product from the fermentation may be difficult.

Isolated and purified enzymes may overcome most, if not all, of these limitations. The most obvious advantages are easier handling, greater specificity in catalytic function, and greater predictability of activity.

The beginning of modern enzyme technology is marked by the contributions of 5 men:

1894 Takamine was issued a United States patent to cover the production of diastatic enzymes from molds.

1907 Röhm in Germany obtained a patent for using pancreatic proteases in bating hides. Some years later he started the use of these enzymes as laundry aids.

1911 Wallerstein in the United States obtained a patent for chillproofing beer by use of proteolytic enzymes.

1917 Boidin and Effront in France introduced bacterial enzymes for textile desizing.

During the first decades of this century the scientific foundation of enzymology had been laid and thus it is not surprising that during this period progress in enzyme technology was poor. However, after about World War II a rapid development began. Several reasons may be responsible for this: (1) the discovery of numerous new enzymes in the course of investigations on metabolic processes; (2) the increased knowledge of enzyme properties and, as a result, the detection of the potentiality of utilizing enzymes as industrial catalysts; (3) the finding that all enzymes of industrial interest can be produced from microorganisms; and (4) the progress in fermentation technology. Nevertheless, at present only a relatively small number of microbial enzymes have found commercial application. This may be due to legal limitations, particularly in the food industry where most (in number and amount) of the enzymes are used, as well as to difficulties in enzyme processing.

PRINCIPLES OF INDUSTRIAL ENZYMOLOGY

Enzymes are proteins produced by living cells and utilized by these cells to catalyze specific chemical reactions. Their biological function is to bring about and regulate, by their concerted action, the metabolic processes in the organism. It is a characteristic feature of enzymes that they are highly specific in their action (in catalyzing only one reaction step), work at high conversion rates, and exhibit their action under physiological conditions of low pressure and temperature and in aqueous solutions. The normal locus of action is the interior of the living cell, but extracellular enzymes are excreted in order to prepare nutrients which otherwise could not enter the cell.

Commercial production and utilization of enzymes are based on two facts:

(1) Enzymes are produced by living cells.
(2) Enzymes can exert their specific action independent of living cells.

It is well known that enzymes possess a number of properties, the knowledge and consideration of which are of great importance both for their manufacture and for their application.

Substrate Specificity

This term refers to the characteristic property of enzymes to selectively activate certain chemical compounds or types of compounds. The appearance of the large number of enzymes necessary for metabolism can be attributed to this property. At present, more than 1000 enzymes are known. According to their action they are classified into 6 main groups (IUPAC 1972):

(1) Oxidoreductases comprise enzymes which catalyze oxidation-reduction reactions.
(2) Transferases are enzymes that transfer groups (e.g., methyl-, glycosyl-groups) from one substrate (donor) to another substrate (acceptor).
(3) Hydrolases catalyze the splitting of bonds, such as C-O, C-N, C-C, etc., by hydrolytic action.

(4) Lyases cleave bonds, such as C-C, C-O, C-N, etc., with the formation of a double bond, or catalyze the binding of groups to double bonds.
(5) Isomerases catalyze conformational changes of a molecule.
(6) Ligases (synthetases) catalyze coupling of two molecules in conjunction with hydrolysis of an energy-rich triphosphate.

These main groups are further subdivided according to a more specific characterization of the catalyzed reaction.

Enzyme specificity is important when selecting an enzyme for a process. Some industrial applications demand enzymes with low specificity. For example, laundries prefer proteases which are able to split as large numbers as possible of various peptide bonds. In other cases highly specific enzymes are required. Glucose oxidase may be mentioned as an example. However, specificity may also be utilized for purification of enzymes. This method, called "affinity chromatography" (Cuatrecasas and Anfinsen 1971), makes use of substrate analogs which are attached to inert carriers. By means of chromatographic procedures it is possible to selectively separate from a mixture of enzymes the desired one which has the binding site that recognizes the substrate analog.

Enzyme Activity

In 1913 Michaelis and Menten outlined a general theory of enzymatic catalysis, the basis of which was the postulate of an enzyme-substrate complex (ES) formed by a reversible action between enzyme (E) and substrate (S):

$$E + S \overset{k_{+1}}{\underset{k_{-1}}{\rightleftharpoons}} ES$$

The proper enzymatic conversion can only proceed from the enzyme-substrate complex:

$$ES \overset{k_{+2}}{\rightarrow} P + E$$

where P is the product (k_{+1}, k_{-1}, and k_{+2} are reaction rate constants). In the majority of cases this reaction is the rate determining step of the process. Therefore, the rate of dissociation of the enzyme-substrate complex equals the overall rate of the enzymatic process:

$$v = k_2 \, [ES]$$

The actual concentration of the enzyme-substrate complex depends on the concentration of the components E and S. With [E] constant, increase in concentration of [S] leads to an increase in v until the total amount of enzyme is present in the form of the complex. Then the reaction rate has reached its maximum (V) and the enzyme shows substrate saturation.

The Michaelis constant

$$K_M = \frac{k_{-1} + k_{+2}}{k_{+1}} = \frac{[E]\ [S]}{[ES]}$$

is the substrate concentration at which the reaction rate is half of its maximum: It is approximately equal to the dissociation constant (K_s) of the enzyme-substrate complex. In most enzyme reactions K_M is in the range of 10^{-2} to 10^{-5} mole per liter. Low K_M values indicate high affinity between enzyme and substrate.

The rate of enzymic conversion is given by the Michaelis-Menten equation:

$$v = \frac{k_{+2}\ [E^\circ]\ [S]}{K_M\ [S]} = \frac{V\ [S]}{K_M\ [S]}$$

This equation relates to the initial velocity, i.e., the velocity at zero time when the concentration of the product is zero. This may be useful in the study of the properties of enzymes. Industrial use of enzymes, however, is concerned with the time necessary for a reaction to proceed to equilibrium or to completion. In these cases advantage is taken of the form of the Michaelis-Menten equation integrated with respect to time:

$$Vt = K_M\ \ln\frac{[S_o]}{[S_t]} - ([S_o] - [S_t])$$

where t is the time of reaction, $[S_o]$ the initial substrate concentration, $[S_t]$ the substrate concentration after time t, and $([S_o] - [S_t])$ the product concentration after time t. [For an introduction to the subject of enzyme kinetics, the reader is referred to Gutfreund (1965).]

It is also of practical importance that the enzyme activity is affected by a number of variables. With respect to pH, activity response of enzymes is mostly in the form of a bell-shaped curve. The pH of incubation that produces the greatest activity is referred to as optimum pH. It is often believed to be a characteristic property of a particular enzyme, but it depends on ions in the reaction mixture, on the substrate, etc. The position of the temperature optimum is very greatly influenced by the time of incubation. For very short incubation times, temperatures as high as 70°–80°C may yield most product. Increase in incubation time causes inactivation of the enzyme and, therefore, greater activity is found with lower temperatures.

Inhibitors.—Compounds that reduce the rate of an enzymic action are called inhibitors. According to their mode of action several types of inhibitors can be distinguished:

Competitive Inhibitors.—These are compounds that act on the binding site of the enzyme molecule. They need to be similar in size, shape, and charge distribution to the substrate. Inhibition is reversible and depends on concentration, i.e., it can be overcome by excess substrate.

Noncompetitive Inhibitors.—In contrast, noncompetitive inhibition depends only on the concentration of the inhibitor, but not on substrate concentration. From this fact, a potential use can be derived. They may be added to a reaction mixture to rapidly reduce the enzyme activity to zero when the reaction has proceeded to the desired stage.

Uncompetitive Inhibitors.—Uncompetitive inhibition results when the inhibitor reacts only with the intermediate enzyme-substrate complex. It has only been found with enzymes that have at least two substrates.

High Substrate Concentration.—High concentrations of substrates can also reduce the rate of enzyme reactions (substrate inhibition).

End Product Inhibitors.—End product inhibition means affecting activity by a metabolite that has no similarity to the substrate, coenzyme, or product of the enzyme; i.e., it is allosteric. Such allosteric effectors are mostly the end product of a synthetic chain and act through feedback inhibition.

Enzyme inhibitors may be useful in enzyme purification and utilization. For example, specific irreversible inhibitors may be used for inactivating a contaminating enzyme which otherwise cannot be readily removed. Other enzyme inhibitors should be avoided. For instance, ethanol inhibits the action of pectic enzymes. Therefore, these enzymes should be employed prior to fermentation, when used for producing clear wines.

Enzyme Assay

The amount of enzyme is not determined on the base of its chemical constitution but on the basis of its catalytic activity, the rate of substrate conversion serving as a measure of the activity. In principle, the laws of the classical reaction kinetics hold good. However, the formation of an intermediate enzyme-substrate complex in the course of the reaction makes many enzymic reactions a special case for which the Michaelis-Menten equation is valid. Although these reactions are very often mono-, di-, or trimolecular in nature, they can be simplified to a zero order reaction. This is possible by choice of suitable conditions of assay, such as excess substrate ($[S] > K_M$), and slow progress of the reaction. The latter can then be described by the relationship

$$\frac{dx}{dt} = K$$

or

$$v = k \cdot [E] = v_{max}$$

where $\dfrac{dx}{dt}$ is the velocity of the increase of the amount of product. This means that the dependence of enzyme activity is linear.

In order to obtain comparable values for a given enzyme the Enzyme Commission of the International Union of Biochemistry defined the "international unit" (U) as the amount of enzyme that catalyzes the conversion of 1 μM substrate per minute under standardized conditions of substrate concentration, optimum pH, absence of inhibitors, presence of activators.

However, this definition does not apply to most commercial enzyme processes because the assays used do not even resemble the proposed conditions of enzyme application. For this reason enzyme producer and user sometimes agree on an assay that meets the requirement of the user. In most cases, however, assays are chosen for their convenience to the enzyme manufacturer and not to the enzyme user.

From this outline, it is clear that for the majority of commercial enzymes the choice of substrate to be used in an assay may be problematic. To illustrate: Proteins are well-defined molecules, but the substrates on which commercial proteases act usually contain several, and even many, different proteins. A list of substrates that have been used for the assay of hydrolases is presented (see also Collier 1970):

For amylases and/or amyloglucosidase	raw starch, Lintner starch, Zulkowsky starch, modified starch of known dextrose equivalent (D.E.), blue starch polymer, amylose, amylodextrin, phenylmaltoside
For cellulases	ground bran, cotton fibers, cellulose powder, filter paper, dyed insoluble cellulose, CM cellulose, hydroxyethyl cellulose, cellulose phosphate
For pectinase	potato tuber disk, fruit puree or juice (fresh or freeze-dried), pectin, pectic acid, pectinic acid
For proteases	hide powder, azocoll, raw meat (e.g., beef muscle tissue), milk powder, casein, azocasein, N,N'-dimethylcasein, gelatin (also photographic film), acetyl-gelatin, hemoglobin, peptone, peptides

LARGE SCALE PRODUCTION: FERMENTATION

Sources of Enzymes

Industrial enzymes are produced from plants, animals, and microorganisms, but manufacture from the first two groups is limited for several reasons. Cultivation of plants is restricted to areas where climate is suitable. It is generally seasonal, impeding steady enzyme production. As the concentration of enzymes in plant tissues is generally low, processing of large amounts of plant material is necessary. For example, approximately the annual yield of a tree is required for the production of 0.45 kg (1lb) papaya latex. On the other hand, enzymes of animal origin are by-products of the meat industry and for this reason limited in supply. Moreover, they often compete with other end users for the supply of suitable glands.

In contrast, microbial enzymes can be produced in amounts meeting all demands of the market. Seasonal fluctuations of raw material do not count and there are possibilities for genetic and environmental manipulation of bacteria and fungi to give increased yields of desired enzymes in a way not possible with higher organisms. Moreover, the diversity of enzymes available from microorganisms is very great. Lastly, microbial enzymes present a wide spectrum of characteristics that makes them utilizable for quite specific applications.

Selection of Microorganisms

The first step in the manufacture of an enzyme involves the selection of an organism suitable to produce the desired enzyme in amounts as large as possible. The general aspects of this procedure can be outlined as follows:

(1) Extracellular enzymes are preferred, because difficult and costly methods of cell disruption are not necessary. As compared with intracellular enzymes, they are present in a relatively pure form in the culture liquor. Intracellular enzymes are industrially used to a lesser extent because of difficult procedures of cell disruption and separation of contaminating cell components.

(2) High yields of enzymes should be obtained with an economical time required for culture production.

(3) The strain must be stable with respect to productivity, requirement for culture conditions, and sporulation.

(4) The organism should be able to grow on cheap substrates.

(5) Synthetic activity should be as far as possible in the direction of the desired enzyme. Formation of interfering by-products should be low.

(6) Clarification of the culture liquor or extract should be possible without difficulties.

(7) The strain must not produce toxic substances and should be free of antibiotic activities. It should not belong to related strains that synthesize toxins.

Mostly, enzymes with particular properties, e.g., with respect to stability and activity, are desired. This requires special screening programs. As has been demonstrated, the technique of screening is also influenced by the method of cultivation of the organisms at the industrial production stage. This is particularly valid for fungi. If the submerged culture technique is employed, an appropriate selection of strains should be done at an early stage of the program. It is worth mentioning that shaken cultures behave differently from deep cultures. For this reason shaken culture screens only permit a first rough selection. Further selection in deep culture is necessary.

Detection of mutants with increased productivity is difficult. Methods based on checking for halo formation are possible with extracellular enzymes but often worthless for industrial purposes (see also Aunstrup 1974). Therefore, attempts have been undertaken to find out correlations between production of a particular enzyme and one, or several, distinct physiological or morphological characteristics. For example, Nasuno (1972) correlated formation of *Aspergillus oryzae*-type alkaline protease with smooth conidia and production of *Aspergillus sojae*-type alkaline protease with echinulate or tuberculate conidia. This correlation was valid for a large number of *Aspergillus* species tested. For detection of mutants with altered regulation mechanisms of enzyme biosynthesis see Demain (1971).

Mechanisms of Enzyme Biosynthesis

Metabolism is principally regulated by a change of the rate of enzyme reactions. Therefore, regulation of metabolism is mainly a problem of kinetics. As outlined previously, the rate of most enzyme reactions can be described by the Michaelis-Menten equation:

$$v = \frac{k_{-2} [E_o] [S]}{K_M - [S]}$$

where [S] (substrate concentration), $[E_o]$ (total enzyme concentration), k_{-2} (rate constant), and K_M (Michaelis-Menten constant) are independent variables. The enzyme concentration is varied by two mechanisms, namely, by controlled protein synthesis and controlled protein degradation.

For control of enzyme synthesis the microbial cells bring into action the mechanisms of induction and repression. Since Jacob and Monod (1961) outlined their theory of enzyme regulation, these mechanisms have become increasingly understandable, and practical applications have led to sometimes drastic increases of enzyme production by environmental and genetic manipulations.

In this chapter only some practical aspects will be outlined. For a basic treatment of the regulation of enzyme biosynthesis see, e.g., Clarke (1971) and Demain (1972A).

Inducible Enzymes.—There are only few enzymes synthesized in substantial concentration under all conditions of growth. These "constitutive" enzymes include, for example, the enzymes of the hexose-monophosphate pathway. Many of the enzymes used commercially fall into the inducible group. Their biosynthesis requires the presence of substrate in the medium. For example, starch acts as an inducer for amylase. Dextrin, a degradation product of starch, was found to give 16% higher amylase production than did starch in *Bacillus polymyxa*, whereas maltose induced the synthesis of only 50% of the activity induced by starch. Often inducers are analogs or derivatives of the substrate, e.g., isopropyl-β-D-thiogalactoside for β-galactosidase. In other cases compounds structurally similar to the substrate may serve as inducer, e.g., sophorose for cellulase. For polymer substrates which cannot enter the microbial cell, it has been shown that the dimer is the true inducer, such as cellobiose for cellulase. However, the dimers are active as inducers only when they are present in very low concentrations. At higher levels catabolite repression occurs. Thus, it appears that the inductive effects of the polymers result from their slow hydrolysis to dimers which are consumed by the organism as rapidly as they are formed. The same result can be achieved when slowly metabolizable derivatives of the dimers are used, e.g., sucrose monopalmitate for invertase synthesis in yeasts and molds. Table 15.1 is a representation of inducers active toward commercial enzymes.

Repression Mechanisms.—*Feedback Repression.*—Repression mechanisms are of the feedback or of the catabolite type. Feedback repression means that biosynthesis of an enzyme is inhibited when end products of a pathway are accumulated or added to the growth medium. For example, protease production in many bacilli is repressed in certain amino acid-containing media, and protease formation by *Aspergillus niger* is apparently sensitive to repression by sulfur-containing amino acids.

Catabolite Repression.—Catabolite repression is very important in commercial enzyme production since in many of the strains employed this type of regulation is effective. It occurs when cells are grown rapidly on readily metabolizable carbon sources. The classical example of catabolite repression is the repression of β-galactosidase in *Escherichia coli* by glucose ("glucose effect," Eppes and Gale 1942). The repressed enzymes can be of the constitutive or of the inducible type, but in most cases inducible enzymes are involved. By the intervention of catabolite repression it is ensured that when several substrates are present, only the enzymes acting on the best substrate will be formed and wasteful production of other enzymes avoided. Some examples of repressible enzymes are shown in Table 15.2.

The mechanism of carbon catabolite repression in bacteria is now largely understood (Buettner *et al.* 1973; Brickman *et al.* 1973; Magasanik 1970). However, comparatively little is known about the mechanism(s) by which readily utilizable C sources repress the synthesis of many enzymes involved in the carbon metabolism of fungi; and still less is known about the way in which carbon catabolite repression is integrated with ammonium repres-

TABLE 15.1.   INDUCIBLE ENZYMES

Enzyme	Organism	Inducer
α-Amylase	*Bacillus* spp.	Starch, dextrin, maltose
Catalase	*Aspergillus niger* *Candida tropicalis*	H_2O_2, O_2 H_2O_2, O_2, hydrocarbons

Cellulase	*Pestalotiopsis westerdijkii*
	
	Trichoderma lignorum
	Trichoderma viride
		Cellulose, cellobiose,
		    cellobiose octoacetate
		Lactose, sucrose monopalmitate
		Cellulose, cellobiose,
		    cellobiose tripalmitate, sophorose

Glucose isomerase	*Bacillus coagulans*	D-Xylose
Glucose oxidase	*Aspergillus niger* *Penicillium vitale*	Glucose, sucrose Sucrose
Invertase	*Pullularia pullulans*	Sucrose, sucrose monopalmitate
Lactase	*Aspergillus nidulans* *Escherichia coli*	Lactose Lactose, isopropyl-β-D- thiogalactoside
Lipase	*Candida cylindracea* *Candida lipolytica* *Candida paralipolytica*	Tripalmitin Sorbitan monooleate Cholesterol
Pectinmethyl- esterase	*Rhizopus stolonifer*	Pectin
Pectin *trans*- eliminase	*Rhizoctonia solani*	Citrus pectin
Polygalacturonase	*Aspergillus niger* *Rhizoctonia solani*	Pectin Na-Polypectate
Pullulanase	*Aerobacter* sp.	Maltose

sion of enzymes involved in nitrogen metabolism (for genetic basis of ammonium repression see Arst and Cove 1973).

Cyclic AMP.—Investigations by Perlman and Pastan (1969) indicate that inhibition of cyclic 3',5'-adenosine monophosphate formation holds a key position in catabolite repression. In *Escherichia coli*, its intracellular concentration is depressed 1000-fold by growth on glucose, whereas the addition of this nucleotide reverses catabolite repression of many enzymes.

Manipulation of Enzyme Biosynthesis

A number of methods are available to overcome any one of the control mechanisms which may exert an inhibiting effect on the production of large amounts of a given enzyme. These techniques can be divided into two main categories: manipulation of the genetic function of the organism, and manipulation of the environment of the organism.

The methods of genetic manipulation include the classical techniques of mutant formation and a class of novel techniques which is often termed

TABLE 15.2.   CATABOLITE REPRESSION-SENSITIVE ENZYMES

Enzyme	Organism	Repressor
α-Amylase	*Bacillus stearothermophilus*	Fructose
Amyloglucosidase	*Endomycopsis bispora*	Glucose, maltose, starch, glycerol
Catalase	*Rhodotorula mucilaginosa*	Glucose
Cellulase	*Trichoderma viride*	Glucose, cellobiose, starch, glycerol
C_1-Cellulase	*Trichoderma* sp.	Glucose, sucrose
C_x-Cellulase	*Rhizoctonia solani*	Glucose, cellobiose
Glucose isomerase	*Streptomyces phaeochromogenes*	Glucose
Invertase	*Aspergillus nidulans*	Glucose
Lactase	*Escherichia coli*	Glucose
Pectin *trans-eliminase*	*Penicillium expansum*	Glucose, arabinose, mannose, galactose, sucrose, raffinose, galacturonic acid
Polygalacturonic acid *trans-eliminase*	*Aeromonas liquefaciens*	Glucose, polygalacturonic acid
Polygalacturonase	*Aspergillus niger*, *Penicillium expansum*, *Rhizoctonia solani*	Glucose, galacturonic acid Glucose, cellobiose
Protease	*Bacillus megaterium*, *B. subtilis*, *Candida lipolytica*	Glucose
Alkaline protease	*Aspergillus nidulans*, *Neurospora crassa*	Low molec. wt. sources of C, N, and S
Neutral protease	*Aspergillus nidulans*, *Neurospora crassa*	Low molec. wt. sources of C, N, and S

"genetic engineering." While mutant formation is quite usual in enzyme manufacture, genetic engineering techniques are restricted to research laboratories, unless easier handling permits their introduction into industrial practice.

There are two ways in which mutations can cause overproduction of enzymes. The first one is concerned with an alteration in the regulation mechanisms. Such mutational events effect removal of inducer requirement, resistance to end product repression, and resistance to catabolite repression. The second group of mutations leads to an increase in copies of the gene responsible for the production of the enzyme.

Genetic engineering means transfer of genes from one strain to another. Terms such as "plasmid transfer," "phage escape synthesis," etc., may be cited to characterize the methods employed.

Manipulations of the environment enable the biochemical engineer to overcome inhibition of enzyme biosynthesis as caused by regulatory mechanisms, by selection of suitable medium composition, or culture conditions.

In the following, methods of genetic and environmental manipulations, which lead to substantial increases in enzyme production, are enumerated. For more details see Demain (1972B).

For inducible enzymes the application of two methods is possible: (1) mutation to constitutivity, or (2) incorporation of inducers into the medium. Often the most potent inducers are nonmetabolizable substrate analogs. For industrial practice it is important for a particularly expensive or not readily available inducer to be successfully replaced by compounds which can be converted by the organism to the required inducer (see Table 15.1).

End product repression of enzyme biosynthesis can be avoided by several means:

(1) Avoiding presence of end products as medium constituents. For instance, protease production by *Aspergillus niger* is derepressed under conditions of sulfate limitation.
(2) Limitation of end product accumulation is generally possible by adding an inhibitor of the pathway to the medium. External accumulation can further be limited by using mixed cultures with a second organism that metabolizes the repressive substance and by applying dialysis or ultrafiltration techniques in the fermentor system. Internal buildup of end product co-repressors can be limited by starving an auxotroph mutant of the end product required for growth. There are several means: limited feeding of the end product, using slowly utilized derivatives of the required end product, growing partial auxotrophs ("leaky mutants") in the absence of their end product requirement.
(3) Selection of regulatory mutants which are not repressed by end products (constitutive mutants). A commonly used method is selection for resistance to a toxic analog of the end product, but other methods are also available.

As mentioned, catabolite repression is of great importance since many enzymes produced in industry are subject to this type of regulation. Catabolite repression can be avoided by the following means:

(1) Avoidance of the use of repressing carbon sources in the medium. For example, replacement of fructose by glycerol increases α-amylase production of *Bacillus stearothermophilus* more than 25-fold.
(2) Derepression of the enzyme synthesis by growth limitation. Suitable means are slow feeding of the repressive substrate and use of slowly metabolizable analogs or derivatives of the substrate. For example, application of sucrose monopalmitate instead of sucrose was found to increase invertase production 80-fold. Cultivation at lower tempera-

tures, addition of toxic substances to the medium, etc., are further methods of growth restriction.

(3) Mutation to resistance against catabolite repression may increase productivity considerably. For example, a yeast mutant has been obtained that produced 2% of its cellular protein in the form of invertase.

Kinetics of Enzyme Biosynthesis

Rates of fermentation processes are desirable to know for both engineering and fundamental scientific reasons. Bioengineers are concerned with microbial dynamics from the point of view of process design and optimization, building on the experience in the classical chemical industry. Bioscientists, on the other hand, are interested in the dynamic response of microorganisms as a tool for gaining insight into the mechanisms of microbial physiology.

However, biochemical processes are extremely complex and sensitive to a number of factors and rendering their mathematical modeling is most difficult. The available models for predicting the causal effects of changes of control variables are not simple and not accurate. This is the reason why commercial fermentation processes have not been significantly optimized in an engineering sense.

Excellent contributions to the development of kinetic models of enzyme formation stem from Terui and his associates (1967). The presented models refer to commercially produced hydrolases. They are based on the assumptions that the rate-limiting ability of the enzyme-forming system (EFS) corresponds to mRNA and that the specific rate of enzyme production (ϵ, units $\cdot$ mg^{-1} $\cdot$ hr^{-1}) is proportional to the quantity per cell of mRNA (r, quantity $\cdot$ mg^{-1}), i.e., $\epsilon \propto a\mu$.

For growth-associated enzyme production ($\epsilon = a\mu$) the following hypothetical relationship was proposed:

$$\frac{d\epsilon}{dt} = a\mu - b \frac{d\mu}{dt} - k$$

where μ is the specific growth rate (hr^{-1}), k the monomolecular decay rate constant of the specific mRNA (hr^{-1}), and a and b are the system constants. The second term on the right-hand side of the equation is based on the negative correlation of the change of ϵ with that of μ. It expresses the rate of growth-associated repression exerted at the level of transcription. The third term represents the decay rate of mRNA or EFS.

The preceding model has been shown to be in accord with the actual fermentation processes for the production of α-amylase by *Bacillus subtilis*, and amyloglucosidase, acid protease, and polygalacturonase by *Aspergillus niger*. In other cases, such as production of amyloglucosidase, acid protease, polygalacturonase, and C_x-cellulase by *Aspergillus niger* and C_x-cellulase by *Penicillium variabile*, where enzyme formation in the stationary growth phase plays a major role in enzyme accumulation, another model was proposed. It concerns remaining mRNA formed in the preceding growing phase and turnover of RNA to mRNA in the nongrowing phase. This model has the form

$$\epsilon = \epsilon_m e^{-k(t - t_m)} + K_1 (e^{-\lambda(t - t_m)} - e^{-k(t - t_m)})$$

where ϵ_m is the maximum rate of enzyme production at time t_m, when growth has just ceased; λ is the monomolecular decay constant for cell RNA; and K_1 is a system constant. To comment upon the terms on the right-hand side of the equation: The first term represents ϵ due to the mRNA carried over from the growing phase and the second term is the change of ϵ due to the turnover synthesis and degradation of the mRNA.

Cultivation Techniques

Solid Substrate Cultivation.—This method plays an important role in commercial enzyme production from fungal sources, especially in Japan. Advantages and disadvantages of this method, as compared with the submerged culture technique, have often been analyzed. Considering modern deep bed processes, the following advantages can be stated as a matter of fact: (1) enzyme yield per unit volume of incubator is high; (2) power requirement is low; (3) minimum control is necessary; (4) extraction yields highly concentrated enzyme solutions; (5) only small equipment for enzyme recovery is required due to the small amounts of extracts obtained; (6) scaling-up is easy.

Problems which can be solved, if need be, may be listed as follows: (1) continuous operation is possible; (2) feeding substrates during cultivation is possible; (3) defined media can be applied by using suitable inert carriers (see Meyrath 1965). Caused by the nature of the complex media used, the extracts contain considerable amounts of fungal pigments, the removal of which is difficult and costly. Their formation might be avoided by using the previously mentioned "solidified" synthetic media.

The methods of solid substrate cultivation can be divided into two groups: thin layer and deep bed processes.

The thin layer techniques, also called tray processes, work with substrate layers of 2 to 4 cm height spread on wooden or metallic trays. These are incubated in air-conditioned rooms or cabinets (hence cabinet method). The process is described in detail by Underkofler and Hickey (1954). Figure 15.1 shows a flow sheet of the tray cultivation. Jeffreys (1948) has tried to mechanize this process, but it does not seem that manufacturers have been very successful in fully mechanized tray methods. Usually the heat produced by the growing culture is removed by moistened cool air which passes over the surfaces of the trays or is pressed through the bran mass. Kalashnikov *et al.* (1960) recommended a water-cooling system for the trays.

The deep bed process, developed by Terui and co-workers (1958, 1959A,B) in order to meet the enormous demand for enzymes needed for the tradi-

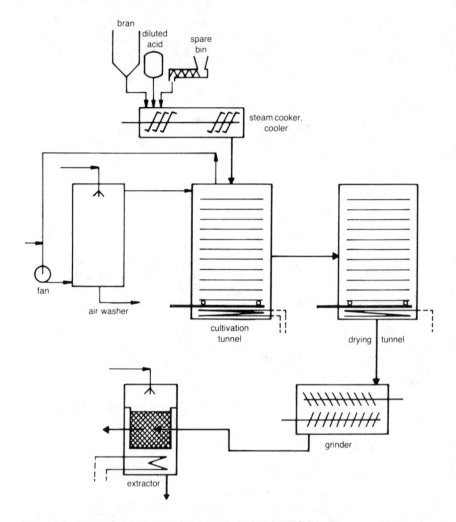

FIG. 15.1.   TRAY CULTIVATION PROCESS: FLOW SHEET

tional soybean fermentations, uses substrate layers usually as deep as 0.6 m (2 ft), but beds as deep as 1.5–1.8 m (5–6 ft) have also been reported. The dimension of rectangular beds is of the order of 5.5 × 61 m (18 × 200 ft). Circular beds are also operated. In general, the equipment used in deep bed processes is quite similar to that known in the malting industry. As mentioned above, the deep bed plants are fully automated. Continuously operated deep bed techniques may, for example, guide the culture mass through air-conditioned tunnels by means of conveyor belts, as claimed by Christensen (1940). Figure 15.2 presents the scheme of a small scale deep bed cultivator.

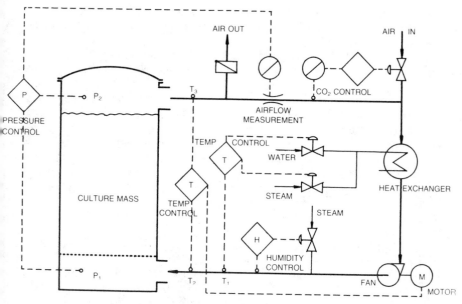

FIG. 15.2. DEEP BED CULTIVATOR

The use of rotating drums, as recommended by Takamine (1913) for the production of mold bran on an industrial scale, was only of temporary importance. With this method manufacturers experienced many difficulties and the desired results were not always obtained. Some workers attributed this to damage of the mold mycelium by mechanical action, but it is not really clear why this method does not work.

Media used in solid substrate fermentations are mainly based on wheat bran. This material is particularly suitable because of its high content of nutrients and its large surface. Other basic substrates are rice bran and soybean or sweet potato flakes, as well as grains or soybeans, etc. Kernels should be selected for optimal size or cracked to give particles of the desired size. On bran, superficial growth of the fungus is sufficient for utilization of nutrients. Kernels, however, must permit the inoculum to penetrate. This is not difficult with polished rice, but corn or soybeans must be cracked or freed of the hulls. The amount of water needed for moistening the substrates is in the range of 40 to 70%. In the case of grains and soybeans the optimum content of moisture is about 30%.

Pressure sterilization of large masses of wet bran in bulk presents serious problems. A convenient method to ensure thorough sterilization is by direct steam injection into the bran. During this process the mass is agitated so that each particle of the moist bran is in constant direct contact with the steam. Experience, however, has shown that when acidic solutions are employed in place of water for moistening the bran it was quite sufficient to sterilize the bran at 95°C for 15 to 30 min. In some cases decontamination of the bran was achieved by means of bactericides, e.g., formaldehyde or β-propiolactone.

Inoculation of the sterilized medium is carried out by use of spores in a dry or suspended form. The amount of inoculum varies from process to process depending on a number of factors. The actual amount must be determined empirically. Underkofler et al. (1947) found that even as low an inoculation ratio as 0.04% of dry spore culture was satisfactory in fungal amylase production. Attempts had been undertaken to use an inoculum in the mycelial form which can be easily produced in large quantities by submerged culture. However, this method is not convenient, resulting in non-uniform growth of the fungus throughout the bran mass. Conidia can be produced in large quantities in special cultures, whereby the method of fermentation is quite similar to that applied in enzyme production. In order to promote spore formation it may be beneficial to add a balanced solution of trace elements (Fe^{3+}, Zn^{++}, C^{++}, Mn^{++}).

Among the conditions of incubation, moisture and temperature present the most serious problems. It was soon recognized that these two factors are dependent variables. Growing cultures produce heat which tends to evaporate the water of the substrate. This effect is intensified by warming the air when it passes the culture and thereby enlarging its capacity to dissolve water vapor. These processes inevitably cause a drying out of the culture. However, the water lost is partially replaced by water that is formed by the metabolic activity of the organism. The difference must be made up by application of sprayed water (this method, of course, is unfit for nonagitated cultures). The microbiological approach to this problem is the selection of slowly growing strains with high productivity. Short-time fermentations may also be helpful.

Heat production is, indeed, considerable. Its magnitude can be readily appraised by the loss in dry weight of the culture. It has been demonstrated that almost all this loss is due to the oxidation of organic compounds to CO_2. In many cases it was found that approximately half of the dry weight of the bran disappears. Kalashnikov et al. (1960) observed with Aspergillus niger that under industrial conditions up to $380\ J \cdot h^{-1} \cdot kg^{-1}$ of fermented bran were liberated at maximum heat production. Removal of this amount of heat required $20\ m^3$ air with 20°–28°C temperature and 100% moisture.

The problem of heat removal is complicated by the heat-insulating properties of the bran. Attempts to minimize this problem by diluting the bran through incorporation of inert materials, such as grain husks, only cause other problems, e.g., a greater requirement of space.

Regarding the moisture content of the medium, it has been found in many cultures that careful observation of the tolerated limits has a decisive influence on the production of the enzyme. In this respect it is very important to keep in mind the previously mentioned formation of water by metabolic processes. Initial pH and the course of the pH during the development of the culture also play a great role. Several means are known to influence the pH development. For instance, this can be done by incorporation of suitable inorganic or organic salts into the medium (see later section on "Particular Technical Enzyme Preparations").

Submerged Cultivation Techniques.—Treatment of this subject may be limited to some considerations of enzyme production.

The fermentation equipment used for the large-scale production of enzymes is the same as that used in the production of other microbial metabolites. As far as is known, processes are batch-operated; however, attempts have been made to introduce continuous processes to enzyme production. Continuous fermentation may be employed if the optimal conditions for the process are known. For inducible enzymes, however, optimal conditions for cell growth are often different from those required for induction and synthesis of the enzymes (Pardee 1969). In such cases a 2-stage, continuous-culture system can be used efficiently for production of the enzyme. By using this method, the growth stage can be operated with an optimal set of culture conditions. It may be useful to insert a washing of the cells between the first and the second stage in order to decrease the requirement of inducer and to produce less impure preparations. Two-stage, continuous-culture processes can also be used when reduction or elimination of catabolite repression due to a high medium concentration is desired (Ryu *et al.* 1972).

Addition of a substrate to a batch process under controlled conditions is a well-known technique in fermentation technology for prolonging growth and increasing product accumulation. Edwards *et al.* (1970) have called this method "extended culture." Since it permits prolonged maintenance of a constant environment with respect to the added substrate, the system is more similar to a continuous process than to a batch fermentation. In many cases drastic increases in enzyme production result from the use of extended culture.

Quite similar to solid substrate fermentations, the pH also plays an important role in submerged cultivations. Generally, the culture starts with a certain pH which depends on the strain employed and the enzyme desired. The course of pH during the fermentation is often manipulated by addition of suitable agents. In some processes the pH is "fixed" at a certain level, within limits. This is the case in the production of amyloglucosidase from *Aerobacter aerogenes*. In a number of batch processes it is convenient to change the pH because of differences in optimum pH for growth and enzyme formation. For example, in the case of neutral metalloproteinase from a *Bacillus* species, the pH is allowed to change "naturally." The course of pH may, however, be programmed. As an example, the acid stable α-amylase of *Aspergillus niger* may be mentioned.

LARGE SCALE PRODUCTION: ENZYME RECOVERY

In enzyme production there is a very unfavorable ratio between input of raw material and output of product. This requires the installation of concentration procedures. For economic reasons of enzyme application a concentration up to 10-fold is usually satisfactory for industrial enzyme preparations. For example, enzyme products employed in detergents contain about 5–10% protease while amylase preparations for use in flour treatment contain only about 0.1% pure α-amylase. However, in applications where high purity enzymes are required, e.g., in enzymic analysis, 1000-fold purification is quite common.

In some applications, such as baking and dextrose manufacture, the presence of contaminating enzymes must be very low or rigidly controlled. Moreover, the raw enzyme solutions obtained from microbial cultures contain—independent of their source—different types of by-products. Separation of all these substances may be necessary because of the possibility of undesired effects.

Considering enzyme stability there is another reason for treatment of crude enzyme preparations. Since the trend in enzyme applications is toward use of liquid preparations, stabilization is an important procedure.

Figure 15.3 is a diagrammatic presentation of some treatments which are used in the preparation of enzymes on a commercial scale. Techniques for the large-scale isolation and (partial) purification of enzymes from microbial sources make use mainly of traditional procedures. Most of the equipment can be found in food-processing plants. Large-scale equipment specific for enzyme isolation is not marketed.

Nearly all process operations are carried out at low temperatures (preferably 0°–10°C), with the exception of drying.

Separation processes are usually conducted in batches rather than continuously. However, the scale-up of batch operations inherently causes extended processing times which for many enzymes result in increased losses of activity due to denaturation of the enzyme protein. For this reason the application of continuous operations seems to be useful, but the necessity for highly reliable machines and ingenious process control delays introduction of continuous methods. In addition the value of continuous processing is lost when a single process step is conducted batchwise, perhaps during precipitation.

Extraction Methods

The first step in the isolation of enzymes is their extraction. Techniques that fall into this group are employed either to separate enzymes from solid substrate culture or to release enzymes from the interior of microbial cells.

Extraction of Solid Substrate Cultures.—Enzymes produced by solid substrate cultivation used to be of the extracellular type. It is therefore easily conceived that extraction of mold brans is rather a washing out process. Countercurrent techniques of percolation are the most frequently used unit operation.

In many cases the mold bran is dried prior to extraction. This is convenient when the utilization of the particular enzyme preparation is seasonal. The cultures can be produced in relatively small equipment all the year round, while the extraction is conducted in times of enzyme demand. On the other hand, it is easily seen that extraction from dried bran will yield

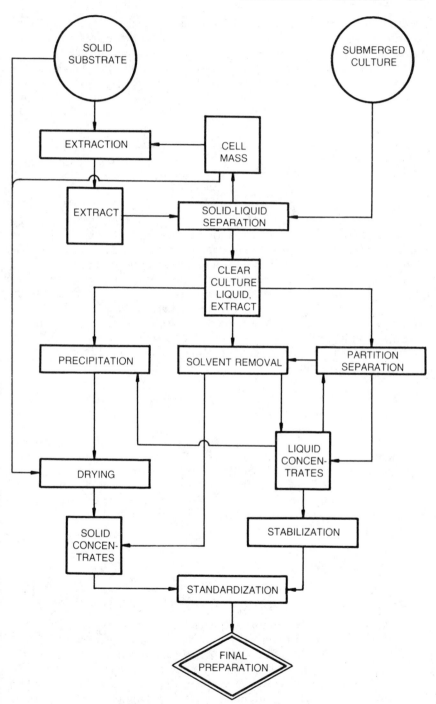

FIG. 15.3.   CONCENTRATION AND PURIFICATION OF ENZYMES

solutions with higher enzyme concentrations. And last, drying avoids interference caused by the activity of living cells of fresh cultures. This argument, however may not apply in continuously operated culture plants.

In all cases the extractant is water which, however, may contain acids (inorganic or organic), salts, buffer, or other substances to facilitate solubilization of the enzyme or to improve its stability in solution, or to exclude or minimize undesired effects caused by contaminating by-products or microorganisms.

Extraction of Cells.—The decision on whether to employ whole cells for a biochemical process or to use isolated enzymes depends on many factors (Lilly and Dunnill 1971). Technical difficulties and the related cost of large-scale isolation play an important role.

There are a number of methods for cell disruption, as reviewed by Hughes *et al.* (1971). Chemical and biochemical methods, such as autolysis, treatment with solvents, detergents, or lytic enzymes, have the disadvantage of being in principle batch operations. Their conduct is difficult to standardize and optimize. More recommendable are mechanical techniques.

At present, the APV-Manton-Gaulin homogenizer seems to be the most versatile type for cell disintegration. In this machine the cell suspension passes a homogenizing valve at the selected operating pressure and impinges on an impact ring. The strong shearing forces combined with the sudden decompression lead to a disruption of the cell wall. Dunnill and Lilly (1972), who examined the disruption of yeast, found that release of protein can be described by a first order rate equation:

$$\log \frac{R_m}{R_m - R} = KNP^{2.9}$$

where R is the amount of soluble protein released in g per kg cell mass, R_m the maximum amount of soluble protein released, K a temperature-dependent rate constant, N the number of times the cell suspension has passed the homogenizer, and P the operating pressure. With industrial models, between 50 and 9000 liters of bacterial suspension per hr can be treated, depending on the size of the machine. Ball mills available on the market have a volume capacity of 0.6–250 liters.

Separation Methods

It is possible here to give only the barest outline of methods that find wider application in the large-scale production of enzymes.

Solids Separation Techniques.—Such methods are involved in the clarification of culture liquors and extracts, in the separation of precipitates, and in the sterilization of liquid enzyme preparation by mechanical methods.

The solids to be separated may have a number of properties which make separation processes difficult. For instance, they may be greasy, sometimes colloidal, and often density differences between solid particles and liquid phase are very small. Therefore, pretreatment of the liquor is usually inevitable, as conducted by acidification, addition of water miscible solvents or liquid polyions, mild heating, etc.

The problems of large-scale solid-liquid separation are complex and diverse. There are two approaches: centrifugation and filtration. Industrial centrifuges are not ideal for removal of finely divided biological solids. Disc type centrifuges without solid discharge have proved most efficient for separation of easily settling suspensions of greasy particles. Decanters are used in cases where solids content is high but easily settling, e.g., in the production of dried acetone precipitates. In cases of poorly settling protein precipitates, hollow bowl centrifuges are employed for separation from low solids suspensions as obtained during fractionated enzyme precipitation. In all cases flow rates must be determined empirically. Sometimes (e.g., with precipitates) the throughput is reduced to less that 10% of the nominal capacity. This requires the integration of cooling devices.

Most frequently filters are more suitable for separation of biological particles. Generally large proportions of filter aids are required. In continuous processes vacuum drum filters are used, with diatomaceous earth, wood-meal, or starch as precoat materials. Batch operations are conducted with filter presses.

Membrane Separation Techniques.—Membrane processes allow separation of solutes from one another or from a solvent, with no phase change or interphase mass transfer. There are many different kinds of membrane processes, the classification of which is based on the driving forces that cause the transfer of solutes through the membrane. Such a force may be transmembrane differences in concentrations, as in dialysis; electric potential, as in electrodialysis; or hydrostatic pressure, as in microfiltration, ultrafiltration, and reverse osmosis.

At present, ultrafiltration is the only membrane process of importance in large-scale enzyme production (Porter and Michaels 1971, 1972). From normal filtration processes it differs just by the size range of the particles to be separated (molecular weight cutoffs between 500 and 300,000). Two types of ultrafiltration membranes are used, which differ in their transport mechanisms and their separation properties (Michaels 1968).

Isotropic porous membranes are the type most similar to conventional filters. They possess a spongy structure with extremely small random pores the average size of which is in the range of 0.05 to 0.5 microns. Molecules with a diameter smaller than that of the smallest pore will pass the membrane quantitatively, whereas particles larger than the largest pore will be retained at the filter surface. Molecules of intermediate size, however, will only pass to some extent. Another proportion of these particles will be retained within the structure of the membrane. This leads, first, to a decrease in retention (or vice versa, in permeation) with a resulting fouling of the membrane, and secondly, to a reduced discrimination among solutes of different size. In order to minimize fouling it is useful to use membranes with a mean pore size well below that of the solute to be retained. Therefore, porous membranes are advisable for the concentration of high molecular weight solutes (molec. weight $< 1 \times 10^6$).

Diffusive membranes are capable of more selective molecular discrimination. They are essentially homogeneous hydrogel layers, through which the solvent as well as the solute is transported by molecular diffusion under the driving force of a concentration or a chemical potential gradient. The transportation of a molecule through the membrane requires considerable kinetic energy. This depends, of course, on the dimensions of the diffusing molecule and on the mobility of the single polymer chains within the membrane matrix. As a rule, the rate of diffusion is high when the polymer segments of the matrix are only loosely interlaced, i.e., when the gel matrix is highly hydrated. For this reason all membranes made from hydrophilic polymers and capable of swelling in water to a certain degree are principally suited as pressure filtration membranes for aqueous solutions.

In addition to the retention potential of the membrane the flux is important for economical reasons. Since in both the porous and the diffusive filter types the flux depends largely on the thickness of the membrane, it is necessary to keep the membrane as thin as possible. This requirement has been fulfilled by the construction of anisotropic membranes. They consist of a highly consolidated but very thin (0.1–5 μm) active layer on a comparatively thick (1 to 20 microns) highly porous support. The advantage of these anisotropic membranes is that there is no reduction in solvent permeability at constant hydrostatic pressure because there is no blockage within the membrane.

In any ultrafiltration system, accumulation of solutes at the membrane surface occurs, which leads to formation of a "slime" that increasingly impedes solvent flow through the membrane, until convective transport of solute toward the membrane is equal to the rate of back diffusive transport away from the membrane. This phenomenon is called "concentration polarization" (Blatt *et al.* 1970). Proteins and colloidal particles build up solid or thixotropic gels when concentrated beyond a certain point. The solute concentration on the membrane surface reaches an upper limit which is typically between 20 and 70% solute by volume. In order to reduce the polarization effect, in industrial ultrafiltration equipment the feed solution passes the membrane surface at high flow rate.

When a macromolecular solution is ultrafiltered, flux of solvent is described by the relationship (de Filippi and Goldsmith 1970):

$$J = A (\Delta P = \Delta\pi)$$

where A is the membrane constant (dependent on temperature, independent of pressure over the normal operating range), ΔP is the hydrostatic pressure driving force, and $\Delta\pi$ is the osmotic pressure difference across the membrane. For macromolecular solutions with concentrations over 1% w/v, osmotic pressures exceeding 10 to 50 psi are not uncommon.

There are several basic types of ultrafilters: thin channel, tubular, helical tubes, spiral wrapped, and hollow fiber systems. Suitability of a single type depends on the properties of the system to be treated.

Gel Filtration.—The technique of ultrafiltration has the advantage of combining both separation of impurities and concentration of the desired enzyme. However, due to its principle, it is a rather nonselective process. Better results, regarding separation of molecules from each other, can be obtained by gel filtration, but in many cases its application is not economically feasible.

In principle, gel filtration is the diffusional partitioning of solute molecules between the readily mobile solvent phase and that confined in spaces within the porous gel particles that make up the stationary phase (Ackers 1970). Diffusional exchange of solutes takes place between the stationary and mobile phases. The extent to which a molecule penetrates the stationary phase is represented by the partition coefficient K_{av}, according to the equation

$$K_{av} = \frac{V_e - V_o}{V_t - V_o}$$

where V_e is the volume of solvent required to elute solute from the gel column, V_o is the void volume (i.e., the volume of liquid external to the gel particles), and V_t is the total volume of column bed. K_{av} is inversely proportional to the log of the molecular weight, as shown empirically.

Gel filtration works rapidly and preservingly, without mechanical stress as in ultrafiltration. However, precipitation of proteins within the gel column may occur as a consequence of desalting. In some cases it has been observed that the metal bridges of enzyme quaternary structures were uncoupled.

Adsorption Techniques.—Because of the possibility of highly selective separations, adsorption processes are increasingly used. Properties of enzyme molecules as different as, e.g., lipophily, electric charge, specificity, etc., are the basis of separation. This results in a great number of adsorbents, such as active carbon, hydroxyapatite, ion exchangers, carrier fixed substrate analogs, and so forth. A common feature of all adsorption techniques is the principle of adsorption followed by desorption or elution. Separation is achieved by adsorption and elution of either enzyme or impurities. Two methods are available, batchwise adsorption or column chromatography. The latter process has greater separation efficiency and, in addition, offers the possibility of semicontinuous operation.

Among the different adsorption techniques, affinity chromatography is of very great interest, but far from being applicable on a large scale. Ion exchangers available for large-scale processes are of either the resin, large-pore gels, or cellulose types. In particular, ion exchange resins exhibit useful properties for industrial production of enzymes. It must, however, be taken into consideration that the proximity of a resin matrix with a high charge density can affect the structural integrity of enzymes. Large-pore ion exchangers and cellulose exchangers have a number of properties which make them very suitable for enzyme separation processes, but the former are very costly. Obviously, they are only suitable for batch processes because they are compressed in columns.

Precipitation Techniques.—Separation from solution by salting out is one of the oldest and yet most important procedures of concentration and purification of enzymes. The logarithm of the decrease in protein solubility in concentrated electrolyte solutions is a linear function of increasing salt concentration (ionic strength), as described by the equation

$$\log s = B^1 - K^1 s \; \frac{\tau}{2}$$

where s is the solubility of the protein in g/liter solution; τ the ionic strength in moles per liter; B^1, an intercept constant, is dependent on pH, temperature, and the nature of the protein in solution; K^1 is the salting out constant which is independent of pH and temperature, but varies with the protein in solution and the salt used (Charm and Matteo 1971). From the preceding relationship it can be derived that precipitation of protein of known concentrations will occur when the ionic strength satisfies the equation

$$\frac{\tau}{2} = \frac{B^1 - \log s}{K^1 s}$$

This means that the electrolyte concentration required for protein precipitation varies with protein concentration.

The influence of the most important precipitation parameters can be outlined shortly as follows: Higher valency salts produce higher ionic strength than lower valency salts. At a constant ion strength, protein solubility increases with increasing distance (in both directions) from its isoelectric point. As a result, lower ionic strength is required for precipitation when carried out at the isoelectric point of the protein.

The commonly used salt for precipitation is ammonium sulfate. The reasons can be found in the high solubility of this salt and in its low price. In addition, ammonium sulfate is nontoxic for most enzymes and in many cases it acts as a stabilizing agent (Dixon and Webb 1961). In ammonium sulfate solutions precipitated enzymes are often storable for years without significant loss when kept at low temperatures.

In contrast to neutral salts, solvents are less customary for large-scale precipitation of enzymes. The reason is higher costs of raw materials and equipment. Explosion proof equipment and recycling of the solvents are inevitable requirements.

Solvent precipitation is based on the fact that the solubility of enzymes decreases with the decreasing dielectric constant (ϵ) of the solvent. The concentration required is lower the less hydrophilic the solvent is. Thus, an increasing precipitating effect can be achieved in the series methanol (ϵ_{25} = 33), ethanol (ϵ_{25} = 24), isopropanol (ϵ_{25} = 18). Besides aliphatic alcohols, acetone (ϵ_{25} = 20) is often used as a precipitant.

Solvent precipitates are distinguished from salt precipitates by the ease with which they settle. However, at temperatures above 4°C denaturation of the enzyme protein can occur. Therefore, it is quite normal to work at temperatures below zero. This, however, requires large cooling capacity in industrial manufacture.

All these difficulties can be avoided by using polyethyleneglycol with a molecular weight of 6000. This precipitant does not effect enzyme denaturation and is relatively independent of temperature and electrolyte concentration. However, there is a strong dependence on hydrogen ion concentration. The best results are obtained at the isoelectric point of the enzyme to be precipitated.

As the solubility of a protein molecule is lowest at its isoelectric point, successive precipitation of different enzymes from a solution can be achieved by changing the pH. These precipitates settle easily and can easily be separated from the solution by centrifugation.

Conversion to Storage Form

Storability of an enzyme requires the preparation of a suitable storage form. Commercial enzyme products are available either in solution or in solid state. Generally, users prefer solutions because of their easier handling, but enzymes are usually very unstable in aqueous solution. For this reason stabilization of dissolved enzymes is a very important step in the manufacture of liquid enzyme preparations. The storage stability is affected by the following two factors: microbial deterioration of the enzyme solution and denaturation of the enzyme protein. These two problems seem to be closely related to each other.

Many treatments have been tried in order to prevent growth of microorganisms. The methods include, for instance, incorporation of chemical preservatives, pasteurization, addition of salts and polyhydric alcohols, and irradiation. But some of those treatments are undesirable due to legal aspects. Therefore, the most suitable method to repress microbial growth is to dissolve the enzyme in a highly concentrated solution of salts and sugars.

With liquid preparations, storage at low temperatures and at suitable pH is essentially inevitable. It is well known that substrates almost invariably protect the corresponding enzyme against physical, chemical, or physicochemical agents. This can be attributed to either conformational stabilization or steric or competitive protection (Grisolia 1964). From a number of publications it can be seen that almost any effect on enzyme stability may *a priori* be an allosteric one due to attack at sites other than the active site of the enzyme (McKinley-McKee *et al.* 1971). For example, in thermolysin, a bacterial protease, Ca ions stabilize the enzyme molecule, while Zn ions are required for activity. The enormous value of Ca ions in stabilizing bacterial α-amylase has long been known.

A number of techniques are available for stabilization. Some of them are presented in the following list (see Wiseman 1973), immobilization methods excluded:

(1) Conformational or charge stabilization and/or protection from dilution-dissociation by using buffers, glycerol, substrates, or inhibitors.
(2) Protection of active site thiol via disulfide exchange by thiols, redox dyes, oxygen-binding agents, or chelating agents.
(3) Miscellaneous methods include, e.g., inhibition or removal of proteolytic enzymes; protection from light by photosensitive dyes; lowering activity of water by viscosity effectors, salts, or sugars; lowering surface energy by antifoams; cooling and crystallization protection by antifreeze; removal of harmful agents; and sterilization for protection against microbial attack.

Commercially available solid enzyme preparations are dried mold brans, dried precipitates, or dried solutions. Spray drying is the preferred method for removal of water from enzyme solutions due to economic reasons. However, it is only applicable to enzymes sufficiently resistant to the temperature conditions of this process. On the other hand, freeze drying is most preserving, but its use is limited by cost considerations as well as by the fact that unless the salt concentrations of the enzyme solution are sufficiently reduced, eutectic mixtures may be formed. This may lead to incomplete drying or to severe foaming and protein denaturation. A specific method of drying sometimes used is granulation in a fluidized bed with milk sugar or maltodextrin as carrier. In this case, of course, sufficiently high specific activity of the enzyme is required in order to ensure satisfactory activity of the commercial preparation.

Enzyme Immobilization

In commercial applications enzymes are used commonly in the soluble or "free" form. This practice, however, is very wasteful, because the enzyme is discharged at the end of the reaction, although its activity is scarcely lessened in reactions carried out under optimum conditions. Immobilization prevents diffusion of the enzymes in the reaction mixture and permits their recovery from the product stream by simple solid-liquid separation methods. As a consequence, reaction products are free of enzyme and reuse of the enzyme is possible. Another advantage of immobilized enzymes is that they can be used in continuously operated reactors.

Methods of Immobilization.—In principle, immobilization of an enzyme can be achieved by fixing it on the surface of a water-insoluble material, by trapping it inside a matrix that is permeable to the enzyme's substrate and products, and by cross-linking it with suitable agents to give insoluble particles. Schematic representation of the available immobilization methods is given in Fig.15.4. For general review of immobilization see Zaborsky (1973).

A

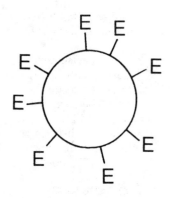

A1

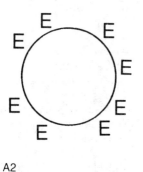

A2

B

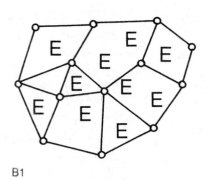

B1

B2

C

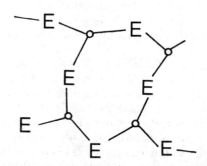

FIG. 15.4.   METHODS OF ENZYME IMMOBILIZATION
A—Bonding. A1—Covalent bonding. A2—Adsorption.
B—Inclusion. B1—Entrapment. B2—Microencapsulation.
C—Cross-linking.

Bound enzymes may be prepared by covalent coupling to active matrices or by heteropolar and/or van der Waals binding to adsorbents or ion exchangers.

Covalent coupling to activated carrier materials is achieved by methods known in peptide and protein chemistry. Some examples of enzymes immobilized in this pattern are given in Table 15.3. The formation of covalent bonds has the advantage of an attachment which is not reversed by pH, ionic strength, or substrate. However, covalent binding offers the possibility that the active site of enzyme may be blocked through the chemical reaction used in the immobilization reaction and the enzyme rendered inactive.

TABLE 15.3.   ENZYMES IMMOBILIZED BY COVALENT COUPLING

Enzyme	Carrier Matrix	Binding Agent/Reaction
α-Amylase	DEAE-cellulose	Direct coupling
Amyloglucosidase	DEAE-cellulose	Cyanuric chloride
Cellulase	Polyurethane	Isocyanate
Glucose isomerase	Polyurethane	Isocyanate
Glucose oxidase	Porous glass	Isothiocyanate
Invertase	DEAE-cellulose Polyaminostyrene Porous glass	Direct coupling Polydiazonium salt Polydiazonium salt
Lactase	Cellulose Polyurethane Sephadex	Cyanuric chloride Isocyanate Cyanogen bromide activation
Pectinase	Polyurethane	Isocyanate
Pronase	CM-Sephadex	Carbodiimide activation

There are a large number of methods of covalent attachment. The groups of enzymes that take part in the formation of the chemical bond are: amino, imino, amide, hydroxyl, carboxy, thiol, methylthiol, guanidyl, imidazole groups, and the phenol ring. Methods have been developed for covalently attaching enzymes to inorganic carriers such as alumina, glass, silica, stainless steel, etc. (Weetall 1972).

Adsorption of enzymes at solid surfaces (Table 15.4) offers the advantage of extreme simplicity (McLaren and Packer 1970). It is carried out according to the principles of chromatography. The conditions of adsorption involve no reactive species and thus do not result in modification of the enzyme. The binding of enzymes, however, is reversible and for this reason adsorbed enzymes present the problem of desorption in the presence of substrate or increased ionic strength. Commonly used adsorbents include many organic and inorganic materials such as alumina, carbon, cellulose, clays, glass (including controlled-pore glass), hydroxyapatite, metal oxides, and various siliceous materials. Ion exchange resins bind enzyme by elec-

TABLE 15.4.   ENZYMES IMMOBILIZED BY ADSORPTION

Enzyme	Carrier Matrix
α-Amylase	Calcium phosphate
Amyloglucosidase	Agarose gel, DEAE-Sephadex
Catalase	Charcoal
Glucose oxidase	Cellophane (followed by cross-linking with glutaraldehyde), inorganic adsorbents
Invertase	Charcoal, DEAE-Sephadex
Subtilisin	Cellulose

trostatic interactions. The first successful commercial application of immobilized aminoacylase (for resolution of DL-amino acids) involved fixing of the enzyme by adsorption to DEAE-Sephadex as carrier (Tosa *et al.* 1966).

Inclusion of enzymes in polymer gels, microcapsules, or filamentous structures has the advantage of relatively mild reaction conditions. This method is free from the risk of blocking active site groups on the enzyme molecule by chemical bonds; the enzyme is retained in its native state. The major drawbacks of this immobilization technique are two: retardation of the enzymic reaction due to diffusional control of the transport of substrate and products (particularly with high molecular weight substrates and/or products); and continuous loss of enzyme due to the distribution of pore sizes. Materials used for entrapment include silicone rubber, silica gel, starch, and, preferably, polyacrylamides (Bernfeld and Wan 1963). Examples of enzymes immobilized by entrapment are given in Table 15.5.

TABLE 15.5.   ENZYMES IMMOBILIZED BY ENTRAPMENT

Enzyme	Polymer Matrix
α-Amylase (fungal)	Polyacrylamide gel
Amyloglucosidase	Cellulose triacetate, polyacrylamide gel, polyvinyl alcohol
Catalase	Cellulose triacetate, polyacrylamide gel
Glucose isomerase	Cellulose triacetate
Glucose oxidase	Cellulose triacetate, polyacrylamide gel
Invertase	Cellulose triacetate, polyacrylamide gel, polyvinyl alcohol
Lactase	Cellulose triacetate
Acid protease	Polyacrylamide gel
Alkaline protease	Polyacrylamide gel
Neutral protease	Polyacrylamide gel

A variation of the inclusion method is encapsulation within semipermeable membranes (Chang 1964, 1972). Materials such as collodion polystyrene, cellulose derivatives, and, most commonly, nylon have been used to form thin, spherical, semipermeable membranes shaped into microcapsules which include the enzyme to be immobilized. The size of the capsules can range from 1 μm to many μm. As has been demonstrated by Kitajima and Kondo (1971) with yeast, it is possible to encapsulate multienzyme systems from cell extracts and to carry out fermentation in such artificial cells.

Enzymes can be polymerized by cross-linking with low molecular weight multifunctional agents (see Table 15.6). This method leads to the formation of a three-dimensional network of enzyme molecules when the reaction is carried out in the absence of a support. However, usually it results in a considerable loss of activity (Levin et al. 1964; Silman et al. 1966). Commonly, enzymes are cross-linked after adsorption onto a suitable carrier. Cross-linking agents most commonly used include diazobenzidine and its derivatives and particularly glutaraldehyde. On the other hand, enzymes can become immobilized by copolymerization, i.e., covalent incorporation into polymers (Bar-Eli and Katchalski 1960). The methods most often employed involve copolymerization with maleic anhydride and ethylene. As with entrapped and microencapsulated enzymes, these derivatives show little or no activity toward macromolecular substrates.

TABLE 15.6.   ENZYMES IMMOBILIZED BY CROSS-LINKING

Enzyme	Cross-linking Agent
Catalase	Glutaraldehyde
Glucose oxidase	Glutaraldehyde and cellophane membrane
Rennin	Glutaraldehyde and aminoethylcellulose
Subtilisin	Glutaraldehyde

An alternative to the immobilization of isolated enzymes is immobilization of whole microbial cells (Mosbach and Mosbach 1966). This method provides a means of avoiding expensive enzyme purification operations. Entrapment of enzymes within whole cells may also be useful when various enzymes are involved in a given process. And, finally, immobilized intact cells have proved effective in processes involving enzymes that require cofactors for mediating their catalytic action. Some examples of whole cell immobilization are shown in Table 15.7. Cells can be immobilized by fixing them to carriers, such as fibers or granular materials, or by entrapment.

Properties of Immobilized Enzymes.—After immobilization of an enzyme, its properties can be changed significantly. Such alterations may be attributed to (1) the physical and chemical nature of the carrier used, (2) the chemical and/or conformational changes in the enzyme structure, and (3) the "heterogeneous nature" of catalysis caused by immobilization.

TABLE 15.7. SOME IMMOBILIZED WHOLE CELL SYSTEMS

Organism	Enzyme	Carrier Matrix	Immobilization Technique
Mold	Amino acid acylase	Cellulose nitrate	Entrapment
Fungal spores	Invertase	Cellulose	Adsorption
Mold	Glucose isomerase	Collagen	Complexation

Effects of altered reactivity include kinetic constants (resulting from a change in activation energy), optimum pH, Michaelis constant, and substrate specificity. A matrix charge can affect the hydrogen ion concentration in the locus of the attached enzyme and thus change its apparent pH optimum in one direction or the other, depending on the use of either cationic or anionic carriers, and the apparent Michaelis constant if the substrates are also charged; it is increased if the matrix and the substrate charges are alike and decreased if they are opposite. Changes in enzyme specificity can result from conformational changes in the enzyme molecule caused by the attachment itself.

One of the more important results of enzyme immobilization is the retention of activity for considerable periods of time under suitable conditions of storage. The stability of immobilized enzymes to storage, heat, and pH basically depends on the nature of the carrier surface to which the enzyme is bound. Among 50 immobilized enzymes, as compared with their soluble counterparts, Melrose (1971) found 30 more stable and 8 less stable than the soluble forms; 12 showed no difference from the free systems. The reasons for the observed increase in stability is not clear. It may be attributed, for example, to prevention of conformational inactivation or to shielding of active groups on the enzyme from reactive groups in solution.

ENZYME UTILIZATION IN INDUSTRIAL PROCESSES

From the standpoint of chemical process engineering, enzymes have some properties which make them ideal catalysts.They are able to carry out chemical reactions at high rates and with high specificity at ambient temperatures and normal pressure in aqueous solutions. Thus enzymes provide a means of eliminating many of the operational difficulties of high temperatures and pressures often encountered in chemical catalytic processes. Practical application of bulk enzymes, however, is rather limited to industries which process natural materials. The reasons for this comparatively mediocre utilization are complex, but the following ones may be mentioned:

(1) Until recently, the development of enzyme applications was founded on an empirical rather than a rational basis.
(2) Enzymes are almost exclusively used as processing materials in other industries. Consequently, the growth of the enzyme market is determined by the growth in the market of the final product.

(3) Strong government regulations, required to protect health and safe-ty, tend to inhibit development and introduction of potential enzyme applications in many industries.

Economic considerations of the use of enzymes are governed by a number of variables and are not always simple to determine. Justification of utiliza-tion is often calculated on a plant-by-plant basis rather than on an industry-wide basis.

Fields of Application

At present, the bulk enzyme industry still relies almost entirely on the production of relatively simple enzymes, primarily for application in the food and other industries. By far the majority of these enzymes are hydro-lases, such as amylases, cellulases, pectinases, proteases, etc. They are used mostly as additives or processing aids in the baking, dairy, fruit juice, and other industries. In contrast to the bulk enzymes, the market for the indi-vidual enzymes glucose isomerase, glucose oxidase, etc., is expanding rapid-ly. This fits also for highly purified enzymes used in the pharmaceutical industry, therapeutic application, and clinical and chemical analysis. A brief abstract of applications is presented in Table 15.8.

Use of microbial enzymes in food processing falls roughly into 3 catego-ries (Fox 1974):

(1) Those in which enzymes form an essential part of a process, e.g., production of cheese, beer, spirits
(2) Those in which enzymes are used to improve the economics of a process, e.g., extraction of fruit juices and essential oils
(3) Those in which enzymes are used to improve product quality, e.g., meat tenderization, loaf volume, flexibility of glucose syrups and dextrins, etc.

Application of immobilized enzymes in food industries is restricted to a very few cases. For instance, aminoacylase (for resolution of racemic mix-tures of amino acids) and glucose isomerase (for converting glucose to fructose) may be mentioned.

Legal Aspects

Large amounts of enzymes are presently used in the food industry. There-fore, manufacturers must conform to food laws which, however, differ from country to country. The Codex Committee of Food Additives has published a list of enzymes derived from plant, animal, and microbial sources, limiting their application in food products. This has to be in accordance with the general principles regulating the use of "Food Additives."

In the United States the following enzymes are "Generally Recognized as Safe" (GRAS), when manufactured with sound processing practice: proteases and carbohydrases from *Aspergillus oryzae*; carbohydrases, cellulases, glucose oxidase, pectic enzymes, and lipases from *Aspergillus niger*; invertase from *Saccharomyces cerevisiae*; lactase from *Saccharomyces fragilis*; and carbohydrases and proteases from *Bacillus subtilis*. The use of enzymes from other microbial sources is limited by the "Food Additive Amendment." This is true, for example, for carbohydrases from *Rhizopus oryzae*; milk clotting enzymes from *Mucor miehei, Mucor pusillus, Bacillus cereus,* and *Endothia parasitica*; and for catalase from *Micrococcus lysodeikticus.*

Care must be taken that the organisms used for enzymes that will be applied in the food industry are known or shown to be nonpathogenic and do not produce toxic substances, β-nitropropionic acid included. New strains must be screened also for variability; it must be certain that the properties of the organism show no difference from those originally tested and approved.

Safety, furthermore, is dependent on freedom from the organism used as well as on the materials used in the concentration and purification processes. Where carryover of materials or residues might persist, these must be of food grade. Fillers and additives must conform to normal standards.

Harmful effects on process workers must also be considered. Suitable means for circumventing these effects are careful handling and change of the physical form of the enzyme preparation. Thus, enzymes are now usually in granular or, less often, encapsulated form.

PARTICULAR TECHNICAL ENZYME PREPARATIONS

Of the 1000 and more enzymes isolated, fewer than 20 are now commercially used on a scale that has significant impact on either the enzyme industry or the user industries. Only the more important will be treated here.

It is worth remembering that commercial enzyme preparations represent mixtures in which the principal enzyme activity is only one among many others. In enzyme applications this fact may imply that desired effects are interfered with by contaminating enzymes, or that effects observed trace back to such activities. The situation becomes even more complex when one considers the varying qualities and quantities of accompanying enzymes depending on the microbial source, the cultivation method, and the recovery process. This is one of the reasons for variable suitability of different products of the market or even of different prices for the same product. In some applications, it is invariably necessary to eliminate the interfering activity unless a mutant unable to produce this activity has been found. The success of such a procedure can decisively influence the development of a special enzyme market.

Amylolytic Enzymes

Amylolytic enzymes represent a group of catalytic proteins of great importance to the food industry. They were also one of the first enzymes to be

TABLE 15.8.   COMMERCIAL APPLICATIONS OF MICROBIAL ENZYMES.

Industry	Application	Enzyme	Source
Baking and milling	Reduction of dough viscosity, acceleration of fermentation process, increase in loaf volume, improvement of crumb score and softness, maintenance of freshness and softness	Amylase	Fungal
	Improvement of dough texture, reduction of mixing time, increase in loaf volume	Protease	Fungal/bacterial
Beer	Mashing	Amylase	Fungal/bacterial
	Chillproofing	Protease	Fungal/bacterial
	Improvement of fine filtration	β-Glucanase	Fungal/bacterial
Cereals	Precooked baby foods, breakfast foods	Amylase	Fungal
	Condiments	Protease	Fungal/bacterial
Chocolate, cocoa	Manufacture of syrups	Amylase	Fungal/bacterial
Coffee	Coffee bean fermentation	Pectinase, hemicellulase	Fungal
	Preparation of coffee concentrates	Pectinase, pectinase	Fungal
Confectionery, candy	Manufacture of soft center candies and fondants	Invertase, pectinase	Fungal/yeast
Corn syrup	Sugar recovery from scrap candy	Amylase	Fungal/bacterial
	Manufacture of high-maltose syrups	Amylase	Fungal
	Production of low D.E. syrups	Amylase	Bacterial
	Production of glucose from corn syrup	Amyloglucosidase	Fungal
	Converting corn syrup to a sweeter fructose-containing product	Glucose isomerase	Bacterial
Dairy	Residual H_2O_2 removal from milk (subsequent to sterilization by H_2O_2)	Catalase	Fungal
	Manufacture of protein hydrolysates	Protease	Fungal/bacterial
	Stabilization of evaporated milk	Protease	Fungal

		Lactase	Yeast
	Production of whole milk concentrates, whey concentrates, and ice cream and frozen desserts	Lactase	
	Curdling milk	Protease	Fungal/bacterial
Distilled beverages	Mashing	Amylase	Fungal/bacterial
Eggs, dried	Glucose removal	Glucose oxidase	Fungal
Feeds, animal	Pig starter rations	Amylase, protease	Fungal
Flavors	Clarification (starch removal)	Amylase	Fungal
	Oxygen removal	Glucose oxidase	Fungal
Fruit juices	Clarification, preventing gelling of concentrates, improvement of juice extraction yield	Pectinases	Fungal
	Oxygen removal	Glucose oxidase	Fungal
Laundry	Detergents	Protease	Bacterial
Leather	Dehairing, bating	Protease	Fungal/bacterial
Meat	Tenderization	Protease	Fungal
	Preparation of fish protein concentrates	Protease	Fungal/bacterial
Pharmaceutical and clinical	Digestive aids	Amylase, protease	Fungal
	Injection for bruises, inflammation, etc.	Streptokinase	Bacterial
	Various clinical tests	Numerous	Fungal/bacterial
Photography	Recovery of silver from spent film	Protease	Bacterial
Protein hydrolysates	Preparation of protein hydrolysates	Protease	Fungal/bacterial
Soft drinks	Stabilization of citrus terpenes from light-catalyzed oxidation	Glucose oxidase and catalase	Fungal
Textiles	Desizing of fabrics	Amylase	Bacterial
Vegetables	Preparation of hydrolysates	Pectinase, cellulase	Fungal
	Liquefying purees and soups	Amylase	Fungal
Wine	Clarification of must	Pectinase	Fungal

produced commercially by microorganisms. Since industrial starch processing requires certain physical properties or a specific carbohydrate composition of the final starch hydrolysate, great efforts have been made to find microbial enzymes with the specific characteristics required. Consequently, a remarkable number of new starch-degrading enzymes have been discovered, most of which have characteristics and properties that clearly distinguish them from all amylolytic enzymes previously known.

Starch-degrading enzymes can be divided into two main groups, α-1,4-glucanases and α-1,6-glucanases (debranching enzymes). The more important members of the class are shown in Table 15.9.

α-Amylase.—This enzyme (α-1,4-glucan-glucanohydrolase, EC 3.2.1.1) acts on starch components, which contain at least three α-1,4-linked glucose units, as an endoase, i.e., in an essentially random manner, with the production of reducing sugars. Mode of action, properties, and products of hydrolysis differ somewhat, depending on the source of the enzyme. Two types of microbial α-amylases have been recognized, termed "liquefying" and "saccharifying" α-amylases. The main difference between them is that the saccharifying enzyme produces a higher yield of reducing sugars than the liquefying enzyme.

Bacterial α-amylase might be produced by a number of *Bacillus* species, *Pseudomonas saccharophila*, and *Clostridium* species, but on an industrial scale specially selected strains of *Bacillus subtilis* seem to be preferred. Fungal α-amylases for commercial purposes are derived from *Aspergillus oryzae*. Certain strains of *Aspergillus niger* and particularly of *Aspergillus oryzae* produce large amounts of the saccharifying enzyme.

In industrial fermentation the cultivation of bacterial α-amylase producers is principally conducted by the submerged culture technique. The media employed are generally based on the use of natural raw materials, because of their cheapness and stimulatory effect. The latter effect is attributed to enzyme inducers or certain growth factors, such as trace elements, vitamins, or suitable combinations of amino acids. A careful balance of carbohydrate and nitrogen ingredients of the medium is most important. From reports in the literature it appears that the concentration of the N source should normally be higher than necessary for meeting full growth of the organism. Mostly, organic N is supplied, but inorganic N can also serve as the major nitrogenous component: it is, however, usually combined with small amounts of proteinaceous compounds.

Fungal α-amylase was originally and is still produced in significant amounts in solid substrate culture. Wheat bran serves as the basic component of the medium. Most publications on this procedure mention weakly acidic solutions as a means of moistening the bran, but acid has been shown to be harmful to amylase production (Meyrath 1966). The moisture content of the culture depends on the height of the culture and on incubation temperature. The incubation time required to reach maximum yield of enzyme varies between "overnight" and about 4 days.

While Dunn *et al.* (1959) succeeded in selecting strains capable of producing good yields of α-amylase under submerged culture conditions, fungal

TABLE 15.9. CLASSIFICATION OF STARCH-DEGRADING ENZYMES OF INDUSTRIAL INTEREST

			Source of Enzyme (Examples)
starch-degrading enzyme	α-1,4-glucanase	endo-α-1,4-glucanase — α-amylase	Bacillus, Aspergillus
		exo-α-1,4-glucanase — exo-maltohexaohydrolase	Aerobacter
		exo-maltopentaohydrolase	Bacillus
		exo-maltotetraohydrolase	Pseudomonas
		β-amylase	Bacillus, Streptomyces
		amyloglucosidase	certain bacteria, Aspergillus, Rhizopus
		isopullulanase	Aspergillus
	α-1,6-glucanase	endo-α-1,6-glucanase — pullulanase	Aerobacter
		isoamylase	Pseudomonas
		exo-α-1,6-glucanase — exo-pullulanase	Cladosporium

α-amylase is also produced in deep tank fermentation. Usually flour or starch serves as raw material, supplemented with inorganic salts. Addition of stillage, corn steep liquor, or yeast extract contributes stimulating agents for some strains. In most cases of fungal α-amylase fermentation, the enzyme accumulates largely in the stationary phase, and quite often during the phase of autolysis. The amount of α-amylase that is formed during these growth phases is very strongly dependent on environmental and internal factors.

Published work indicates the existence of common features in both bacterial and fungal α-amylase fermentations. It has been shown quite generally that in both cases a proper course of pH change toward alkalinity during cultivation is of great importance. Salts of organic acids, e.g., citrate, gluconate, or acetate, can serve as pH regulators acting in the desired sense. The necessary amounts depend on the carbohydrate concentration or on the kind of concentration of organic and inorganic N sources. The same effect can be achieved by alkalizing nitrogenous compounds, such as nitrates, urea, and proteinaceous matter. When ammonium salts are used as the N source, organic salts have to be given preference due to their capability of maintaining the pH of the culture at neutral or of resulting in a rising pH, as shown by Horvath and Inczefi (1972). Proteinaceous matter, e.g., peptone, exerts the same effect. Fully synthetic media must not contain ammonium salts, when lacking salts of organic acids, but should contain nitrates or urea. In contrast to the acidifying effect of inorganic NH_4^+ salts, ammonium acetate tends to maintain the pH of the culture at neutral or to result in a rising pH. This can be attributed to the fact that the organisms produce nonutilizable organic acids.

The question of the type of mechanism that regulates the formation of α-amylase can not be answered clearly or generally. While Coleman (1967) stated that in *Bacillus subtilis* α-amylase is constitutive and controlled by the size of the pool of nucleic acid precursors, Schaeffer (1969) and Meers (1972) found in bacilli that biosynthesis of this enzyme is governed by catabolite repression. Using a carbon-limited medium it was demonstrated that the presence of an inducer was not necessary for α-amylase biosynthesis.

Dunn *et al.* (1959) found that addition of Ca phytate to natural and synthetic media increased the yield of dextrinogenic α-amylase from a *Bacillus* strain. Similarly, Yamada (1961) observed an increase in production of both dextrinogenic and saccharogenic amylase as a response to the incorporation of 0.01–0.05% phytic acid in the medium used for *Aspergillus oryzae* or *Aspergillus awamori*.

Commerical preparations of bacterial α-amylase are commonly produced with a minimum of purification. Highly active or purified preparations are obtained by precipitation and/or adsorption techniques. For some applications the absence of other enzymes, especially proteinase, is essential. For this purpose several methods are available, e.g., adsorption procedures, fractional precipitation (Anderson and Keay 1968), and selective inactivation (Miller and Johnson 1954).

Bacterial α-amylase has a molecular size on the order of 50,000, each molecule containing 1 gram atom of Ca^{++} (Fischer and Stein 1960). In the presence of zinc, a dimer is formed containing 1 atom of zinc (Kakiuchi *et al.* 1964). The calcium maintains the enzyme molecule in the optimum conformation for maximum activity and stability (Whitaker 1972); it does not participate directly in the reaction. The addition of Ca salts is generally recommended to achieve maximum heat stability of the enzyme.

The maximum activities of α-amylase are in the slightly acidic region between pH 4.5 and 7.0, with differences depending on the enzyme source. The α-amylases from *Bacillus subtilis*, and especially *Bacillus stearothermophilus*, are particularly heat stable (Manning and Campbell 1961). In contrast, thermal stability of fungal α-amylase is relatively low.

Bacterial α-amylases preparations are mainly used in the continuous process for desizing of textile fabrics. Other applications include modification of starches suitable for preparation of adhesives, sizes and coatings for the paper industry, as well as manufacture of glucose and glucose syrups and brewing processes. Fungal α-amylases are extensively used in flour treatment for supplementing the diastatic activity of flour.

Amyloglucosidase.—The enzyme amyloglucosidase (glucoamylase, α-1, 4-glucan glucohydrolase, EC 3.2.1.3) acts as an exoase that liberates the α-1,4-linked glucose units consecutively from the nonreducing ends of the starch chains. Terminal α-1,6-bonds are also cleaved, but much more slowly than α-1,4-linkages (Pazur and Kleppe 1962). Maltose is only slowly attacked; increasing the chain length up to 5 or 6 glucose units gives faster attack.

Amyloglucosidase occurs in many microorganisms, particularly in starch degrading molds (*Aspergillus, Mucor, Rhizopus, Endomyces*) and certain bacteria (*Aerobacter, Clostridium*). The mold species commonly used for large-scale production is *Aspergillus niger*. However, this species also produces transglucosidase, which transfers glucose and higher saccharides to oligosaccharides, resulting in synthesis of polysaccharide. These products cause serious reduction in glucose yield and impede its crystallization. Strains are chosen which produce low levels of transglucosidase. Some organisms, e.g., members of the *Mucor, Rhizopus*, and *Aspergillus phoenicis* groups, usually produce amyloglucosidase without simultaneous formation of transglucosidase.

When *Aspergillus niger* is used as the production strain, the submerged fermentation is the cultivation method of choice. *Mucor* and *Rhizopus* strains are used in solid substrate fermentation because they are obviously unsuitable for submerged culture. The reason is that when the nonseptate hyphae of these species are damaged by the strong shearing action of high-speed impellers at one single point the protoplasm of the whole filament will be extruded.

Media employed for amyloglucodsidase production in submerged fermentation contain usually high solid contents of organic matter (of the order of 12–20%). Starchy material, such as maize, wheat, barley, rye, and sorghum, are the raw materials of choice. These cereals provide also the N source, the content of which is usually sufficient. Inorganic nitrogen can be utilized, ammonium salts being preferred. Stimulation of amyloglucosidase formation may be obtained by incorporation of yeast extract or stillage into the medium (Van Lanen and Smith 1965). An acid reaction of the mash seems to be necessary for maximal production of amyloglucosidase. In *Aspergillus niger* fermentations the pH of the culture decreases to values as low as 2.8 at the end of the cultivation, caused by the excretion of organic acids.

Removal of transglucosidase activity is an important step in the processing of amyloglucosidase. One of the oldest methods to free the culture filtrate from transglucosidase takes advantage of the fact that amyloglucosidase is highly acid resistant (Drews *et al.* 1954). Adsorption techniques (Croxall 1964) using synthetic water-insoluble hydrous magnesium silicate or clay minerals (e.g., attapulgite and bentonite) are said to be either unreliable or not sufficiently selective. Barton (1967) claimed coprecipitation with both maleic anhydride copolymers and heteropolyacids to be commercially practicable for removal of transglucosidase. Another claim of Sternberg (1967) suggests the precipitation of transglucosidase by chloroform, but the conditions of the precipitation process are very critical.

Amyloglucosidase has been bound covalently to a variety of matrices (Zaborsky 1973) and also has been physically adsorbed on or bound to various materials. Further, ultrafiltration reactors have been used for the continuous conversion of starch to glucose but obviously there is little inclination in industry to adopt processes based on immobilized amyloglucosidase.

Amyloglucosidase has largely replaced acid hydrolysis in glucose production (Denault and Underkofler 1963; Underkofler *et al.* 1965). Following the solubilization of starch by acid or, preferably by a heat stable bacterial α-amylase, further degradation is achieved using amyloglucosidase.

Pullulanase.—The enzyme pullulanase (amylopectin 6-glucanohydrolase, EC 3.2.1.41) splits α-1,6-glucosidic linkages at a branch point as well as in a linear chain of a number of oligo- and polysaccharides. It degrades pullulan completely to maltotriose and is active on amylopectin, glycogen, and their β-dextrins. The minimum chain which pullulanase liberates is maltose; a 3 unit chain (maltotriose) is optimal.

Pullulanase has been found in *Aerobacter aerogenes* (Bender and Wallenfels 1961) and other bacteria, *Streptomyces* included (Ueda *et al.* 1971). Production of the enzyme is inducible by growth on α-glucans ranging from maltose to glycogen. Yokobayashi *et al.* (1968) found that starch or liquefied starch with a low dextrose equivalent (D.E. 5–10) was particularly inducive. Highest yields occurred when this low D.E. syrup was used in all steps of inoculum buildup as well as in the production stage (CPC Intern. 1969). Further, for optimal synthesis, a comparatively high pH is required. Ueda *et al.* (1971) found that in *Streptomyces* the pH of the culture rose up to 8.5 and the enzyme was abundantly produced in the autolytic phase.

Isoamylase.—This enzyme (amylopectin 6-glucanohydrolase, EC 3.2.1. 68) is able to completely debranch glycogen, but unable to act on pullulan. It differs from pullulanase in that it acts more readily on the native polymers. The α-1,6-linkages are only split when present at branch points in oligo- and polysaccharides. Isoamylase requires a minimum of 3 glucose units on the A-chain of amylopectin, while pullulanase requires only 2.

Isoamylase occurs in yeast and bacteria. It is produced commercially from *Cytophaga* (Gunja-Smith 1972) and *Pseudomonas* (Harada *et al.* 1968), and is an inducible enzyme.

Other Starch-degrading Enzymes.—Among these, the following are worthy of mention: An exo-amylase from *Pseudomonas stutzeri* which splits off maltotetraose from the ends of starch chains (Robyt and Ackerman 1971); an exo-amylase releasing maltohexaose, which is produced by *Aerobacter aerogenes* (Kainuma *et al.* 1972); an enzyme from *Bacillus licheniformis* that forms maltopentaose as the principal product (Saito 1973); isopullulanase produced by certain species of *Aspergillus niger* (Sakano *et al.* 1972). This enzyme degrades pullulan to isopanose. It shows no action on amylopectin but is able to hydrolyze terminal isomaltosyl groups as produced by amyloglucosidase.

Cellulase

The name "cellulase" is given to all enzymes which cleave β-1,4-glucosidic linkages in cellulose and chemically or physically modified cellulose, in cellodextrin, and in cellobiose.

It is well established that cellulase is a multi-enzyme complex, the different components of which bring about the complete degradation of cellulose to monosaccharide residues (Gilligan and Reese 1954). A classification scheme of cellulolytic enzymes is given in Table 15.10. The terms C_1-cellulase and C_x-cellulase were originally proposed by Reese *et al.* (1950) for two components of the cellulase complex that differed in their substrate specificities against cotton fiber. According to this concept, C_1 can attack native cellulose of higher crystallinity (e.g., cotton fiber), while C_x can not attack such cellulose but can split in turn the cellulose fragments which have been produced by the action of C_1. The mode of action of the C_1-enzyme has long been questioned, but now many investigators tend to identify it with β-1,4-glucan cellobiohydrolase, which liberates cellobiose units from the nonreducing end of the cellulose chain (Halliwell *et al.* 1972).

Many bacteria and fungi are cellulolytic, but preparations marketed for industrial applications are derived only from *Aspergillus niger*, *Trichoderma viride*, *Neurospora*, and some other organisms. The *Aspergillus* enzyme exerts good activity on carboxymethylcellulose (CMC), but fails to attack solid cellulose because it lacks C_1-cellulase. In contrast, *Trichoderma viride* produces an enzyme complex with high levels of C_1-cellulase, which extensively degrades insoluble cellulose.

TABLE 15.10.   CLASSIFICATION OF CELLULOLYTIC ENZYMES

			Source of Enzyme (Examples)
cellulolytic enzyme	β-1,4-glucanase	endo-β-1,4-glucanase (= C_x-cellulase)	Many fungi
		exo-β-1,4-glucanase	
		β-1,4-glucan cellobiohydrolase (= C_1-cellulase)	*Trichoderma, Penicillium, Fusarium*
		β-1,4-glucan glucohydrolase (= cellobiase)	*Trichoderma*
	β-1,4-glucosidase		Many fungi

Cellulase in fungi is an inducible enzyme (Mandels and Reese 1957). It is only produced when the cells are grown on cellulose, on glucans of mixed linkages including the β-1,4 bond, and on a few oligosaccharides. The inducing effect of cellulose is due to soluble hydrolysis products of the cellulose, in particular cellobiose (Mandels and Reese 1960). Lactose, a β-1,4-galactoside, and sophorose, a β-1,2-glucoside, are the only known cellulase inducers that do not have a β-1,4-glucoside bond. The inductive action of sophorose is limited to *Trichoderma viride*. However, in spite of the impressive inductive power of this rare sugar, the levels of enzyme produced are not equal to those on cellulose.

Cellobiose plays a complex role: in low concentrations (0.1%) it serves as an inducer of cellulase; in high concentrations (0.5–1.0%) it represses cellulase formation, and, in addition, it can also act as an inhibitor of cellulase action.

Cellulase yields can be increased by various additives to the medium. Reese and Maguire (1971) observed that Tween 80 and Tween 40 doubled the cellulase yield in *Trichoderma* cultures. The mechanism of the action of these surfactants is not understood but may be related to increased permeability of the cell membrane. Nevertheless, Tween 80 has proven useful in the fermentation industry and is routinely incorporated into the culture medium. Further, enhancement of enzyme production can be achieved by supplying peptone at one-tenth the cellulose concentration. This leads to a decrease in the lag of growth and cellulase synthesis.

In actual large-scale fermentation *Aspergillus niger* is mostly cultured by the wheat bran-tray method. This process has no problems and leads to high yields of cellulase. The extraction of cellulase from solid substrate cultures is performed by percolation of the dried mold bran with 0.02 to 0.1 M lactic acid.

Neurospora and *Trichoderma* are grown by submerged culture. For continuous culture it is advantageous that *Trichoderma viride* produces a suspension of short mycelial threads, rarely forming pellets. Mitra and Wilke (1975) proposed a 2-stage operation in continuous stirred tank reactors. The first stage utilizes glucose for biomass production and the second stage utilizes pure spruce wood cellulose for enzyme formation. A significant increase in enzyme productivity was obtained.

Bran, straw, and other plant materials pretreated with alkali serve as cellulose-containing raw materials for submerged cultivations. Ammonium ions can be used as suitable N source. A correct pH profile is necessary to give optimum enzyme yields in batch culture (Nystrom and Allen 1976; Brown *et al.* 1975). It has been shown that a drop to the range of pH 3.5–3.0 is optimal for *Trichoderma viride*. This is usually achieved by an empirical procedure which involves medium composition and initial pH.

Length of cultivation affects the relative amount of the various cellulase components present in the medium. This has been demonstrated for *Myrothecium verrucaria* with CMC and with swollen substrates. In both cases C_1 appeared prior to C_x.

Concentration and purification of the enzyme is carried out by precipitation, adsorption, or gel filtration techniques. Granulated preparations can be obtained by mixing with salt hydrates (e.g., $Na_2SO_4 \cdot H_2O$) and subsequent vacuum drying.

Current use of cellulase is limited to improving texture and palatability of poor quality vegetables. It is also useful for accelerating drying of vegetables. A potential application of cellulase is the conversion of cellulosic materials to glucose and other sugars which in turn can be used as microbial substrates to produce single cell protein or a variety of fermentation chemicals (alcohol, etc.).

Pectolytic Enzymes

Many plant pathogenic bacteria and fungi have long been known to produce pectolytic enzymes and it is widely accepted that the production of these enzymes is a major means by which microorganisms invade the host tissue (Bateman and Millar 1966). Moreover, pectolytic enzymes are essential in the decay of dead plant material by nonpathogenic microorganisms and thus assist in recycling carbon compounds in the biosphere. Lastly, these enzymes play a decisive role in the microbial spoilage of fruits and vegetables.

Several types of enzymes are involved in the degradation of pectic materials. They are divided into 2 main groups, depolymerizing enzymes and saponifying enzymes or pectinesterases. According to the scheme of Neukom (1963), the depolymerizing pectolytic enzymes are further classified by applying the following 3 criteria: preference for pectic acid or pectin as substrate, hydrolytic or transeliminative cleavage of the glycosidic linkages, and endo- or exo-types of the action mechanism. By various combinations of these characteristics, 8 groups of depolymerizing enzymes can be listed, but the existence of the exo-polymethylgalacturonase and exo-polygalacturonate lyase types is doubtful. In 1971 Hatanaka and Ozawa proposed a scheme which included only those enzymes whose existence had been demonstrated (Table 15.11).

Occurrence of pectolytic enzymes has been reported in a large number of bacteria and fungi (Bhat et al. 1968). Commercial enzymes are generally obtained from fungal sources since the pH optima of these enzymes are in the range found naturally in materials to be processed. Most potent strains are selected from *Aspergillus niger*. Japanese enzyme manufacturers also use *Sclerotinia libertiana* and *Coniothyrium diplodiella* as producers of pectolytic enzymes.

In the majority of cases microorganisms produce a variety of pectolytic enzymes and, for this reason, commercial preparations are mixtures of these enzymes. The relative amount of the single components varies considerably with the particular strain employed, medium composition, and culture conditions. Careful observation of the factors responsible for the promoted synthesis of certain enzyme fractions or limited formation of others

TABLE 15.11. ALTERED HATANAKA AND OZAWA SCHEME OF THE CLASSIFICATION OF DEPOLYMERIZING PECTOLYTIC ENZYMES

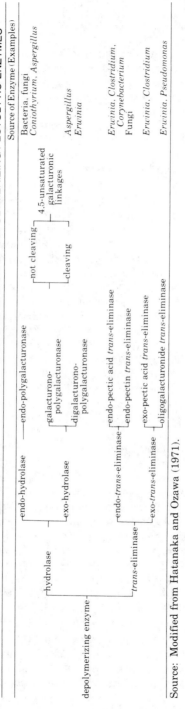

Source: Modified from Hatanaka and Ozawa (1971).

enables the manufacturer to "control" the composition of the preparation and to meet the need for specific formulations.

Biosynthesis of pectolytic enzymes is constitutive or controlled by the mechanisms of induction or catabolite repression. No uniformity exists among the various organisms and the various components of the enzyme complex. Phaff (1947) found pectolytic enzymes to be adaptive in *Penicillium chrysogenum*, but constitutive in *Aspergillus foetidus*. Saito (1955) showed that in *Aspergillus niger* endopolygalacturonase (endo-PGase) is adaptive. For *Clostridium felsineum*, a plant retting organism, Osman *et al.* (1969) demonstrated pectinmethylesterase (PME) to be adaptive and PGase constitutive.

On an industrial scale pectolytic enzymes are produced by the solid substrate method as well as by submerged culture (Beckhorn 1960). Wheat bran or defatted rice bran have been recognized as satisfactory basic substrates in solid substrate cultures. It is well known that some by-products of the food industry, such as beet pulp, apple pulp, or grape pulp, exert a promoting effect on enzyme formation. Other ingredients, e.g., nutrient salts, acid, or buffers, are also incorporated to regulate the pH during the growth of fungi. The time of cultivation can extend up to 7 days, but when *Aspergillus niger* is used, the desired enzyme level is normally reached within 36 to 72 hr. After fermentation the mold bran is dried and can be used as such. For obtaining concentrates the dried mold bran is extracted with suitable aqueous solutions and concentrated under vacuum or by ultrafiltration. Crude or refined solid concentrates are obtained by spray drying or precipitation with neutral salts or solvents.

Submerged cultures, in contrast to solid substrate cultures, seem to have the disadvantages of poor yields and undesirable composition. Brooks and Reid (1955), for example, found that *Aspergillus foetidus* produced endo-PGase and exo-PGase in surface culture, but only endo-PGase in submerged culture.

The production of pectolytic enzymes by submerged fermentation has been described by Nyiri (1968, 1969). As an example, the method reported for *Aspergillus alliaceus* can be cited. The fungus is grown in a liquid medium composed of 2% wheat bran, 2% $(NH_4)_2SO_4$, 0.25% KH_2PO_4, 0.25% yeast extract, 0.1% pectin (degree of esterification = 59%). The initial pH is 3.8, adjusted with HCl. Traces of silicon serve as antifoam. After inoculation with conidia, the medium is agitated and aerated, with the pressure inside the fermentor maintained at 141.8 kPa (1.4 atm). The fermentation is completed within 72 hr and the mycelium separated by filtration. The filtrate is cooled to 0°–1°C and the enzyme precipitated by addition of $(NH_4)_2SO_4$ during a period of 4 hr. After standing for 12 hr, the precipitate is washed and dried to give a solid concentrate.

Initial pH of the medium and pH development in the growing culture play an important role with regard to both enzyme composition and yield of the enzyme fractions (Tuttobello and Mill 1961; Tejerina and Fernandez 1966; Perley and Page 1971). Low pH values, in particular a decreasing pH during cultivation, are favorable for the production of pectinases used in the fruit juice industry, according to Hauptmann (1951), pH 2–3 at culture maturity of *Aspergillus niger* being desirable. In contrast, Tuttobello and Mill (1961) allowed the pH to rise up to 4 at the end of the culture following a drop to about 3 on the fifth day of cultivation.

Tuttobello and Mill (1961) found an aqueous extract of nondefatted peanut flour strongly stimulated the production of pectolytic enzymes. They also observed a strong influence of inoculum on enzyme production. The kind of inoculum and, particularly, its size have to be standardized carefully. With *Aspergillus niger* Tuttobello and Mill (1961) found that the production of pectolytic enzymes was strongly influenced by inoculum size in the range of 10^4 to 2×10^5 conidia per ml, while mycelium formation remained unchanged.

From many reports in the literature it can be concluded that there is a strong variation in relative activity of the various components of the pectolytic system in the course of fermentation (Wang and Pinckard 1971); Edstrom and Phaff 1964; Bateman 1966), indicating their sequential production.

Extensive use of pectolytic enzymes is made in processing fruit juices for increasing juice yields on pressing, as aid in clarification of juices, and for depectinizing in order to obtain high density fruit juice concentrates. Fungal enzymes are widely used in producing apple juice, grape juice, and wine. In the production of coffee beans the residual mucilaginous coating surrounding the bean can be liquefied by commercial pectolytic enzyme preparations, thus offering an alternative to the usually used fermentation process. The curing or fermentation of cocoa, tea, and tobacco also can involve pectolytic enzymes. One of the oldest applications of these enzymes is the process of retting, in which textile fibers, such as flax, hemp, and jute, are loosened from their plant stems. The enzyme system of *Clostridium felsineum*, an organism that is involved in aerobic retting, contains endopolygalacturonate *trans*-eliminase, but not pectinesterase. Recently, pectolytic enzymes have been proposed as a means to make commercial softwoods, such as Sitka and Norway spruce, more permeable to preservatives. It has been demonstrated that treatment with enzyme preparations as well as with the specific bacteria that produce them is possible.

Hemicellulase

Plant cell wall polysaccharides other than cellulose and pectic substances are referred to as hemicelluloses. They are complex compounds and very few of their chemical structures have been clarified. During recent years many enzymes have been recognized which specifically act on different types of hemicelluloses. By far the best known group is that of the xylanases. Other groups of hemicellulases are, for example, mannanases, galactanases, etc.

Many strains of bacteria and fungi are known to produce hemicellulases inducibly or constitutively, but on an industrial scale only fungal strains seem to be used as enzyme sources. Even in these cases hemicellulases are mostly obtained as side activities in the production of other enzymes such as cellulase (for commercial enzyme preparations containing hemicellulases see Sinner and Dietrichs 1975). Therefore, as a rule, the hemicellulase activity of the commercially available enzyme systems is low. This fact reflects either an inherent instability of these enzymes or lack of knowledge of how to produce them.

The potential application for hemicellulases is great. For example, β-glucanase is used in the brewing industry to degrade the barley β-glucans for solving pumping and filtering problems (Enkenlund 1972).

Invertase

Sucrase and invertase are two of the older names of the enzyme β-fructofuranosidase (EC 3.2.1.26). It catalyzes the hydrolysis of the terminal nonreducing β-fructofuranoside residues in β-fructofuranosides. The name "invertase" was derived from the action in splitting sucrose, which is optically dextrorotatory, to form glucose and fructose, a mixture that is levorotatory:

$$\underset{[\alpha]_D \ = \ +66.5°}{\text{sucrose}} \quad + \ H_2O \ \xrightarrow{\text{invertase}} \ \underset{[\alpha]_D \ = \ +52.5°}{\text{D\,(+)-glucose}} \quad + \ \underset{[\alpha]_D \ = \ -92°}{\text{D\,(-)-fructose}}$$

$$[\alpha]_D \ = \ -20°$$

This reaction can also be carried out by α-D-glucosidase (so-called glucosidoinvertase), but this enzyme is unable to split off fructose from the trisaccharide raffinose as is β-fructosidase (so-called fructosidoinvertase).

β-Fructosidase can be prepared from a variety of microbial sources, but only the enzymes from *Saccharomyces cerevisiae* and *Saccharomyces carlsbergensis* have industrial importance.

Biosynthesis of invertase is controlled by a catabolite repression mechanism in *Saccharomyces fragilis* and by repression through unknown effectors in *Saccharomyces cerevisiae*. For a long time invertase was considered to be totally an intracellular enzyme. It has, however, been established that in derepressed cells only a small proportion of the invertase is located inside the cytoplasmic membrane (Demis *et al.* 1954; Sutton and Lampen 1962), most of it being retained externally within the cell wall or between the wall and the cell membrane. In fully repressed cells all the enzyme is intracellular (Friis and Ottolenghi 1959).

The release of the invertase from yeast is achieved by destruction of the structures responsible for the retention of the enzyme. There are various ways in which the separation from the cells can be accomplished. One method is autolysis with chloroform, toluene, or ethylacetate at 30°C for not over 3 hr. Following extraction from yeast, comparatively high purification of invertase is necessary for its application in foods because the enzyme preparation usually has an undesirable, irritating taste originating from yeast. These procedures include common methods for purification of enzymes such as ultrafiltration, precipitation, and adsorption techniques.

The commercial preparation of invertase usually starts with an accumulation step. For this purpose pressed bottom yeast is suspended in a 20-fold amount of nutrient broth containing 4 parts $(NH_4)_2 \cdot HPO_4$, 4 parts KH_2PO_4, 1 part $Mg(NO_3)_2$, and 1 part KNO_3. The mixture is aerated for 3–8 hr, while the temperature is maintained at 28° to 30°C and at a pH of 4.5. During the same period, 3 to 20% sucrose in solution is added continuously, a procedure that ensures reduced catabolite repression. At the end of the process the invertase activity of the yeast is increased up to 15-fold (Underkofler and Hickey 1954).

The intracellular yeast invertase has a molecular weight of 135,000 and is free of carbohydrate, whereas the external enzyme has a molecular weight of 270,000. Approximately half of the external β-fructofuranoside consists of mannan. The pH activity curve of invertase is rather broad between pH 3.5 and 5.5, with an optimum between 4 and 4.5. This is the same range within which the enzyme exhibits its highest stability. Yeast invertase is strongly inhibited by heavy metal ions (especially Ag^+); they combine with the histidine side chains of the enzyme molecule, not with its thiol groups (Myrbaeck 1967).

Invertase has a number of interesting uses in the confectionery industry for soft center candies, fondant, and chocolate coatings. Its use in the preparation of invert sugar by hydrolyzing sucrose has been restricted by glucose isomerase which permits production of invert sugar from cheaper sources.

Lactase

Lactose or milk sugar is enzymically split to glucose and galactose by the action of enzymes called β-galactosidases or, more commonly, lactases (β-D-galactoside galactohydrolase, EC 3.2.1.23). Lactase is very specific for the galactose residue but much less specific for the aglycone moiety of β-galactosides. The enzyme is also responsible for transfer activities which occur with the formation of oligosaccharides (Pazur et al. 1958).

Lactase is widely distributed in microorganisms. Some strains of Escherichia coli are very potent producers, but are not suitable for food purposes. Available commercial preparations are derived from lactose fermenting yeasts such as Saccharomyces fragilis, Zygosaccharomyces lactis, and Candida pseudotropicalis (Myers and Stimpson 1956; Young and Healey 1957; Wendorff and Amundson 1971) or from fungi like Aspergillus niger, and particularly a mutant strain of Aspergillus foetidus (Borglum and Sternberg 1972).

The biosynthesis of β-galactosidase has been extensively investigated, largely in Escherichia coli in connection with studies on the biosynthesis of proteins and its genetic control. The enzyme is of the inducible type, with lactose serving as an inducer. In Fusarium oxysporum and Verticillium alboatrum the lactase can be induced by D-galacturonic acid and, to a lesser extent, by D-galactose.

On an industrial scale the enzyme is obtained, for example, by growing yeast on a lactose medium or on whey. The separated yeast is autolyzed or extracted, and a cell-free extract is obtained by centrifugation or filtration. The enzyme may then be further processed by salt or solvent precipitation (Connors and Sfortunato 1956). Another procedure, described by Stimpson (1954), involved spray drying of the washed yeast at temperatures which destroy any residual alcoholic fermentation activity, thus leading to crude products which can immediately be used as lactase preparations.

The lactases from various microbial sources differ in properties such as pH optima, etc. For example, the pH optimum of the bacterial enzymes is around 7.0; that of the fungal preparations near 5.0; and that of the yeast enzymes near 6.0; the lactase from *Corticium rolfsii* is distinguished by its unusual maximum activity and stability at pH 1.8–2.0.

Yeast lactase is activated by potassium and ammonium ions and is inhibited by certain metals such as copper and iron. Metal-chelating agents do not stimulate the enzyme, indicating little sensitivity to trace heavy metals. The addition of reducing compounds, e.g., cysteine, sodium sulfide, or potassium metasulfite, is able to overcome the effect of metal inactivators and to activate the enzyme (Stimpson and Stamberg 1956).

Hereditary intolerance to lactose precludes use of milk as a valuable protein source in large areas of Asia and Africa. In addition, lactose causes a number of problems in the dairy and allied industry because of its poor solubility, resulting in crystallization in concentrated dairy products. Enzymic hydrolysis of the milk sugar is helpful in overcoming these problems. Moreover, lactase is used in the production of sweet syrups from sources of lactose such as cheese whey and in making waste whey a better substrate for growing microorganisms for single cell protein.

When treating milk it is preferred to employ lactase from yeast because of legal reasons, although this enzyme is less stable than the bacterial one. During past years the use of skim milk powders in bread has been considerably reduced. This has reduced the interest in lactase for the formation of fermentable sugars in baking.

Proteases

The microbial proteases which are of interest for application in the food industry are all of the endopeptidase type and are all extracellular enzymes. There are many different types of proteases produced by an extraordinarily large number of microorganisms, but in actual practice the enzymes prepared commercially are of a very limited number of types and they are derived from very few organisms (see Table 15.12).

The proteolytic enzymes from microorganisms are classified into 4 main groups according to the scheme of Hartley (1960) and based on the mechanism of their action: serine proteinases, thiol proteinases, metalloproteinases, and acid proteinases. Further subgroupings refer to the side-chain specificity of the proteinases and to the properties of their active centers (see Morihara 1974). A classification scheme of microbial endopeptidases is given in Table 15.13.

TABLE 15.12.   ORGANISMS CURRENTLY USED FOR PROTEASE PRODUCTION

Organism	Enzyme Type	Product[1]
Bacillus subtilis	Metallo, serine	Montase, Milezyme
	Serine	Alcalase, Maxatase
B. thermoproteolyticus	Metallo	Thermoase
Streptomyces griseus	Metallo, serine	Pronase
Aspergillus oryzae	Acid	Rhozyme A-4
A. saitoi	Acid	Molsin
Mucor pusillus	Acid	Microbial Rennet

[1]Trade names, except for "Microbial Rennet."

Industrial production of microbial proteases is carried out on a large scale by a number of companies in Europe, Japan, and the United States. For cultivation of the microorganisms the submerged fermentation is the preferred method; with bacteria it is the exclusively used process. However, fungi usually give higher yields when cultured on solid media so this method continues to play a role. As in most fermentations there is a trend to use highly concentrated media. The reason for this is that one can expect higher enzyme yields per unit volume with a larger cell concentration, although there is no direct correlation between growth and protease production. With regard to serine and metalloproteinases it seems that low concentrations of purely carbonaceous substrates and high concentrations of proteinaceous N sources stimulate production.

Many of the organisms excrete more than one kind of protease. The type of proteolytic enzyme formed may depend on the composition of the medium. For example, Bacillus NRRL B-3411 produces the preferable neutral protease when grown on a grain medium, but mainly alkaline protease when cultured on a fishmeal-enzose-cerelose medium (Keay et al. 1972). The biosynthesis of proteases is often correlated with particular growth phases of the microbial culture. Under most growth conditions, Bacillus species produce extracellular protease during the postexponential growth phase (Schaeffer 1969; Dawson and Kurz 1969). Mandelstam (1958) attributed this behavior to an increased need for turnover of cell proteins at the slower growth rate. Other bacilli synthesize proteases during the exponential growth phase (Chaloupka and Křečková 1966; Chaloupka 1969). However, these kinetics depend on the composition of the medium.

For all protease preparations the degree of purification depends on the intended use. A number of purification procedures are in existence, which follow the general description of recovery discussed on page 651. Of course, various combinations are possible. Particular care is necessary during the drying process in order to avoid the formation of dust. For this reason protease preparations are pelleted or coated with some suitable material.

Bacterial proteases are used on a large scale in enzyme-containing washing powders, but they are not widely used in food processing. Minor uses are in the chillproofing of beer, in the production of protein hydrolysates, in the production of condensed fish solubles, and as feed supplement. In contrast to bacterial preparations fungal proteases are the more interesting group for

TABLE 15.13.   CLASSIFICATION OF MICROBIAL PROTEINASES (ENDO-TYPE)

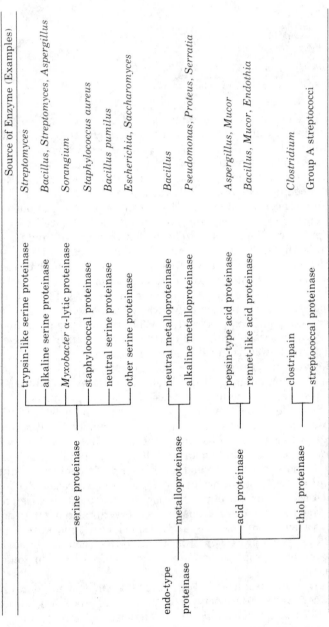

endo-type proteinase			Source of Enzyme (Examples)
serine proteinase	trypsin-like serine proteinase		*Streptomyces*
	alkaline serine proteinase		*Bacillus, Streptomyces, Aspergillus*
	Myxobacter α-lytic proteinase		*Sorangium*
	staphylococcal proteinase		*Staphylococcus aureus*
	neutral serine proteinase		*Bacillus pumilus*
	other serine proteinase		*Escherichia, Saccharomyces*
metalloproteinase	neutral metalloproteinase		*Bacillus*
	alkaline metalloproteinase		*Pseudomonas, Proteus, Serratia*
acid proteinase	pepsin-type acid proteinase		*Aspergillus, Mucor*
	rennet-like acid proteinase		*Bacillus, Mucor, Endothia*
thiol proteinase	clostripain		*Clostridium*
	streptococcal proteinase		Group A streptococci

the food industry. They are used, for example, for the modification of wheat proteins in bread doughs, in meat tenderizing, and in several less important applications.

Serine Proteinases.—These proteases are widespread in bacteria and fungi. They show maximum activity at neutral to alkaline pH and are inhibited by diisopropyl fluorophosphate (DFP) or phenylmethane sulfonylfluoride (PMSF). They can be classified into at least 5 groups: trypsin-like proteinases, alkaline proteinases, *Myxobacter* α-lytic proteinase, staphylococcal proteinases, and serine neutral proteinases. In this article only the serine alkaline proteinases will be treated because of their superior economic importance.

Proteases of this type are most active at pH 9.5–10.5; they are sensitive to DFP and potato inhibitor, but not to tosyl-L-lysine chloromethyl ketone. Their specificity is similar to that of an α-chymotrypsin but somewhat broader. All alkaline serine proteases show specificity toward aromatic or hydrophobic amino acid residues, such as tyrosine, phenylalanine, or leucine, at the carboxyl side of the cleavage point. The molecular weights are 26,000–34,000, slightly below the range of neutral metalloproteases. The isoelectric points are about pH 9. Most of the alkaline proteases are stable from pH 5 to 10 at low temperatures, but show rapid loss of activity at 65°C. Certain strains of *Bacillus*, showing alkalophilic properties, synthesize a serine alkaline proteinase that is most active at pH 11–12 (Aunstrup *et al.* 1972).

Serine alkaline proteinase is produced by numerous species of bacteria and fungi. The best known representatives of this type are the subtilisins, which are produced by *Bacillus subtilis* and related species.

Due to their great economic importance, the biosynthesis of *Bacillus* alkaline proteases has been well investigated. Keay and Moser (1969) have proposed that alkaline serine proteinases produced by different bacilli or different strains of *Bacillus subtilis* can be divided into two groups: subtilisin Carlsberg and subtilisin Novo. These enzymes are quite distinct from each other, but possess many similar properties. It may be mentioned that a similar situation has also been observed with various alkaline serine proteinases from the genus *Aspergillus* (Turkova *et al.* 1972).

The synthesis of these enzymes is linked to particular phases of development of the microbial culture. Some strains, e.g., those of *Bacillus megaterium*, produce the protease during the log phase of growth, while others, like those of *Bacillus subtilis* and *Bacillus cereus*, produce it in the stationary phase. However, as mentioned above, the relationship between growth cycle and enzyme formation depends on the ingredients of the substrate. It is generally valid that the time of biosynthesis is genetically determined and can be changed or extended by selecting proper mutants.

In most species production can be inhibited by certain components in the growth media, such as free ferric ions, amino acids, carbon sources, or several of these. Catabolite repression and availability of nucleic acid precursors are also thought to play a role in alkaline protease synthesis. The concentration of purely carbonaceous medium components should normally be kept on a low level. This can be achieved by incremental feeding of the C source, e.g., glucose, keeping its concentration at a range of 0.4 to 1% (Güntelberg 1954). High concentrations of C sources yield excess organic acids leading to a decrease in pH, which is accompanied by a decrease in alkaline protease production. On the other hand, Keay *et al.* (1972) have shown that the enzyme yield can be largely enhanced when the synthesis of protease is accompanied by an increase in pH of the culture. Such an increase in pH can be reached, for example, by using an organic acid (or its salts) as major C source. Niwa *et al.* (1971) observed good yields of protease under this condition, and the results of Kline *et al.* (1944), Dion (1950), and Maxwell (1952) confirm this assumption.

Bacterial alkaline proteases are produced exclusively by the submerged culture methods. Amounts of more than 1 g protease per liter culture liquor are quite usual. With specially selected strains markedly higher yields are possible; e.g., *Bacillus subtilis* strain AJ 3266 can produce more than 10 g enzyme per liter.

Continuous fermentation techniques do not seem to have been employed in the industrial production of alkaline protease, but there are several publications describing continuous culture on the laboratory scale. Heineken and O'Connor (1972) observed that a continuous fermentation of *Bacillus subtilis* yielded mutants with lower protease productivity.

Fungal alkaline proteases are mainly produced from *Aspergillus* species, in both solid substrate and deep tank fermentations. Solid substrate cultures, extensively used in Japan, are carried out with wheat or rice bran or whole grains as the basic substrate. It has been shown that NH_4^+ ions strongly inhibit production of the enzyme, while nitrates and Na salts of aspartic and glutamic acids promote its formation. Na salts of organic acids had the same effect. Probably the effect of all these compounds on protease synthesis is produced through their influence on pH development during cultivation, as described previously for α-amylase.

The processing of culture filtrates or clarified extracts follows the general description of recovery.

Serine alkaline proteases of bacterial origin are used in large amounts in laundering and to a lesser extent in leather tanning and the food industry.

Metalloproteinases.—Enzymes of this type play a less important role in commercial applications than the serine and acid proteinases. This is mainly due to their relatively poor stability. Metalloproteinases exhibit maximum activity at pH 7 to 8. In the majority of cases they contain a Zn atom in their active center. They are inhibited by metal chelating agents such as ethylenediaminetetraacetate (EDTA) or *o*-phenanthroline (OP), but not by DFP or thiol reagents.

Regarding their pH activity metalloproteinases are divided into neutral and alkaline types. The neutral enzymes all have pH optima around pH 7. Their molecular weights are in the range of 35,000–45,000. The isoelectric points of the proteinases from *Bacillus subtilis, Pseudomonas aeruginosa,* and *Streptomyces* have been determined to be at pH 9.0, 5.9, and 4.2, respectively. Neutral metalloproteinases of bacterial and fungal origin are specific toward hydrophobic or bulky amino acid residues on the amino side of the cleavage point. In general, these enzymes are the least stable of the microbial proteases. The enzyme from *Bacillus subtilis* retains only about 10% of its activity after treatment at 60°C and pH 7 for 15 min. *Bacillus thermoproteolyticus* produces a very stable neutral protease (thermolysin) that retains 50% of its activity after 60 min at 80°C. Neutral proteases tend to undergo very rapid autolysis, which makes their recovery and application difficult.

Neutral proteases are widespread in microorganisms, both fungi and bacteria. Strains used for industrial production belong to the genera *Aspergillus, Bacillus,* and *Streptomyces.* Many of the organisms used for commercial production of neutral proteases also produce alkaline or acid proteases. Only few strains have been found which synthesize neutral protease free of accompanying serine and acid proteases. Such strains are, for example, *Bacillus cereus* ATCC 14579 and NCTC 945, *Bacillus megaterium* ATCC 14581 and MA, and *Bacillus polymyxa* ATCC 842.

The formation of neutral proteases by bacteria does not seem to be correlated with sporulation. In *Bacillus subtilis* enzyme synthesis is subject to catabolite repression. Fogarty and Griffin (1973) found that the enzyme was produced irrespective of the C source used, but that the nature of the peptone had a marked effect on protease accumulation. Without pH adjustment during the fermentation, the culture produced the neutral protease parallel to growth, and enzyme formation reached its maximum toward the end of the log phase. The yield was 15 times that obtained when the culture was run with a "fixed" pH of 6.8. According to Kalunyants *et al.* (1974), the best medium pH (6.9) can be achieved by separate sterilization of carbohydrate and N-containing compounds of the medium. Variable temperature during fermentation (45°C at the beginning and lowering it to 43°, 40°, and 37°C, respectively during the 1st, 2nd, and 3rd 2-hr period) was preferable to a constant temperature. Zn^{++}, Ca^{++}, and Mn^{++} exert a beneficial effect on the level of the metalloproteinase, as described by several authors (Stockton and Wyss 1946; Povilaitiene *et al.* 1976).

In cultures of *Aspergillus* species, biosynthesis and externalization of neutral protease are repressed by low molecular weight sources of C, N, and S. Protease production and release occur when the medium is deficient for any of these elements (Cohen 1973). For *Aspergillus terricola* it has been shown that the enzyme accumulation in the medium was maximal when the N/C ratio was 0.5 (Aleeva *et al.* 1973).

Due to their high instability, processing of metalloproteinases may lead to high activity losses. Therefore, the main problem in the concentration and purification of the enzyme is its stabilization. This can be achieved by strictly observing the tolerated range of pH, by the presence of metal ions (Zn^{++} for activity, Ca^{++} for stability), and by elimination of alkaline protease activity (Feder and Kochavi 1971). One has also to take into consideration that the pigment complex produced by an organism can act as an inhibitor of the neutral protease of this organism, as shown for *Bacillus mesentericus* by Velcheva and Kolev (1975).

The application of neutral metalloproteinases is very limited because of the mentioned instability of these enzymes. Actual and potential uses are: treatment of beer, application in bakeries, and reduction of dental plaque in humans.

Acid Proteinases.—These enzymes are without doubt the most interesting group of proteases with respect to use in the food industry. They are characterized by maximum activity and stability at pH 2.0–5.0. The molecular weight is around 35,000. Acid proteinases are low in basic amino acid content and have low isoelectric points. They are insensitive to SH-reagents, metal chelators, heavy metals, and DFP and are generally stable in the acid pH range (pH 2–6), but are rapidly inactivated at higher pH values. The acid proteases exhibit limited esterase activity, but split a wide range of peptide bonds.

Acid proteinases of commercial importance are prepared exclusively from fungal sources and are tentatively divided into two subgroups by their physiological characteristics: pepsin-like acid proteinases and rennin-like proteinases.

Pepsin-like acid proteinases have usually been reported in the group of black aspergilli, such as *Aspergillus niger, Aspergillus awamori, Aspergillus usamii,* and *Aspergillus saitoi,* but also occur in species of *Penicillium, Rhizopus,* and others. To a large extent they are produced in solid substrate cultures. Biosynthesis of these enzymes is favored by high C/N ratios (Shibata and Nakanishi 1967). Inorganic N sources show an inhibiting effect on the production of acid proteinases, whereas peptone was found highly effective in inducing this enzyme in a strain of *Aspergillus niger* (Shinmyo *et al.* 1968). The inducing action of peptone was much more remarkable when this material was added to a culture during the growth phase than in the stationary phase. From this finding it is evident that the increase in acid proteinase activity observed when growth has ceased represents de novo synthesis. It has been shown to be due to exhaustion of adenine-group growth substances. Adipic and glutaric acids were also highly effective in supporting enzyme formation, but amino acids tested and some dipeptides were less effective than peptone.

Acid proteinases of the pepsin type play an important role in the production of fermented foods by molds from soybeans, rice, and other cereals. They are further used in the baking industry for the modification of wheat proteins in bread doughs.

Rennin-like acid proteinases are produced by strains of *Mucor miehei, Mucor pusillus, Endothia parasitica,* and *Trametes sanguinea.* The enzyme from the *Mucor* species has been, and is now, produced by the solid substrate culture method (Arima *et al.* 1967). However, Aunstrup (1974) isolated a strain of *Mucor miehei,* which he succeeded in growing in submerged culture for rennin production.

Microbial rennet substitutes must be freed of lipase to avoid rancidity of the cheese. This can be achieved by controlled heating or by adjusting to a low pH (Schleich 1970). Unspecific proteolytic enzymes, which may cause bitter taste of the cheese, must also be removed. Their separation is obtained by adsorption on aluminosilicates (Organon 1970). These adsorbents are particularly advantageous because they can be mixed into the culture liquid at the end of the fermentation, even in the presence of the medium, with a good separation effect and without any loss of milk clotting activity. Bentonite, permutite, and attapulgite are also suitable.

Because of their particular properties, rennet-like microbial proteases are used for clotting of milk in cheese manufacture. The process is based on the coagulation of casein under the influence of the rennet-like protease. It is known that the casein in milk is mainly composed of α_s-, β-, and κ-casein. In particular, κ-casein plays an important role in the coagulation process, because it keeps the casein micelles present in milk in solution and protects them against flocculation by calcium ions. The clotting effect of rennins consists of the destabilization of the casein complex. Two phases can be distinguished:

(1) The primary or enzymatic phase, in which the protective colloid (κ-casein) of the casein micelle is broken down and a glycomacropeptide is split off as follows:

$$\kappa\text{-casein} \xrightarrow{\text{rennin}} \underset{\text{insoluble}}{para\text{-}\kappa\text{-casein-}}\underset{\text{soluble}}{\text{glycomacropeptide}}$$

(2) The secondary or nonenzymatic phase, in which the coagulum is formed under the influence of calcium ions

The primary phase has a temperature coefficient, Q_{10}, of about 2 (like most enzyme reactions), whereas the secondary phase has a Q_{10} of about 15. Therefore, it is reasonable to develop a system for continuous clotting of milk employing immobilized enzymes. Consequently, in a 2-stage enzyme reactor the enzymatic phase is conducted in the first stage at low temperatures in order to inhibit the nonenzymatic phase. In the second stage subsequent warming clots the milk by the action of calcium ions. The whole process of cheese manufacturing is schematically illustrated in Fig. 15.5.

Lipases

The enzyme lipase catalyzes the reaction:

$$\text{triacylglycerol-}H_2O \xrightarrow{\text{lipase}} \text{diacylglycerol} + \text{fatty acid anion}$$

This reaction goes to completion, i.e., until glycerol and free fatty acids are formed.

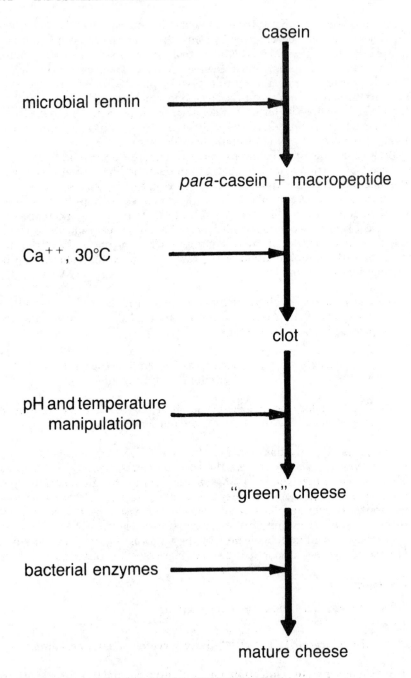

FIG. 15.5.   DIAGRAM OF CHEESE MANUFACTURING

Many, perhaps most, bacteria and fungi produce lipase. Potent producers are among the fat-producing microorganisms. However, no distinct relationship between capacity of fat production and lipase production has been found. The enzyme from *Candida cylindracea* is commercially available. Other producers of lipase are, e.g., *Geotrichum candidum, Rhizopus arrhizus,* and *Aspergillus niger. Geotrichum candidum* lipase is unique with respect to its specificity properties (Brockerhoff and Jensen 1974). The enzyme from the thermophile, *Humicola lanuginosa,* exhibits better thermostability (Liu *et al.* 1972).

The production of lipase is markedly affected by many factors. Generally, synthetic media produce lower yields of lipase than complex media. Lipase formation is highly dependent on nutrient and physical conditions. In a number of cases, addition of lipid material or fatty acids to the culture medium was found to enhance lipase production (Yoshida *et al.* 1968). In contrast, Smith and Alford (1966) observed inhibition of lipase formation in media supplied with lard, sodium oleate, or salts of other unsaturated fatty acids. These authors also reported that the inhibition was prevented, but not reversed, for example, by some divalent cations and Tweens. Glucose is unsuitable as the C source and ammonium ions seem to be unsuitable as the N source. Incorporation of $CaCO_3$ acts differently, depending on the strain used. In some cases promotion of lipase production was observed, whereas in other cases investigators found an inhibitory effect using $CaCO_3$.

Sometimes lipase preparations prove to be unstable. This can be due to the presence of proteases, the removal of which will lead to stable products.

Despite the fact that there is a considerable industry based on fats there is little industrial application of lipases. The main use is as digestive aid.

Glucose Oxidase

Glucose oxidase (β-D-glucose:oxygen oxidoreductase, EC 1.1.3.4), also known as notatin, acts in the presence of molecular oxygen to convert glucose to gluconic acid and hydrogen peroxide:

$$C_6H_{12}O_6 + O_2 + H_2 \xrightarrow{\text{glucose oxidase}} C_6H_{12}O_7 + H_2O_2$$

It is highly specific for β-D-glucose, although slight activity is found with 2-deoxyglucose (Gibson *et al.* 1964).

At present, glucose oxidase is commercially prepared from *Aspergillus niger* and *Penicillium amagasakiense* in submerged culture. It has also been reported that *Penicillium notatum* and *Penicillium chrysogenum* synthesize glucose oxidase on liquid media in surface culture, but not in submerged culture.

During the growth of the fungal culture, the enzyme occurs in the phase following the lag phase. By feeding glucose this phase can be extended and thus the enzyme yield enhanced. The special culture conditions, however, depend markedly on microbial strains used. For example, beet molasses has proved to be a suitable carbon source in *Penicillium purpurogenum*, but not suitable in *Penicillium chrysogenum*; high aeration rates supported enzyme synthesis in *Penicillium purpurogenum*, but did not in *Penicillium chrysogenum* (Nakamatsu *et al.* 1975).

For the purpose of concentration and purification, glucose oxidase must be separated from cells by extraction. The crude solutions also contain catalase which may interfere with glucose oxidase in some applications. In these cases separation is conducted by adsorption of the catalase on alumina or kaolin (Gulyi *et al.* 1967; Lebedeva *et al.* 1966). For preparation of solid products, glucose oxidase can be precipitated by neutral salts or solvents, but liquid preparations are preferred.

Glucose oxidase is a good glycoprotein. The enzyme from *Aspergillus niger* contains 10.5% carbohydrate, which is believed to contribute to the stability and not to affect the overall structure. Two FAD molecules per molecule of enzyme act as the prosthetic group. The molecular weight of the *Aspergillus niger* enzyme is 186,000, that of *Penicillium notatum* is 152,000. The optimum pH of glucose oxidase is about 5.5. The enzyme is fully stable between pH 4 and 6 at 40°C for 2 hr. Specially stabilized preparations for use at pH 2.5 are available. Use above pH 8.0 may be possible, but requires a high glucose concentration. Glucose oxidase is very unstable above 50°C, although glucose has some protective effect. Normal increase in activity caused by increased temperatures is counteracted by decrease in dissolved oxygen concentration at higher temperatures.

According to the reaction equation, glucose oxidase can be used in order to remove glucose or oxygen or to form hydrogen peroxide or gluconic acid. Indeed, in food processing glucose oxidase finds application for removal of residual glucose prior to the preparation of dried eggs or to remove it from other products in order to reduce nonenzymatic browning. It is highly effective in removing residual oxygen from beer, wine, fruit juices, high fat products (mayonnaise), or packaged dehydrated foods. In the treatment of flour, when the formation of peroxide is desired, only catalase-free preparations can be used.

Catalase

Catalase (EC 1.11.1.6) splits hydrogen peroxide to water and oxygen:

$$2H_2O_2 \xrightarrow{\text{catalase}} 2H_2O + O_2$$

The enzyme is widely distributed in microorganisms. Its biological role has been studied by a number of investigators. In methanol-utilizing yeasts, it is generally accepted that catalase must be involved in the metabolism of methanol, since hydrogen peroxide is liberated during methanol oxidation by alcohol dehydrogenase (Sahm and Wagner 1973). This suggestion is supported by the fact that catalase is markedly induced when the yeast cells are grown on methanol.

Commercially, catalase is prepared from *Aspergillus niger, Penicillium vitale*, or *Micrococcus lysodeikticus*. It is a hemo-protein containing 4 ferri-protoporphyrin prosthetic groups per molecule of enzyme, with a molecular weight of 250,000. The optimum pH of *Aspergillus niger* catalase is at pH 6.0; 75% of its optimum activity occurs between pH 3.0 and 9.0. The enzyme is inactivated by cyanides, phenols, alkali, urea, freezing, and by sunlight under aerobic conditions.

Production of catalase is conducted in deep tank cultivation. Biosynthesis occurs simultaneously with glucose oxidase formation. The ratio of these enzymes is controlled by quality and quantity of the inoculum, the composition of the medium, and by aeration conditions.

With *Penicillium vitale*, Nikolskaya et al. (1972) found that greater amounts of catalase were accumulated when the C:N ratio in the medium was as high as 12:1. Optimum pH was 4.5–5.5 during the first 48–72 hr of growth at 26°–27°C. L-Cysteine and DL-methionine promoted catalase production by the same fungus (Nikolskaya and Sinyavskaya 1971). The stimulating effect observed with $CaCO_3$ was demonstrated to be not the result of neutralization (Pokrovskaya and Kislyakova 1966). Ca^{++} was believed to facilitate the transport of the enzyme from the mycelium into the medium.

Catalase produced by the directed biosynthesis can be separated selectively from extracts containing catalase and glucose oxidase. The recovery of catalase from *Micrococcus lysodeikticus* starts with lysing of the cells in a solution of sodium chloride (0.5 to 2%). The next step is fractionation of the lysate by centrifugation of a mixture of the lysate, an organic solvent (ethanol in an end concentration of 40 to 50% v/v), and a salt (sodium or potassium chloride adjusted to a concentration of 1 to 2% either before or after addition of the solvent). The dissolved catalase can then be precipitated from solution by adding ethanol to a final concentration of 75%.

Separation of catalase from solution can also be conducted by adsorption methods using alumina or kaolin as adsorbents (Gulyi *et al.* 1967; Lebedeva *et al.* 1966). For commercial application, liquid preparations are preferred.

Catalase finds application wherever the removal of hydrogen peroxide is required or the controlled release of oxygen from hydrogen peroxide is desired. Therefore, in the food industry catalase is employed to remove the excess of hydrogen peroxide used for cold sterilization in milk and cheese processing. Catalase may also be employed in cake baking as well as in irradiated foods, in the process of which hydrogen peroxide is formed.

Glucose Isomerase

This enzyme converts D-glucose to D-fructose. The main substrate of this enzyme, however, is xylose and, indeed, the glucose isomerizing enzyme is a D-xyloseketoisomerase (EC 5.3.1.5) with side activities to D-glucose and D-ribose (Yamanaka 1968), as shown in Fig. 15.6.

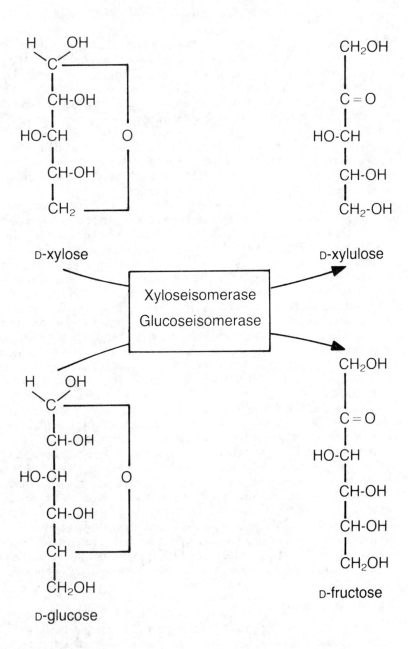

FIG. 15.6. ACTION OF GLUCOSE ISOMERASE ON D-XYLOSE AND D-GLUCOSE

A large number of genera of bacteria and some yeasts have been found to produce a glucose isomerizing enzyme. But the strains most widely used as sources for commercial production are members of the genus *Streptomyces* (Takasaki *et al.* 1969). Outtrup (1974) found thermophilic atypical variants of *Bacillus coagulans* to be particularly suited as the enzyme source due to the properties of its glucose isomerase.

The isomerase in *Streptomyces* is an inducible enzyme which requires the presence of D-xylose in the culture medium for its production (Tsumura and Sato 1965). Sanchez and Quinto (1975) selected a double mutant strain of *Streptomyces phaeochromogenes* which was able to produce glucose isomerase constitutively and was insensitive to catabolite repression. Diers (1976), who worked with *Bacillus coagulans* in chemostat cultures, found that glucose isomerase production of this strain was regulated mainly by catabolic repression. The latter occurred when inorganic compounds limited growth, whereas carbon limitation and particularly carbon-oxygen limitation were advantageous.

Media for the commercial production of glucose isomerase are based on xylan or xylan-containing raw materials such as wheat bran, maize husks, sulfite liquor, etc. As the enzyme is of the intracellular type, it can be used in the form of whole cells. In 1969 Takasaki *et al.* described the immobilization of *Streptomyces* cells by heating them to over 60°C for about 10 min. This procedure prevents autolysis of the cells and "fixes" the glucose isomerase. Purified preparations with higher activities can be obtained by application of the usual methods of cell rupture and solubilizing of the enzyme. After discarding the cellular material, the glucose isomerase is then adsorbed on DEAE-cellulose or a similar material, recovered, and washed.

Co and Mg ions are well-known activators of glucose isomerase and essential for obtaining maximum activity, whereas copper, nickel, and zinc are strong inhibitors. The other characteristics such as pH and temperature activity and stability vary with the enzyme source and depend on whether the enzyme is in the native or an immobilized form. In order to prevent alkaline conversion of fructose produced during the isomerization process it is necessary that the employed glucose isomerase be highly active at pH 6.5.

Principal producers and users of glucose isomerase are found in the corn wet milling industry, where the enzyme is used to convert glucose in corn syrups to fructose. Crystalline D-glucose can be used as a substrate for D-fructose production. However, in industrial practice, high D.E. starch hydrolysates are a more economical source of D-glucose. Such a hydrolysate can be prepared by the combined action of bacterial α-amylase, amyloglucosidase, and isoamylase on starch to yield a glucose syrup of about 95–98 D.E. Subsequent isomerization is carried out in agitated vessels at 65°C and pH 7 for 18 to 24 hr using, for example, immobilized cells. Following conversion to D-fructose, the immobilized enzyme is removed and the liquor further treated to give an invert sugar syrup of about 45% fructose and 55% glucose. The process is illustrated in Fig. 15.7. It will probably replace invertase in the manufacture of invert sugar.

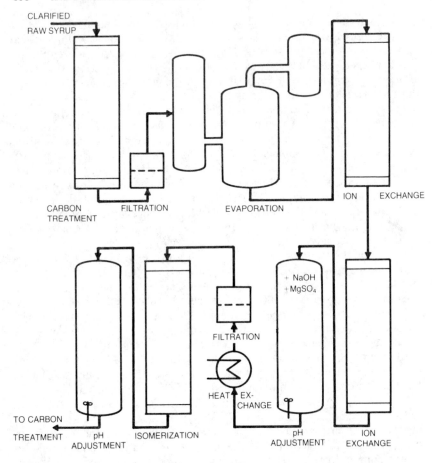

FIG. 15.7.   GLUCOSE ISOMERIZATION: FLOW SHEET

REFERENCES

ACKERS, G.K.   1970.   Analytical gel chromatography of proteins. Adv. Prot. Chem. *24*, 343–446.

ALEEVA, V.D., FEDOTOVA, A.I. and ARAVINA, L.A.   1973.   Biosynthesis of proteases by *Aspergillus terricola* under various cultivation conditions. Mikrobiologiya *42*, 428–433. (Russian)

ANDERSON, R.G. and KEAY, L.   1968.   Purification and fractionation of enzyme mixture from aqueous solution. U.S. Pat. 3,616,232. Aug. 14.

ARIMA, K., IWASAKI, S. and TAMURA, G.   1967.   Milk clotting enzymes from microorganisms. Part. I. Screening test and the identification of the potent fungus. Agric. Biol. Chem. *31*, 540–545.

ARST, H.N., JR. and COVE, D.J.   1973.   Nitrogen metabolite repression in *Aspergillus nidulans*. Mol. Gen. Genet. *126*, 111–141.

AUNSTRUP, K. 1974. Industrial production of proteolytic enzymes. FEBS Proc. Meet. *30* (Ind. Aspects Biochem., Pt. 1) 23–46.

AUNSTRUP, K., OUTTRUP, H., ANDRESEN, O. and DAMBMANN, C. 1972. Proteases from alkalophilic *Bacillus* species. Ferment. Technol. Today, Proc. 4th Intern. Ferment. Symp., 1972. G. Terui (Editor). Soc. Ferment. Technol., Osaka, Japan, 299–305.

BAR-ELI, A. and KATCHALSKI, E. 1960. A water-insoluble trypsin derivative and its use as a trypsin column. Nature *188*, 856–857.

BARTON, R. 1967. Separation of amyloglucosidase from transglucosidase. U.S. Pat. 3,483,085. May 1.

BATEMAN, D.F. 1966. Hydrolytic and trans-eliminative degradation of pectic substances by extracellular enzymes of *Fusarium solani* f. *phaseoli*. Phytopathology *56*, 238–244.

BATEMAN, D.F. and MILLAR, R.L. 1966. Pectic enzymes in tissue degradation. Annu. Rev. Phytopathol. *4*, 119–146.

BECKHORN, E.J. 1960. Production of industrial enzymes. Wallerstein Lab. Commun. *23*, 201–212.

BENDER, H. and WALLENFELS, K. 1961. Investigations on pullulan. I. Specific degradation by a bacterial enzyme. Biochem. Z. *334*, 79–95.

BERNFELD, P. and WAN, J. 1963. Antigens and enzymes made insoluble by entrapping them into lattices of synthetic polymers. Science *142*, 678–679.

BHAT, J.V., JAYASANKAR, N.P., AGATE, A.D. and BILIMORIA, M.H. 1968. Microbial degradation of pectic substances. J. Sci. Ind. Res. *27*, 196–203.

BLATT, W.F., DRAVID, A., MICHAELS, A.S. and NELSEN, L. 1970. Solute polarization and cake formation in membrane ultrafiltration: Causes, consequences, and control techniques. *In* Membrane Science and Technology. J.E. Flinn (Editor). Plenum Press, New York.

BOIDIN, A. and EFFRONT, J. 1917. Process of manufacturing diastase and toxins by oxidizing ferments. U.S. Pat. 1,227,525. May 22.

BORGLUM, G.B. and STERNBERG, M.Z. 1972. Properties of a fungal lactase. J. Food Sci. *37*, 619–623.

BRICKMAN, E., SOLL, L. and BECKWITH, J. 1973. Genetic characterization of mutations which affect catabolite-sensitive operons in *Escherichia coli*, including deletions of the gene for adenyl cyclase. J. Bacteriol. *116*, 582–587.

BROCKERHOFF, H. and JENSEN, R.G. 1974. Lipolytic Enzymes. Academic Press, New York, San Francisco, London.

BROOKS, J. and REID, W.H. 1955. The complex nature of the polygalacturonase of *Aspergillus foetidus*. Thom and Raper. Chem. Ind. (London), 325–326.

BROWN, D.E., HALSTED, D.J. and HOWARD, P. 1975 Biosynthesis of cellulase by *Trichoderma viride* QM 9123. *In* Proceedings of the Symposium on Enzymatic Hydrolysis of Cellulose. M. Bailey, T.M. Enari, and M. Linko (Editors). Tech. Res. Cent. Finland, Helsinki.

BUETTNER, M.J., SPITZ, E. and RICKENBERG, H.V. 1973. Cyclic adenosine 3′,5′-monophosphate in *Escherichia coli*. J. Bacteriol. *114*, 1068–1073.

CHALOUPKA, J. 1969. Dual control of megateriopeptidase synthesis. Ann. Inst. Pasteur, Paris *117*, 631–636.

CHALOUPKA, J. and KŘEČKOVÁ, P.   1966.   Regulation of the formation of protease in *Bacillus megaterium*. I. The influence of amino acids on the enzyme formation. Folia Microbiol. (Prague) *11*, 82−88.

CHANG, T.M.S.   1964.   Semipermeable microcapsules. Science *146*, 524−525.

CHANG, T.M.S.   1972.   Artificial Cells. Charles Thomas Publisher, Springfield, Ill.

CHARM, S.E. and MATTEO, C.C.   1971.   Scale-up of protein isolation. *In* Methods in Enzymology, Vol. 22. S.P. Colowick (Editor). Academic Press, New York.

CHRISTENSEN, L.M.   1940.   Apparatus for promoting mold growth. U.S. Pat. 2,325,368. Oct. 7.

CLARKE, P.H.   1971.   Methods for studying enzyme regulation. *In* Methods in Microbiology, Vol. 6A. J.R. Norris (Editor). Academic Press, London.

COHEN, B.L.   1973.   Regulation of intracellular and extracellular neutral and alkaline proteases in *Aspergillus nidulans*. J. Gen. Microbiol. *79* (Pt. 2) 311−320.

COLEMAN, G.   1967.   Studies on the regulation of extracellular enzyme formation by *Bacillus subtilis*. J. Gen. Microbiol. *49*, 421−431.

COLLIER, B.   1970.   How active are commercial enzymes? Process Biochem. *5* (8) 39−42.

CONNORS, W.M. and SFORTUNATO, T.   1956.   Purification of lactase enzyme and spray drying with sucrose. U.S. Pat. 2,773,002. Dec. 4.

CPC INTERN.   1969.   Pullulanase enzyme from *Aerobacter aerogenes (Enterobacter aerogenes)*. Br. Pat. 1,232,130. May 19.

CROXALL, W.J.   1964.   Process for removing transglucosidase from amyloglucosidase. U.S. Pat. 3,254,003. Jan 2.

CUATRECASAS, P. and ANFINSEN, C.B.   1971.   Affinity chromatography. Annu. Rev. Biochem. *40*, 259−278.

DAWSON, P.S.S. and KURZ, W.G.W.   1969.   Continuous phased culture—A technique for growing, analyzing, and using microbial cells. Biotechnol. Bioeng. *11*, 843−851.

DE FILIPPI, R.P. and GOLDSMITH, R.L.   1970.   Application and theory of membrane processes for biological and other macromolecular solutions. *In* Membrane Science and Technology. J.E. Flinn (Editor). Plenum Press, New York.

DEMAIN, A.L.   1971.   Overproduction of microbial metabolites and enzymes due to alteration of regulation. Adv. Biochem. Eng. *1*, 113−142.

DEMAIN, A.L.   1972A.   Theoretical and applied aspects of enzyme regulation and biosynthesis in microbial cells. Biotechnol. Bioeng. Symp. *3*, 21−32.

DEMAIN, A.L.   1972B.   Increasing enzyme production by genetic and environmental manipulations. *In* Methods in Enzymology, Vol. 22. Enzyme Purification and Related Techniques. W.B. Jakoby (Editor). Academic Press, New York.

DEMIS, D.J., ROTHSTEIN, A. and MEIER, R.   1954.   The relation of the cell surface to metabolism. X. The location and function of invertase in the yeast cell. Arch. Biochem. Biophys. *48*, 55−62.

DENAULT, L.J. and UNDERKOFLER, L.A. 1963. Conversion of starch by microbial enzymes for production of syrups and sugars. Cereal Chem. *40*, 618–629.

DIERS, I. 1976. Glucose isomerase in *Bacillus coagulans*. *In* Continuous Culture 6: Applications and New Fields. A.C.R. Dean, D.C. Ellwood, C.G.T. Evans and J. Melling (Editors). Ellis Horwood Publisher, Chichester, England.

DION, W.M. 1950. The proteolytic enzymes of microorganisms. II. Factors affecting the production of proteases in submerged culture. Can. J. Res. *28C*, 586–599.

DIXON, M. and WEBB, E.C. 1961. Enzyme fractionation by salting-out: a theoretical note. Adv. Prot. Chem. *16*, 197–219.

DREWS, B., SPECHT, H. and OLBRICH, H. 1954. Amyloglucosidase of *Aspergillus oryzae*. Branntweinwirtschaft *76*, 21–23. (German)

DUNN, C.G. *et al.* 1959. Production of amylolytic enzymes in natural and synthetic media. Appl. Microbiol. *7*, 212–218.

DUNNILL, P. and LILLY, M.D. 1972. Continuous Enzyme Isolation. Biotechnol. Bioeng. Symp. *3*, 97–113.

EDSTROM, R.D. and PHAFF, H.J. 1964. Purification and certain properties of pectin trans-eliminase from *Aspergillus fonsaceus*. J. Biol. Chem. *239*, 2403–2408.

EDWARDS, V.H. *et al.* 1970. Extended culture: The growth of *Candida utilis* at controlled acetate concentrations. Biotechnol. Bioeng. *12*, 975–999.

ENKENLUND, J. 1972. Externally added beta-glucanase. Process Biochem. *7* (7) 27–29.

EPPES, H.M.R. and GALE, E.F. 1942. The influence of the presence of glucose during growth on the enzymic activities of *Escherichia coli*: Comparison of the effect with that produced by fermentation acids. Biochem. J. *36*, 619–623.

FEDER, J. and KOCHAVI, D. 1971. Stabilization of neutral protease in alkaline protease containing enzyme mixtures by utilization of alkaline protease inhibitors. Ger. Offen. 2,112,730. Mar. 17. (German)

FISCHER, E.H. and STEIN, E.A. 1960. α-Amylases. *In* the Enzymes, Vol. 4. P.D. Boyer, H. Lardy and K. Myrbaeck (Editors). Academic Press, New York.

FOGARTY, W.M. and GRIFFIN, P.J. 1973. Production and purification of the metalloprotease of *Bacillus polymyxa*. Appl. Microbiol. *26*, 185–190.

FOX, P.F. 1974. Enzymes in food processing. FEBS Proc. Meet. *30* (Ind. Aspects Biochem., Pt. 1) 213–239.

FRIIS, J. and OTTOLENGHI, P. 1959. Localization of invertase in a strain of yeast. C.R. Trav. Lab. Carlsberg *31*, 259–271.

GIBSON, A.H., SWOBODA, B.E.P. and MASSEY, V. 1964. Kinetics and mechanism of action of glucose oxidase. J. Biol. Chem. *239*, 3927–3934.

GILLIGAN, W. and REESE, E.T. 1954. Evidence for multiple components in microbial cellulases. Can. J. Microbiol. *1*, 90–107.

GRISOLIA, S. 1964. Catalytic environment and biological results. Physiol. Rev. *44*, 657–712.

GULYI, M.F. *et al.* 1967. Catalase and glucose oxidase production from fungi. U.S.S.R. Pat. 204,962. Nov. 13. (Russian)

GUNJA-SMITH, Z. 1972. Cytophaga isoamylase. Ger. Offen. 2,162,923. June 29. (German)

GÜNTELBERG, A.V. 1954. A method for the production of the plakalbumin-forming proteinase from *Bacillus subtilis*. C.R. Trav. Lab. Carlsberg *29*, 27–35.

GUTFREUND, H. 1965. An Introduction to the Study of Enzymes. John Wiley & Sons, London.

HALLIWELL, G., GRIFFIN, M. and VINCENT, R. 1972. The role of component C_1 in cellulolytic systems. Biochem. J. *127*, 1–43.

HARADA, T., YOKOBAYASHI, K. and MISAKI, A. 1968. Formation of isoamylase by *Pseudomonas*. Appl. Microbiol. *16*, 1439–1444.

HARTLEY, B.S. 1960. Proteolytic enzymes. Annu. Rev. Biochem. *29*, 45–72.

HATANAKA, C. and OZAWA, J. 1971. Pectolytic enzymes of exo-types. Part I. Oligogalacturonide trans-eliminase of a *Pseudomonas*. Agric. Biol. Chem. *35*, 1617–1624.

HAUPTMANN, K.H. 1951. Process for clarification of musts, fruit juices, etc. Ger. Pat. 837,644. Aug. 24. (German)

HEINEKEN, F.G. and O'CONNOR, R.J. 1972. Continuous culture studies on the biosynthesis of alkaline protease, neutral protease, and α-amylase by *Bacillus subtilis* NRRL-B 3411. J. Gen. Microbiol. *73*, 35–44.

HORVATH, E.F.M. and INCZEFI, I. 1972. Regulated biosynthesis of protease and alpha-amylase in *Bacillus subtilis*. Acta Microbiol. *19*, 77–85.

HUGHES, D.E., WIMPENNY, J.W.T. and LLOYD, D. 1971. The disintegration of microorganisms. *In* Methods in Microbiology, Vol. 5B. J.R. Norris and D.W. Ribbons (Editors). Academic Press, London, New York.

IUPAC. 1972. Recommendation of the International Union of Pure and Applied Chemistry and the International Union of Biochemistry on Enzyme Nomenclature. Elsevier, Amsterdam.

JACOB, F. and MONOD, J. 1961. Genetic regulatory mechanisms in the synthesis of proteins. J. Mol. Biol. *3*, 318–356.

JEFFREYS, G.A. 1948. Mold enzymes produced by continuous tray method. Food Ind. *20*, 688–690, 825–826.

KAINUMA, K., KOBAYASHI, S., ITO, T. and SUZUKI, S. 1972. Isolation and action pattern of maltohexaose-producing amylase from *Aerobacter (Enterobacter) aerogenes*. FEBS Lett. *26*, 281–285.

KAKIUCHI, K., KATO, S., IMANISHI, A. and ISEMURA, T. 1964. Association and dissociation of *Bacillus subtilis* α-amylase molecule. II. Monomer-dimer transformation by gel filtration. J. Biochem. (Tokyo) *55*, 102–109.

KALASHNIKOV, E.Ya., LIFSHITS, D.B. and TRAĬNINA, T.I. 1960. Heat liberation by *Aspergillus oryzae* mold as cultured in commercial enzyme production. Mikrobiologiya *29*, 899–905. (Russian)

KALUNYANTS, K.A. 1974. Effect of various factors on the biosynthesis of neutral proteinase by cultures of *Bacillus subtilis*-103. Tr. Vses. Nauchno-Issled. Inst. Biosint. Belkovykh Veshchestv *2*, 64–71. (Russian)

KEAY, L. and MOSER, P.W. 1969. Differentiation of alkaline proteases from *Bacillus* species. Biochem. Biophys. Res. Commun. *34*, 600–604.

PRODUCTION OF ENZYMES   703

KEAY, L. *et al.* 1972. Production and isolation of microbial proteases. Biotechnol. Bioeng. Symp. *3*, 63–92.

KITAJIMA, M. and KONDO, A. 1971. Fermentation without multiplication of cells using microcapsules that contain zymase complex and muscle enzyme extract. Bull. Chem. Soc. Jpn. *44*, 3201–3202.

KLINE, L., MacDONNELL, L.R. and LINEWEAVER, H. 1944. Bacterial proteinase from waste asparagus butts. Ind. Eng. Chem. *36*, 1152–1158.

LEBEDEVA, E.I. *et al.* 1966. Processing a catalase preparation. U.S.S.R. Pat. 195,414. May 4. (Russian)

LEVIN, Y., PECHT, M., GOLDSTEIN, L. and KATCHALSKI, E. 1964. A water-insoluble polyanionic derivative of trypsin. I. Preparation and properties. Biochemistry *3*, 1905–1913.

LILLY, M.D. and DUNNILL, P. 1971. Biochemical reactors. Process Biochem. *6* (8) 29–32.

LIU, W.H., BEPPU, T. and ARIMA, K. 1972. Cultural conditions and some properties of the lipase of *Humicola lanuginosa* S-38. Agric. Biol. Chem. *36*, 1919–1924.

MAGASANIK, B. 1970. Glucose effects: inducer exclusion and repression. *In* The Lactose Operon. J.R. Beckwith (Editor). Cold Spring Harbor Lab., Cold Spring Harbor, N.Y.

MANDELS, M. and REESE, E.T. 1957. Induction of cellulase in *Trichoderma viride* as influenced by carbon sources and metals. J. Bacteriol. *73*, 269–278.

MANDELS, M. and REESE, E.T. 1960. Induction of cellulase in fungi by cellobiose. J. Bacteriol. *79*, 816–826.

MANDELSTAM, J. 1958. Turnover of protein in growing and nongrowing populations of *Escherichia coli*. Biochem. J. *69*, 110–119.

MANNING, G.B. and CAMPBELL, L.L. 1961. Thermostable α-amylase of *Bacillus stearothermophilus*. I. Crystallization and some general properties. J. Biol. Chem. *236*, 2953–2957.

MAXWELL, M.E. 1952. Enzymes of *Aspergillus oryzae*. I. The development of a culture medium yielding high protease activity. Aust. J. Sci. Res. Ser. B *5*, 43–55.

McKINLEY-McKEE, J.S., MORRIS, D.L. and REYNOLDS, C.H. 1971. Conformation changes in chemically modified alcohol dehydrogenase. Biochem. J. *125*, 111–112.

McLAREN, A. and PACKER, L. 1970. Enzyme reactions in heterogeneous systems. Adv. Enzymol. Relat. Areas Mol. Biol. *33*, 245–308.

MEERS, J.L. 1972. Regulation of α-amylase production in *Bacillus licheniformis*. Antonie van Leeuwenhoek J. Microbiol. Serol. *38*, 585–590.

MELROSE, G.J.H. 1971. Insolubilized enzymes: Biochemical applications of synthetic polymers. Rev. Pure Appl. Chem. *21*, 83–119.

MEYRATH, J. 1965. Production of amylase on vermiculite by *Aspergillus oryzae*. J. Sci. Food Agric. *16* (1) 14–18.

MEYRATH, J. 1966. Fungal amylase production. Process Biochem. *1* (4) 234–238.

MICHAELIS, L. and MENTEN, M.L. 1913. The kinetics of invertin action. Biochem. Z. *49*, 333–369. (German)

MICHAELS, A.S.   1968.   Ultrafiltration. *In* Progress in Separation and Purification. E.S. Perry (Editor). John Wiley & Sons, New York.

MILLER, B.S. and JOHNSON, J.A.   1954.   Differential inactivation of enzymes. U.S. Pat. 2,683,682. July 13.

MITRA, G. and WILKE, C.R.   1975.   Continuous cellulase production. Biotechnol. Bioeng. *17*, 1–13.

MORIHARA, K.   1974.   Comparative specificity of microbial proteinases. Adv. Enzymol. *41*, 179–243.

MOSBACH, K. and MOSBACH, R.   1966.   Entrapment of enzymes and microorganisms in synthetic cross-linked polymers and their application in column techniques. Acta Chem. Scand. *20*, 2807–2810.

MYERS, R.P. and STIMPSON, E.G.   1956.   Lactase-active, zymase-inactive yeast. U.S. Pat. 2,762,749. Sept. 11.

MYRBAECK, K.   1967.   Studies on yeast β-fructofuranosidase (invertase). XXII. The reaction with the zinc ion, Zn^{2+}. Ark. Kemi *27*, 507–514.

NAKAMATSU, T., AKAMATSU, T., MIYAJIMA, R. and SHIIO, I.   1975. Microbial production of glucose oxidase. Agric. Biol. Chem. *39*, 1803–1811.

NASUNO, S.   1972.   Electrophoretic studies of alkaline proteinases from strains of *Aspergillus flavus* group. Agric. Biol. Chem. *36*, 684–689.

NEUKOM, H.   1963.   On the degradation of pectic substances. Schweiz. Landwirtsch. Forsch. *2*, 112–122.

NIKOLSKAYA, E.A. and SINYAVSKAYA, O.I.   1971.   Amino acids as a source of nitrogen nutrition for *Penicillium vitale*, a producer of glucose oxidase and catalase. *In* Metabolism of Micromycetes. N.M. Pidoplichko (Editor). Naukova Dumka, Kiev, U.S.S.R. (Russian)

NIKOLSKAYA, E.A., ZAKORDONETS, L.A. and SINYAVSKAYA, O.I. 1972.   Formation of glucose oxidase and catalase during growth of *Penicillium vitale* on media with different ratios of nitrogen and carbon. Mikrobiol. Zh. (Kiev) *34*, 718–723. (Russian)

NIWA, K. *et al.*   1971.   Microbial protease by fermentation in an acid medium. Ger. Offen. 2,048,811. Apr. 8. (German)

NYIRI, L.   1968.   Manufacture of pectinases. Process Biochem. *3* (8) 27–30.

NYIRI, L.   1969.   Manufacture of pectinases. II. Process Biochem. *4* (8) 27–30.

NYSTROM, J.M. and ALLEN, A.L.   1976.   Pilot scale investigations and economics of cellulase production. Biotechnol. Bioeng. Symp. *6*, 55–74.

ORGANON, N.V.   1970.   Purified microbial rennins for use in cheese production. Fr. Pat. 1,592,965. June 26. (French)

OSMAN, H.G., ABDEL-FATTAH, A.F. and ABDEL-SAMIE, W.   1969.   Bacterial pectolytic enzymes. I. Isolation of *Clostridium felsineum* from the retting water of Egyptian flax, separation and pectolytic activity of its polygalacturonase. J. Chem. U.A.R. *12*, 543–550.

OUTTRUP, H.   1974.   Novel glucose isomerase, process for its production and utilization of same. Ger. Offen. 2,400,323. Jul. 25. (German)

PARDEE, A.B.   1969.   Enzyme production by bacteria. *In* Fermentation Advances. D. Perlman (Editor). Academic Press, New York.

PAZUR, J.H. and KLEPPE, K.   1962.   The hydrolysis of α-D-glucosides by amyloglucosidase from *Aspergillus niger*. J. Biol. *237*, 1002–1006.

PAZUR, J.H., MARSH, J.M. and TIPTON, C.L. 1958. Reversible transgalactosylation. J. Am. Chem. Soc. 80, 1433–1435.

PERLEY, A.F. and PAGE, O.T. 1971. Differential induction of pectolytic enzymes of Fusarium roseum (Lk.) emend. Snyder and Hansen. Can. J. Microbiol. 17, 415–420.

PERLMAN, R.L. and PASTAN, I. 1969. Pleiotropic deficiency of carbohydrate utilization in adenyl cyclase deficient mutant of Escherichia coli. Biochem. Biophys. Res. Commun. 37, 151–157.

PHAFF, H.J. 1947. The production of exocellular pectic enzymes by Penicillium chrysogenum. I. On the formation and adaptive nature of polygalacturonase and pectinesterase. Arch. Biochem. 13, 67–81.

POKROVSKAYA, N.V. and KISLYAKOVA, O.V. 1966. Mechanism of the stimulatory effect of calcium carbonate on glucose oxidase and catalase activity of Penicillium vitale. Mikrobiologiya 34, 793–800. (Russian)

PORTER, M.C. and MICHAELS, A.S. 1971. Membrane ultrafiltration. Chem. Technol. 1, 56–63, 248–254, 440–445, 633–637.

PORTER, M.C. and MICHAELS, A.S. 1972. Membrane ultrafiltration, Part 5. Chem Technol. 2, 56–61.

POVILAITIENE, J., CIURLYS, T. and UZKURENAS, A. 1976. Effect of some metal ions on the synthesis of proteinases by a Bacillus mesentericus culture. (2. Effect of the calcium ion). Liet. TSR Mokslu Akad. Darb. Ser. C, 141–148. (Russian)

REESE, E.T. and MAGUIRE, A. 1971. Increase in cellulase yields by addition of surfactants to cellobiose cultures of Trichoderma viride. Dev. Ind. Microbiol. 12, 212–224.

REESE, E.T., SIU, R.G.H. and LEVINSON, H.S. 1950. The biological degradation of soluble cellulose derivatives and its relationship to the mechanism of cellulose hydrolysis. J. Bacteriol. 59, 485–497.

ROBYT, J.F. and ACKERMAN, R.J. 1971. Isolation, purification and characterization of a maltotetraose-producing amylase from Pseudomonas stutzeri. Arch. Biochem. Biophys. 145, 105–114.

RÖHM, O. 1907. Process for bating hides. Ger. Pat. 200,519. June 7. (German)

RYU, D.Y., LEE, B.K. and THOMA, R.W. 1972. Production of 3-ketosteroid-delta-1-dehydrogenase by a two-stage, continuous-culture system. Proc. IV IFS: Ferment. Technol. Today, Proc. 4th Intern. Ferment. Symp., 1972. G. Terui (Editor). Soc. Ferment. Technol., Osaka, Japan.

SAHM, H. and WAGNER, F. 1973. Microbial assimilation of methanol. Ethanol- and methanol-oxidizing enzymes of the yeast Candida boidinii. Eur. J. Biochem. 36, 250–256.

SAITO, H. 1955. Pectic glycosidases of Aspergillus niger. J. Gen. Appl. Microbiol. 1, 38–60.

SAITO, N. 1973. Thermophilic extracellular α-amylase from Bacillus licheniformis. Arch. Biochem. Biophys. 155, 290–298.

SAKANO, Y., HIGUCHI, M. and KOBAYASHI, T. 1972. Pullulan 4-glucanohydrolase from Aspergillus niger. Arch. Biochem. Biophys. 153, 180–187.

SANCHEZ, S. and QUINTO, C.M. 1975. D-Glucose isomerase: Constitutive and catabolite repression-resistant mutants of Streptomyces phaeochromogenes. Appl. Microbiol. 30, 750–754.

SCHAEFFER, P. 1969. Sporulation and the production of antibiotics, exo-enzymes, and exotoxins. Bacteriol. Rev. *33*, 48–71.

SCHLEICH, H. 1970. Eliminating esterase from microbial rennin produced by fermentation. Fr. Pat. 2,031,141. Nov. 13. (French)

SHIBATA, Y. and NAKANISHI, K. 1967. Effect of carbon:nitrogen ratio and cultivation temperature on the composition of protease system and mycelial weight in asbestos koji. Chomi Kagaku *14*, 16–21. (Japanese)

SHINMYO, A., OKAZAKI, M. and TERUI, G. 1968. Kinetic studies on enzyme production by microbes. IV. Physiological bases for kinetic studies on acid protease production by *Aspergillus niger*. Hakko Kogaku Zasshi *46*, 733–742.

SILMAN, I.H., ALBU-WEISSENBERG, M. and KATCHALSKI, E. 1966. Some water-insoluble papain derivatives. Biopolymers *4*, 441–448.

SINNER, M. and DIETRICHS, H.H. 1975. Enzymatic hydrolysis of hardwood xylans. I. Investigation of commercial enzyme preparations of fungal origin with respect to xylanases and other polysaccharide-decomposing enzymes. Holzforschung *29*, 123–130. (German)

SMITH, J.L. and ALFORD, J.A. 1966. Inhibition of microbial lipases by fatty acids. Appl. Microbiol. *14*, 699–705.

STERNBERG, M. 1967. Removal of transglucosidase from amyloglucosidase preparations by chloroform. U.S. Pat. 3,483,084. May 29.

STIMPSON, E.G. 1954. Drying of yeast to inactivate zymase and preserve lactase. U.S. Pat. 2,693,440. Nov. 2.

STIMPSON, E.G. and STAMBERG, O.E. 1956. Conversion of lactose to glucose, galactose and other sugars in the presence of lactase activators. U.S. Pat. 2,749,242. June 5.

STOCKTON, J.R. and WYSS, O. 1946. Proteinase produced by *Bacillus subtilis*. J. Bacteriol. *52*, 227–228.

SUTTON, D.D. and LAMPEN, J.O. 1962. Localization of sucrose and maltose fermenting systems in *Saccharomyces cerevisiae*. Biochim. Biophys. Acta *56*, 303–312.

TAKAMINE, J. 1894. Preparing and making taka-koji. U.S. Pat. 525,820. Sept. 11.

TAKAMINE, J. 1913. Process for producing diastatic product. U.S. Pat. 1,054,324. Feb. 25.

TAKASAKI, Y., KOSUGI, Y. and KANBAYASHI, A. 1969. *Streptomyces* glucose isomerase. *In* Ferment. Adv. Pap. 3rd Intern. Ferment. Symp. D. Perlman (Editor). Academic Press, New York.

TEJERINA, G. and FERNANDEZ, P. 1966. Effect of pH on the production of pectolytic enzymes. An. Bromatol. *18*, 67–72. (Spanish)

TERUI, G., OKAZAKI, M. and KINOSHITA, S. 1967. Kinetic studies on enzyme production by microbes. I. On kinetic models. J. Ferment. Technol. *45*, 497–503.

TERUI, G., SHIBASAKI, I, and MOCHIZUKI, T. 1958. The high-heap aeration process as applied to some industrial fermentations. II. General description of the improved process. J. Ferment. Technol. *36*, 109–116.

TERUI, G., SHIBASAKI, I. and MOCHIZUKI, T. 1959A. High-heap aeration process as applied to some industrial fermentations. IV. Some industrial

TERUI, G., SHIBASAKI, I. and MOCHIZUKI, T. 1959A. High-heap aeration process as applied to some industrial fermentations. IV. Some industrial heap cultures with wheat bran as main raw materials. J. Ferment. Technol. *37*, 479–494.

TERUI, G., SHIBASAKI, I. and TAKANO, M. 1959B. Problems in the process control of high-heap culture. J. Ferment. Technol. *37*, 534–540.

TOSA, T., MORI, T., FUSE, N. and CHIBATA, I. 1966. Studies on continuous enzyme reactions. 2. Preparation of DEAE-cellulose-aminoacylase column and continuous optical resolution of acetyl-DL-methionine. Enzymology *31*, 225–238.

TSUMURA, N. and SATO, T. 1965. Enzymatic conversion of D-glucose to D-fructose. VI. Properties of the enzyme from *Streptomyces phaeochromogenes*. Agric. Biol. Chem. *29*, 1129–1134.

TURKOVA, J. *et al.* 1972. Alkaline proteinases of the genus *Aspergillus*. Biochim. Biophys. Acta *257*, 257–263.

TUTTOBELLO, R. and MILL, P.J. 1961. The pectic enzymes of *Aspergillus niger*. The production of active mixtures of pectic enzymes. Biochem J. *79*, 51–57.

UEDA, S., YAGISAWA, M. and SATO, Y. 1971. Production of isoamylase by *Streptomyces* species No. 28. J. Ferment. Technol. *49*, 552–558.

UNDERKOFLER, L.A., DENAULT, L.J. and HORE, E.F. 1965. Enzymes in the starch industry. Stärke *17*, 179–184.

UNDERKOFLER, L.A. and HICKEY, R.J. 1954. Industrial Fermentations, Vol. 2. Chemical Publishing Co., New York.

UNDERKOFLER, L.A., SEVERSON, G.M., GOERING, K.J. and CHRISTENSEN, L.M. 1947. Commercial production and use of mold bran. Cereal Chem. *24*, 1–12.

VAN LANEN, J.M. and SMITH, M.B. 1965. Glucamylase production from *Aspergillus niger*. U.S. Pat. 3,418,211. Dec. 13.

VELCHEVA, P. and KOLEV, D. 1975. Inhibitory effect of the pigmentary complex of *Bacillus mesentericus*, strain 90, on the neutral protease from the same strain. Prilozhna Mikrobiol. *6*, 31–39. (Bulgarian)

WALLERSTEIN, L. 1911. Method of treating beer or ale. U.S. Pat. 995,824 and 995,826. June 20.

WANG, S.C. and PINCKARD, J.A. 1971. Pectic enzymes produced by *Diplodia gossypina* in vitro and in infected cotton bolls. Phytopathology *61*, 1118–1124.

WEETALL, H.H. 1972. Insolubilized enzymes on inorganic materials. *In* Chemistry of Biosurfaces, Vol. 2. M. Hair (Editor). Marcel Dekker, New York.

WENDORFF, W.L. and AMUNDSON, C.H. 1971. Characterization of beta-galactosidase from *Saccharomyces fragilis*. J. Milk Food Technol. *34*, 300–306.

WHITAKER, J.R. 1972. Principles of Enzymology for the Food Sciences. Food Science Series, Vol. 2. Marcel Dekker, New York.

WISEMAN, A. 1973. Industrial enzyme stabilization. Process Biochem. *8* (4) 14–15.

YAMADA, K. 1961. Starch-hydrolyzing enzymes. U.S. Pat. 2,976,219. Mar. 21.

YAMANAKA, K. 1968. Purification, crystallization and properties of the

D-xylose isomerase from *Lactobacillus brevis*. Biochim. Biophys. Acta *151*, 670–680.

YOKOBAYASHI, Y., SUGIMOTO, K. and SATO, Y. 1968. Process for production of isoamylase. Ger. Offen. 1,767,653. May 31. (German)

YOSHIDA, F., MOTAI, H. and ICHISHIMA, E. 1968. Effect of lipid materials on the production of lipase by *Torulopsis ernobii*. Appl. Microbiol. *16*, 845–847.

YOUNG, H. and HEALY, R.P. 1957. Production of *Saccharomyces fragilis* with an optimum yield of lactase. U.S. Pat. 2,776,928. Jan. 8.

ZABORSKY, O.R. 1973. Immobilized Enzymes. Chemical Rubber Co., Cleveland.

16

Citric Acid

K.K. Kapoor, K. Chaudhary, and P. Tauro

Citric acid (CH_2COOH COH COOH CH_2COOH), a tricarboxylic acid, was first isolated from lemon juice and crystallized in 1784 by Scheele. It is found as a natural constituent of a variety of fruits. However, members of the citrus family are especially rich in this organic acid. Citric acid extracted from fruits is commercially known as "natural citric acid" in contrast to the citric acid produced by microbial fermentations. Until the early part of this century, citric acid was produced mostly from lemon juice although Wehmer (1893) had described this organic acid as a metabolic product of molds of the genera *Penicillium* and *Mucor*. These fungi were reported to produce citric acid from a nutrient solution containing sucrose as a carbon source. Today, most of the citric acid used in food and other industries comes from fungal fermentations. Although chemical synthesis of this organic acid is possible, as yet there is no competitive synthetic process developed that is superior to fungal fermentations.

Citric acid has a variety of uses. About 70% of the citric acid produced is used in the food and beverage industry, about 12% in pharmaceuticals, and about 18% in other industrial applications. The food and beverage industry uses citric acid mostly as an acidulant because of its high solubility, extremely low toxicity, and a pleasant sour taste. Table 16.1 summarizes the major end uses of citric acid or its esters and salts.

The increasing use of citric acid in a variety of industries has demanded a steady increase in citric acid production. Recently, because of its easy biodegradability, this organic acid has also found a ready acceptance in the detergent industry in place of phosphates and in the removal of sulfur in stack gases, a process applicable to power stations and other facilities where sulfur must be removed. These additional uses have placed greater stress on increasing citric acid production and on the search for more efficient, newer processes. Current estimates suggest that the worldwide demand for this acid is about 220,000 MT per year and this is bound to increase in coming years if this organic acid finds use in other industries.

TABLE 16.1.   USES OF CITRIC ACID

Industry	Uses
Beverages	As flavor enhancer; as preservative. Eliminates haze due to trace metals; prevents color and flavor deterioration In wines—prevents turbidity; inhibits oxidation; adjusts pH In soft drinks—gives a cool taste; maintains carbonation
Food and candy	In confectionery—enhances flavor; inverts sucrose; prevents oxidation. Produces darker colors in hard candies, jams, and jellies; and adjusts pH In frozen foods—neutralizes residual lye; protects ascorbic acid from oxidation; inactivates trace metals. Inactivates oxidative enzymes by lowering the pH to prevent changes in color and flavor In dairy products—acts as an antioxidant; and as emulsifier in cheese, ice cream, etc.
Pharmaceutical	As a solvent and flavoring agent; produces effervescence when combined with bicarbonate
Cosmetics	As antioxidant and synergist
Other industrial uses	For treatment of boiler water and in metal plating In detergents—as a builder In tanning and in textiles

The estimated production in the major citric acid producing countries is shown in Table 16.2. Among the citric acid producing countries, the United States is by far the largest producer. Based on 1976 estimates, the world production in 1977 was about 180,000 MT (200,000 tons) (Messing and Schmitz 1976). Some of the major citric acid producers of the world are listed below.

(1) Joh. A. Benckshiser, GmbH, Ludwigshafen/Rhein, West Germany.
(2) C.H. Boehringer Sohn, Ingelheim/Rhein, West Germany.
(3) Citrique Belge, Tienen, Belgium.
(4) Miles Laboratories, Inc., Elkhart, Indiana.
(5) Pfizer, Inc., New York, N.Y.
(6) Rhone-Poulenc S.A., Paris, France.
(7) San Fu Chemical Company, Ltd., Taipei, Republic of China.
(8) John and E. Sturge, Ltd., Birmingham, England.
(9) Tai Nan Fermentation Industrial Company, Ltd., Taipei, Republic of China.

In the United States, Miles Lab., Inc. and Pfizer, Inc. are the two main producers of citric acid. Because of their dominant position in this area, no specific information on capacities, production, or marketing of citric acid is available. However, it is estimated the production capacity of Miles Lab., Inc. is about 54,000 MT (60,000 tons)/year while that of Pfizer, Inc. is about 112,500 MT (125,000 tons)/year. Details of fermentation processes used by

TABLE 16.2.   ESTIMATED PRODUCTION OF CITRIC ACID BY VARIOUS CITRIC ACID-PRODUCING COUNTRIES

Countries	Estimated Production (MT/Year)	(Tons/Year)
Western European countries, United Kingdom, France, Netherlands, Belgium, Austria, West Germany, and Ireland	90,000	100,000
U.S.S.R.	18,000	20,000
Canada	9,000	10,000
Japan	6,300	7,000
Czechoslovakia	3,600	4,000
Australia	2,700	3,000
Poland	2.250	2.500
Developing countries	10,800	12,000
Israel	3,600	4,000
Others	14,400	16,000

Source: Atticus (1975).

these companies are closely guarded secrets and therefore not much information is available about the processes used for large-scale production of this organic acid.

Over the years, a large number of organisms including fungi, yeast, and bacteria have been screened for citric acid production. Yet, *Aspergillus niger* (Fig. 16.1), first used by Currie (1917), still remains the organism of choice for industrial production of citric acid. Other organisms such as yeast and bacteria appear to have the potential to produce substantial amounts of citric acid but are yet to be exploited fully for commercial production.

The basic methodology for citric acid production using fungi and easily fermentable carbon sources has been modified over the years. Since the publication of the book *Industrial Microbiology* by Prescott and Dunn in 1959, the number of reports available up to the beginning of 1978 has been over 500. It is likely that in preparation of this manuscript we have missed a few. Nonetheless, this large number and the fact that these have come from different parts of the world testifies to the worldwide interest in and importance of this organic chemical. Although brief chapters on citric acid production have appeared in a few books (Perlman and Sih 1960; Casida 1968; Peppler 1967), no comprehensive report on this subject has been published since the publication of the book by Prescott and Dunn (1959). Since these authors adequately covered the work done prior to 1959, as far as possible, no attempt has been made to review the work done prior to that date. Also, since the number of reports that have appeared during the past two decades is very large, only developments which in our opinion are important have been reviewed here.

CITRIC ACID PRODUCTION BY FUNGI

Fungal Strains

Since the early demonstration by Wehmer (1893) of the presence of citric acid in culture media containing sugars and inorganic salts with species of

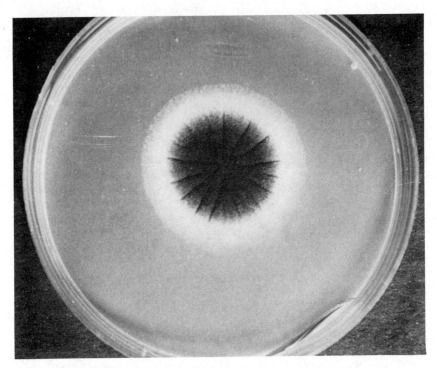

FIG. 16.1.   *ASPERGILLUS NIGER* ON POTATO DEXTROSE AGAR PLATE

Penicillium, a variety of fungi have been screened for citric acid production. The various fungi which have been found to accumulate citric acid in their culture media include strains of *Aspergillus niger, A. awamori, A. fonsecaeus, A. luchensis, A. wentii, A. saitoi, A. usami, A. fumaricus, A. phoenicus, A. lanosus, A. flavus, Penicillium janthinellum, P. restrictum, Trichoderma viride, Mucor piriformis, Ustulina vulgaris*, and species of *Botrytis, Ascochyta, Absidia, Talaromyces, Acremonium*, and *Eupenicillium*.

Currie (1917) first reported that strains of *A. niger*, when cultured in a sugar-containing medium of low pH, produced citric acid. Prior to that, this organism was known to produce only oxalic acid. In fact, even today, this organism when cultured in sugar-containing media at a higher pH produces oxalic acid. Since this finding, strains of *A. niger* have dominated others both in laboratory and industrial scale production of citric acid. The major advantages in using this organism for producing citric acid are (1) the ease with which it can be handled, (2) the cheap raw materials that it can utilize for citric acid production, and (3) high and consistent yields, thereby making the process economical.

Types of Fermentations

The development of processes for citric acid fermentation can be divided into 3 phases. In the first phase, citric acid production was confined to species of *Penicillium* and *Aspergillus* under stationary or surface culture conditions. The second phase, beginning in the 1930s, consisted of the development of submerged fermentation processes for citric acid production using *A. niger*. The third stage, which is of recent origin, involves the development of solid state culture, continuous culture, and multi-stage fermentation techniques for citric acid production. Thus, there are at least 3 principal methods which are today available for producing citric acid using fungi.

In the surface culture technique, sterile nutrient medium containing sugar is allowed to flow into stainless steel or high grade aluminum trays which are arranged in tiers in sterile fermentation chambers. The size and the number of trays and the volume per tray vary from plant to plant. Most fermentation chambers have controls over temperature, relative humidity, and circulation of purified air. The fermentation medium is then inoculated with spores of *A. niger* and the temperature is maintained around 28°–30°C and the relative humidity between 40 and 60% for 8–12 days. As the organism grows and spreads over the surface, the medium becomes acidic and the course of fermentation can be followed by determining either the pH or the total acid content of the broth. After the fermentation is finished, the fermented liquor is drained off and further processed for recovery of citric acid. In some cases, the preformed mycelium is reused for one or two more rounds of fermentation. The fermentation chambers and trays are then sterilized using water, dilute formaldehyde, and sulfur dioxide.

Although the surface process is still being used, most of the newly built plants have adopted the submerged or the deep fermentation process. In this, the nutrient media after inoculation are subjected to vigorous, controlled aeration and agitation in large fermentors. The time interval involved is much shorter (3–5 days) at temperatures of 25°–30°C. The mother liquor after fermentation is drained off and citric acid is extracted. As in the surface culture, sometimes the mycelial pellets are reused for a second fermentation.

A 2-stage submerged fermentation process involving a "growth stage" and a "production stage" has also been developed. In this, the growth medium is first inoculated with the spores and after 3–4 days of growth, the mycelium is separated from the solution and added to the fermentation medium. The fermentation is then carried out for 3–4 days at 25°–30°C while oxygen is dispersed through the medium.

The third process, namely, the solid state fermentation, was first described by Cahn (1935). Despite its potential, this technique has not drawn much attention mainly because it is a labor intensive technique. In this process, the fermentation medium is impregnated in porous solid materials such as sugarcane bagasse, potato or beet pulp, pineapple pulp, etc., in an appropriate ratio, sterilized, and then inoculated with a suspension of fungal spores. The mash is then incubated in trays at $25°-30°C$ for $6-7$ days (Hisanaga and Nishimura 1968; Yo 1975; Lakshminarayana et al. 1975). The height of the mash during incubation can be variable. After fermentation, the mash is extracted with water, concentrated, and then processed for citric acid precipitation.

In recent years semicontinuous, continuous, or multi-stage processes have been patented for citric acid production. However, full details of these processes are not available. Zhuravskii et al. (1974) have patented a continuous multi-stage process for citric acid production. In this process culture medium is added at a rate providing for its complete replacement in 24 hr. The inflow coefficient of the medium is kept at 4%. The medium from the first stage then flows to a second fermentor. It is then aerated gradually with increasing amounts of air from 20 to 30 $m^3/m^3/hr$. During the fermentation, caustic alkali is also added in such a way that one-third of the citric acid formed is neutralized.

In another Russian patent, Brehznoi et al. (1976) have described a continuous process for citric acid production. In this, the nutrient medium containing molasses is added at a constant rate and the final product is removed at a rate that maintains sugar concentration and total acidity at a constant level.

The semicontinuous processes described in literature or patents are mostly replacement culture processes. In these, after the fermentation under batch culture condition is over, the mother liquor is withdrawn and replaced with fresh sterile medium. This process has been partially successful and makes use of the fact that citric acid production occurs by cells which are not in the active stages of growth. Thus, the use of mycelia for subsequent fermentation, although it does not represent a continuous process in the correct sense, is yet close to a semicontinuous culture technique. Although this process is economical, subsequent yields are generally lower.

Despite various advances in process development, the surface and submerged batch culture techniques are still being used as the exclusive processes for large-scale production of citric acid. Depending upon the strain of fungus, appropriate cultural conditions have to be developed.

Cultural Conditions

Carbon, Nitrogen, and Phosphate.—Cultural conditions for citric acid production by fungi vary from strain to strain and also depend on the type of process. This has been so since a strain that produces citric acid efficiently under surface culture conditions fails to do so under submerged culture conditions. Also, strains that can use one carbon source efficiently fail to show good acid production when cultured in a medium containing another. Because of these difficulties, the best substrate for each organism has to be determined.

During the early periods of citric acid fermentation with *Penicillium* species, culture media contained glucose in the range of $10-12\%$ concentration with incubation times of 4–6 weeks at 15°–20°C. Under these conditions about 50% of the glucose was converted into citric acid.

Efficient citric acid production by fungi requires simple synthetic media. In fact, it has been found that yields are much higher when fungi are cultivated in simple synthetic media than in complex media (Sanchez-Marroquin et al. 1963). As seen later, this may be due to the influence of metals and other components in citric acid production. These days a variety of carbon sources, such as sucrose, citrus molasses, cane juice, starch from various sources, cane and beet molasses, have been used for citric acid production, yet sucrose and molasses have remained mostly the substrates of choice. The use of beet or cane molasses has made the process economical although yields may be slightly low.

The initial sugar concentration has been found to determine the amount of citric acid and also the amount of other organic acids produced by A. niger. Normally, strains of A. niger need a fairly high initial concentration (15–18%) of sugars in the medium. A concentration higher than 15–18%, however, leads to greater amounts of residual sugars, making the process uneconomical; while on the other hand, a lower concentration of sugar leads to lower yields of citric acid as well as to the accumulation of oxalic acid (Kovats 1960).

Molasses, a by-product of the sugar industry, is the syrupy liquid left after the removal of sugar from the mother syrup. It contains 50–60% sucrose which cannot be removed by simple crystallization. The average composition of molasses is given in Table 16.3.

TABLE 16.3.   AVERAGE COMPOSITION OF BEET AND CANE MOLASSES

Constituents	Beet Molasses (%)	Cane Molasses (%)
Water	16.5	20.0
Sugars	53.0	62.0
Nonsugars	19.0	10.0
Inorganics (ash)	11.5	8.0

Source: Anon. (1972).

There are three types of molasses, the blackstrap, refinery, and invert or high test molasses. The blackstrap molasses is molasses from the sugar factory obtained from the last stages of crystallization while refinery molasses is that molasses obtained at a second stage of refining sugar. Invert or high test molasses is partly inverted cane juice syrup from which no sugar has been extracted. Refinery molasses contains about 48–50% sugar and has relatively less ash content. The sucrose contained in molasses is well suited as a raw material for fermentation processes. However, one major disadvantage in using this material for citric acid production is its high ash content, which inhibits efficient citric acid production.

When molasses is used, it is diluted to a sugar concentration of 15–20% with dilute sulfuric acid and to a pH 5.5–6.5. Other nutrients required for fungal growth are then added and the mixture is sterilized for 30 min. To reduce or eliminate the inhibitory action of metals, appropriate pretreatment is given to the molasses solution.

Citric acid production by *A. niger* involves both a "growth stage" and a "fermentation stage." The organisms therefore need major elements such as carbon, nitrogen, phosphorus, and sulfur in addition to various trace elements for growth and citric acid production. The concentration of all these constituents has a profound effect on the yield of citric acid. The nitrogen requirement for citric acid production is generally met by the addition of inorganic nitrogen sources such as $(NH_4)_2SO_4$, NH_4NO_3, $NaNO_3$, KNO_3, urea, etc. However, the type of nitrogen source and its concentration affect the performance of the fungus considerably. For example, ammonium sulfate prolongs vegetative growth while ammonium nitrate favors a shorter period of vegetative growth. However, a concentration of ammonium nitrate greater than 0.25% leads to the accumulation of oxalic acid (Naguchi and Bando 1960; Gupta *et al.* 1976).

Among the nitrates, sodium and potassium nitrates have been found to be superior to ammonium nitrate for citric acid production by *A. niger*. Dhankar *et al.* (1974) found that sodium nitrate at a concentration of 0.4% was superior to ammonium nitrate. In general, a high concentration of nitrogen leads to greater vegetative growth and delays the onset of the production phase. It is, therefore, necessary to correctly determine the nitrogen source and the concentration essential for maximal citric acid production by different fungal strains under different fermentation conditions.

In addition to carbon and nitrogen, phosphate constitutes the third major essential element for fungal growth. The concentration of phosphate in the fermentation medium has a profound effect on the amount of citric acid produced. A high concentration of phosphate promotes more growth and less acid production (Khan *et al.* 1970). Earlier reports suggested that citric acid production begins only after the available phosphorus compounds are assimilated by the mold (Szucs 1944). Although there are not too many reports on this aspect, in general a phosphate concentration of about 0.1–0.2% in the fermentation medium appears to be adequate.

Trace Elements.—*A. niger* needs a variety of divalent trace elements such as Fe^{++}, Cu^{++}, Zn^{++}, Mn^{++}, and Mg^{++}, etc., for growth and citric acid production. However, citric acid production is very sensitive to the concentration of these metals in the fermentation media. In fact, successful citric acid production depends to a great extent on the control of the concentration of trace elements.

Magnesium is essential for a variety of enzyme reactions in the cell and is required both for growth as well as for citric acid production. The optimum concentration of $MgSO_4$ for maximum citric acid production varies from 0.02 to 0.025%.

Among the other trace elements, Fe^{++} and Zn^{++} have a critical role to play in determining the efficiency of fermentation. There are conflicting reports regarding the exact concentration of these two metals essential for optimal citric acid production, although it is generally agreed that the concentration should be very low. A high concentration of these metals allows vegetative growth at the cost of acid excretion.

The optimum concentration of Fe^{++} required for maximal citric acid production has been found to vary with the strain of fungus. Schweiger (1961) has reported that citric acid production by *A. niger* under submerged fermentation conditions using molasses as the substrate is severely affected by the presence of iron at a concentration as low as 0.2 ppm; however, the addition of copper at 0.1–500 ppm at the time of inoculation or during the first 50 hr of fermentation was found to counteract the deleterious effect of iron. Such beneficial effect of Cu^{++} in counteracting the Fe^{++} effect has also been reported by Fedoseev *et al.* (1970), who found that the addition of $CuSO_4$ at 4.7 mg/100 g molasses resulted in better conversion of sugar into citric acid.

The effect of Zn^{++} concentrations on citric acid production by *A. niger* has been a subject of much investigation. Like Fe^{++}, the concentration of this metal drastically influences the outcome of the fermentation. Zn^{++} at a concentration of 1–2 μM allows continuation of the growth phase, but at a concentration of less than 1 μM restricts growth (Wold and Suzuki 1976A). In fact, addition of excess of Zn^{++} to a citric acid-producing culture has been found to reverse the acid production phase. It is suggested that Zn^{++} has an important role in the regulation of growth and citric acid production and that Zn^{++} deficiency during growth apparently signals the transition from the growth phase to the acid production phase.

It has also been suggested that Zn^{++} has an indirect role in the functioning of cyclic AMP (Wold and Suzuki 1976B). While addition of cAMP during the production phase enhances citric acid production, addition of Zn^{++} retards acid production.

Other trace elements such as Mn^{++}, Ba^{++}, Al^{+++}, etc., have been reported to have an effect on fungal morphology and citric acid production, at concentrations that generally do not inhibit growth (Kumamoto and Okamura 1972; Tondon and Srivastava 1971). The exact role of some of these metals in citric acid production is not known.

Pretreatment of Raw Materials.—Since the concentration of trace elements affects citric acid production profoundly, various techniques have been used to minimize the concentration of these metals in fermentation media. Complete elimination of trace elements is practically impossible, especially when raw materials such as molasses are used. In recent years therefore, two approaches have been used to overcome this difficulty: firstly, pretreatment of raw material with chemicals, ion exchange resins, etc., to reduce the trace element concentration; and secondly, the development of strains of fungi that have the ability to produce citric acid in the presence of a high concentration of trace elements.

The use of chemicals such as potassium ferrocyanide to reduce the concentration of Fe^{++} in fermentation medium was first described by Mazzadroli (1938). Since then, this method has been extensively used for the reduction of Fe^{++} concentration in raw materials such as molasses, etc.

There are two methods by which ferrocyanide can be used for reducing Fe^{++} concentration: (1) by addition of ferrocyanide directly into the fermentation media at a low, nongrowth inhibitory concentration, or (2) by treatment of molasses media with a high concentration of ferrocyanide before inoculation.

Ferrocyanide at high concentrations is toxic to fungi, and a high level in the growth medium will precipitate Fe^{++} and Zn^{++} and cause a deficiency of these metals. The concentration of ferrocyanide required for molasses clarification depends on the type of molasses. Kovats (1960) has reported the use of 0.04–0.6% of ferrocyanide at pH 2.2 while others (Clark 1962; Leopold and Valtr 1969) have suggested a concentration of 0.005–0.020%. These concentrations can not, however, be taken as a rule and will depend on the raw material used.

Leopold and Valtr (1967) suggest the following pretreatment of molasses with ferrocyanide to reduce the trace metal content. To 8 kg of molasses in 6.5 liters of water, 150 g of potassium ferrocyanide is added and boiled for 30 min. To this, H_2SO_4 is added to adjust the pH to 6.5–7.2. Twenty-five ml of phosphoric acid (45%) is then added to the mixture and again boiled for 20 min. The mixture is cooled and diluted to adjust sugar concentration to 15%. A nitrogen source is added and the medium is used for citric acid fermentation.

In the alternate method, ferrocyanide is added to the fermentation medium at the time of inoculation. Hustede and Rudy (1976) added about 1.2 g of ferrocyanide per liter of molasses medium containing $NH_4H_2PO_4$ adjusted to pH 5.0, before inoculation with A. niger spores. The yield of citric acid under these conditions was nearly 89%. Clark (1962) has also reported that adding ferrocyanide to the fermentation medium at a concentration of 20–40 ppm was adequate. It is, however, suggested that during the time of high acid production, the concentration of ferrocyanide should not exceed more than 20 ppm.

The concentration of ferrocyanide tolerated by A. niger depends on the stage of fermentation; for example, during the initial growth period, a concentration of ferrocyanide as high as 10–200 mg/ml is tolerated, while during the acid production stage, the concentration must be below 20 µg/ml (Clark 1964).

The addition of potassium ferrocyanide does not affect the carbon, nitrogen, or phosphorus content of the substrate but only reduces the ash content. Metals known to interfere with citric acid production, mainly Fe^{++} and Zn^{++}, are precipitated. Whether citric acid production can be increased by adding ferrocyanide to the fermentation in parts before and after sterilization of the medium has also been examined. Leopold and Valtr (1969) have found that when two-thirds of ferrocyanide is added before and one-third after boiling, the yield of citric acid by A. niger is increased.

Other chemicals that are used for reducing the metal content of molasses are the chelating agents such as EDTA (Chaudhary and Pirt 1966), activated charcoal (Dhankar 1972), and polyethylene amine (Garbataya et al. 1975).

Use of quaternary ammonium compounds, such as diisobutylphenoxy-ethyl dimethylbenzylammoniumchloride, to a molasses medium has been reported to increase the yield of citric acid by 92% (Miles Lab. 1969). Mints *et al.* (1976) have reported that addition of trilon B at 100–500 mg/liter to a molasses medium in addition to ferrocyanide improved the yield of citric acid by *A. niger.*

The use of ion exchange resins for reduction of the metal content of molasses media or sugar solutions has been commonly practiced (Miles Lab. 1962; Sanchez-Marroquin *et al.* 1970) in the production of citric acid, and it appears to be superior to the chemical treatment. However, clarification of complex materials such as molasses may pose certain problems.

pH.—Generally, a neutralizing agent such as $CaCO_3$ is not used in citric acid production by fungi since these organisms can tolerate substantial amounts of acidity. The maintenance of a favorable pH is, however, very essential for the successful production of citric acid. The initial pH required depends on the carbon source used. For example, when sucrose or glucose or clarified molasses or other relatively pure materials are used, a low pH (3.0) is desirable. A higher initial pH leads to the accumulation of oxalic acid (Shadafza *et al.* 1976). When crude molasses is used as a carbon source, a higher pH is desirable. In fact, a low pH in cane molasses medium has been found inhibitory for the growth of *A. niger* (Chaudhary *et al.* 1972). However, when decationized molasses is used, the initial pH can be 1.4–2.8 (Miles Lab. 1962). A low initial pH has the main advantage of preventing contamination, suppressing oxalic acid formation, and making the sterilization operation more efficient. Inorganic acids such as HCl or H_2SO_4, and NaOH are used to adjust the pH of the fermentation media.

A typical composition of basal media used for citric acid production by *A. niger* is as follows:

Total reducing sugars (molasses, sucrose)	14–15%
N source	0.25%
KH_2PO_4	0.10–0.15%
$MgSO_4 \cdot 7H_2O$	0.02–0.025%
pH: molasses medium	5.0–6.0
sucrose medium	2.0–3.0

Inoculum Development.—Citric acid-producing fungal strains are generally maintained on standard media for fungi although stocks may be preserved in soil, silica gel, glycerol, etc. The inoculum for fermentation can be either the spores or the pregrown mycelia. When spores are used as the inoculum, a suspension of spores is freshly prepared in sterile water or sterile water containing Tween 80 and this is used to inoculate the fermentation medium. The initial concentration of spores may vary between 1×10^5 and 1×10^7/ml of fermentation media. For practical reasons, in surface and solid state culture fermentations, a spore inoculum is generally used. In surface culture fermentation, after the spores are mixed with the fermentation medium, the fungus grows on the surface of the liquid. This mycelial mat can be reused as in the "replacement culture technique."

In submerged fermentations, spores as well as mycelial pellets are used as inoculum. When mycelial pellets are used, spores are first inoculated into a growth medium and allowed to grow under submerged conditions for 2–3 days. The mycelial pellets are then used as inoculum for larger fermentations. When pregrown mycelia are to be used as inoculum as in a large fermentation, the inoculum and fermentation media are of the same composition. In fact, the use of complex media such as wort or koji has been found to be inferior as media for inoculum development (Usami and Takatomi 1958).

Spore viability varies with the age of the spores and, as a consequence, the germination properties also change with age (Sussman and Halvorson 1966). There have been very few studies made to assess the effect of the age of spores on fermentation abilities of *A. niger*. Chaudhary *et al.* (1978) have found that the spores from a 3-day-old slant culture are as good as spores from a 7–8-day-old slant culture for citric acid production. This suggests that the age of the spores may not have a bearing on citric acid-producing ability. However, it is difficult to determine the exact age of a population of spores derived from slant cultures since a culture of *A. niger* sporulates in 24–36 hr. Whether spores formed very early are as effective as those formed late is not known. Also, 7–8 days is too short a period to assess the effect of spore development on efficiency.

The composition of the sporulation medium and the source of spores has been found to affect subsequent performance. For example, spores produced under submerged fermentation were found to be poor acid producers, while spores from a surface culture were found satisfactory (Usami and Takatomi 1958). Miles Lab. (1958) has patented a process for citric acid production in which the inoculum for submerged fermentation was prepared by inoculating moist sterile maize bran with *A. niger* spores. The mixture was incubated for 7–14 days at 28°–30°C and this when used as inoculum was found to be superior to the spores grown on other media.

Aeration, Temperature, and Time of Incubation.—The citric acid fermentation is essentially an aerobic fermentation and the organism needs an abundant supply of oxygen beyond that required for growth. The organism depends on the terminal oxidation of the reduced coenzymes of the respiratory chain by oxygen. Therefore, it is absolutely necessary to ensure an adequate supply of oxygen during the fermentation.

Aeration and agitation are primarily employed in liquid fermentations to ensure an appropriate oxygen supply as well as to maintain ventilation and prevent contamination by providing a positive pressure inside the fermentor. The degree of agitation and aeration will depend upon the organism, the medium, and the size of fermentor. It is generally agreed that the oxygen demand of a fermenting culture is so high that the amount of oxygen in a saturated aqueous medium is inadequate. This has made it essential to maintain an appropriate supply of aeration and agitation in a fermentor during the fermentation. In recent years, this aspect of aeration in fermentation has received considerable attention, but there are problems yet to be solved both in the theory of oxygen transfer and in the application of this theory to tanks containing thousands of liters (gallons) of fermentation material.

Three ways in which aeration can be ensured are (1) surface cultivation where oxygen diffuses through the surface which is maintained in a static condition; (2) by submerged aeration in which sterile air or oxygen is pumped from beneath the surface of the culture; and (3) shake culture, which is a process of both surface as well as submerged aeration. This process is, however, not feasible on an industrial scale.

Citric acid fermentation by surface culture is carried out in shallow pans of high grade aluminum or stainless steel containers. The volume of media is determined by the strain and other conditions. Under these conditions growth occurs on the surface of the medium and conversion of sugar to acid occurs intracellularly before it is excreted. Thus, the rate of the diffusion process will determine the time of citric acid accumulation. Agitation of the medium by gentle or moderate shaking under surface culture conditions has been found to retard citric acid production in strains suitable for surface fermentation. However, the flow of a small amount of air over the fungal mats has no deleterious effect. Although in surface culture humidified air is blown over the trays, the amount of the air supply has to be predetermined.

Under submerged culture conditions, either oxygen or air is continuously supplied to the growing culture and a limitation in the supply of oxygen reduces citric acid production. As in surface culture conditions, the oxygen supply should continue at an optimal rate throughout the fermentation. Generally, strains of fungi suitable for surface fermentation do not perform equally well in submerged culture suggesting that oxygen requirements are different for different culture conditions. This necessitates the development of cultures for different processes. Usami et al. (1960) have determined the effect of aeration on citric acid production by A. niger in 5 liter jar fermentors. With an increase in the aeration rate, the fermentation time was decreased and the yield of citric acid was increased. The effective aeration rate was found to be 3×10^{-6} g moles O_2/ml/min or higher. Under these conditions the citric acid yield was 68%.

In submerged fermentors either purified compressed air or oxygen along with agitation is used. The cost of production may not be significantly affected when air is used, but when purified oxygen is used, the cost of production increases. In order to reduce the cost of production but at the same time retain the efficiency, the possibility of reusing fermentation gases has been examined. Clark and Lentz (1971) recirculated fermentation gases from a submerged fermentation of ferrocyanide-treated beet molasses using O_2 for aeration after removal of carbon dioxide and found that this had no effect on citric acid production.

Aeration and agitation in submerged culture fermentation not only provide necessary oxygen for the growth of the organism but also promote efficient excretion of citric acid into the fermentation medium. Wendel (1967) has shown by electrical conductivity measurements during citric acid production that in the initial stages of fermentation citric acid diffuses slowly into the medium. The stratification that occurs on the mycelium during the period inhibits fungal metabolism and decreases the rate of citric acid production. By a combination of stirring and circulatory pumping applied to the fermentation medium, such stratification has been prevented, allowing metabolism and efficient citric acid production.

Interruption of aeration during fermentation under submerged conditions affects acid production; however, the extent of damage depends upon the duration of the interruption and the phase of fermentation. For example, a 30 min interruption of a 24-hr-old fermentation decreased the acidity of the medium by 13%, a 60 min interruption by 20%, and a 7.5 hr interruption by 60% (Kovats and Gackowska 1976). This suggests that the older the fermentation, the more sensitive it is to interruption of aeration. Apparently, in later stages when the oxygen supply becomes limiting, citric acid is remetabolized by the organism. Our current knowledge of the actual site of dislocation of the fermentation during breakdown in aeration is not known. It is therefore safer and more economical to devise preventive measures rather than experimenting on recovery conditions for recovering dislocated fermentations.

In the solid state fermentation, the fungus is allowed to grow in a medium impregnated in a carrier. Under optimal conditions, because of the increased surface area exposed, this system provides an ideal supply of oxygen for the growth and citric acid production.

As stated earlier, aeration conditions and oxygen requirements for each culture under different sets of conditions have to be determined. The amount of oxygen required for efficient conversion has to be determined for the strain of fungus and the cultural conditions, based on the ratio of surface area to volume at which maximum conversion of sugar into citric acid occurs.

A. niger and other fungi used in citric acid fermentation have an optimum temperature between 25° and 30°C. An increase in the temperature of incubation beyond 30°C has been found to decrease the citric acid yield and increase the oxalic acid accumulation (Doelger and Prescott 1934), irrespective of the fermentation conditions; therefore, maintenance of optimum temperature is necessary for optimum citric acid production.

The optimum time of incubation for maximal citric acid production varies both with the organism and the fermentation conditions. In the surface culture, the fermentation is usually complete in 7–10 days while under submerged conditions the incubation period is much shorter (4–5 days). Attempts have been made to decrease the fermentation time under both surface and submerged culture by altering the cultural conditions. Meyrath and Ahmed (1966) have attempted to reduce the fermentation time by the addition of vermiculite and found that the time can be reduced from 10 to 5 days. In submerged fermentation conditions with the current methodology, 4–5 days appears to be the minimum required. As in other fermentations, the incubation period has to be adequate for allowing growth and product formation. Since the fungi are relatively slow-growing organisms, it is doubtful if incubation periods less than 3–4 days will be adequate. Under solid state fermentation conditions, a period of 6–7 days is recommended (Lakshminarayana et al. 1975). However, this process has not been rigorously tested on a large scale to arrive at precise conclusions.

Additives and Stimulants.—A variety of stimulants have been tested for improving citric acid yields by A. niger. The most important of these is methanol (CH_3OH). The use of methanol in citric acid production was first reported by Moyer (1953). Since then it has been used in surface, submerged, and solid culture fermentations for increasing yields of citric acid by A. niger. There have been several reports which are consistent with Moyer's findings, that among alcohols, methanol and, to a smaller extent, ethanol are beneficial in increasing citric acid yield.

The effect of methanol in increasing citric acid yield appears to be a general phenomenon in strains of A. niger. This chemical at a concentration of 3–4% has been found to retard growth, delay sporulation, and increase citric acid yields. For methanol to be beneficial, it is essential that this be added to the medium before inoculation. Addition of methanol after 24 hr or later has not been found to be beneficial (Chaudhary et al. 1978). This suggests that methanol perhaps has some role in conditioning the mycelia without impairing their metabolism. Earlier, it was suggested that methanol perhaps increases the tolerance of fungi to trace elements such as Fe, Mn, Zn, etc. The derivation of mutants that tolerate a high concentration of trace metals but still respond to methanol addition is inconsistent with this assumption (Tauro 1977). It is likely that methanol also affects the permeability properties and enables greater excretion of citric acid.

The effect of other chemicals with no nutritional value in improving citric acid production has also been examined. For example, inhibitors of metabolism such as CaF, NaF, KF (Takami 1970) at a concentration of 10^{-4} M have been found to accelerate citric acid production. Similarly, malic hydrazide (1 g/2.5 kg molasses) under submerged culture conditions has been found to increase citric acid production by A. niger. In the absence of this chemical, the yield of citric acid was 30% while in its presence, it was 70% (Ohtsuka et al. 1975). Addition of mild oxidizing agents such as hydrogen peroxide or naphthoquinone or methylene blue to the fermentation medium has also been reported to stimulate acid production (Bruchmann 1966). The exact mechanism by which these affect citric acid production is not known.

Other chemical additives, such as aromatic amides, esters of dichloroacetic acid, sodium sulfite, and crysyllic acid, have also been tested and their effect appears to be more in controlling contamination from other microorganisms than in controlling citric acid fermentation (Maslova and Khaskin 1976; Lockwood and Batti 1965).

Short chain carbohydrates such as glycerol or lipid materials or other metabolizable complex compounds have been examined for their effect on increasing citric acid yields by A. niger. Addition of glycerol to a molasses medium at a rate of 30–50 g/liter has been found to increase citric acid yield by nearly 30% (Leopold 1971). Similar results were also obtained when sorbitol, mannitol, or erythritol were added to the medium. Glycerol when added was found to be metabolized and utilized for citric acid production.

The use of lipid materials such as various vegetable oils, fatty acids, etc., has been reported to be beneficial for citric acid production. Millis et al. (1963) reported that addition of fatty acids of chain length of 15 carbon atoms or fewer or natural oils containing a high amount of unsaturated fatty acids such as corn oil, almond oil, linseed oil, peanut oil, etc., increased citric acid yield by 20% (Table 16.4). It is suggested that unsaturated fatty acids may also serve as alternate hydrogen acceptors during citric acid fermentation in addition to serving as antifoam agents and stimulants. Similar results on the use of refined peanut oil in molasses medium or cane juice medium for improved citric acid yields have also been reported by Dhankar (1972) and Kumar and Ethiraj (1976).

TABLE 16.4.   EFFECT OF DIFFERENT AMOUNTS OF VEGETABLE OILS ON THE YIELD OF CITRIC ACID BY A. NIGER, MUTANT 72-74

Amount of Oil Added (% v/v)	Maize	Oil Added Peanut Citric Acid Yield (g/100 ml)	Olive
0	4.6	7.0	5.2
0.1	6.7	—	5.7
0.2	5.7	—	5.2
0.5	6.8	7.3	9.0
1.0	7.5	9.0	7.2
2.0	8.1	9.9	8.5
3.0	8.0	9.2	8.5
4.0	8.5	9.0	8.4
5.0	9.4	10.0	—

Source: Millis et al. (1963).

Antifoam agents such as octadecanol (0.75% solution) or Antifoam AE (silicone oil) have been tested and found to be effective in controlling foaming and increasing the yields of citric acid (Usami et al. 1960). Apparently, the presence of antifoam agents allows increased aeration and agitation.

Complex components such as mycelial digests, bakers' yeast, rice bran, etc., have been tested for their usefulness in increasing citric acid yields. A Belgian patent describes the use of a mycelial digest from A. niger for increasing citric acid yields (Czech. Acad. Sci. 1963, 1964). The stimulatory component from the digest has been isolated by chromatography on anion exchange resin or cellulose. The yield of the stimulant is increased by the addition of sulfonamides or streptomycin. The stimulant is dialyzable and thermostable, but unstable in alkali. Addition of this stimulant to a sucrose medium increases citric acid production by 60–70% while without the stimulant the yield is 40–50%. Suprisingly, the stimulant can be prepared from A. terreus and A. niger as well as from Escherichia coli.

Bakers' yeast when added to the fermentation medium has been found to improve the yield of citric acid by *A. niger*. Leopold (1965) has patented a process in which the addition of 1.92 mg of pressed bakers' yeast to a culture medium and cultivation for 9 days at 32°C increased the yield of citric acid from 61.5 to 72.1%.

Cyclic AMP, which has been of interest in recent years, has also drawn the attention of investigators in citric acid production. Wold and Suzuki (1973) have reported that a strain of *A. niger* accumulated citric acid in the medium when cAMP concentration was 10^{-6} M or higher (Table 16.5). The rate of citrate synthesis under these conditions was found to be increased. Adenosine, ATP, and/or cGMP also stimulated citric acid production at a concentration of 10^{-3} M, but were ineffective at 10^{-4} M. AMP had no effect while GMP and guanosine inhibited citric acid production slightly. ADP on the other hand strongly inhibited citric acid accumulation. Addition of thiophylline along with cAMP was found to increase the effect of cAMP. It is suggested that citric acid production in fungi perhaps results from an abnormal cAMP metabolism.

Normally the fermentation medium is sterilized prior to inoculation. When sucrose or deionized molasses is used as the carbon source, the pH of the medium is also kept low and this is adequate to prevent bacterial

TABLE 16.5. INFLUENCE OF ADENINE AND GUANINE NUCLEOTIDES ON CITRATE ACCUMULATION

Effector	Specific Activity	% of Control
Experiment 1		
Control	1372 ± 79^{1}	100
10^{-3} M 3',5'-cyclic AMP (9)	2454 ± 195	179
10^{-3} M 2',3'-cyclic AMP (9)	1363 ± 131	99
10^{-3} M adenosine (9)	1481 ± 122	108
10^{-3} M 5'-AMP (9)	1263 ± 105	92
10^{-3} M ADP (9)	792 ± 125	58
10^{-3} M ATP (9)	1736 ± 122	127
Experiment 2		
Control (9)	1875 ± 158	100
10^{-3} M 3',5'-cyclic AMP (4)	3082 ± 74	163
2×10^{-4} M 3',5'-cyclic AMP (4)	2374 ± 89	127
10^{-3} M 3',5'-cyclic GMP (4)	2831 ± 349	151
2×10^{-4} M 3',5'-cyclic GMP (5)	2067 ± 201	110
10^{-3} M 5'-GMP (8)	1694 ± 156	90
10^{-3} M guanosine (9)	1485 ± 196	79
Experiment 3		
Control (8)	1155 ± 203	100
10^{-3} M 3',5'-cyclic AMP (9)	1808 ± 278	156
10^{-3} M 5'-AMP	1229 ± 169	106
10^{-3} M adenosine (9)	1756 ± 244	152

Source: Wold and Suzuki (1973).
[1]Standard deviation of mean.
Note: Citrate was determined by a couple of citratases (prepared from *Aerobacter aerogenes*) and malic dehydrogenase. The number in brackets represents the number of flasks used.

contamination. However, when crude molasses media are used, necessitating a higher pH, the chances of bacterial contamination are greater. To prevent this, antiseptics such as pentachlorophenolate, formic acid, 5-nitro-3-furaldehyde semicarbazone, tetracyclines, etc., have been used. It is claimed that these chemicals effectively control bacterial growth without affecting the citric acid producing ability of A. niger (Halama 1974).

CITRIC ACID PRODUCTION BY YEASTS

Until recently, almost all the citric acid produced by fermentation was manufactured using A. niger and a few other fungi. It is now well known that many kinds of yeasts also accumulate citric acid in their growth media along with relatively large amounts of isocitric acid. The variety of yeasts known to produce citric acid from various sources are species of Candida, Hansenula, Pichia, Debaryomyces, Torulopsis, Kloeckera, Trichosporon, Torula, Rhodotorula, Sporobolomyces, Endomyces, Nocardia, Nematospora, Saccharomyces, and Zygosaccharomyces. Out of these yeast genera, the species of Candida (Saccharomycopsis) are the ones that are widely used for studies on citric acid production. These species include C. lipolytica, C. tropicalis, C. zeylanoides, C. fibrae, C. intermedia, C. parapsilosis, C. petrophilum, C. subtropicalis, C. oleophila, C. hitachinica, C. citrica, C. guilliermondii, and C. sucrosa. A variety of substrates have been used for producing citric acid by yeast. These include glucose, acetate, hydrocarbons, molasses, alcohols, fatty acids, and natural oils.

Citric acid production from glucose is carried out in media containing glucose (10–18%), NH_4Cl, KH_2PO_4, and $MgSO_4 \cdot 7H_2O$, and a neutralizing agent such as $CaCO_3$. The concentration of various nutrients in the medium depends on the yeast strain used. Fermentation is always carried out under aerobic conditions, and a temperature between 22° and 30°C is optimum. During fermentation, isocitric acid is accumulated along with citric acid and the amount of isocitric acid produced depends on the cultural conditions and the strain of yeast. Generally, yeast strains producing the least amount of isocitric acid but high levels of citric acid are desirable. The time of fermentation varies from 3 to 6 days.

The metabolic pattern of these yeasts makes them ideal for use in citric acid production. These yeasts are highly oxidative and therefore need a considerable amount of aeration for their growth and metabolism. Fermentations using these yeasts are therefore carried out with vigorous agitation.

Oh et al. (1973) have reported citric acid production by a strain of Hansenula anomala in a 10% glucose medium containing 3% $CaCO_3$ at a temperature of 30°C with agitation at 110 rpm. Potassium dihydrogen phosphate (0.05%) and magnesium sulfate (0.025%) are also added. Ammonium chloride at 0.1% concentration was most effective as a N source. In a 6 day fermentation, 46% citric acid was produced.

Ishi et al. (1972) have developed a process for the production of citric acid by fermentation of waste glucose (32.6% glucose, 3% fructose) left after separation of fructose from a fructose-glucose mixture, after the addition of urea, corn steep liquor, KH_2PO_4, $MgSO_4$, $MnSO_4$, and $CaCO_3$. The medium was inoculated with *C. oleophila* and fermentation was carried out at 28°C under aeration for 5 days to yield 70.5% calcium citrate based on the initial carbon content.

Citric acid production from molasses by species of *Candida* has also been reported. Using *C. guilliermondii* and *C. lipolytica*, a continuous process for citric acid production from molasses has been described (Miall and Parker 1975). The medium contains 50–250 g/liter sugarcane or beet molasses; a nitrogen source such as $(NH_4)_2SO_4$, NH_4Cl, or NH_4NO_3; minerals; and vitamin B complex. Fermentation is carried out for 189 hr with 0.3–1.5 vol. air/min. Initial pH of the medium is 5.5–6.5 which is stabilized at 2.8–4.0 during citrate formation. From 3145 kg sugar, 1119 kg of citrate monohydrate was produced. Similarly, Liu (1975) has also reported citric acid production using *Candida* spp. in a medium containing molasses with yields as high as 50%.

Various alcohols such as methanol, butanol, ethanol, and C_{12-16} alcohols can serve as carbon sources for citric acid production by strains of *C. fibrae, C. subtropicalis, Pichia farinosa, C. lipolytica*, and *Torulopsis xylinus* (Tabuchi *et al.* 1969; Yoshinaga *et al.* 1972; Ikeno *et al.* 1975). The initial concentration of alcohol in the medium varies from 1 to 2% and additional alcohol is added during the fermentation to maintain the concentration. Besides alcohol, the medium should contain a nitrogen source, a phosphate source, and Fe^{++}, Mg^{++}, Zn^{++}, Mn^{++}, and Cu^{++} as trace elements. Fermentation is carried out for 7 days and yields of 27.9 g citric acid/liter are obtained (Ikeno *et al.* 1975).

Production of citric acid from acetate by yeast has been described by various workers (Yoshinaga *et al.* 1972; Takayama and Tomiyama 1973; Tabuchi *et al.* 1973; and Uchio 1976). Species of *Candida* such as *C. zeylanoides, C. parapsilosis, C. fibrae*, and *C. subtropicalis* are cultivated using acetic acid or calcium acetate as a carbon source. Fermentation is carried out for about 3 days at a pH of 5–6 with shaking.

Fatty acids, natural oils, and fats have been tested for citric acid production utilizing various genera of yeasts like *Candida, Hansenula*, and *Pichia* (Tabuchi *et al.* 1969; Nubel and Fitts 1974; Masuda *et al.* 1975; Ikeno *et al.* 1975). Tallow, coconut oil, palm oil, olive oil, soybean oil, linseed oil, rapeseed oil, fish oil, corn oil, and free fatty acids have been tested for this purpose. A typical medium for citric acid production from coconut oil is given below (Ikeno *et al.* 1975):

Coconut oil	5.0%
$(NH_4)_2SO_4$	0.2%
Corn steep liquor	0.26%
KH_2PO_4	0.0125%
$MgSO_4 \cdot 7H_2O$	0.01%
$CaCO_3$	0.5%
Antifoam	0.5%
Thiamin hydrochloride	100 µg/ml

Fermentation is carried out at 30°C with shaking for 84 hr to give a 146% yield of citric acid from palm oil with a mutant strain of *C. lipolytica*. Various strains of *C. lipolytica* have been extensively investigated for citric acid production from hydrocarbons such as n-paraffins, n-alkanes, and alkenes (Ikeno *et al*. 1975; Tabuchi *et al*. 1969). Strains of *C. lipolytica* yielding a high amount of citric acid without the formation of isocitric acid have been developed by mutagenic treatment (Suzuki *et al*. 1971; Fukuda *et al*. 1971; Nakanishi *et al*. 1972; Akiyama *et al*. 1972, 1973A,B; Hustede and Siebert 1975; Hustede 1975; and Maldonado *et al*. 1975).

Several processes for citric acid production from n-paraffins have been described (Kimura and Nakanishi 1971; Akiyama *et al*. 1972; Nakanishi *et al*. 1972; Iizuka *et al*. 1973; Takahashi and Ikeno 1974; Fukuda *et al*. 1975). The process involves cultivation of mutants of *Candida* with high citric acid-producing ability in an n-paraffin-containing medium containing nitrogen, phosphate, $MgSO_4$, $CaCO_3$, and vitamins. Fermentation is carried out with shaking at $28°-30°C$ for $4-6$ days. A typical medium for citric acid production from n-paraffins includes hydrocarbons, $40-60$ g; NH_4Cl, 2 g; KH_2PO_4, 0.5 g; $MgSO_4$, 0.5 g; cornsteep liquor, 1 g; and $CaCO_3$, 30 g/liter (Tabuchi *et al*. 1969).

As with *A. niger* fermentations, the concentration of ferric ion in the medium affects the accumulation of citric and isocitric acids. An increase in citric and a decrease in isocitric acid can be achieved by adjusting the iron content to very low concentrations. Since the maintenance of a low concentration of iron in complex media is difficult, fluoroacetate-sensitive strains of *C. lipolytica* with low aconitase activity have been developed which produce increased amounts of citric acid from n-paraffins (Akiyama *et al*. 1972).

Small amounts of polyhydric alcohols such as mannitol, sorbitol, or erythritol are detected temporarily at early stages of the fermentation in an n-paraffin culture medium containing a neutralizing agent such as $CaCO_3$. In a culture medium containing no neutralizing agent, the amount of citrates is decreased (Tabuchi and Hara 1973).

The concentration of thiamin in the fermentation medium has a marked effect on the amount of citric and isocitric acid produced. With sufficient thiamin in the medium a large amount of citric acid is produced, while in a thiamin restricted medium a large amount of α-ketoglutarate is produced with a reduced amount of citrate (Tabuchi and Hara 1973).

Citric acid production from n-alkanes is similar to that from n-paraffins. A typical medium for citric acid production from alkanes using strains of *C. lipolytica* includes hydrocarbon, $4-6\%$; NH_4Cl, 0.2%; KH_2PO_4, 0.05%; $MgSO_4 \cdot 7H_2O$, 0.05%; cornsteep liquor, 0.1%; and $CaCO_3$, 3%. The fermentation time was $6-8$ days under shake culture conditions and the yield of citric acid was $17-56\%$, depending upon the alkane (Tabuchi *et al*. 1969).

A unique medium containing a lead salt has been described for citric acid production by yeast. In a French patent (Pfizer 1971), citric acid production by *Candida* in aqueous nutrient media containing $0.5-1.5$ g lead oxide or lead salts has been described. A medium containing cerelose, 150 g; $CaCO_3$,

10 g; NaCl, 4 g; yeast extract, 5 g; peptone, 15 g; CaCO$_3$, 10 g; and lead acetate, 1.5 g/liter was seeded with an inoculum of *C. guilliermondii* and incubated for 72 hr at 30°C under aeration. It yielded 53 g citric acid/liter. Without lead acetate the yield was only 29 g. The role of the lead acetate in this fermentation is not clear.

One advantage of using yeasts for citric acid production, as compared to aspergilli, is the tremendous potential for developing a continuous process. In fact, a continuous process for citric acid production using *Candida* species has been patented by Miall and Parker (1975).

CITRIC ACID PRODUCTION BY BACTERIA

Despite the availability of a large amount of information regarding the biochemical activities of bacteria, this group of organisms has not been vigorously exploited for the production of citric acid. In the past 10 years only about 10 references have appeared and most of these are patents. Hence, few details are available regarding this aspect of citric acid production.

Some bacteria such as *Bacillus licheniformis*, *Bacillus subtilis*, and *Brevibacterium flavum* have been found to possess the ability to produce citric acid from either glucose, isocitric acid, or from hydrocarbons. Sardinas (1972) has patented a process for producing citric acid using *B. licheniformis* from glucose-containing media. Fermentation is carried out under aerobic conditions at 30°–37°C for 36–120 hr at a pH of 7.0. Besides glucose, the medium is supplemented with a nitrogen source such as urea, ammonium sulfate, or glutamate in addition to salts and calcium carbonate. Yields up to 42 g/liter have been reported. Kyowa Fermentation Industries (Kyowa Ferment. Ind. 1970) have patented a process for the production of citric acid by *Arthrobacter paraffinens*. The organism is cultured in an aqueous medium containing dodecane or a mixture of $C_{12}-C_{14}$ paraffins in addition to salts after preculturing in an inoculum medium. The fermentation medium was inoculated at a 5% level and aerated with 3 v/v/min at 28°C for 72 hr to yield 28 mg citric acid/ml.

Using n-paraffins as a carbon source, citric acid is also produced by *Corynebacterium* sp. and *Brevibacterium* sp. Fukuda *et al.* (1970) have patented a process for citric acid production which involves culturing various species of *Corynebacterium* in paraffin-containing media. Yields of 41.4 mg of citric acid/ml have been obtained by culturing for 64 hr at 32°C.

Alternate methods of producing citric acid by growing bacteria such as *Klebsiella, Aerobacter, Pseudomonas, Micrococcus, Bacillus, Brevibacterium, Corynebacterium,* and *Arthrobacter* in media containing isocitric acid have also been reported (Hitachi Chem. Co. 1973; Ohmori and Ikeno 1973; Takayama and Adachi 1974). The organisms are cultured in media containing 5–6% isocitric acid at a pH of 5–8 for about 2 days at 32°–37°C under shake culture conditions to convert isocitric acid into citric acid.

A mutant of *Brevibacterium flavum* requiring L-glutamate for growth has been described to produce citric acid when cultured in a medium containing glucose, 3.6%; urea, 0.2%; KH_2PO_4, 0.01%; $MgSO_4$, 0.04%; sodium L-glutamate, 0.05%, and $CaCO_3$, 1%, in addition to soybean hydrolysate (3 ml/liter) and minor amounts of Vitamin B_{12}, $FeSO_4$, and $MnSO_4$. After incubation at 30°C for 4 days at pH 7.4, the yield of citric acid was 50 mg/ml.

These studies on bacteria, although few in number compared to the reports on fungi, have opened a new avenue for citric acid production.

RECOVERY METHODS

There have been no substantial departures from the methods reported before 1959 for recovery of citric acid from culture filtrates. The crude fermented liquor containing citric acid, obtained either from surface or submerged culture, and the extract from solid state culture, is filtered to remove mycelia or cells and other suspended impurities. The waste mycelia are pressed to recover most of the fermented liquor. The mother liquor is heated to 80°–90°C by the addition of small amounts of hydrated lime to allow precipitation of oxalic acid. Citric acid is then precipitated as calcium citrate using 1 part of hydrated lime for every 2 parts of liquor added over a 1 hr period while the temperature is raised to 95°C. Although this is a generally accepted procedure, Heding and Gupta (1975) have reported a precipitation temperature of 50°C for 20 min to be optimum for calcium citrate precipitation and recovery of citric acid (Table 16.6). According to these investigators, the product obtained by precipitation at 50°C is easy to filter. The precipitation rate, although slightly slower, is complete at this temperature.

TABLE 16.6.   INFLUENCE OF TEMPERATURE ON THE PRECIPITATION OF CALCIUM CITRATE AT pH 5.0

Calcium Carbonate Addition Temp (°C)	Time Taken for Precipitation (min)	Yield of Citric Acid (%)
20	120	70
40	40	90
50	20	~100
70	2	100
80	0.5	100
100	0.5	~100

Source: Heding and Gupta (1975).

The precipitated calcium citrate is filtered and washed with water several times. It is then transferred to acidulators and treated with H_2SO_4. The solution is again filtered to remove $CaSO_4$. The mother liquor containing citric acid is decolorized by charcoal and passed through ion exchange resin columns. The liquor is concentrated in vacuum and finally run into low temperature crystallizers where citric acid crystallizes as citric acid monohydrate.

The quality of citric acid can be improved by using $Ca(OH)_2$ free of contaminating metals such as Mg^{++}, Fe^{++}, Al^{+++}, etc. Also, by the addition of 9–12% calcium ferrocyanide to a mother liquor at 95°–97°C for 5–8 min, the recovery of citric acid has been found to improve (Savchenko *et al.* 1976).

In large-scale production the preceding recovery method is commonly used (Fig. 16.2), but each manufacturer has his own modifications. In recent years, another procedure, the solvent extraction process, has been introduced. Rieger and Kioustelidis (1975) have patented a process in which citric acid is separated from aqueous fermentation solutions by 2-phase extraction using tridecylamine or triisononylamine and a water insoluble ester, ketone, or alcohol. Citric acid salts are then precipitated by extracting the mixed solvent extracts with aqueous solutions of ammonia, alkalihydroxides, carbonates, or bicarbonates. In another solvent extraction process recommended by the U.S. Food and Drug Administration (Anon. 1975), a mixture of n-octyl alcohol, synthetic isoparaffinic petroleum hydrocarbons, and tridodecylamine are used for the recovery of citric acid from fermented liquors. Citric acid recovered by this method is recommended for food, drug, and cosmetic uses.

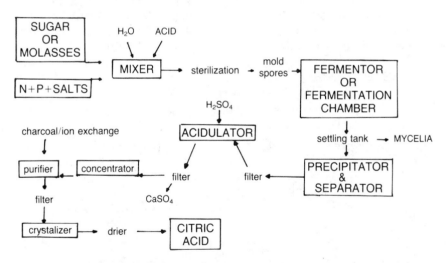

FIG. 16.2.   FLOW SHEET FOR CITRIC ACID PRODUCTION

In yeast fermentations using *Candida*, isocitric acid is produced along with citric acid, and this needs to be separated. Nara *et al.* (1971) have patented a process for separation of these acids from the fermented medium. Citric acid is first precipitated by mixing the culture broth with 3.3 parts of $Ca(OH)_2$ and maintaining the temperature at 85°–90°C for 2 hr. Isocitrate is then precipitated by the addition of 2 more parts of $Ca(OH)_2$ to the mother liquor and again maintaining the temperature at 85°–90°C for 2 hr. These salts are then dissolved in sulfuric acid and processed further. To obtain sodium or potassium citrate, the calcium citrate is treated with a 15%

excess of NaOH or KOH at 95°–98°C, and the sodium or potassium citrates are crystallized from the supernatant (Hentschel 1975).

BIOCHEMISTRY OF CITRIC ACID FERMENTATION

The functioning of the citric acid cycle (TCA cycle) in fungi has been well documented (Fig. 16.3), and citric acid is produced as an overflow product due to the faulty operation of the TCA cycle. The presence of enzymes of the TCA cycle has also been demonstrated in *A. niger* (Ramakrishnan and Martin 1955; Niederpruem 1965; Mueller and Frosch 1975). Studies on enzyme content of *A. niger* in relation to citrate accumulation and incorporation of ^{14}C from labelled substrates into citrate have indicated the central role of this cycle in this fermentation. Two key enzymes that have been examined in detail in relation to citric acid fermentation are aconitase and isocitrate dehydrogenase. The activities of these enzymes have been shown to decrease to very low levels during the period of citric acid accumulation while the activity of the condensing enzyme has been found to increase (Ramakrishnan *et al.* 1955; Bruchmann 1961; Bertrand and Wolf 1962; La Nauze 1966; Tabuchi *et al.* 1973). The final step in the synthesis of citric acid is the condensation of acetyl CoA and oxaloacetate and this condensation of C_2 and C_4 components is the major route of citrate synthesis.

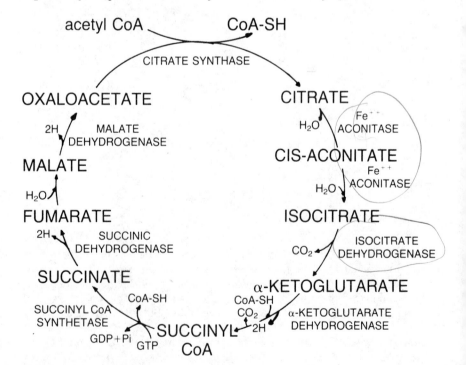

FIG. 16.3.   THE TRICARBOXYLIC ACID CYCLE

Direct carboxylation of pyruvate catalyzed by the malic enzyme provides malate which is readily used for citric acid synthesis after being converted into oxaloacetate through malic dehydrogenase. The incorporation of radioactive CO_2 into citrate has been determined (Bhattacharya et al. 1962) and the fixation involves two different enzyme systems. One system involves the condensation of CO_2 with pyruvate catalyzed by pyruvate carboxylase (Woronick and Johnson 1960; Bloom and Johnson 1962; Fier and Suzuki 1969).

$$\text{pyruvate} + CO_2 + H_2O + ATP \underset{}{\overset{Mg^{++}}{\rightleftharpoons}} \text{oxaloacetate} + ADP + Pi$$

The second system appears to be identical to the phosphoenol pyruvate carboxylase kinase system found in plants and yeasts. It utilizes phosphoenol pyruvate and CO_2 but does not require ATP (Woronick and Johnson 1960).

$$\text{phosphoenol pyruvate} + ADP + CO_2 \rightleftharpoons \text{oxaloacetate} + ATP$$

The glyoxylate cycle acts as another source of oxaloacetate for citrate synthesis (Shah and Ramakrishnan 1963; Mueller 1975). Acetyl CoA condenses with glyoxylate and the reaction is catalyzed by malate synthetase.

$$\text{glyoxylate} + \text{acetyl CoA} \rightleftharpoons \text{malate} + \text{coenzyme A}$$

The glyoxylate required for the synthetase reaction is supplied by the isocitritase reaction as shown:

$$\text{isocitrate} \rightarrow \text{succinate} + \text{glyoxylate}$$

Aconitase, one of the two key enzymes in citric acid production, is sensitive to a high concentration of metallic ions such as Fe^{++}. Restricting the activity of this enzyme during the production stage appears to be the key to success in citric acid production.

Citric acid production and excretion are apparently two different phenomena. Mutants with altered aconitase need not excrete citric acid. There is in fact evidence to suggest this although direct experimental proof is lacking. A mutant of A. niger which is not sensitive to trace metals but sporulates normally, yet produces citric acid, has been described (Chaudhary et al. 1978).

The overall success in citric acid fermentation using fungi depends to a large extent on the regulation and functioning of the TCA cycle. The TCA cycle is complex and involves several enzymes. The regulation of synthesis of each one of these enzymes and their activities is perhaps under various control mechanisms that are known for controlling enzyme synthesis and function. We have little information about the control of these enzymes except that the activity of some, such as aconitase, can be regulated by controlling trace element concentration.

Yeasts such as *Candida* are highly aerobic organisms and have a well developed TCA cycle. The production of citric acid by *Candida* is generally accompanied by a simultaneous accumulation of isocitric acid. According to Tabuchi *et al.* (1973), the activities of aconitase and NAD- or NADP-linked isocitrate dehydrogenase do not decrease much throughout the period of citrate fermentation when *C. lipolytica* is grown in a medium containing glucose or hexadecane. As in *A. niger*, addition of Fe^{++} increases the aconitase activity with decreased citrate and increased isocitrate accumulation. Addition of inhibitors of aconitase such as fluoroacetate leads to iron deprivation and to citrate accumulation. Also, in monofluoroacetate-sensitive mutants an increase in the accumulation of citrate and a decrease in the accumulation of isocitrate has been reported (Akiyama *et al.* 1973A; Tabuchi *et al.* 1974).

The activity of isocitrate dehydrogenase in this organism also determines the amount of isocitrate and citrate produced. Addition of potassium ferrocyanide or thiamin deficiency in a glucose medium leads to a decrease in isocitrate dehydrogenase activity and to an increased citrate accumulation (Nakanishi *et al.* 1972; Tabuchi *et al.* 1973).

Recently, Marchal *et al.* (1977) have also examined the regulation of the central metabolism in relation to citric acid and isocitric acid production in *Saccharomycopsis lipolytica (Candida lipolytica)*. In this organism, when grown on n-paraffins, citric and isocitric acid accumulation begins when growth ceases due to nitrogen limitation. At the same time, the concentrations of AMP and ADP decrease to a low level. The activity of NAD-dependent isocitric dehydrogenase which requires AMP for its activity also decreases significantly. This prevents the oxidation of citrate while isocitrate lyase activity is not inhibited. Citric and isocitric acid accumulation occurs since the inhibition of citrate synthase by ATP is inadequate to stop n-paraffin degradation. One can, therefore, conclude that the excretion of citric and isocitric acids probably occurs due to other physiological changes brought about by nitrogen limitation and also by the alteration of cell permeability to these acids.

The involvement of the glyoxylate cycle and CO_2 fixation has also been demonstrated in *C. lipolytica* during citrate accumulation (Tabuchi and Hara 1974). In this yeast, it appears that isocitrate lyase and malate synthetase, the key enzymes of the glyoxylate cycle, are inducible. When n-paraffins and acetate are used as carbon sources, oxaloacetate for citrate synthesis is supplied through the glyoxylate cycle. While growing in a glucose-containing medium, oxaloacetate is generated by the CO_2 fixation reaction. During active citric acid synthesis, the activity of citrate synthetase is high as compared to other enzymes of the TCA cycle (Glazunova and Finogenova 1976).

Tabuchi and Hara (1974) have reported the production of a large amount of threo-D_s-methyl isocitric acid and trace amounts of 1-methyl citric acid and 2-methyl-*cis*-aconitic acid by a mutant of *C. lipolytica* when grown in a mixture of n-alkanes. Methylisocitric acid is produced mainly from the odd carbon alkanes from the terminal C_3-residues left after successive removal of C_2-fragments by β-oxidation. Tabuchi *et al.* (1974) have proposed a hypothetical pathway for partial oxidation of propionyl CoA (formed during β-oxidation of odd carbon alkanes) to pyruvate via C_7-tricarboxylic acids in yeast. This proposed hypothetical pathway (methylcitric acid cycle) for citric acid accumulation is given in Fig. 16.4.

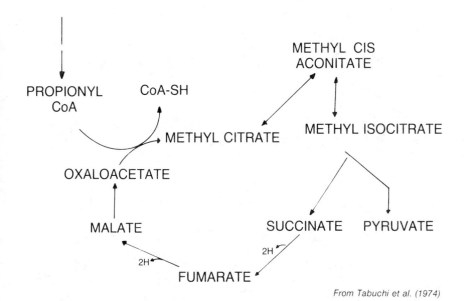

From Tabuchi et al. (1974)

FIG. 16.4.   METHYL CITRIC ACID CYCLE

According to this scheme, propionyl-CoA formed through the β-oxidation of n-alkanes condenses with oxaloacetate to yield methyl citric acid. This is then isomerized to methyl-isocitric acid. This is further cleaved to yield pyruvate and succinate. Succinate is then oxidized through the TCA cycle to yield oxaloacetate which is then recycled. A methyl citrate condensing enzyme and a methyl isocitrate cleaving enzyme have been demonstrated in *C. lipolytica* (Tabuchi and Uchiyama 1975); Tabuchi and Satoh 1976; Uchiyama and Tabuchi 1976; Tabuchi and Satoh 1977). On the basis of the methyl citric acid cycle, the glyoxylate pathway, and the CO_2 fixation reaction, the hypothetical pathway for citric acid production from n-alkanes as proposed by Tabuchi and Serizawa (1975) is shown in Fig. 16.5.

According to this pathway, 1 mole of odd carbon alkane having $(2n + 1)$ carbon atoms yields $(n - 1)$ moles of acetyl-CoA and 1 mole of propionyl CoA through β-oxidation. Propionyl CoA is then oxidized to pyruvate via the methyl citric acid cycle. Carboxylation of pyruvate to oxaloacetate and

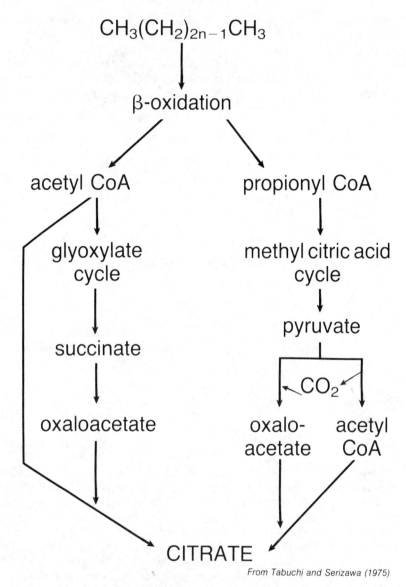

FIG. 16.5.   HYPOTHETICAL PATHWAY OF CITRATE PRODUCTION FROM n-ALKANES BY YEASTS

condensation with acetyl CoA yields citric acid. Also, acetyl CoA formed by β-oxidation enters the glyoxylate cycle to yield oxaloacetate, which on condensation with acetyl CoA again yields citrate. Thus 1 mole of n-alkane having $(2n + 1)$ carbon atoms gives rise to $(2n + 1)/6$ moles of citrate.

GENETIC IMPROVEMENT IN CITRIC ACID-PRODUCING MICROORGANISMS

Despite great advances in our knowledge of the genetics and biochemistry of microorganisms, the application of genetics to the improvement of microorganisms producing citric acid has not been very significant. The main technique used for genetic improvement of citric acid-producing fungi has been the age-old technique of mutation and selection. Normally, a spore population is suspended in a liquid medium or a buffer and subjected to mutagenesis, and the survivors are screened for their ability to produce citric acid.

One difficulty in screening mutants for better citric acid producers is the lack of a precise and quick method by which citric acid-producing strains can be easily selected. Agar plates with nutrient media containing an indicator have been used for preliminary screening of survivors for citric acid-producing mutants. This method is, however, not very precise. It is well known that the type of acid produced by fungi is pH dependent and when agar medium of pH 6–7 containing sucrose is used for detecting acid producers, normally oxalic acid-producing mutants are detected (Tauro 1977).

To obtain mutants that produce more citric acid in molasses medium, agar medium containing smaller concentrations of molasses and an indicator have been used (Chaudhary 1977). By this method it has been possible to isolate mutants producing more citric acid in molasses medium.

As discussed in the earlier section on "Trace Elements," citric acid production is extremely sensitive to the concentration of trace elements, especially iron. Citric acid production in a molasses medium is more economical than in a sucrose medium. However, the control of the concentration of trace elements in such complex media is difficult and therefore attempts have been made to isolate mutants that would tolerate a high concentration of trace elements and produce large quantities of citric acid when cultured in crude cane molasses medium (Chaudhary et al. 1978).

Mutants of A. niger that have better ability to produce citric acid have been derived by mutagenesis and extensive screening. Trumphy and Millis (1963) reported the isolation of one mutant of A. niger, after screening nearly 40,000 survivors, which produced 4 times more citric acid than the parent culture. Shcherbakova (1964) has reported the isolation of A. niger mutants by UV irradiation. These mutants produce equal amounts of citric acid in synthetic media, but in a medium containing molasses, they produce 1.5 times more citric acid than the parent. Das and Nandi (1969) have also reported the isolation of mutants of A. niger using UV, gamma-rays, and nitrogen mustard. The most promising mutant was the gamma-ray mutant which produced about twice the amount of citric acid as the parent culture. This mutant was further subjected to UV irradiation and a second stage mutant was produced which gave 3 times more citric acid than the parent culture. Kahlon and Vyas (1971) have reported the screening of 800 survivors of nitrous acid and UV mutagenesis. Of these, one mutant produced a large amount of citric acid in cane sugar medium while another produced good amounts of citric acid only in molasses medium. Banik (1975) has reported the derivation of mutants of *A. niger* by sequential mutagenesis using UV and ethyleneamine, which had the ability to produce about 70.5 mg of citric acid/ml in a sucrose medium while the wild type produced only 11.6 mg/ml. Similarly, Sharma (1973) has reported the isolation of a mutant of *A. niger* by sequential mutagenesis of *A. niger* spores using nitrous acid, nitrosoguanidine, and UV light. This mutant has been found to produce more citric acid than the parent when cultured in a molasses medium.

Besides induction of mutants using various mutagens, strain improvement in *A. niger* by somatic recombination has been attempted, but without much success.

Yeasts, particularly the members of the genus *Candida*, are increasingly being used for citric acid production because of their ability to produce citric acid from a variety of substrates such as glucose, molasses, hydrocarbons, acetate, alcohols, lipids, etc. In these fermentations citric acid is produced along with isocitric acid. The development of mutants that accumulate only citric acid in high concentration without concomitant accumulation of isocitric acid has been attempted. Also, mutants with either low aconitase activity or those with an enzyme sensitive to fluoroacetate have been isolated and used for citric acid production (Akiyama *et al.* 1972). Ikeno *et al.* (1975) have studied various *Candida* mutants derived by nitrosoguanidine mutagenesis for citric acid production in glucose, n-paraffin, or coconut oil-containing media (Table 16.7). These mutants produce higher amounts of citric acid in glucose, n-paraffin, or in coconut oil-containing media as compared to the parent culture. Sriprakash (1977) has isolated a mutant of *C. lipolytica* which accumulates 140 g of citric acid/liter, representing a conversion of about 70% on the basis of the carbohydrate input. The ease with which yeasts can be handled and the high yields reported make them ideal test systems for genetic improvement studies.

TABLE 16.7.   RELATION BETWEEN MAIN CARBON SOURCES AND YIELD OF CITRIC ACID BY VARIOUS MUTANT STRAINS OF *C. LIPOLYTICA*

| | Yield of Citric Acid | | |
C. lipolytica	Glucose (%)	n-Paraffin (%)	Coconut Oil (%)
Parent Strain 281	36.8	71.1	32.6
Mutants N-9	45.2	92.6	38.2
N-157	46.0	112	46.8
N-1236	52.4	143	63.3
N-2487	64.8	155	58.5
N-1813	82.0	—	—
N-2508	57.4	150	65.0
N-5704	55.5	158	73.5

Source: Ikeno *et al.* (1975).

CONCLUSIONS

Although the need for citric acid has been steadily increasing because of its use in industries other than food, there has been no significant improvement or major breakthrough in the past two decades in producing this organic acid using aspergilli and other fungi. The basic methodology of culturing this organism has not undergone any drastic change. Also the organisms commonly used, as far as information is available, do not appear to be much different from those used two decades ago. It appears that the capacity of the aspergilli to produce citric acid has perhaps been fully exploited. This thinking is supported by the fact that mutants of *A. niger* with abilities to convert sucrose into citric acid at almost theoretical limits have been reported. For example, with an initial concentration of 14% sucrose, it has been possible to convert nearly 80–85% of this into citric acid (Miles Lab. 1962; Chaudhary *et al.* 1978). It is likely that the major industrial producers have also reached these limits in citric acid production.

One improvement that is wanting is the development of cultures that will produce citric acid in unclarified molasses medium in a shorter period of time. Normally, crude fermentation materials such as molasses, starch hydrolysates, etc., are purified to remove trace elements before use in fermentation. This step has increased the cost of citric acid production. Secondly, under surface culture conditions, an incubation time of 8–10 days, and in submerged culture, a minimum of 4–5 days is required. Improvements in this area should involve the development of strains which will produce citric acid from crude raw materials without any pretreatment, and, secondly, mutants that will grow faster and produce citric acid in a shorter time without loss in efficiency.

Generally, the use of methanol has become a common practice in citric acid production. Although the use of this alcohol ensures consistent and high yields, the use of this chemical has added to the cost of production. Also, since methanol has more important alternate uses these days, it is imperative that strains that do not need methanol be developed.

Despite a great deal of understanding of the basic genetic mechanisms existing in microorganisms, there have been no substantial advances made to use this knowledge for improving fungi useful in citric acid production. This may be due to the absence of a concerted effort by those interested seriously in this area of industrial microbiology. Generally, genetic improvement of industrial microorganisms has remained with the industries, and therefore progress has mostly remained a secret.

The discovery that yeasts and bacteria can produce substantial amounts of citric acid in a shorter period of time than fungi and from a variety of raw materials holds much promise for the future. Information available on these aspects and reviewed herein supports this line of thinking. These organisms because of their versatile nature are ideally suited for citric acid production and therefore need further exploitation.

ACKNOWLEDGEMENTS

The assistance of Mr. K.M. Sharma in typing this manuscript is gratefully acknowledged.

REFERENCES

AKIYAMA, S. *et al.* 1972. Production of citric acid from n-paraffins by fluoroacetate sensitive mutant strains of *Candida lipolytica.* Agric. Biol. Chem. *36,* 339–341.

AKIYAMA, S. *et al.* 1973A. Production of citric acid from n-paraffins by fluoroacetate sensitive mutants. 1. Induction and citric acid productivity of fluoroacetate sensitive mutant strain of *Candida lipolytica.* Agric. Biol. Chem. *37,* 879–884.

AKIYAMA, S. *et al.* 1973B. Production of citric acid from n-paraffins by fluoroacetate sensitive mutants. II. Relation between aconitate hydratase activity and citric acid productivity in fluoroacetate sensitive mutant strain of *Candida lipolytica.* Agric. Biol. Chem. *37,* 885–888.

ANON. 1972. The Market for Sucrose-based Chemicals. International Trade Centre UNCTAD/GATT. U.N. Conf. on Trade and Dev., Gen. Agreement on Tariff and Trade, New York.

ANON. 1975. Food additives: Solvent extraction process for citric acid. U.S. Food Drug Admin., Fed. Regist. *40* (204) 49080–49082. Oct. 21.

ATTICUS. 1975. Citric acid forges ahead as an industrial chemical. Chem. Age India *26,* 49–54.

BANIK, A.K. 1975. Fermentative production of citric acid by *Aspergillus niger.* Strain selection and optimum cultural conditions for improved citric acid production. J. Food Sci. Technol. *12,* 111–114.

BERTRAND, D. and WOLF, A.D. 1962. Influence of iron and manganese as trace elements in the synthesis of aconitase in *Aspergillus niger.* C.R. Acad. Fr. (Paris) *254,* 4381–4383.

BHATTACHARYA, P.K., VAKIL, J.R., DASGUPTA, A.K. and DAMODARAN, M. 1962. Role of carbondioxide in citric acid fermentation. 1. Incorporation of carbondioxide into citrate. J. Sci. Ind. Res. *21,* 193–201.

BLOOM, S.J. and JOHNSON, M.J. 1962. The pyruvate carboxylase of *Aspergillus niger.* J. Biol. Chem. *237,* 2718–2720.

BREHZNOI, Y.D., NOVIKOVA, L.A., IGNATOV, Y.L. and KACHANOV, Y.E. 1976. Citric acid. U.S.S.R. Pat. 510,509. April 15.

BRUCHMANN, E.E. 1961. Enzymatic studies on mold fermentation. II. Inhibition of aconitase and its significance for citric acid accumulation by *Aspergillus niger* in submerged culture. Biochem. Z. *335,* 199–211.

BRUCHMANN, E.E. 1966. Action of hydrogen peroxide on the accumulation of citric acid by *Aspergillus niger.* Naturwissenschaften *53,* 226–227.

CAHN, F.J. 1935. Citric acid fermentation on solid materials. Ind. Eng. Chem. 27, 201–203.

CASIDA, L.E., JR. 1968. Industrial Microbiology. John Wiley & Sons, New York.

CHAUDHARY, A.Q. and PIRT, S.J. 1966. The influence of metal complexing agents on citric acid production of Aspergillus niger. J. Gen. Microbiol. 43, 71–81.

CHAUDHARY, K. 1977. Unpublished. Haryana Agric. Univ., Hissar, India.

CHAUDHARY, K., CHAKARVORTY, S.C. and TAURO, P. 1972. Inhibition of fungal growth by sugarcane molasses. J. Res. Haryana Agric. Univ. 1, 48–52.

CHAUDHARY, K., LAKSHMINARAYANA, K., DEV, I.K. and VYAS, S.R. 1974. Nitroguanidine induced mutation of Aspergillus niger for obtaining high citric acid producing strains. Indian J. Microbiol. 14, 42–43.

CHAUDHARY, K., LAKSHMINARAYANA, K., ETHIRAJ, S. and TAURO, P. 1978. Citric acid production by Aspergillus niger from Indian cane molasses by solid state fermentation. J. Ferment. Technol. Jpn. 56, 554–557.

CLARK, D.S. 1962. Submerged citric acid fermentation of ferrocyanide treated cane molasses. Biotechnol. Bioeng. 4, 17–21.

CLARK, D.S. 1964. Citric acid. U.S. Pat. 3,118,821. Jan. 21.

CLARK, D.S. and LENTZ, C.P. 1971. Submerged citric acid fermentation of sugarbeet molasses: Effect of pressure and recirculation of oxygen. Can. J. Microbiol. 7, 447–453.

CURRIE, J.N. 1917. The citric acid fermentation of Aspergillus niger. J. Biol. Chem. 31, 15–37.

CZECH. ACAD. SCI. 1963. Stimulant for the production of acids of TCA cycle by fungi. Belg. Pat. 600,537. Aug. 1.

CZECH. ACAD. SCI. 1964. Influence of a mycelial extract of Aspergillus niger on the formation of citric acid. Fr. Pat. 1355,922. March 20.

DAS, A. and NANDI, P. 1969. New strain of Aspergillus niger producing citric acid. Experientia 25, 1211–1213.

DHANKAR, H.S. 1972. Production of citric acid by soil fungus (Aspergillus niger) from pretreated sugarcane molasses. M.Sc. Thesis. Haryana Agric. Univ., Hissar, India.

DHANKAR, H.S., ETHIRAJ, S. and VYAS, S.R. 1974. Effect of methanol on citric acid production from sugarcane molasses by Aspergillus niger. Indian J. Technol. 12, 316–317.

DOELGER, W.P. and PRESCOTT, S.C. 1934. Citric acid fermentation. Ind. Eng. Chem. 26, 1142.

FEDOSEEV, V.F. et al. 1970. Effect of copper on the fermentative conversion of molasses to citric acid. Khlebopek. Konditer. Promst. 14, 33–35.

FIER, H.A. and SUZUKI, I. 1969. Pyruvate carboxylase of *Aspergillus niger*. Kinetic study of a biotin containing carboxylase. Can. J. Biochem. *47*, 697–710.

FUKUDA, H., SUZUKI, T., AKIYAMA, S. and SUMINO, Y. 1971. Citric acid fermentation using *Candida* yeasts. Ger. Pat. 2,108,094. Sept. 2.

FUKUDA, H., SUZUKI, T., AKIYAMA, S. and SUMINO, Y. 1975. Citric acid from hydrocarbons by *Candida* yeasts. Can. Pat. 973,824. Sept. 2.

FUKUDA, H., SUZUKI, T., SUMINO, Y. and AKIYAMA, S. 1970. Microbial preparation of citric acid. Ger. Pat. 2,003,221. Dec. 3.

GARBATAYA, E.I., MINTS, E.S. and SMIRNOV, V.A. 1975. Initial treatment of molasses before fermentation for the production of citric acid. U.S.S.R. Pat. 469,743. May 5.

GLAZUNOVA, L.M. and FINOGENOVA, Y.V. 1976. Activity of the enzymes of citrate glyoxylate and pentose phosphate cycles during synthesis of citric acid by *Candida lipolytica*. Mikrobiologiya *45*, 444–449.

GUPTA, J.K., HEDING, L.G. and JARGENSEN, O.B. 1976. Effect of sugars, pH and ammonium nitrate on formation of citric acid by *Aspergillus niger*. Acta Microbiol. Acad. Sci. Hung. *23*, 63–67.

HALAMA, D. 1974. Combination of thermal and chemical sterilization. Biotechnol. Bioeng. Symp. *4* (Pt. 2) 891–898.

HEDING, L.G. and GUPTA, J.K. 1975. Improvement of conditions for precipitation of citric acid from fermentation mash. Biotechnol. Bioeng. *17*, 1363–1364.

HENTSCHEL, G.O. 1975. Potassium or sodium salts of citric acid and tartaric acid. Swed. Pat. 380,516. Nov. 10.

HISANAGA, S. and NISHIMURA, Y. 1968. Citric acid production by fungal fermentation. Jpn. Pat. 68,20,708. Sept. 5.

HITACHI CHEM. CO. 1973. Microbiologically producing citric acid. Br. Pat. 1,306,986. Feb. 14.

HUSTEDE, H. 1975. Yeast mutants with high citric acid producing ability. Ger. Pat. 2,264,764. March 20.

HUSTEDE, H. and RUDY, H. 1976. Citric acid by submerged fermentation. U.S. Pat. 3,941,656. March 2.

HUSTEDE, H. and SIEBERT, B. 1975. Citric acid by fermentation. Ger. Pat. 2,264,763. May 22.

IIZUKA, H. et al. 1973. Citric acid by fermentation. Jpn. Pat. 7,326,981. April 9.

IKENO, Y. et al. 1975. Citric acid production from various raw materials by yeasts. Hakko Kogaku Zasshi *53*, 752–756.

ISHI, K., NAKAJIMA, Y. and IWAKURA, T. 1972. Citric acid by fermentation of waste glucose, Ger. Pat. 2,157,847. May 25.

KAHLON, S.S. and VYAS, S.R. 1971. Production of citric acid by mutants of *Aspergillus niger*. J. Res. Punjab Agric. Univ. *8*, 356–359.

KHAN, M.A.A., HUSSAIN, M.M., KHALIQUE, M.A. and RAHMAN, M.A. 1970. Methods of citric acid fermentation from molasses by *Aspergillus niger*. Pak. J. Sci. Ind. Res. *13*, 439–444.

KIMURA, K. and NAKANISHI, T. 1971. Citric acid by fermentation of microorganisms in the presence of alcohols. Ger. Pat. 2,040,356. Apr. 29.

KOVATS, J. 1960. Studies on submerged citric acid fermentation. Acta Microbiol. Pol. *9*, 275–287.

KOVATS, J. and GACKOWSKA, L. 1976. Effect of interruption in aeration of liquid in deep citric acid fermentation in molasses solution. Przem. Ferment. Rolny *20*, 22–25. (Polish)

KUMAMOTO, H. and OKAMURA, T. 1972. Fermentative manufacture of citric acid. Jpn. Pat. 7,234,955. Sept. 2.

KUMAR, K. and ETHIRAJ, S. 1976. Influence of methanol and groundnut oil on citric acid production from sugarcane juice by *Aspergillus niger*. Intern. Sugar J. *78*, 13–15.

KYOWA FERMENT. IND. 1970. Citric acid prepared by fermentation. Br. Pat. 1,187,610. Apr. 8.

LAKSHMINARAYANA, K., CHAUDHARY, K., ETHIRAJ, S. and TAURO, P. 1975. A solid state fermentation method for citric acid production using sugarcane bagasse. Biotechnol. Bioeng. *17*, 291–293.

LA NAUZE, J.M. 1966. Aconitase and isocitric dehydrogenase of *Aspergillus niger* in relation to citric acid production. J. Gen. Microbiol. *44*, 73–81.

LEOPOLD, H. and VALTR, Z. 1969. Effect of potassium ferrocyanide on preparation of molasses solution for citric acid fermentation. V. New boiling process. Nahrung *13*, 11–19.

LEOPOLD, J. 1971. Fermentation production of citric acid. Czech. Pat. 142,338. Aug. 15.

LEOPOLD, J. and COISONU, D. 1974. Control of infection by *Aspergillus flavus* in citric acid fermentation. Czech. Pat. 156,327. Dec. 15.

LEOPOLD, J. and VALTR, Z. 1967. Preparing molasses substrates for citric acid fermentation. Czech. Pat. 121,832. Feb. 15.

LEOPOLD, S. 1965. Stimulating the production of TCA cycle acids, especially citric acid. Czech. Pat. 113,640. Feb. 16.

LIU, Y.T. 1975. Studies on the fermentation production of citric acid. II. Screening of the yeasts producing citric acid from cane molasses. T'ai-wan T'ang Yeh Yen Chiu So Yen Chiu Hui Pao *68*, 55–56. (Chinese)

LOCKWOOD, L.B. and BATTI, M.A. 1965. Biosynthesis of citric acid. U.S. Pat. 3,189,527. June 15.

MALDONADO, P., CHARPENTIER, M. and GLIKMANS, G. 1975. Citric acid, isocitric acid and yeast cells by fermentation. Ger. Pat. 2,451,481. May 15.

MARCHAL, R., VANDECASTEELE, J. and METCHE, M. 1977. Regulation of the central metabolism in relation to citric acid production of *Saccharomycopsis lipolytica*. Arch. Microbiol. *113*, 99–104.

MASLOVA, T.I. and KHASKIN, I.G. 1976. Citric acid. U.S.S.R. Pat. 525,749. Aug. 25.

MASUDA, M., IKENO, Y. and MATSUMOTA, Y. 1975. Fermentative production of citric acid. Jpn. Pat. 7,525,789. Mar. 18.

MAZZADROLI, G. 1938. Fermentation of citric acid. Fr. Pat. 833,631. Oct. 26.

MESSING, W. and SCHMITZ, R. 1976. Citric acid from sucrose, an industrially useful microbial process based on molasses. Chem. Exp. Didaktik 2, 309–316. (German)

MEYRATH, J. and AHMED, S.A. 1966. Reduction of incubation time in citric acid fermentation by vermiculite. Experientia 22, 806–808.

MIALL, M. and PARKER, G.F. 1975. Continuous preparation of citric acid by Candida lipolytica. Ger. Pat. 2,429,224. Jan. 16.

MILES LAB. 1958. Cultivation of Aspergillus niger spores. Br. Pat. 885,747. Dec. 28.

MILES LAB. 1962. Citric acid. Br. Pat. 908,024. Oct 10.

MILES LAB. 1969. Citric acid production by Aspergillus niger. Br. Pat. 1,145,520. March 19.

MILLIS, N.F., TRUMPHY, B.H. and PALMER, B.M. 1963. The effect of lipids on citric acid production by Aspergillus niger mutants. J. Gen. Microbiol. 30, 365–379.

MINTS, E.S., IVANOVA, L.F., TERENT'EVA, O.F. and KHMELIVINA, R.G. 1976. Molasses for fermentation during the production of citric acid. U.S.S.R. Pat. 526,659. Aug. 30.

MOYER, A.J. 1953. Effect of alcohols on the mycological production of citric acid in surface and submerged culture. I. Nature of the alcohol effect. Appl. Microbiol. 1, 1–7.

MUELLER, H.M. 1975. Oxalate accumulation from citrate by Aspergillus niger. Biosynthesis of oxalate from its ultimate precursor. Arch. Microbiol. 103, 185–189.

MUELLER, H.M. and FROSCH, S. 1975. Oxalate accumulation from citrate by Aspergillus niger. II. Involvement of tricarboxylic acid cycle. Arch. Microbiol. 104, 159–162.

NAGUCHI, Y. and BANDO, Y. 1960. Formation of oxalic acid in the citric acid fermentation in the methanol added molasses medium. Hakko Kogaku Zasshi 38, 485–488. (Japanese)

NAKANISHI, T., YAMAMOTO, M., KIMURA, K. and TANAKA, K. 1972. Fermentative production of citric acid from n-paraffins by yeasts. Hakko Kogaku Zasshi 50, 855–867. (Japanese)

NARA, H. et al. 1971. Separation of citric acid and isocitric acid. Ger. Pat. 2,046,576. Apr. 1.

NIEDEPRUEM, D.J. 1965. Carbohydrate metabolism. II. Tricarboxylic acid cycle. In The Fungi, Vol. 1. G.C. Ainsworth and A.S. Sussman (Editors). Academic Press, New York.

NUBEL, R.C. and FITTS, R.A.   1974.   Citric acid manufacture by fermentation.   Ger. Pat. 2,419,605. Nov. 14.

OH, M.J., PARK, Y.J. and LEE, S.K.   1973.   Citric acid production by *Hansenula anomala* var. *anomala.*   Han'guk Sikp'um Kwahakhoe Chi 5, 215–223. (Korean)

OHMORI, I. and IKENO, Y.   1973.   Citric acid by fermentation.   Jpn. Pat. 7,443,157. Nov. 19.

OHTSUKA, M., OHZAKI, Y. and ARAMIYA, H.   1975.   Citric acid by fermentation.   Jpn. Pat. 7,530,158. Sept. 29.

PEPPLER, H.J.   1967.   Microbial Technology.   Reinhold Publishing Corp., New York.

PERLMAN, D. and SIH, C.J.   1960.   Fungal synthesis of citric, fumaric and itaconic acids.   *In* Progress in Industrial Microbiology, Vol. 2.   Interscience Publishers, New York.

PFIZER.   1971.   Citric acid by *Candida* fermentation.   Fr. Pat. 2,051,108. May 7.

PRESCOTT, S.C. and DUNN, C.G.   1959.   The citric acid fermentation.   *In* Industrial Microbiology.   McGraw-Hill Book Co., New York.

RAMAKRISHNAN, C.V. and MARTIN, S.M.   1955.   Isocitric dehydrogenase in *Aspergillus niger.*   Arch. Biochem. Biophys. 55, 403–407.

RAMAKRISHNAN, C.V., STEEL, R. and LENTZ, C.P.   1955.   Mechanism of citric acid formation and accumulation in *Aspergillus niger.*   Arch. Biochem. Biophys. 55, 270–273.

RIEGER, M. and KIOUSTELIDIS, J.   1975.   Obtaining alkali metal or ammonium citrates.   Ger. Pat. 2,355,039. May 7.

SANCHEZ-MARROQUIN, A., CARRENO, R. and LEDEZMA, M.   1970.   Effect of trace elements on citric acid fermentation by *Aspergillus niger.*   Appl. Microbiol. 20, 888–892.

SANCHEZ-MARROQUIN, A. *et al.*   1963.   Citric acid fermentation of sucrose.   Rev. Soc. Quim. Mex. 7, 191–198.

SARDINAS, J.L.   1972.   Fermentation production of citric acid.   Fr. Pat. 2,113,668. July 28.

SAVCHENKO, N.I., NOVOTEL'NOVA, N.Y. and GOMA, I.G.   1976.   Separation of citric acid from calcium citrate.   U.S.S.R. Pat. 510,508. Apr. 15.

SCHWEIGER, L.B.   1961.   Citric acid by fermentation.   U.S. Pat. 2,970,084. Jan. 31.

SHADAFZA, D., OGAWA, T. and FAZELI, A.   1976.   Comparison of citric acid production from beet molasses and date syrup with *Aspergillus niger.*   Hakko Kogaku Zasshi 54, 65–75.

SHAH, V.K. and RAMAKRISHNAN, C.V.   1963.   Acid metabolism of *Aspergillus niger*: Metabolic changes during citric acid utilization in *A. niger.*   Enzymologiya 26, 33–43.

SHARMA, S.K.   1973.   Production of citric acid by *Aspergillus niger* using cane molasses.   M.Sc. Thesis. Haryana Agric. Univ., Hissar, India.

SHCHERBAKOVA, E.Y.   1964.   Characteristics of biochemically active *Aspergillus niger* variants produced through UV.   Mikrobiologiya *33*, 49–55.

SRIPRAKASH, K.S.   1977.   Personal communication.   Sarabhai Res. Cent. Baroda, India.

SUSSMAN, A.S. and HALVORSON, H.O.   1966.   Spores: Their Dormancy and Germination.   Academic Press, New York.

SUZUKI, T., SUMINO, Y., AKIYAMA, S. and FUKUDA, H.   1971.   Biosynthesis of citric acid.   Ger. Pat. 2,115,514. Oct. 21.

SZUCS, J.   1944.   Method of producing citric acid by fermentation.   U.S. Pat. 2,353,771. July 18.

TABUCHI, T. and HARA, S.   1973.   Organic acid fermentation by yeasts. VII. Conversion of citrate fermentation to polyol fermentation in *Candida lipolytica*.   Nippon Nogei Kagaku Kaishi *47*, 485–490. (Japanese)

TABUCHI, T. and HARA, S.   1974.   Organic acid fermentation of yeasts. X. Mechanism of citrate fermentation in *Candida lipolytica*.   Nippon Nogei Kagaku Kaishi *48*, 417–424. (Japanese)

TABUCHI, T. and SATOH, T.   1976.   Distinction between isocitrate lyase and methylisocitrate lyase in *Candida lipolytica*.   Agric. Biol. Chem. *40*, 1863–1869.

TABUCHI, T. and SATOH, T.   1977.   Purification and properties of methylisocitrate lyase, a key enzyme in propionate metabolism, from *Candida lipolytica*.   Agric. Biol. Chem. *41*, 169–174.

TABUCHI, T. and SERIZAWA, N.   1975.   A hypothetical cyclic pathway for the metabolism of odd-carbon n-alkanes or propionyl-CoA via seven-carbon tricarboxylic acids in yeasts.   Agric. Biol. Chem. *39*, 1055–1061.

TABUCHI, T., SERIZAWA, N. and UCHIYAMA, H.   1974.   A novel pathway for the partial oxidation of propionyl-CoA to pyruvate via seven-carbon tricarboxylic acids in yeasts.   Agric. Biol. Chem. *38*, 2571–2572.

TABUCHI, T., TAHARA, Y., TANAKA, M. and YANAGIUCHI, S.   1973.   Organic acid fermentation by yeasts. IX. Preliminary experiments on the mechanism of citrate fermentation in yeasts.   Nippon Nogei Kagaku Kaishi *47*, 617–622. (Japanese)

TABUCHI, T., TANAKA, M. and AEE, M.   1969.   Studies on organic acid fermentation in yeasts. II. Production of citric acid by *Candida lipolytica* strain No. 228.   Nippon Nogei Kagaku Kaishi *43*, 154–158. (Japanese)

TABUCHI, T. and UCHIYAMA, H.   1975.   Methylcitrate condensing and methylisocitrate cleaving enzymes; evidence for the pathway of oxidation of propionyl-CoA to pyruvate via C_7-tricarboxylic acids.   Agric. Biol. Chem. *39*, 2035–2042.

TAKAHASHI, N. and IKENO, Y.   1974.   Citric acid production by *Candida*.   Jpn. Pat. 7,430,589. Mar. 19.

TAKAMI, W.   1968.   Effect of sodium fluoride and its related agents on the citric acid fermentation in *Aspergillus niger*. II. Effect of related agents of sodium fluoride in citric acid fermentation in *A. niger*.   Hakko Kyokaishi *26*, 77–82.

TAKAMI, W. 1970. Citric acid fermentation by *Aspergillus niger.* Jpn. Pat. 7,009,830. Apr. 9.

TAKAYAMA, K. and ADACHI, T. 1974. Citric acid. Jpn. Pat. 7,426,483. Mar. 8.

TAKAYAMA, K. and TOMIYAMA, T. 1973. Citric acid and isocitric acid by fermentation. Ger. Pat. 2,301,079. July 19.

TAURO, P. 1977. Unpublished data. Haryana Agric. Univ., Hissar, India.

TONDON, S.P. and SRIVASTAVA, A.S. 1971. The influence of some trace elements on the growth of *Aspergillus niger* and on the citric acid fermentation. Sydowia Ann. Mycol. Ser. II *25*, 1–6.

TRUMPHY, B.H. and MILLIS, N.F. 1963. Nutritional requirement of an *Aspergillus niger* mutant for citric acid production. J. Gen. Microbiol. *30*, 381–394.

UCHIO, R. 1976. Fermentative production of citric acid. Jpn. Pat. 7,638,488. Mar. 3.

UCHIYAMA, H. and TABUCHI, T. 1976. Properties of methylcitrate synthetase from *Candida lipolytica.* Agric. Biol. Chem. *40*, 1411–1418.

USAMI, S., SUZUKI, H. and TAKATOMI, N. 1960. Production of citric acid in five liter jar fermentor equipped with sparger and agitator. Kogyo Kagaku Zasshi *63*, 1766–1768.

USAMI, S. and TAKATOMI, N. 1958. Spore cultivation of mold in production of citric acid by fermentation. Kogyo Kagaku Zasshi *61*, 1494–1497.

WEHMER, C. 1893. As quoted *In* Industrial microbiology. S.C. Prescott and C.G. Dunn. 1959. McGraw-Hill Book Co., New York.

WENDEL, D.G. 1967. Diffusion of citric acid in molasses solutions, possibilities of technical use for surface culture fermentation. Zucker *20*, 238–246.

WOLD, W.S.M. and SUZUKI, I. 1973. Cyclic AMP and citric acid accumulation by *Aspergillus niger.* Biochem. Biophys. Res. Commun. *50*, 237–244.

WOLD, W.S.M. and SUZUKI, I. 1976A. The citric acid fermentation by *Aspergillus niger*: Regulation by zinc of growth and acidogenesis. Can. J. Microbiol. *22*, 1083–1092.

WOLD, W.S.M. and SUZUKI, I. 1976B. Regulation by zinc and adenosine cyclic 3′,5′-monophosphate of growth and citric acid accumulation in *Aspergillus niger.* Can. J. Microbiol. *22*, 1093–1101.

WORONICK, C.L. and JOHNSON, M.J. 1960. Carbon dioxide fixation by cell free extracts of *Aspergillus niger.* J. Biol. Chem. *235*, 9–15.

YO, K. 1975. Citric acid fermentation. Jpn. Pat. 75,154,487. Dec. 12.

YOSHINAGA, F., TSUCHIDA, T., NAKASE T. and OKUMURA, S. 1972. Fermentative production of citric acid by *Candida.* Jpn. Pat. 72,25,383. Oct. 20.

ZHURAVASKII, G.I. *et al.* 1974. Citric acid. U.S.S.R. Pat. 432,186. June 15.

17

Amino Acids

K. Nakayama

Fermentative production of amino acids has become an industrial reality by the discovery of an efficient glutamic acid producer, *Corynebacterium glutamicum* (synonym *Micrococcus glutamicus*), by Kinoshita *et al.* (1957). The bacterium was found during a time of increasing demand for monosodium glutamate as a flavoring agent, and following this, much research activity has been focused on microbial amino acid production. The primary reason for these efforts was the hope of improving the nutritional value of low-cost vegetable proteins by enrichment with essential amino acids. Once *C. glutamicum* was discovered by screening isolates from nature, similar efforts led to the isolation of bacteria producing DL-alanine or L-valine. However, it was found that most wild type strains isolated from nature could not produce industrially significant amounts of other amino acids except the few amino acids already referred to. One of the main reasons is that regulation of cellular metabolism avoids oversynthesis. The existence of these regulatory phenomena was just being clarified at the time of the isolation of *C. glutamicum* and is now well recognized.

An auxotrophic mutant which cannot produce the regulatory effector or corepressor (usually the end product or a derivative of the end product) overproduces and excretes the precursor or the related metabolite of a blocked reaction when grown on a limiting supply of the required nutrient. This is the principle of the application of an auxotrophic mutant to the microbial production of amino acids. Active attempts to utilize this phenomenon for the industrial production of microbial metabolites were launched in the 1950s and many amino acids are now produced with auxotrophic mutants. It is obviously useless to accumulate the end product of an unbranched pathway such as arginine and histidine with an auxotrophic mutant. The production of such a metabolite depends on the use of a regulatory mutant. A mutant which has lost some biosynthetic regulation can be obtained by selection of an analog-resistant and prototrophic revertant from the auxotroph having a deficiency in a regulatory enzyme.

To improve the yield of an amino acid, mutants having multiple markers, including auxotrophy and analog-resistance contributing to production of the designated amino acid, are selected. Multiple markers also contribute to the yield by stabilizing the productivity against back mutation during fermentation (Nakayama 1972A). With metabolic regulation, the permeability barrier is another mechanism which protects microorganisms from leaking organic compounds to the environment. It allows cells to retain intermediates and macromolecules necessary for the life of the microorganisms.

Permeability is another important factor for amino acid production. In fact, excess production of L-glutamic acid by *C. glutamicum* was found to be due mainly to the permeability change induced by limiting the supply of biotin required by the bacterium. Although certain specific environmental conditions make it possible to exploit a single enzymic process or a process using a precursor, most amino acids can now be produced by the so-called "direct fermentation" process, i.e., microbial production from a cheap carbon source by fermentation. Examples of each type and process will be described in the later sections of this chapter.

Many processes for production of various amino acids have been developed. Total world production of L-glutamic acid is considered to be in excess of 150,000 MT (165,000 tons) per year. It is used mainly as a flavoring agent. L-lysine has also been produced on a large scale by fermentation and is used mainly as a feed-supplement. Total world production of it is probably in excess of 35,000 MT (38,500 tons) per year. Methionine, alanine, glycine, and cysteine are not produced commercially by fermentation. The racemic forms of methionine, alanine, and glycine are produced by chemical synthesis, and L-methionine and L-alanine can be made from the racemic form by an enzymatic process. Some other amino acids are produced by fermentation on a scale of less than 1000 MT (1100 tons) per year.

L-GLUTAMIC ACID AND L-GLUTAMINE

Glutamic Acid Production from Carbohydrate

Production of L-glutamic acid and L-glutamine has been reviewed by Kinoshita and Tanaka (1972). Production of glutamic acid from carbohydrate in high yield is carried out by a group of bacteria represented by *Corynebacterium glutamicum* (synonym *Micrococcus glutamicus*). These bacteria are classified as species in different genera: they include *Corynebacterium glutamicum* (*Micrococcus glutamicus*), *Brevibacterium flavum*, *B. lactofermentum*, *B. divaricatum*, *B. thiogenitalis*, *Corynebacterium callunae*, *C. herculis*, *Microbacterium ammoniaphilum*, and others. The guanine + cytosine (G-C) content of the DNA of these bacteria falls into a narrow range from 51.2 to 54.4 moles %. Morphological and physiological properties also support their close affinities, and they could be classified into a group of bacteria belonging to the Corynebacteriaceae. For convenience these bacteria will be referred to as "glutamic acid bacteria" hereafter.

The special conditions which allow these bacteria to excrete large amounts of glutamate are a nutritional requirement for biotin and the lack, or very low content, of α-ketoglutarate dehydrogenase. The biotin requirement is the major controlling factor in the fermentation. When enough biotin is supplied for optimal growth, the organism produces lactate. Glutamate is excreted under conditions of suboptimal growth.

Under optimal culture conditions, glutamic acid bacteria convert about 50% of the supplied carbohydrate into L-glutamic acid with little formation of by-products. Various carbohydrate materials can be used as the carbon source. Glucose and sucrose are particularly suitable. For industrial purposes, hydrolyzed starch solutions, cane molasses, and beet molasses are preferred. Other carbon sources such as acetic acid and ethanol are also used. Carbon sources, such as cane molasses, with a high content of biotin are used with the addition of penicillin during logarithmic growth or of fatty-acid derivatives such as polyoxyethylene sorbitan-monooleate (Tween 60) before or during logarithmic growth.

Ammonium sulfate, ammonium chloride, ammonium phosphate, aqueous ammonia, ammonia gas, and urea have been used as nitrogen sources. Although a large amount of ammonium ion is necessary, a high concentration of it is inhibitory to growth of the organism as well as to production of glutamic acid. Therefore, ammonium ions are added as the fermentation progresses. Ammonia water, or gaseous ammonia, is generally used industrially.

Other ions supplied include K^+, Mg^{2+}, Fe^{2+}, Mn^{2+}, PO_4^{3-}, SO_4^{2-}, and Cl^-. These are usually supplied by the following inorganic salts (% w/v): 0.05-0.2 KH_2PO_4, 0.05-0.2 K_2HPO_4, 0.025-0.1 $MgSO_4 \cdot 7H_2O$, 0.0005-0.01 $FeSO_4 \cdot 7H_2O$, 0.0005-0.005 $MnSO_4 \cdot 4H_2O$ and 0.5-4 $CaCO_3$.

The most important factor in the medium for the glutamic acid fermentation is biotin, which is an essential growth factor for glutamic acid bacteria. The concentration of biotin must be suboptimal for growth. The best concentration of biotin for the glutamic acid fermentation depends on the strain, kind, and concentration of the carbon source, but it is generally somewhat below 5 μg per liter of medium. Some strains require thiamin or cystine in addition to biotin.

Certain iron-chelating compounds are necessary for growth of glutamic acid bacteria, but their presence is not necessary when the carbohydrate is autoclaved simultaneously with other ingredients of the medium because iron-chelating compounds are formed by autoclaving.

The pH value optimal for growth and glutamic acid production is 7.0–8.0. Continuous feeding of NH_4^+ can adjust the pH value and also supply ammonium ions to the medium. Urea can replace the ammonium ion in those glutamic acid bacteria which possess urease activity. The optimal value of Kd (the overall coefficient of oxygen transfer) for glutamic acid fermentation is considered to be 3–5 × 10^{-6} (mol O_2) atm^{-1}min^{-1}ml^{-1}. Under conditions of insufficient oxygen, production of glutamic acid is poor and large amounts of lactic acid and succinic acid accumulate, while excess oxygen increases the amount of lactic acid and α-ketoglutaric acid. The optimal temperature for the glutamic acid fermentation is usually 30° to 35°C. The time course of the glutamic acid fermentation by *C. glutamicum* No. 541 is shown in Figure 17.1. The medium used had the following composition (w/v): 10% glucose, 0.05% K_2HPO_4, 0.05% KH_2PO_4, 0.025% $MgSO_4 \cdot 7H_2O$, 0.001% $FeSO_4 \cdot 7H_2O$, 0.001% $MnSO_4 \cdot 4H_2O$, 0.5% urea, and 2.5 µg/liter biotin. The fermentation was carried out in 5 liter jar fermentors at 28°C. The pH value of the medium was kept between 7 and 8 by feeding of urea.

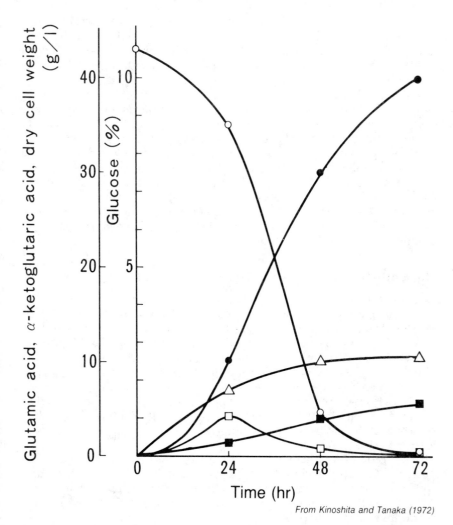

From Kinoshita and Tanaka (1972)

FIG. 17.1.   TIME COURSE OF GLUTAMIC ACID FERMENTATION BY *CORYNEBACTE-RIUM GLUTAMICUM* NO. 541
Symbols: ● Glutamic acid. □ Lactic acid. ○ Glucose. ■ 2-Ketoglutaric acid. △ Dry cell weight.

Glutamic Acid Production from Noncarbohydrate Materials

The availability of acetic acid at a reasonable price and the waste water problem associated with the use of cane molasses as the carbon source for the glutamic acid fermentation prompted the search for a process using acetic acid. With *Brevibacterium flavum*, L-glutamic acid production reached 98 g per liter (48% on the basis of acetic acid) in 48 hr (Tanaka *et al.* 1971). Adding 25 to 250 μg of Cu^{2+} per liter to a medium increased L-glutamate yield from acetate by *B. thiogenitalis*. The poor yield of L-glutamate from acetate by copper-deficient cells seems to be due to a decrease in energy supply which is caused by the low efficiency of oxidative phosphorylation (Sugiyama *et al.* 1973). The productivity of an oleic acid auxotroph of *B. thiogenitalis* was superior to the parent strain (Kanzaki *et al.* 1972). *Brevibacterium* sp. B 136 converted ethanol to L-glutamic acid with a yield of 60%.

A group of bacteria represented by *Nocardia erythropolis* (synonym *Corynebacterium hydrocarboclastus*) (Kōmura *et al.* 1973) produces L-glutamic acid from n-paraffins in media containing suboptimal concentrations of thiamin. The glutamic acid yield is stimulated by addition of penicillin to an exponentially growing culture with excess thiamin in the medium. Cupric ions stimulate growth and glutamic acid production from n-paraffins by *Arthrobacter paraffineus* (Suzuki *et al.* 1971). Production of glutamic acid by *A. paraffineus* reached 82 g per liter at 48 hr. A penicillin-resistant mutant of *N. erythropolis* produced 84 g of L-glutamic acid per liter (Kobayashi *et al.* 1971). A glycerol auxotroph of *C. alkanolyticum* produced 72 g of L-glutamic acid per liter (Kikuchi *et al.* 1972).

Biosynthetic Pathway from Glucose

The pathway of glutamate biosynthesis from glucose is shown in Fig. 17.2. The major route, shown with the heavy arrows, involves at least 16 enzymic steps. α-Ketoglutarate is converted to glutamate by reductive amination. The enzyme catalyzing this conversion is the NADP-specific glutamate dehydrogenase. The presence of this enzyme is essential for glutamate formation. When resting cells are incubated with glucose in the absence of ammonia, α-ketoglutarate accumulates rather than glutamate. The NADPH required for the action of glutamate dehydrogenase is supplied by the preceding isocitrate dehydrogenase reaction. Glutamate dehydrogenase, in turn, provides the NADP required for isocitrate dehydrogenase activity.

A low content of α-ketoglutarate dehydrogenase favors glutamate production. The importance of the lack of this enzyme has been demonstrated with *E. coli*, which does not require biotin and is not a glutamate excreter. Even without the biotin requirement, a mutant which lacked α-ketoglutarate dehydrogenase was found to excrete 2.3 g of glutamate per liter while its parent excreted none.

In addition to the EMP pathway, glutamic acid bacteria also use the HMP pathway to convert glucose to 3-carbon and 2-carbon compounds (shown by

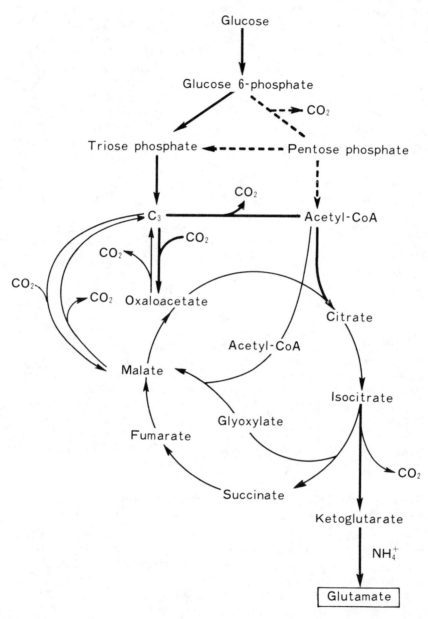

FIG. 17.2.   THE METABOLIC PATHWAYS OF THE BIOSYNTHESIS OF GLUTAMIC ACID FROM GLUCOSE

broken lines in Fig. 17.2). These compounds can then feed into the TCA cycle. Estimates of the relative utilization of glucose by the two pathways have shown the predominance of the EMP pathway under fermentation conditions.

When glutamate fermentation is carried out in the presence of $^{14}CO_2$, the radioactivity is fixed into the α-carboxyl group of glutamate. Two enzymes have been found in glutamate excreters which participate in the fixation of carbon dioxide, namely, oxalacetate carboxylase and the NADP-linked malic enzyme which catalyzes fixation of carbon dioxide to pyruvate to yield malate. Malate is oxidized to oxaloacetate by malate dehydrogenase, then oxaloacetate is converted to citrate. Because of a deficiency in the TCA cycle, glutamic acid bacteria use the glyoxylate pathway shown by thin lines in Fig. 17.2. Therefore, the two competing reactions of isocitrate are important. During the growth phase, the isocitritase reaction is needed for energy production and to produce intermediates for biosynthetic reactions. However, after the growth phase, glutamate production would be better without operation of the isocitritase reaction. This would indicate that optimal conditions for the growth phase and for the glutamate production phase should be different. After growth, the ideal fermentation would proceed by the following reaction:

$$C_6H_{12}O_6 + NH_3 + 1.5\ O_2 \longrightarrow C_5H_9O_4N + CO_2 + 3H_2O$$

This represents a 100% molar conversion or 81.7% weight conversion of sugar to glutamic acid. On the other hand, the poorest fermentation would be represented by complete oxidation of glucose as occurs during growth in media containing an optimal concentration of biotin and resulting in no conversion. What is actually found lies between these limits, i.e., 50–75% molar conversion by resting cells. This indicates that carbon is being lost as carbon dioxide by reversal of action of the malic enzyme and of oxaloacetate carboxylase during operation of the glyoxylate bypass. Therefore, assimilation of ammonium ions in this organism is almost totally dependent upon action of NADP-linked glutamate dehydrogenase.

Alteration of Permeability Relevant to Glutamic Acid Production

Alteration of permeability in relation to glutamic acid production was reviewed by Demain and Birnbaum (1968). Production of a large amount of glutamic acid in bacteria is attained by the alteration of the permeability barrier. Increased permeability can be induced in glutamic acid bacteria by either biotin deficiency, oleic acid deficiency in an oleic acid auxotroph, glycerol deficiency in a glycerol auxotroph (Nakao et al. 1972), treatment with fatty acid derivatives, or by addition of penicillin, cephalosporin C, or T-125 (tunicamycin-like antibiotic) to the growth medium (Kitano et al. 1976). Biotin deficiency and treatment with fatty acid derivatives cause an aberration in the normal synthesis and distribution of cellular fatty acids.

The cell membrane from such cells contains an abnormal ratio of saturated to unsaturated fatty acids. This abnormality directly correlates with increased permeability of the cell and with the excretion of high concentrations of glutamic acid.

Biotin auxotrophs and oleic acid auxotrophs can not be used for the production of L-glutamic acid from n-paraffins. Excretion of L-glutamic acid by addition of penicillin or cephalosporin was accompanied by excretion of both phospholipid and N-acetylglucosamine, which are known to be components of the cell membrane and the cell wall, respectively (Nakao et al. 1973). Based on this observation, a glycerol auxotroph of Corynebacterium alkanolyticum was isolated. In the auxotroph, cellular phospholipid synthesis was regulated by the amount of glycerol supplied. The mutant produced about 40 g of L-glutamic acid per liter from n-paraffins in the culture in the presence of 0.01% of glycerol but in the absence of penicillin (Nakao et al. 1972). Studies with the auxotroph suggest that the permeability of L-glutamic acid is not always controlled by the cellular content of unsaturated fatty acyl residues and that phospholipid content controls the membrane permeability of L-glutamic acid (Kikuchi and Nakao 1973).

Addition of penicillin to log phase cultures of C. glutamicum results in a rapid 97–99.5% decrease in viability accompanied by a rapid rate of glutamate excretion. Cell mass represented by optical density and total cell count remain fairly constant after a slight initial increase. The packed cell volume, however, decreases by 60–80% during the phase of decreasing viability indicating a change in the surface properties of the cells. The cells continue to produce glutamate for 40–50 hr after penicillin addition with no lysis throughout the entire fermentation. Apparently, the lack of an overabundance of potent mucopeptidases allows the cells to retain their form and to metabolize as permeable "resting" entities. During growth in a medium containing a high concentration of biotin, C. glutamicum synthesizes glutamate until the cell becomes saturated at 25–35 μg per mg of dry weight. Cessation of production is assumed to be due to some feedback control of glutamate toward its own synthesis. When the cells become more permeable, glutamate passes from the cells into the medium, thus releasing the feedback effect and allowing further synthesis.

L-Glutamine and N-Acetyl-L-glutamine

L-Glutamine and N-acetyl-L-glutamine are produced along with glutamic acid under certain conditions by glutamic acid bacteria. Production of glutamine is increased by maintaining the medium at a weakly acidic pH value, and production of N-acetylglutamine is increased by maintaining a neutral to weakly acidic pH value in the medium. Under optimal conditions, a 20% conversion of glucose to glutamine and over 20% to N-acetylglutamine are observed. High concentrations of ammonium ion retard growth and glutamic acid production. Increase in the biotin supply and addition of natural nutrients such as corn-steep liquor and meat extract to the medium stimulate growth and increase glutamine production. These nutrients or Zn^{++} at a concentration over that required for growth re-presses formation of N-acetylglutamine (Nakanishi 1975). Selection of a suitable strain (KY 9003) and limiting the Zn^{++} concentration in the medium permitted preferential N-acetylglutamine production. L-Gluta-mine and N-acetylglutamine are used in the treatment of gastric ulcers.

L-LYSINE

L-Lysine Production by Fermentation

Microbial production of L-lysine was reviewed by Nakayama (1972B). A microbial process for L-lysine production was first developed by a combina-tion of diaminopimelate production by a lysine auxotroph of *Escherichia coli* and decarboxylation of the compound by *Aerobacter aerogenes* or wild type *E. coli*. Direct production of L-lysine from carbohydrate was developed first with a homoserine (or threonine plus methionine)-auxotroph of *Cory-nebacterium glutamicum*. The same type of process was reported with a homoserine auxotroph of *Brevibacterium flavum*. The leaky homoserine auxotroph was recognized as a threonine-sensitive mutant because growth was inhibited by excess threonine and the inhibition was released by addi-tion of methionine. This phenomenon is due to feedback inhibition of resid-ual homoserine dehydrogenase by threonine. Homoserine (or threonine plus methionine)-auxotrophs of other bacteria were also found to produce L-lysine, but the yields were lower than that from the homoserine auxo-troph of coryneform bacteria. Threonine auxotrophs and leucine auxo-trophs of *C. glutamicum* produce fairly large amounts of L-lysine, but they are inferior to the homoserine auxotroph. Other auxotrophs of *C. glutam-icum* and other bacteria were also inferior to the homoserine auxotroph of *C. glutamicum*. Double auxotrophs, which require, in addition to homo-serine, at least one of the amino acids, threonine, isoleucine, or methionine, for growth, have been found to be highly stabilized, showing little tendency to revert to homoserine independence. It is possible not only to prevent reversion of the cultures to a wild type state, but also to produce lysine in higher yields since many of the microorganisms are double mutants in the homoserine pathway.

Cane molasses is now generally used as a carbon source in the industrial production of lysine, although other carbohydrate materials, acetic acid, and ethanol can be used. The pH value of the medium is maintained near neutrality during the fermentation by feeding ammonia or urea. Ammo-nium salts are generally good nitrogen sources, and urea can be used for organisms having urease activity. An example of a fermentation using cane molasses in a 2 kl fermentor is as follows. The medium for first seed culture contained 2% glucose, 1% peptone, 0.5% meat extract, and 0.25% NaCl in tap water. For the second seed culture, the medium contained 5% cane molasses, 2% $(NH_4)_2SO_4$, 5% corn-steep liquor, and 1% $CaCO_3$ in tap water. For the fermentation, the medium contained 20% reducing sugars ex-pressed as invert (as cane molasses) and 1.8% soybean meal hydrolysate (as weight of meal before hydrolysis with 6 N H_2SO_4 and neutralization with ammonia water) in tap water. The fermentation was carried out at 28°C. Figure 17.3 shows the time course of the fermentation using *C. glutamicum* No. 901 (a homoserine auxotroph), which produced 44 g of L-lysine per liter

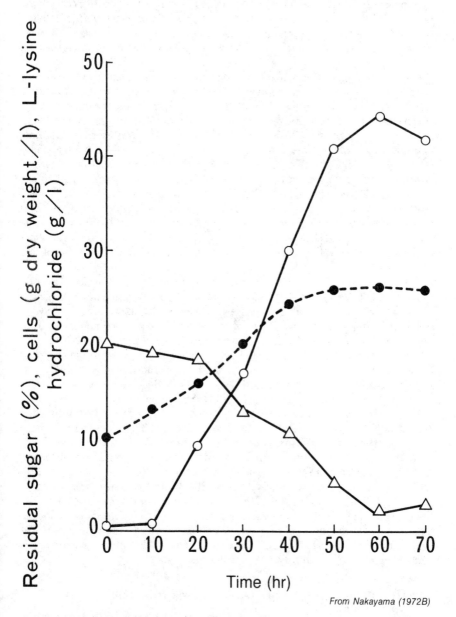

From Nakayama (1972B)

FIG. 17.3.   TIME COURSE OF LYSINE FERMENTATION
Symbols: ○ L-Lysine. △ Residual sugar: ● Dry cell weight.

in 60 hr. Foaming in the aerated culture can be repressed by addition of proper antifoaming agents. The amount of the growth factors (homoserine or threonine and methionine) should be appropriate for the production of L-lysine. It is supplied in limited amounts and is suboptimal for growth. The biotin concentration in the medium must generally be greater than 30 μg per liter. With a limited supply of biotin, L-glutamate accumulates in place of L-lysine. Cane molasses usually supplies enough biotin. Yields of L-lysine as the monohydrochloride reach 30–40% in relation to the initial sugar concentration.

Coryneform glutamic acid-producing bacteria can utilize acetic acid as a carbon source for growth and lysine production. L-Lysine production from acetic acid by a homoserine-leaky (threonine-sensitive) threonine auxo-troph mutant of *Brevibacterium flavum* reached 75 g (as monohydro-chloride) per liter or 29% on the basis of acetic acid and glucose supplied (Tanaka *et al.* 1971). The medium contained 0.7% acetic acid, 0.2% KH_2PO_4, 0.04% $MgSO_4\cdot7H_2O$, 0.001% $FeSO_4\cdot7H_2O$, 0.001% $MnSO_4\cdot H_2O$, 3.5% hydro-lysate of soybean protein, 3.0% glucose, 50 μg biotin per liter, and 40 μg thiamin·HCl per liter (pH 6.0). Fermentation was carried out at 33°C with feeding of a solution of acetic acid. The feeding solution contained 60% acetic acid composed of a mixture of acetic acid and ammonium acetate having a molar ratio of 100:25, and 3% glucose. Feeding was controlled automatically until the end of the fermentation, keeping the pH value of the medium at 7.4.

A mutant of *Brevibacterium flavum* resistant to 5-(β-aminoethyl)-L-cysteine (AEC), a lysine analog, produced fairly large amounts of L-lysine (Sano and Shiio 1970). The increase in lysine yield (more than 10%) was obtained using a mutant of *C. glutamicum*, which requires homoserine and leucine and is resistant to AEC. It produced 39.5 g of L-lysine per liter in a medium containing 10% reducing sugars expressed as invert (as cane mo-lasses) while the homoserine plus leucine auxotroph produced 34.5 g of L-lysine per liter (Nakayama and Araki 1973). Some patents have been issued for the process to produce L-lysine from n-paraffin.

Regulation of lysine biosynthesis is represented in Fig. 17.4. A similar regulatory pattern was also observed in *B. flavum*. The blocking of homo-serine synthesis at homoserine dehydrogenase results in the release of the concerted feedback inhibition by threonine and lysine on aspartokinase, and the aspartic semialdehyde produced proceeds to lysine through the lysine synthetic pathway on which no feedback inhibition is found, a situa-tion which differs from that in *E. coli*. Resistance to AEC brought about by the desensitization of aspartokinase also releases the concerted feedback inhibition. The conversion of aspartic semialdehyde to threonine is feed-back-inhibited by L-threonine. Thus the overproduced aspartic semialde-hyde is channelled into L-lysine production.

L-Lysine Production from DL-α-Aminocaprolactam

L-Lysine production from DL-α-aminocaprolactam was first found with *Aspergillus ustus*. However, the yield was low. More recently, a very effi-

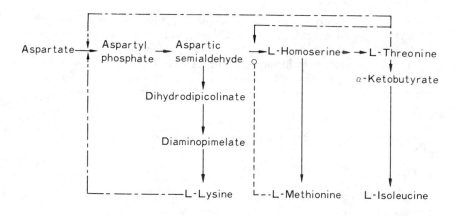

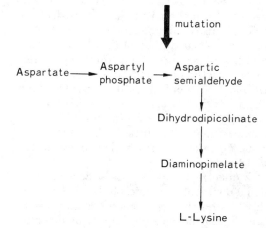

FIG. 17.4.   DEREGULATION OF LYSINE BIOSYNTHESIS IN HOMOSERINE AUXO-
TROPH OF *CORYNEBACTERIUM GLUTAMICUM*
Symbols: ← ——— *Feedback inhibition.* ○— — — — — —*Repression.*

cient process has been developed for the conversion (Fukumura 1976A,B).
Incubation of a mixture of 100 ml of 10% DL-α-aminocaprolactam (adjusted
to pH 8.0 with HCl), 0.1 g acetone-dried cells of *Cryptococcus laurentii*, and
0.1 g acetone-dried cells of *Achromobacter obae* nov. sp. with gentle shaking
at 40°C for 24 hr resulted in the conversion of DL-α-aminocaprolactam to
L-lysine in 99.8% yield.

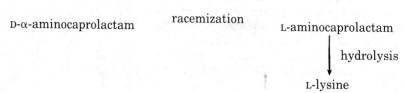

Cryptococcus laurentii produces L-aminocaprolactam hydrolase inductively in a medium containing L-α-aminocaprolactam, glucose, and other ingredients. *Achromobacter obae* produces aminocaprolactam racemase using both D- and L-α-aminocaprolactam as an inducer (Sato *et al.* 1974). A similar optimal pH value of both enzymes allows the efficient conversion in what appears to be a single step.

L-THREONINE, L-HOMOSERINE, AND L-SERINE

L-Threonine

L-Threonine production from L-homoserine has been studied by several groups (Shimura 1972A), but it has been unsuccessful as an industrial process because of the high cost of homoserine. Direct production of L-threonine from carbohydrate was pioneered by Huang (1961), but its industrial establishment was delayed until Nakayama's group established the process using an *E. coli* auxotroph. Using a diaminopimelate auxotroph and diaminopimelate plus methionine double auxotroph of *E. coli*, Huang obtained L-threonine production at 2–4 g per liter. Triple auxotrophs of *E. coli*, which require diaminopimelate, methionine, and isoleucine, and their isoleucine revertants, produced L-threonine in increased yields. One of the isoleucine revertants, KY 8280, produced 13.5 g of L-threonine per liter. The cultivation was carried out in 5 liter jar fermentors containing 3 liters of a medium having the following composition: 7.5% fructose, 1.4% $(NH_4)_2SO_4$, 0.3% KH_2PO_4, 0.3% $MgSO_4·7H_2O$, 2.0% $CaCO_3$ (pH 7.8). The time course is shown in Fig. 17.5 (Kase *et al.* 1971).

The process using the *E. coli* auxotroph has also been developed by Hirakawa's group. They isolated a methionine auxotroph (No. 15) from *E. coli* C-6. It produced 4.3 g of L-threonine per liter in a medium containing 5% reducing sugars expressed as invert (as cane molasses). A methionine plus valine-leaky double auxotroph (No. 234), derived from strain No. 15, produced 13.6 g of L-threonine per liter in a medium containing 10% glycerol (Hirakawa *et al.* 1973). In strain No. 15, the lysine- or methionine-sensitive aspartokinase, which is insensitive to feedback inhibition, was derepressed about 5-fold when the auxotroph was cultured in the presence of a limited concentration of methionine (Hirakawa and Watanabe 1974). The production of L-threonine by *E. coli* auxotrophs was greatly increased by the presence of borrelidin. Addition of aspartic acid with borrelidin further increased the production. The maximum amount of L-threonine accumulated by *E. coli* No. 234 was 15 g per liter in the medium containing 50 g of glucose and 5 g of sodium aspartate per liter (Hirakawa *et al.* 1974). Borrelidin, an antibiotic, is known to selectively inhibit the activity of the threonyl-tRNA synthetase of some microorganisms.

L-Threonine production by auxotrophic mutants is also encountered in other members of the Enterobacteriaceae (Kase *et al.* 1971; Komatsubara *et al.* 1978) and *Candida guilliermondii* var. *membranaefaciens* (Tsukada and

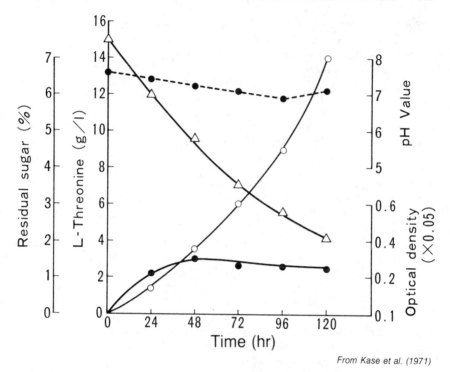

From Kase et al. (1971)

FIG. 17.5.   TIME COURSE OF L-THREONINE FERMENTATION IN *ESCHERICHIA COLI* KY 8280

Symbols: ○ L-Threonine. △ Residual sugar. − − − ● − − pH. ——●— Optical density (OD) at 660 nm.

Sugimori 1971). A mutant deficient in threonine dehydrogenase and threonine dehydratase was isolated from *Serratia marcescens*. A mutant resistant to α-amino-α-hydroxyvaleric acid (AHV, an analog of L-threonine), derived from the above mutant, produced 14 g of L-threonine per liter (Komatsubara *et al.* 1978).

The regulation of biosynthesis in glutamic acid-producing bacteria as represented by *C. glutamicum* is different from that in *E. coli*. Therefore, the organism could not be used successfully for the production of L-threonine. But L-threonine production could be obtained by using a regulatory mutant. An AHV-resistant mutant of *Brevibacterium flavum* BB 82 produced 13.5 g of L-threonine per liter in a medium having the following composition: 10% glucose, 3% $(NH_4)_2SO_4$, 1.5% KH_2PO_4, 0.04% $MgSO_4 \cdot 7H_2O$, 2 ppm Fe^{2+}, 2 ppm Mn^{2+}, 200 μg biotin per liter, 300 μg thiamin·HCl per liter, 4 ml Mieki (an HCl-hydrolysate of soybean protein) per liter, 5% $CaCO_3$, pH 7.2 (adjusted with KOH) (Shiio and Nakamori 1970). Strain BBM-21, a methionine auxotroph derived from another AHV-resistant mutant of *B. flavum*, produced about 18 g of L-threonine per liter (Nakamori and Shiio 1972). A methionine auxotroph of *C. glutamicum* resistant to

AHV and thialysine produced 14 g and 10 g, respectively, of L-threonine per liter in a medium containing 10% or 5% reducing sugars expressed as invert (as cane molasses). Another AHV- and thialysine-resistant methionine auxotroph was found to produce both 9 g of L-methionine per liter and 5.5 g of L-lysine (Kase and Nakayama 1972). Production of L-threonine with an AHV-resistant mutant was reported with other bacteria including *Proteus rettgeri* and *Corynebacterium acetoacidophilum.*

An example of L-threonine production from acetic acid by an AHV-resistant mutant of *B. flavum* was reported by Tanaka *et al.* (1971). The fermentation medium had the following composition: 0.4% ammonium acetate, 0.41% sodium acetate, 1.0% $(NH_4)_2SO_4$, 0.2% urea, 0.3% KH_2PO_4, 0.04% $MgSO_4·7H_2O$, 0.001% $FeSO_4·7H_2O$, 0.001% $MnSO_4·4H_2O$, 50 μg biotin per liter, 5 mg thiamin·HCl per liter, 1.5% hydrolysate of soybean protein, and 0.3% glucose (pH 7.2). After inoculation of the seed culture, the fermentation was conducted at 31°C with automatic maintenance of the pH value at 7.7 by feeding a mixed solution of acetic acid and ammonium acetate having a molar ratio of 100:17. After 48 hr culture, L-threonine production reached 27 g per liter (14% on the basis of acetic acid and glucose).

An isoleucine-leaky auxotroph of *Arthrobacter paraffineus* produced L-threonine and L-valine each at 9 g per liter. Besides these amino acids, 2 g per liter each of L-serine and L-leucine were produced. Some of the double auxotrophs derived from the isoleucine auxotroph, and some of their revertants with respect to isoleucine requirement, produced more threonine than the original isoleucine auxotroph (Kase and Nakayama 1973). A revertant derived from a methionine plus isoleucine double auxotroph produced 12 g of L-threonine per liter in 7 days incubation with the medium containing 10% n-paraffin (C_{12}–C_{14} rich). Formation of valine as a by-product was decreased in this strain.

Regulation of threonine biosynthesis in *E. coli* and *C. glutamicum* is shown in Fig. 17.6. *Corynebacterium glutamicum* has one homoserine dehydrogenase which is threonine-sensitive and methionine-repressible, and one aspartokinase which is concerted feedback-inhibited by threonine plus lysine. On the other hand, *E. coli* has two homoserine dehydrogenases, one of which is methionine-repressible and threonine-insensitive while the other is threonine-sensitive and threonine-repressible. Furthermore, phosphorylation of aspartate in *E. coli* is catalyzed by 3 isoenzymes of aspartokinase, one of which is repressible by methionine and insensitive to threonine and is a complex with the homoserine dehydrogenase repressible by methionine; the second one is multivalent, repressible by threonine and isoleucine, and inhibited by threonine, and forms a complex with the homoserine dehydrogenase sensitive to threonine; the third aspartokinase is repressible and inhibited by lysine. These regulatory processes explain the contribution of methionine, lysine, and isoleucine deficiencies to threonine overproduction in *E. coli*. Isoleucine auxotrophy contributes also to blocking metabolism of the threonine produced. Figure 17.6 also explains the reason why the auxotrophic mutant of *C. glutamicum* could not overproduce L-threonine.

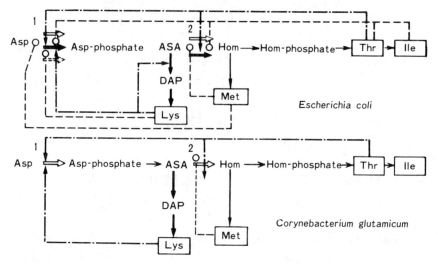

FIG. 17.6.   REGULATION OF THREONINE BIOSYNTHESIS IN *ESCHERICHIA COLI* AND *CORYNEBACTERIUM GLUTAMICUM*
Symbols: 1—Aspartokinase. 2—Homoserine dehydrogenase. ←————Feedback inhibition.
○——————Repression.

The genealogy of the threonine-producing strains of *C. glutamicum* is shown in Fig. 17.7 (Kase and Nakayama 1974A). The activities of homoserine dehydrogenase in the mutant which produced L-threonine or L-threonine together with L-lysine were slightly less susceptible to inhibition by L-threonine than the activity in the original methionine auxotroph (KY 9159) from which the threonine producers were derived. The genetic alteration of the enzyme may be one of the causes of L-threonine production in these mutants. The aspartokinase in the threonine-producing mutants (KY 10484 and 10230), which are resistant to AHV and more sensitive to AEC than the parent, were sensitive to concerted feedback inhibition by L-lysine and L-threonine to the same degree as KY 9159. The aspartokinase activity in KY 10440 was less susceptible to concerted feedback inhibition than KY 10484 or KY 9159. In KY 10251, which produces both L-threonine and L-lysine, the simultaneous addition of L-threonine and L-lysine hardly inhibited the activity of aspartokinase. The difference in L-lysine production between strains KY 10440 and KY 10251 could be mainly due to the difference in the susceptibility of the aspartokinase to concerted feedback inhibition by L-lysine and L-threonine. Furthermore, higher levels of L-threonine production in AHV- and AEC-resistant mutants, as compared with the parent AHV-resistant mutants, could be brought about by the genetically determined desensitization of aspartokinase to end product inhibition.

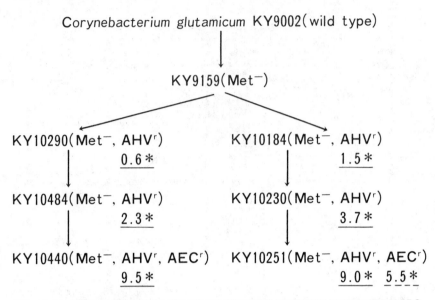

FIG. 17.7.   GENEALOGY OF MUTANTS OF *CORYNEBACTERIUM GLUTAMICUM* PRO-
DUCING L-THREONINE AND L-LYSINE
Solid underline—Amount of L-threonine in g per liter.   ·
Dashed underline—Amount of L-lysine in g per liter.
Abbreviations: AHV—α-Amino-β-hydroxyvaleric acid. AEC—S-(β-Aminoethyl)-L-cysteine.
r—Resistant.

L-Homoserine

Nara (1972) has reviewed homoserine production. A threonine auxotroph
of *C. glutamicum* produced 13–15 g of L-homoserine per liter and 9 g of
L-lysine per liter in a chemically defined medium consisting of 10% glucose,
2% $(NH_4)_2SO_4$, 400 mg L-threonine per liter, 30 µg biotin per liter, 0.1%
K_2HPO_4, 0.03% $MgSO_4 \cdot 7H_2O$, and 2% $CaCO_3$. In the presence of excess
methionine, the yield of homoserine was lowered and that of lysine increased without any change in the total amount produced of both amino
acids. The effect of methionine explained the low yield in natural media.
Excess threonine lowered the yield of both amino acids. Homoserine production by threonine auxotrophs has been reported also for *E. coli* and *B.
flavum*. Homoserine production from n-paraffins has also been reported in a
threonine auxotroph of *Corynebacterium* sp. KY 4403.

L-Serine

Corynebacterium glycinophilum ATCC 21341, produced 10 g of L-serine
per liter with a medium containing 2% glycine under favorable conditions
(Kubota *et al.* 1972). This strain was isolated from a putrefied banana and
resistant to a high concentration of glycine. A leucine-methionine double
auxotroph, AJ 3414, derived from *C. glycinophilum* ATCC 21341 produced

14 g of L-serine per liter in a medium containing 3% glycine. The serine dehydratase activity of these mutants was reduced to 31 and 1.3%, respectively, of the parent strain (Kubota *et al.* 1975). The serine hydroxymethyl transferase is induced by glycine, and serine dehydratase is induced by serine formed during fermentation. Chloramphenicol (10 μg per liter) addition at 16 and 48 hr in the culture of the parent strain increased serine production to 11 g per liter compared with 4.6 g per liter in the control culture. *Pseudomonas* 3 ab, a facultative methylotropic organism, was incubated for 1 day in a medium containing 1% methanol. After supplementation of this medium with methanol (8 g per liter) and glycine (20 g per liter) and incubation at pH 8.5 for 3 days, 4.7 g of serine per liter were produced. At pH 8.5–9.0, there was only a small degradation of L-serine and glycine (Keune *et al.* 1976).

An attempt to obtain a mutant which produced L-serine directly from cheap carbon sources was not successful although some mutants produced a small amount of L-serine (Yoshida and Nakayama 1974). Formation of L-serine as a by-product by threonine producers was noted (Kase and Nakayama 1974B). Threonine, homoserine, and glycine contributed to L-serine production by some bacteria. L-Serine production by some bacteria with a medium containing DL-glycerate has been patented.

L-ISOLEUCINE, L-LEUCINE, AND L-VALINE

L-Isoleucine

Figure 17.8 illustrates the regulation of biosynthesis of branched amino acids in bacteria. It consists of (1) feedback inhibition of L-threonine dehydratase by isoleucine, (2) inhibition of α-acetolactate (ALA) synthetase by valine, (3) multivalent repression of isoleucine-valine biosynthetic enzymes by all 3 end products of this sequence, i.e., valine, leucine, and isoleucine, (4) feedback inhibition of α-isopropylmalate (IMP) synthetase by leucine, and (5) regulation of intracellular levels of threonine, a precursor of isoleucine. Thus wild type strains of bacteria do not excrete appreciable amounts of isoleucine in a salts-sugar medium. Microbial production of isoleucine was, therefore, first established by addition of precursors such as α-aminobutyrate (α-AB), D-threonine, and α-hydroxybutyric acid which all escape the regulation.

Fermentation methods using these precursors have been established using various bacteria by several groups and have been utilized industrially. These processes have been reviewed in detail (Shimura 1972B). DL-α-Bromobutyric acid could also be utilized as a precursor for isoleucine production (Matsushima *et al.* 1974). In addition to its role as a precursor by deamination to α-ketobutyrate, α-AB stimulates isoleucine production by releasing the feedback inhibition of threonine of threonine dehydratase by isoleucine, and acts as a regulatory factor on acetohydroxyacid (AHA) synthetase, stimulating synthesis of acetohydroxybutyrate (isoleucine pre-

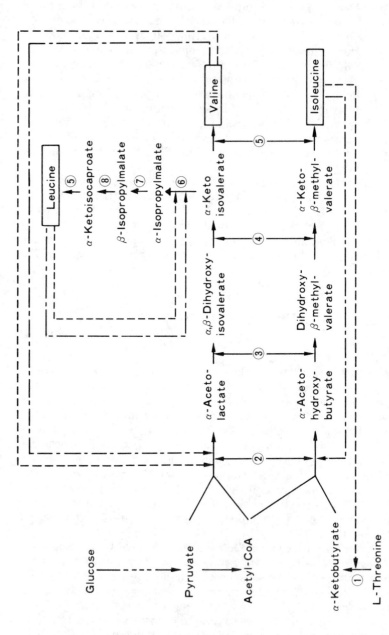

FIG. 17.8. REGULATION OF THE BIOSYNTHETIC PATHWAYS OF ISOLEUCINE, LEUCINE, AND VALINE
1—Threonine dehydratase. 2—Acetohydroxyacid synthetase. 3—Dihydroxyacid reductoisomerase. 4—Dihydroxyacid dehydratase. 5—
Branched chain amino acid transaminase. 6—α-Isopropylmalate synthetase. 7—α-Isopropylmalate isomerase. 8—β-Isopropylmalate dehydrogenase.
Symbols: ◄─────── Feedback inhibition. ◄─ ─ ─ ─ ─ Repression.

cursor) and inhibiting that of acetolactate (valine precursor) in *Bacillus subtilis*. D-Threonine induces synthesis of D-threonine dehydratase which is resistant to end-product inhibition by L-threonine, in contrast to L-threonine dehydratase in the species *Serratia* and *Pseudomonas*. Hence, D-threonine supplies α-ketobutyrate continuously for isoleucine biosynthesis, avoiding feedback control by isoleucine. In addition, α-AB formed from excess α-ketobutyrate exerts a stimulating effect on the enzyme systems in these bacteria in favor of isoleucine biosynthesis. In *Serratia marcescens* No. 1, D-threonine derepresses synthesis of D-threonine dehydratase and AHA synthetase. In an α-AB-resistant mutant derived from *S. marcescens* No.1, D-threonine dehydratase and AHA synthetase are genetically derepressed. This strain produced 15–16 g of L-isoleucine per liter in the presence of D-threonine.

Recently, processes for L-isoleucine production directly from carbohydrates have been established. Isoleucine hydroxamate (IH) shows false feedback inhibition of L-threonine dehydratase activity (Kisumi *et al.* 1971C). A mutant of *Serratia marcescens* which is resistant to IH and has desensitized L-threonine dehydratases, and α-AB-resistant mutants which are genetically derepressed for the isoleucine-valine biosynthetic enzyme, produce no isoleucine in a salts-sugar medium in the absence of isoleucine precursors (Kisumi *et al.* 1971B,D). A double mutant (IHA$_v$ 818) resistant to both IH and α-Ab, isolated from the IH-resistant strain, produced 6–7 g of L-isoleucine per liter in a medium in the absence of threonine. The double mutant lacked both feedback inhibition and repression of isoleucine biosynthesis. Increasing the α-AB resistance led strain GIHVLA$_v$ 2795 to produce 12 g of L-isoleucine per liter in a glucose medium (Kisumi *et al.* 1977A).

In *Brevibacterium flavum*, growth inhibition by α-amino-β-hydroxyvaleric acid (AHV), an analog of threonine, was reversed not only by L-threonine but also partially by L-isoleucine. Threonine-producing mutants, which were isolated as analog-resistant mutants, produced a small amount of L-isoleucine (Shiio and Nakamori 1970). Strain ARI-129, which is resistant to AHV and which was isolated on a medium supplemented with 2 mg of AHV per ml, produced 11 g of L-isoleucine per liter (Shiio *et al.* 1973B). Homoserine dehydrogenase from strain ARI-129 was insensitive to feedback inhibition by L-isoleucine and threonine, while no difference was observed in the activity of threonine dehydratase between strains ARI-129 and 2247A, a wild type parent strain. Another AHV-resistant mutant (ARI-199), which was selected on a plate containing 3 g per liter of AHV, produced 15.1 g of L-isoleucine per liter. The medium had the following composition: 10% glucose, 5% $(NH_4)_2SO_4$, 0.15% KH_2PO_4, 0.05% $MgSO_4\cdot 7H_2O$, 2 ppm Fe^{2+}, 2 ppm Mn^{2+}, 200μg biotin per liter, 400 μg thiamin·HCl per liter, 4 mg Mieki (protein hydrolysate) per liter, and 5% $CaCO_3$. Maximum production of L-isoleucine was obtained after 66 hr. Strain ARI-129 grew slightly better than ARI-199, and the maximum production of L-isoleucine was obtained after 44 hr. The productivity of strain ARI-129 was more stable than that of ARI-199. The difference between the isoleucine and L-threonine producing mutants was attributed to differences in permeability to L-threonine. In isoleucine-producing mutants, intracellular accumulation of threonine overcomes feedback inhibition of threonine dehydratase by isoleucine because inhibition of threonine dehydratase by isoleucine is competitive with respect to L-threonine. An O-methyl-threonine-resistant mutant (AORI-126), which was derived from ARI-129, produced 14.5 g of L-isoleucine per liter. The specific activity of threonine dehydratase from strain AORI-126 increased about 2-fold over that of strains No. 2247A and ARI-129, whereas the degree of inhibition of the enzyme by L-isoleucine was the same (Shiio et al. 1973B).

An ethionine (20 mg per ml)-resistant mutant (No. 168) of B. flavum was isolated after mutagenic treatment of strain BB-69, an AHV-resistant mutant which produces L-threonine from glucose in a yield of 13%. Mutant strain No. 168 produced L-isoleucine in a yield of 10% (Ikeda et al. 1976). Another ethionine (15 mg per ml)-resistant mutant (No. 14083) was derived from B. flavum FAB-3-1, which was resistant to both thialysine (AEC), a lysine analog, and AHV, and produced L-threonine in 11% yield. The mutant (No. 14083) produced L-isoleucine in 12% yield. Production of L-isoleucine by this process using acetic acid as the carbon source reached 33.5 g per liter and a 10% yield (Ikeda et al. 1976).

An L-isoleucine producing mutant (ASAT 372) of Corynebacterium glutamicum was isolated as a thiaisoleucine-resistant mutant from a threonine-producing strain having 3 markers: methionine-less, AHV resistance, and thialysine resistance by mutation. ASAT 372 produced 5 g of isoleucine and 5 g of threonine per liter. This strain was further improved by adding as markers: ethionine resistance, 4-azaleucine resistance, and α-AB resistance. The strain thus obtained, RAM-55, produced 10.6 g of L-isoleucine (Kase and Nakayama 1977).

L-Leucine

A leucine-producing mutant was obtained as a leaky-type isoleucine auxotroph from an α-AB-resistant mutant of Serratia marcescens. The isoleucine auxotroph (No. 149) produced L-leucine by long incubation in a medium lacking isoleucine. S-13 and S-11, partial revertants of strain No. 149, produced 13 g per liter of L-leucine in 48 hr in a medium lacking isoleucine (Kisumi et al. 1973). The medium had the following composition: 2% glucose, 10% dextrin, 1% urea, 0.1% K_2HPO_4, 0.05% $MgSO_4·7H_2O$, 2% $CaCO_3$. The original α-AB-resistant mutant (Ar 130-1) was derepressed for isoleucine-valine biosynthetic enzymes. Acetohydroxy acid synthetase and transaminases of strains S-3 and S-11 were found to be derepressed 10–20-fold compared with the parent strain No. 149. Furthermore, α-isopropylmalate (α-IPM) synthetase, the first enzyme on the leucine biosynthetic pathway of the α-AB-resistant mutant, was also found to be constitutive. Reversion of the isoleucine auxotrophy was accompanied by desensitization of the leucine biosynthetic enzyme. On the other hand, α-IPM synthetase in the wild type strain is repressed by leucine. L-Threonine dehydratase activity was not detected in leucine-producing revertants by the assay method used. These results indicate that leucine production was due to a lack of both feedback inhibition and repression. In the L-leucine-producing revertant, leucine enzymes catalyzed α-ketobutyrate formation from pyruvate via citramalate, citraconate and β-methylmalate (Kisumi *et al.* 1977B) (Fig. 17.9).

Tsuchida *et al.* (1974B) found an L-leucine producer in the 2-thiazolealanine-resistant mutants derived from an isoleucine-methionine double auxotroph of *Brevibacterium lactofermentum* 2256. One of the mutants, No. 218, produced 19 g of L-leucine per liter in a medium having following composition: 8% glucose, 4% $(NH_4)_2SO_4$, 0.1% KH_2PO_4, 0.04% $MgSO_4 \cdot 7H_2O$, 50 μg biotin per liter, 300 μg thiamin·HCl per liter, 2 ppm Fe^{2+}, 2 ppm

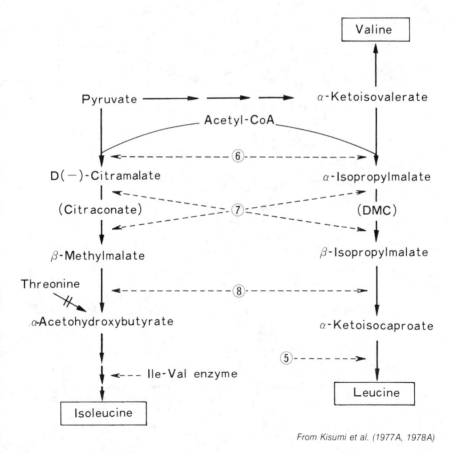

From Kisumi et al. (1977A, 1978A)

FIG. 17.9. PATHWAY OF ISOLEUCINE FORMATION IN A LEUCINE-ACCUMULATING STRAIN OF *SERRATIA MARCESCENS*

Mn^{2+}, 40 mg DL-methionine per liter, 20 mg L-isoleucine per liter, and 5% $CaCO_3$.The time course of the reaction is shown in Fig. 17.10.

Strain No. 218 produced 28 g of L-leucine per liter after 72 hr cultivation when 13% glucose was supplied as a carbon source (Tsuchida *et al.* 1975B). Maximum production of L-leucine was obtained with a certain degree of oxygen deficiency. When oxygen deficiency degree is defined as r_{ab}/KrM, where r_{ab} is respiration rate (moles of O_2/ml min) of microorganism and KrM (moles of O_2/ml min) is maximum oxygen demand rate of microorganism, maximum leucine production by *Brevibacterium lactofermentum* was attained at an oxygen deficiency degree between 0.8 and 0.9 (Akashi *et al.* 1977, 1978) (Fig. 17.11). This oxygen deficiency was attained at the redox potential of the medium (Eh, mv) between -200 mv and -220 mv. In a wild

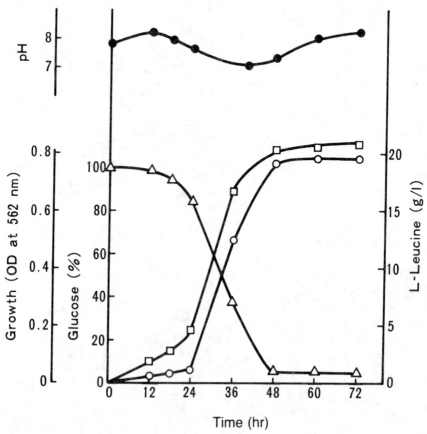

From Tsuchida et al. (1974B)

FIG. 17.10.   TIME COURSE OF L-LEUCINE PRODUCTION
Symbols: ○ L-Leucine. □ Growth. △ Residual glucose. ● pH.

type strain of *Brevibacterium flavum*, α-IPM synthetase is sensitive to inhibition and repression by L-leucine, and AHA synthetase is weakly sensitive to inhibition by valine, leucine, or isoleucine and is strongly repressible by a combination of the 3 amino acids. On the other hand, in the leucine producer No. 218, the regulatory property of AHA synthetase was

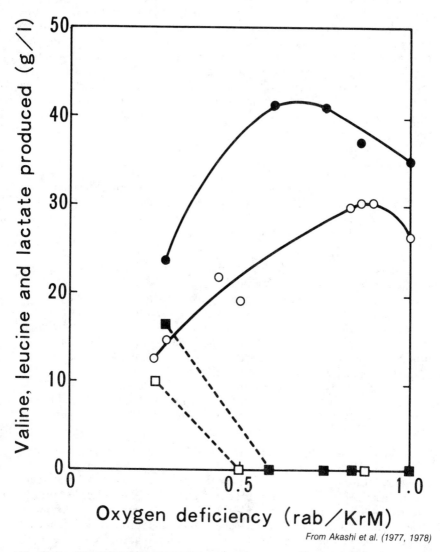

From Akashi et al. (1977, 1978)

FIG. 17.11. RELATIONSHIP BETWEEN MICROBIAL PRODUCTS AND DEGREE OF OXYGEN DEFICIENCY *(r$_{ab}$/KrM)* IN VALINE AND LEUCINE FERMENTATION

Symbols: ● Valine. ○ Leucine. ■ Lactate in valine fermentation. □ Lactate in leucine fermentation.

almost the same as that of a wild type strain, while α-IPM synthetase was found to be derepressed 3-fold and resistant to feedback inhibition. From these results, both genetic release of feedback inhibition by the mutation causing the 2-thiazolealanine resistance and derepression of AHA synthetase brought about by the mutation leading the isoleucine auxotrophy seem to be the cause of overproduction of leucine (Tsuchida and Momose 1975).

An L-leucine production process using an auxotrophic mutant of *Corynebacterium glutamicum* has also been developed (Araki *et al.* 1974C). Mutant No. 190 was found to produce a large amount of L-leucine in the culture medium. The nutritional requirements of the mutant are rather complex but its growth was remarkably stimulated by L-phenylalanine. Acetate (1.5–3.0%) or pyruvate (3%) stimulated L-leucine production. A histidine auxotrophic derivative, Pα-129, produced twice as much L-leucine as the parent strain, i.e., L-leucine production by the strain reached 16 g per liter in a medium containing 12% glucose, 2.5% CH_3COONH_4, and other ingredients.

L-Valine

Uemura *et al.* (1972) reviewed L-valine production by a fermentation process. Valine-producing bacteria are often found in nature, but processes using auxotrophic or regulatory mutants have also been developed. The addition of certain drugs to the culture medium will induce valine production.

Valine Production by Wild Type Bacteria.—Most of the wild type organisms capable of copious valine production belong to the family Enterobacteriaceae, especially to the genera *Aerobacter* and *Escherichia*.

Optimal carbon:nitrogen ratios for valine production seem to be approximately 100:7 to 100:4. It is noteworthy that heavy metal ions profoundly influence valine production. Ferrous ions seem to be essential for valine production largely because α-acetolactate synthetase (ALSase) and β-hydroxyacid dehydratase are Fe^{2+}-requiring enzymes. Control of the oxygen supply to the culture medium is another important factor for valine production. Using a selected wild type culture of a strain of Enterobacteriaceae, 12–13 g of L-valine per liter, equivalent to 20% conversion of consumed glucose, could be produced.

Mechanism of Valine Production in Wild Type Bacteria.—Since the early studies, most prototrophic valine-producing bacteria have been shown to belong to the group of bacteria which have 2 α-acetolactate synthetases (ALSases), namely, pH 8 ALSase and pH 6 ALSase. The former enzyme has optimum activity at pH 8 and functions as the initial enzyme in the biosynthesis of L-valine. The second enzyme has optimum activity at pH 6 and performs a biodegrading function leading to acetoin formation, although it can function as a biosynthetic enzyme at lower pH values. In the bacteria described above, pH 6 ALSase, which is valine-insensitive, is induced by growth in the fermentation medium and functions in the overproduction of valine. This was clearly demonstrated in *Paracolobactrum coliforme*. Production of valine seemed to be further enhanced due to a lack of ALA decarboxylase in this bacterium. In *Aerobacter aerogenes* No. 19-35, studied by Uemura *et al.* (1972), the pH 6 enzyme was formed inductively in the presence of an optimal concentration of inorganic phosphate or 6-thiol-purines. In addition to *P. coliforme* and *A. aerogenes, Aerobacter cloacae* and some *E. coli* strains belong to the group of bacteria described above.

Another group of valine-producing bacteria has only pH 8 ALSase. *Escherichia freundii, Serratia marcescens,* some *E. coli* strains, *Bacillus subtilis,* and *Brevibacterium ammoniagenes* belong to this group. In these bacteria, the pH 8 ALSase was insensitive or less sensitive to feedback inhibition by valine.

When altered regulatory mechanisms are participating, valine overproduction may be understood as a consequence of the extent to which pyruvate accumulates as a catabolic intermediate, the extent to which ALA is efficiently synthesized from pyruvate as a first precursor leading to valine, and to which an amino-donor in the form of glutamic acid is available for a transamination reaction. Each of these processes is influenced by several external factors, e.g., the presence of amino acids, drugs, and heavy metal ions, partial pressure of oxygen, pH value, and quality and quantity of carbon and nitrogen sources. Hence valine production is dependent on environmental changes, often leading to another mode of fermentation.

Acetohydroxy acid (AHA) synthetase is sensitive to catabolite repression in wild type *E. coli* B. The synthetase in a streptomycin-dependent mutant of *E. coli* B which excretes L-valine was relatively resistant to catabolite repression (Coukell and Polglase 1969).

Valine Production by Auxotrophic and Regulatory Mutants.—Isoleucine and leucine auxotrophic mutants of *C. glutamicum* produce a large amount of L-valine in culture broths. Some other auxotrophs produce L-valine in smaller amounts. L-Valine production by an isoleucine auxotroph reached 11 g per liter in a medium containing 7.5% glucose. Addition of leucine and valine markedly increased valine production in a synthetic medium. Antagonisms among L-isoleucine, L-leucine, and L-valine were observed in growth and valine production. It was speculated that L-leucine and L-valine compete with isoleucine, reversing the inhibition of isoleucine on valine production although permitting protein formation to a certain extent to enable overproduction of valine. Permeation into the cell was considered as a probable step for antagonism by the 3 amino acids.

A mutant of *Serratia marcescens* resistant to α-AB produced a remarkable amount of L-valine (Kisumi *et al.* 1971 B). This overproduction of valine by the mutant was correlated with genetic release from multivalent repression of the formation of isoleucine-valine enzymes. This was particularly true for the pH 8 ALSase which was sensitive to feedback inhibition by valine to the same extent as the enzyme in the parent strain. In a number of analog-resistant mutants, the best valine producer, strain No. 9, showed high activity of pH 8 ALSase, being also less sensitive to valine inhibition.

A thiazolealanine-resistant mutant of *Brevibacterium lactofermentum* produced 23.1 g of L-valine per liter in a medium containing 8% glucose (Tsuchida *et al.* 1975A). Maximum valine production was attained at an oxygen deficiency degree between 0.5 and 0.7 (Akashi *et al.* 1977,1978) (Fig. 17.11). This oxygen deficiency degree was attained at the redox potential of the medium (Eh,mv) between −210 mv and −260 mv.

L-TRYPTOPHAN AND OTHER AROMATIC AMINO ACIDS

L-Tryptophan

Microbial production of L-tryptophan using precursors has been reviewed by Terui (1972). With *Hansenula anomala*, L-tryptophan production from anthranilic acid reached 5.7 g per liter. By feeding anthranilic acid to a derepressed anthranilic acid auxotroph of *Bacillus subtilis*, 5.5 g of L-tryptophan per liter was produced from 5 g of anthranilic acid per liter (Arima *et al.* 1971). *Candida utilis* (synonym *Torulopsis utilis*) 295-t produced 6.4 g of L-tryptophan per liter from 4.2 g of anthranilic acid per liter in 36 hr (Bekers *et al.* 1971). By feeding indole, a 5-methyltryptophan-resistant mutant of *B. subtilis* ATCC 21336 produced 10.4 g of L-tryptophan per liter in 96 hr with a medium containing 7% glucose (Thieman and Pagani 1972). A strain carrying an *F Try*-episome in addition to the chromosomal *try* operon was obtained by sexduction from a feedback-resistant and derepressed mutant of *E. coli* K12 which is resistant to 5-methyltryptophan (5-MT) and 5-fluorotryptophan (5-FT). This strain produced 5 g of L-tryptophan per liter by feeding indole (1.5 g per liter) and L-serine (7 g per liter) (Sahm and Zähner 1971).

Tryptophanase, which catalyzes synthesis of L-tryptophan by reversal of the α,β-elimination reaction at rates similar to the forward reaction, was utilized for production of L-tryptophan and related compounds such as 5-hydroxytryptophan (Nakazawa *et al.* 1972). The culture broth of *Proteus rettgeri* (AJ 2770) was used as the enzyme for the reaction. For the synthesis of L-tryptophan, a reaction mixture contained 6.0 g of indole in 10 ml of methanol, 8.0 g of sodium pyruvate, 8.0 g of ammonium acetate, 0.001 g of pyridoxal phosphate, 0.1 g of Na_2SO_4, and 100 ml of the cultured broth in a total volume of 120 ml. After the pH value of the mixture was adjusted to 8.8 with 6 N KOH, it was incubated at 34°C for 48 hr. Under these conditions, 7.5 g of L-tryptophan were synthesized. Similarly, 5-hydroxy-L-tryptophan was synthesized from 5-hydroxyindole, pyruvate, and ammonia.

$$\text{indole} + CH_3COCOOH + NH_3 \longrightarrow \text{L-tryptophan} + H_2O$$

The preceding methods using precursors are not advantageous since the precursor compounds are expensive at present. Recently, direct production of L-tryptophan from carbohydrate by fermentation has been developed to the practical level.

Figure 17.12 shows the genealogy of L-tryptophan-producing mutants of *Corynebacterium glutamicum*. Mutants producing a large amount of L-tryptophan were derived from a phenylalanine and tyrosine double auxotroph of *C. glutamicum* KY 9456 which produced only a trace amount of L-tryptophan and anthranilate (Nakayama *et al.* 1976). A mutant (4MT-11), which stepwise acquired resistance to 5-MT, tryptophan hydroxamate (TrpHx), 6-fluorotryptophan (6-FTF), and 4-methyltryptophan (4-MT), produced L-tryptophan to a concentration of 4.9 g per liter in a cane molasses medium containing 10% reducing sugar as invert. L-Tryptophan production with this mutant was inhibited by L-phenylalanine and L-tyrosine. Accordingly, mutants resistant to phenylalanine and tyrosine analogs such as *p*-fluorophenylalanine (PFP), *p*-aminophenylalanine (PAP), tyrosine hydroxamate (TyrHx), and phenylalanine hydroxamate (PheHx), were derived from this mutant. One of the mutants thus obtained (Px-115-97) produced 12 g of L-tryptophan per liter in the molasses medium. The medium used had the following composition: 10% reducing sugars as invert (as cane molasses), 0.05% KH_2PO_4, 0.05% K_2HPO_4, 0.025% $MgSO_4 \cdot 7H_2O$, 2% $(NH_4)_2SO_4$, 1% corn-steep liquor, 2% $CaCO_3$ (pH 7.2). Production of L-tryptophan with the mutant was still sensitive to L-phenylalanine and L-tyrosine. Hence, further genetic improvement of the strain may be possible.

L-Tryptophan produced
(g per litre) *

KY9456 Phe⁻, Tyr⁻   0.15

  ↓ 5MTʳ, TrpHxʳ, 6FTʳ, 4MTʳ

4MT-11   4.9

  ↓ PFPʳ

PFP-2-32   5.7

  ↓ PAPʳ

PAP-126-50   7.1

  ↓ TyrHxʳ

Tx-49   10.0

  ↓ PheHxʳ

Px-115-97   12.0

From Nakayama et al. (1976)

FIG. 17.12. GENEALOGY OF L-TRYPTOPHAN PRODUCING MUTANTS AND THEIR L-TRYPTOPHAN PRODUCTIVITY
*With a cane molasses medium containing 10% sugar (as glucose).
Abbreviations: 5MT—5-Methyltryptophan. TrpHx—Tryptophan hydroxamate. 6FT—6-Fluorotryptophan. 4MT—4-Methyltryptophan. PFP—*p*-Fluorophenylalanine. PAP—*p*-Aminophenylalanine. TyrHx—Tyrosine hydroxamate. PheHx—Phenylalanine hydroxamate.

A 5-FT-resistant mutant derived from *B. subtilis* produced 4 g of L-tryptophan and L-phenylalanine per liter. A leucine auxotroph derived from the mutant produced 6.15 g of L-tryptophan per liter in a medium containing 300 μg of L-leucine per ml (Shiio *et al.* 1973A). Suppressor prototrophic revertants from No. 149 *rec⁻* histidine auxotroph of *B. subtilis* produced L-tryptophan in the medium. Elimination of *rec⁻* mutation from the gene using genetic transformation resulted in an increase in tryptophan productivity by 60%. This strain produced 5 – 6 g of L-tryptophan per liter (Alikhanian 1972).

A 5-FT-resistant histidine auxotrophic mutant of *Brevibacterium flavum* produced 2.4 g of L-tryptophan per liter. Production was increased to 3.8 g per liter using an *m*-fluorophenylalanine (MFP) resistant mutant derived from the 5-MT resistant one. A phenylalanine auxotrophic mutant from the latter mutant produced 6.2 g of L-tryptophan (Shiio *et al.* 1975).

Flavobacterium aminogenes nov. sp. has an intracellular enzyme system which degrades aromatic amino acid hydantoins into corresponding L-amino acids. Molar conversion of DL-tryptophan hydantoin into L-tryptophan was 100% when 5% each of DL-tryptophan hydantoin and inosine were treated at 40°C for 100 hr with cells of a mutant of *F. aminogenes* in which the activity to break down tryptophan had been reduced (Yokozeki *et al.* 1976). In this system inosine removes L-tryptophan from the reaction system by forming a complex with it. Spontaneous racemization of the substrate allows the conversion of the D-form hydantoin into L-tryptophan.

L-Phenylalanine

L-Phenylalanine production has been much improved using a regulatory mutant after the review of Oishi (1972). A prototrophic mutant resistant to *p*-fluorophenylalanine produced 5.5 g of L-phenylalanine per liter and trace amounts of L-tyrosine in a medium containing 10% reducing sugars as invert (as cane molasses) (Hagino and Nakayama 1974). A tyrosine auxotrophic mutant, resistant to PFP and PAP, produced 9.5 g of L-phenylalanine per liter in the molasses-containing medium. The medium used had the following composition: 10% reducing sugars as invert (as cane molasses), 2% (NH₄)₂SO₄, 0.05% KH₂PO₄, 0.05% K₂HPO₄, 0.025% MgSO₄·7H₂O, 0.25% NZ-amine (enzymic digest of casein), and 2% CaCO₃ (pH 7.2). L-Phenylalanine production in these mutants was inhibited by L-tyrosine and was stimulated by L-tryptophan (Hagino and Nakayama 1974).

A mutant of *B. subtilis* resistant to 5-FT produced 6.0 g of L-phenylalanine per liter in addition to 4.0 g of L-tryptophan per liter (Shiio *et al.* 1973A). Maximum L-phenylalanine production by a mutant of *Brevibacterium lactofermentum* was attained at an oxygen deficiency degree between 0.45 and 0.65 (Akashi *et al.* 1979). This oxygen deficiency degree was attained at the redox potential of the medium (Eh, mv) between −200 mv and −275 mv. A tyrosine auxotroph of a *Corynebacterium* species, an *n*-paraffin-utilizing glutamic acid producer, has also been reported to produce L-phenylalanine (Tokoro *et al.* 1970).

L-Tyrosine and L-DOPA

The tyrosine productivity of a phenylalanine auxotroph was improved by endowing a certain analog resistance (Hagino and Nayakama 1973). Figure 17.13 shows the genealogy of the finally selected L-tyrosine producers. A combination of auxotrophy and multiple resistance to analogs of aromatic amino acids was necessary to yield a large amount of L-tyrosine. A phenylalanine auxotroph (91-1-x-71), which became multiply resistant to the analogs of phenylalanine and tyrosine [PFP, PAP, 3-aminotyrosine (3-AT), and tyrosine hydroxamate (TyrHx)], produced 13.5 g of L-tyrosine per liter in a cane molasses medium containing 10% reducing sugars as invert. This strain is a so-called leaky mutant. It has some L-phenylalanine synthesizing activity similar to the original strain; in fact, it sometimes excreted trace amounts of L-phenylalanine. The L-phenylalanine pool of this mutant may reach a value high enough to inhibit synthesis of L-tyrosine due to deviation of its regulation for L-phenylalanine synthesis, in addition to its deviation of the regulation of L-tyrosine synthesis. Therefore, mutants which are definitely defective in L-phenylalanine synthesis were expected to produce higher amounts of L-tyrosine. Such a mutant may be selected as an L-tyrosine-sensitive strain. It grows slowly in the minimal medium supplemented with excess L-tyrosine because L-tyrosine antagonizes the entrance of L-phenylalanine into the cells and inhibits growth of a phenylalanine auxotroph in proportion to the degree of its requirement for L-phenylalanine. Colonies grown slowly in the presence of tyrosine were selected as mutants sensitive to L-tyrosine. Some mutants thus obtained produced larger amounts of L-tyrosine than the parent strain, notably strains Pr-20 and Pr-102, which produced L-tyrosine at a concentration of 17.6 g and 17.3 g per liter, respectively. The medium had the following composition: 10% reducing sugars as invert (as cane molasses), 2% $(NH_4)_2SO_4$, 0.05% K_2HPO_4, 0.05% KH_2PO_4, 0.025% $MgSO_4 \cdot 7H_2O$, and 2% $CaCO_3$ (pH 7.2). An increase in L-tyrosine production was also noted in many auxotrophic mutants derived from a phenylalanine auxotroph of C. glutamicum. Among them, LM-96, a phenylalanine and purine double auxotrophic strain, produced L-tyrosine at a concentration of 15.1 g per liter in a medium containing 20% sucrose (Hagino et al. 1973).

A reversion of the α,β-elimination reaction catalyzed by β-tyrosinase was utilized for preparation of L-tyrosine and L-DOPA (Yamada and Kumagai 1975).

$$\text{pyruvic acid} + NH_3 + \text{phenol} \longrightarrow \text{l-tyrosine} + H_2O$$
$$\text{pyruvic acid} + NH_3 + \text{pyrocatechol} \longrightarrow \text{l-DOPA} + H_2O$$

Cells of *Erwinia herbicola* prepared by growing at 28°C for 28 hr in an appropriate medium were used as a source of enzyme. A reaction mixture (100 ml) containing 0.5 g sodium pyruvate, 1.0 g phenol or 0.8 g pyrocatechol, 5 g ammonium acetate, and cells was incubated. At intervals, sodium pyruvate and phenol or pyrocatechol were added. Under these conditions, 6.05 g of L-tyrosine or 5.85 g L-DOPA were synthesized.

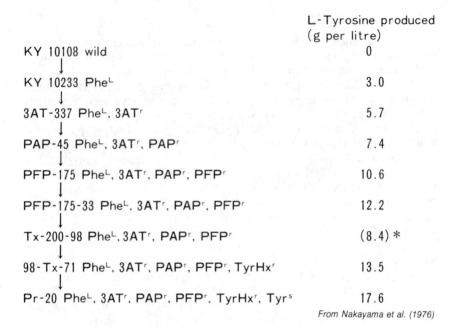

From Nakayama et al. (1976)

FIG. 17.13.   GENEALOGY OF L-TYROSINE-PRODUCING MUTANTS
*The value was obtained when the parent, PFP-175-33, gave the value 6.2 g per liter.
Abbreviations: 3AT—3-Aminotyrosine. PAP—p-Aminophenylalanine. PFP—p-Fluorophenyl-alanine. TyrHx—Tyrosine hydroxamate. L—Leaky. r—Resistant. s—Sensitive.

A 3-AT-resistant mutant of *Pseudomonas maltophila* (synonym *P. melanogenum*) produced 14–15 g of L-DOPA per liter from 26 g of L-tyrosine per liter (68% molar conversion ratio). The high L-DOPA productivity of the improved mutants was found to be due to the increased tyrosinase activity of the mutants (Tanaka *et al.* 1974).

Mechanism of Overproduction of Aromatic Amino Acids

Regulatory properties of the enzyme involved in aromatic amino acid biosynthesis in *C. glutamicum* wild and mutant strains were investigated (Nakayama *et al.* 1976). The overall control pattern (Fig. 17.14) is a new addition to the list of control patterns in aromatic amino acid biosynthesis in microorganisms. A phenylalanine and tyrosine double auxotrophic L-tryptophan producer, Px-115-97, has anthranilate synthetase partially released from the inhibition by L-tryptophan and DAHP synthetase of a wild type. L-Tryptophan production by the mutant appeared to be caused by the release from the feedback inhibition of anthranilate synthetase by L-tryptophan and blockage of chorismate mutase.

Deregulation for L-tyrosine overproduction can be understood by Fig. 17.15 which shows the control of aromatic amino acid biosynthesis in *C. glutamicum* mutant Pr-20.

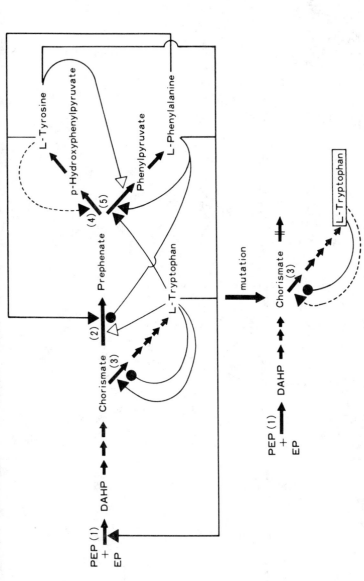

FIG. 17.14. CONTROL OF AROMATIC AMINO ACID BIOSYNTHESIS IN *C. GLUTAMICUM* AND DEREGULATION IN TRYPTO-PHAN-PRODUCING MUTANT

1—DAHP synthetase. 2—Chorismate mutase. 3—Anthranilate synthetase. 4—Prephenate dehydrogenase. 5—Prephenate dehydra-tase.

Abbreviations: PEP—Phosphoenolpyruvate. EP—Erythrose-4-phosphate. DAHP—3-Deoxy-D-arabinoheptulosonic acid-7-phosphate.
Symbols: ▼—Feedback inhibition. ----Partial inhibition. ▽—Activation. ●—Repression. ☐ Overproduced metabolite. ╪ Blocked reaction.

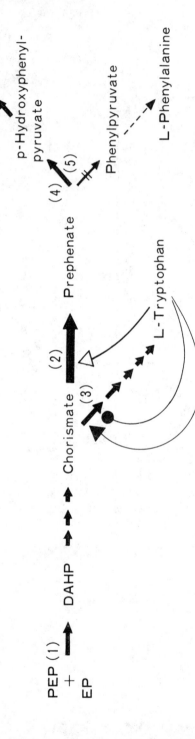

FIG. 17.15.   CONTROL OF AROMATIC AMINO ACID BIOSYNTHESIS IN *CORYNEBACTERIUM GLUTAMICUM* MUTANT Pr-20

Symbols are the same as those of Fig. 17.14.

L-ARGININE, L-ORNITHINE, AND L-CITRULLINE

L-Arginine

Efficient production of L-arginine was obtained by using regulatory mutants of *Bacillus subtilis* (Kisumi *et al.* 1971A), *Corynebacterium glutamicum* (Nakayama and Yoshida 1972), *Brevibacterium flavum* (Kubota *et al.* 1973), and *Serratia marcescens* (Kisumi *et al.* 1978B). An L-arginine hydroxamate-resistant mutant of *B. subtilis* produced 4.5 g of L-arginine per liter in shaken culture. This mutant formed N^δ-acetylornithine as a by-product and its ornithine carbamoyltransferase was strongly derepressed. Therefore, a shortage of carbamoyltransferase was thought to be the cause of N-acetylornithine formation. A 6-azauracil-resistant mutant derived from the arginine and hydroxamate-resistant mutant produced 28 g of L-arginine per liter without forming N^δ-acetylornithine in a medium containing 8% glucose and 3.5% glutamic acid.

Corynebacterium glutamicum DSS8, isolated as a D-serine-sensitive mutant from an isoleucine auxotroph KY 10150, was found to be sensitive to D-arginine and arginine hydroxamate. Furthermore, strain DSS8 produced L-arginine in a culture medium. Most of the L-arginine analog-resistant mutants derived from DSS8 produced large amounts of L-arginine. An isoleucine revertant from one of these mutants produced 19.6 g of L-arginine per liter in a medium containing 15% reducing sugars as invert (as cane molasses). Strain DSS8 seems to be a mutant with increased permeability to D- and L-arginine.

An efficient L-arginine producer was obtained through several mutation and selection steps from strain No. 33038, a guanine auxotroph of *B. flavum* (Fig. 17.16). Strain No. 352, a final, isolated mutant, produced 25.3 g of L-arginine per liter. Maximum L-arginine production of 28.4 g per liter was found at a concentration of 0.1 mg of guanine per ml which was a suboptimal concentration for growth of the mutant. Increasing the concentration of $(NH_4)_2SO_4$ in the medium increased L-arginine production. Maximum production of 29.4 g per liter was attained with 7% $(NH_4)_2SO_4$. L-Histidine markedly retarded growth of strain No. 352 and also that of the parent strain ATCC 14067. The time course of L-arginine production is presented in Fig. 17.17.

Arginine-producing strains of *Serratia marcescens* were obtained by a rather complex process owing to the insensitivity of the bacterium to arginine analogs (Kisumi *et al.* 1978B). The process is shown in Fig. 17.18. First, a mutant (ArD^-) which does not utilize arginine as a sole nitrogen source was isolated from the wild strain by mutagenic treatment. A lysine auxotroph, PA 2028 derived from the mutant (ArD^-) can grow on acetyllysine owing to the substrate specificity of acetylornithinase which allows degradation of acetyllysine to lysine. But it (ArD^- $lysA^-$) cannot grow on acetyllysine in the presence of arginine because arginine represses the formation of acetylornithinase. By mutagenic treatment of PA 2028, a

Brevibacterium flavum **A.T.C.C.14067**

| X-ray irradiation

No.33038(gu⁻),

| NG treatment(TA:5g per litre)

No.112(gu⁻, TAʳ),

| L-histidine producer(2g per litre)

| NG treatment(TA:10g per litre)

No.179(gu⁻, TAʳ),

| L-arginine producer(14.3g per litre)

| Diethyl sulphate treatment

No. 352(gu⁻, TAʳ),

| L-arginine producer(25.3g per litre)

From Kubota et al. (1973)

FIG. 17.16.   GENEALOGY OF L-ARGININE-PRODUCING MUTANTS OF *BREVIBACTE-RIUM FLAVUM*
Abbreviations: TA—2-Thiazolealanine. NG—N-Methyl-N′-nitro-N-nitrosoguanidine.

mutant strain which is able to grow in the presence of both acetyllysine and arginine is obtained. This mutant, PA 3179 (*ArD⁻ lys*A *arg*R), produced 3.2 g of L-arginine per liter of the medium.

On the other hand, a revertant of *pro*A/B was found to be sensitive to arginine, and inhibition by arginine was reversed by proline. This phenomenon is due to indirect suppression. N-Acetylglutamate-γ-semialdehyde accumulated due to *arg*D mutation and was transformed to glutamate-γ-semialdehyde by acetylornithinase and finally to proline (Fig. 17.19). Thus it grows without proline but is sensitive to arginine because arginine represses acetylornithinase.

A mutant resistant to arginine derived from the mutant (*pro*A/B, *arg*D) excreted proline because of derepression of the arginine pathway including acetylornithinase. But proline excretion was inhibited by arginine because of the inhibition of N-acetylglutamate synthetase by arginine. A regulatory mutant in which N-acetylglutamate synthetase became desensitized to the inhibition by arginine was isolated as one which grows in the presence of arginine and 3,4-dehydro-DL-proline. The latter compound competes with proline and inhibits the growth of wild type strains and *pro*A/B, *arg*D, *arg*R. The desensitized mutant (RA 4240) can grow by overcoming the inhibitory action of 3,4-dehydro-DL-proline by overproduction of proline. Finally,

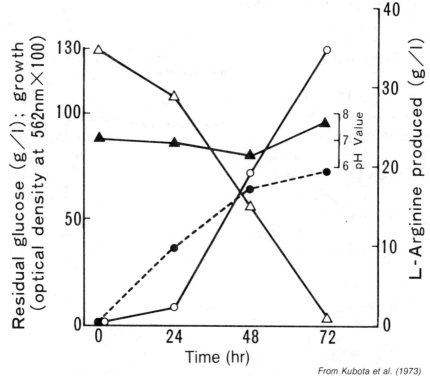

From Kubota et al. (1973)

FIG. 17.17. TIME COURSE OF L-ARGININE PRODUCTION WITH *BREVIBACTERIUM FLAVUM* NO. 532 IN A MEDIUM CONTAINING 13% GLUCOSE
Symbols: △ Glucose. ● Growth. ○ L-Arginine. ▲ pH.

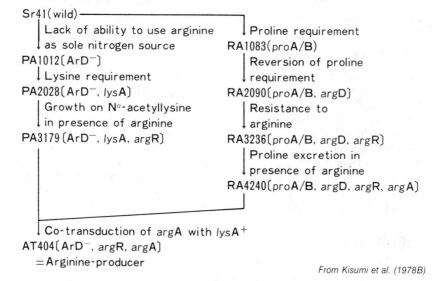

From Kisumi et al. (1978B)

FIG. 17.18. OUTLINE OF THE CONSTRUCTION OF AN ARGININE-PRODUCING MUTANT

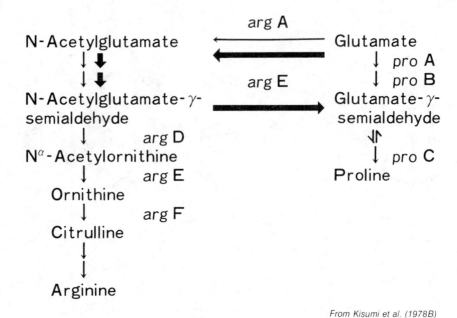

From Kisumi et al. (1978B)

FIG. 17.19.   BIOSYNTHETIC PATHWAYS OF ARGININE AND PROLINE IN ENTERIC BACTERIA
Gene symbols as used in *Escherichia coli* K12.

AT404 (ArD^-, argR, argA) was obtained by cotransduction of argA with $lysA^+$ form RA 4240 into PA 3179 (ArD^-, lysA, argR). AT404 produced 25.2 g of L-arginine per liter in a medium containing 10% glucose.

L-Ornithine

L-Ornithine production from carbohydrate is now known among arginine or citrulline auxotrophs of many microorganisms such as species of *Corynebacterium, Brevibacterium, Arthrobacter, Bacillus,* and *Escherichia* (Udaka 1972) and of *Streptomyces* (Kondo et al. 1970). L-Ornithine production from hydrocarbons by the same type of auxotroph of *C. hydrocarboclastus* and *Arthrobacter paraffineus* is also known. Among these, a *C. glutamicum* mutant is the first reported efficient L-ornithine producer. The molar yield was as high as 36%. The fermentation conditions are similar to those for glutamic acid production except that the medium contained an appropriate concentration of arginine and a large concentration of biotin.

L-Citrulline

Arginine auxotrophs of *C. glutamicum* and *B. subtilis* produced 10.7 g and 16.5 g of L-citrulline per liter in a medium containing 10 and 13% glucose, respectively. An arginine auxotroph of *Corynebacterium* sp. produced 8 g of L-citrulline per liter in a medium containing 10% (v/v) n-paraffin (Udaka 1972). Production of L-citrulline by an arginine auxotroph resistant to arginine hydroxamate was not influenced by the concentration of arginine in the medium (Kato *et al.* 1977). L-Citrulline production reached 26 g per liter using an arginine auxotroph derived from a mutant resistant to both arginine hydroxamate and 6-azauracil (Kato *et al.* 1977).

Mechanism of Overproduction of Arginine, Ornithine, and Citrulline

Arginine is synthesized from glutamic acid via ornithine and citrulline in microorganisms as shown in Fig. 17.20. Eight different enzymes participate in the chain of reactions. The first and fifth step reactions are different depending on the microorganism. In the bacteria of the Enterobacteriaceae and species of *Bacillus*, ornithine is formed by hydrolytic cleavage of acetylornithine with an acylase. In this group of microorganisms, end product inhibition acts on the first enzyme in the pathway. On the other hand, in species of *Pseudomonas, Corynebacterium, Streptomyces,* and in yeast, transacetylase catalyzes the transfer of an acetyl residue from acetylornithine to glutamic acid forming ornithine and acetylglutamic acid. In this group of microorganisms, arginine regulates the activity of the second enzyme, and probably the first enzyme as well. If the second enzyme is not regulated, arginine synthesis proceeds without control because acetyl-

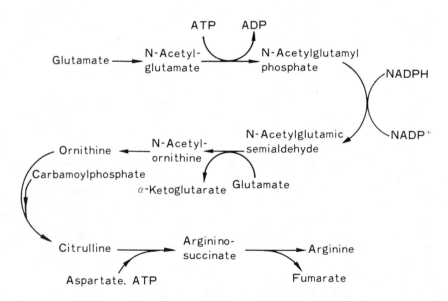

FIG. 17.20.   BIOSYNTHETIC PATHWAY OF L-ARGININE FORMATION

glutamic acid will be formed by cyclic utilization of the acetyl residue regardless of the activity of the first enzyme. The end product (arginine) has the role of corepressor in some bacteria, i.e., enzyme formation is repressed only in the presence of arginine. Other bacteria also have a repression mechanism since a much larger amount of enzyme is synthesized after a single mutation. Overproduction of ornithine becomes possible because of an increase in the amount of enzyme and the weak or nonexistent inhibition of enzyme activity by arginine when arginine is supplied in limited concentrations to cultures of an arginine auxotroph. A similar mechanism also explains citrulline production by arginine auxotrophs. In Brevibacterium flavum No. 352, N-acetylglutamokinase activity is 9 times greater than with the wild strain. It is not repressed by addition of 1% arginine, which represses the enzyme in the wild strain (Morioka et al. 1977).

L-HISTIDINE

L-Histidine Production

L-Histidine is an essential amino acid for some animals though not for humans.

Before the establishment of L-histidine production by a "direct fermentation" method, histidinol production by a histidine auxotrophic mutant and its conversion to L-histidine had been studied.

A 1,2,4-triazole-3-alanine (TRA)-resistant mutant (KY 10260) and a 2-thiazolealanine (TA)-resistant mutant derived from *C. glutamicum* ATCC 13761, a wild type strain, produced several grams of L-histidine per liter in a cane molasses medium. The histidine productivity of the TRA-resistant mutant could be improved stepwise by successively endowing resistance to 6-mercaptoguanine, 8-azaguanine, 4-thiouracil, and 6-mercaptopurine, and increased resistance to triazolealanine and 5-methyl-tryptophan resistance (Araki *et al.* 1974B). Improvement of L-histidine productivity in each step was rather minor but, as a total, the finally selected mutant strain, KY 10522, produced about twice the amount of L-histidine as the original L-histidine producer. Among the steps, insertion of 4-thiouracil resistance resulted in a most significant increase in L-histidine productivity. The rationale of the improvement was an increased supply of 5-phosphoribosyl pyrophosphate and adenine nucleotide for L-histidine biosynthesis by releasing the supposed feedback regulation of their biosynthesis. This is based on the speculation that the regulatory mechanism of L-histidine biosynthesis and related biosynthesis known in some other microorganisms is applicable to this bacterium. An improvement caused by increased resistance to TRA could be explained by a further release of end product regulation of the histidine pathway. This could be the result of an additional mutation.

Strain KY 10522 produced 15 g of L-histidine per liter equivalent to 10% (w/w) of the initial sugar. The medium used had the following composition:

6% reducing sugars as invert (as cane molasses), 9% sucrose, 4% $(NH_4)_2SO_4$, 0.2% KH_2PO_4, 0.1% K_2HPO_4, 0.05% $MgSO_4 \cdot 7H_2O$, 0.2% urea, 0.75% meat extract, 1 mg thiamin·HCl per liter, 80 μg biotin per liter, and 3% $CaCO_3$. Among the auxotrophic derivatives of strain KY 10260, a leucine auxotroph produced L-histidine at a concentration of 11 g per liter equivalent to 5.8% (w/w) of the initial sugar (Araki et al. 1974A).

L-Histidine-producing mutants were also isolated from Brevibacterium flavum (Kamijo et al. 1973; Mihara et al. 1973). A TA-resistant mutant of B. flavum (No. 2247) produced 4 g of L-histidine per liter. Production was improved by adding the following markers successively; sulfa drug-resistance; resistance to α-amino-β-hydroxyvaleric acid, a threonine analog; and to 2-aminobenzothiazole. An L-histidine-producing mutant was also found among sulfisoxazole-resistant mutants of B. flavum No. 2247. The productivity was improved by successive insertion of the following markers: resistance to sulfadiazine; S-(β-aminoethyl)-L-cysteine, a lysine-analog; and ethionine. The mutant thus derived produced 9.7 g of L-histidine per liter. An isoleucine-valine double auxotroph of Proteus rettgeri produced 3.8 g of L-histidine per liter in a medium containing 10% glucose (Nakamura et al. 1973). A deamination product of L-histidine, urocanic acid, has a "sun-screening" effect and is used in cosmetics. Hence an efficient enzymic process for urocanic acid production from L-histidine has been developed (Kajiwara et al. 1971; Shibatani et al. 1974).

A 1,2,4-triazole-3-alanine-resistant mutant derived from histidase-deficient Serratia marcescens produced 13 g of L-histidine per liter (Kisumi et al. 1977B). A strain producing urocanic acid was obtained by transducing the triazolealanine-resistance from the mutant into a urocanase-deficient mutant (Kisumi et al. 1977B). A urocanase-deficient mutant, KY-10550, of Brevibacterium ammoniagenes produced 7.2 g of urocanic acid per liter from 10 g of L-histidine per liter. A mutant resistant to 2-thiazole-DL-alanine derived from KY 10550 produced urocanic acid in a medium containing no histidine. The productivity was improved by a sequence of mutations which endowed resistance to 2-fluoroadenine, 6-mercaptopurine, 8-azaguanine, and streptomycin. The mutant thus derived produced 7.2 g of urocanic acid per liter in a medium containing 15% glucose (Kobayashi et al. 1977).

Mechanism of L-Histidine Production

Genetic and enzymic studies on Salmonella typhimurium have provided much information on the biosynthetic pathway of histidine and control mechanisms of the pathway. The first step on the pathway is a condensation of 5-phosphoribosyl-1-pyrophosphate (PRPP) and ATP to form N-1-(5'-phosphoribosyl) adenosine triphosphate (phosphoribosyl-ATP) and pyrophosphate. The reaction is catalyzed by phosphoribosyl-ATP pyrophosphorylase and is subject to feedback inhibition by L-histidine. Certain mutants of S. typhimurium, resistant to TA (2-thiazolealanine) were found to have a pyrophosphorylase resistant to feedback inhibition. This property offered an explanation for L-histidine excretion by TA-resistant mutants of this bacterium and *E. coli.* The histidine pathway is also under feedback repression control. Mutants of *S. typhimurium* resistant to TRA (1,2,4-triazole-3-alanine) have been found to be relieved of the repression control. Phosphoribosyl-ATP pyrophosphorylases in 2 L-histidine producers of *C. glutamicum,* each selected as a TA-resistant and a TRA-resistant strain, were found to be 100-fold more resistant to L-histidine inhibition than the wild type enzyme (Araki and Nakayama 1974). It was also resistant to inhibition by TA, but was still as sensitive as the wild type enzyme to inhibition by α-methylhistidine. Formation of the pyrophosphorylase in these mutants was not significantly derepressed. However, 2-fold derepression was noted with a further improved L-histidine producer, namely, strain KY 10522. Phosphoribosyl-ATP pyrophosphorylase of strain KY 10522 was found to be resistant to feedback inhibition like its parent strain. Thus, loss of feedback inhibition and derepression of phosphoribosyl-ATP pyrophosphorylase synthesis offer an explanation for L-histidine production in the above described mutants of *C. glutamicum.*

OTHER AMINO ACIDS

Alanine

Direct Fermentation.—Many bacteria, fungi, yeasts, and actinomycetes isolated from natural sources produce alanine in culture media (Kitai 1972). Prominent producers are *Corynebacterium gelatinosum, Brevibacterium monoflagellum, B. alanicum, B. amylolyticum, B. pentoso-aminoacidicum, Bacillus coagulans,* and a mutant strain (13W) of *Brevibacterium* 22. In contrast to other amino acids, the alanine produced in these cultures is usually the racemic form. *Pseudomonas* sp. No. 483 and *Micrococcus sodonensis* are rare examples of L-alanine producers. D-Alanine is produced by *Corynebacterium fascians* under certain conditions (Yamada *et al.* 1973). An arginine hydroxamate-resistant mutant of *Microbacterium ammoniaphilum,* a glutamic acid producer, also produced a large amount of alanine in a cane molasses medium (Yoshida and Nakayama 1975).

The most common substrate for alanine production is D-glucose, although other carbohydrates are also used. Some alanine producers which utilize n-paraffins as substrate have been reported, but the amounts of alanine produced are small. The concentration of the nitrogen source and the degree of aeration are important factors in producing alanine. Manganese ions (*Brevibacterium pentoso-alanicum*) and Zn^{2+} (*Fusarium moniliforme*) stimulate L-alanine production. Pyruvate and ammonium lactate increase alanine production with *Pseudomonas* sp. No. 483. Yields of 40% by a *C. gelatinosum* strain have been reported.

In *Pseudomonas* sp. No. 483, alanine is formed from pyruvate by the action of alanine dehydrogenase. Alanine biosynthesis by *Arthrobacter* sp. 19d in a molasses medium also involves reductive amination (Setty and

Bhat 1973). On the other hand, in *C. gelatinosum*, transamination from glutamate to pyruvate was suggested as the mechanism of alanine formation. Alanine racemase catalyzes conversion of L-alanine to D-alanine yielding an equilibrium racemate mixture.

Enzymic Conversion of L-Aspartic Acid to L-Alanine.—The difficulty in obtaining optically active alanine led to the development of an enzymic process for producing L-alanine from L-aspartic acid (Kitai 1972). *Pseudomonas dacunhae* and *Xanthomonas oryzae* have been selected for the purpose. With *P. dacunhae*, high L-aspartic-β-decarboxylase activity was obtained by shaking a culture at 30°C in a medium containing ammonium fumarate, sodium fumarate, corn-steep liquor, peptone, and inorganic salts. For the enzymic conversion of L-aspartic acid to L-alanine, the whole culture broth was employed as an enzyme source. Large amounts of L-aspartic acid (as much as 40% of the broth) were converted stoichiometrically to L-alanine in 72 hr at 37°C. Control of the pH value at around 5.0 was essential for achieving a yield of more than 90%, since otherwise decomposition and racemization of the accumulated L-alanine occurred.

L-Aspartic Acid

Aspartase activity was exploited for the production of L-aspartic acid from ammonium fumarate. *Escherichia coli, E. freundii,* and *Pseudomonas fluorescens* are good sources of aspartase (Kitahara 1972). *Escherichia coli* is a facultative aerobe, so that cell yield is strongly affected by culture conditions such as the degree of aeration. To avoid glucose repression of aspartase formation, the glucose concentration in the medium should be low, though a small amount of glucose sometimes gives good results causing improved growth of the bacterium. Kitahara's procedure is as follows:

A large concentration of ammonium fumarate, corresponding to 50 g fumaric acid, was suspended in 100 ml water and subjected to the action of aspartase. Crystals of ammonium fumarate gradually disappeared and were converted to a solution of acid ammonium aspartate. Hydrochloric acid was added to bring the pH value of the solution to 2.8, the isoelectric point of aspartic acid. The solubility of aspartic acid being only 0.6% at this point, the greater part immediately crystallizes out.

$$
\begin{array}{ccc}
\underset{\text{solid phase}}{\overset{\displaystyle CHCOONH_4}{\underset{\displaystyle CHCOONH_4}{\|}}} & \underset{\substack{\text{liquid phase}\\\text{solubility 20\%}}}{\overset{\displaystyle CHCOONH_4}{\underset{\displaystyle CHCOONH_4}{\|}}} & \underset{\substack{\text{solubility}\\60\%}}{\overset{\displaystyle CHNH_2COOH}{\underset{\displaystyle CH_2COONH_4}{|}}}
\end{array}
$$

$$
\xrightarrow[\text{pH 2.8}]{\text{HCl}} \quad \overset{\displaystyle CHNH_2COOH}{\underset{\displaystyle CH_2COOH}{|}} \quad + \quad NH_4Cl
$$

Later *Alcaligenes* and *Pseudomonas ovalis* were selected as bacteria which produce L-aspartate from ammonium maleate. The activity of *A. faecalis* was high when it was grown in an acidic medium due to the permeation of maleate, an inducer of maleate *cis-trans*-isomerase. Malonic acid was found to be a gratuitous inducer.

Continuous production of L-aspartic acid from ammonium fumarate was attained by employing an enzyme column packed with the immobilized aspartase, a preparation of partially purified aspartase from *E. coli* entrapped in a polyacrylamide gel lattice (Chibata *et al.* 1974). Furthermore, *E. coli* was used in place of the enzyme. When a solution of 1 M ammonium fumarate (pH 8.5) containing 1 mM Mg^{2+} was passed through the immobilized cell column at a flow rate of space velocity 0.8 at 37°C, the highest rate of reaction was attained. L-Aspartic acid was obtained in good yield from the column effluent. The half-life of the immobilized column was 120 days (Tosa *et al.* 1974; Sato *et al.* 1975).

L-Proline

L-Proline, which has a characteristically sweet taste, is not an essential amino acid, but recently medical and food industries have begun to make use of it.

Production of L-proline by fermentation was reviewed by Okumura (1972). L-Proline productivity of an isoleucine auxotroph and a histidine auxotroph of *Brevibacterium flavum* increased significantly by making them resistant to sulfaguanidine by mutation (Tsuchida *et al.* 1974A). Some auxotrophic mutants of coryneform glutamic acid-producing bacteria, represented by *C. glutamicum*, produced large amounts of L-proline. These include isoleucine, histidine, and ornithine auxotrophs. According to Araki *et al.* (1975), a certain base auxotroph of *C. glutamicum* produced 31 g of L-proline per liter in a medium containing 15% reducing sugars as invert (as cane molasses). A tyrosine-phenylalanine double auxotroph of *Corynebacterium melassecola* also produced a fairly large amount of L-proline (Kanamitsu 1971). Some wild type strains of *C. glutamicum* also produce significant amounts of L-proline under certain conditions (Nakanishi *et al.* 1973). Production of L-proline by *Kurthia catenaforma* increased in a serine auxotrophic mutant.

One of the characteristic conditions for L-proline production is a high concentration of ammonium ions. An excessive supply of biotin and unusually high concentrations of magnesium ions were also required for proline production by auxotrophic mutants of coryneform glutamic acid-producing bacteria. In the case of *K. catenaforma*, the presence of L-aspartic acid and a high concentration of potassium ions in the medium markedly stimulated L-proline production. L-Glutamic acid was also effective when some surfactants were added to the medium to increase its transport (Kato *et al.* 1972A,B). The nutrients required for auxotrophic mutants should also be limited to an appropriate concentration which is suboptimal for their growth. Addition of high concentrations of both ammonium and chloride ions to the medium was effective for L-proline production by wild type strains of *C. glutamicum*. Highly aerobic conditions, a temperature near 30°C, and an initial pH value of 7.0–8.0 were also required for good production of L-proline by *C. glutamicum*. Adding alcohols increased L-proline production by *C. glutamicum* KY 9003 (Nakanishi *et al.* 1973).

L-Proline production by *B. flavum* No. 14-5, an isoleucine auxotroph, can be explained on the following basis: An intracellular accumulation of threonine resulting from a blockage of threonine dehydratase activity; availability of a high level of ATP resulting from inhibition of aspartate kinase and of homoserine kinase by the threonine; promotion of glutamate kinase activity by the ATP; insensitivity of one of the two isozymic kinases to proline; and the availability of intracellular glutamate under biotin-rich conditions (Okumura 1972). However, this explanation may be questioned because many isoleucine auxotrophs deficient in threonine dehydratase are unable to produce a large amount of proline. The mechanism of L-proline production is still waiting to be clarified.

L-Cysteine

Very recently, two new processes for cysteine production have been reported. High activity of L-cysteine desulfhydrase was found in the bacteria of Enterobacteriaceae. The bacterial cells produced 25 g of L-cysteine per liter from β-chloroalanine and sodium sulfide. The yield increased to 48.5 g per liter (molar conversion from β-chloroalanine is 80.2%) by adding up to 8% acetone to the reaction mixture. L-Cysteine could be easily recovered as precipitate of cystine by aeration of the mixture at pH 5 after removing hydrogen sulfide in the mixture (Ōgishi *et al.* 1977). *Pseudomonas thiazolinophilum* AJ 3854 quantitatively converted DL-2-aminothiazoline-4-carboxylate (DL-ATC), an intermediate in the chemical synthesis of cysteine, into L-cysteine. A part of the L-cysteine produced was spontaneously oxidized to L-cystine by air. In a mutant which lost cysteine desulfhydrase, 95% of DL-ATC was converted into L-cysteine and the concentration of L-cysteine in the reaction mixture reached 31.4 g per liter. ATC-racemase, ATC-hydrolase, and S-carbamylcysteine hydrolase are involved in the conversion (Sano and Mitsugi 1978).

L-Methionine

L-Methionine production of *Corynebacterium glutamicum* KY 9276 (Thr⁻) was improved by sequential addition of resistance to 5 methionine-analogs. The finally selected strain produced 2 g of L-methionine per liter in a medium containing 10% glucose (Kase and Nakayama 1975). Increase of L-methionine production was accompanied by increased levels and reduced repressibility of methionine-forming enzymes.

AERATION, AGITATION, AND RECOVERY

Aeration and Agitation

The importance of aeration and agitation can not be overestimated for the practical operation and equipment design of aerobic fermentations. In leucine, valine, and phenylalanine production by mutants of *Brevibacterium lactofermentum*, maximum production is obtained at conditions of oxygen deficiency as described in each section. On the other hand, maximum proline and glutamine production (each 42 g and 42.6 g per liter) by the same species requires oxygen sufficiency. The optimal oxygen deficiency degree (defined as r_{ab}/KrM, where r_{ab} is the respiration rate of the cell, mol of O_2/ml min, and KrM is maximum oxygen requirement rate, mole of O_2/ml min) for these fermentations was 1.0. This oxygen deficiency degree occurs at the redox potentials of the medium (Eh, mv) of -130 mv and -150 mv for each fermentation (Akashi *et al.* 1979). At an oxygen deficiency degree of less than 0.3, lactic acid is predominantly formed.

Valine and leucine are produced through pyruvic acid and phenylalanine is produced through phosphoenolpyruvic acid. Therefore, for the efficient production of these amino acids, oxidation of pyruvate or phosphoenolpyruvate through the TCA cycle must be halted, and the microorganism must be incubated at an oxygen supply less than the critical dissolved oxygen level for the cell's respiration. On the other hand, for maximum production of amino acids produced through the TCA cycle, such as glutamine and proline, the oxygen demand of the cells must be satisfied.

Recovery

Methods for extraction and purification of amino acids from fermentation broths were described by Samejima (1972). Crystallization of glutamic acid hydrochloride from concentrated broths has often been employed. Aspartic acid is usually crystallized from broths or enzymic mixtures at its isoelectric point, pH 2.8. Ion exchange resins have been widely used for the extraction and purification of amino acids from fermentation broths.

SUMMARY

Amino acids produced by microbial processes are generally L-forms, except alanine, which accumulates in a medium usually as the racemic form because of the wide distribution of alanine racemase. The sterospecificity of the amino acid produced by fermentation makes the process advantageous compared with the synthetic process. Glycine and DL-methionine are now produced by chemical synthesis. DL-Alanine is also produced mostly by chemical synthesis. Absence of isomerism in glycine and the fact that stereospecificity is not required for most uses of the latter two amino acids make the synthetic process advantageous.

Most amino acids are now produced by "direct fermentation" from cheap carbon sources such as carbohydrate materials or acetic acid. L-Aspartic acid, L-alanine, L-serine, and L-cysteine are exceptions. These amino acids are produced by one-step enzymic reactions or use of a precursor. Microorganisms employed in direct fermentation processes are divided into 4 classes; wild type strains, auxotrophic mutants, regulatory mutants, and auxotrophic regulatory mutants. Examples of processes using *Corynebacterium glutamicum* strains are shown in Table 17.1. The main contributions to overproduction are also listed in the table. To increase the yield of amino acid the strain is improved by the addition of many supplementary markers. An example is L-lysine production in which a homoserine auxotroph was further improved by addition of a requirement for leucine and S-(β-aminoethyl)-L-cysteine.

TABLE 17.1.   AMINO ACID PRODUCTION USING *CORYNEBACTERIUM GLUTAMICUM* STRAINS

Amino Acid Produced	Type of Strain	Main Contributors to Production
L-Glutamic acid	Wild type	Change of permeability and low level of α-ketoglutarate dehydrogenase
L-Valine	Auxotrophic mutant	Ile$^-$, Leu$^-$
L-Homoserine		Thr$^-$
L-Lysine		Homoserine$^-$, Thr$^-$
L-Ornithine		(Arg/Cit)$^-$
L-Citrulline		Arg$^-$
L-Leucine		Amino acids$^-$
L-Proline		Base$^-$, Arg$^-$, Ile$^-$, His$^-$
L-Arginine	Regulatory mutant	Arg-analog[R]
L-Histidine		His-analog[R]
L-Threonine	Auxotrophic regulatory mutant	Met$^-$, Thr-analog[R], Lys-analog[R]
L-Isoleucine		Met$^-$, Thr-analog[R], Lys-analog[R], Ile-analog[R]
L-Tryptophan		Phe$^-$, Tyr$^-$, Phe-analog[R], Tyr-analog[R], Trp-analog[R]
L-Tyrosine		Phe$^-$, Tyr-analog[R], Phe-analog[R]
L-Phenylalanine		Tyr$^-$, Phe-analog[R], Tyr-analog[R]
L-Lysine		Homoserine$^-$, Leu$^-$, Lys-analog[R]

REFERENCES

AKASHI, K., IKEDA, S., SHIBAI, H., KOBAYASHI, K. and HIROSE, Y. 1978. Determination of redox potential levels critical for cell respiration and suitable for L-leucine production. Biotechnol. Bioeng. *20*, 27−41.

AKASHI, K., SHIBAI, H. and HIROSE, Y. 1977. Effect of oxygen supply on L-valine fermentation. J. Ferment. Technol. *55*, 364−368.

AKASHI, K., SHIBAI, H. and HIROSE, Y.   1979.   Effect of oxygen supply on L-phenylalanine, L-proline, L-glutamine and L-arginine fermentations. J. Ferment. Technol *57*, 321–327.

ALIKHANIAN, S. I.   1972.   Genetic methods for the improvement of microbiological fermentation. *In* Fermentation Technology Today. G. Terui (Editor). Society of Fermentation Technology, Japan, Osaka.

ARAKI, K., KATO, F., ARAI, Y. and NAKAYAMA, K.   1974A.   Histidine production by auxotrophic histidine analog-resistant mutants of *Corynebacterium glutamicum*. Agric. Biol. Chem. *38*, 189–194.

ARAKI, K. and NAKAYAMA, K.   1974.   Feedback-resistant phosphoribosyl-ATP pyrophosphorylase in L-histidine-producing mutants of *Corynebacterium glutamicum*. Agric. Biol. Chem. *38*, 2209–2218.

ARAKI, K., SHIMOJO, S. and NAKAYAMA, K.   1974B.   Histidine production by *Corynebacterium glutamicum* mutants, multiresistant to analogs of histidine, tryptophan, purine and pyrimidine. Agric. Biol. Chem. *38*, 837–846.

ARAKI, K., TAKASAWA, Y. and NAKAJIMA, J.   1975.   Fermentative production of L-proline with auxotrophs of *Corynebacterium glutamicum*. Agric. Biol. Chem. *39*, 1193–1200.

ARAKI, K., UEDA, H. and SAIGUSA, S.   1974C.   Fermentative production of L-leucine with auxotrophic mutants of *Corynebacterium glutamicum*. Agric. Biol. Chem. *38*, 565–572.

ARIMA, K., NOGAMI, I. and YONEDA, M.   1971.   L-Tryptophan production from anthranilic acid bacteria. Abstr. Meet. Agric. Chem. Soc. Jpn., Tokyo, 1971, 153. (Japanese)

BEKERS, M., ABOLINS, T., VIESTURS, U., SELGA, S., RAMINA, L. and OZOLINS, S.   1971.   Effect of aeration on L-tryptophan biosynthesis from anthranilic acid by *Candida utilis (Torulopsis utilis)* 295-t. Prikl. Biokhim. Mikrobiol. *7*, 103–106. (Russian)

CHIBATA, I., TOSA, T. and SATO, T.   1974.   Immobilized aspartase-containing microbial cells: Preparation and enzymatic properties. Appl. Microbiol. *27*, 878–885.

COUKELL, M.B. and POLGLASE, W.J.   1969.   Relaxation of catabolite repression in streptomycin-dependent *Escherichia coli*. Biochem. J. *111*, 279–286.

DEMAIN, A.L. and BIRNBAUM, J.   1968.   Alteration of permeability for the release of metabolites from the microbial cells. Curr. Top. Microbiol. Immunol. *46*, 1–25.

FUKUMURA, T.   1976A.   Screening, classification and distribution of L-α-amino-ε-caprolactam-hydrolyzing yeasts. Agric. Biol. Chem. *40*, 1687–1693.

FUKUMURA, T.   1976B.   Hydrolysis of L-α-amino-ε-caprolactam by yeasts. Agric. Biol. Chem. *40*, 1695–1698.

HAGINO, H. and NAKAYAMA, K.   1973.   L-Tyrosine production by analog-resistant mutants derived from a phenylalanine auxotroph of *Corynebacterium glutamicum*. Agric. Biol. Chem. *37*, 2013–2023.

HAGINO, H. and NAKAYAMA, K.   1974.   L-Phenylalanine production by analog-resistant mutants of *Corynebacterium glutamicum*. Agric. Biol. Chem. *38*, 157–161.

HAGINO, H., YOSHIDA, H., KATO, F., ARAI, Y., KATSUMATA, R. and NAKAYAMA, K.   1973.   L-Tyrosine production by polyauxotrophic mutants of *Corynebacterium glutamicum*. Agric. Biol. Chem. *37*, 2001–2005.

HIRAKAWA, T., MORINAGA, H. and WATANABE, K.   1974.   Effect of the antibiotic, borrelidin, on the production of L-threonine by *E. coli* auxotrophs. Agric. Biol. Chem. *38*, 85–89.

HIRAKAWA, T., TANAKA, T. and WATANABE, K.   1973.   L-Threonine production by auxotrophs of *E. coli*. Agric. Biol. Chem. *37*, 123–130.

HIRAKAWA, T. and WATANABE, K.   1974.   Mechanism of L-threonine production in *E. coli* auxotrophs. Agric. Biol. Chem. *38*, 77–84.

HUANG, H.T.   1961.   Production of L-threonine by auxotrophic mutants of *Escherichia coli*. Appl. Microbiol. *9*, 419–424.

IKEDA, S., FUJITA, I. and HIROSE Y.   1976.   Culture conditions of L-isoleucine fermentation from acetic acid. Agric. Biol. Chem. *40*, 517–522.

KAJIWARA, M., YASUNAGA, M., NAGANUMA, F., SHIBUYA, M., SHIRO, T. and OKUMURA, S.   1971.   Production of urocanic acid by enzymic method. Abstr. Meet. Agric. Chem. Soc. Jpn., Tokyo, 1971, 124. (Japanese)

KAMIJO, H., MIHARA, O. and KUBOTA, K.   1973.   L-Histidine production by a histidine-analog resistant mutant of glutamic acid-producing bacteria. Abstr. Meet. Agric. Chem. Soc. Jpn.,Tokyo, 1973, 112. (Japanese)

KANAMITSU, O.   1971.   L-Proline production by phenylalanine-tyrosine double auxotroph of *Corynebacterium melassecola*. Abstr. Meet. Agric. Chem. Soc. Jpn., Tokyo, 1971, 156. (Japanese)

KANZAKI, T., KITANO, K., SUMINO, Y. and OKAZAKI, Y.   1972.   L-Glutamic acid fermentation with acetic acid by an oleic acid-requiring mutant. I. Cultural characteristics. Nippon Nogei Kagaku Kaishi *46*, 95–101. (Japanese)

KASE, H. and NAKAYAMA, K.   1972.   Production of L-threonine by analog-resistant mutants. Agric. Biol. Chem. *36*, 1611–1621.

KASE, H. and NAKAYAMA, K.   1973.   L-Threonine production by a mutant of *Arthrobacter paraffineus*. Agric. Biol. Chem. *37*, 1643–1649.

KASE, H. and NAKAYAMA, K.   1974A.   Mechanism of L-threonine and L-lysine production by analog-resistant mutants of *Corynebacterium glutamicum*. Agric. Biol. Chem. *38*, 993–1000.

KASE, H. and NAKAYAMA, K.   1974B.   L-Serine production by mutants of *Corynebacterium glutamicum* and *Arthrobacter paraffineus*. Nippon Nogei Kagaku Kaishi *48*, 209–213.

KASE, H. and NAKAYAMA, K.   1975.   L-Methionine production by methionine analog-resistant mutants of *Corynebacterium glutamicum*. Agric. Biol. Chem. *39*, 153–160.

KASE, H. and NAKAYAMA, K. 1977. L-Isoleucine production by analog-resistant mutants derived from threonine-producing strain of *Corynebacterium glutamicum*. Agric. Biol. Chem. *41*, 109–116.

KASE, H., TANAKA, H. and NAKAYAMA, K. 1971. Studies on L-threonine fermentation. I. Production of L-threonine by auxotrophic mutants. Agric. Biol. Chem. *35*, 2089–2096.

KATO, J., FUKUSHIMA, H., KISUMI, M. and CHIBATA, I. 1972A. Mechanism of proline production by *Kurthia catenoforma*. Appl. Microbiol. *23*, 699–703.

KATO, J., KISUMI, M. and CHIBATA, I. 1972B. Effect of L-aspartic acid and L-glutamic acid on production of L-proline. Appl. Microbiol. *23*, 758–764.

KATO, J., KISUMI, M., TAKAGI, T. and CHIBATA, I. 1977. Increase in arginine and citrulline production by 6-azauracil-resistant mutants of Bacillus subtilis. Appl. Environ. Microbiol. *34*, 689–694.

KEUNE, H., SAHM, H. and WAGNER, F. 1976. Production of L-serine by the methanol utilizing bacterium, *Pseudomonas* 3ab. Eur. J. Appl. Microbiol. *2*, 175–184.

KIKUCHI, M., DOI, M., SUZUKI, M. and NAKAO, Y. 1972. Culture conditions for the production of glutamic acid from *n*-paraffins by glycerol auxotroph GL-21. Agric. Biol. Chem. *36*, 1141–1146.

KIKUCHI, M. and NAKAO, Y. 1973. Relation between cellular phospholipids and the excretion of L-glutamic acid by a glycerol auxotroph of *Corynebacterium alkanolytium*. Agric. Biol. Chem. *37*, 515–519.

KINOSHITA, S. and TANAKA, K. 1972. Glutamic acid. *In* The Microbial Production of Amino Acids. K. Yamada, S. Kinoshita, T. Tsunoda and K. Aida (Editors). Kōdansha, Tokyo.

KINOSHITA, S., TANAKA, K., UDAKA, S. and AKITA, S. 1957. Glutamic acid fermentation. Proc. Intern. Symp. Enzyme Chem. *2*, 464–468.

KISUMI, M., KATO, J., SUGIYAMA, M. and CHIBATA, I. 1971A. Production of L-arginine by arginine hydroxamate-resistant mutants by *Bacillus subtilis*. Appl. Microbiol. *22*, 987–991.

KISUMI, M., KOMATSUBARA, S. and CHIBATA, I. 1971B. Valine accumulation by α-aminobutyric acid-resistant mutants of *Serratia marcescens*. J. Bacteriol. *106*, 493–499.

KISUMI, M., KOMATSUBARA, S. and CHIBATA, I. 1973. Leucine accumulation by isoleucine revertants of *Serratia marcescens* resistant to α-aminobutyric acid: lack of both feedback inhibition and repression. J. Biochem. (Tokyo) *73*, 107–115.

KISUMI, M., KOMATSUBARA, S. and CHIBATA, I. 1977A. Enhancement of isoleucine hydroxamate-mediated growth inhibition and improvement of isoleucine-producing strains of *Serratia marcescens*. Appl. Environ. Microbiol. *34*, 647–653.

KISUMI, M., KOMATSUBARA, S. and CHIBATA, I. 1978A. Pathway of isoleucine formation from pyruvate by leucine biosynthetic enzymes in leucine-accumulating isoleucine revertants of *Serratia marcescens*. J. Biochem. *82*, 95–103.

KISUMI, M., KOMATSUBARA, S., SUGIYAMA, M. and CHIBATA, I. 1971C. Isoleucine hydroxamate, an isoleucine antagonist. J. Bacteriol. *107*, 741–745.

KISUMI, M., KOMATSUBARA, S., SUGIYAMA, M. and CHIBATA, I. 1971D. Properties of isoleucine hydroxamate-resistant mutants of *Serratia marcescens*. J. Gen. Microbiol. *69*, 291–297.

KISUMI, M., NAKANISHI, N., TAKAGI, T. and CHIBATA, I. 1977B. L-Histidine production by histidase-less regulatory mutants of *Serratia marcescens* constructed by transduction. Appl. Environ. Microbiol. *34*, 465–472.

KISUMI, M., TAKAGI, T. and CHIBATA, I. 1978B. Construction of L-arginine-producing mutant in *Serratia marcescens*. Use of wide substrate specificity of acetylornithinase. J. Biochem. *84*, 881–890.

KITAHARA, K. 1972. Aspartic acid. *In* The Microbial Production of Amino Acids. K. Yamada, S. Kinoshita, T. Tsunoda and K. Aida (Editors). Kōdansha, Tokyo.

KITAI, A. 1972. Alanine. *In* The Microbial Production of Amino Acids. K. Yamada, S. Kinoshita, T. Tsunoda and K. Aida (Editors). Kōdansha, Tokyo.

KITANO, K., KANETAKA, K., KATAMOTO, K., NARA, K. and NAKAO, Y. 1976. Membrane-permeability of L-glutamic acid—induction of release of L-glutamic acid from cell by tunicamycin-like antibiotic T-125. Abstr. Meet. Soc. Ferment. Technol. Jpn., Osaka, 1976, 203. (Japanese)

KOBAYASHI, K., IKEDA, S., TAKAHASHI, K., HIROSE, Y. and SHIRO, T. 1971. Production of L-glutamic acid from hydrocarbon by penicillin-resistant mutants of *Corynebacterium hydrocarboclastus*. Agric. Biol. Chem. *35*, 1241–1247.

KOBAYASHI, S., ARAKI, K. and NAKAYAMA, K. 1977. Urocanic acid-producing mutants of *Brevibacterium ammoniagenes*. Nippon Nogei Kagaku Kaishi *51*, 543–550.

KOMATSUBARA, S., KISUMI, M., MURATA, K. and CHIBATA, I. 1978. Threonine production by regulatory mutants of *Serratia marcescens*. Appl. Environ. Microbiol. *35*, 834–840.

KŌMURA, I., KOMAGATA, K. and MITSUGI, K. 1973. A comparison of *Corynebacterium hydrocarboclastus* Iizuka and Komagata 1964 and *Nocardia erythropolis* (Gray and Thornton) Waksman and Henrici 1948. J. Gen. Microbiol. *19*, 161–170.

KONDO, E., MITSUGI, T. and HANEI, F. 1970. Genetic control of microbial metabolites. I. Accumulation of L-ornithine and its related compounds by an auxotrophic mutant of *Streptomyces*. Amino Acid Nucleic Acid *22*, 151–155.

KUBOTA, K., KAGEYAMA, K., MAEYASHIKI, I., YAMADA, U. and OKUMURA, S. 1972. Fermentative production of L-serine. Production of L-serine from glycine by *Corynebacterium glycinophilum* nov. sp. J. Gen. Appl. Microbiol. *18*, 365–375.

KUBOTA, K., ONDA, T., KAMIJO, H., YOSHINAGA, F. and OKAMURA, S. 1973. Microbial production of L-arginine. I. Production of L-arginine by mutants of glutamic acid-producing bacteria. J. Gen. Appl. Microbiol. *19*, 339–352.

KUBOTA, K., YOKOI, M. and HIROSE, Y. 1975. L-Serine production from glycine by *Corynebacterium glycinophilum* and its mechanism. Abstr. Meet. Soc. Ferment. Technol. Jpn., Osaka, 1975, 80. (Japanese)

MATSUSHIMA, H., MURATA, K. and MASE, Y. 1974. Culture conditions for the microbial production of L-isoleucine from DL-α-bromo-butyric acid. Hakko Kogaku Zasshi *52*, 20–27. (Japanese)

MIHARA, O., KAMIJO, H. and KUBOTA, O. 1973. L-Histidine fermentation. III. L-Histidine production by mutants of glutamic acid-producing bacteria resistant to chemicals other than histidine-analogs. Abstr. Meet. Agric. Chem. Soc. Jpn., Tokyo, 1973, 112.

MORIOKA, H., ISHII, K. and TAKINAMI, H. 1977. L-Arginine fermentation. II. Mechanism of arginine accumulation in arginine-producing bacterium. Abstr. Meet. Agric. Chem. Soc. Jpn., Yokohama, 1977, 213. (Japanese)

NAKAMORI, S. and SHIIO, I. 1972. Microbial production of L-threonine. III. Production by methionine and lysine auxotrophs derived from α-amino-β-hydroxyvaleric acid resistant mutants of *Brevibacterium flavum*. Agric. Biol. Chem. *36*, 1209–1216.

NAKAMURA, T., EBIHARA, M. and SHIRAI, T. 1973. Method of L-histidine production. Jpn. Pat.-Kokai 48–92588. Nov. 30. (Japanese)

NAKANISHI, T. 1975. Effect of inorganic ions on the conversion of L-glutamic acid fermentation to L-glutamine fermentations by *Corynebacterium glutamicum*. Hakko Kogaku Zasshi *53*, 551–558.

NAKANISHI, T., YOKOTE, Y. and TAKETSUGU, Y. 1973. Conversion of L-glutamic acid fermentation to L-proline fermentation by *Corynebacterium glutamicum*. Hakko Kogaku Zasshi *51*, 742–749.

NAKAO, Y., KANAMARU, T., KIKUCHI, M. and YAMATODANI, S. 1973. Extracellular accumulation of phospholipids, UDP-N-acetylhexosamine derivatives and L-glutamic acid by penicillin-treated *Corynebacterium alkanolytikum*. Agric. Biol. Chem. *37*, 2399–2404.

NAKAO, Y., KIKUCHI, M., SUZUKI, M. and DOI, M. 1972. Microbial production of L-glutamic acid by glycerol auxotrophs. I. Induction of glycerol auxotrophs and production of L-glutamic acid from n-paraffins. Agric. Biol. Chem. *36*, 490–496.

NAKAYAMA, K. 1972A. Microorganisms in amino acid fermentation. *In* Fermentation Technology Today. G. Terui (Editor). Society of Fermentation Technology, Japan, Osaka.

NAKAYAMA, K. 1972B. Lysine and diaminopimelic acid. *In* The Microbial Production of Amino Acids. K. Yamada, S. Kinoshita, T. Tsunoda and K. Aida (Editors). Kōdansha, Tokyo.

NAKAYAMA, K. and ARAKI, K. 1973. Process for producing L-lysine. U.S. Pat. 3,708,395. Jan. 2.

NAKAYAMA, K., ARAKI, K., HAGINO, H., KASE, H. and YOSHIDA, H. 1976. Amino acid fermentations using regulatory mutants of *Corynebacterium glutamicum*. *In* Genetics of Industrial Microorganisms. K.D. MacDonald (Editor). Academic Press, London and New York.

NAKAYAMA, K. and YOSHIDA, H. 1972. Fermentative production of L-arginine. Agric. Biol. Chem. *36*, 1675–1684.

NAKAZAWA, H., ENEI, H., OKUMURA, S., YOSHIDA, H. and YAMADA, H. 1972. Enzymatic production of L-tryptophan and 5-hydroxy-L-tryptophan. FEBS Lett. *25*, 43–45.

NARA, T. 1972. Homoserine. *In* The Microbial Production of Amino Acids. K. Yamada, S. Kinoshita, T. Tsunoda and K. Aida (Editors). Kōdansha, Tokyo.

ŌGISHI, H., NISHIKAWA, D., KUMAGAI, H. and YAMADA, H. 1977. Synthesis of L-cysteine and its derivatives using cysteine desulfhydrase. Abstr. Meet. Agric. Chem. Soc. Jpn., Yokohama, 1977, 90. (Japanese)

OISHI, K. 1972. Phenylalanine and tyrosine. *In* The Microbial Production of Amino Acids. K. Yamada, S. Kinoshita, T. Tsunoda and K. Aida (Editors). Kōdansha, Tokyo.

OKUMURA, S. 1972. Proline. *In* The Microbial Production of Amino Acids. K. Yamada, S. Kinoshita, T. Tsunoda and K. Aida (Editors). Kōdansha, Tokyo.

SAHM, H. and ZÄHNER, H. 1971. Metabolic products of microorganisms. 90. Studies on the formation of tryptophan by *Escherichia coli* K12. Arch. Microbiol. *76*, 223–251.

SAMEJIMA, H. 1972. Methods for extraction and purification. *In* The Microbial Production of Amino Acids. K. Yamada, S. Kinoshita, T. Tsunoda and K. Aida (Editors). Kōdansha, Tokyo.

SANO, K. and MITSUGI, K. 1978. Enzymatic production of L-cysteine from DL-2-amino-Δ^2-thiazoline-4-carboxylic acid by *Pseudomonas thiazolinophilum*: optimal conditions for the enzyme formation and enzymatic reaction. Agric. Biol. Chem. *42*, 2315–2321.

SANO, K. and SHIIO, I. 1970. Microbial production of L-lysine. III. Production by mutants resistant to S-(2-aminoethyl)-L-cysteine. J. Gen. Appl. Microbiol. *16*, 373–391.

SATO, E., KAWABATA, Y. and FUKUMURA, T. 1974. Production of L-aminocaprolactam hydrolase and aminocaprolactam racemase. Abstr. 23rd Symp. Amino Acid Nucleic Acid Assoc., Jpn., Tokyo, 1974, 2. (Japanese)

SATO, T., MORI, T., TOSA, T., CHIBATA, I., FURUI, M., YAMASHITA, K. and SUMI, A. 1975. Engineering analysis of continuous production of L-aspartic acid by immobilized *Escherichia coli* cells in fixed beds. Biotechnol. Bioeng. *17*, 1797–1804.

SETTY, T.M.R. and BHAT, J.V. 1973. Microbial production of amino acids. IV. Studies on the pathway of alanine biosynthesis in *Arthrobacter* sp. $C_{19}d$. Curr. Sci. *42*, 484–489.

SHIBATANI, T., NINIMURA, N., ABE, R., KAKIMOTO, T. and CHIBATA, I. 1974. Enzymatic production of urocanic acid by *Achromobacter liquidum*. Appl. Microbiol. *27*, 688–694.

SHIIO, I., ISHII, K. and YOKOZEKI, K. 1973A. Production of L-tryptophan by 5-fluorotryptophan resistant mutants of *Bacillus subtilis*. Agric. Biol. Chem. *37*, 1991–2000.

SHIIO, I. and NAKAMORI, S. 1970. Microbial production of L-threonine. II. Production of α-amino-β-hydroxyvaleric acid resistant mutants of glutamate producing bacteria. Agric. Biol. Chem. *34*, 448–456.

SHIIO, I., SASAKI, A., NAKAMORI, S. and SANO, K. 1973B. Production of L-isoleucine by AHV resistant mutants of *Brevibacterium flavum*. Agric. Biol. Chem. *37*, 2053–2061.

SHIIO, I., SUGIMOTO, S. and NAKAGAWA, M. 1975. Production of L-tryptophan by mutants of *Brevibacterium flavum* resistant to both tryptophan and phenylalanine analogues. Agric. Biol. Chem. *39*, 627–635.

SHIMURA, K. 1972A. Threonine. *In* The Microbial Production of Amino Acids. K. Yamada, S. Kinoshita, T. Tsunoda and K. Aida (Editors). Kōdansha, Tokyo.

SHIMURA, K. 1972B. Isoleucine. *In* The Microbial Production of Amino Acids. K. Yamada, S. Kinoshita, T. Tsunoda and K. Aida (Editors). Kōdansha, Tokyo.

SUGIYAMA, Y., KITANO, K. and KANZAKI, T. 1973. The role of copper ions in the regulation of L-glutamate biosynthesis. Agric. Biol. Chem. *37*, 1837–1847.

SUZUKI, T., YAMAGUCHI, K. and TANAKA, K. 1971. Effects of cupric ion on the production of glutamic acid and trehalose by a n-paraffin-grown bacterium. Agric. Biol. Chem. *35*, 2135–2137.

TANAKA, K., SUZUKI, T. and OKUMARA, S. 1971. Production of sugars and amino acids from hydrocarbons and petrochemicals by microorganisms. 5th World Pet. Congr. Proc. *5*, 165–170. Applied Science Publication, London.

TANAKA, Y., YOSHIDA, H. and NAKAYAMA, K. 1974. Production of L-DOPA by mutants of *Pseudomonas melanogenum*. Agric. Biol. Chem. *38*, 633–639.

TERUI, G. 1972. Tryptophan. *In* The Microbial Production of Amino Acids. K. Yamada, S. Kinoshita, T. Tsunoda and K. Aida (Editors). Kōdansha, Tokyo.

THIEMAN, J.E. and PAGANI, H. 1972. Production of L-tryptophan by fermentation. U.S. Pat. 3,700,558. Oct. 24.

TOKORO, Y., OSHIMA, K., OKII, M., YAMAGUCHI, K., TANAKA, K. and KINOSHITA, S. 1970. Utilization of hydrocarbons by microorganisms. XII. Microbial production of L-phenylalanine from hydrocarbons. Amino Acid Nucleic Acid *21*, 49–55.

TOSA, Y., SATO, T., MORI, T. and CHIBATA, I. 1974. Basic studies for continuous production of L-aspartic acid by immobilized *Escherichia coli* cells. Appl. Microbiol. *27*, 886–889.

TSUCHIDA, T., KUBOTA, K., YOSHINAGA, F. and HIROSE, Y. 1974A. L-Proline fermentation. VI. Increase of L-proline production capacity in sulfaguanidine-resistant mutants. Abstr. 23rd Symp. Amino Acid Nucleic Acid Assoc., Jpn., Tokyo, 1974, 4. (Japanese)

TSUCHIDA, T. and MOMOSE, H. 1975. Genetic changes of regulatory mechanisms occurred in leucine and valine producing mutants derived from *Brevibacterium lactofermentum* 2256. Agric. Biol. Chem. *39*, 2193-2198.

TSUCHIDA, T., YOSHINAGA, F., KUBOTA, K. and MOMOSE, H. 1975A. Production of L-valine by 2-thiazolealanine resistant mutants derived from glutamic acid producing bacteria. Agric. Biol. Chem. *39*, 1319-1322.

TSUCHIDA, T., YOSHINAGA, F., KUBOTA, K., MOMOSE, H. and OKUMURA, S. 1974B. Production of L-leucine by a mutant of *Brevibacterium lactofermentum* 2256. Agric. Biol. Chem. *38*, 1907-1911.

TSUCHIDA, T., YOSHINAGA, F., KUBOTA, K., MOMOSE, H. and OKUMURA, S. 1975B. Cultural conditions for L-leucine production by strain No. 218, a mutant of *Brevibacterium lactofermentum* 2256. Agric. Biol. Chem. *39*, 1149-1153.

TSUKADA, Y. and SUGIMORI, T. 1971. Induction of auxotrophic mutants from *Candida species* and their application to L-threonine fermentation. Agric. Biol. Chem. *35*, 1-7.

UDAKA, S. 1972. Ornithine, citrulline and arginine. *In* The Microbial Production of Amino Acids. K. Yamada, S. Kinoshita, T. Tsunoda and K. Aida (Editors). Kōdansha, Tokyo.

UEMURA, T., SUGISAKI, Z. and TAKAMURA, T. 1972. Valine. *In* The Microbial Production of Amino Acids. K. Yamada, S. Kinoshita, T. Tsunoda and K. Aida (Editors). Kōdansha, Tokyo.

YAMADA, H. and KUMAGAI, H. 1975. Synthesis of L-tyrosine-related amino acids by β-tyrosinase. Adv. Appl. Microbiol. *19*, 249-288.

YAMADA, S., MAESHIMA, H., WADA, H. and CHIBATA, I. 1973. Production of D-alanine by *Corynebacterium fascians*. Appl. Microbiol. *25*, 630-640.

YOKOZEKI, K., SANO, K., EGUCHI, A., YASUDA, N., NODA, I. and MITSUGI, K. 1976. L-Amino acid production from the corresponding hydantoin of aromatic amino acid by enzymic method. Abstr. Meet. Agric. Chem. Soc. Jpn., Kyoto, 1976, 238. (Japanese)

YOSHIDA, H. and NAKAYAMA, K. 1974. Production of L-serine by L-serine analog-resistant mutants from various bacteria and the effect of L-threonine and L-homoserine on the production of L-serine. Nippon Nogei Kagaku Kaishi *48*, 201-208. (Japanese)

YOSHIDA, H. and NAKAYAMA, K. 1975. Production of alanine by arginine hydroxamate-resistant mutants of *Microbacterium ammoniaphilum*. Nippon Nogei Kagaku Kaishi *49*, 527-532. (Japanese)

18

Vinegar

Heinrich Ebner

DEFINITION

In the United States vinegar is the product obtained by acetic acid fermentation of alcohol-containing solutions. It must contain at least 4 g of acetic acid per 100 ml (at 20°C) and may not contain more than 0.5 vol. % of ethanol. Since the vinegar fermentation is carried out at higher concentrations, the appropriate acid level can be obtained by dilution with water. However, diluted acetic acid cannot be called vinegar in the United States although it may be used for the preparation of foods if it is properly labelled. Similar definitions apply to most of the countries of the world. In Europe the use of diluted acetic acid for the preparation of foods is sometimes permitted and sometimes prohibited.

PROPERTIES

Vinegar is an aqueous clear liquid which is either colorless or has the color of the raw material. The content of solutes depends on the compounds of the raw material used for the fermentation. The density, boiling point, freezing point, surface tension, and viscosity of vinegar may deviate more or less from that of pure water depending on the concentration of acetic acid and the raw material used. The pH is between 2 and 3.5.

PRODUCTION

Basics of Vinegar Fermentation

The vinegar fermentation is an oxidative fermentation in which diluted solutions of ethanol are oxidized by *Acetobacter* with the oxygen of air to acetic acid and water. The oxidation proceeds according to the basic equation:

$$C_2H_5OH + O_2 \longrightarrow CH_3COOH + H_2O \qquad H = -493\,kJ$$

The alcohol-containing solution is called a mash. Its alcohol concentration is given in volume percent. Usually it also contains some acetic acid which is determined by titration with 1 N NaOH and expressed as grams acetic acid per 100 ml. The sum of the vol. % ethanol and the weight percent (g/100 ml) of acetic acid is called "total concentration." The sum of these rather incommensurable values gives the maximal concentration of acetic acid which can be obtained by complete fermentation. The quotient of the "total concentration" of the produced vinegar over the "total concentration" of the mash gives the concentration yield. The quotient of the acetic acid concentration of the produced vinegar over the "total concentration" of the mash gives the acid yield (for details see Hromatka and Ebner 1959).

Raw Material

All mashes must contain ethanol, water, and nutrients for the acetic acid bacteria. By far the largest percentage of vinegar is distilled vinegar which is produced from diluted, purified ethanol or from fusel oil containing crude spirit. Other common names for the same product are white vinegar, spirit vinegar, alcohol vinegar, or grain vinegar. It is customary in almost all countries to denature the ethanol which serves as raw material for the vinegar industry. In the United States this is mostly done with ethyl acetate, which is split during the fermentation to ethanol and acetic acid. In most European countries denaturation is carried out with distilled vinegar. Mashes obtained by the alcoholic fermentation of natural sugar-containing liquids also serve as a raw material. The vinegar is designated according to the particular raw material used. For instance, wine vinegar is produced by vinegar fermentation of grape wine. In some countries with large wine production the use of ethanol as raw material for vinegar production is not permitted.

Cider vinegar is produced from fermented apple juice. It is particularly well known in the United States, Switzerland, and Austria because of its desirable aroma.

Malt vinegar is the product made by the alcoholic and subsequent acetous fermentation, without distillation, of an infusion of barley malt or cereals whose starch has been converted by the malt. It is well known in the United States, in England, and in South Africa.

Whey vinegar is produced by the alcoholic and subsequent acetous fermentation of concentrated whey. Fruit vinegar is made from fruits which are available in surplus in some countries. Its production from dates, citrus fruit, or bananas, for instance, is gaining in importance.

Sugar vinegar is made by the alcoholic and subsequent acetous fermentation of sugar syrup, molasses, or refiners' syrup.

Glucose vinegar is made by the alcoholic and subsequent acetous fermentation of glucose solutions.

Rice vinegar is made by the saccharification of rice starch, followed by alcoholic and acetous fermentation.

Water for Processing

The water used for the preparation of mashes must be clear, colorless, odorless, and without the presence of sediment or suspended particles. It may be hard or soft but frequent changes in hardness as they occur in municipal water supplies may interfere with the fermentation. In extreme cases the water must be demineralized followed by the addition of minerals. The water must be bacteriologically clean. Therefore, care has to be used when well water or water previously used for cooling purposes is used. If municipal water is used, it must be free from chlorine.

Nutrients

Most natural raw materials do not require the addition of extra nutrients. Apple cider is usually low in nitrogenous materials which can be corrected by the addition of 100 g ammonium phosphate to 1000 liters of mash. Some grape wines require the addition of up to 400 g ammonium phosphate for a satisfactory fermentation. In rare cases the same nutrients must be added as those for the production of distilled vinegar which are listed later.

For the production of distilled vinegar a mixture of the required nutrients had to be developed. *Acetobacter* definitely require between 500 and 1000 g of glucose per 1000 liters mash. Further, a total of 300 g of the following substances are required per 1000 liters mash: potassium, magnesium, calcium, ammonium as phosphate, sulfate, and chloride salts. The following trace minerals are required: iron, manganese, cobalt, copper, molybdenum, and vanadium. Demineralized water requires the addition of 100 g of calcium carbonate and 100 g of sodium chloride per 1000 liters.

These additions are sufficient for a satisfactory acetous fermentation. Commercial mixtures containing supplements such as malt extract or dried yeast are available. These are used to restart a fermentation more quickly if it has been stopped by a disturbance, for instance, by an interruption of the power supply. The amount added is not more than 200 g per 1000 liters mash. The amounts of nutrients mentioned are required for the submerged fermentation process. For the trickle process about one-third of the listed amounts are needed. In principle, nutrients should be added sparingly in order to exert a selection pressure in the direction of a low requirement for nutrients.

Acetobacter

Classification.—Since the time of Pasteur's work, it has been known that acetic acid bacteria cause the souring of stored wine. In 1900 Beijerinck suggested the designation *Acetobacter* for the genus, a designation which is still current today. The *Acetobacter* belong to the family Pseudomonadaceae, but this classification is in dispute (Asai 1968).

The first attempt to classify acetic acid bacteria was made by Hansen in 1894 (Asai 1968), followed by classifications of other investigators. The number of isolated strains became larger and larger and their classification more complicated. But already Visser't Hooft (1925) had indicated briefly that acetic acid bacteria of prior collections did not retain their properties.

For a while it appeared as if Frateur (1950) had succeeded in the final classification of the group by the designation of 4 main groups: peroxydans, oxydans, mesoxydans, and suboxydans. But subsequently new strains with different properties were discovered, and these had to be accommodated in ever more complex schemes of classification.

As the result of an extensive study, Shimwell (1959) found that a strain of *Acetobacter* changes its properties right at the time of its isolation. The identification of a strain of *Acetobacter* reflects therefore only its properties at the time of isolation and gives no information on its properties at an earlier or later date. The mutants which arise immediately give rise to new mutants. However, it is not certain that the word "mutation" can properly be applied to this change. Other concepts such as cytoplasmic variation, segregation, recombination, or other more complex types of reproduction are under consideration. The variability of *Acetobacter* goes so far that the property of oxidizing ethanol to acetic acid may be lost. Hence, *Acetobacter* really defy classification (for details see Asai 1968).

The vinegar industry is only interested in using a strain of *Acetobacter* which tolerates high concentrations of acetic acid and of "total concentration," which requires small amounts of nutrients, and which does not overoxidize the formed acetic acid. By 1950 it had already been observed that with each cultivation of a strain in Petri dishes there was a danger of a change in properties. Therefore, vinegar fermentations are carried out continuously in the laboratories of the firm Heinrich Frings in Bonn in order to supply the strain for fermentations in all parts of the world. A newly installed fermentor is directly inoculated by means of a transportable fermentor. If this is not possible, a vinegar from the mentioned laboratory fermentation is poured into bottles (partially filled) which are transported in the pressurized cabin of a plane. At the destination the vinegar is aerated until the fermentation begins again; and this is used to inoculate the new commercial fermentor. In this manner variations in the *Acetobacter* strain can be avoided. In spite of the propensity of *Acetobacter* for variation, it is possible to perform vinegar fermentations year-in year-out without interruption. This can be done if one does not give mutants which require more nutrients or which are more sensitive to acetic acid a chance to survive. This is accomplished by keeping the total concentration high and the concentration of nutrients low.

Biochemistry of Vinegar Fermentation.—In a satisfactory vinegar fermentation, ethanol is almost quantitatively oxidized to acetic acid. Yields between 95 and 98% are normal, and the remainder is mainly lost in the effluent gas. Acetic acid should not be oxidized further to water and CO_2 (overoxidation). From Wieland and Bertho (1929) it is known that the oxidation of ethanol takes place in two steps with acetaldehyde as the intermediate product. Other investigators have studied this reaction in detail. However, this work has been carried out with varying strains of

Acetobacter. This means that the properties of the *Acetobacter* at the time of the investigation varied, and that they often did not correspond to the properties of the strains during the vinegar fermentation.

King and Cheldelin (1954) successfully purified an alcohol dehydrogenase from *Acetobacter suboxydans* which is NAD-dependent and which does not act on acetaldehyde. The same authors also purified an acetaldehyde dehydrogenase which required NADP as coenzyme. Prieur (1968) found two enzyme systems in the oxidation of ethanol by *Acetobacter xylinum*. One of these with maximum activity at pH 5.7 requires no NAD; the other one with maximum activity at pH 8.1 is NAD-dependent.

The most exhaustive investigations have been carried out by Nakayama (1959, 1960, 1961A,B). He isolated a highly purified ethanol oxidizing enzyme from an *Acetobacter* sp. with some similarity to *A. suboxydans*. It contains a heme protein and has an absorption spectrum similar to that of cytochrome C and an absorption maximum at 553 nm. In the presence of ethanol the enzyme reduces several oxidation-reduction dyes, but not TPN and DPN. Its optimum pH is at 3.8. Its temperature inactivation curve parallels that of the denaturation of the heme protein. With ferricyanide as electron acceptor, the enzyme shows broad substrate specificity by oxidation of many saturated and unsaturated straight chain mono-alcohols. The heme protein is also reduced by acetaldehyde in presence of an aldehyde dehydrogenase which can also be isolated from *Acetobacter* and which is not coenzyme-dependent. Further, an aldehyde dehydrogenase has been isolated which is TPN-dependent and has broad substrate specificity. The author arrives at the following scheme of oxidation:

Ethanol is oxidized by the alcohol-cytochrome-553-reductase (E_1) to acetaldehyde. The electrons are transferred to the iron of the heme protein of E_1. Acetaldehyde is further oxidized by the coenzyme-independent aldehyde dehydrogenase (E_2) or by the TPN-dependent aldehyde dehydrogenase (E_3). For oxidation by E_2 the freed electrons are also transferred to the heme iron of E_1. The reduced cytochrome 553 is oxidized by a cytochrome oxidase present in the cell. Electrons liberated by the oxidation of acetaldehyde by E_3 reduce TPN to $TPNH_2$. It is assumed that the presence of $TPNH_2$ interferes with the further oxidation of acetic acid by the tricarboxylic acid cycle. The acid pH optima of E_1 and E_2 favor the accumulation of acetic acid by the microorganism. TPN must be available for these reactions through the metabolism of the cell.

Loitsyanskaya (1955) was the first to show that *Acetobacter schutzenbachii, A. curvum,* or *A. aceti* does not oxidize glucose to gluconic acid in a medium containing both glucose and ethanol (which is not true in the absence of ethanol). Apparently the glucose is used for cell metabolism. It is not absolutely necessary to use a microorganism classified as *A. suboxydans* according to Frateur in order to avoid overoxidation. However, it must be assumed that the *Acetobacter* used commercially in concentrated alcohol vinegar fermentations approaches the properties of this group. For instance, it was possible to adapt a lyophilized strain of *A. aceti* (ATCC 15937) to the production of 13.5% alcohol vinegar. This strain could not then be distinguished in its fermentation properties from a previously used *A.*

suboxydans strain which had been slowly adapted to such high concentrations (Frings 1974). Therefore, it seems appropriate to consider primarily investigations with A. *suboxydans* if one wishes to gather information on the carbohydrate metabolism of acetic acid bacteria during the vinegar fermentation.

King and Cheldelin (1952) were the first to point out that glucose may be oxidized directly or after phosphorylation. The metabolism of phosphorylated glucose may occur along 3 pathways: the glycolytic path of Embden-Meyerhof-Parnas, the pentose cycle, or the Entner-Doudoroff pathway. The presence of enzymes of the EMP pathway does not mean that this glycolytic path actually operates since each of the enzymes of this scheme also participates in the pentose cycle or the Entner-Doudoroff pathway (Cheldelin 1960). The active participation of the pentose cycle and the presence of the corresponding enzymes in A. *suboxydans* have been demonstrated by Hauge *et al.* (1955).

A quantitative study of the pentose cycle as respiration mechanism in A. *suboxydans* has been done by Kitos *et al.* (1958) with aerated resting cells. In pure oxygen, 100 molecules of glucose were metabolized as follows: 28 molecules were oxidized to 2-ketogluconic acid, and of the remaining 72 molecules 63 entered the pentose cycle. All of the CO_2 produced from glucose was derived from the pentose cycle. During oxidation with air the pentose cycle delivered the major portion of the CO_2; a minor portion of the oxidation of trioses followed a different path. The tricarboxylic acid cycle as well as the Entner-Doudoroff pathway seems to be of minor importance.

According to Kitos *et al.* (1958), acetate is not oxidized by A. *suboxydans*. If $CH_3{}^{14}COOH$ is added to respiring cells, only a negligible part of the ^{14}C appears in respiratory CO_2, while 25% of it can be recovered in the lipid fraction of the cell wall. Kitos *et al.* (1957) showed that the latter occurs only in the presence of glucose. Raghavendra Rao and Stokes (1953) found that A. *suboxydans* and A. *melanogenum* require sugar to initiate growth; but then ethanol may be used as an additional carbon source and incorporated into cell material although most of the ethanol is oxidized to acetic acid.

Razumovskaya and Belousova (1952) were the first to show that A. *schutzenbachii* is inhibited in its growth if CO_2 is removed from the air, particularly in mineral media in the presence of ethanol. Loitsyanskaya (1958) showed that A. *schutzenbachii* and A. *aceti* fix CO_2, particularly during growth, and that the uptake rate depends on the nitrogen source. Inorganic nitrogen causes a higher CO_2 uptake than peptone or yeast extract. An increase in the CO_2 concentration in the air to 1% stimulates growth, while higher concentrations are inhibitory. Hromatka *et al.* (1962) confirmed that the growth of A. *suboxydans* is also inhibited if CO_2 is removed from the air. A. *suboxydans* uses some CO_2 carbon for incorporation into cell substance as shown by experiments with $^{14}CO_2$ so that approximately 0.1% of the cell carbon is derived from CO_2 (Hromatka and Gsur 1962). A very small but measurable portion of acetic acid is derived from CO_2 metabolism. Claus *et al.* (1969) showed that dialyzed extracts of A. *suboxydans* catalyze the assimilation of $^{14}CO_2$ in the presence of phosphoenolpyruvate and divalent cations.

In contrast to *A. suboxydans*, strains of *Acetobacter* which belong to other groups (as classified by Frateur) prefer different paths of glucose metabolism. King *et al.* (1956) found that *A. pasteurianum* which belongs to the oxydans group showed strong activity of the tricarboxylic acid cycle. In strains of the mesoxydans group, use of the EMP pathway has been shown (Bourne and Weigel 1954; Prieur 1968), while for *A. peroxydans* the function of the TCA cycle is still in doubt.

It is likely that the enzymes for the different pathways are always present in the *Acetobacter* cells and that repression of one or the other pathway, for instance that of the TCA cycle, is caused by particular conditions of growth, such as, for instance, a high acid concentration.

The oxidation of ethanol to acetic acid is not entirely dependent on cell multiplication. Even after cell growth stops, for instance, when a high concentration of acetic acid has been reached, the cells are capable of oxidizing ethanol to acetic acid for a certain period of time. After this period the cells die quickly and oxidation ceases (Ebner 1976). In confirmation Vera and Wang (1977) found that the formation of acetic acid follows growth kinetics up to an acetic acid concentration of 3 g/100 ml, while a mixed kinetic model is required for concentrations between 4 and 7 g per 100 ml in which the formation of acetic acid depends both on growth rate and on actual cell concentration.

The exact connection of ethanol oxidation with the energy balance of cell metabolism, the mechanism which enables cells to withstand high concentrations of acetic acid, and the reasons for the unusual properties of *Acetobacter* during vinegar fermentations have not yet been clarified.

Special Behavior of Acetobacter During Vinegar Fermentation.—*Sensitivity Toward Lack of Oxygen.*   During commercial use of many aerobic microorganisms, it is common practice to permit the inoculum to remain in a fermentor for several days without aeration before inoculation into the main fermentor. This is also true for the oxidation of sorbitol to sorbose by *A. suboxydans*. However, this procedure is unthinkable with a vinegar fermentation.

Even today the work of Hromatka and Ebner (1949) forms the basis of all commercial submerged vinegar fermentations. According to these authors it is customary to express productivity by the Eta value, that is, the increase in acetic acid in g per 100 ml per 24 hr. If the log of the Eta value is plotted against time, a straight line is obtained during exponential growth of the bacteria. Its slope is a function of the growth rate. A change of the slope or any curvature of the line indicates a change in the conditions of the fermentation. The use of *Acetobacter* strains which are adapted to the highest "total concentration" and the highest acetic acid concentration occurring in an experiment is essential if one wishes to draw valid conclusions from experimental work. One can not generalize from the results if these conditions are not fulfilled.

Extensive tests have been conducted by Hromatka *et al.* (1951) and Hromatka and Exner (1962) on the damage to *Acetobacter* due to an interruption of aeration as a function of several variables. The percentage of killed bacteria was calculated from the change of the log Eta curve. Table 18.1 shows the effect of "total concentration," acetic acid concentration, ethanol concentration, rate of fermentation, and length of interruption of aeration. First one notices a dependence of damage on "total concentration." At a "total concentration" of 5% an interruption of aeration from 2 to 8 min leads to the same damage as an interruption from 15 to 60 sec at a "total concentration" of 11–12%. The degree of damage is between 10 and 100%. It increases sharply with increasing "total concentration" for equal periods of interruption of aeration. At constant total concentration, damage increases with increasing concentrations of acetic acid and with an increasing fermentation rate which is proportional to the number of bacteria. At Eta = 1 there are about 50×10^9 cells per liter, corresponding to about 20 mg/liter of dry substance. Commercial vinegar fermentations have an Eta = 4, that is, they have about 80 mg/liter of dry bacterial solids. Today up to 150 g of acetic acid can be produced per liter, which means that only 0.6 mg of bacterial solids is needed for each g of final product.

The productivity of a commercial vinegar fermentation is kept at about 1.7 g acetic acid per hr per liter. Higher values can be obtained but presuppose higher additions of nutrients which are costly and which may interfere with the maintenance of the strain. Vera and Wang (1977) found in laboratory experiments that the rate of a continuous fermentation at a "total concentration" of 7% could be increased by recycling bacterial cells to the fermentor. The organisms were recovered by ultrafiltration, but the death rate of the cells was so high that 50% of the cells in the fermentor were dead in spite of the aeration with oxygen during separation and reinoculation of the cells. Therefore, it seems that the high rate of productivity, 11.5 g acetic acid per liter per hr, can not yet be attained commercially.

Experiments with ^{14}C-ethanol and *A. rancens* showed that the uptake of oxygen corresponded to the theoretical value of the oxidation of ethanol. The cell material in grams produced per gram atom of oxygen which was taken up is only about one-tenth that of other microorganisms. It was concluded that the cell yield based on ATP was very small (Mori and Terui 1972B).

Meyrath (1973) considered the reason for the extreme sensitivity of *Acetobacter* during the fermentation to be a lack of oxygen. He assumes that *Acetobacter* has high apyrase activity so that ATP which accumulates during the oxidation of ethanol is rapidly hydrolyzed and therefore only poorly available for other metabolic activities of the cell. When aeration is interrupted, the ATP pool disappears so quickly that the cell does not have the ability to adjust to the changed conditions. Experiments to substantiate this theory have not been published. Also, it does not account for the dependence of the degree of cell damage on acetic acid concentration and "total concentration." It may be assumed that the ATP pool is required to prevent the entry of ethanol or acetic acid into the cell interior.

The effect of air bubble size on the achievement of maximal Eta values has been studied by Hromatka and Ebner (1951). At a "total concentration" of 9% in a submerged vinegar fermentation, decreasing bubble size leads to higher Eta values. The utilization of oxygen in the airstream can be raised

TABLE 18.1.   DAMAGE TO *ACETOBACTER* FROM INTERRUPTION OF AERATION AS A FUNCTION OF FERMENTATION PARAMETERS

Experiment No.	Acetic Acid g/100 cm³	Alcohol % by Vol.	Total Concentration %	Efficiency Eta at Beginning of Interruption g/100 ml/24 hr	Duration of Interruption Sec	Degree of Damage of the *Acetobacter* %
1	2.51	2.29	4.80	1.10	120	34.0
2	2.42	2.29	4.71	1.38	300	42.5
3	3.16	1.71	4.87	6.03	480	99.5
4	7.90	3.80	11.70	2.63	15	10.8
5	8.05	3.70	11.75	5.25	30	74.8
6	6.17	5.20	11.37	1.00	45	43.1
7	8.62	2.80	11.42	8.14	45	80.0
8	8.05	3.30	11.35	4.47	60	99.9

Source: Hromatka *et al.* (1951); Hromatka and Exner (1962).

to the point where only 4 vol.% residual oxygen remains in the effluent gas without adverse effects on the fermentation. Such a high utilization of oxygen, up to 80%, can also be achieved commercially. A sparing use of air is quite important because of the volatility of ethanol and acetic acid. The use of pure oxygen or highly oxygen-enriched air damages the acetic acid bacteria.

Sensitivity to Lack of Ethanol.—Acetic acid bacteria are damaged if a vinegar fermentation is carried on to the point where all of the ethanol has been oxidized and if the addition of fresh ethanol-containing mash is delayed beyond that point. This is analogous to the damage resulting from an interruption of the oxygen supply and depends also primarily on "total concentration" and the duration of the interruption. The interruption of the ethanol supply can, therefore, cause severe damage to the bacteria.

Sensitivity to Changes in Temperature.—The fermentation is not affected if the temperature is cycled every 2 hr between 32° and 26°C. However, the maximal Eta value decreases markedly if the change in temperature takes place every 30 min (Hromatka et al. 1953). If cooling is stopped during a vinegar fermentation, the temperature rises higher and higher. Damage to the cells of Acetobacter increases with the duration of the interruption of cooling, with higher temperatures, and with higher concentrations of acetic acid. Ultimately the fermentation ceases.

Specific Growth Rate.—The specific growth rates were calculated from the slopes of the log-Eta curves for fermentations which were carried out with varying "total concentrations" and at varying temperatures (Hromatka et al. 1953). The values are shown in Table 18.2. For such semicontinuous fermentations the specific growth rate did not depend on the acetic acid concentration but it decreased rapidly at higher "total concentrations." At the same time the optimal fermentation temperature decreased with increasing "total concentration."

TABLE 18.2.   INFLUENCE OF THE TOTAL CONCENTRATION ON SPECIFIC GROWTH RATE OF *ACETOBACTER* IN SEMICONTINUOUS FERMENTATIONS

Total Concentration %	Specific Growth Rate hr^{-1}	Generation Time hr	Optimum Fermentation Temperature °C
5	0.49	1.4	35
7	0.36	1.9	33
9	0.21	3.4	30
11	0.16	4.4	28

Source: Hromatka et al. (1953).

There is a trend toward higher fermentation temperatures in commercial fermentations in order to permit the reuse of water from evaporative coolers in warmer climates. The trend toward higher "total concentrations," which is also present in order to save nutrients, storage, and transportation costs, makes the solution of this problem more difficult. But fermentation temperatures between 31° and 33°C at a total concentration of 12% are possible.

In a continuous culture with a "total concentration" of 12% in the feed, the dilution rate was kept constant until a constant concentration of ethanol and acetic acid was reached in the effluent (Ebner 1976). This showed that the specific growth rate of 0.027 hr^{-1} at 7.5 g/100 ml acetic acid and 4.5 vol.% ethanol decreased in linear fashion to a growth rate of 0.006 hr^{-1} at 11.0 g acetic acid per 100 ml and 1.0 vol.% ethanol. It should be mentioned that the values obtained with continuous fermentations, that is, with constant acetic acid and ethanol concentrations, are less favorable than those obtained under semicontinuous conditions with variable acetic acid and ethanol concentrations. Additionally, there is a decrease in the specific growth rate with decreasing ethanol concentrations. Therefore, it makes sense to run commercial continuous fermentations, which have to be carried out with low alcohol concentrations, only up to a "total concentration" betweeen 9 and 10%.

Vera and Wang (1977) carried out batch fermentations with A. suboxydans at a "total concentration" of 7.5%. Yeast extract was used as nutrient. The value for the specific growth rate of 0.30 hr^{-1} agrees with that of Hromatka et al. (1953).

The specific growth rate of A. aceti rises sharply with the rate of aeration at an acetic acid concentration of 2% (Alian et al. 1963). At an increased acetic acid concentration of 5%, the dependence of the growth rate on aeration becomes rather slight. The same dependence could be observed for the oxidation capacity of the Acetobacter cells. Mori et al. (1970) did batch experiments with A. rancens at a "total concentration" of 7%, starting with an acetic acid concentration of 2 g/100 ml. Acetic acid formation paralleled cell growth. In confirmation of Alian et al. (1963), the specific growth rate depended on oxygen concentration at low acetic acid concentrations but not at higher concentrations. Increasing concentrations of acetic acid inhibited the uptake of oxygen by the bacterial cells, which was explained by inhibition of the cytochrome system. Most likely this is really due to insufficient adaptation of the Acetobacter strain to acetic acid concentrations between 5 and 7%.

Mori and Terui (1972A) worked with A. rancens in continuous fermentations. The specific growth rate was strongly dependent on the acetic acid and ethanol concentrations. It rose from 0.05 hr^{-1} at 6 g/100 ml acetic acid to 0.4 hr^{-1} at 2.5 g/100 ml acetic acid. They also found a strong dependence of the specific product formation on the acetic acid concentration at a "total concentration" of 7%. Maximum specific product formation was 15 g/g·hr at 3.0 g/100 ml acetic acid, and dropped in approximately linear fashion to 2 g/g·hr at 7.0 g/100 ml acetic acid.

In contrast to these results Hromatka and Ebner (1949) found a specific product formation of 21 g/g·hr in semicontinuous tests and at a "total concentration" of 10%. Specific product formation was independent of the concentration of acetic acid between 4.5 and 7.2 g per 100 ml. The explanation may be in part that their strain was better adapted to industrial conditions. This is supported by the fact that the respiratory activity was high, namely 7750 ml O_2 per g of bacterial dry substance, and independent of acetic acid concentration between 4.5 and 7.2 g per 100 ml. The strain used by Mori and Terui (1972A) had a maximal respiratory activity of only 3160 ml O_2/g·hr with strong dependence on the concentration of acetic acid (Mori et al. 1970). Again, it appears that a change in process conditions (continuous vs. semicontinuous) has influenced the results.

Vera and Wang (1977) found a maximal value for product formation of only 5.5 g/g·hr in a continuous fermentation and at a "total concentration" of 7%. This shows the effect of the particular strain and of the process technique.

Mori et al. (1972) reported an inhibition of the specific growth rate as a function of the starting concentration of acetic acid for A. rancens. However, successive inoculations of these bacteria during the logarithmic growth phase permitted adaptation and the disappearance of this dependence.

According to Hromatka and Ebner (1959), ethanol concentrations above 6 vol.% should be avoided in commercial fermentations since such higher ethanol concentrations reduce the specific growth rate.

Mori and Terui (1972C) attempted to determine the effect of the ethanol concentration at the beginning of the fermentation on the specific growth rate of A. rancens. This included experiments with starting concentrations of up to 10 vol.% ethanol, but none of the experiments gave a higher yield of acetic acid than 5.5 g/100 ml since the strain of Acetobacter was only adapted to a "total concentration" of 6%. That means that fermentations started at higher "total concentrations" became stuck. This was primarily due to lack of adaptation to higher "total concentrations" and not —as the authors concluded—to the starting concentration of ethanol. It is clear that ethanol as well as acetic acid and the "total concentration" influence cell metabolism. But we are far from a detailed understanding since these concentrations can not be varied independently of each other.

Sensitivity to Changes in Concentration.—Commercial vinegar fermentations are generally conducted as semicontinuous fermentations. Shortly before the concentration of ethanol reaches zero, about 40% of the fermentor contents are pumped out of the fermentor which may contain at this point 13.3 g/100 ml acetic acid and 0.2 vol.% ethanol (Frings 1974). Without interruption of the aeration and at a constant temperature of the fermentation, the fermentor is refilled with new mash containing, for instance, 1.0 g/100 ml acetic acid and 13.0 vol.% ethanol. Strong concentration gradients of acetic acid and ethanol are formed locally if the fresh mash is merely pumped through a pipe ending near the wall of the fermentor. The cells of Acetobacter are unavoidably exposed to this concentration gradient and damaged to the point where the cells die. The new fermentation starts then, for instance, with only 10% of the number of bacteria of the preceding fermentation period instead of the 60% which would have been otherwise available. The dead cells cause considerable foaming. In practice, damage to the cells is avoided by adding fresh mash in such a way that it is rapidly mixed with the fermentor contents by the aerator (Hromatka and Ebner 1959).

There is no lag phase at all if the time for pumping out of the fermentor and the method of adding fresh mash is chosen correctly, not even with very high "total concentrations." This presupposes, however, that the "total concentration" of the new mash does not differ greatly from the "total concentration" in the fermentor. If this is not the case, then the changes in concentration will have a detrimental effect on the fermentation.

Overoxidation.—Overoxidation is the undesirable oxidation of acetic acid to carbon dioxide and water. Continued aeration in the absence of ethanol may trigger a change in cell metabolism which occurs more rapidly at lower "total concentrations" than at higher ones. Further, overoxidation is always correlated with bacterial growth; and the reaction is faster at lower concentrations. It is well known in industrial practice that overoxidation once started can proceed simultaneously with the oxidation of ethanol to acetic acid. It can be recognized by lower yields which are caused by the drop in "total concentration." An abundance of nutrients in the mash favors overoxidation. According to Hitschmann and Meyrath (1972), overoxidation is stimulated by the addition of 10^{-6} molar DPN or 10^{-3} molar pyruvate. Divies *et al.* (1969) showed that addition of succinic acid causes overoxidation. On the basis of extensive experiments, Hitschmann and Meyrath stated that overoxidation takes place only at acetic acid concentrations below 6%, and that it does not proceed at all in the presence of ethanol. This is not in accord with the experience gained in industrial practice.

It is most important to make sure that the fermentation does not proceed until ethanol is used up if one wishes to avoid overoxidation. In the submerged process this can be achieved by appropriate automation. Natural mashes which are low in total concentration and high in nutrients, such as apple cider, are best fermented continuously but with automatic control of the ethanol concentration. Distilled vinegar should be produced semicontinuously, and preferably at high "total concentrations" and at low concentrations of nutrients.

It is more difficult to avoid overoxidation in trickle fermentations. In such fermentations the bacterial cells adhere to a carrier, usually beechwood shavings which are aerated and through which the fermentor liquid trickles. The unavoidable formation of clumps (due in part to the formation of slime), uneven aeration, and uneven development of heat often lead to zones in which ethanol has been completely used up, while other parts still contain a sufficient concentration. This alone can cause overoxidation. Once overoxidation has been recognized in a commercial fermentor, it becomes necessary to stop the fermentation and to clean and sterilize the fermentor and all tanks, lines, and pumps ahead of the fermentor.

Submerged Vinegar Fermentation

The Frings Acetator.—The first commercial equipment for the submerged process was described by Hromatka (1952). The basic patent (Hromatka and Ebner 1955) emphasizes the need for uninterrupted aeration of the fermentation with fine bubbles as the necessary conditions for the production of satisfactory vinegar of high concentration. The process was further improved by the firm of Heinrich Frings, Bonn, for the production of vinegar with more than 12% acetic acid in a semicontinuous manner (Ebner 1969). In this process each fermentation cycle takes the same time course, as preceding and following periods. The starting concentration of each cycle is about 7.5 g acetic acid per 100 ml and 5.5 vol.% ethanol. In order to obtain higher concentrations of acetic acid, the concentration at the beginning of the cycle may slowly be raised to 8 to 10 g/100 ml. This can be achieved by discharging a smaller amount of vinegar than 40% at the end of the cycle and feeding mash of a higher "total concentration." The concentration of ethanol at the start of the new fermentation cycle remains the same, i.e., about 5.5 vol.%. This process is already practiced commercially in factories producing up to 15 g per 100 ml acetic acid. The process requires excellent performance by the fermentor (called Acetator ®) (Ebner 1967). The important parts of the equipment are:

The Frings Aerator.—It consists of a hollow body turbine surrounded by a stator as shown in Fig. 18.1 (Enenkel and Maurer 1955; Ebner 1973). The hollow body (A) has 6 openings for the escape of air, which are arranged radially and which are open against the direction of rotation (B). The openings are preceded in the direction of rotation by vertical surfaces (C). These force liquid entering from above and below in intimate mixture with the sucked-in air through the stator (D) to the outside. The rotor has an upper and a lower ring (E, F). The stator also consists of an upper and lower ring (G, H). The rings are connected with vertical baffles (I) which form an angle of 30° with the radius. These baffles receive the air-liquid emulsion very close to the rotor and direct it outward. The whole aerator works by "self aspiration," that is, without need for compressed air. The rotor sits on the shaft of a motor mounted below the fermentor and turns at 1450 or 1750 rpm. The hollow body of the rotor is connected with an air suction pipe. Proper dimensions of the rotor and stator cause a sufficient amount of air to be aspirated, and cause the air-liquid emulsion to be thrown outward radially with a given speed. This speed is chosen in such a manner that the turbulence of the stream causes a uniform distribution of the air over the whole cross section of the fermentor. Rotation of the entire fermentor contents is prevented by appropriate baffles. The particular behavior of *Acetobacter* cells during the fermentation makes an even distribution of aeration an absolute necessity. The maximum fermentation rate of an operating fermentor is limited by conditions prevailing at that point of the fermentor having the least aeration. Because of the rapid mixing in the

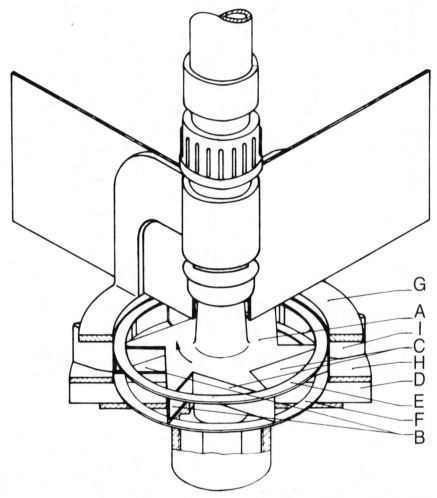

Courtesy of Heinrich Frings Co., Bonn

FIG. 18.1.   FRINGS AERATOR
A—Hollow body. B—Openings for air escape. C—Working surfaces. D—Stator. E—Upper rotor ring. F—Lower rotor ring. G—Upper stator ring. H—Lower stator ring. I—Stator blades.

fermentor, each *Acetobacter* cell must eventually pass through that part of the fermentor which is inadequately aerated and suffers damage which can not be compensated by increased aeration in other parts of the fermentor. This damage can increase to the point where the death rate reaches or surpasses the rate of multiplication of cells, causing a slowdown or even a complete cessation of the fermentation.

The extreme requirements for uniform aeration are rarely fulfilled by aeration devices commonly used in other fermentors. An instrument with which bubble size can be measured has been developed by Ebner (1973).

This is suitable for determining a spectrum of air bubble sizes at many different points of the fermentor. Optimal fermentation conditions can only be obtained if the size of bubbles (converted to equivalent pressures) and number of bubbles show the same average values at all measuring points lying at different heights and at different radial distances from the fermentor axis. The mean bubble diameter is about 1 mm with only small deviation from this value.

The Frings Defoamer.—Foam which forms occasionally must be prevented from leaving the fermentor. Its bulk must be removed at the end of the fermentation cycle in order to prevent its accumulation in the fermentor. The Frings aerator is equipped with a mechanical defoamer (Ebner 1966). As shown in Fig. 18.2, it consists of a spiral housing (E) in which a rotor (B) turns at 1000 or 1450 rpm. The rotor is equipped with radial wings (F). At the gas exit side the rotor has a peripheral ring which runs tightly against a counter flange of the housing. Foam enters the housing with the exit gas at (A) and from there it enters the rotor axially. Liquid and foam particles are thrown centrifugally to the outside and are pumped back into the fermentor through the return pipe (C). The air freed from foam flows axially from the inner part of the rotor into the exit duct (D). The defoamer adjusts automatically to varying requirements for foam removal and is activated by a foam electrode. In order to remove the foam from the Acetator at the end of the fermentation cycle, the return pipe is connected to the discharge pipe through which the product is pumped out of the fermentor at the end of a cycle. This arrangement permits complete removal of foam so that there is no need for chemical defoamers. This is advantageous because vinegar is used as a food.

The Frings Alkograph.—An automatic instrument for measuring the percentage of alcohol for control of the fermentation has been introduced (Ebner and Enenkel 1966, 1969). It is called Alkograph®. Small amounts of fermentor liquid flow continuously through the analyzer; first through a vessel which adjusts the temperature and then through 2 boiling vessels. The temperature of the boiling point of the incoming liquid is measured in the first boiling vessel. While alcohol is distilled off continuously, the higher boiling point of the liquid from which ethanol has been removed is measured in the second boiling vessel. The difference in temperature is a measure of the ethanol content and is recorded automatically. It is independent of other components of the liquid and of variations in barometric pressure.

Other portions of the process can also be controlled automatically (Els and Martens 1961). The pump which pumps the vinegar out of the fermentor is actuated by a contact in the Alkograph and stopped when a certain level in the Acetator has been reached. A pump adding fresh mash is started at the same time. It is controlled in such a manner that the Acetator is filled slowly and with maintenance of a constant temperature. When the desired liquid level in the Acetator has been reached, the pump is stopped. The device for automatic cooling stays in operation until the alkograph shows

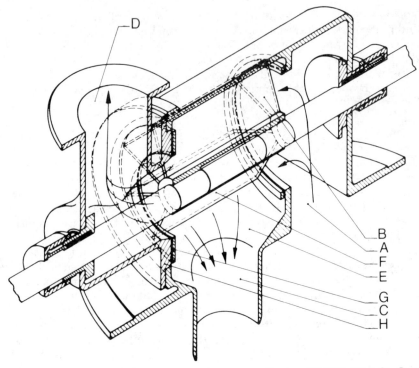

FIG. 18.2.   FRINGS DEFOAMER
A—Foam entry. B—Rotor. C—Recycling pipe. D—Exhaust gas pipe. E—Spiral housing.
F—Radial wings. G—Rotor ring. H—Counter flange of housing.

again the end of the fermentation cycle and activates the discharge pump. The length of a fermentation cycle is on the average about 36 hr.

Continuous fermentation can be carried out at low "total concentrations" in two ways. First, one may maintain a semicontinuous operation but reduce the discharged amounts drastically and create a sequence of short semicontinuous cycles, which approach the conditions of a continuous operation. Secondly, one can choose a truly continuous operation. The Alkograph contact set at an ethanol concentration of 0.2 vol.% regulates by means of a controller a valve in the feed pipe. Continuous addition of mash is regulated so that the alcohol concentration is kept constant. Vinegar may be removed continuously through an overflow pipe or it may be pumped out at short intervals.

At this time the Frings Acetator is the technically most proficient equipment capable of handling all "total concentrations" from 5 to 15% automatically and all raw materials. Energy consumption is about 1260 kJ per liter of fermented alcohol.

Figure 18.3 is a schematic drawing of the Acetator with the aerator (A), the defoamer (B), the Alkograph (C), level control switches (D), mash pump (E), vinegar pump (F), cooler (G), thermometer for temperature control (H), cooling water valve (I), airflow meter (K), and return pipe (L). Acetator vessels are made from stainless steel or wood and the instrumentation from stainless steel and polyvinylchloride. Commercial sizes are for the fermentation of 75, 150, 300, 600, and 1200 liters of pure ethanol in 24 hr. At the end of 1980, 428 Acetators were in operation in the world with a total capacity of 715 million liters of vinegar (at 10% acetic acid) per year. Of this capacity, 133 million liters are in the United States, 90 million in France, 44 million in Japan, and the rest is distributed over 50 other countries.

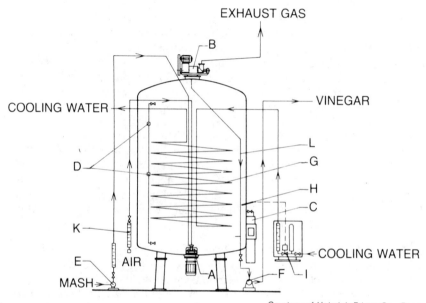

Courtesy of Heinrich Frings Co., Bonn

FIG. 18.3.   FRINGS ACETATOR

A—Aerator. B—Defoamer. C—Alkograph. D—Level switch. E—Charging pump. F—Discharging pump. G—Cooler. H—Temperature gage. I—Cooling water valve. K—Air meter. L—Recycling pipe.

All other fermentation processes mentioned subsequently have less efficient aeration systems, and therefore they can operate only at "total concentrations" of up to 10%. They are either not automated or only partly automated and do not have mechanical defoamers. Therefore, the occasional use of chemical defoamers is required. Energy consumption is generally higher than with the Acetator.

The Yeomans Cavitator.—In the United States, the Cavitator of the firm Yeomans Brothers Co., Melrose Park, Ill., was sold between 1959 and 1970 (Burgoon et al. 1960; Mayer 1961, 1963; Rowse et al. 1970). The aerator consists of a simple hollow body turbine with arms and paddles. It is driven from above by a long, hollow shaft through which it sucks air. A cylinder which surrounds this shaft sucks liquid and foam from the top of the fermentor. The liquid air emulsion is ejected in an uncontrolled and uneven fashion. A cooling coil with automatic temperature control completes the equipment. The Cavitator was used mainly for continuous cider vinegar fermentations with manual control (Beaman 1963, 1967) and was not suitable for fermentations at higher concentrations. At the end of 1980 such Cavitators were still in operation in the United States and Japan with an annual capacity of about 50 million liters of vinegar.

The Bourgeois Process.—The process of the firm Bourgeois Frères & Cie. in Ballaigues, Switzerland, is based on the use of 2 fermentors with alternating fermentation cycles. In one of the fermentors the fermentation proceeds until the ethanol concentration has dropped to zero. A thermostat stops the aeration at this time. In the meantime fermentation proceeds in the other fermentor whose contents are used for the inoculation of the first fermentor. A simple turbo mixer with air compressor and an automatic device for temperature control complete the equipment (Bourgeois Frères 1956). The lack of ethanol in the finished product lessens its aroma. Therefore, the addition of ethanol after completion of the fermentation is recommended. Total production per fermentor volume is not favorable. The system is not suitable for fermentations at higher "total concentrations" since, apart from lack of aeration, the cells of *Acetobacter* cannot be adapted to such higher concentrations. At the end of 1980 this type of equipment was mainly operated in Italy and Spain with a total capacity of about 20 million liters of vinegar per year.

The Fardon Process.—Several of the Fardon installations are operated for the production of malt vinegar (Fardons Vinegar Co. and White 1961, 1964). Aeration is provided by continuous pumping of the fermentor liquid through a Venturi tube which aspirates air, returning a liquid-air mixture to the fermentor. Cooling is done automatically. Low productivity, high energy consumption, and restriction to mashes of low "total concentration" are the principal disadvantages of this system. At the end of 1980 several units were in operation, mainly in South Africa, with a total capacity of 10 million liters of vinegar per year.

Other Processes.—The following processes have not been expanded beyond a single installation and have not reached practical significance at the present time.

R.N. Greenshields (1972) uses a high, aerated fermentation tower. Malt mash is added continuously at the bottom, and malt vinegar (fully or not fully fermented) flows out at the top of the tower. Devices which largely prevent vertical mixing in the tower have been installed.

British Vinegars (Br. Vinegars 1955) propose aeration at increased pressure as the fermentation progresses.

Patentauswertung Vogelbusch GmbH (1964) suggest control of the fermentation through measurement of the ethanol concentration of the effluent gas.

Richardson (1959), Simonin and Bernard (1956), and Dothey (1952) provide aeration by pumping the liquid through air aspirating jets. Fuchs (1963) uses a rotating self-priming impeller.

Trickling Process

Older Trickling Processes.—Detailed descriptions of the history of the vinegar fermentation have been published by Haeseler (1955), Allgeier *et al.* (1974) and Connor and Allgeier (1976). The equipment used for the surface fermentations of the *New Orleans* method can only be seen in museums of factories. The next step in the development of equipment for the fermentation of vinegar was the so-called trickling process in which the bacteria adhere to the large surface areas of a carrier material which is surrounded by air. The carrier material up to this date is still beechwood shavings, birch twigs, or corn cobs. Installations which operate on the Schutzenbach principle are today very rare. In such factories a number of smaller fermentors with approximately 2 m³ of carrier material are joined in a battery. Periodically mash is poured over the carrier material. The mash has a relatively high concentration of acetic acid and such a low concentration of ethanol that the fermentation is completed after the liquid has passed 1, 2, or 3 times through the column of carrier material. The fermentors have openings in the lower portion through which air is sucked in because of the evolved heat.

This equipment can be enlarged by the addition of a collection tank from which the liquid is pumped continuously over the carrier column, with cooling of the liquid and with openings for the aspiration of air, or by blowing air in by means of ventilators. This scheme leads to the *generator process*. The process has some basic disadvantages in comparison with equipment using the submerged fermentation process. It is impossible to distribute the liquid trickling over the carrier material so uniformly that the ethanol content is everywhere the same. There is, therefore, always the danger that the ethanol content will drop to zero at some points of the fermentor and lead to losses of fermentation capacity and to overoxidation. Mashes high in nutrient content and low in ethanol concentration aggravate this problem by formation of slimy deposits on the carrier material which may plug up the column. Therefore, generators filled with birch twigs are still in use today for the production of malt vinegar. The nonhomogeneity of the carrier column makes it impossible to distribute the air, which has been sucked in or blown in, evenly. This results in variations in temperature within the column which cannot be corrected. The interruption of aeration affects the fermentation less rapidly than with the submerged system because of the reserves of air within the carrier material.

However, once the fermentation rate has been reduced, it can not be as quickly reestablished as with the submerged process, since the recolonization of those parts of the column where the *Acetobacter* have died is a slow process. It is difficult to stop the fermentation completely; and switching from one raw material to another produces mixed types of vinegar for some time.

Yields are also lower than with the submerged process. Today quite a number of generators are still in operation which were designed and built by vinegar factories.

The Frings Generator.—The problem of automatic temperature control within the carrier column has been solved, as far as that is possible, by the Frings generator process (Frings 1932; Meynen 1935). Figure 18.4 shows such a generator. Mash is pumped from the collecting chamber of the fermentor (B) by pump (D) continuously through the heat exchanger (H) into the feed tank (I). From there it returns to the collection chamber either indirectly through the distributing wheel (K) and the beechwood shavings (A) or directly through the overflow pipe (C). The column of shavings contains 3 contact thermometers (M) which activate valves in the feed tank (E) for automatic control of the circulating liquid. A contact thermometer (G) in the mash pipe controls the flow of cooling water to the heat exchanger (H) by means of a magnetic valve (F). The mash is cooled in this fashion to 28°C and at very high fermentation rates to 26°C. The contact points of the thermometers which protrude into the column of shavings are set between 28° and 33°C. Air is blown into the carrier column from the bottom and escapes through the exit gas pipe (N).

When a residual ethanol content of 0.3 vol.% has been reached, the collection tank is largely emptied. It is subsequently refilled in 2 to 3 steps over a period of several days. There is a decided drop in rate of fermentation at the beginning of each fermentation period which is caused by the dying off of bacteria in the upper part of the carrier column due to the rapid change in alcohol and acetic acid concentration.

Once the newly added mash has been mixed with the vinegar which was absorbed by the shavings, favorable conditions (e.g., 8% acetic acid and 4% ethanol) are established for the production of vinegar with 11% acetic acid. The yield depends on the age of the shavings column. It is between 85 and 90%. Oxygen utilization during the peak fermentation period is about 50%. The length of the fermentation period depends on the volume ratio of the column of shavings to the collection chamber. It is generally between 4 and 10 days. With beechwood shavings as carrier material, a productivity of 5 liters of acetic acid per m^3 of shavings per day can be obtained.

The durability of such installations is such that at the end of 1980 Frings generators with capacities of 20, 40, and 60 m^3 for the shavings column were still in operation. These generators were first delivered more than 25 years before. Total output of such installations is estimated at 400 million liters per year.

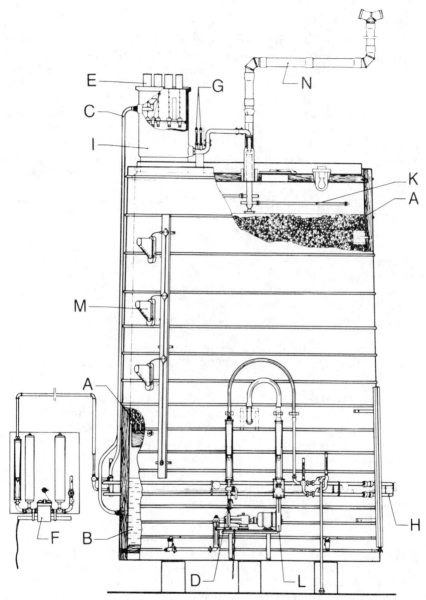

FIG. 18.4.   FRINGS GENERATOR
A—Beechwood shavings. B—Collecting chamber. C—Overflow. D—Sprinkling pump. E—Valves for control of circulating liquid. F—Magnetic valve for cooling water. G—Contact thermometer in mash pipe. H—Mash cooler. I—Sprinkling jar. K—Sprinkling wheel. L—Ventilator. M—Contact thermometer in shavings column. N—Exhaust pipe.

Treatment of Raw Vinegar

The pH of the mash drops during the vinegar fermentation. With mashes from natural raw materials there is a certain lability after completion of the fermentation with regard to the solubility of previously dissolved compounds. This lability is greater the smaller the drop in the pH during the fermentation. For instance, insoluble materials will precipitate over a period of several months if a cider vinegar of 5% is produced from fresh apple juice soon after its alcoholic fermentation. But the process of precipitation is completed much faster if apple juice concentrate is fermented to 10% cider vinegar. A rest period of several months is, therefore, recommended for all kinds of vinegar produced from natural raw materials. This does not apply to distilled vinegar.

It is a well recognized fact that the quality of vinegar improves on storage. To a lesser extent this is also true for distilled vinegar. Residual ethanol which forms esters plays a part in this improvement of quality. In general there is no difference in quality between vinegar produced by submerged fermentation and by trickling processes. Poor quality of vinegar produced in generators usually indicates that the column of shavings has been used too long or that slime has built up on the shavings. Poor quality of vinegar from the submerged process indicates problems with the fermentation, such as a low concentration of acetic acid, a high concentration of nutrients, poor selection of bacteria, or poor aeration. Vinegar should have a clean aroma which is related to the raw material that has been used.

During storage of vinegar, precipitated materials settle. This facilitates the subsequent operations.

Refining.—Raw vinegar contains acetic acid bacteria which make it opaque. This is also true, although to a lesser extent, for vinegar produced in generators. A portion of the acetic acid bacteria settles during storage. Subsequent filtration is facilitated if the vinegar is fined with bentonite, but this is not absolutely essential. For this purpose an aqueous suspension of bentonite is added to the vinegar and mixed with it intimately. The suspended particles are permitted to settle for several hours, so the supernate is usually clear and easy to filter.

Filtration.—Filtration is carried out with suspensions of diatomaceous earth whether the vinegar has been stored for some time or not, and whether it has been fined or not. Filtration must remove all suspended material such as vinegar bacteria or the occasionally appearing "vinegar eels." The latter occur often during production by the trickling process, but rarely during submerged fermentations. For the filtration of distilled vinegar produced by the trickling process, simple plate and frame filters are also satisfactory.

A membrane ultrafiltration process has been developed by Ebner and Enenkel (1976) which eliminates the fining process and simplifies the filtration. This process permits the continuous and automatic production of a bacteria-free filtrate from unrefined vinegar from the submerged process.

At low hydrostatic pressure, the flux through the filter modules is so rapid that there is no concentration polarization at the surface of the membranes, which means that the pores of the filter always remain open. The vinegar, which contains only about 0.04% of filter aid, is recirculated through the system and the amount of filtrate is replaced with raw vinegar. The filtrate is automatically checked for cloudiness and pumped off. The addition of raw vinegar is stopped after 4 weeks of continuous operation.

Filtration is continued until a concentrate is obtained which constitutes only about 0.5% of the filtered volume. The ultrafilter is now emptied, cleaned, and the process started again. This ecologically sound process requires only about 10 g of filter aid per 1000 liters of filtered vinegar. It is assumed that this continuous method of ultrafiltration will replace other methods of filtration because of its considerable advantages. The labor-saving process may also eliminate the use of sterilizing filters directly before bottling and avoid the undesirable presence of bentonite and diatomaceous earth in the plant effluents.

Pasteurization and Addition of SO_2

Vinegar is often pasteurized by short heating just before bottling, since entirely reliable methods of sterile filtration did not exist until very recently. This is particularly true for fruit vinegars which are low in total concentration and high in nutrient content. In some countries, sulfiting of vinegar up to 50 mg SO_2 per liter is permitted. In this case, sulfiting replaces pasteurization. However, it is necessary to add sulfite immediately before bottling, since SO_2 oxidizes rapidly in vinegar with high extract values and thereby loses its effectiveness.

Bottling and Keeping Quality

Vinegar for domestic use is sold exclusively in bottles of glass, polyvinyl-chloride, or polyethylene with plastic screw top closures. The closure must be air tight. For institutional consumers, such as restaurants, vinegar is sold in 5 to 25 liter plastic containers. Industrial users of vinegar use tank trucks of stainless steel.

The keeping quality of bottled vinegar must be guaranteed. This presents no problems with distilled vinegar. However, vinegar of low "total concentration" and high extract value is more readily subject to spoilage. The best guarantees for a long shelf life are a sufficient storage period to complete the precipitation of cloud-forming materials, ultrafiltration, or fining followed by filtration, absence of bacteria, and tight closures of the bottles. Pasteurization provides additional protection but can cause new problems with after-precipitation due to the heating. Heavy metals can also cause problems and their presence should be eliminated.

Concentration.—Vinegar is used in large quantities by the canning and other food industries. For some purposes, vinegar with an acetic acid concentration higher than can be obtained by fermentation is required. The producer of vinegar also prefers a product with high acetic acid content because of the savings in freight and in storage capacity.

Vinegar of 20 to 30% acetic acid is produced by freezing of vinegar with 10 to 13% acetic acid. The ice which forms during this process contains very little acetic acid. It is removed by centrifuging, leaving vinegar with the desired high concentration of acetic acid. The ice can be thawed and reused for the preparation of mashes. Unfortunately, the process of concentration is rather costly.

It is advantageous to reduce the cost of the freezing process by producing vinegar with the highest possible concentration during the fermentation. It is possible to produce vinegar with up to 15 g/100 ml of acid in the Acetator (Ebner 1969). This limit is set by the ability of the *Acetobacter* to multiply. Recently a 2-step production sequence has been developed which permits the manufacture of vinegar at higher percentages of acetic acid (Ebner and Enenkel 1978). At a certain stage of the fermentation the "total concentration" of a fermentor is increased by the addition of ethanol. A portion of the contents is pumped into a second fermentor where the process of acidification is completed in a second fermentation stage. The remaining portion of the fermentor contents of the first stage is replenished with fresh mash until its "total concentration" has dropped to the original value. This permits the acetic acid bacteria to multiply again so that a sufficiently large number of bacteria is again available for the second stage. The first fermentor is thus used for semicontinuous fermentations at varying "total concentrations." The second fermentor completes the fermentation as a batch process. It is then emptied and refilled from the first fermentor. At the end of 1979 the first unit (Acetator 1200) was in commercial operation and produced vinegar with 18.5% acetic acid per 100 ml with this procedure. It is anticipated that the acetic acid concentration can be further increased.

COMPONENTS OF VINEGAR

A consideration of the individual components of vinegar is important for two reasons. First, it is important for the distinction between vinegar and diluted acetic acid. Second, it permits a distinction among the various types of vinegar. Analytical values and limits have been compiled by Herrmann (1970) and are useful in the context of food regulations.

Distilled Vinegar

Schanderl and Staudenmayer (1956) identified 8 amino acids in alcohol vinegar by means of paper chromatography, as well as several vitamins of the B group. The vinegar had been produced by submerged fermentation and only glucose and inorganic salts had been added as nutrients. These substances must, therefore, have been formed during the fermentation. Haeseler et al. (1957) found in distilled vinegar 8–38 µg/liter riboflavin,

60–100 μg/liter nicotinamide, and 12–38 μg/liter pantothenic acid. Bourgeois (1957) detected the following amino acids in distilled vinegar: alanine, arginine, cystine, glycine or serine, isoleucine, leucine, lysine, methionine, phenylalanine, proline, tryptophan, and tyrosine. The amounts found in vinegar with 4.5% acetic acid varied from 0.016 to 4.4 mg/liter. Bergner and Petri (1959, 1960) increased the number of detected amino acids to 18 and found values for total amino acids between 3.1 and 11.4 mg/liter.

An investigation of the volatile components of vinegar showed only the presence of ethyl acetate (Suomalainen and Kangasperko 1963). However, Aurand et al. (1966) identified acetaldehyde, acetone, ethyl acetate, and ethanol in all samples of distilled vinegar. They found another 14 compounds which could not be identified. Finally Kahn et al. (1972) increased the number of identified compounds to 27. They neutralized and filtered distilled vinegar and extracted the filtrate with ether-pentane. After removal of the solvents from the extract, the material was analyzed with a gas-liquid chromatograph.

Wine Vinegar

The presence of acetoin in wine vinegar has long been established. Galoppini and Rotini (1956) as well as Rotini and Galoppini (1957) showed that it is formed in varying amounts during the fermentation. According to Federico (1948) and Federico and Gobis (1949), it is formed by oxidation of 2,3-butanediol, but may also be formed by different pathways. Hadorn and Beetschen (1965) found as a result of extensive experiments with wine vinegar that there is no direct correlation between the increase in acetoin and the decrease in 2,3-butanediol during the fermentation of wine vinegar, although the presence of some 2,3-butanediol seems to be indispensable. The concentration of acetoin may vary between 10 and 800 mg/liter. Haeseler et al. (1957) found 64.5 μg riboflavin, 266 μg nicotinamide, and 167 μg pantothenic acid per liter in wine vinegar.

Bourgeois (1957) found the same amino acids as in distilled vinegar and additionally glutamic acid and threonine with variations between 1 and 540 mg/liter. Suomalainen and Kangasperko (1963) found ethyl acetate, diacetyl, 2-butanol, isobutanol, isobutyl acetate, acetoin, isoamyl alcohol, and isoamyl acetate in ether-pentane extracts of wine vinegar by chromatographic methods. Using gas-liquid chromatography, Aurand et al. (1966) found 17 volatile components and Kahn et al. (1972) found 42.

Cider Vinegar

An ether-pentane extract of a cider vinegar contained ethyl acetate, diacetyl, 2-butanol, isobutanol, isobutyl acetate, acetoin, isoamyl alcohol and isoamyl acetate (Suomalainen and Kangasperko 1963). Kahn et al. (1972) increased the number of compounds in ether-pentane extracts of cider vinegar to 31.

Malt Vinegar

Jones and Greenshields (1969, 1970A,B, 1971) investigated the volatile materials which are formed during the alcoholic fermentation of malt mashes and the following acetic acid fermentation. Acid aldehyde, ethanol, isoamyl alcohol, acetic acid, n-propanol, 2-butanol, isobutyl alcohol, n-amyl alcohol, 2-butyl acetate, isobutyl acetate, and n-amyl acetate were formed during the alcoholic fermentation and propionic acid, isobutyric acid, and acetoin were formed additionally during the acetic acid fermentation. Kahn *et al.* (1972) increased the number of compounds identified in an ether-pentane extract of neutralized malt vinegar to 48.

Whey Vinegar

Extensive investigations of whey vinegar were carried out by Hadorn and Zürcher (1973) to arrive at specifications for regulatory purposes. They determined the concentration of lactic acid, lactose, extract, acetoin, and the metal ions potassium, sodium, calcium, magnesium, and iron. Bourgeois (1957) found the following amino acids: alanine, arginine, cystine, glutamic acid or threonine, glycine or serine, isoleucine, leucine, lysine, methionine, phenylalanine, proline, tryptophan, and tyrosine. The total concentrations varied between 2 and 380 mg/liter.

USES

Vinegar is used in the home for the preparation of salads and vegetables. In the food industry it is used for the production of pickles, other vegetables, fish, mustard, mayonnaise, and salad dressings.

PRODUCTION VOLUME

The Vinegar Industry Survey 1980 (Anon. 1981) gives the following figures for production in the United States: 372 million liters distilled vinegar (10% acetic acid); 65 million liters cider vinegar (5% acetic acid); 11 million liters wine vinegar; 14 million liters corn vinegar; and 7 million liters of other vinegar types; for a total production of 469 million liters vinegar (not all of the same strength). The end use was distributed as follows: bottled for home use 29.0%; industrial use for pickles 14%, for salad dressings 20.7%, for tomato products 14%, for mustard 9%, and for other processed foods 13%. The European Economic Community in 1979 produced 273 million liters of distilled vinegar, 112 million liters of wine vinegar, and 49 million liters of other vinegar types; a total of 434 million liters of vinegar (all at 10% acetic acid) (Anon. 1980).

REFERENCES

ALIAN, A., RABATNOVA, I.L., NIKOLAEV, P.I. and IVANOV, V.A.   1963. Submerged culture of vinegar bacteria under different aeration conditions. Mikrobiologiya *32* (4) 703–710. (Russian)

ALLGEIER, R.J., NICKOL, G.B. and CONNER, H.A.   1974.   Vinegar: History and development. Food Prod. Dev. *8* (5) 69–77; *8* (6) 50–53, 56.

ANON.   1981.   Vinegar Industry Survey 1980. Robert H. Kellen Co., Atlanta.

ANON.   1980.   The Vinegar Industry of the European Community in the Years 1978 and 1979. EEC Statistics of the Assoc. of German Vinegar Manufacturers, Bonn.

ASAI, T.   1968.   Acetic Acid Bacteria. Univ. of Tokyo Press, Tokyo.

AURAND, L.W., SINGLETON, J.A., BELL, T.A. and ETCHELLS, J.L.   1966. Volatile components in the vapors of natural and distilled vinegars. Food Sci. *31*, 172–177.

BEAMAN, R.G.   1963.   A continuously operated submerged fermentation process for the production of vinegar. Meet. Am. Chem. Soc., New York, 1963.

BEAMAN, R.G.   1967.   Vinegar fermentation. *In* Microbial Technology. H.J. Peppler (Editor).   Reinhold Publishing Corp., Subsidiary of Chapman Reinholt, New York.

BERGNER, K.G. and PETRI, H.   1959.   Amino acids in alcohol vinegar. Angew. Chem. *71*, 31–32. (German)

BERGNER, K.G. and PETRI, H.   1960.   Amino acids of spirit vinegar. I. Detection of amino acids in spirit vinegar and in vinegar bacteria. Z. Lebensm. Unters. Forsch. *111*, 319–333. (German)

BOURGEOIS FRÈRES.   1956.   Process and apparatus for the acetous fermentation of a liquid, especially for the production of vinegar. Fr. Pat. 1,132,093. Oct. 29. (French)

BOURGEOIS, J.   1957.   Chromatographic and microbiologic determination of amino acids in different kinds of vinegar. Branntweinwirtschaft *79*, 250. (German)

BOURNE, E.J. and WEIGEL, H.   1954.   C^{14}-cellulose from *Acetobacter acetigenum*. Chem. Ind., 132.

BR. VINEGARS.   1955.   Manufacture of vinegar. Br. Pat. 727,039. Mar. 30.

BURGOON, D.W., CIABATTARI, E.J. and YEOMANS, C.   1960.   Mixing apparatus. U.S. Pat. 2,966,345. Dec. 27.

CHELDELIN, V.H.   1960.   Metabolic Pathways in Microorganisms.   John Wiley & Sons, New York.

CLAUS, G.M., ORCUTT, M.L. and BELLY, R.T.   1969.   Phosphoenolpyruvate carboxylation and aspartate synthesis in *Acetobacter suboxydans*. J. Bacteriol. *97*, 691–696.

CONNER, H.A. and ALLGEIER, R.J.   1976.   Vinegar: Its history and development. Adv. Appl. Microbiol. *20*, 81–133.

DIVIES, C., DUPUY, P. and SCHNEIDER, C.   1969.   Oxidation of acetate by *Acetobacter rancens*. Ann. Technol. Agric. *18* (4) 339. (French)

DOTHEY, G.   1952.   Process for the production of vinegar. Belg. Pat. 509,550. Mar. 15. (French)

EBNER, H.   1966.   Apparatus for separating gas from foam. U.S. Pat. 3,262,252. July 26.

EBNER, H.   1967.   Latest development in the technical verification of the submerged vinegar fermentation. Zentralbl. Bakteriol. Parasitenk. Infektionskr. Hyg. Abt. 1, Suppl. *2*, 65–72. (German)

EBNER, H.   1969.   Process for acetic acid fermentation. U.S. Pat. 3,445,245. May 20.

EBNER, H.   1973.   About the use of self-aspirating aerators in large fermentors. H. Dellweg (Editor). 3. Symp. Tech. Microbiol., Berlin, 1973.   Publisher Institute for Industrial Fermentations and Biotechnology, W. Berlin.

EBNER, H.   1976.   Vinegar. *In* Ullmanns Encyclopedia of Technical Chemistry, Vol. 11.   Verlag Chemie, Weinheim, W. Germany. (German)

EBNER, H. and ENENKEL, A.   1966.   Process and apparatus for the analysis of mixtures of liquids. U.S. Pat. 3,290,924. Dec. 13.

EBNER, H. and ENENKEL, A.   1969.   The Frings Alkograph. GIT Fachz. Lab. *13*, 651–654. (German)

EBNER, H. and ENENKEL, A.   1974.   Device for aerating liquids. U.S. Pat. 3,813,086. May 28.

EBNER, H. and ENENKEL, A.   1976.   Ultrafiltration process and apparatus using low hydrostatic pressure to prevent concentration polarization. U.S. Pat. 3,974,068. Aug. 10.

EBNER, H. and ENENKEL, A.   1978.   Two-stage process for the production of vinegar with high acetic acid concentration. U.S. Pat. 4,076,844. Feb. 28.

ELS, H. and MARTENS, F.   1961.   Method for producing acetic acid from alcohol containing fermentation medium. U.S. Pat. 3,014,804. Dec. 26.

ENENKEL, A. and MAURER, R.   1955.   Apparatus for the aeration of liquids. Br. Pat. 724,791. Feb. 23.

FARDONS VINEGAR CO. and WHITE, J.   1961.   Process for the aeration of malt liquors. Br. Pat. 878,949. Oct. 4.

FARDONS VINEGAR CO. and WHITE, J.   1964.   Process and apparatus for the manufacture of vinegar. Br. Pat. 963,481. July 8.

FEDERICO, L.   1948.   About the origin of acetylmethylcarbinol in vinegar fermentation. Ann. Chim. Appl. *38*, 619–624. (Italian)

FEDERICO, L. and GOBIS, L.   1949.   About the origin of acetylmethylcarbinol in vinegar fermentation. II. Ann. Chim. Appl. *39*, 278–282. (Italian)

FRATEUR, J.   1950.   Trial on classification of *Acetobacter*. Cellule *53*, 287–392. (French)

FRINGS, H.   1932.   Manufacture of vinegar. U.S. Pat. 1,880,381. Oct. 4.

FRINGS, H.   1974.   Process for the production of vinegar with more than 12% acetic acid by submerged fermentation of alcohol containing mashes. Ger.

Pat. 1,517,898. Oct. 10. (German)

FUCHS, J.C.P.   1963.   Apparatus for introduction of a gas into a liquid and for rigorous movement of the latter. Fr. Pat. 1,366,173. June 1. (French)

GALOPPINI, C. and ROTINI, O.T.   1956.   Further investigations on the formation of acetylmethylcarbinol in vinegar fermentation. Ann. Fac. Agrar. Univ. Pisa *17*, 99−111. (Italian)

GREENSHIELDS, R.N.   1972.   Improvements in and relating to the fermentation of alcohol to product vinegar. Br. Pat. 1,263,059. Feb. 9.

HADORN, H. and BEETSCHEN, W.   1965.   About true vinegar with extremely low content of acetoin. Mitt. Geb. Lebensmittelunters. Hyg. *56* (1) 46−62. (German)

HADORN, H. and ZÜRCHER, K.   1973.   Production analysis and valuation of whey vinegar. Mitt. Geb. Lebensmittelunters. Hyg. *64*, 480−503. (German)

HAESELER, G.   1955.   Vinegar. *In* Ullmanns Encyclopedia of Technical Chemistry, Vol. 6.   Urban and Schwarzenberg, Munich, Berlin. (German)

HAESELER, G., HERBST, A.M. and JUST, F.   1957.   Quantitative analysis of a few representatives of the vitamin B-complex in different vinegars. Branntweinwirtschaft *79*, 156−157. (German)

HAUGE, J.G., KING, T.E. and CHELDELIN, V.H.   1955.   Oxidation of dihydroxyacetone via the pentose cycle in *Acetobacter suboxydans*. J. Biol. Chem. *214*, 11−26.

HERRMANN, K.   1970.   Vinegar. *In* Handbook of Food Chemistry, Vol. 6. L. Acker *et al.* (Editors).   Springer Verlag, Berlin, Heidelberg, New York. (German)

HITSCHMANN, A. and MEYRATH, J.   1972.   Utilization of acetic acid by *Acetobacter*. Mitt. Versuchsstn. Gaerungsgewerbe Wien *3*, 48−55. (German)

HROMATKA, O.   1952.   The submerged vinegar fermentation. Chem. Ztg. *76*, 776−778, 815−817. (German)

HROMATKA, O. and EBNER, H.   1949.   Investigations of the vinegar fermentation. I. Trickling method and submerged method. Enzymologia *13*, 369−387. (German)

HROMATKA, O. and EBNER, H.   1951.   Investigations of the vinegar fermentation. III. About the influence of the aeration on submerged fermentation. Enzymologia *15*, 57−69. (German)

HROMATKA, O. and EBNER, H.   1955.   Method for the production of vinegar acids by oxidative fermentation of alcohols. U.S. Pat. 2,707,683. May 3.

HROMATKA, O. and EBNER, H.   1959.   Vinegar by submerged oxidative fermentation. Ind. Eng. Chem. *51*, 1279−1280.

HROMATKA, O., EBNER, H. and CSOKLICH, C.   1951.   Investigations of the vinegar fermentation. IV. About the influence of a total interruption of the aeration. Enzymologia *15*, 134−153. (German)

HROMATKA, O. and EXNER, W.   1962.   Investigation of the vinegar fermentation. VIII. Further knowledge on interruption of aeration. Enzymologia *25*, 37−51. (German)

HROMATKA, O. and GSUR, H.   1962.   Investigations about the vinegar fermentation. XII. The influence of radioactive $^{14}CO_2$ on *Acetobacter*. Enzymologia *25*, 81–86. (German)

HROMATKA, O., KASTNER, G. and EBNER, H.   1953.   Investigations about the vinegar fermentation. V. About the influence of temperature and total concentration on the submerged fermentation. Enzymologia *15*, 337–350. (German)

HROMATKA, O., KASTNER, G., GSUR, H. and GRUBER, T.   1962.   Investigations about the vinegar fermentation. IX. The influence of CO_2 on the submerged vinegar fermentation. Enzymologia *25*, 52–64. (German)

JONES, D.D. and GREENSHIELDS, R.N.   1969.   Volatile constituents of vinegar. I. A survey of some commercially available malt vinegars. J. Inst. Brew. London *75* (5) 457–463.

JONES, D.D. and GREENSHIELDS, R.N.   1970A.   Volatile constituents of vinegar. II. Formation of volatiles in a commercial malt vinegar process. J. Inst. Brew. London *76* (1) 55–60.

JONES, D.D. and GREENSHIELDS, R.N.   1970B.   Volatile constituents of vinegar. III. Formation and origin of volatiles in laboratory acetifications. J. Inst. Brew. London *76* (3) 235–242.

JONES, D.D. and GREENSHIELDS, R.N.   1971.   Volatile constituents of vinegar. IV. Formation of volatiles in the Frings Process and a continuous process of malt vinegar manufacture. J. Inst. Brew. London *77* (2) 160–163.

KAHN, J.H., NICKOL, G.B. and CONNER, H.A.   1972.   Identification of volatile components in vinegars by gas chromatography-mass spectrometry. Agric. Food Chem. *20* (2) 214–218.

KING, T.E. and CHELDELIN, V.H.   1952.   Phosphorylative and nonphosphorylative oxidation in *Acetobacter suboxydans*. J. Biol. Chem. *198* (1) 135–141.

KING, T.E. and CHELDELIN, V.H.   1954.   Oxidations in *Acetobacter suboxydans*. Biochim. Biophys. Acta *14*, 108–116.

KING, T.E., KAWASAKI, E.H. and CHELDELIN, V.H.   1956.   Tricarboxylic acid cycle activity in *Acetobacter pasteurianum*. J. Bacteriol. *72* (3) 418–421.

KITOS, P.A., KING, T.E. and CHELDELIN, V.H.   1957.   Metabolism of fructose-1,6-diphosphate and acetate in *Acetobacter suboxydans*. J. Bacteriol. *74*, 565–571.

KITOS, P.A., WANG, C.H., MOHLER, B.A., KING, T.E. and CHELDELIN, V.H.   1958.   Glucose and gluconate dissimilation in *Acetobacter suboxydans*. J. Biol. Chem. *233* (6) 1295–1298.

LOITSYANSKAYA, M.S.   1955.   Bacterial utilization of glucose in vinegar production. Mikrobiologiya *24*, 598–607. (Russian)

LOITSYANSKAYA, M.S.   1958.   The significance of carbon dioxide for *Acetobacter*. Vestn. Leningr. Univ. Ser. Biol. Geogr. Geol. *13* (9) (2) 73–82. (Russian)

MAYER, E.   1961.   Process for making vinegar. U.S. Pat. 2,997,424. Aug. 22.

MAYER, E.   1963.   Historic and modern aspects of vinegar making (acetic fermentation). Food Technol. *17*, 582–584.

MEYNEN, H.L.K.   1935.   Apparatus for manufacturing vinegar. U.S. Pat. 2,022,970. Dec. 3.

MEYRATH, J.   1973.   Cellularenergetic considerations on submerged vinegar fermentation. Mitt. Versuchsstn. Gaerungsgewerbe Wien 11, 180–183. (German)

MORI, A., KONNO, N. and TERUI, G.   1970.   Kinetic studies on submerged acetic acid fermentation. I. Behavior of Acetobacter rancens cells toward dissolved oxygen. J. Ferment. Technol. 48 (4) 203–212. (Japanese)

MORI, A. and TERUI, G.   1972A.   Kinetic studies on submerged acetic acid fermentation. II. Process kinetics. J. Ferment. Technol. 50 (2) 70–78. (Japanese)

MORI, A. and TERUI, G.   1972B.   Kinetic studies on submerged acetic acid fermentation. III. Efficiency of energy metabolism in acetic fermentation using Acetobacter rancens. J. Ferment. Technol. 50 (8) 510–517. (Japanese)

MORI, A. and TERUI, G.   1972C.   Kinetic studies on submerged acetic acid fermentation. V. Inhibition by ethanol. J. Ferment. Technol. 50 (11) 776–786. (Japanese)

MORI, A., YOSHIKAWA, H. and TERUI, G.   1972.   Kinetic studies on submerged acetic acid fermentation. IV. Product inhibition and transient adaptation of cells to the product. J. Ferment. Technol. 50 (8) 518–527. (Japanese)

NAKAYAMA, T.   1959.   Studies on acetic acid bacteria. I. Biochemical investigation of ethanol oxidation. J. Biochem. 46 (9) 1217–1225.

NAKAYAMA, T.   1960.   Studies on acetic acid bacteria. II. Intracellular distribution of enzymes related to acetic acid fermentation and some properties of a highly purified TPN-dependent aldehyde dehydrogenase. J. Biochem. 48 (6) 812–830.

NAKAYAMA, T.   1961A.   Studies on acetic acid bacteria. III. Purification and properties of coenzyme-independent aldehyde dehydrogenase. J. Biochem. 49 (2) 158–163.

NAKAYAMA, T.   1961B.   Studies on acetic acid bacteria. IV. Purification and properties of a new type of alcohol dehydrogenase, alcohol-cytochrome-553-reductase. J. Biochem. 49 (3) 240–251.

PATENTAUSWERTUNG VOGELBUSCH GMBH.   1964.   Process for control of feed and outflow in biological oxidation of alcohol to acetic acid. Austrian Pat. 237,560. Dec. 28. (German)

PRIEUR, P.   1968.   Repression of 6-phosphofructokinase and phosphoglyceromutase biosynthesis in Acetobacter xylinum by fructose. Bull. Soc. Chim. Biol. 50 (10) 1769–1782. (French)

RAGHAVENDRA RAO, M.R. and STOKES, J.L.   1953.   Utilization of ethanol by acetic acid bacteria. J. Bacteriol. 66 (6) 634–638.

RAZUMOVSKAYA, Z.G. and BELOUSOVA, T.Z.   1952.   Relations of Acetobacter to carbon dioxide in vinegar production. Mikrobiologiya 21, 403–407. (Russian)

RICHARDSON, A.C.   1959.   Process for the production of vinegar. U.S. Pat. 2,913,343. Nov. 17.

ROTINI, O.T. and GALOPPINI, C.   1957.   The acetylmethylcarbinol acetous fermentation and the origin of commercially available vinegars. Ann. Sper. Agrar. *11* (6) 1355–1372. (Italian)

ROWSE, J.A., ROWSE, D.F. and BRODY, J.   1970.   Advanced vinegar production. Food Eng. *42* (6) 86–87.

SCHANDERL, H. and STAUDENMAYER, T.   1956.   Differentiation between vinegar and diluted acetic acid in aminoacids content. Z. Lebensm. Unters. Forsch. *104* (1) 26–28. (German)

SHIMWELL, J.L.   1959.   A re-assessment of the genus *Acetobacter*. Antonie van Leeuwenhoek J. Microbiol. Serol. *25*, 49–67.

SIMONIN, R.F. and BERNARD, M.   1956.   Dynamic process and apparatus for aerobic fermentations in liquid phase and products of these fermentations. Fr. Pat. 1,114,988. Apr. 18. (French)

SUOMALAINEN H. and KANGASPERKO, J.   1963.   Aroma substances of vinegar. Z. Lebensm. Unters. Forsch. *120* (5) 353–356. (German)

VERA, F.M. and WANG, D.I.C.   1977.   Increasing productivity in the acetic acid fermentation by continuous culture and cell recycle. 174th Am. Chem. Soc. Meet., Chicago, Aug. 28–Sept. 2, 1977.

VISSER'T HOOFT, F.   1925.   Biochemical investigations on *Acetobacter*. Thesis Techn. Univ. Delft. (Dutch)

WIELAND, H. and BERTHO, A.   1929.   About the mechanism of oxidations. XV. The acetic acid fermentation. Ann. Chem. (Liebig) *467*, 95–157. (German)

19

Production of Fermentation Alcohol as a Fuel Source

Gerald Reed

INTRODUCTION

This chapter on alcohol production for use as a source of energy has been included because of the growing interest in the use of fermentation alcohol as a fuel for motor cars. Analogous fermentation processes have been treated in the chapter on distilled beverages and on biomass production. Anderson (1978) has reviewed the political and economic questions of this application, and Weisz and Marshall (1979) have reviewed the technological potential and constraints.

Approximately 80% of the world supply of alcohol is produced by fermentation, although in countries with advanced technology, such as the United States, almost all of the industrial ethanol is made from ethylene derived from petroleum sources. It takes about 1.9 kg (4.2 lb) of ethylene to make 3.8 liters (1 gal.) of ethanol, and prices of industrial alcohol have risen sharply with the increased cost of imported petroleum. This and the critical dependence on imported oil have stimulated the search for other sources of energy. While the United States has ample supplies of coal, these can not be used directly in motor cars as fuel, and the conversion of coal to liquid fuel appears to be costly and not very energy efficient.

This chapter does not deal with the energy crisis or with alternate energy sources but merely with the technology of producing ethanol by fermentation of agricultural raw materials. The distillation process and the use of ethanol in motor cars have been treated only briefly.

USE OF ETHANOL IN MOTOR CARS

Ethanol (or methanol) has been used as such or in blends with gasoline as motor car fuel since the beginning of the century although this use has never been widespread. With present combustion engines and present carburetors, blends containing 10% ethanol can be used without any modification, and blends up to 20% ethanol can be used without major modifications in present automobiles. Combustion engines can be built to run on straight ethanol, or on ethanol of lower concentration (80% ethanol and 20% water).

Current interest in the United States centers on a mixture usually called "gasohol," which is a blend of 90% unleaded gasoline and 10% ethanol. Anhydrous ethanol is preferred for this application but 96.5% ethanol can be used if a mutually miscible solvent such as isopropyl alcohol is added.

The effect of a gasoline-ethanol blend for an unmodified motor car engine is a leaner fuel-air mixture. This is apt to give somewhat better fuel economy, but may give a somewhat poorer driving performance as shown by hesitation, stalling, and other motor malfunctions. Specifically, a 10% ethanol blend increases the octane rating, but the extent of the increase depends on the octane rating of the base gasoline and on its hydrocarbon composition. Emissions of nitrogen oxides and carbon monoxide are reduced with 10% ethanol-gasoline blends (Brinkman et al. 1975), but it is not entirely clear whether this effect is due to a leaner fuel-air mixture or whether it is a specific effect of ethanol.

An extensive discussion of the use of ethanol as a fuel for passenger cars is that of Freeman et al. (1976). Rao and Murthy (1963) and Wrage and Goering (1979) have treated the use of ethanol in diesel engines.

Extensive use of fermentation alcohol as a motor car fuel would have far-reaching economic and political effects. For the United States a goal of a 10% replacement of gasoline with ethanol has been proposed (Anon. 1979A). At a present annual use of 416 billion liters (110 billion gal.) of gasoline, this would require 41.6 billion liters (11 billion gal.) of ethanol. If corn is used as the raw material, this would require between 102 and 115 million MT (4 and 4.5 billion bu) of corn or about 70% of the current United States crop. It is obvious that a shift to the production of large amounts of ethanol from corn would seriously affect corn prices, the availability of concentrated protein feeds (from distillers' dried grains) and United States export markets for corn. For a discussion of these effects and their political implications the reader is referred to Anderson (1978). An indirect government subsidy in the form of a 4¢ per 3.8 liter (1 gal.) tax remission for "gasohol" is currently in effect. This amounts to a subsidy of about 40¢ per 3.8 liters (1 gal.) of ethanol. Additional tax incentives have been provided by some midwestern states.

Brazil has made extensive efforts to lower its dependence on imported oil by producing ethanol from cane sugar molasses, cane juice, and cassava starch. The current goal is the replacement of gasoline with an 80/20 blend of gasoline and ethanol in motor cars. This program with extensive government support is well on its way (Jackson 1976; Anon. 1979B).

RAW MATERIALS FOR ETHANOL FERMENTATIONS

As a general rule it can be assumed that 45 kg (100 lb) of fermentable sugar (as glucose) yield from 18 to 23 kg (40 to 50 lb) of ethanol [or 23 to 28 liters (6 to 7.5 gal.)]. For starchy materials, the yield is about the same, that is, between 40 and 50% based on the dry weight of carbohydrate. Complete hydrolysis of 45 kg (100 lb) of starch yields about 50 kg (110 lb) of glucose, but conversion is never complete, and with a 90% conversion the yields will be as indicated. For cellulosic raw materials, the yields of ethanol are substantially less because α-cellulose is quite resistant to enzymatic attack. Yields of 50% fermentable sugar (based on cellulose) can be obtained, and yields as high as 90% have been reported for various experimental conditions. Since the enzymatic conversion of cellulose is not yet practiced on a commercial scale, it is difficult to say what the ethanol yields would be in practice.

Types of raw materials can be usefully divided into sugar-containing materials, which can be fermented directly, starchy materials, which can be easily hydrolyzed by enzymes or acids to fermentable sugars, and cellulosic materials, which are difficult to hydrolyze.

Sugar-containing Raw Materials

Ethanol may be produced from any sugar-containing fruits, fruit juices, or extracts, such as grape juice, apple juice, honey, date syrup, or sugar-containing effluents of canneries. Such sources are usually too costly in comparison with sugar beets, sugar cane, or sweet sorghum. The production of crystalline sucrose yields a by-product, molasses, which until recently has been the cheapest source of fermentable sugar. Molasses of about 80° Brix containing about 50–55% fermentable sugar can be fermented without difficulty. This fermentation is the basis of the production of rum. The composition of molasses is shown in Chapter 14. The high osmotic pressure of molasses protects it from microbial spoilage, and it can be easily transported by barge, tank truck, or railroad. Molasses can be pumped easily, although in northern climates it may have to be heated before pumping to reduce its viscosity. For production of ethanol in nonsugar-producing areas, and particularly in areas with suitable waterways for barge shipment, molasses may be the best available raw material for fermentation.

In sugar-producing areas, economic considerations favor production from the juice of sugar beets or sugar cane. In the United States, production from beet juice is not likely because government support of sugar prices makes the production of beet sugar more attractive. Besides, beet sugar molasses finds a ready market in the feed industry, as a nutrient for fermentations in the pharmaceutical industry, and for the production of glutamic acid and bakers' yeast. However, in Brazil the production of ethanol from cane juice is already practiced (Jackson 1976; Coutinho 1976). The sugar cane is preferably crushed within 6 hr after harvest or the juice is obtained by high temperature extraction. The juice can be clarified without liming. Cane stalks contain about an equal weight of bagasse and of sucrose. The bagasse (the residue after extraction of the juice) is burned and can supply the energy needs of the fermentation plant.

Total production of sugar in developing countries in 1975 is estimated at 36 million MT (40 million tons), and projected production in 1985 is 47 million MT (52 million tons). At a yield of 0.9 MT (1 ton) of ethanol from 2.02 MT (2.25 tons) of sugar, the increased production could yield about 4.5 million MT (5 million tons) of ethanol or about 6 billion liters (1.6 billion gal.). The major sugar-producing countries are Brazil, Cuba, Mexico, and the Philippines, in that order.

Total production of cane molasses in 1975 was 24 million MT (27 million tons) and projected production in 1985 is 30 million MT (33 million tons). Yields of ethanol are 0.9 MT (1 ton) from 3 to 3.6 MT (3.3 to 4 tons) of cane molasses. The amount of ethanol available from 5.8 million MT (6.4 million tons) of surplus molasses (potentially available in 1985) is 1.7 million MT (1.9 million tons) or 2.4 million kl (630 million gal.) (Hepner 1979). In the United States, sugar cane is available in the southern states, principally in Louisiana. A more recent development is the proposal to use sweet sorghum in the United States.

Cheese whey contains about 6% solids, of which three-fourths is lactose, a fermentable disaccharide. While the production of ethanol from cheese whey is practical and appears to be economical, the overall impact of this raw material on ethanol production is relatively small. In the United States, about 16 billion kg (35 billion lb) of cheese whey are available from the annual production of about 1.6 billion kg (3.5 billion lb) of cheese. This could provide a total of no more than 378.5 million liters (100 million gal.) of ethanol annually.

Starchy Raw Materials

Starch which has been gelatinized by heating can be readily hydrolyzed to fermentable sugars by enzymes. Such starches occur in cereal grains (rice, wheat, corn), root crops (cassava), or tubers (potatoes). All of these materials have been used for the production of distilled alcoholic beverages; the use of wheat or corn for production of whiskey, and the use of potatoes for production of vodka are well known.

World production of starchy roots was 117 million MT (130 million tons) in 1975 and is estimated to reach 144 million MT (160 million tons) in 1985. Cassava accounts for about 78% of the total and yams for about 17%. The major producing areas are Brazil, Nigeria, Indonesia, and India, in that order. The projected increase in production of about 25 million MT (28 million tons) per year could yield 2.5 million MT (2.8 million tons) of ethanol or about 3.4 million kl (900 million gal.) (Hepner 1979).

In the United States, corn is the most abundant source of starch. Corn can be processed by customary methods of wet corn milling with the recovery of prime starch. This method is presently used for the production of glucose syrups, dextrose (glucose), and high fructose syrups (MacAllister et al. 1975; Allen et al. 1979). For the production of ethanol, mashing of corn (as described in Chapter 11) is frequently used. This follows the procedures used in the production of distilled beverages from grains, such as wheat, corn, rye, and barley malt. The principal materials for brewers' mashes are barley malt, corn grits, corn syrup, and rice. Both industries use barley malt for the enzymatic conversion of the carbohydrate of the grains. In brewing, at least 50% of the mash bill must consist of malt. In the distilling industry about 10% of malt is used because this is sufficient to obtain the desired conversion to fermentable sugars. For the production of industrial alcohol the use of bacterial and fungal amylases (instead of barley malt) is more efficient and less costly.

There are important differences between the two industries with regard to the formation of fermentable sugars. In brewing, the conversion of mashes is carried out until all available carbohydrate has been converted to fermentable sugars or to soluble oligosaccharides. The liquid portion of the mash is drawn off before all of the carbohydrate is in the form of fermentable sugars. This liquid wort is then boiled which arrests enzyme action and sterilizes the wort. Fermentation is then carried on until the fermentable sugars have been converted to ethanol and carbon dioxide, leaving a considerable fraction of unfermentable oligosaccharides in the beer.

In contrast the distiller converts his grain mashes with malt until about 30% of the carbohydrate is converted to fermentable sugars. At this point yeast is added and fermentation proceeds simultaneously with the conversion of the rest of the carbohydrate. The distiller is not interested in retaining unconverted carbohydrate. His aim is the conversion of all of the carbohydrate to fermentable sugars since he wishes to obtain as much ethanol as possible. It is obvious from what has been said that the production of industrial alcohol patterns itself after procedures used by the distilled beverage industry.

Cellulosic Raw Materials

The utilization of cellulosic materials for the production of ethanol is ultimately the most promising endeavor, but it also presents the most difficult technological and economic problems. The supply of cellulosic agricultural residues is extremely large and estimated to be 7 quads (7 × 10^{15} BTU) in the year 2000 (Anon. 1979C). The Food and Agricultural Organization of the United Nations (Kapsiotis 1979) has made a comprehensive survey of such residues on a worldwide basis, as well as estimates of current and proposed utilization. The following discussion deals only with cellulosic residues since the utilization of sugar-containing residues and of starchy materials has already been discussed. Such cellulosic residues contain variable proportions of α-cellulose, hemicellulose, and lignin, and comprise such varied materials as, for example, saw mill residue, paper mill residue, newsprint, potato peelings, rice straw, corn stover, peanut shells, cocoa and coffee husks, tobacco stalks, wheat straw, etc. Of these materials, paper mill waste, which has already been partly delignified, is probably the most promising.

The use of acids for the conversion of cellulosic materials is certainly not new. It has been widely practiced in Eastern Europe and current processes have been reviewed by Wenzel (1970) and more recently by Ladish (1979). During the past 10 years enzymatic processes have been employed in most of the experimental work thanks to the development of more efficient and less costly cellulases (β-1,4-glucan glucanohydrolases). Nevertheless, hydrolysis is slow compared to the hydrolysis of starches with α-1,4-amyloglucosidases. This is due to the partial crystallinity of cellulose and to lignin "seals" which make the polymer less accessible to enzymatic hydrolysis. Therefore, pretreatment of the cellulosic materials with acid, alkali, or mechanical disruption (ball milling) is essential. A promising approach is the pretreatment of cellulosic materials with Kadoxen (Ladish *et al.* 1978).

Cellulases from *Aspergillus niger* have been used in the past but more effective enzymes have been isolated from the genus *Trichoderma*, particularly *T. reesei* (Andren *et al.* 1976; Ryu *et al.* 1979). A pilot plant for the enzymatic conversion of cellulosic materials has been in operation at the Natick Laboratory of the U.S. Quartermaster Corps for several years. Microbial cellulases have been reviewed by Kulp (1975).

One of the more promising developments is the 2-step hydrolysis of cellulosic materials with (1) dilute acid to hydrolyze the hemicellulose to pentoses, followed by (2) acid or enzymatic hydrolysis of the separated cellulose. Several such processes have been proposed. One of these will serve as an example. It is of particular interest because it combines acid hydrolysis for the production of fermentable sugars with fermentation by yeasts in a fixed film reactor (Sitton *et al.* 1979).

Corn stover containing 15% pentosans, 35% hexosans, and 15% lignin is subjected to hydrolysis by dilute sulfuric acid (4.4%) at 100°C for 50 min. This treatment hydrolyzes the pentosan fraction with a yield of 94% of fermentable sugars from the hemicellulose. The insoluble residue is dried and impregnated with 85% H_2SO_4 and then diluted to an 8% concentration of H_2SO_4. It is then hydrolyzed for 10 min at 110°C with a yield of 89% glucose based on the hexosan content. The acid is recovered by electrodialysis.

A fixed film (immobilized yeast cell) reactor is prepared by coating small, ceramic rings with gelatin and nutrient salts and cross-linking with glutaraldehyde (Sitton and Gaddy 1978). For production of ethanol from pentose solutions, *Candida utilis* is used, and glucose is fermented with immobilized cells of *S. cerevisiae*. The ceramic rings (0.6 cm or 1/4 in. diameter) are placed into a cylindrical Plexiglass reactor and occupy about two-thirds of its volume. The substrate solution is fed continuously into the immobilized cell reactor. Dilution rates of 0.455 hr^{-1} have been achieved with this system on an experimental scale. Figure 19.1 shows the conversion of fermentable sugar to ethanol determined at 5.08 cm (2 in.) intervals over the length of the 91.4 cm (36 in.) reactor.

A different approach is the use of a single organism for the enzymatic conversion of cellulose and for the fermentation of the formed sugars. Such organisms are thermophilic and produce both ethanol and acetic acid from cellulosic residues, an example being *Clostridium thermocellum* (Wang *et al.* 1979). Stutzenberger (1979) has used *Thermonospora curvata* for this purpose.

Simultaneous enzymatic hydrolysis by isolated enzymes and fermenta-

FLOW RATE = 250 ML/HR, S_0 = 30 G/LITER, T = 22°C

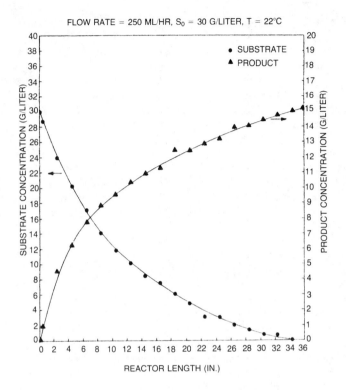

REACTOR LENGTH (IN.)

From Sitton et al. (1979)

FIG. 19.1.   PERFORMANCE OF IMMOBILIZED CELL REACTOR
AFTER 1 DAY OF CONTINUOUS OPERATION
Flow rate = 250 ml/hr. Dilution rate = 0.455 hr^{-1}. S = Substrate in
g/liter. P = Ethanol in g/liter. Temperature = 22°C. 1 in. = 2.54 cm.

tion of the formed sugars has also been suggested. Cellobiose, a disaccharide
containing 2 glucose molecules in the β-1,4-linkage, and glucose are feedback inhibitors of cellulases. Therefore, the removal of glucose by fermentation increases the rate of cellulose and cellobiose hydrolysis. The principle is
analogous to the simultaneous conversion of grain mashes by malt and
alcoholic fermentation of the formed maltose in the distilled beverage
industry. The principle has been applied to cellulose in the Gulf Oil process
(Emmert and Katzen 1979). Enzymatic hydrolysis is carried out with a
cellulase from *T. reesei* and fermentation with either *S. cerevisiae, S. uvarum,* or *Candida brassicae* at a temperature of 40°C.

MICROORGANISMS

Most of the experimental work, pilot scale work, and commercial production is carried out with *Saccharomyces cerevisiae*. This is the organism

traditionally used for the leavening of doughs and for the production of ale and distilled beverages. *S. uvarum* (formerly *S. carlsbergensis*), the lager beer yeast, has also been used. This species is closely related to *S. cerevisiae* and may have a certain advantage because it is more flocculent and sediments faster than *S. cerevisiae*. *Candida utilis* is used for the fermentation of waste sulfite liquor since it also ferments pentoses. Recent experimentation with *Schizosaccharomyces pombe* has shown promise (Rose 1976). For *S. cerevisiae* the optimum temperature for growth and ethanol fermentation is 30°C but somewhat higher temperatures (35°–38°C) are tolerated. However, at such higher temperatures growth rate, yield of ethanol, and the death rate may be affected. It would be very desirable to obtain strains which perform well at higher temperatures because it would reduce the energy required for cooling of the fermentors, and it might increase productivity. At this time it is doubtful whether this can actually be achieved with *S. cerevisiae* strains.

Yeast growth and ethanol formation are inhibited by solutions of high osmotic pressure and by high concentrations of ethanol. Many strains of *S. cerevisiae* will readily produce 10–13% of ethanol (by vol.) in batch fermentations of grape musts, malt worts, or molasses solutions with initial fermentable sugar concentrations of 20–25%. Higher concentrations up to 16–18% can be achieved by incremental feeding. However, productivity[1] of ethanol declines rapidly at alcohol concentrations above 5–7% by vol. Therefore, traditional batch fermentations for the production of ethanol from molasses, for instance, have been carried out with an initial sugar concentration of about 12% (20° Brix molasses). The effect of osmotic pressure on the fermentation rate of *S. cerevisiae* has been treated in more detail in Chapter 14.

It has been known for some time that certain sterols increase fermentation in grape musts (see Chapter 9). However, there has been no serious study of the mechanism of this effect. Recently, A.H. Rose (1979) has tested various sterols and unsaturated fatty acids to determine their incorporation into the yeast cell wall and their effect on alcohol tolerance. He reported that ergosterol and stigmasterol provided better ethanol tolerance than other sterols and that linoleic acid was more effective than oleic acid. Finally, he found that yeasts with higher alcohol tolerance contained less alcohol inside the cells. This promising approach will provide better insight into the biochemical problems of alcohol tolerance and may lead to the selection of improved strains.

Strains of *S. cerevisiae* used in the distilling industry must ferment maltose rapidly because enzymatic hydrolysis of cereal starches by malt produces predominantly maltose. This need not be of concern if cereal mashes or starches are hydrolyzed by fungal amylases which produce mainly glucose. And, of course, it is of no concern for the fermentation of molasses, cane juice, glucose syrups, or cellulose hydrolysates. For commercial

---

[1] Productivity means g ethanol produced by 1 fermentor vol. per hr. Specific productivity means g ethanol produced per g yeast solids per hr.

fermentation of distillers' mashes to produce industrial alcohol the use of compressed or active dry bakers' yeast is customary. Recently, an active dry distillers' yeast has become available which is used at a rate of 0.24 kg per kl (2 lb per 1000 gal.) of mash (Christensen 1978). The fermentation of whey requires the use of a "dairy" yeast. Suitable species are *Kluyveromyces fragilis* and *K. lactis* (formerly classified in the genus *Saccharomyces*). The rate of fermentation of lactose with these yeasts does not differ substantially from the rate of fermentation of glucose by *S. cerevisiae*. Hydrolysis of lactose to its 2 constituent monosaccharides, glucose and galactose, followed by fermentation with *S. cerevisiae* has been suggested. This is not advantageous since *S. cerevisiae* strains ferment galactose at a slow rate.

Literature references for the various yeast species will be found in the chapters to which reference has already been made.

FERMENTATION NUTRIENTS

In many traditional fermentations the use of yeast nutrients is not required. However, in some instances the supply of such nutrients may be limiting. For instance, in the production of fruit wines (other than grape) the supply of nitrogen is often inadequate. Or, if yeast is grown on beet molasses, as in the production of bakers' yeast, the supply of biotin is limiting. Grain mashes usually contain a sufficient amount of the required nutrients, but glucose syrups made from cornstarch are greatly deficient.

In batch fermentations there is usually a 5- to 10-fold multiplication of yeast cells. With an inoculum of 6–8 million cells per ml or about 0.24 kg (2 lb) of yeast solids per kl (1000 gal.) one would expect the growth of about 1.2 to 2.2 kg (10 to 18 lb) of yeast solids in 1 kl (1000 gal.) of wort or mash. Newer processes of continuous fermentation with some aeration during the fermentation period may result in the production of larger quantities of yeast.

Yeast requires the following nutrients for growth per 45 kg (100 lb) of yeast solids: About 3–3.6 kg (7–8 lb) of nitrogen from an assimilable nitrogen source such as ammonia, ammonium salts, urea, or amino acids; about 1.1 kg (2.5 lb) of phosphorus (as P_2O_5), usually in the form of orthophosphate; and minor quantities of potassium, magnesium, and calcium salts, as well as trace minerals. The addition of vitamins is rarely required but thiamin often accelerates the rate of fermentation. *S. cerevisiae* requires the presence of biotin; *Candida utilis* does not.

Nitrogen can be supplied most economically as ammonia. Phosphate is usually supplied as phosphoric acid or orthophosphate salts. The best source of all required nutrients including the trace minerals and vitamins is often yeast extract. This nutrient is used extensively by the pharmaceutical industry in growth media. It is also used in the production of cheese starter cultures, and to some extent in the wine industry. Details on nutrients required for the growth of *S. cerevisiae* will be found in Chapter 14.

FERMENTATION

Batch Fermentation

The traditional alcoholic fermentation of distillers' mashes has been described in Chapter 11. The total fermentation time varies between 48 and 72 hr, and final alcohol concentrations of 6–8% by vol. are obtained. The process is based on the use of ground and mashed grain. The proteinaceous residue which is a valuable feed product is economically important to the operation of the process. Usual yields are 8.7 to 9.8 liters (2.3 to 2.6 gal.) of ethanol and about 7.7 kg (17 lb) of distillers' dried grains from 25.5 kg (1 bu) of corn.

Alternately, starch may be recovered from the grain in the wet milling process. The starch is cooked (gelatinized) and converted by a 2-step enzymatic process to a glucose syrup. Such syrups, preferably of a D.E. of 96, can be fermented directly but require the addition of yeast nutrients. Fermentation periods are generally shorter than with grain mashes because all of the carbohydrate is already present in the form of fermentable sugars. For batch processes, fermentation periods may vary from 15 to 30 hr. Glucose syrups are also suitable for continuous fermentation processes with or without yeast recycle. The same applies also to other sugar-containing, reasonably clarified solutions, such as molasses or cane juice.

Batch fermentations are carried out without the need for establishing pure culture conditions, that is, without the need for complete sterilization of media (although pasteurization is advantageous) and without maintenance of complete sterility of equipment. However, this presupposes a rapid start of the yeast fermentation. This fermentation inhibits the growth of other microbes by depleting the available nutrients, by a lowering of the pH, and most importantly by the formation of ethanol. Industrial alcohol fermentations require cooling either through internal cooling coils or through external heat exchangers. The temperature is usually maintained between 30° and 35°C and the pH between 4.5 and 6. Fermentors are not stirred and not aerated. The rapid development of carbon dioxide gas during the fermentation provides some mixing.

Most of the currently practiced alcohol fermentations are based on the traditional processes which have been outlined above, and so are many of the projected installations. However, the need for reducing capital investment, better utilization of energy, and increased productivity have led to the development of advanced methods. Such advances are the use of continuous fermentations, the increase of the yeast population by recycling, and the removal of ethanol during the fermentation.

Continuous Fermentation

Most of the experimental work has been carried out with homogeneous fermentations, that is, with a single stirred fermentor. Fresh medium is

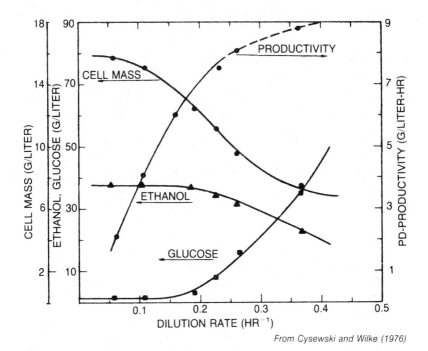

From Cysewski and Wilke (1976)

FIG. 19.2.   EFFECT OF DILUTION RATE ON PRODUCTIVITY, AND CONCENTRA-
TIONS OF ETHANOL, GLUCOSE, AND CELL MASS
Continuous fermentation with adapted yeast; glucose conc. 8.9%; oxygen tension 0.07
mm Hg.

continuously pumped into the fermentor and an equal volume of the fermentor liquid is continuously pumped out for recovery of ethanol and yeast.
It is assumed that the concentration of all substances is uniform in all parts
of the fermentor and that the same concentration prevails in the withdrawn
(harvested) liquid. The rate at which medium is added (or at which fermentor liquid is withdrawn) is usually expressed as the dilution rate, D.
This is the ratio of withdrawn liquid to the volume of total liquid in the
fermentor. Hence, a dilution rate of D = 0.1 indicates that one-tenth of the
fermentor liquid is withdrawn per hour (or that the "residence time" in the
fermentor is 10 hr).
    Figure 19.2 shows the results of an experimental continuous fermenta-
tion with sugars from the enzymatic hydrolysis of cellulose and with C.
utilis yeast. Results are shown for different dilution rates. Productivity
increases dramatically with increased dilution rates, but at D values higher
than 0.17 the concentration of glucose in the fermentor also increases
rapidly. This means that unfermented glucose is constantly withdrawn
from the fermentor, and the yield of ethanol declines. Therefore, it would

not be practical to run continuous fermentations at dilution rates which do not permit almost complete utilization of the substrate.

Cell Recycle

Cell recycle is a method for maintaining a high yeast cell population in a continuous fermentation. Yeast cells are separated from the withdrawn fermentor liquid by means of a centrifuge (or by a process of settling) and returned to the fermentor while the supernate essentially free from yeast cells is sent to the still. For the continuous fermentation depicted in Fig. 19.2, that is, without cell recycle, the cell mass declines above a dilution rate of 0.06. Figure 19.3 shows the effect of cell recycle in a continuous fermenta-

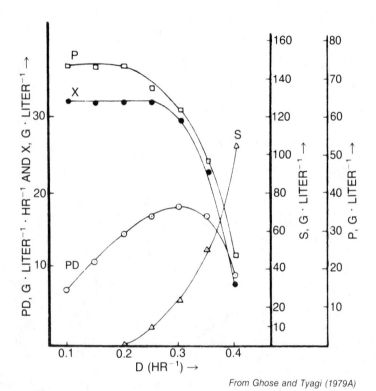

From Ghose and Tyagi (1979A)

FIG. 19.3.   CONTINUOUS FERMENTATION OF BAGASSE HYDROLY-
SATE WITH CELL RECYCLE (GLUCOSE CONC. 15%)
(●) X—Cell mass conc. in g/liter. (△) S—Glucose conc. in g/liter. (□) P—
Ethanol conc. in g/liter. (○) PD—Ethanol productivity in g/liter/hr. D—Dilution
rate/hr.

tion with a bagasse hydrolysate and *S. cerevisiae* as the fermenting organism. In this particular fermentation the cell mass does not decline until the dilution rate is greater than 0.2. Productivity of ethanol increases until D = 0.3 and declines at higher values of D. However, sugar concentration in the fermentor begins to rise above a dilution rate of 0.2. This is unfermented glucose which is withdrawn from the fermentor, and will be reflected in a decline of alcohol yields.

There is an additional reason for cell recycle which is not apparent from Fig. 19.1 and 19.2. A loss of cell viability usually occurs at high cell concentrations in alcoholic fermentations. For instance, Del Rosario *et al.* (1979) report a 16% loss in viability at lower yeast concentrations in 5 hr and a 35% loss in viability at high yeast cell concentrations. This loss in viability of the yeast population can be counteracted by cell recycle and by providing oxygen during the fermentation.

A decline in yeast activity (and presumably viability) has also been reported for batch fermentations with recycle of yeast and maintenance of high cell populations (Espinosa *et al.* 1978). This could be alleviated by working with lower sugar concentrations by diluting sugar cane juice from 16 to 13% of glucose.

Most authors provide information on the concentration of yeast in terms of cell numbers per ml or in terms of dry cell weight per liter, rarely in both. This makes comparison of results often difficult, since the ratio of cell number to cell weight varies with the size of the cells. For a particular culture of bakers' yeast the cell number is about 20–30×10^9 per g of dry cell weight (for cells about 4–5 microns in width and 6–7 microns in length). For other species of yeast and even for other strains of bakers' yeast this ratio may differ greatly. Del Rosario *et al.* (1979) give a figure of 36×10^9 cells per g of yeast solids for a strain of *S. uvarum*. In their work with continuous fermentation of molasses and cell recycle they used between 2 and 3×10^9 cells per ml. Bernstein *et al.* (1977) described the commercial production of ethanol from whey. With a continuous fermentation and cell recycle they used about 1×10^9 cells of *K. fragilis* per ml. These values are in sharp contrast with conventional batch fermentations in which the cell population is usually 5–10×10^6 cells per ml at the start and 50–150×10^6 cells per ml at the end of the fermentation (Reed 1974).

Other workers indicate cell mass values of 30–60 g/liter for continuous fermentations with cell recycle (Del Rosario *et al.* 1979; Cysewski and Wilke 1976, 1977, 1978; Ghose and Tyagi 1979A,B). If the cell mass is given in g per liter, one can calculate the specific productivity of ethanol (g ethanol/g yeast solids/hr). Maximum values in the range of 0.5 to 1 g/g/hr have been reported (Aiba *et al.* 1968; Bazua and Wilke 1977; Del Rosario *et al.* 1979). This compares to a specific productivity of 0.7 to 1.4 g/g/hr for the fermentation of bread doughs, beer, and wine (Reed 1974). It is clear that the continuous fermentations with cell recycle show high productivities because of the very much larger number of yeast cells, and not because of an improved performance of the individual cells.

Aeration

Continuous fermentations definitely require the introduction of air into the fermentor to maintain viability of cells and an efficient fermentation. The extent of aeration by sparging is often expressed as volumes of air per fermentor volume per minute (VVM). For continuous alcoholic fermentations, a VVM of 0.1 to 0.2 is commonly used. However, it must be kept in mind that the VVM by itself does not determine the concentration of dissolved oxygen in the fermentor. Oxygen tension in the fermentor also depends on the shape of the fermentor, fermentor height, and on other factors which affect the rate of oxygen transfer (see Chapter 14).

Vacuum Fermentation

The effect of higher alcohol concentrations in inhibiting fermentation rate has already been mentioned. This inhibition makes it impractical to run continuous fermentations (or batch fermentations) at alcohol levels exceeding 6–8% by vol. An interesting way of solving this problem is the constant removal of alcohol during the fermentation by boiling it off under vacuum. This process has been described by Cysewski and Wilke (1977, 1978) and by Ramalingham and Finn (1977). The former authors conducted fermentations at a temperature of 35°C and at 50 mm Hg pressure which keeps the fermentation broth boiling. The latter authors used a temperature of 30°C to prevent damage to the yeast cells, and consequently a lower pressure of 32 mm Hg.

Productivities were 3 to 4 times higher than with continuous operation without vacuum removal of ethanol. In addition, more concentrated sugar solutions could be fermented. This is the direct consequence of the low alcohol concentration which could be maintained at all times. The major disadvantage of the system is the need for removal of large volumes of CO_2 to the atmosphere, and the construction of fermentors which can be operated under vacuum. Under such conditions, sparging with air is impractical but sparging with oxygen may be feasible.

Conclusion

In conclusion it is reasonably certain that traditional batch operations will be replaced by continuous operations with or without cell recycle. This applies to raw materials in which the feed is a reasonably clear solution of fermentable sugars. It does not seem practical to apply the concepts of continuous fermentation to grain mashes since the presence of insoluble solids and a high viscosity make it difficult to pump the material efficiently and to obtain uniform conditions throughout the fermentor.

Table 19.1 shows a comparison of batch and continuous processes under the optimum experimental conditions established by Cysewski and Wilke (1978). The data given are useful for purposes of comparison, but it is not certain whether the indicated productivities can be obtained in commercial practice.

TABLE 19.1. LABORATORY FERMENTATION SYSTEMS[1]

Fermentation System	Oxygen Tension (Optimal) in mm Hg	Sugar Conc. (Optimal) in %	Cell Mass Conc. (Optimal) in g Solids/liter	Ethanol Productivity (Maximum) in g/liter/hr
Batch	[2]	—	5.6[3]	2.2[4]
Continuous	0.07	10.0	12.0	7.0
Continuous, cell recycle	0.07	10.0	50.0	29.0
Vacuum, cell recycle	[5]	—	124.0	82.0

Source: Cysewski and Wilke (1978).
[1]Optimum pH = 4.0; optimum temp. = 35°C in all cases.
[2]Optimum procedure: Saturate fermentation medium initially with air; no air sparging during fermentation; inoculum 2.0% by vol., grown aerobically.
[3]At end of the batch fermentation.
[4]Assumes 6 hr fermentor down time between 16 hr batch fermentations.
[5]Oxygen tension could not be determined. Optimum procedure: Sparge with pure oxygen at 0.10 VVM.

DISTILLATION

A discussion of the principles of distillation and of distillery practice is outside the scope of this volume; but reference will be made to the terminology used in the industry, and the production of anhydrous ethanol will be outlined.

The distilling process may be divided into the distillation proper, which separates the volatile components of the fermentor liquid from the insoluble solids and which concentrates the ethanol to a distillate containing 30–96% ethanol by weight, and the rectification, which separates the ethanol from other volatile components. Ethanol has a lower boiling point than water and the vapor phase in equilibrium with an alcoholic solution has a higher concentration of ethanol. For instance, a solution containing 10% of ethanol by weight is in equilibrium with a vapor containing about 50% of ethanol by weight. Therefore, the concentration of ethanol is greatly enriched in the distillate. Water and ethanol form an azeotrope. This is a mixture of volatile substances which at a given concentration has identical liquid and vapor compositions. For water and ethanol this concentration is 96% by weight of ethanol; and the boiling point for the azeotrope is 78.2°C. Therefore, ethanol can be concentrated by distillation to 96% by weight, but not to higher concentrations.

Anhydrous ethanol may be obtained from a 96% ethanol solution by forming a ternary azeotrope with benzene. This azeotrope contains 74% benzene, 18.5% ethanol, and 7.5% water. The boiling point is 68°C. Since the ternary azeotrope is richer in water than the ethanol-water azeotrope, it can be distilled overhead and leaves an anhydrous ethanol behind. Hexane may also be used for this purpose. The process requires a 2-column system consisting of a stripping column and a dehydration column. The process is fairly expensive and current research suggests the possibility of alternate methods of removing water from the ethanol-water azeotrope (Hartline 1979).

Water may be removed by the use of adsorbents, such as cellulose, or other dry plant material, such as cracked grain. Inorganic desiccants have also been suggested for that purpose. An extension of this method is the use of an endless loop of yarn fibers which is pulled through the ethanol-water mixture. In all of these methods it is necessary to remove water from the adsorbent or desiccant by heating, except in the case of cracked grain, which may be mashed with additional water as substrate for the next fermentation.

It is also possible to use solvents for the ethanol which do not mix readily with water. Among these, dibutyl phthalate and carbon dioxide (in the critical fluid state) have been suggested. Work is presently going on with membranes permeable to ethanol and impermeable to water. Zeolites used as molecular sieves in the production of n-paraffins have also been suggested to separate ethanol from water.

Finally, it has been suggested that the water-ethanol azeotrope, which can be separated by extractive distillation with benzene, can be subjected to extractive distillation with gasoline. In this case the organic solvent does not have to be removed if the alcohol is used as a source of motor fuel.

General descriptions of the principles of distillation will be found in Considine (1974) and Shinskey (1977).

ENERGY

The evaluation of the energy requirements of the fermentation process and of energy inputs and outputs is one of the more controversial questions of the use of ethanol as a motor fuel. Table 19.2 shows the energy analysis of some authors who are proponents or opponents of the use of corn as a source of fuel alcohol. Some calculations had to be performed to obtain data which could be compared in a single table, and, therefore, the reader is encouraged to check the source of the data. For purposes of comparison all data have been expressed on the same basis, namely, as BTU per gal. of 100% ethanol (1 BTU/gal. = 277.63 J/liter).

There are some obvious discrepancies among the individual values shown by various authors, but these are not necessarily critical for the overall evaluation of the process. However, there are two considerations which are decisive for the overall energy balance. The first of these is the energy value of the raw material, corn. In columns A and D it has been assumed that the energy used to grow and harvest the corn is suitable for calculating the energy input for that raw material. In contrast, in columns B, C, and E the energy content of the corn is used for the calculation. Obviously, this includes the solar energy captured by the growing crop.

The second critical consideration is the fuel value of the corn stalks, leaves, and husks, etc. In column A it has been assumed that 75% of this material can be transported to the alcohol plant and used as a source of fuel for the generation of steam. In the other columns no allowance is made for

TABLE 19.2. ENERGY BALANCE FOR THE PRODUCTION OF ETHANOL FROM CORN
(Expressed as BTU per gal. of ethanol produced)[1]

Materials, Processing, Products	Scheller A	Midwest Solvents B	Tong (A) C	Tong (B) D	Reilley E
Input					
Raw materials:					
grow crop	46,000	—	—	55,000	—
transport stalks	1,200	—	—	—	—
energy in grain	—	172,000	147,000	—	145,000
Process:					
mash, ferment, distill	108,000[2]	—	101,500	101,500	—
dry distilled grains	63,200[2]	—	46,000	46,000	—
total process	171,200	131,150	147,500	147,500	120,000–130,000
Total input	218,400	303,150	294,500	202,500	270,000
Output					
ethanol	75,600	84,000	84,000	84,000	84,000
fusel oils	1,100	—	—	—	—
distilled dried grains	37,200[3]	85,720	45,000	45,000	50,000
stalks, etc.	124,000	—	—	—	—
Total output	247,900	169,720	129,000	129,000	134,000
Gain in energy	29,500	—	—	—	—
Loss in energy	—	133,400	165,500	73,500	136,000

Sources: Data adapted from: Columns A and B: Anderson (1978); Scheller and Mohr (1976, 1977); Scheller (1979A,B).
Columns C and D: Tong (1979A,B).
Column E: Reilley (1979).

[1] 1 BTU/gal. = 277.63 J/liter.
[2] Figures shown are for a traditional beverage distillery. For a late 1979 fuel alcohol plant, the figures are 53,000 BTU and 17,000 BTU, respectively (Scheller 1979B).
[3] Value calculated by subtraction of other items from total output.

the fuel value of this material, and it must be assumed that the authors did not consider it practical to collect and transport this material to the plant.

With the exception of data shown in column A, there is a substantial loss in overall energy for the conversion of corn to ethanol.

Data shown in Table 19.2 are based on conventional fermentations and processes for the production of ethanol from corn. Newer processes, which, however, may be more capital-intensive, will reduce the energy required for the fermentation, distillation, and drying of the distillers' dried grains. For instance, a process which separates not only the bran and germ but also the protein meal prior to fermentation of corn, wheat, or rye and which recycles the thermal energy spent on heating the starch and in the distillation, is claimed to require only 15,270 kJ (55,000 BTU) (as steam) per liter (1 gal.) of 100% ethanol (Loser 1979). With thermal compression this figure is reduced to 8606.5 kJ/liter (31,000 BTU/gal.) ethanol; but these figures do not include electricity for motors, etc. An additional energy saving can be made by use of the methane generated by the process. The general energy relationships in the production of fuels from biomass have been discussed by Lewis (1976).

The practicability of the use of ethanol as a motor car fuel can not be judged simply on the basis of the energy balance. We can compare the energy value of various fuels in a fairly exacting manner by such units as BTU's, calories, joules. But the energy that can be usefully applied may differ considerably. The combustion engine in our motor cars requires a liquid fuel, and an amply available fuel such as coal or the plant material we grow on our land can not be used directly. This consideration does not apply only to the current efforts to produce ethanol from corn but also to the production of other liquid fuels from coal.

Secondly, it is not entirely clear how the energy value of ethanol should be considered for use in gasohol. Many authors have assumed that car mileage (distance in kilometers) obtained with gasohol is the same as that obtained with gasoline. On the other hand, the results of the "Nebraska 2 million mile road test" (3.2 million km) indicate a 3–5% improvement in mileage (distance in kilometers) for cars using gasohol vs. cars using gasoline. This would indicate that 3.78 liters (1 gal.) of ethanol can replace 4.9 or 5.7 liters (1.3 or 1.5 gal.) of gasoline (since "gasohol" contains 10% ethanol). But in either case the efficiency of ethanol in "gasohol" is appreciably better than one could calculate from its energy (joules or BTU) value.

Finally, it is not easy to assign a realistic value to the important by-product, distillers' dried grains, on the basis of its energy value. This by-product is currently available from the production of distilled alcoholic beverages. It contains about 27% protein, 9% crude fat, and 13% crude fiber. That means that kilogram for kilogram (pound for pound) the material is superior as feed to the corn from which it is derived (corn contains about 8.9% protein, 3.5% crude fat, and 2.9% crude fiber). Obviously, the value of distillers' dried grains is not properly reflected in its energy (joules or BTU) value. It has been claimed that the improvement in feed value of the higher protein material (for instance, 2.8 kl or 80 bu of corn plus distillers' dried grains from 0.7 kl or 20 bu of corn) more than compensates for the partial loss of carbohydrates from the total 35.2 kl or 100 bu of corn (Scheller 1979 A,B).

ECONOMICS

A discussion of the economics of the various processes for the production of fermentation alcohol is outside the scope of this chapter. However, some factors pertaining specifically to alcoholic fermentations which affect the economy of the operation will be mentioned.

For most raw materials, with the possible exception of molasses or glucose syrups, it is essential that the plant be located close to the source of the raw material. For Brazil, 3 types of installations have been considered: (1) a distillery built at the site of an existing sugar factory, (2) a distillery built in close proximity to several existing sugar refineries, and (3) a new plant built especially for the utilization of sugar cane juice or cassava.

In general, transportation costs have to be judged on the basis of fermentable sugars in a particular raw material. On that basis and for equal distances the cost of transporting molasses with 50% fermentable sugar will be only one-tenth that of transportation of cheese whey with 5% fermentable sugar. In addition, the keeping quality of the raw materials must be considered.

The conduct of the fermentation is important for the overall cost of the process. For dilute media, the rate of fermentation may be high but fermentor productivity may be relatively low; and the cost of distillation will be high because of the low concentration of ethanol in the fermentor liquid. For media containing more than 10–15% fermentable sugar, productivity in batch fermentations will also be low because of the inhibiting effect of ethanol, but distilling costs will be lower. For continuous fermentations with cell recycle, fermentation rates will be high and productivity will be excellent, but at higher dilution rates yields may suffer. Some of these relationships are shown in Fig. 19.4. This figure shows the cost of fermentation, which depends largely on productivity, and the cost of distillation, which depends on ethanol concentration as a function of the percentage sugar in the feed. For this particular system a 10% concentration of sugars in the feed is optimal. The figure is reproduced for purposes of illustration only. Actual costs and cost relationships may differ for different processes.

The cost of distillation is high with low ethanol concentrations in the fermentor liquid. Finn (1975) cites the following figures for various alcohol concentrations in the fermentor liquid (figures in parentheses are kg or lb of steam per Imperial gal. of ethanol; 1 U.S. gal. = 1.2 Imperial gal. = 3.8 liters): 5% (15.4 kg or 33.9 lb); 6% (10.05 kg or 27.6 lb); 7% (11.7 kg or 25.9 lb); 8% (10.5 kg or 23.2 lb).

A considerable part of the cost of fermentation is associated with the operation of pumps and centrifuges. This is particularly true for processes involving cell recycle. Meyrath (1979) points out that the fermentor liquid that has to be centrifuged is the sum of the volume of fresh medium plus the

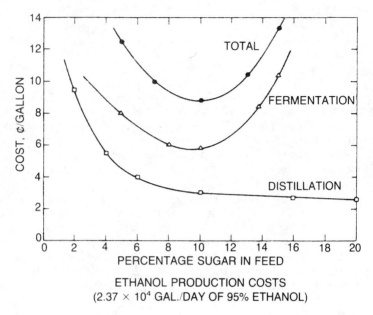

ETHANOL PRODUCTION COSTS
(2.37 × 10⁴ GAL./DAY OF 95% ETHANOL)

From Cysewski and Wilke (1976)

FIG. 19.4. ETHANOL PRODUCTION COSTS EXCLUDING RAW MATE-
RIALS AS A FUNCTION OF FEED SUGAR CONCENTRATION

amount of recirculated yeast cream and says "with increase of productivity
(reduction of fermentation time or fermentor volume) the proportion of the
re-cycle stream increases much faster than that of the fresh mash added;
that is, centrifuge costs rise very sharply."

Finally, one must pay close attention to costs associated with disposal of
still residues. For any given raw material it is potentially easier to transport (or to concentrate and dry) the still residue the higher the concentration of the raw material in the fermentor liquid.

The preceding examples have been mentioned because they relate to cost
factors which are often overlooked. A general review of the economics of
fermentation processes has been published by Bartholomew and Reisman
(1979).

COMMERCIAL PLANTS

Most of the plants producing industrial ethanol are designed to operate
conventional processes which have been successful for the production of
distilled beverages such as whiskey, rum, etc. These processes which have
been described by Underkofler and Hickey (1954) have undergone little

change during the past 50 years. This is not surprising since aging of distilled beverages over a period of several years makes experimentation costly and tampering with the process risky. Some innovation has taken place within the industry. For instance, for the production of industrial alcohol, starch hydrolysis is carried out with bacterial and fungal amylases in the place of malt.

Commercial plants customarily operate slow batch processes. For the fermentation of grain mashes, the rate-limiting step until about the thirtieth hour is the fermentation. After that point the rate-limiting step is the enzymatic conversion of residual dextrins. This is shown in Fig. 19.5 (Pan *et al.* 1950). For both malt- and fungal-converted mashes, the solid line represents total carbohydrate, the dashed line represents dextrin, and the distance between the 2 lines indicates the concentration of fermentable sugars. After about 30 hr the quantity of fermentable sugars has become minimal,

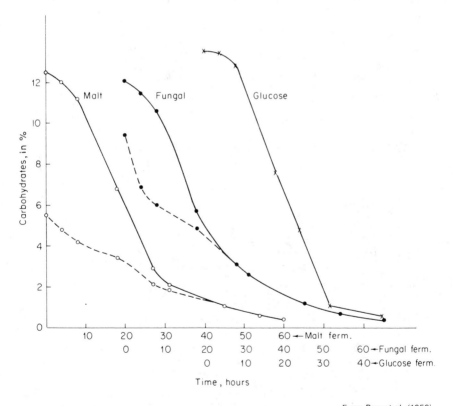

From Pan et al. (1950)

FIG. 19.5.   CHANGE OF CARBOHYDRATE CONCENTRATIONS DURING FERMENTATION OF MALT AND FUNGAL ENZYME-CONVERTED CORN MASHES AND DURING A GLUCOSE FERMENTATION

and the newly converted sugar is fermented as fast as it is formed. For the glucose fermentation, the rate of the alcoholic fermentation is the limiting step throughout.

There is no doubt that the fermentation of worts containing fermentable sugars can be accomplished more rapidly, even in batch processes, by using larger cell concentrations. Continuous fermentation processes with high cell counts can reduce the fermentation time to 10 hr or less (dilution rate of 0.1 hr^{-1}). One such process with cheese whey has been described by Reesen and Strube (1978). A United States plant for the continuous fermentation of cheese whey has been in operation for several years. The operating parameters are shown in Table 19.3 (Tzeng *et al.* 1979).

TABLE 19.3.   OPERATING CONDITIONS OF A 2-STAGE CONTINUOUS FERMENTATION OF CHEESE WHEY

Operating Condition	First Stage Fermentor	Second Stage Fermentor
Temperature, °C	30	30
pH	4.5	4.5
Aeration rate, VVM	1.0	0.3
Lactose conc. in feed, %	12–16	4–7
Flow rate, liters/hr	6435–7571	6435–7571
Av. yeast solids conc., g/liter	10–15	10–15

Source: Tzeng *et al.* (1979).

REFERENCES

AIBA, S., SHODA, M. and NAGATANI, M.   1968.   Kinetics of product inhibition in alcoholic fermentations. Biotechnol. Bioeng. *10*, 845–864.

ALLEN, B.R., CHARLES, M. and COUGHLIN, R.W.   1979.   Fluidized bed immobilized-enzyme reactor for the hydrolysis of corn starch to glucose. Biotechnol. Bioeng. *21*, 689–706.

ANDERSON, E.A.   1978.   Gasohol: Energy mountain or molehill. Chem. Eng. News *56* (31) 8–15.

ANDREN, R.K., MENDELS, M. and MODEIROS, J.E.   1976.   Production of sugars from waste cellulose by enzymatic hydrolysis: Primary evaluation of substrates. Process Biochem. *11* (8) 2–11.

ANON.   1979A.   U.S. Senate alcohol fuels legislation. U.S. Senate Bill *932*, Nov. 8, 1979, Washington, D.C.

ANON.   1979B.   Mandioca and sugar cane fuel alcohol. Gasohol (1) 33–34.

ANON.   1979C.   Biomass potential in 2000 put at 7 quad. Chem. Eng. News *57* (7) 20–22.

BARTHOLOMEW, W.H. and REISMAN, H.B.   1979.   Economics of fermentation processes. *In* Microbial Technology, 2nd Edition, Vol. 2. H.J. Peppler and D. Perlman (Editors). Academic Press, New York.

BAZUA, C.D. and WILKE, C.R.   1977.   Ethanol effects on the kinetics of continuous fermentations with *Saccharomyces cerevisiae*. Biotechnol. Bioeng. Symp. 7, 105–118.

BERNSTEIN, S., TZENG, C.H. and SISSON, D.   1977.   Commercial fermentation of cheese whey for the production of protein and/or alcohol. Biotechnol. Bioeng. Symp. 7, 1–9.

BRINKMAN, D., GALLOPOULIS, N.E. and JACKSON, M.W.   1975.   Exhaust emissions, fuel economy and driveability of vehicles fueled with alcohol-gasoline blends. Automot. Eng. Congr. Exposition, Soc. Automot. Eng., Detroit, Feb. 1975, Pap. *750120*.

CHRISTENSEN, G.R.   1978.   Personal communication. Universal Foods Corp., Milwaukee, Wis.

CONSIDINE, D.A.   1974.   Chemical and Process Technology Encyclopedia. McGraw-Hill Book Co., New York.

COUTINHO, N.   1976.   Economic and political considerations of alcohol production. Bras. Acucareiro, 19–41.

CYSEWSKI, G.R. and WILKE, C.R.   1976.   Utilization of cellulosic materials through enzymatic hydrolysis. I. Fermentation of hydrolysate to ethanol and single cell protein. Biotechnol. Bioeng. *18*, 1297–1313.

CYSEWSKI, G.R. and WILKE, C.R.   1977.   Rapid ethanol fermentation using vacuum and cell recycle. Biotechnol. Bioeng. *19*, 1125–1143.

CYSEWSKI, G.R. and WILKE, C.R.   1978.   Process design and economic studies of alternate fermentation methods for the production of ethanol. Biotechnol. Bioeng. *20*, 1421–1444.

DEL ROSARIO, E.J., LEE, K.J. and ROGERS, P.L.   1979.   Kinetics of alcoholic fermentation at high yeast levels. Biotechnol. Bioeng. *21*, 1477–1482.

EMMERT, G.H. and KATZEN, R.   1979.   Chemicals from biomass by improved enzyme technology. Pap. Am. Chem. Soc., Chem. Soc. Jpn. Joint Chem. Congr., Honolulu, April 1979.

ESPINOSA, R., COJULUN, V. and MARROQUIN, F.   1978.   Alternatives for energy savings at plant level for the production of alcohol for use as automotive fuel. Biotechnol. Bioeng. Symp. *8*, 69–74.

FINN, R.K.   1975.   The prospects for fermentation alcohol from hydrolysed cellulose. Biotechnol. Bioeng. Symp. *5*, 279–283.

FREEMAN, J.H. *et al.*   1976.   Alcohols. A technical assessment of their application as fuels. Am. Pet. Inst. *4261*.

GHOSE, T.K. and TYAGI, R.D.   1979A.   Rapid ethanol fermentation of cellulose hydrolysate. I. Batch versus continuous system. Biotechnol. Bioeng. *21*, 1387–1400.

GHOSE, T.K. and TYAGI, R.D.   1979B.   Rapid ethanol fermentation of cellulose hydrolysate. II. Product and substrate inhibition and optimization of fermenter design. Biotechnol. Bioeng. 21, 1401–1420.

HARTLINE, F.F.   1979.   Lowering the cost of alcohol. Science 206 (4414) 41–42.

HEPNER, L.   1979.   Technology and economics of industrial alcohol production. Pap. Intern. Microbiol. Food Ind. Congr., Paris, Oct. 1979.

HUMPHREY, A.E.   1977.   Studies on cellulose degradation by Thermoactinomyces. Univ. of Pa. Studies, U.S. Dep. Energy COO/4070-2.

JACKSON, E.A.   1976.   Brazil's national alcohol programme. Process Biochem. 11 (5) 29–36.

KAPSIOTIS, G.D.   1979.   Agricultural residues: Quantitative survey. FAO Agric. Serv. Bull., Food Agric. Organ. U.N., Rome.

KULP, K.   1975.   Carbohydrases. In Enzymes in Food Processing, 2nd Edition. G. Reed (Editor). Academic Press, New York.

LADISH, M.R.   1979.   Fermentable sugars from cellulosic residues. Process Biochem. 14 (1) 21–25.

LADISH, M.R., LADISH, C.M. and TSAO, G.T.   1978.   Cellulose to sugars: New path gives quantitative yield. Science 201 (4357) 743–745.

LEWIS, C.   1976.   Energy relationships of fuel from biomass. Process Biochem. 11 (9) 29.

LOSER, R.   1979.   Personal communication. Chemapec, Inc. Woodbury, N.Y.

MacALLISTER, R.V., WARDRIP, E.K. and SCHNYDER, B.J.   1975.   Modified starches, corn syrups containing glucose and fructose, and crystalline dextrose. In Enzymes in Food Processing, 2nd Edition. G. Reed (Editor). Academic Press, New York.

MEYRATH, J.   1979.   Industrial production of alcohol by continuous fermentation. Pap. Intern. Microbiol. Food Ind. Congr., Paris, Oct. 1979.

PAN, S.C., ANDREASEN, A.A. and KOLACHOV, P.   1950.   Rate of secondary fermentation of corn mashes converted by Aspergillus niger. Ind. Eng. Chem. 42, 1783–1789.

RAMALINGHAM, A. and FINN, R.K.   1977.   The Vacuferm process: A new approach to fermentation alcohol. Biotechnol. Bioeng. 19, 583–589.

RAO, M.R.K. and MURTHY, N.S.   1963.   Alcohol as fuel for Diesel engines. Symp. New Dev. Chem. Ind. Relating to Ethyl Alcohol, Its By-products Wastes, New Delhi, Oct. 1963.

REED, G.   1974.   Comparison of the use of commercial yeasts. Process Biochem. 9 (9) 11–12, 32.

REESEN, L. and STRUBE, R.   1978.   Complete utilization of whey for alcohol and methane production. Process Biochem. 13 (11) 21–24.

REILLEY, P.J.   1979.   Gasohol—future prospects. Distill. Feed Res. Counc. Conf. Proc. 34, 4–14.

ROSE, A.H.   1979.   Ethanol tolerance in S. cerevisiae. Pap. Annu. Meet., Am. Soc. Microbiol., Los Angeles, May 1979.

ROSE, D.   1976.   Yeasts for molasses alcohol. Process Biochem. 11 (2) 10–12, 36.

RYU, A.D., ANDREOTTI, R., MANDELS, M., GALLO, B. and REESE, E.T. 1979. Studies on quantitative physiology of *Trichoderma reesei* with 2 stage continuous culture for cellulase production. Biotechnol. Bioeng. *21*, 1887–1903.

SCHELLER, W.A. 1979A. Energy and ethanol. Gasohol (4) 33–38.

SCHELLER, W.A. 1979B. Gasohol, ethanol and energy. Pap. Natl. Gasohol Comm. Meet., San Antonio, Tex., Dec. 1979.

SCHELLER, W.A. and MOHR, B.J. 1976. Net energy analysis of ethanol production. Pap. Am. Chem. Soc., Fuel Chem. Div., Prep. *21* (2) 29.

SCHELLER, W.A. and MOHR, B.J. 1977. Gasoline does, too, mix with alcohol. Chemtec *7*, 616–623.

SHINSKEY, F.G. 1977. Distillation Control. McGraw-Hill Book Co., New York.

SITTON, O.C., FOUTCH, G.L., BOOK, N.L. and GADDY, J.L. 1979. Ethanol from agricultural residues. Process Biochem. *14* (9) 7–10.

SITTON, O.C. and GADDY, J.L. 1978. Ethanol production in an immobilized cell reactor. Proc. Eur. Congr. Biotechnol., Interlaken, Switzerland, 1978.

STUTZENBERGER, F.J. 1979. Degradation of cellulosic substances by *Thermonospora curvata*. Biotechnol. Bioeng. *21*, 909–913.

TONG, G.E. 1979A. Industrial chemicals from fermentation. Enzyme Microb. Technol. *1* (3) 173–179.

TONG, G.E. 1979B. Anaerobic fermentation for chemicals and fuel. Pap. Div. Microb. Biochem. Technol., Am. Chem. Soc., Miami, Sept. 1978.

TZENG, C.H., BERNSTEIN, B.A., CHU, S.W. and BERNSTEIN, S. 1979. Commercial production of ethanol by the continuous, two-stage fermentation of cheese whey. Pap. U.S.-Jpn. Intersoc. Microbiol. Congr., Honolulu, May 1979, Am. Soc. Microbiol.

UNDERKOFLER, L.A. and HICKEY, R.J. 1954. Industrial Fermentations, Vol. 2. Chemical Publishing Co., New York.

WANG, D.I.C., BIOCIC, I., FONG, H.Y. and WANG, S.D. 1979. Direct microbiological conversion of cellulosic biomass to ethanol. Proc. 3rd Annu. Biomass Energy Syst. Conf., Golden, Col., June 1979.

WEISZ, P.W. and MARSHALL, J.F. 1979. High-grade fuels from biomass farming: Potentials and constraints. Science *206* (4414) 24–29.

WENZEL, H.F. 1970. The Chemical Technology of Wood. Academic Press, New York.

WRAGE, K.E. and GOERING, C.E. 1979. Technical feasibility of Diesohol. Agric. Eng. *60* (10) 34–36.

Appendix

TABLE A.1.   SOURCES OF INDUSTRIALLY IMPORTANT CULTURES

Collection	Type of Microbes
Argentina	
Departamento de Microbiologia	Bacteria
Centro de Investigaciones en Ciencias Agronomicas	Fungi
Instituto Nacional de Tecnologia Agropecuaria	Yeasts
Casilla de Correo No. 25	
Castelar—F.C.D.F.S.	
Buenos Aires, Argentina	
Belgium	
Collection of the Laboratory for Microbiology	Bacteria
Laboratorium voor Microbiologie, Fac. Wetenschappen	
Rijksuniversiteit	
Ledeganckstraat 35	
B-9000 Gent, Belgium	
Canada	
Mold Herbarium and Culture Collection (UAMH)	Fungi
University of Alberta	
Edmonton, Alberta T6G 2H7	
Canada	
Division of Biological Sciences	Bacteria
National Research Council of Canada	Fungi
100 Sussex Drive	Yeasts
Ottawa, Ontario K1A 0R6	
Canada	
Prairie Regional Laboratory	Bacteria
National Research Council of Canada	Fungi
Saskatoon, Saskatchewan S7N 0W9	Yeasts
Canada	
France	
Institut Pasteur de Lyon: IPL	Yeasts
77 rue Pasteur	Bacteria
69365 Lyon CEDEX 2, France	Animal viruses
Collection de Microorganismes Associés aux Invertébrés	Mycoplasma
Station de Recherches de Pathologie Comparée	Rickettsia
I.N.R.A.-C.N.R.S.	Bacteria
Montpellier 34060 Saint-Christol 30380, France	Fungi
	Animal and plant viruses
Laboratoire de Cryptogamie M.N.H.N.	Fungi
Mycothèque	
12, rue de Breffon	
75005 Paris, France	
Service des Anaérobies	Bacteria
Institut de Pasteur de Paris	
25, rue du Docteur Roux	
Paris XVième, France	
West Germany	
Deutsche Sammlung für Mikroorganismen,	Yeasts
Teilsammlung Hefen,	

TABLE A.1.   *(Continued)*

Collection	Type of Microbes
Versuchs- und Lehranstalt für Spiritusfabrikation und Fermentationstechnologie in Berlin, Institut fur Gärungsgewerbe und Biotechnologie Seestr. 13 D-1000 Berlin 65, West Germany	
Instituts Sammlung Biologische Bundesanstalt für Land- und Forstwirtschaft, Institut fur Biologische Schädlingsbekampfung, Heinrichstr. 243 D-61 Darmstadt, West Germany	Bacteria Fungi Animal viruses
Japan	
Department of Fermentation Technology (HUT) Faculty of Engineering Hiroshima University Hiroshima, 730 Japan	Bacteria Fungi Yeasts
Culture Collection Institute for Fermentation Osaka 17-85 Juso-honmachi 2-chome, Yodogawa-ku Osaka, 532 Japan	Bacteria Fungi Yeasts Bacteriophages
ATU Culture Collection Department of Agricultural Chemistry Faculty of Agriculture University of Tokyo 1-chome, Yayoi, Bunkyo-ku 113 Tokyo, Japan	Bacteria Fungi Yeasts Actinomycetes
Institute of Applied Microbiology (IAM) University of Tokyo 1-chome, Yayoi, Bunkyo-ku 113 Tokyo, Japan	Bacteria Fungi Yeasts Algae Actinomycetes
The Research Institute of Fermentation Faculty of Engineering, Yamamashi University Kitashin 1-chome, Kofu Yamamashi, Japan	Bacteria Fungi Yeasts
Netherlands	
Culture Collection Laboratory of Microbiology University of Technology Julianalaan 67a, Delft, Netherlands	Bacteria
Centraalbureau voor Schimmelcultures Oosterstraat 1—Post Office Box 273 Baarn, Netherlands	Fungi Yeasts Actinomycetes
Switzerland	
Mikrobiologisches Institut Eidg. Techn. Hochschule ETH-Zentrum CH-8092 Zürich Switzerland	Bacteria Fungi Yeasts Actinomycetes

TABLE A.1.   *(Continued)*

Collection	Type of Microbes
United Kingdom	
The Wellcome Bacterial Collection Wellcome Foundation Limited Wellcome Research Laboratories Langley Court Beckenham, Kent BR3 3BS United Kingdom	Bacteria
Commonwealth Mycological Institute Ferry Lane Kew, Surrey United Kingdom	Fungi
British National Collection of Yeast Cultures Brewing Industry Research Foundation Nutfield, Surrey RH1 4HY United Kingdom	Yeasts
National Collection of Industrial Bacteria Ministry of Agriculture, Fisheries and Food Torry Research Station P.O. Box 31, 135 Abbey Rd. Aberdeen AB9 8DG Scotland, United Kingdom	Bacteria Bacteriophages
National Collection of Marine Bacteria Ministry of Agriculture, Fisheries and Food P.O. Box 31, 135 Abbey Rd. Aberdeen AB9 8DG Scotland, United Kingdom	Bacteria Bacteriophages
United States	
AR Culture Collection (NRRL) Northern Regional Research Laboratory Agricultural Research Service U.S. Department of Agriculture 1815 N. University St. Peoria, IL 61604 U.S.A.	Bacteria Fungi Yeasts Actinomycetes
IMRU Collection Waksman Institute of Microbiology Rutgers University P.O. Box 759 Piscataway, NJ 08854 U.S.A.	Actinomycetes *Bacillus* sp.
American Type Culture Collection (ATCC) 12301 Parklawn Drive Rockville, MD 20852 U.S.A.	Bacteria Fungi Yeasts Actinomycetes Algae Protozoa Animal viruses Bacteriophages Cell lines

TABLE A.2. ETIOLOGIC AGENT LABELS AND SHIPPING CONTAINERS REQUIRED UNDER THE AUTHORITY OF THE INTERSTATE QUARANTINE REGULATIONS (42 CFR, SECTION 72.25) ARE AVAILABLE FROM A NUMBER OF SOURCES INCLUDING THOSE LISTED BELOW

Labels

Interex Corporation
66 Woerd Ave.
Waltham, MA   02154

Marion Manufacturing Co.
P.O. Box 17704
Atlanta, GA   30316

Nuclear Associates, Inc.
35 Urban Ave.
Westbury, NY   11590

Shamrock Scientific Specialty Systems, Inc.
34 Davis Drive
Bellwood, IL   60104

Shipping Containers

Arthur H. Thomas Co.
Vine and Third Sts.
Philadelphia, PA   19105

Harold Liner
58 Warren St.
New York, NY   10007

Index

Other AVI Books

BASIC FOOD MICROBIOLOGY
  Banwart
ENCYCLOPEDIA OF FOOD SCIENCE
  Vol. 3   *Peterson and Johnson*
ENCYCLOPEDIA OF FOOD TECHNOLOGY
  Vol. 2   *Johnson and Peterson*
SOURCE BOOK OF FLAVORS
  Heath
FOOD ANALYSIS: THEORY AND PRACTICE
  Pomeranz and Meloan
FOOD AND BEVERAGE MYCOLOGY
  Beuchat
FOOD MICROBIOLOGY: PUBLIC HEALTH AND
SPOILAGE ASPECTS
  DeFigueiredo and Splittstoesser
FOOD PROCESSING WASTE MANAGEMENT
  Green and Kramer
FOOD QUALITY ASSURANCE
  Gould
FOOD SCIENCE
  3rd Edition   *Potter*
FUNDAMENTALS OF FOOD MICROBIOLOGY
  Fields
FUNDAMENTALS OF FOOD PROCESS ENGINEERING
  Toledo
HANDBOOK OF SUGARS
  2nd Edition   *Pancoast and Junk*
MICROBIOLOGY OF FOOD FERMENTATIONS
  2nd Edition   *Pederson*
SOURCE BOOK FOR FOOD SCIENTISTS
  Ockerman
TECHNOLOGY OF WINE MAKING
  4th Edition   *Amerine et al.*
THE TECHNOLOGY OF FOOD PRESERVATION
  4th Edition   *Desrosier and Desrosier*